Advances in Experimental Medicine and Biology

Volume 1504

Advances in Experimental Medicine and Biology provides a platform for scientific contributions in the main disciplines of the biomedicine and the life sciences. This series publishes thematic volumes on contemporary research in the areas of microbiology, immunology, neurosciences, biochemistry, biomedical engineering, genetics, physiology, and cancer research. Covering emerging topics and techniques in basic and clinical science, it brings together clinicians and researchers from various fields.

Advances in Experimental Medicine and Biology has been publishing exceptional works in the field for over 40 years, and is indexed in SCOPUS, Medline (PubMed), EMBASE, Reaxys, EMBiology, the Chemical Abstracts Service (CAS), and Pathway Studio.

2023 CiteScore: 5.9

Margarida Gama-Carvalho
Andreia Figueiredo
Francisco Rodrigues Pinto
Editors

Omics Approaches in Biomedicine and Biotechnology

From Technologies to Data to Biological Knowledge

Springer

Editors
Margarida Gama-Carvalho
BioISI – BioSystems and Integrative Sciences Institute
Faculty of Sciences
Universidade de Lisboa
Lisbon, Portugal

Andreia Figueiredo
BioISI – BioSystems and Integrative Sciences Institute
Faculty of Sciences
Universidade de Lisboa
Lisbon, Portugal

Francisco Rodrigues Pinto
BioISI – BioSystems and Integrative Sciences Institute
Faculty of Sciences
Universidade de Lisboa
Lisbon, Portugal

ISSN 0065-2598 ISSN 2214-8019 (electronic)
Advances in Experimental Medicine and Biology
ISBN 978-3-032-18965-3 ISBN 978-3-032-18966-0 (eBook)
https://doi.org/10.1007/978-3-032-18966-0

This Springer imprint is published by the registered company Springer Nature Switzerland AG
The registered company address is: Gewerbestrasse 11, 6330 Cham, Switzerland

Preface

The past two and a half decades have brought a profound transformation to the life sciences through technological innovation, ushering in the era of omics approaches. Emerging in the wake of the Human Genome Project, these methods were developed to systematically interrogate genomic diversity and gene expression and have since expanded to encompass nearly all layers of biological complexity. They mark the age of big data in biology, reshaping how questions are formulated and how results are interpreted.

This rapid methodological evolution has also exposed a gap. The increasing centrality of omics technologies in biomedicine and biotechnology has not been matched by adequate training of human resources able to fully exploit their potential, from experimental design to computational analysis and biological interpretation. Today, there is a pressing need for a new generation of professionals who are both biologically literate and computationally skilled, capable of designing and applying new analytical approaches to complex biological problems.

Omics approaches span a wide spectrum of applications, from mapping environmental biodiversity to probing molecular interactions and small molecules. Despite increasing accessibility, most students and researchers still encounter the field only superficially. As a result, many remain unaware of the technological opportunities and limitations, or of the critical importance of tailored experimental designs and informed data analysis.

This volume, shaped by our experience in teaching an eponymous postgraduate course, seeks to address this gap. It aims to equip readers with a comprehensive understanding of omics technologies and to provide a resource for applying them effectively to diverse biological contexts. Contributions by experts in the field cover technological platforms, computational approaches, and applications, reflecting the multidimensional nature of omics.

The book is further enhanced by the illustrations of Filipa Santos, whose work builds on the authors' concepts to create a coherent visual language throughout the volume. We hope it will serve both as an entry point for those new to the field and as a reference for experienced researchers wishing to expand their perspective on the transformative role of omics in the life sciences.

The editors gratefully acknowledge the support of the BioSystems and Integrative Sciences Institute (BioISI) in the preparation of this book.

Lisbon, Portugal
December 2025

Margarida Gama-Carvalho
Andreia Figueiredo
Francisco Rodrigues Pinto

Contents

The Era of Large-Scale Data in Biological Sciences

1

Margarida Gama-Carvalho, Andreia Figueiredo, and Francisco Rodrigues Pinto

Abstract

We are living in the era of large-scale data in the biological sciences. The completion of the Human Genome Project (HGP) marked a pivotal moment, not only for its monumental achievement but also for introducing the concept of Big Science to the life sciences. This project catalysed a technological revolution, shifting biology from a primarily descriptive and hypothesis-driven discipline to a data-intensive, interdisciplinary field capable of systematic and quantitative exploration of biological systems at an unprecedented scale.

This new paradigm is embodied by omics approaches, defined as experimental methods based on technologies that enable the systematic, qualitative, quantitative, and unbiased characterization of all molecular components of a given type within a biological system, or the interactions between them. From genomics and transcriptomics to proteomics and metabolomics, these technologies have moved us from studying individual molecules to capturing global views of biological processes. This has been fundamental for the emergence of Systems Biology, a unifying approach that seeks to understand complex biological systems by identifying their components and interactions and predicting systems behaviour through mathematical and computational models.

In this chapter we provide an overview of the rise of the "-omes" and their corresponding omics approaches, their conceptual foundations, and the implications they carry for research in the life sciences, including controversies, challenges, and opportunities. We conclude by discussing how the convergence of omics and artificial intelligence is reshaping the epistemological foundations of biological research, bridging data accumulation and inference of molecular mechanisms.

Keywords

Omics approaches · Genomics history · Big data in biology · High-throughput technologies · Scientific method · AI-driven biological inference

1.1 Omics Approaches: What's in a Word?

Both everyday language and scientific terminology follow rules that allow the derivation of new words whose meanings become intuitively graspable. The German botanist Hans Winkler is

M. Gama-Carvalho (✉) · A. Figueiredo · F. R. Pinto
Faculdade de Ciências, Universidade de Lisboa, Lisbon, Portugal

BioISI – Biosystems and Integrative Sciences Institute, Faculty of Sciences, Universidade de Lisboa, Lisbon, Portugal
e-mail: mhcarvalho@fc.ul.pt

M. Gama-Carvalho et al. (eds.), *Omics Approaches in Biomedicine and Biotechnology*, Advances in Experimental Medicine and Biology 1504, https://doi.org/10.1007/978-3-032-18966-0_1

credited with proposing in 1920 that the haploid chromosome set of a species should be termed the genome (Winkler 1920). There is some dispute about the origin of this word. Although some sources suggest it derives from the conjunction of "gene" and "chromosome", others note that Winkler was likely familiar with the botanical use of the suffix "-ome", as in *rhizome* or *tracheome,* to designate the collective set of roots or vascular tissue in plants (Lederberg and McCray 2001). The suffix "-ome" may therefore have been originally used in the sense of "collectivity".

The term genome later became the root for genomics, the first "-omics" word adopted in the life sciences. Genomics was coined in 1987 as the name for both a new discipline and a new journal established by Victor McKusick and Frank Ruddle, following a suggestion by T. H. Roderick. The term was meant to represent collective efforts aimed at mapping and sequencing genomes, just as discussions of the Human Genome Project (HGP) were beginning. In their launch editorial, McKusick and Ruddle highlight how the suffix "-ics", as in mathematics or economics, conveys the sense of an active, systematic pursuit of problems and solutions, in contrast with the more scholarly tradition of "-ology", which also denotes organized knowledge (McKusick and Ruddle 1987). As McKusick later noted, although "both the enterprise that aimed to map the human genome completely and even the new term genomics were under attack" at the time (McKusick 1997), within a decade genomics was established as a widely accepted, vigorous new scientific field (Botstein and Cherry 1997; McKusick 1997).

However, the acceptance of this new term, and the path toward "big data" in the life sciences that followed the HGP, also opened a Pandora's box of new names and disciplines (Fields and Johnston 2002). By 2000, roughly 100 different "-omes" and "-omics" were mentioned in the literature, a number that more than doubled over the following decade (McDonald et al. 2012). At their core, these terms were intended to convey the concept of *comprehensiveness*—the whole set of a certain type of constituents in a system, not necessarily biological, as the nomenclature spread into other fields. Yet, as Jonathan Eisen argued more than a decade ago, this "badomics" proliferation often resembled a marketing exercise that contributed little to science (Eisen 2012). It also obscured the perspective we and others subscribe to: *that omics should be understood as a methodological approach to support research in specific fields, rather than as a field in itself* (Petsko 2002).

For this reason, we prefer the expression "omics approaches" to refer to technologies and methods that—in an ideal setting—enable the unbiased identification and quantification, in a single experimental assay, of all biological molecules of a specific class within the system under study, whether a single cell, a whole organism, or a community of organisms (Fig. 1.1). This definition also encompasses methods to identify the physical and functional interactions established by these molecules. At present, omics approaches broadly fall into three categories: nucleic acid sequencing–based methods, mass spectrometry–based methods, and genome-wide functional screens (e.g., CRISPR or siRNA screens). As new technological breakthroughs emerge, additional omics-scale methodologies will likely appear.

The application of these approaches has profoundly changed how we address biological problems. It has also sparked debates about the scientific method and reshaped key aspects of scientific practice, including data sharing and publishing. These issues will be discussed in the following sections.

1.2 The Omics Revolution: A Paradigm Shift in Life Sciences

1.2.1 From Genomics to Proteomics: The Birth of Omics Technologies

As with many scientific movements, the growth of the omics concept required momentum before it could reshape research practice. The second "-ome" to be defined was the proteome,

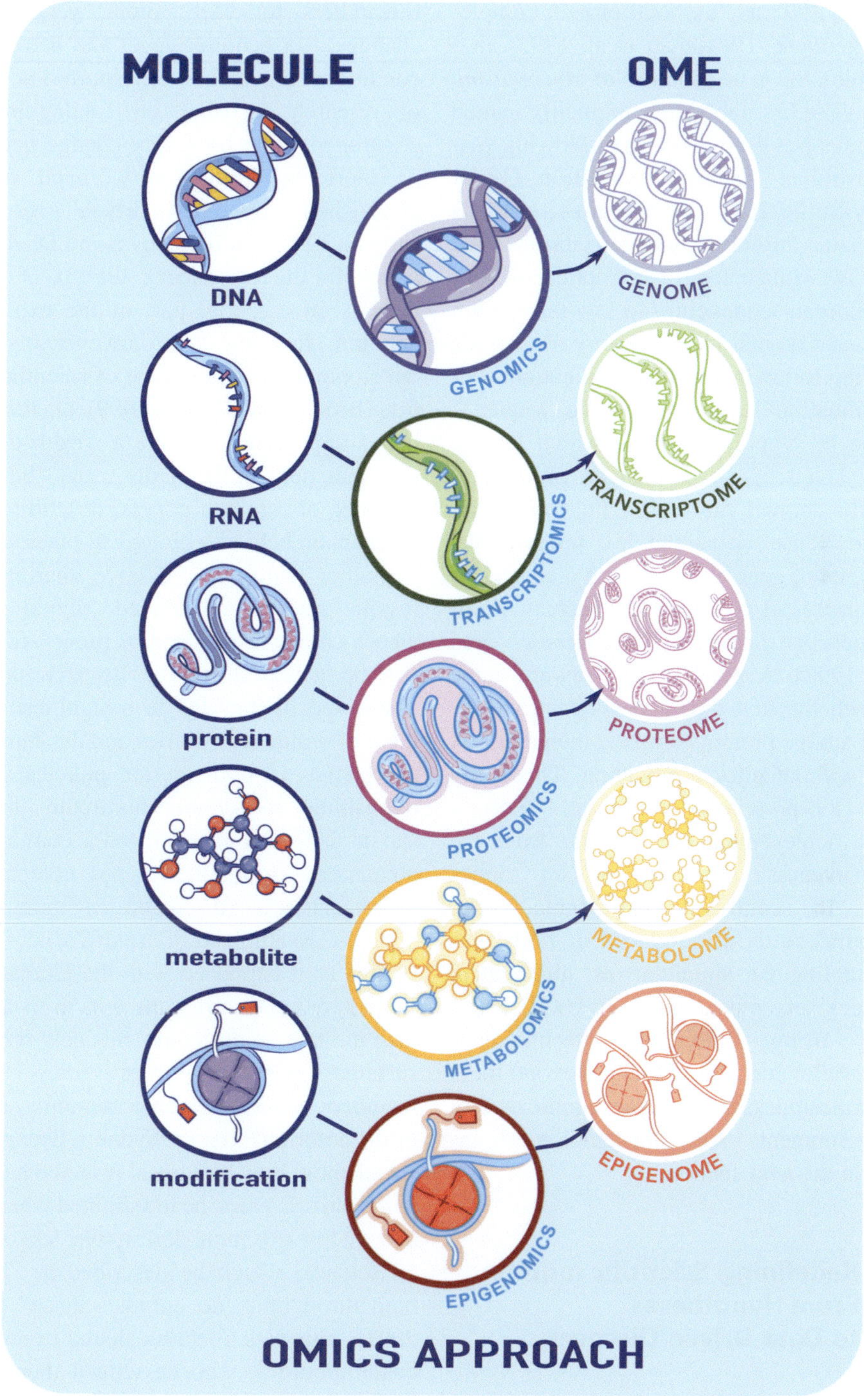

Fig. 1.1 From molecules to "omes" using omics approaches. Biological systems contain multiple classes of molecules that together define distinct "-omes", representing the set of molecules of a given type present at a specific time in a cell, tissue, organ, organism, or species. Omics technologies enable the unbiased quantitative and qualitative characterization of these molecular collectives, providing detailed insights into the composition and organization of biological systems

introduced in 1995 as "the total protein complement of a genome" (Wasinger et al. 1995), in a study mapping the gene products of *Mycoplasma genitalium*. The first laboratory explicitly named for proteomics was established the following year at the Australian National Innovation Center (Humphery-Smith 2015). At that time, methods for systematic protein characterization relied largely on two-dimensional electrophoresis coupled with Edman sequencing—a low-throughput approach—and on mass spectrometry for peptide fingerprinting following protein purification. Protein identification required genome sequence information to support mapping, which limited throughput and scope. Accordingly, proteomics was initially defined more as a field of study, aiming to "use quantitative protein-level measurements of gene expression to characterize biological processes and decipher mechanisms of gene expression control" (Anderson and Anderson 1998). A notable step forward came in 1999 with the first direct mass-spectrometric analysis of a large protein complex, the yeast 80S ribosome, without prior purification. This study identified 189 proteins (Link et al. 1999), a remarkable achievement but still far from the comprehensiveness expected of an omics approach. In contrast, that same year, transcriptomics, a term introduced in 1996, was already enabling the simultaneous, quantitative measurement of more than 100,000 target sequences using microarray technologies (Gama-Carvalho and Sousa 2026). Beyond these technical milestones, the broader significance of these developments was epistemological, as discussed in the next section.

1.2.2 Redefining Scientific Inquiry: From Hypotheses to Data-Driven Discovery

The sudden generalization of microarray technologies created a paradigm shift that sparked significant discussion around the scientific method and the concept of "true science" as opposed to "fishing expeditions", which continues to this day (Weinstein 1998; Torson et al. 2020; Guo et al. 2025). By allowing researchers to easily probe gene expression changes at a genomic level and across multiple conditions, this technology spurred new avenues of research focusing on cataloguing system components and how they change between two or more conditions—very much unlike the established "hypothesis-driven" research. As early as 1999, Patrick Brown and David Botstein argued for the relevance of this type of descriptive data as an essential part of the exploration of genomic function that, however, required the development of a new type of scientific publication (Brown and Botstein 1999). On the one hand, omics approaches provide a wealth of information that detracts from the usual "bottom line" message of research papers, describing one key mechanism behind a biological phenomenon. On the other hand, the effective sharing of these results requires dedicated digital platforms where complete raw and/or processed data files can be accessed by the research community. These specificities led to a significant development of online repositories and databases, as well as changes in both journal policies and online publishing to define minimum information standards for submission, and accommodate supplementary information types compatible with omics datasets (discussed in more detail in (Gama-Carvalho and Sousa 2026)).

Just as it happened with the Human Genome Project, or even earlier attempts to systematically map the human proteome, this new type of work suffered pushback from some prominent members of the research community, one of the most notorious being Sydney Brenner. When asked about how biological research had changed in the past 50 years, he highlighted what he saw as a move towards more descriptive, less experimental science, which he described as "low input, high throughput, no output science" (Friedberg 2008). This was a feeling shared by many of his contemporaries, who developed their careers at the height of molecular biology, the science that pushed biology away from the old days of Natural History and specimen collecting to the investigation and discovery of causal mechanisms.

The scepticism against the value of these approaches was deeply rooted in the modern philosophy of science, built on Karl Popper's

Table 1.1 Examples of large international projects applying omics approaches to characterize different biological systems

Omics domain	Representative projects
Genomics/functional genomics	1000 Genomes (Auton et al. 2015), ENCODE (Dunham et al. 2012), Vertebrate Genomes Project (Rhie et al. 2021), FAANG (Giuffra et al. 2018), 10KP plan (Cheng et al. 2018)
Transcriptomics	GTEx (Consortium Gte 2013), TCGA (Network et al. 2013), The Human Cell Atlas (Regev et al. 2017), One Thousand Plant Transcriptomes Initiative (Leebens-Mack et al. 2019)
Proteomics	Human Proteome Project (Legrain et al. 2011)
Epigenomics	International Human Epigenome Consortium (Stunnenberg et al. 2016)
Metagenomics	Earth Microbiome Project (Gilbert et al. 2014), Tara Oceans (Turnbaugh et al. 2007; Sunagawa et al. 2020), Human Microbiome Project (Turnbaugh et al. 2007)

rejection of inductivism and the proposal of falsificationism. This approach was central to defining scientific progress as hypothesis-driven testing rather than mere data accumulation. With time, however, this feeling gradually dissipated, as the value of generating systematic maps of biological components as a starting point for further inquiries became increasingly appreciated (Weinstein 1998). In fact, since the completion of the Human Genome Project, many international consortia have been established to derive high-quality data across a wide range of "-omes" and biological systems, marking a new era in the Life Sciences (Table 1.1). The large omics datasets acquired through combined independent research projects or the organized activity of large international consortia contribute to a detailed mapping of the biological world. Shared across Open Science platforms, they can serve to identify novel and unsuspected patterns and create a systems-level perspective that supports the investigation of complex, emergent properties, which cannot be assessed through the study of individual components. These descriptive observations, in turn, can fuel the proposal of new hypotheses and models that will have to be tested by traditional deductive approaches to achieve mechanistic explanations. As discussed in detail by Fulvio Mazzocchi, "both strategies are necessary and can complement each other", opening new opportunities that were not accessible prior to the advent of these technologies (Mazzocchi 2015).

Notwithstanding these advancements, hidden tensions remain in the community surrounding the place for big data approaches in biology and they occasionally resurface to question the increasingly widespread use of omics technologies and large-scale projects. One noteworthy discussion surrounded a famous article in the *Wired* magazine by editor-in-chief Chris Anderson published in 2008 with the title "The End of Theory: the data deluge makes the scientific method obsolete" (https://www.wired.com/2008/06/pb-theory/). In this provocative piece, Anderson argued that the generation of petabytes of data in different fields—specifically in biology—together with the advent of high-capacity computational clusters, would allow science to advance simply through the identification of patterns and correlations in data, without the need for models or mechanistic explanations. As expected, there was significant pushback from the community regarding this very narrow perspective of the aims of science. As Massimo Pigliucci (2009) aptly expressed it in his editorial "The End of Theory in Science", although finding patterns is undoubtedly part of the research process, it is not the endgame but rather the discovery of an explanation for those patterns. If research activity cannot produce these explanations, "it becomes akin to stamp collecting" and should not be called science. The big challenge then becomes how can we analyse and integrate such a massive amount of data to effectively generate new knowledge and understanding.

1.3 From Data to Wisdom: The Challenge of Interpretation

The debates regarding the influence of omics approaches on life sciences research will likely continue for a long time. Given the dynamic complexity and diversity of biological systems, the drive to acquire more descriptive data is very strong. However, more data do not necessarily generate more knowledge or understanding. The actual meaning of these words—knowledge and understanding—is part of a profound philosophical debate that lies beyond the scope of this chapter. The idea that there is a hierarchy moving from raw data and information to knowledge and wisdom is not new, although of somewhat uncertain origin. From early literary references, such as T.S. Eliot's poem "The Rock", the so-called DIKW model seems to have been first clearly formulated by Russell Ackoff in 1988, in his presidential address to the International Society for General Systems Research (ISGSR) (Rowley 2007). Widely applied across disciplines, the model captures the challenges raised by the generalized use of omics approaches in the life sciences particularly well.

1.3.1 Omics Limits and Risks: Avoiding Over-Interpretation

High-throughput technologies have transformed biology into a data-intensive discipline, enabling the rapid generation of multidimensional datasets across genomics, transcriptomics, proteomics, and metabolomics. The recent emergence of single-cell, spatial, and multi-omics technologies has further expanded both analytical and computational complexity. These data need to be organized and contextualized through appropriate processing and analysis to become information. Integrating approaches with distinct resolution, scale, and noise profiles requires new statistical frameworks and scalable computational infrastructure. Although data sharing and standardization of raw omics formats were major achievements of the first decade of omics growth, current challenges are now shifting toward ensuring reproducible workflows, transparent model provenance, and privacy-preserving data governance (Powell 2021; Holt and Inouye 2025). As omics approaches expand to clinical and biotechnological applications, their limitations and potential risks must be acknowledged. Practical examples—from basic research to biomedicine and biotechnology—illustrate both the power and the limitations of data-driven inference.

Public release of the ENCODE project results (Dunham et al. 2012) provided an early illustration of the risks of over-interpreting large-scale omics data, igniting a major debate in the genomics community (Doolittle 2013; Eddy 2013; Mazzocchi 2015). Building on the human genome sequence, ENCODE aimed to catalogue functional elements genome-wide by mapping open chromatin, chromatin modifications, protein-binding sites, and actively transcribed regions (Consortium TEP 2011). Media coverage of its findings proclaimed that "80% of the genome serves some purpose" (Pennisi 2012), a statement widely criticized as an over-interpretation of biochemical activity. Detecting transcription, chromatin accessibility, or protein binding does not in itself imply biological function (Doolittle 2013; Eddy 2013). Such claims require testing against a null hypothesis—e.g. whether transcription-factor-bound sequences regulate expression more effectively than random DNA of equivalent length. Such an experiment was performed by Michael White and colleagues using a massively parallel transcriptional enhancer assay in mouse retina, comparing 1300 genomic binding sites for the Crx photoreceptor transcription factor with 3000 random DNA sequences, who failed to find significant differences between the two groups (White et al. 2013). The ENCODE controversy thus illustrates a recurrent pitfall in omics research: equating biochemical activity or statistical association with biological function.

Beyond research, similar challenges emerge across applied domains. The application of omics-based technologies in medical contexts illustrates how their potential to drive important advances is accompanied by significant limitations and ethical concerns. During the COVID-19 pandemic, large-scale viral sequencing enabled near-real-time tracking of variants and transmission dynamics, but mechanistic explanations for immune escape or pathogenicity required targeted experimental validation (Knyazev et al. 2022). In oncology, omics-based stratification can reveal mutational signatures and molecular subtypes with prognostic value, but functional and clinical confirmation remain essential before biomarkers can safely guide treatment decisions (Brenton et al. 2000). In both cases, the availability of individual genomic data is associated with fundamental ethical concerns regarding the full extent to which the acquired data should or not be used (Stoeklé et al. 2018; Hunter and Helft 2023; Jose et al. 2024).

In agriculture, integrated multi-omics has enabled the identification of regulatory networks and candidate genes underlying yield, stress tolerance, and senescence across key crops such as wheat, rice, maize, soybean, and tomato (Choi 2019). However, despite these successes, most discovered associations remain correlative. Predictive models and editing targets derived from omics integration must be validated experimentally, and field trials to confirm causal relationships and environmental robustness are essential. Nevertheless, novel interventions are proposed for yield prediction and smart farming based on omic datasets, from genome to phenotype. As a successful example, in wheat, convolutional and transformer architectures are applied for disease detection, pest classification, and in-field robot guidance, achieving high performance in object detection and classification tasks (Hu and Yu 2024). While such applications can deliver specific outcomes ahead of the full mechanistic understanding, the last remains a black box.

Hence, while automated pattern discovery is an essential step toward hypothesis generation, deductive, hypothesis-driven reasoning and experimental testing remain cornerstones for the advancement of our understanding of biological systems. As the complexity of biological data increases, addressing these interpretative limits increasingly depends on computational methods capable of integrating multiple layers of information—most notably artificial intelligence.

1.3.2 AI and the Future of Knowledge Integration

With the exponential increase in the number, size, and complexity of omics datasets, modern computational and AI platforms have become central to these discovery processes. Recent deep-learning frameworks for genome analysis and molecular design demonstrate remarkable capacity to uncover latent structure and generate correct predictions of systems behaviour without explicit human understanding of underlying mechanisms (Callaway 2025; Jung et al. 2025). Predictive models trained on biased or incomplete datasets can propagate inequities or unsafe recommendations, underscoring the need for independent validation, mechanistic interpretation, and post-deployment monitoring. Automated discovery of correlations and patterns, while indispensable, does not equate to biological knowledge or causal insight.

Contemporary AI approaches are highly dependent on inductive learning, primarily through machine-learning and deep-learning algorithms that identify patterns from large datasets. Interestingly, much of the research-based omics approaches—including genomics, transcriptomics, proteomics, and metabolomics—similarly rely on inductive strategies to gain insight into the molecular functioning of biological systems. This fundamental alignment suggests that AI methods are ideally suited to extract information and knowledge from omics data, creating what appears to be a natural methodological partnership.

This convergence is enthusiastically embraced by many biologists who find themselves overwhelmed by the sheer volume of data generated by omics experiments. Even relatively simple experiments comparing just two experimental conditions can produce datasets of daunting complexity. There is often hope that computational tools will automatically analyse these data and distil results into readily interpretable biological insights. The AI-omics marriage is particularly well suited for identifying molecular signatures associated with different phenotypes, as this challenge can be elegantly formulated as a classification problem—a task at which machine learning excels.

However, knowing which molecular signatures associate with specific phenotypes does not inherently explain why or how those signatures contribute to the observed biology. This limitation becomes especially problematic given the inherent complexity and context-dependency of biological systems. In biology, remarkably small molecular differences can precipitate dramatically different phenotypes, as is the case in many monogenic diseases. Conversely, highly distinct molecular configurations may converge on similar phenotypic outcomes. For example, different individuals may have a clinically similar cancer with distinct molecular origins. These intricate relationships impose a significant practical constraint: achieving high-quality machine learning classification models requires very large and representative datasets encompassing many different biological conditions to adequately capture the full spectrum of biological variability.

Fortunately, the AI field is evolving beyond purely black-box approaches. The development of explainable AI methods (Budhkar et al. 2025), neurosymbolic AI frameworks (Velasquez et al. 2025), and symbolic regression techniques such as genetic programming (Vanneschi and Silva 2023; Makke and Chawla 2024) represents a shift towards more mechanistic understanding. These emerging approaches aim to transcend correlative pattern discovery and instead uncover causal relationships that genuinely explain the underlying data. Such methods hold promise for bridging the gap between molecular signatures and mechanistic biological understanding.

Nevertheless, a fundamental challenge remains: the available data can often be explained by multiple distinct mechanistic models. Neither humans nor AI methods can solve this challenge if no additional data is available. When faced with equally plausible mechanistic explanations, discriminating between them requires the generation of new data through specifically designed experiments to test competing hypotheses. This iterative cycle of computational modelling and targeted experimentation represents the frontier of integrating AI with biological discovery, moving beyond descriptive pattern recognition toward genuine mechanistic insight.

1.3.3 Conclusion

Ultimately, omics should be regarded not as an endpoint but as a powerful engine for hypothesis generation, driving mechanistic exploration. Their greatest contribution lies in revealing complex system behaviour and emergent properties that guide the formulation of new, testable ideas and inform experimental design. The true biological insight will continue to depend on integrating computational scale with causal reasoning, experimental rigor, and interdisciplinary literacy, transforming data not merely into information, but into knowledge and, ultimately, wisdom.

Acknowledgements This work was supported by UID/04046/2025 - Biosystems and Integrative Sciences Institute Centre grant from FCT, Portugal (https://doi.org/10.54499/UID/04046/2025)

References

Anderson NL, Anderson NG (1998) Proteome and proteomics: new technologies, new concepts, and new words. Electrophoresis 19(11):1853–1861. https://doi.org/10.1002/elps.1150191103

Auton A, Abecasis GR, Altshuler DM, Durbin RM, Abecasis GR, Bentley DR, Chakravarti A, Clark AG, Donnelly P, Eichler EE, Flicek P, Gabriel SB, Gibbs RA, Green ED, Hurles ME, Knoppers BM, Korbel JO, Lander ES, Lee C, Lehrach H, Mardis ER, Marth GT, McVean GA, Nickerson DA, Schmidt JP, Sherry ST, Wang J, Wilson RK, Gibbs RA, Boerwinkle E, Doddapaneni H, Han Y, Korchina V, Kovar C, Lee S, Muzny D, Reid JG, Zhu Y, Wang J, Chang Y, Feng Q, Fang X, Guo X, Jian M, Jiang H, Jin X, Lan T, Li G, Li J, Li Y, Liu S, Liu X, Lu Y, Ma X, Tang M, Wang B, Wang G, Wu H, Wu R, Xu X, Yin Y, Zhang D, Zhang W, Zhao J, Zhao M, Zheng X, Lander ES, Altshuler DM, Gabriel SB, Gupta N, Gharani N, Toji LH, Gerry NP, Resch AM, Flicek P, Barker J, Clarke L, Gil L, Hunt SE, Kelman G, Kulesha E, Leinonen R, McLaren WM, Radhakrishnan R, Roa A, Smirnov D, Smith RE, Streeter I, Thormann A, Toneva I, Vaughan B, Zheng-Bradley X,

Bentley DR, Grocock R, Humphray S, James T, Kingsbury Z, Lehrach H, Sudbrak R, Albrecht MW, Amstislavskiy VS, Borodina TA, Lienhard M, Mertes F, Sultan M, Timmermann B, Yaspo M-L, Mardis ER, Wilson RK, Fulton L, Fulton R, Sherry ST, Ananiev V, Belaia Z, Beloslyudtsev D, Bouk N, Chen C, Church D, Cohen R, Cook C, Garner J, Hefferon T, Kimelman M, Liu C, Lopez J, Meric P, O'Sullivan C, Ostapchuk Y, Phan L, Ponomarov S, Schneider V, Shekhtman E, Sirotkin K, Slotta D, Zhang H, McVean GA, Durbin RM, Balasubramaniam S, Burton J, Danecek P, Keane TM, Kolb-Kokocinski A, McCarthy S, Stalker J, Quail M, Schmidt JP, Davies CJ, Gollub J, Webster T, Wong B, Zhan Y, Auton A, Campbell CL, Kong Y, Marcketta A, Gibbs RA, Yu F, Antunes L, Bainbridge M, Muzny D, Sabo A, Huang Z, Wang J, Coin LJM, Fang L, Guo X, Jin X, Li G, Li Q, Li Y, Li Z, Lin H, Liu B, Luo R, Shao H, Xie Y, Ye C, Yu C, Zhang F, Zheng H, Zhu H, Alkan C, Dal E, Kahveci F, Marth GT, Garrison EP, Kural D, Lee W-P, Leong WF, Stromberg M, Ward AN, Wu J, Zhang M, Daly MJ, DePristo MA, Handsaker RE, Altshuler DM, Banks E, Bhatia G, del Angel G, Gabriel SB, Genovese G, Gupta N, Li H, Kashin S, Lander ES, McCarroll SA, Nemesh JC, Poplin RE, Yoon SC, Lihm J, Makarov V, Clark AG, Gottipati S, Keinan A, Rodriguez-Flores JL, Korbel JO, Rausch T, Fritz MH, Stütz AM, Flicek P, Beal K, Clarke L, Datta A, Herrero J, McLaren WM, Ritchie GRS, Smith RE, Zerbino D, Zheng-Bradley X, Sabeti PC, Shlyakhter I, Schaffner SF, Vitti J, Cooper DN, Ball EV, Stenson PD, Bentley DR, Barnes B, Bauer M, Cheetham RK, Cox A, Eberle M, Humphray S, Kahn S, Murray L, Peden J, Shaw R, Kenny EE, Batzer MA, Konkel MK, Walker JA, MacArthur DG, Lek M, Sudbrak R, Amstislavskiy VS, Herwig R, Mardis ER, Ding L, Koboldt DC, Larson D, Ye K, Gravel S, Swaroop A, Chew E, Lappalainen T, Erlich Y, Gymrek M, Willems TF, Simpson JT, Shriver MD, Rosenfeld JA, Bustamante CD, Montgomery SB, Vega FMDL, Byrnes JK, Carroll AW, DeGorter MK, Lacroute P, Maples BK, Martin AR, Moreno-Estrada-A, Shringarpure SS, Zakharia F, Halperin E, Baran Y, Lee C, Cerveira E, Hwang J, Malhotra A, Plewczynski D, Radew K, Romanovitch M, Zhang C, Hyland FCL, Craig DW, Christoforides A, Homer N, Izatt T, Kurdoglu AA, Sinari SA, Squire K, Sherry ST, Xiao C, Sebat J, Antaki D, Gujral M, Noor A, Ye K, Burchard EG, Hernandez RD, Gignoux CR, Haussler D, Katzman SJ, Kent WJ, Howie B, Ruiz-Linares A, Dermitzakis ET, Devine SE, Abecasis GR, Kang HM, Kidd JM, Blackwell T, Caron S, Chen W, Emery S, Fritsche L, Fuchsberger C, Jun G, Li B, Lyons R, Scheller C, Sidore C, Song S, Sliwerska E, Taliun D, Tan A, Welch R, Wing MK, Zhan X, Awadalla P, Hodgkinson A, Li Y, Shi X, Quitadamo A, Lunter G, McVean GA, Marchini JL, Myers S, Churchhouse C, Delaneau O, Gupta-Hinch A, Kretzschmar W, Iqbal Z, Mathieson I, Menelaou A, Rimmer A, Xifara DK, Oleksyk TK, Fu Y, Liu X, Xiong M, Jorde L, Witherspoon D, Xing J, Eichler EE, Browning BL, Browning SR, Hormozdiari F, Sudmant PH, Khurana E, Durbin RM, Hurles ME, Tyler-Smith C, Albers CA, Ayub Q, Balasubramaniam S, Chen Y, Colonna V, Danecek P, Jostins L, Keane TM, McCarthy S, Walter K, Xue Y, Gerstein MB, Abyzov A, Balasubramanian S, Chen J, Clarke D, Fu Y, Harmanci AO, Jin M, Lee D, Liu J, Mu XJ, Zhang J, Zhang Y, Li Y, Luo R, Zhu H, Alkan C, Dal E, Kahveci F, Marth GT, Garrison EP, Kural D, Lee W-P, Ward AN, Wu J, Zhang M, McCarroll SA, Handsaker RE, Altshuler DM, Banks E, del Angel G, Genovese G, Hartl C, Li H, Kashin S, Nemesh JC, Shakir K, Yoon SC, Lihm J, Makarov V, Degenhardt J, Korbel JO, Fritz MH, Meiers S, Raeder B, Rausch T, Stütz AM, Flicek P, Casale FP, Clarke L, Smith RE, Stegle O, Zheng-Bradley X, Bentley DR, Barnes B, Cheetham RK, Eberle M, Humphray S, Kahn S, Murray L, Shaw R, Lameijer E-W, Batzer MA, Konkel MK, Walker JA, Ding L, Hall I, Ye K, Lacroute P, Lee C, Cerveira E, Malhotra A, Hwang J, Plewczynski D, Radew K, Romanovitch M, Zhang C, Craig DW, Homer N, Church D, Xiao C, Sebat J, Antaki D, Bafna V, Michaelson J, Ye K, Devine SE, Gardner EJ, Abecasis GR, Kidd JM, Mills RE, Dayama G, Emery S, Jun G, Shi X, Quitadamo A, Lunter G, McVean GA, Chen K, Fan X, Chong Z, Chen T, Witherspoon D, Xing J, Eichler EE, Chaisson MJ, Hormozdiari F, Huddleston J, Malig M, Nelson BJ, Sudmant PH, Parrish NF, Khurana E, Hurles ME, Blackburne B, Lindsay SJ, Ning Z, Walter K, Zhang Y, Gerstein MB, Abyzov A, Chen J, Clarke D, Lam H, Mu XJ, Sisu C, Zhang J, Zhang Y, Gibbs RA, Yu F, Bainbridge M, Challis D, Evani US, Kovar C, Lu J, Muzny D, Nagaswamy U, Reid JG, Sabo A, Yu J, Guo X, Li W, Li Y, Wu R, Marth GT, Garrison EP, Leong WF, Ward AN, del Angel G, DePristo MA, Gabriel SB, Gupta N, Hartl C, Poplin RE, Clark AG, Rodriguez-Flores JL, Flicek P, Clarke L, Smith RE, Zheng-Bradley X, MacArthur DG, Mardis ER, Fulton R, Koboldt DC, Gravel S, Bustamante CD, Craig DW, Christoforides A, Homer N, Izatt T, Sherry ST, Xiao C, Dermitzakis ET, Abecasis GR, Kang HM, McVean GA, Gerstein MB, Balasubramanian S, Habegger L, Yu H, Flicek P, Clarke L, Cunningham F, Dunham I, Zerbino D, Zheng-Bradley-X, Lage K, Jespersen JB, Horn H, Montgomery SB, DeGorter MK, Khurana E, Tyler-Smith C, Chen Y, Colonna V, Xue Y, Gerstein MB, Balasubramanian S, Fu Y, Kim D, Auton A, Marcketta A, Desalle R, Narechania A, Sayres MAW, Garrison EP, Handsaker RE, Kashin S, McCarroll SA, Rodriguez-Flores JL, Flicek P, Clarke L, Zheng-Bradley X, Erlich Y, Gymrek M, Willems TF, Bustamante CD, Mendez FL, Poznik GD, Underhill PA, Lee C, Cerveira E, Malhotra A,

Romanovitch M, Zhang C, Abecasis GR, Coin L, Shao H, Mittelman D, Tyler-Smith C, Ayub Q, Banerjee R, Cerezo M, Chen Y, Fitzgerald TW, Louzada S, Massaia A, McCarthy S, Ritchie GR, Xue Y, Yang F, Gibbs RA, Kovar C, Kalra D, Hale W, Muzny D, Reid JG, Wang J, Dan X, Guo X, Li G, Li Y, Ye C, Zheng X, Altshuler DM, Flicek P, Clarke L, Zheng-Bradley X, Bentley DR, Cox A, Humphray S, Kahn S, Sudbrak R, Albrecht MW, Lienhard M, Larson D, Craig DW, Izatt T, Kurdoglu AA, Sherry ST, Xiao C, Haussler D, Abecasis GR, McVean GA, Durbin RM, Balasubramaniam S, Keane TM, McCarthy S, Stalker J, Chakravarti A, Knoppers BM, Abecasis GR, Barnes KC, Beiswanger C, Burchard EG, Bustamante CD, Cai H, Cao H, Durbin RM, Gerry NP, Gharani N, Gibbs RA, Gignoux CR, Gravel S, Henn B, Jones D, Jorde L, Kaye JS, Keinan A, Kent A, Kerasidou A, Li Y, Mathias R, McVean GA, Moreno-Estrada A, Ossorio PN, Parker M, Resch AM, Rotimi CN, Royal CD, Sandoval K, Su Y, Sudbrak R, Tian Z, Tishkoff S, Toji LH, Tyler-Smith C, Via M, Wang Y, Yang H, Yang L, Zhu J, Bodmer W, Bedoya G, Ruiz-Linares A, Cai Z, Gao Y, Chu J, Peltonen L, Garcia-Montero A, Orfao A, Dutil J, Martinez-Cruzado JC, Oleksyk TK, Barnes KC, Mathias RA, Hennis A, Watson H, McKenzie C, Qadri F, LaRocque R, Sabeti PC, Zhu J, Deng X, Sabeti PC, Asogun D, Folarin O, Happi C, Omoniwa O, Stremlau M, Tariyal R, Jallow M, Joof FS, Corrah T, Rockett K, Kwiatkowski D, Kooner J, Hiê`n TT, Dunstan SJ, Hang NT, Fonnie R, Garry R, Kanneh L, Moses L, Sabeti PC, Schieffelin J, Grant DS, Gallo C, Poletti G, Saleheen D, Rasheed A, Brooks LD, Felsenfeld AL, McEwen JE, Vaydylevich Y, Green ED, Duncanson A, Dunn M, Schloss JA, Wang J, Yang H, Auton A, Brooks LD, Durbin RM, Garrison EP, Kang HM, Korbel JO, Marchini JL, McCarthy S, McVean GA, Abecasis GR (2015) A global reference for human genetic variation. Nature 526(7571):68–74. https://doi.org/10.1038/nature15393

Botstein D, Cherry JM (1997) Molecular linguistics: extracting information from gene and protein sequences. Proc Natl Acad Sci 94(11):5506–5507. https://doi.org/10.1073/pnas.94.11.5506

Brenton JD, Aparicio SA, Caldas C (2000) Molecular profiling of breast cancer: portraits but not physiognomy. Breast Cancer Res 3(2):77. https://doi.org/10.1186/bcr274

Brown PO, Botstein D (1999) Exploring the new world of the genome with DNA microarrays. Nat Genet 21 (Suppl 1):33–37. https://doi.org/10.1038/4462

Budhkar A, Song Q, Su J, Zhang X (2025) Demystifying the black box: a survey on explainable artificial intelligence (XAI) in bioinformatics. Comput Struct Biotechnol J 27:346–359. https://doi.org/10.1016/j.csbj.2024.12.027

Callaway E (2025) DeepMind's new AlphaGenome AI tackles the 'dark matter' in our DNA. Nature 643(8070):17–18. https://doi.org/10.1038/d41586-025-01998-w

Cheng S, Melkonian M, Smith SA, Brockington S, Archibald JM, Delaux P-M, Li F-W, Melkonian B, Mavrodiev EV, Sun W, Fu Y, Yang H, Soltis DE, Graham SW, Soltis PS, Liu X, Xu X, Wong GK-S (2018) 10KP: a phylodiverse genome sequencing plan. Gigascience 7(3):giy013. https://doi.org/10.1093/gigascience/giy013

Choi H-K (2019) Translational genomics and multi-omics integrated approaches as a useful strategy for crop breeding. Genes Genom 41(2):133–146. https://doi.org/10.1007/s13258-018-0751-8

Consortium Gte (2013) The Genotype-Tissue Expression (GTEx) project. Nat Genet 45(6):580–585. https://doi.org/10.1038/ng.2653

Consortium TEP (2011) A user's guide to the encyclopedia of DNA elements (ENCODE). PLoS Biol 9(4): e1001046. https://doi.org/10.1371/journal.pbio.1001046

Doolittle WF (2013) Is junk DNA bunk? A critique of ENCODE. Proc Natl Acad Sci 110(14):5294–5300. https://doi.org/10.1073/pnas.1221376110

Dunham I, Kundaje A, Aldred SF, Collins PJ, Davis CA, Doyle F, Epstein CB, Frietze S, Harrow J, Kaul R, Khatun J, Lajoie BR, Landt SG, Lee B-K, Pauli F, Rosenbloom KR, Sabo P, Safi A, Sanyal A, Shoresh N, Simon JM, Song L, Trinklein ND, Altshuler RC, Birney E, Brown JB, Cheng C, Djebali S, Dong X, Dunham I, Ernst J, Furey TS, Gerstein M, Giardine B, Greven M, Hardison RC, Harris RS, Herrero J, Hoffman MM, Iyer S, Kellis M, Khatun J, Kheradpour P, Kundaje A, Lassmann T, Li Q, Lin X, Marinov GK, Merkel A, Mortazavi A, Parker SCJ, Reddy TE, Rozowsky J, Schlesinger F, Thurman RE, Wang J, Ward LD, Whitfield TW, Wilder SP, Wu W, Xi HS, Yip KY, Zhuang J, Bernstein BE, Birney E, Dunham I, Green ED, Gunter C, Snyder M, Pazin MJ, Lowdon RF, Dillon LAL, Adams LB, Kelly CJ, Zhang J, Wexler JR, Green ED, Good PJ, Feingold EA, Bernstein BE, Birney E, Crawford GE, Dekker J, Elnitski L, Farnham PJ, Gerstein M, Giddings MC, Gingeras TR, Green ED, Guigó R, Hardison RC, Hubbard TJ, Kellis M, Kent WJ, Lieb JD, Margulies EH, Myers RM, Snyder M, Stamatoyannopoulos JA, Tenenbaum SA, Weng Z, White KP, Wold B, Khatun J, Yu Y, Wrobel J, Risk BA, Gunawardena HP, Kuiper HC, Maier CW, Xie L, Chen X, Giddings MC, Bernstein BE, Epstein CB, Shoresh N, Ernst J, Kheradpour P, Mikkelsen TS, Gillespie S, Goren A, Ram O, Zhang X, Wang L, Issner R, Coyne MJ, Durham T, Ku M, Truong T, Ward LD, Altshuler RC, Eaton ML, Kellis M, Djebali S, Davis CA, Merkel A, Dobin A, Lassmann T, Mortazavi A, Tanzer A, Lagarde J, Lin W, Schlesinger F, Xue C, Marinov GK,

Khatun J, Williams BA, Zaleski C, Rozowsky J, Röder M, Kokocinski F, Abdelhamid RF, Alioto T, Antoshechkin I, Baer MT, Batut P, Bell I, Bell K, Chakrabortty S, Chen X, Chrast J, Curado J, Derrien T, Drenkow J, Dumais E, Dumais J, Duttagupta R, Fastuca M, Fejes-Toth K, Ferreira P, Foissac S, Fullwood MJ, Gao H, Gonzalez D, Gordon A, Gunawardena HP, Howald C, Jha S, Johnson R, Kapranov P, King B, Kingswood C, Li G, Luo OJ, Park E, Preall JB, Presaud K, Ribeca P, Risk BA, Robyr D, Ruan X, Sammeth M, Sandhu KS, Schaeffer L, See L-H, Shahab A, Skancke J, Suzuki AM, Takahashi H, Tilgner H, Trout D, Walters N, Wang H, Wrobel J, Yu Y, Hayashizaki Y, Harrow J, Gerstein M, Hubbard TJ, Reymond A, Antonarakis SE, Hannon GJ, Giddings MC, Ruan Y, Wold B, Carninci P, Guigó R, Gingeras TR, Rosenbloom KR, Sloan CA, Learned K, Malladi VS, Wong MC, Barber GP, Cline MS, Dreszer TR, Heitner SG, Karolchik D, Kent WJ, Kirkup VM, Meyer LR, Long JC, Maddren M, Raney BJ, Furey TS, Song L, Grasfeder LL, Giresi PG, Lee B-K, Battenhouse A, Sheffield NC, Simon JM, Showers KA, Safi A, London D, Bhinge AA, Shestak C, Schaner MR, Kim SK, Zhang ZZ, Mieczkowski PA, Mieczkowska JO, Liu Z, McDaniell RM, Ni Y, Rashid NU, Kim MJ, Adar S, Zhang Z, Wang T, Winter D, Keefe D, Birney E, Iyer VR, Lieb JD, Crawford GE, Li G, Sandhu KS, Zheng M, Wang P, Luo OJ, Shahab A, Fullwood MJ, Ruan X, Ruan Y, Myers RM, Pauli F, Williams BA, Gertz J, Marinov GK, Reddy TE, Vielmetter J, Partridge E, Trout D, Varley KE, Gasper C, Bansal A, Pepke S, Jain P, Amrhein H, Bowling KM, Anaya M, Cross MK, King B, Muratet MA, Antoshechkin I, Newberry KM, McCue K, Nesmith AS, Fisher-Aylor KI, Pusey B, DeSalvo G, Parker SL, Balasubramanian S, Davis NS, Meadows SK, Eggleston T, Gunter C, Newberry JS, Levy SE, Absher DM, Mortazavi A, Wong WH, Wold B, Blow MJ, Visel A, Pennachio LA, Elnitski L, Margulies EH, Parker SCJ, Petrykowska HM, Abyzov A, Aken B, Barrell D, Barson G, Berry A, Bignell A, Boychenko V, Bussotti G, Chrast J, Davidson C, Derrien T, Despacio-Reyes G, Diekhans M, Ezkurdia I, Frankish A, Gilbert J, Gonzalez JM, Griffiths E, Harte R, Hendrix DA, Howald C, Hunt T, Jungreis I, Kay M, Khurana E, Kokocinski F, Leng J, Lin MF, Loveland J, Lu Z, Manthravadi D, Mariotti M, Mudge J, Mukherjee G, Notredame C, Pei B, Rodriguez JM, Saunders G, Sboner A, Searle S, Sisu C, Snow C, Steward C, Tanzer A, Tapanari E, Tress ML, van Baren MJ, Walters N, Washietl S, Wilming L, Zadissa A, Zhang Z, Brent M, Haussler D, Kellis M, Valencia A, Gerstein M, Reymond A, Guigó R, Harrow J, Hubbard TJ, Landt SG, Frietze S, Abyzov A, Addleman N, Alexander RP, Auerbach RK, Balasubramanian S, Bettinger K, Bhardwaj N, Boyle AP, Cao AR, Cayting P, Charos A, Cheng Y, Cheng C, Eastman C, Euskirchen G, Fleming JD, Grubert F, Habegger L, Hariharan M, Harmanci A, Iyengar S, Jin VX, Karczewski KJ, Kasowski M, Lacroute P, Lam H, Lamarre-Vincent N, Leng J, Lian J, Lindahl-Allen M, Min R, Miotto B, Monahan H, Moqtaderi Z, Mu XJ, O'Geen H, Ouyang Z, Patacsil D, Pei B, Raha D, Ramirez L, Reed B, Rozowsky J, Sboner A, Shi M, Sisu C, Slifer T, Witt H, Wu L, Xu X, Yan K-K, Yang X, Yip KY, Zhang Z, Struhl K, Weissman SM, Gerstein M, Farnham PJ, Snyder M, Tenenbaum SA, Penalva LO, Doyle F, Karmakar S, Landt SG, Bhanvadia RR, Choudhury A, Domanus M, Ma L, Moran J, Patacsil D, Slifer T, Victorsen A, Yang X, Snyder M, White KP, Auer T, Centanin L, Eichenlaub M, Gruhl F, Heermann S, Hoeckendorf B, Inoue D, Kellner T, Kirchmaier S, Mueller C, Reinhardt R, Schertel L, Schneider S, Sinn R, Wittbrodt B, Wittbrodt J, Weng Z, Whitfield TW, Wang J, Collins PJ, Aldred SF, Trinklein ND, Partridge EC, Myers RM, Dekker J, Jain G, Lajoie BR, Sanyal A, Balasundaram G, Bates DL, Byron R, Canfield TK, Diegel MJ, Dunn D, Ebersol AK, Frum T, Garg K, Gist E, Hansen RS, Boatman L, Haugen E, Humbert R, Jain G, Johnson AK, Johnson EM, Kutyavin TV, Lajoie BR, Lee K, Lotakis D, Maurano MT, Neph SJ, Neri FV, Nguyen ED, Qu H, Reynolds AP, Roach V, Rynes E, Sabo P, Sanchez ME, Sandstrom RS, Sanyal A, Shafer AO, Stergachis AB, Thomas S, Thurman RE, Vernot B, Vierstra J, Vong S, Wang H, Weaver MA, Yan Y, Zhang M, Akey JM, Bender M, Dorschner MO, Groudine M, MacCoss MJ, Navas P, Stamatoyannopoulos G, Kaul R, Dekker J, Stamatoyannopoulos JA, Dunham I, Beal K, Brazma A, Flicek P, Herrero J, Johnson N, Keefe D, Lukk M, Luscombe NM, Sobral D, Vaquerizas JM, Wilder SP, Batzoglou S, Sidow A, Hussami N, Kyriazopoulou-Panagiotopoulou S, Libbrecht MW, Schaub MA, Kundaje A, Hardison RC, Miller W, Giardine B, Harris RS, Wu W, Bickel PJ, Banfai B, Boley NP, Brown JB, Huang H, Li Q, Li JJ, Noble WS, Bilmes JA, Buske OJ, Hoffman MM, Sahu AD, Kharchenko PV, Park PJ, Baker D, Taylor J, Weng Z, Iyer S, Dong X, Greven M, Lin X, Wang J, Xi HS, Zhuang J, Gerstein M, Alexander RP, Balasubramanian S, Cheng C, Harmanci A, Lochovsky L, Min R, Mu XJ, Rozowsky J, Yan K-K, Yip KY, Birney E (2012) An integrated encyclopedia of DNA elements in the human genome. Nature 489(7414):57–74. https://doi.org/10.1038/nature11247

Eddy SR (2013) The ENCODE project: missteps overshadowing a success. Curr Biol 23(7):R259–R261. https://doi.org/10.1016/j.cub.2013.03.023

Eisen JA (2012) Badomics words and the power and peril of the ome-meme. GigaScience 1(1):6. https://doi.org/10.1186/2047-217x-1-6

Fields S, Johnston M (2002) A Crisis in Postgenomic Nomenclature. Science 296(5568):671–672. https://doi.org/10.1126/science.1070208

Friedberg EC (2008) Sydney Brenner. Nat Rev Mol Cell Biol 9(1):8–9. https://doi.org/10.1038/nrm2320

Gama-Carvalho M, Sousa L (2026) Microarrays and the dawn of omics data analysis. In: Omics approaches in biomedicine and biotechnology, advances in experimental medicine and biology 1504. https://doi.org/10.1007/978-3-032-18966-0_2

Gilbert JA, Jansson JK, Knight R (2014) The earth microbiome project: successes and aspirations. BMC Biol 12(1):69. https://doi.org/10.1186/s12915-014-0069-1

Giuffra E, Tuggle CK, Consortium TF (2018) Functional Annotation of Animal Genomes (FAANG): current achievements and roadmap. Annu Rev Anim Biosci 7(1):1–24. https://doi.org/10.1146/annurev-animal-020518-114913

Guo W, Velleman SG, Strasburg GM (2025) From discovery to application: merging modern omics with traditional hypothesis-driven approaches in muscle myopathy studies. Front Physiol 15:1520196. https://doi.org/10.3389/fphys.2024.1520196

Holt KE, Inouye M (2025) Genomic data sharing: you don't know what you've got (till it's gone). Nat Rev Genet 26(5):293–294. https://doi.org/10.1038/s41576-025-00820-7

Hu H, Yu F (2024) Embracing the omics era for plant breeding. Crop Breed Genet Genom 7(2). https://doi.org/10.20900/cbgg20250002

Humphery-Smith I (2015) The 20th anniversary of proteomics and some of its origins. Proteomics 15(11):1773–1776. https://doi.org/10.1002/pmic.201400582

Hunter CL, Helft PR (2023) Yes, we can, but should we? Ethical considerations in reporting germline findings from paired tumor-normal genomic testing in patients with advanced cancer. J Clin Oncol 41(11):1982–1985. https://doi.org/10.1200/jco.22.00796

Jose S, Geller G, Bollinger J, Mathews D, Kahn J, Garibaldi BT (2024) The ethics of using COVID-19 host genomic information for clinical and public health decision-making: a survey of US health professionals. Hum Genet Genomics Adv 5(1):100255. https://doi.org/10.1016/j.xhgg.2023.100255

Jung V, Vincent-Cuaz C, Tumescheit C, Fournier L, Darsinou M, Xu ZM, Saadat A, Wang Y, Tsantoulis P, Michielin O, Fellay J, Patani R, Ramos A, Frossard P, Hastings J, Riccio A, van der Plas L, Luisier R (2025) Decoding the interactions and functions of non-coding RNA with artificial intelligence. Nat Rev Mol Cell Biol 26(10):797–818. https://doi.org/10.1038/s41580-025-00857-w

Knyazev S, Chhugani K, Sarwal V, Ayyala R, Singh H, Karthikeyan S, Deshpande D, Baykal PI, Comarova Z, Lu A, Porozov Y, Vasylyeva TI, Wertheim JO, Tierney BT, Chiu CY, Sun R, Wu A, Abedalthagafi MS, Pak VM, Nagaraj SH, Smith AL, Skums P, Pasaniuc B, Komissarov A, Mason CE, Bortz E, Lemey P, Kondrashov F, Beerenwinkel N, Lam TT-Y, Wu NC, Zelikovsky A, Knight R, Crandall KA, Mangul S (2022) Unlocking capacities of genomics for the COVID-19 response and future pandemics. Nat Methods 19(4):374–380. https://doi.org/10.1038/s41592-022-01444-z

Lederberg J, McCray AT (2001) Ome Sweet Omics-- A Genealogical Treasury of Words. Scientist 7(15):8

Leebens-Mack JH, Barker MS, Carpenter EJ, Deyholos MK, Gitzendanner MA, Graham SW, Grosse I, Li Z, Melkonian M, Mirarab S, Porsch M, Quint M, Rensing SA, Soltis DE, Soltis PS, Stevenson DW, Ullrich KK, Wickett NJ, DeGironimo L, Edger PP, Jordon-Thaden IE, Joya S, Liu T, Melkonian B, Miles NW, Pokorny L, Quigley C, Thomas P, Villarreal JC, Augustin MM, Barrett MD, Baucom RS, Beerling DJ, Benstein RM, Biffin E, Brockington SF, Burge DO, Burris JN, Burris KP, Burtet-Sarramegna V, Caicedo AL, Cannon SB, Çebi Z, Chang Y, Chater C, Cheeseman JM, Chen T, Clarke ND, Clayton H, Covshoff S, Crandall-Stotler BJ, Cross H, dePamphilis CW, Der JP, Determann R, Dickson RC, Stilio VSD, Ellis S, Fast E, Feja N, Field KJ, Filatov DA, Finnegan PM, Floyd SK, Fogliani B, García N, Gâteblé G, Godden GT, Goh F, Qi Y, Greiner S, Harkess A, Heaney JM, Helliwell KE, Heyduk K, Hibberd JM, Hodel RGJ, Hollingsworth PM, Johnson MTJ, Jost R, Joyce B, Kapralov MV, Kazamia E, Kellogg EA, Koch MA, Konrat MV, Könyves K, Kutchan TM, Lam V, Larsson A, Leitch AR, Lentz R, Li F-W, Lowe AJ, Ludwig M, Manos PS, Mavrodiev E, McCormick MK, McKain M, McLellan T, McNeal JR, Miller RE, Nelson MN, Peng Y, Ralph P, Real D, Riggins CW, Ruhsam M, Sage RF, Sakai AK, Scascitella M, Schilling EE, Schlösser E-M, Sederoff H, Servick S, Sessa EB, Shaw AJ, Shaw SW, Sigel EM, Skema C, Smith AG, Smithson A, Stewart CN, Stinchcombe JR, Szövényi P, Tate JA, Tiebel H, Trapnell D, Villegente M, Wang C-N, Weller SG, Wenzel M, Weststrand S, Westwood JH, Whigham DF, Wu S, Wulff AS, Yang Y, Zhu D, Zhuang C, Zuidof J, Chase MW, Pires JC, Rothfels CJ, Yu J, Chen C, Chen L, Cheng S, Li J, Li R, Li X, Lu H, Ou Y, Sun X, Tan X, Tang J, Tian Z, Wang F, Wang J, Wei X, Xu X, Yan Z, Yang F, Zhong X, Zhou F, Zhu Y, Zhang Y, Ayyampalayam S, Barkman TJ, Nguyen N, Matasci N, Nelson DR, Sayyari E, Wafula EK, Walls RL, Warnow T, An H, Arrigo N, Baniaga AE, Galuska S, Jorgensen SA, Kidder TI, Kong H, Lu-Irving P, Marx HE, Qi X, Reardon CR, Sutherland BL, Tiley GP, Welles SR, Yu R, Zhan S, Gramzow L, Theißen G, Wong GK-S (2019) One thousand plant transcriptomes and the phylogenomics of green plants. Nature 574(7780):679–685. https://doi.org/10.1038/s41586-019-1693-2

Legrain P, Aebersold R, Archakov A, Bairoch A, Bala K, Beretta L, Bergeron J, Borchers CH, Corthals GL, Costello CE, Deutsch EW, Domon B, Hancock W, He F, Hochstrasser D, Marko-Varga G, Salekdeh GH, Sechi S, Snyder M, Srivastava S, Uhlén M, Wu CH, Yamamoto T, Paik Y-K, Omenn GS (2011) The human proteome project: current state and future direction. Mol Cell Proteomics 10(7):M111.009993. https://doi.org/10.1074/mcp.m111.009993

Link AJ, Eng J, Schieltz DM, Carmack E, Mize GJ, Morris DR, Garvik BM, Yates JR (1999) Direct analysis of protein complexes using mass spectrometry. Nat Biotechnol 17(7):676–682. https://doi.org/10.1038/10890

Makke N, Chawla S (2024) Interpretable scientific discovery with symbolic regression: a review. Artif Intell Rev 57(1):2. https://doi.org/10.1007/s10462-023-10622-0

Mazzocchi F (2015) Could Big Data be the end of theory in science? EMBO Rep 16(10):1250–1255. https://doi.org/10.15252/embr.201541001

McDonald D, Clemente JC, Kuczynski J, Rideout JR, Stombaugh J, Wendel D, Wilke A, Huse S, Hufnagle J, Meyer F, Knight R, Caporaso JG (2012) The Biological Observation Matrix (BIOM) format or: how I learned to stop worrying and love the ome-ome. GigaScience 1(1):7. https://doi.org/10.1186/2047-217x-1-7

McKusick VA (1997) Genomics: structural and functional studies of genomes. Genomics 45(2):244–249. https://doi.org/10.1006/geno.1997.5020

McKusick VA, Ruddle FH (1987) A new discipline, a new name, a new journal. Genomics 1(1):1–2. https://doi.org/10.1016/0888-7543(87)90098-x

Network CGAR, Weinstein JN, Collisson EA, Mills GB, Shaw KRM, Ozenberger BA, Ellrott K, Shmulevich I, Sander C, Stuart JM (2013) The cancer genome atlas pan-cancer analysis project. Nat Genet 45(10): 1113–1120. https://doi.org/10.1038/ng.2764

Pennisi E (2012) ENCODE project writes eulogy for junk DNA. Science 337(6099):1159–1161. https://doi.org/10.1126/science.337.6099.1159

Petsko GA (2002) No place like Ome. Genome Biol 3(7): comment1010.1. https://doi.org/10.1186/gb-2002-3-7-comment1010

Pigliucci M (2009) The end of theory in science? EMBO Rep 10(6):534 534. https://doi.org/10.1038/embor.2009.111

Powell K (2021) The broken promise that undermines human genome research. Nature 590(7845):198–201. https://doi.org/10.1038/d41586-021-00331-5

Regev A, Teichmann SA, Lander ES, Amit I, Benoist C, Birney E, Bodenmiller B, Campbell P, Carninci P, Clatworthy M, Clevers H, Deplancke B, Dunham I, Eberwine J, Eils R, Enard W, Farmer A, Fugger L, Göttgens B, Hacohen N, Haniffa M, Hemberg M, Kim S, Klenerman P, Kriegstein A, Lein E, Linnarsson S, Lundberg E, Lundeberg J, Majumder P, Marioni JC, Merad M, Mhlanga M, Nawijn M, Netea M, Nolan G, Pe'er D, Phillipakis A, Ponting CP, Quake S, Reik W, Rozenblatt-Rosen O, Sanes J, Satija R, Schumacher TN, Shalek A, Shapiro E, Sharma P, Shin JW, Stegle O, Stratton M, Stubbington MJT, Theis FJ, Uhlen M, van Oudenaarden A, Wagner A, Watt F, Weissman J, Wold B, Xavier R, Yosef N, Participants HCAM (2017) The human cell atlas. Elife 6:e27041. https://doi.org/10.7554/elife.27041

Rhie A, McCarthy SA, Fedrigo O, Damas J, Formenti G, Koren S, Uliano-Silva M, Chow W, Fungtammasan A, Kim J, Lee C, Ko BJ, Chaisson M, Gedman GL, Cantin LJ, Thibaud-Nissen F, Haggerty L, Bista I, Smith M, Haase B, Mountcastle J, Winkler S, Paez S, Howard J, Vernes SC, Lama TM, Grutzner F, Warren WC, Balakrishnan CN, Burt D, George JM, Biegler MT, Iorns D, Digby A, Eason D, Robertson B, Edwards T, Wilkinson M, Turner G, Meyer A, Kautt AF, Franchini P, Detrich HW, Svardal H, Wagner M, Naylor GJP, Pippel M, Malinsky M, Mooney M, Simbirsky M, Hannigan BT, Pesout T, Houck M, Misuraca A, Kingan SB, Hall R, Kronenberg Z, Sović I, Dunn C, Ning Z, Hastie A, Lee J, Selvaraj S, Green RE, Putnam NH, Gut I, Ghurye J, Garrison E, Sims Y, Collins J, Pelan S, Torrance J, Tracey A, Wood J, Dagnew RE, Guan D, London SE, Clayton DF, Mello CV, Friedrich SR, Lovell PV, Osipova E, Al-Ajli FO, Secomandi S, Kim H, Theofanopoulou C, Hiller M, Zhou Y, Harris RS, Makova KD, Medvedev P, Hoffman J, Masterson P, Clark K, Martin F, Howe K, Flicek P, Walenz BP, Kwak W, Clawson H, Diekhans M, Nassar L, Paten B, Kraus RHS, Crawford AJ, Gilbert MTP, Zhang G, Venkatesh B, Murphy RW, Koepfli K-P, Shapiro B, Johnson WE, Palma FD, Marques-Bonet T, Teeling EC, Warnow T, Graves JM, Ryder OA, Haussler D, O'Brien SJ, Korlach J, Lewin HA, Howe K, Myers EW, Durbin R, Phillippy AM, Jarvis ED (2021) Towards complete and error-free genome assemblies of all vertebrate species. Nature 592(7856):737–746. https://doi.org/10.1038/s41586-021-03451-0

Rowley J (2007) The wisdom hierarchy: representations of the DIKW hierarchy. J Inf Sci 33(2):163–180. https://doi.org/10.1177/0165551506070706

Stoeklé H-C, Mamzer-Bruneel M-F, Frouart C-H, Tourneau CL, Laurent-Puig P, Vogt G, Hervé C (2018) Molecular tumor boards: ethical issues in the new era of data medicine. Sci Eng Ethics 24(1): 307–322. https://doi.org/10.1007/s11948-017-9880-8

Stunnenberg HG, Consortium TIHE, Abrignani S, Adams D, de Almeida M, Altucci L, Amin V, Amit I, Antonarakis SE, Aparicio S, Arima T, Arrigoni L, Arts R, Asnafi V, Esteller M, Bae J-B, Bassler K, Beck S, Berkman B, Bernstein BE, Bilenky M, Bird A, Bock C, Boehm B, Bourque G, Breeze CE, Brors B, Bujold D, Burren O, Bussemakers MJ, Butterworth A, Campo E, Carrillo-de-Santa-Pau E, Chadwick L, Chan KM, Chen W, Cheung TH, Chiapperino L, Choi NH, Chung H-R, Clarke L, Connors JM, Cronet P, Danesh J, Dermitzakis M, Drewes G, Durek P, Dyke S, Dylag T, Eaves CJ, Ebert P, Eils R, Eils J, Ennis CA, Enver T, Feingold EA, Felder B, Ferguson-Smith A, Fitzgibbon J, Flicek P, Foo RS-Y, Fraser P, Frontini M, Furlong E, Gakkhar S, Gasparoni N, Gasparoni G, Geschwind DH, Glažar P, Graf T, Grosveld F, Guan X-Y,

Guigo R, Gut IG, Hamann A, Han B-G, Harris RA, Heath S, Helin K, Hengstler JG, Heravi-Moussavi A, Herrup K, Hill S, Hilton JA, Hitz BC, Horsthemke B, Hu M, Hwang J-Y, Ip NY, Ito T, Javierre B-M, Jenko S, Jenuwein T, Joly Y, Jones SJM, Kanai Y, Kang HG, Karsan A, Kiemer AK, Kim SC, Kim B-J, Kim H-H, Kimura H, Kinkley S, Klironomos F, Koh I-U, Kostadima M, Kressler C, Kreuzhuber R, Kundaje A, Küppers R, Larabell C, Lasko P, Lathrop M, Lee DHS, Lee S, Lehrach H, Leitão E, Lengauer T, Lernmark Å, Leslie RD, Leung GKK, Leung D, Loeffler M, Ma Y, Mai A, Manke T, Marcotte ER, Marra MA, Martens JHA, Martin-Subero JI, Maschke K, Merten C, Milosavljevic A, Minucci S, Mitsuyama T, Moore RA, Müller F, Mungall AJ, Netea MG, Nordström K, Norstedt I, Okae H, Onuchic V, Ouellette F, Ouwehand W, Pagani M, Pancaldi V, Pap T, Pastinen T, Patel R, Paul DS, Pazin MJ, Pelicci PG, Phillips AG, Polansky J, Porse B, Pospisilik JA, Prabhakar S, Procaccini DC, Radbruch A, Rajewsky N, Rakyan V, Reik W, Ren B, Richardson D, Richter A, Rico D, Roberts DJ, Rosenstiel P, Rothstein M, Salhab A, Sasaki H, Satterlee JS, Sauer S, Schacht C, Schmidt F, Schmitz G, Schreiber S, Schröder C, Schübeler D, Schultze JL, Schulyer RP, Schulz M, Seifert M, Shirahige K, Siebert R, Sierocinski T, Siminoff L, Sinha A, Soranzo N, Spicuglia S, Spivakov M, Steidl C, Strattan JS, Stratton M, Südbeck P, Sun H, Suzuki N, Suzuki Y, Tanay A, Torrents D, Tyson FL, Ulas T, Ullrich S, Ushijima T, Valencia A, Vellenga E, Vingron M, Wallace C, Wallner S, Walter J, Wang H, Weber S, Weiler N, Weller A, Weng A, Wilder S, Wiseman SM, Wu AR, Wu Z, Xiong J, Yamashita Y, Yang X, Yap DY, Yip KY, Yip S, Yoo J-I, Zerbino D, Zipprich G, Hirst M (2016) The international human epigenome consortium: a blueprint for scientific collaboration and discovery. Cell 167(5):1145–1149. https://doi.org/10.1016/j.cell.2016.11.007

Sunagawa S, Acinas SG, Bork P, Bowler C, Acinas SG, Babin M, Bork P, Boss E, Bowler C, Cochrane G, de Vargas C, Follows M, Gorsky G, Grimsley N, Guidi L, Hingamp P, Iudicone D, Jaillon O, Kandels S, Karp-Boss L, Karsenti E, Lescot M, Not F, Ogata H, Pesant S, Poulton N, Raes J, Sardet C, Sieracki M, Speich S, Stemmann L, Sullivan MB, Sunagawa S, Wincker P, Eveillard D, Gorsky G, Guidi L, Iudicone D, Karsenti E, Lombard F, Ogata H, Pesant S, Sullivan MB, Wincker P, de Vargas C (2020) Tara Oceans: towards global ocean ecosystems biology. Nat Rev Microbiol 18(8):428–445. https://doi.org/10.1038/s41579-020-0364-5

Torson AS, Dong Y, Sinclair BJ (2020) Help, there are 'omics' in my comparative physiology! J Exp Biol 223(24). https://doi.org/10.1242/jeb.191262

Turnbaugh PJ, Ley RE, Hamady M, Fraser-Liggett CM, Knight R, Gordon JI (2007) The human microbiome project. Nature 449(7164):804–810. https://doi.org/10.1038/nature06244

Vanneschi L, Silva S (2023) Lectures on intelligent systems. Nat Comput Ser;205–257. https://doi.org/10.1007/978-3-031-17922-8_8

Velasquez A, Bhatt N, Topcu U, Wang Z, Sycara K, Stepputtis S, Neema S, Vallabha G (2025) Neurosymbolic AI as an antithesis to scaling laws. PNAS Nexus 4(5):pgaf117. https://doi.org/10.1093/pnasnexus/pgaf117

Wasinger VC, Cordwell SJ, Cerpa-Poljak A, Yan JX, Gooley AA, Wilkins MR, Duncan MW, Harris R, Williams KL, Humphery-Smith I (1995) Progress with gene-product mapping of the Mollicutes: mycoplasma genitalium. Electrophoresis 16(1):1090–1094. https://doi.org/10.1002/elps.11501601185

Weinstein JN (1998) Fishing expeditions. Science 282(5389):627–627. https://doi.org/10.1126/science.282.5389.627g

White MA, Myers CA, Corbo JC, Cohen BA (2013) Massively parallel in vivo enhancer assay reveals that highly local features determine the cis-regulatory function of ChIP-seq peaks. Proc Natl Acad Sci 110(29):11952–11957. https://doi.org/10.1073/pnas.1307449110

Winkler H (1920) Verbreitung und Ursache der Parthenogenesis im Pflanzen- und Tierreiche. Verlag Fisher, Jena

Part I

Capturing the Molecular Universe

Microarrays and the Dawn of Omics Data Analysis

2

Margarida Gama-Carvalho and Lisete Sousa

Abstract

Microarrays have played a pivotal role in the early stages of omics data generation and analysis. For over 15 years, microarray technology dominated the field of transcriptomics, providing high-throughput platforms to measure gene expression levels across thousands of genes simultaneously. Microarrays enabled a wide range of genomic studies, leading to a vast repository of data in public databases such as the Gene Expression Omnibus (GEO). However, with the advent of next-generation sequencing (NGS), the use of microarrays has steadily declined. Despite this, the legacy of microarrays persists, particularly through the wealth of data generated over the years, which continues to fuel genomic research today. This chapter reviews the origins of microarrays and their impact on omics research and discusses the basic principles behind the technology, how it has evolved over the years, and how it lay the groundwork for the field as we know it today.

Keywords

Microarrays · DNA chips · Oligonucleotide arrays · Transcriptomics · Genomics · Data analysis

2.1 Historical Overview: The 15-Year Reign of Microarrays

Microarray technology was the first widely available high-throughput approach for characterizing genome structure and function. At a time when Northern blots, RT-PCR and other low-throughput, gene-by-gene assays dominated, microarrays offered a radically new means of scaling up investigations, enabling thousands of genes to be monitored in parallel. The technology emerged in the mid-1990s to meet a pressing need at a moment when enabling tools (e.g., cDNA banks, oligonucleotide synthesis, robotic spotting and fluorescence detection) had matured. The transition was driven by a convergence of evolving conceptual frameworks (notably the increasing availability of genome sequences), advances in surface chemistry and instrumentation, and growing demand for genome-wide functional insight. Reviewing this history is more than retrospection: it is pivotal for understanding how

M. Gama-Carvalho (✉)
Faculdade de Ciências, Universidade de Lisboa, Lisbon, Portugal

BioISI—Biosystems and Integrative Sciences Institute, Faculty of Sciences, Universidade de Lisboa, Lisbon, Portugal
e-mail: mhcarvalho@fc.ul.pt

L. Sousa
Faculdade de Ciências, Universidade de Lisboa, Lisbon, Portugal

CEAUL—Centro de Estatística e Aplicações, Universidade de Lisboa, Lisbon, Portugal

M. Gama-Carvalho et al. (eds.), *Omics Approaches in Biomedicine and Biotechnology*, Advances in Experimental Medicine and Biology 1504, https://doi.org/10.1007/978-3-032-18966-0_2

new approaches to scientific inquiry, analytic pipelines and data-sharing norms emerged and for recognizing how the microarray era laid the groundwork for the genomics era that followed. The widespread adoption of the technology attests to its added value, and its long period of dominance reflects the absence of comparable alternatives—until the advent of low-cost next-generation sequencing, which built on many of the concepts introduced by microarrays.

2.1.1 Characterizing Genomes and Gene Expression Before Microarrays

The characterization of genomic sequences, transcripts, and their expression patterns has long been a central focus in efforts to understand genome, organism and cell function. The biochemical properties of nucleic acids—particularly their ability to form relatively stable hybrids with complementary sequences, including DNA-DNA, DNA-RNA and RNA-RNA duplexes—enabled the development of early detection and analysis methods based on target-probe hybridization. The introduction of the Southern blot method by Edwin Southern in the 1970s for detecting specific DNA sequences (Southern 1975), followed by the adaptation of this approach into the Northern blot for RNA analysis (Alwine et al. 1977), laid the foundation for detecting immobilized nucleic acid sequences using labelled probes. These techniques involve the initial separation of biological samples by gel electrophoresis, followed by their transfer to a membrane in a process known as blotting. However, while effective for analysing individual genes, these methods were inherently limited in throughput and not well suited for the simultaneous analysis of multiple targets or large sample sets. Given the need to efficiently estimate the relative abundance of several transcripts across many samples, a more efficient option was developed involving the spotting of unlabelled cDNA probes in a grid pattern on a nitrocellulose membrane, which could then be incubated with a radioactively labelled sample of total RNA (Kafatos et al. 1979). The dot hybridization procedure, or "dot blot", began to be widely used in the 1980s, introducing the concept of "arrayed cDNA probes" to parallelize the quantification of dozens of transcripts of interest in a sample.

In the late 1980s, the Human Genome Project (HGP) began to take shape, driving the development of high-density tools for the systematic mapping and screening of genomic libraries (Council 1988). Membrane-based nucleic acid arrays gained prominence for identifying and organizing large-insert clones. These mapped clones would subsequently be subcloned into plasmid vectors for sequencing as part of the second stage of the project. The growing demand for throughput accelerated automation of the spotting process, enabling the production of 22×22 cm membranes carrying up to 10,000 distinct DNA sequences (Nizetić et al. 1991), thereby substantially increasing the scalability of array-based analyses.

2.1.2 Conceptual Advances: From Sequencing-by-Hybridization to Expressed Sequence Tags

Recognizing that existing sequencing methods—largely reliant on electrophoresis—would face significant limitations in scalability and automation, several groups began independently exploring the use of high-density arrays as a more efficient platform for nucleic acid analysis. This led to the development of the sequencing-by-hybridization (SBH) concept, which aimed to leverage array-based approaches not only for genomic "fingerprinting" but also for de novo sequence determination. At the core of this concept was the use of oligonucleotide sets representing all possible sequence combinations of a given length (Dramanac et al. 1989; Khrapko et al. 1989). Two main strategies emerged: the Type I approach took advantage of robotic spotting of nucleic acids onto membranes to map the sequence of thousands of DNA clones by sequential hybridization to oligonucleotides; and the Type II approach proposed the generation of large arrays of oligonucleotides on a solid support, which could then be hybridized with labelled target sequences. This strategy introduced the concept of "DNA chips", anticipating the widespread application of oligonucleotide arrays in genomic analysis (Southern 1996).

While efforts to develop alternative, cost-effective sequencing strategies progressed, parallel advances were enabling the use of arrayed templates for high-throughput gene expression analysis. Although the primary goal of the HGP was to produce a complete map of genomic sequences, as early as 1991 an effort to characterize transcript diversity through partial cDNA sequencing, or expressed sequence tags (ESTs), led to an explosive growth of the available information on gene products (Adams et al. 1991). Within 4 years, the number of cDNA sequences in the GenBank database increased from approximately 10,000 to over 250,000 (Boguski et al. 1993; Boguski 1995). These developments poised the field for the large-scale interrogation of gene expression profiles.

2.1.3 Enabling Technologies

The rapid emergence of high-density microarrays was possible once several key technological components were established. In particular, two critical barriers of earlier approaches had to be surmounted: first, the fabrication of high-density probe arrays on solid surfaces, and second, the development of sensitive, non-radioactive detection systems for reliable monitoring of probe–target hybridization. Once these technical elements matured, multiplexed measurement of gene expression, genomic variation and other molecular events became feasible at scale, setting the stage for the widespread adoption of the technology.

2.1.3.1 Advances in Surface Chemistry and High-Density Probe Synthesis

One major advancement was the shift from porous to solid supports, like chemically modified glass and silicon, which allow nucleic acid coupling, making them reliable substrates. In parallel, adaptations of solid-phase chemistry enabled the in situ synthesis of oligonucleotides directly on these surfaces.

Several independent groups developed distinct approaches to achieve this goal. Edwin Southern's laboratory at Oxford explored the confinement of reagents to defined surface regions using physical barriers such as channels or wells, facilitating the combinatorial synthesis of oligonucleotides at specific array locations (Southern et al. 1992, 1994; Maskos and Southern 1993). Building on the integration of photolabile protecting groups and photolithography (Fodor et al. 1991)—originally developed for semiconductor manufacturing—Fodor and colleagues introduced a light-directed method for oligonucleotide synthesis on solid supports, enabling extreme miniaturization with densities reaching up to 10^5 synthesis sites per cm^2 (Fodor et al. 1993; Pease et al. 1994). A few years later, Leroy Hood's group reported the use of inkjet printing technology to deposit small reagent volumes onto chemically modified silicon dioxide, allowing localized synthesis of oligonucleotides (Blanchard et al. 1996). Due to their technical sophistication and reliance on specialized equipment, these methods were not readily adaptable in standard academic laboratories and instead laid the foundation for commercial microarray platforms, subsequently patented and brought to market by various biotechnology companies.

As an alternative, methods for spotting pre-synthesized probes on glass slides were also developed, with robotic printing supporting a density of ~20,000 targets per array (Guo et al. 1994; Schena et al. 1995). The spotting methods developed by the Smith and Brown labs were to be successfully adapted by many laboratories for local fabrication of microarrays, further facilitating the dissemination of the technology across research settings. These two groups also pioneered another key development in the implementation of microarray technologies, in parallel with work by Fodor and colleagues at the company Affymetrix: the switch from radioactive to fluorescence detection methods.

2.1.3.2 Fluorescence Detection Methods Driving Early Applications of Microarrays

The translation of microarray platforms into high-throughput analytical tools required sensitive detection strategies that could accurately capture and quantify hybridization events across thousands of arrayed probes. The generation of fluorescently labelled samples, either by PCR of genomic DNA (Guo et al. 1994) or reverse transcription of mRNA (Schena et al. 1995), coupled to the use of fluorescent scanners, was shown to allow specific and quantitative detection of target sequences. Guo and colleagues used glass support-bound arrays of allele-specific oligonucleotides coupled to fluorescence detection to analyse genetic variation. By varying the sequence and length of spotted oligonucleotides, they showed that 15-mers with GC contents between 33% and 46% were able to achieve discrimination of single-nucleotide polymorphisms (SNPs) (Guo et al. 1994). In contrast, the Brown Lab at Stanford focused on the production of spotted microarrays on glass slides using PCR-amplified cDNA sequences of ~1 kb in length. In their 1995 study (Schena et al. 1995), they used reverse transcription to generate fluorescently labelled cDNA from *Arabidopsis* mRNA samples and hybridized them to the microarrays, generating a quantitative expression profile for a set of 45 target genes. Notably, they introduced the concept of dual-colour array hybridization for differential gene expression analysis. This was achieved by labelling *Arabidopsis* mRNA samples from root and leaves with distinct fluorophores. By adding a "spike-in" human transcript to the samples and including its complementary cDNA probe in the array, they were able to normalize fluorescence signal intensities and generate relative estimates of gene expression levels.

Although originally conceptualized as a cost-effective alternative to sequencing, the value of microarray platforms for multiple applications was soon realized. By 1996, high-density oligonucleotide arrays with ~10^5 probes were used to map the human mitochondrial genome (Chee et al. 1996) and to detect heterozygous mutations in the BRCA1 gene (Hacia et al. 1996). Spotted cDNA microarrays were applied to monitor the expression of ~1000 human genes using two-colour hybridization assays, identifying stress-responsive genes in T cells (Schena et al. 1996) and genes with altered expression in cancer (DeRisi et al. 1996). This platform was also used to analyse yeast strain karyotypes by generating microarrays with physically mapped genomic clones covering 75% of the yeast genome (Shalon et al. 1996). This study established the use of microarrays for comparative genomic hybridization (CGH), originally developed to detect chromosomal copy number changes on metaphase spreads (Kallioniemi et al. 1992) (see Sect. 3.2.1 for more details).

These foundational studies validated microarrays for genome-wide measurements of gene expression and sequence variation and demonstrated their versatility across diverse biological applications. With the continued growth of information from genome sequencing projects, they set the stage for the broader adoption and refinement of microarray-based assays, as discussed in the following sections.

2.1.4 The Rise of Microarrays

By 1999 the research community was increasingly aware of the promise of microarrays, though access to the technology remained limited to a small number of institutions. In fact, at the time most publications using microarrays originated from Affymetrix, Stanford, or NIH-affiliated groups (Lenoir and Giannella 2006). Reflecting this growing interest, Nature Genetics published a supplement on January 1, 1999, titled "The Chipping Forecast", which aimed to provide a comprehensive overview of available microarray options and their potential applications while also acknowledging the limitations and uncertainties surrounding the technology considering that, as noted in the editorial "Going global" by Bette Phimister, "the limitations and the ultimate utility of the microarray remain(ed) to be determined" (Phimister 1999).

Just 1 year earlier, Affymetrix had launched its GeneChip® technology, marking a key milestone in the commercialization of high-density oligonucleotide arrays and providing access to the technology at increasingly accessible prices. Two types of high-density oligonucleotide arrays were initially developed for distinct applications: gene expression profiling and genotyping (Lipshutz et al. 1999). Affymetrix implemented a single-colour hybridization design based on pairs of perfect-match (PM) and mismatch (MM) probes to improve specificity, background subtraction, robust normalization, and comparison between independent hybridization assays. The analysis of such arrays relied on Affymetrix-developed software for signal preprocessing and background correction (see Sect. 2.2.1 for further discussion of in situ oligonucleotide microarrays).

In parallel, the Brown Lab at Stanford pioneered an open approach to microarray development, sharing protocols and specifications for constructing and using spotted arrays and associated hardware and software (Brown and Botstein 1999). This open-source ethos facilitated the widespread adoption and local adaptation of spotted microarray technologies by other academic laboratories, who contributed to the improvement of the methodologies (Cheung et al. 1999).The emergence of companies offering commercial arrayers and scanners further supported broader access to spotted cDNA arrays (reviewed in Bowtell 1999).

2.1.5 Challenges of a New Type of Data

Researchers realized early on that microarray technology introduced new challenges to biological research, spanning scientific publishing, data management, and analysis tools. The technology enabled a new form of genome-wide exploratory study that did not conform to traditional publication standards while producing large volumes of biologically relevant data that should be made accessible to the research community (Brown and Botstein 1999). Moreover, the diversity of microarray platforms and experimental applications highlighted the need for standardized descriptions of array design and experimental parameters to ensure data comparability and facilitate reuse. This need was especially evident in gene expression profiling, which rapidly became the predominant application of microarray technology.

2.1.5.1 Early Challenges in Data Sharing and Standardization

Many early adopters of the technology embraced open data sharing, providing data tables associated with published manuscripts on laboratory websites and creating institutional databases to disseminate their results (Brown and Botstein 1999; Brazma et al. 2000; Sherlock et al. 2001; Bono et al. 2002). However, the absence of consistent standards made it difficult to interpret or reuse these datasets effectively.

The final data for a gene expression study are a simple table—a gene expression matrix—in which rows represent genes, columns represent various samples and each position in the table describes the measurement for a particular gene in each sample. However, the raw data from which these measurements are obtained are quite different in nature: they are scanned images of fluorescence intensities in one or two channels, depending on the type of array used. For these data to be transformed into a gene expression matrix, several processing steps must be undertaken. First, quantification matrices that reflect the fluorescence intensity per microarray spot must be generated by an image analysis software, together with other measurements like background intensities. These data will be the basis for generating the final gene expression matrix following normalization steps that can vary between different microarray platforms. Therefore, there are at least three levels of data relevant for a microarray experiment connected by a series of transformations.

To be effective, the sharing of this type of data required the development of a set of accepted standards, including experimental descriptions (with controlled vocabularies and ontologies for sample annotation), appropriate data representation formats (e.g., microarray XML exchange files) and conventions for data normalization, quality control and cross-platform comparison (Brazma et al. 2000). As early as 1999, an intercontinental group of experts formed the Microarray Gene Expression Data Society to coordinate the development of these standards, pressuring scientific journals to adopt them as part of the publishing process (Kellam 2001; Ball et al. 2002a, b). Their work was instrumental in shaping community-wide practices and facilitating data sharing and reuse.

2.1.5.2 The MIAME Guidelines and Their Impact

A key outcome of the MGED Society's efforts was the publication of the MIAME (Minimum Information About a Microarray Experiment) guidelines in 2001 (Brazma et al. 2001). Focusing specifically on microarray-generated gene expression data, the most popular application of the technology at the time, the MIAME guidelines aim to specify the minimum information required to ensure that microarray data can be easily interpreted and that results derived from its analysis can be independently verified. Gene expression data are more complex than sequence data because there are as many transcriptomes as there are tissue or cell types responding to specific environmental conditions, and they are only meaningful if these are well described. Additional complexity arose from the relative nature of gene expression measurements, platform-specific detection strategies and differing data processing pipelines.

MIAME guidelines provide a very clear insight into the experimental use of microarrays, establishing that a published microarray-based gene expression experiment should include six sections, with information provided through a combination of controlled vocabulary and free-text format (Brazma et al. 2001):

1. Experimental design: overall structure of the hybridization experiment.
2. Array design: features and layout of the microarray used.
3. Samples: origin, preparation and labelling of biological samples.
4. Hybridizations: conditions and parameters of the hybridization process.
5. Measurements: experimental results including raw images and quantification data.
6. Normalization controls: types, values and specifications.

In a typical microarray experiment, multiple samples are hybridized to arrays to identify relative changes in expression levels. As such, many experiments use a "common reference sample" to which all others are compared. These comparisons require normalizing the data to compensate for technical variations between and within single hybridizations, for which a variety of different schemes were developed. To support the independent validation and re-analysis of microarray-derived gene expression results, MIAME stipulates that experimental results should be presented in three parts, progressing from raw to processed data: raw images; image analysis results; and the final gene expression matrix.

2.1.5.3 Establishing Public Microarray Data Repositories

The effort to define broadly accepted standards for microarray data also aimed to facilitate the establishment of public repositories for microarray data, which the research community recognized as a critical development (Brazma et al. 2000). The European Bioinformatics Institute led these initiatives with the creation of the ArrayExpress database (Brazma et al. 2003), in parallel with the launch of the Gene Expression Omnibus (GEO) database (Edgar et al. 2002) by the US National Centre for Biotechnology Information. GEO became the first open public repository for microarray data and, along with BioStudies (Sarkans et al. 2020), the designated successor to ArrayExpress since 2020, remains a key reference database for this type of data.

As the first public microarray repository, GEO provides a unique view of the use of microarray technology from 2000 onwards. The progressive

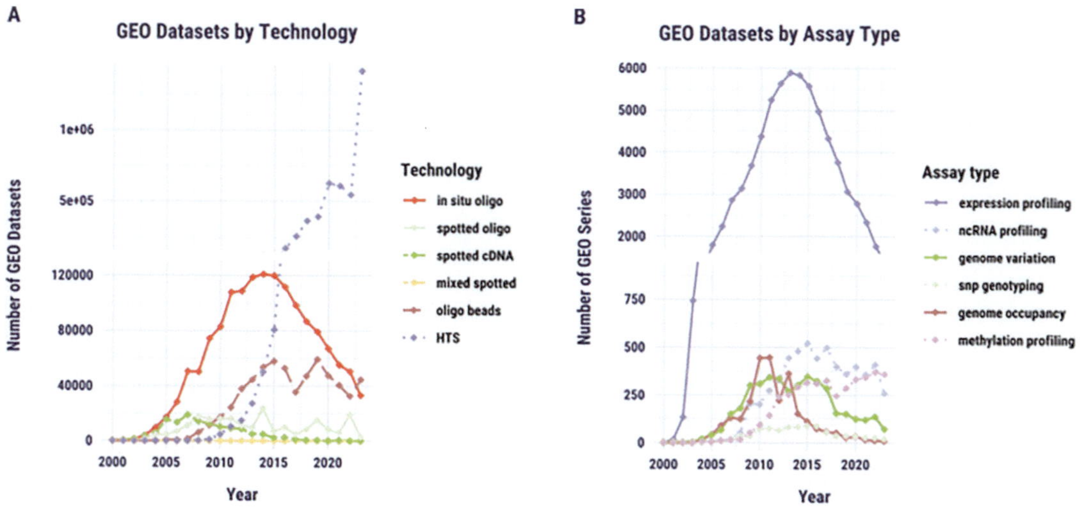

Fig. 2.1 The 15-year dominance of microarray technology. (**a**) Number of GEO datasets for major microarray platforms and high-throughput sequencing (HTS) between 2000 and 2025. Oligonucleotide arrays dominated genomic and transcriptomic studies until approximately 2015, after which HTS rapidly became the prevailing approach. Oligonucleotide bead arrays remain in use, particularly in clinical contexts, although this is not fully captured in GEO because of data-privacy restrictions. (**b**) Number of GEO series using microarray technology by assay type from 2000 to 2025. Expression profiling was the main application, with a marked decline as HTS became widespread. The use of microarrays for non-coding RNA (ncRNA) and methylation profiling studies has remained comparatively more stable over time. Counts for plots were obtained through programmatic queries to the GEO DataSets database (Edgar et al. 2002) using the R rentrez package, with platform types retrieved via [PTYP] field terms and assay classes via [GTYP] terms. (Code is available at https://github.com/GamaPintoLab/microarray_trends_geo)

replacement of this technology by high-throughput sequencing (HTS) after a reign of 15 years is also clear from the steady decline in deposited data (Fig. 2.1). GEO microarray datasets are organized by different types of assays that can be performed using the technology, allowing an insight into the types of studies they are used for: expression (mRNA) profiling, ncRNA profiling, genome variation, SNP genotyping, genome occupancy, and methylation profiling. Expression profiling has been by far the most widely used application for microarrays, and the first to reach significant levels of use, with all other applications representing a much smaller fraction of the deposited data (Fig. 2.1a). Of note, GEO entries represent predominantly the use of these technologies within the research community and in association with published work. They do not provide a good estimate for the use of different technologies by private companies or in the context of medical diagnostics, where microarray use has become increasingly significant as the technology matured (see below).

2.2 Microarray Technology Platforms

Given the broad potential of microarrays to systematically analyse nucleic acid sequences across diverse applications, the maturation of the technology led to the development of a wide range of platforms. These platforms differed in fabrication methods, probe types, detection and quantification strategies, array designs and, consequently, their experimental setups. This diversity is reflected in public GEO datasets, which categorize microarray platforms into five main types: in situ oligonucleotide, mixed spotted oligonucleotide/cDNA, spotted DNA/cDNA, spotted oligonucleotide and oligonucleotide bead arrays.

These categories correspond to the main fabrication technologies that gained traction as the field expanded. Except for spotted cDNA microarrays, most platforms were primarily developed and commercialized by a limited number of companies.

2.2.1 In Situ Oligonucleotide Microarrays

In situ oligonucleotide microarrays refer to platforms in which oligonucleotide probes are synthesized directly on the array surface. Commercially available implementations were developed by Affymetrix, which pioneered high-density oligonucleotide arrays (Fodor et al. 1991) (see Sect. 2.1.4), and NimbleGen Systems, later acquired by Roche. Both companies used proprietary photolithographic methods for high-density oligonucleotide synthesis. The NimbleGen platform originated from research at the University of Wisconsin in the 1990s aiming to develop a maskless array synthesizer based on a digital micromirror system as a more versatile and cost-effective alternative to traditional photolithographic masks (Blanchard and Friend 1999; Singh-Gasson et al. 1999). NimbleGen arrays were discontinued in 2012, following Roche's strategic decision to exit the array market.

For gene expression monitoring, Affymetrix developed a complex system of probe sets composed of 15–20 probe pairs targeting different regions of a given RNA. Each pair included a 25-mer perfect-match (PM) probe and a mismatch (MM) probe with a single internal nucleotide difference, enabling background correction and specificity assessment. Probes were designed based on a set of empirically derived, composition-dependent rules, to ensure similar hybridization kinetics and low levels of cross-hybridization to similar genes (Lipshutz et al. 1999). The same design principles were applied to genotyping arrays (SNP chips), where sets of PM–MM pairs targeted known polymorphisms. A total of eight probe pairs would be used to interrogate each annotated SNP, with one pair centred on the polymorphic position and the remaining pairs offset one and four bases to either side (Lipshutz et al. 1999).

Agilent Technologies developed an alternative in situ microarray platform using SurePrint inkjet technology (Hughes et al. 2001), which synthesizes 60-mer oligonucleotides with high specificity and hybridization sensitivity. This manufacturing technology enabled the company to easily provide customized arrays for specific experiments. The arrays could be used both for single- and dual-colour experimental designs, as further discussed in Sect. 2.2.3.

In situ oligonucleotide arrays dominated the microarray field until the widespread adoption of HTS around 2015, when they suffered a sharp decline in their use (Fig. 2.1b). Affymetrix in particular held a leading market position throughout much of this period. In 2016, Affymetrix was acquired by Thermo Fisher Scientific, and its products are now marketed under the Applied Biosystems brand.

2.2.2 Oligonucleotide Bead Arrays

Oligonucleotide bead arrays were introduced later than other microarray platforms but have remained in significant use, particularly for genome variation and DNA methylation studies (see Sect. 2.3.3.1). Their persistence beyond the rise of HTS can be attributed to cost-effectiveness, high reproducibility and robust data analysis pipelines (Fig. 2.1b).

This platform represents a distinct technological approach. Originally developed by David Walt's group at Tufts University and later commercialized by Illumina (Michael et al. 1998; Walt 2000), bead arrays are based on the random deposition of oligonucleotide-coated beads onto etched microwells on a solid substrate—initially optical fibre bundles and later glass slides. This fabrication method confers several advantages: (i) bead arrays support the highest feature density among array platforms; (ii) they are produced by bulk methods, reducing costs and enabling flexible probe updates; and (iii) each probe is represented by multiple beads, increasing the accuracy of results. However, this random layout requires a decoding step to assign probe identities to specific bead positions, a technical challenge that was only resolved several years after the initial publication of the bead array manufacturing method (Gunderson et al. 2004). Decoding involves the use of 23-mer barcoded oligonucleotides attached to each bead. These are identified through a combinatorial decoding strategy, which sequentially hybridizes fluorescently labelled oligonucleotide "decoders" to each bead. The resulting fluorescent patterns reveal the barcode sequence, which is mapped back to a probe identity from the manufacturing library. Following this decoding process, the array is ready for sample hybridization.

Given their very high density, bead arrays were originally designed in a way that supported the parallel testing of multiple samples and were shown to be efficient for both SNP analysis and gene expression profiling (Fan et al. 2003, 2004; Kuhn et al. 2004). A key technical distinction of this platform lies in the sample labelling and detection system, which contributes to its high reliability, particularly for genotyping applications. Unlike other microarray technologies, bead arrays were initially designed to hybridize unlabelled genomic DNA, followed by a detection step involving a polymerase extension reaction (Gunderson et al. 2005). This approach evolved into the Infinium assay—still in current use—which uses a single-nucleotide extension of the bead-bound probe, with the captured DNA serving as template (Gunderson et al. 2006). The extension incorporates a hapten-labelled nucleotide, which is then detected by fluorescently labelled proteins. This method allows for precise discrimination between homozygous and heterozygous genotypes on the same bead, identified by distinct fluorophores in a dual-channel assay.

2.2.3 Spotted Arrays

As mentioned in Sect. 2.1.3, spotting technologies were developed early as a cost-effective alternative to in situ synthesis methods. This approach requires the selection and organization of a set of pre-synthesized antisense oligonucleotide probes (Guo et al. 1994) or, alternatively, of purified PCR products derived from cDNA clones, ESTs or even genomic DNA (Duggan et al. 1999) into microtiter dishes.

These probe collections are then spotted onto chemically modified solid surfaces, like glass slides coated with poly-L-lysine, aminosilane or other reactive chemistries, using high-precision robotic printing platforms equipped with multiple pins or capillaries (Duggan et al. 1999). The relative affordability and modularity of these systems enabled their widespread adoption in academic laboratories, allowing researchers to develop custom arrays tailored to their specific experimental needs. However, a significant limitation of spotted arrays was the high variability in probe deposition across slides and batches, which hampered quantitative inter-sample comparisons. To compensate for this problem, spotted arrays were commonly used in dual-colour hybridization assays, using a test and a reference sample on the same array. In this setup, each sample was labelled with a different fluorophore (typically Cy3 and Cy5), and fluorescence signals were acquired for both channels. The hybridization data are reported as a ratio of the Cy5/Cy3 fluorescent signals at each probe, enabling relative quantification of differential gene expression. This contrasts with the single-channel approach, where an absolute value of fluorescence directly reflecting target abundance is obtained for each probe.

While this approach facilitated within-array normalization and reduced some batch effects, it introduced new sources of bias, particularly those related to dye-specific properties. Indeed, early studies reported systematic intensity differences between dyes, depending on labelling efficiency and signal stability and affecting the measurements that were obtained depending on the dye that was associated with each sample (Churchill and Oliver 2001; Jin et al. 2001). To compensate for this effect, dye-swap designs were commonly used in early two-channel microarray experiments, not only for custom-spotted cDNA arrays but also with platforms such as Agilent dual-colour oligonucleotide arrays. In this design, the fluorophore assignment is reversed in replicate hybridizations, mitigating dye-specific biases through balanced experimental design (Yang and Speed 2002). This strategy contrasts with conventional dual-channel designs, in which fluorophore assignments remain fixed and post-hybridization normalization methods are used to adjust for dye effects without additional compensatory labelling.

The diversity of spotted microarrays developed in academic laboratories required flexible solutions for data submission and annotation. To facilitate the data submission process, GEO developed generic templates that support different array configurations, including specific formats for dual-channel or merged dye-swap designs, as well as generic templates for single-channel and tiling array formats (see below). These templates were designed to capture the essential metadata associated with home-made platforms, facilitating data standardization and public dissemination. Additionally, GEO also offers predefined submission formats for commercial platforms such as Affymetrix, NimbleGen, Agilent and Illumina, as well as the option to submit platform descriptions independently to ensure full documentation of probe content and array design.

2.3 Applications of Microarray Technology

2.3.1 Transcriptome Profiling

As discussed in Sect. 2.1, although the movement that led to the emergence of microarray technologies was associated with the need for more efficient sequencing and genotyping methods, the fact is that gene expression profiling quickly became the most prominent application of the technology (Fig. 2.1b) and led to the generalization of transcriptomics as the first widespread omics approach. Historically, the term transcriptomics was coined in 1996 by Charles Auffray (Piétu et al. 1999) and first used in a publication applying Serial Analysis of Gene Expression (SAGE) (Velculescu et al. 1995), to characterize the universe of transcripts produced from the yeast genome (Velculescu et al. 1997). As a sequencing-based method, SAGE had the advantage of not requiring the availability of a hybridization probe for each transcript. The approach essentially consisted of the concatenation of short expressed tags of 9–11 nucleotides such that sequencing of a limited number of concatemers could provide an insight into the population of expressed genes in a similar way to current short-read sequencing methods. This method played an important role in generating the first insights into genome-wide expression patterns in different models, despite its limitations.

2.3.1.1 Gene Expression Analysis

The explosive growth of information regarding expressed sequences from projects running in parallel to the human and model organism genome sequencing projects (see, e.g. Hillier et al. 1996) quickly established a solid basis for the development of probe-dependent, high-density, expression arrays as a tool to efficiently analyse the known universe of expressed genes. These were not limited to arrays produced by in situ synthesis of probes based on available sequence information. Indeed, the growing availability of arrayed cDNA libraries, like the one developed by the IMAGE consortium (Lennon et al. 1996), facilitated the widespread production of spotted cDNA arrays.

Expression profiling using microarrays has been used to characterize biological systems in several different ways. Firstly, it was used to investigate genome-wide tissue or cellular responses. In this type of study, the transcriptome of a specific biological sample (organism, cell line, tissue sample) was characterized as part of time-course assays or dose-response experiments allowing for the identification of genes whose expression is turned on, off or fine-tuned in response to specific stimuli compared to a control condition. Early examples include the characterization of the yeast temporal program of gene expression accompanying the metabolic shift from fermentation to respiration (DeRisi et al. 1997); the characterization of the gene expression changes in B and T cells in response to activation or stimulation with cytokines (Alizadeh et al. 1998) or the transcriptional response of human fibroblasts to serum exposure (Iyer et al. 1999) or the identification of sex-dependent transcriptional profiles in *Drosophila* (Jin et al. 2001). As discussed in Sect. 2.4, the identification of these changes required the development of novel analytic methods that could find correlations and significant variation—or differential expression—between a very large number of variables measured across a small number of samples.

2.3.1.2 Unravelling Gene Structure and Alternative Splicing

In addition to gene expression profiling, microarray technology was adapted to support the characterization of transcript isoform expression, enabling global studies of alternative splicing regulation (Moore and Silver 2008). The prevalence of alternative splicing in human tissues was first revealed by systematic sequencing of ESTs during the early Human Genome Project (Mironov et al. 1999; Brett et al. 2000; Croft et al. 2000). Effective characterization of isoform expression using microarrays requires careful probe design capable of detecting different types of splicing (e.g., exon skipping, intron retention, alternative 5′/3′ sites), a task achievable only with oligonucleotide arrays. An early study using oligonucleotide arrays with 20 probe sets targeting the 3′ UTR of 1600 alternatively spliced rat genes demonstrated the potential of microarrays for genome-wide analysis of alternative splicing (Hu et al. 2001).

Different strategies were developed to detect known splicing events or to perform unbiased surveys of transcript architecture. Exon and tiling microarrays, initially designed as tools to refine human genome annotation, proved useful for identifying splicing isoforms (Shoemaker et al. 2001). The first exon arrays consisted of 50–60-bp-long printed oligonucleotide probes, with two probes per predicted exon, and were used in dual-colour hybridization assays. Using appropriate analysis algorithms, these arrays allowed detection of alternatively spliced exons between samples, although other splicing events remained undetectable. This concept was later implemented by Affymetrix as a single-colour commercial platform covering all annotated human exons (Okoniewski et al. 2007), supporting detailed annotations of exon usage across tissues (Clark et al. 2007).

In contrast to the focused probe design of exon arrays, tiling arrays employ overlapping oligonucleotide probes to cover entire genomic regions (Shoemaker et al. 2001)—or even a whole genome (Yamada et al. 2003; Juneau et al. 2007). Probe lengths range from 25 nt to >1000 nt, with varying spacing between successive probes, which determines the assay resolution and the ability to map precise boundaries of expressed regions (Johnson et al. 2005). Although exacting in terms of data analysis, tiling arrays are particularly interesting because they do not rely on a priori gene models, enabling the detection of novel splicing patterns, including intron retention, having been used to annotate transcript architecture across multiple genomes (Johnson et al. 2005; Moore and Silver 2008).

Despite their ability to detect transcript isoforms, exon and tiling arrays cannot distinguish between spliced and unspliced RNAs (Moore and Silver 2008). This requires small probes designed to span annotated exon junctions. When combined with probes detecting introns and adjacent exons, these arrays support computation of gene-specific indices—the splice-junction (SJ) and intron accumulation (IA) indices—providing a quantitative perspective on splicing changes independent of transcriptional or decay effects (Clark et al. 2002). This pioneer study in yeast provided the first genome-wide view of splicing, revealing functional relationships among mRNA processing factors (Clark et al. 2002). A similar exon-junction array platform, comprising a total of five arrays with ~125,000 36-nt probes to monitor splicing at every exon-exon junction in >10,000 multi-exon human genes, was used in 2003 to chart transcriptome diversity across 52 tissues and human cell lines at an unprecedented level of detail (Johnson et al. 2003).

Paramount to the ability of these methods to faithfully capture the landscape of alternatively spliced isoforms was the development of a novel RNA sample preparation strategy that ensured whole-transcript coverage, rather than the traditional 3′ UTR-focused analysis used for gene expression profiling (Castle et al. 2003). Notwithstanding these advances, microarrays remain limited in their ability to map the actual mRNA isoform combinations present in a sample, given their inability to determine whether two splicing events occur in the same or separate transcripts.

2.3.1.3 Tumour Classification

An important application of large-scale gene expression analysis has been the characterization of tumours. From the onset, the power of microarray technology to provide insights into pathological processes was widely explored by performing comparative transcriptomics between normal and diseased tissues (Zhang et al. 1997). Early analysis of tumour samples revealed that transcriptome data were highly effective at distinguishing tumour subtypes related to different disease prognoses and often difficult to discriminate by standard pathology analysis (Golub et al. 1999; Alizadeh et al. 2000; Perou et al. 2000). The insights derived from such studies were used to successfully personalize the therapeutic approach to the patient's tumour (Golub et al. 1999). Additionally, further investigation of genes with high expression levels in tumours revealed novel driver oncogenes that were shown to be relevant targets for precision therapies (Armstrong et al. 2003). As a consequence, gene expression profiling became a major approach for biomarker discovery and molecular classification of tumours, with increasing application in clinical settings.

2.3.2 Genomic Profiling

As discussed in Sect. 2.1, the initial conceptualization of oligonucleotide arrays was driven by the need for cost-effective platforms to characterize genomic variation. Although microarray applications for gene expression analysis developed earlier and rapidly became the dominant use of the technology, microarray-based genomic profiling has also achieved substantial relevance, particularly in clinical and diagnostic settings, where it continues to serve as a reference method.

2.3.2.1 Copy Number and Structural Variation Analysis

Numerical chromosome abnormalities, aneuploidy, hypodiploidy, hyperdiploidy and polyploidy, structural variations of the genome, deletions, duplications, copy number variations (CNVs), insertions, inversions and chromosomal translocations, are a pervasive source of genetic variation with known pathological consequences in humans (Collins et al. 2020; Comaills and Castellano-Pozo 2023). Their relevance is well documented in the context of unexplained infertility, miscarriages, congenital abnormalities, developmental delays, intellectual disabilities and autism (Grayton et al. 2012; Nagaoka et al. 2012). In addition to this, genomic instability has been known for a long time to play an important role in the diagnosis, prognosis and therapy of many neoplastic disorders (Comaills and Castellano-Pozo 2023). Comparative genomic hybridization (CGH) was developed in the early 1990s to detect and map relative DNA copy number differences between malignant and normal cells in a single assay by co-hybridizing labelled total genomic DNA from both samples to normal metaphase chromosomes (Kallioniemi et al. 1992). With the development of microarray technologies, this method was quickly adapted to different platforms, initially using large, overlapping, BAC-derived clones to represent portions of the human genome (Solinas-Toldo et al. 1997; Pinkel et al. 1998) and later reaching full genome coverage at sub-megabase tiling resolution (Ishkanian et al. 2004). Array CGH (aCGH) variations built using cDNA (Pollack et al. 1999), exon (Dhami et al. 2005) or oligonucleotide-based (IJssel et al. 2005) approaches were also developed and shown to be effective at detecting amplifications, deletions and chromosomal aberrations at increased resolution. The use of these different formats of tiling arrays to perform aCGH analysis supported not only the identification of novel disease gene associations but also the genetic diagnosis of patients by reliably detecting and quantifying whole chromosome gains, high-level deletions and amplifications and single copy-alterations in developmental syndromes and cancer.

To this date, microarrays remain a gold standard for genome-wide detection of chromosomal aberrations in clinical contexts and are extensively used in pre- and postnatal testing, preimplantation genetic screening and cancer research and are considered the first-tier clinical diagnostic test for individuals with developmental disabilities or congenital anomalies (Miller et al. 2010). Dual-colour CGH arrays enable true comparative hybridization and fast delivery of results, with genome-wide coverage and resolutions down to exon level, supporting the detection of aneuploidies, microdeletions, microduplications and regions of copy-neutral loss of heterozygosity (cnLOH) (Miller et al. 2010; Shao et al. 2021). Commercial array platforms optimized for different cytogenetic contexts are currently available and continue to be preferred in laboratories where cost, speed and sample constraints limit the use of next-generation sequencing.

2.3.2.2 Genotyping Arrays and Single-Nucleotide Variation Analysis

Understanding genetic variation and how it correlates with phenotype is one of the major challenges following genome sequencing projects, which typically focus on DNA samples from a limited number of individuals and thus do not mirror the existing sequence diversity (Kruglyak and Nickerson 2001). Single-nucleotide variants are the main source of genome sequence variation and the predominant type of disease-causing mutations. Once the reference genome sequence for a species is available, identification of these variations does not require sequencing-dependent approaches. Indeed, hybridization with oligonucleotides has been known for a long time to be able to discriminate single mismatches from fully complementary sequences.

The first array-based assays for genotyping exploited this principle by using oligonucleotide probes designed to detect single base mismatches within short DNA sequences (Southern et al. 1992). Early commercial implementations used high-density oligonucleotide arrays to interrogate thousands of genomic loci in parallel, laying the groundwork for systematic cataloguing of single-nucleotide polymorphisms (SNPs) across populations (Wang et al. 1998). Platforms such as the Affymetrix GeneChip (Lipshutz et al. 1999) and Illumina BeadArray (Fan et al. 2003) introduced scalable genotyping formats in which allele-specific hybridization or primer extension allowed accurate discrimination of single base variants at hundreds of thousands of loci per assay. These developments supported a dramatic increase in throughput, reproducibility and cost-efficiency of SNP genotyping, replacing labour-intensive gel- or PCR-based techniques and enabling routine processing of large sample sets for both research and clinical applications.

The generation of dense SNP datasets supported the reconstruction of haplotype blocks—regions of the genome where genetic variants are inherited together—and provided the basis for fine-scale linkage analyses (Manolio and Collins 2009). International initiatives such as the HapMap Project systematically characterized common patterns of human genetic variation by genotyping millions of SNPs across diverse populations (Manolio and Collins 2009). These efforts produced comprehensive maps of linkage disequilibrium, facilitating the identification of chromosomal regions associated with Mendelian and complex traits. Microarray-based SNP panels became indispensable in family-based linkage studies and population genetics, supporting both the identification of disease loci and the inference of population structure and ancestry.

The combination of high-density SNP arrays with large, well-phenotyped cohorts enabled the advent of genome-wide association studies (GWAS), which systematically evaluate the association between genetic variants and traits across the entire genome (Uffelmann et al. 2021). Before the widespread adoption of next-generation sequencing, microarray-based GWAS represented the most powerful strategy to detect common variants contributing to complex diseases. The scalability and decreasing cost of SNP arrays made possible studies involving cohorts of thousands of individuals, leading to the discovery of multiple genetic loci associated with traits ranging from metabolic and cardiovascular diseases to neurological and psychiatric disorders (Grant and Hakonarson 2008). The standardization of array formats and genotyping pipelines, coupled with rigorous statistical frameworks and imputation methods, established GWAS as a cornerstone of modern human genetics. Although sequencing-based approaches have since expanded the scope of variant detection, microarray genotyping remains widely used for large-scale association studies, polygenic risk scoring and quality control in population genomics (Uffelmann et al. 2021).

2.3.3 Functional Genomics Beyond Gene Expression

In addition to the characterization of genome sequences and their transcripts, microarray platforms also began to be explored for characterizing regulatory and functional aspects at a genome-wide scale. This became possible essentially through smart adaptations of sample preparation protocols that allowed researchers to ask distinct biological questions using the same technological platforms coupled to adapted analytic pipelines. This allowed, for the first time, the systematic exploration of processes as distinct as the location of open chromatin regions, the presence of DNA methylation or chromatin modifications, transcription factor binding or active transcriptional sites, RNA-protein interaction profiles, global changes of mRNA stability or ribosome occupancy, creating global maps of genome activity, as discussed in the following sections.

2.3.3.1 Mapping Epigenetic Changes and Genome Regulation

Beyond the actual sequence, DNA methylation—specifically, the presence of 5-methylcytosine bases—plays a key role in regulating gene expression processes. Although these bases cannot be directly detected by hybridization-based methods, microarrays can be used to perform genome-wide profiling of this modification through sample pre-treatments that reveal the presence or absence of the methyl group at cytosine residues. These include methylation-sensitive restriction enzymes, bisulphite conversion and affinity-based capture using methyl-binding proteins or 5-methylcytosine-specific antibodies (Laird 2010). The identification in the 1970s of endonucleases like HpaII and SmaI, which are inhibited by the presence of methyl groups, laid the groundwork for the first systematic investigation of genomic DNA methylation, which was early on adapted for use together with microarrays, namely, through hybridization of restriction fragments to CpG-island or promoter-focused microarrays (e.g., Gitan et al. 2002). This method, however, requires high-molecular-weight DNA, is limited by the sequence context of the enzyme and lacks nucleotide-level resolution. The gold standard for single-base methylation analysis was based on pre-treatment of genomic DNA with bisulphite, which converts unmethylated cytosine residues to uracil, altering the hybridization profile of the treated DNA. Originally used together with Sanger sequencing (Frommer et al. 1992), this approach was adapted to support genome-wide profiling of methylated cytosines through hybridization with methylation-specific microarrays (Adorján et al. 2002), including the commercial Infinium HumanMethylation27 BeadChip platform by Illumina (Bibikova et al. 2009). The complexity of bisulphite treatment justified the development of alternative methods based on affinity capture of methylated DNA segments (Weber et al. 2005; Keshet et al. 2006). These methods (MeDIP or mDIP) were based on the use of antibodies against 5-methylcytosine, which are used to immunoprecipitate (IP) DNA fragments, with both input and bound DNA subsequently hybridized to a spotted tiling array in a dual-colour assay. These studies were instrumental in revealing the widespread nature of methylation changes associated with cancer.

The conceptual approach of immunoprecipitating DNA-protein complexes for sequence analysis emerged in the 1980s, with early applications targeting histone-bound DNA (Solomon et al. 1988). With the advent of microarrays, these protocols were adapted for genome-wide profiling of histone modifications and DNA-protein interactions, known as ChIP-on-chip (or ChIP-chip) assays. Pioneer studies by the Young and Brown labs established the first genomic maps of yeast transcription factor (TF) binding sites (Ren et al. 2000; Iyer et al. 2001). The first of these studies combined TF location analysis with expression profiling to map the direct targets of Gal4 and Ste12, providing an initial systems-level view of gene regulatory networks. The Encyclopaedia of DNA Elements (ENCODE) and ModENCODE projects (Celniker et al. 2009), focusing on human and model organism genomes (particularly *D. melanogaster* and *C. elegans*), respectively, pioneered the standardization of high-quality methods and analytic pipelines for these approaches and generated a detailed map of functional elements across a large number of systems, made publicly available through dedicated repositories (Consortium 2011). A key challenge for ChIP-chip methods was the development of appropriate analysis methods able to deal with the signal-to-noise ratios of pull-down samples and generate positional mapping information (Buck and Lieb 2004). Of note, the resolution of these methods was highly dependent on array design—tiling density, probe length and genomic coverage.

The development of genome-wide methods for mapping DNA and chromatin modifications and TF binding enabled integrative analyses of gene expression programs and the reconstruction of regulatory circuits and it was further expanded by the characterization of post-transcriptional and translational regulatory layers, as addressed in the next section.

2.3.3.2 Global View of Gene Expression and Post-transcriptional Regulation

While transcript quantification using microarrays has provided broad functional insights into gene regulation and cellular programs, it captures only steady-state RNA levels and their variation across conditions. These levels reflect the combined effects of transcription and degradation, but the underlying kinetic processes cannot be resolved using standard microarray assays. Moreover, mRNA abundance correlates only partially with protein levels because additional regulatory layers—including ribosome recruitment, translation initiation and elongation and protein turnover—shape the final proteome. Even so, the capacity of microarrays to deliver comprehensive transcriptome profiles, together with the technical limitations of proteomics, established them as a central tool for characterizing cellular functional states. To extend their use beyond steady-state transcriptome measurements, early methods for studying mRNA dynamics and post-transcriptional regulation were rapidly adapted for genome-wide application on microarray platforms.

Identification of actively transcribing genes can be achieved through pulse-labelling of nascent RNA, which was originally done with radioactive nucleotides. Adaptations of this method for use with microarray technology required the use of nucleotide analogues that allowed purification of recently synthesized RNA molecules. These include 4-thiouridine, coupled with isolation of thiolated RNA using an organomercurial affinity matrix (Kenzelmann et al. 2007) or, alternatively, by streptavidin beads following biotinylation (Dölken et al. 2008). Another approach relied on bromouridine, which can be affinity purified using antibodies (Ohtsu et al. 2008). Following purification, newly synthesized RNA can be used as a sample for microarray analysis, allowing for the direct quantification of ongoing transcription. By comparing newly synthesized RNA with total RNA, or pre-existing (unlabelled) RNA with total RNA from pre-stimulus conditions, depending on whether conditions are steady-state or not, Dölken et al. (2008) showed this method could be used to estimate mRNA decay and half-lives with significant precision. Previous studies of mRNA turnover using microarrays were an adaptation of the classic approach of performing a time-course assay following the inhibition of transcription using actinomycin D (Raghavan et al. 2002). By profiling mRNA abundance at three or four time points following treatment (e.g., 1 h, 2 h and 3 h), decay rates for individual RNA molecules can be inferred from the microarray data. However, this method tends to be imprecise for medium- to long-lived transcripts, in contrast with half-life determination from newly transcribed/total RNA ratios (Friedel et al. 2009).

Finally, translation efficiency of mRNA molecules can be assessed at a genome-wide level by quantifying the population of mRNAs that is associated with ribosomes and polysomes (Arava et al. 2003). These complexes can be fractionated by gradient centrifugation and the associated mRNA populations quantified using microarrays. Careful fractionation allows the separation of RNAs associated with a distinct number of ribosomes, supporting estimates of translation initiation efficiency by comparing to the population of free RNAs. An early adaptation of this method was used to identify human mRNAs that exhibit cap-independent translation initiation (Johannes et al. 1999).

Coupled with modifications of the experimental design and sample preparation methods, microarrays supported the genome-wide characterization of the dynamics of gene expression. Many of these processes are influenced by post-transcriptional mechanisms acting on the mRNA, such as poly(A) tail dynamics, interactions with RNA-binding proteins or microRNAs and subcellular localization. These interactions were also explored early on through the development of targeted RNA capture methods, namely, focused on poly(A) tail length (Meijer et al. 2007), immunoprecipitation of ribonucleoprotein complexes (RIP-Chip) (Tenenbaum et al. 2000; Gama-Carvalho et al. 2006; Wang et al. 2010) or targeted cellular fractionation (McGeachie et al. 2005).

Paralleling their use to study chromatin regulation, adaptations in sample-preparation strategies and dedicated analytical pipelines extended microarray applications beyond steady-state transcript profiling to the investigation of dynamic post-transcriptional processes. The breadth and heterogeneity of the resulting datasets were instrumental in driving the development of modern bioinformatics, paving the way for the data-intensive approaches that define current genomics.

2.4 A New Age in the Generation and Analysis of Biological Data

The emergence of microarrays marked the beginning of large-scale molecular profiling, introducing a new statistical paradigm in which thousands of variables could be measured in parallel for the first time. This technological shift reshaped not only experimental biology but also the methodological foundations of data analysis. A series of statistical and computational challenges emerged immediately, from transforming raw scanner images into usable data, to developing valid inference in high dimensions, to integrating biological knowledge through ontologies and pathway analyses. Many of the methodological solutions pioneered during this period remain central to modern omics technologies and continue to underpin contemporary transcriptomics, proteomics and multi-omics.

2.4.1 From Raw Images to Structured and Reproducible Data

Turning images produced by laser scanners into gene expression measurements required sophisticated image processing pipelines. Early software such as GenePix and ScanAlyze had to identify printed spots, segment them from background and quantify fluorescence intensities while addressing various sources of technical variability (Schadt et al. 2002). Background correction alone was a major statistical challenge: local estimators reduced spatial noise but increased variance, while global estimators risked systematic bias for low-intensity spots.

Technical artefacts were pervasive. Dye bias in two-colour arrays caused systematic differences between Cy3 and Cy5 channels (Yang et al. 2002), and print-tip effects introduced local intensity distortions. Normalization methods rapidly became central to microarray analysis. For two-colour arrays, non-parametric Lowess (locally weighted scatterplot smoothing) normalization removed intensity-dependent imbalances (Yang et al. 2002), whereas Affymetrix arrays benefited from the modular pipeline of robust multi-array average (RMA), which combined model-based background adjustment, quantile normalization and median-polish summarization (Irizarry et al. 2003; Bolstad et al. 2003). These innovations transformed heterogeneous raw intensities into consistent, comparable estimates of gene expression.

Importantly, the early 2000s saw the rise of R as the primary environment for statistical computing in genomics. Its open-source nature, extensibility and strong community support made it uniquely suited to rapidly evolving methodological needs. This period also witnessed the creation of Bioconductor (Gentleman et al. 2004), which provided standardized data structures, reproducible workflows and high-quality implementations of cutting-edge methods—including limma, affy and marray. R and Bioconductor became the community standard platform for microarray data analysis, enabling reproducibility and accelerating methodological innovation at a pace unmatched by proprietary software.

To support calibration and quality control, microarray platforms incorporated spike-in controls, i.e. known transcripts added at predefined concentrations. The Affymetrix Latin Square spike-in dataset provided a benchmark for assessing sensitivity, dynamic range and linearity of response and influenced later practices in RNA-seq via ERCC spike-ins (Jiang et al. 2011). Pre-processing thus became not a peripheral step but a foundational determinant of downstream validity.

The heterogeneity of microarray platforms made data organization a major barrier to reproducibility. Early expression datasets were often shared without sufficient metadata, preventing reanalysis. As discussed in Sect. 2.1.5, the community responded by developing standard file formats (e.g., CEL, CHP, GPR) and by launching public repositories such as GEO and ArrayExpress. The pivotal advance was the MIAME standard, which defined the essential components required for a microarray dataset to be interpretable and reproducible (Brazma et al. 2001). MIAME enabled the establishment of structured metadata, from sample annotations to preprocessing steps, and became the conceptual template for later standards in proteomics, metabolomics and sequencing.

Microarray experiments were expensive and technically variable, making robust experimental design critical. Key considerations included biological vs. technical replication; dye swaps in two-colour arrays to counter dye bias (Churchill 2002); blocking and randomization to mitigate batch effects; and sample pooling, sometimes used for cost reasons but discouraged due to loss of variance information.

2.4.2 Statistical Challenges in High Dimensions: The $p \gg n$ Paradigm

Microarrays introduced a statistical regime in which $p >> n$, i.e. the number of genes far exceeded the number of samples. Typical studies measured 10,000–50,000 transcripts using as few as 6–20 biological replicates. Under these conditions, classical statistical methods failed: variances were poorly estimated, t-tests were unstable, and thousands of simultaneous hypotheses created severe multiple testing problems.

One of the key breakthroughs was the adoption of false discovery rate (FDR) control (Benjamini and Hochberg 1995), which allowed researchers to manage error rates in high-dimensional settings without the excessive conservatism of family-wise error control. Complementary methods such as significance analysis of microarrays (SAM) employed permutation-based null distributions to provide robust inference even with small sample sizes (Tusher et al.

2001). Perhaps the most influential development was empirical Bayes shrinkage of gene-wise variances, introduced through the limma framework (Smyth 2004). By borrowing statistical strength across all genes, limma dramatically stabilized variance estimates and improved power, becoming the dominant method for differential expression. These strategies formed the conceptual basis for later approaches in RNA-seq and multi-omics, where high dimensionality and limited replication remain the norm.

2.4.3 Classification, Clustering, and the Emergence of Gene Signatures

One of the earliest and most influential studies demonstrated that acute leukaemias could be accurately distinguished using supervised learning applied to microarray data (Golub et al. 1999). This work introduced the idea of gene signatures, small sets of transcripts capable of predicting phenotypes such as disease subtype, prognosis or treatment response. The small *n*, large *p* setting made models prone to overfitting, and small changes in the training data often produced entirely different sets of selected genes (Simon et al. 2003). Penalized regression (e.g., Lasso), support vector machines, nearest-centroid classifiers and random forests became widely used, but best practices emphasized cross validation, independent test sets, and careful estimation of model performance. Despite instability issues, gene signatures helped establish the idea that molecular phenotypes could be quantitatively predicted, a concept that persists in modern precision oncology.

Unsupervised learning also gained prominence. Hierarchical clustering and heatmaps (Eisen et al. 1998) became iconic tools for visualizing global gene expression patterns. These methods made it possible to detect biologically coherent groups of samples, such as new cancer subtypes that were not identifiable using conventional diagnostic approaches. Choice of distance metric (correlation vs. Euclidean), linkage method and noise handling strongly influenced results, motivating later methods including k-means, self-organizing maps, biclustering and consensus clustering (Monti et al. 2003) to improve stability.

2.4.4 From Gene Lists to Biology: Ontologies and Enrichment Analysis

As microarray datasets produced long gene lists, researchers sought methods to extract biological meaning beyond single-gene interpretation. The Gene Ontology (GO) (Ashburner et al. 2000) and pathway databases provided formal representations of biological processes, enabling the first systematic enrichment analyses. Two major classes of methods emerged: (i) Over-Representation Analysis (ORA), which tests whether predefined biological categories contain more differentially expressed genes than expected by chance, and (ii) Gene Set Enrichment Analysis (GSEA), which evaluates whether gene sets show coordinated shifts in ranked expression values without requiring arbitrary significance thresholds (Subramanian et al. 2005). These approaches transformed interpretation from individual genes to pathways and processes, an essential evolution given the noise, redundancy and instability of single-gene results. Later refinements addressed redundancies in GO annotations and introduced topology aware and network-based enrichment tools.

2.4.5 Cross-Study Integration and the Legacy of Microarray Statistics

Given the small sample sizes characteristic of early studies, meta-analysis emerged as a powerful tool to improve robustness and reproducibility. However, integrating data across platforms was difficult due to differences in probe design, preprocessing and annotation. Strategies included combining p-values using Fisher's or Stouffer's methods, effect size aggregation and rank-based approaches such as RankProd (Hong et al. 2006). Meta-analytic frameworks increased power and revealed consistent expression patterns across diseases, tissues and conditions (Rhodes et al. 2004).

Many statistical innovations developed for microarrays continue to influence the analysis of modern omics data. Normalization, variance shrinkage, multiple testing control, enrichment analysis, metadata standards and concerns about batch effects remain central themes in contemporary RNA-seq, single-cell genomics and multi-omics. Although the technology has evolved, the conceptual advances of the microarray era laid the foundation for today's quantitative biology.

2.5 Conclusion: Legacy and Future Directions

Although microarrays no longer dominate the omics landscape, their contributions to high-throughput biological data generation and analysis remain foundational. They enabled quantitative transcriptomics, functional genomics and early personalized medicine studies, generating vast datasets that continue to be valuable in repositories such as GEO. The development of microarrays also drove the creation of standardized databases, data formats and analytic pipelines—including exploratory data analysis, clustering, differential expression and functional enrichment methods—that now form the core of modern NGS workflows. The technological innovations established during the rise of microarrays, namely, solid-support oligonucleotide synthesis and high-sensitivity fluorescence scanning, laid the groundwork for many current high-throughput sequencing methods. Looking forward, hybrid approaches that integrate microarray principles with sequencing, including spatial transcriptomics, suggest that microarray-derived technologies will retain a role in specialized applications, demonstrating their versatility and enduring impact on omics research.

References

Adams MD, Kelley JM, Gocayne JD, Dubnick M, Polymeropoulos MH, Xiao H, Merril CR, Wu A, Olde B, Moreno RF, Kerlavage AR, McCombie WR, Venter JC (1991) Complementary DNA sequencing: expressed sequence tags and human genome project. Science 252(5013):1651–1656. https://doi.org/10.1126/science.2047873

Adorján P, Distler J, Lipscher E, Model F, Müller J, Pelet C, Braun A, Florl AR, Gütig D, Grabs G, Howe A, Kursar M, Lesche R, Leu E, Lewin A, Maier S, Müller V, Otto T, Scholz C, Schulz WA, Seifert H-H, Schwope I, Ziebarth H, Berlin K, Piepenbrock C, Olek A (2002) Tumour class prediction and discovery by microarray-based DNA methylation analysis. Nucleic Acids Res 30(5):e21–e21. https://doi.org/10.1093/nar/30.5.e21

Alizadeh A, Eisen M, Botstein D, Brown PO, Staudt LM (1998) Probing lymphocyte biology by genomic-scale gene expression analysis. J Clin Immunol 18(6): 373–379. https://doi.org/10.1023/a:1023293621057

Alizadeh AA, Eisen MB, Davis RE, Ma C, Lossos IS, Rosenwald A, Boldrick JC, Sabet H, Tran T, Yu X, Powell JI, Yang L, Marti GE, Moore T, Hudson J, Lu L, Lewis DB, Tibshirani R, Sherlock G, Chan WC, Greiner TC, Weisenburger DD, Armitage JO, Warnke R, Levy R, Wilson W, Grever MR, Byrd JC, Botstein D, Brown PO, Staudt LM (2000) Distinct types of diffuse large B-cell lymphoma identified by gene expression profiling. Nature 403(6769):503–511. https://doi.org/10.1038/35000501

Alwine JC, Kemp DJ, Stark GR (1977) Method for detection of specific RNAs in agarose gels by transfer to diazobenzyloxymethyl-paper and hybridization with DNA probes. Proc Natl Acad Sci 74(12):5350–5354. https://doi.org/10.1073/pnas.74.12.5350

Arava Y, Wang Y, Storey JD, Liu CL, Brown PO, Herschlag D (2003) Genome-wide analysis of mRNA translation profiles in Saccharomyces cerevisiae. Proc Natl Acad Sci 100(7):3889–3894. https://doi.org/10.1073/pnas.0635171100

Armstrong SA, Kung AL, Mabon ME, Silverman LB, Stam RW, Boer MLD, Pieters R, Kersey JH, Sallan SE, Fletcher JA, Golub TR, Griffin JD, Korsmeyer SJ (2003) Inhibition of FLT3 in MLL validation of a therapeutic target identified by gene expression based classification. Cancer Cell 3(2):173–183. https://doi.org/10.1016/s1535-6108(03)00003-5

Ashburner M, Ball CA, Blake JA, Botstein D, Butler H, Cherry JM, Davis AP, Dolinski K, Dwight SS, Eppig JT, Harris MA, Hill DP, Issel-Tarver L, Kasarskis A, Lewis S, Matese JC, Richardson JE, Ringwald M, Rubin GM, Sherlock G (2000) Gene ontology: tool for the unification of biology. Nat Genet 25(1):25–29. https://doi.org/10.1038/75556

Ball CA, Sherlock G, Parkinson H, Rocca-Sera P, Brooksbank C, Causton HC, Cavalieri D, Gaasterland T, Hingamp P, Holstege F, Ringwald M, Spellman P, Stoeckert CJ, Stewart JE, Taylor R, Brazma A, Quackenbush J, Data MGE (2002a) An open letter to the scientific journals. Bioinformatics 18(11):1409–1409. https://doi.org/10.1093/bioinformatics/18.11.1409

Ball CA, Sherlock G, Parkinson H, Rocca-Sera P, Brooksbank C, Causton HC, Cavalieri D, Gaasterland T, Hingamp P, Holstege F, Ringwald M, Spellman P, Stoeckert CJ, Stewart JE, Taylor R, Brazma A, Quackenbush J, Society MGED (MGED) (2002b) Standards for microarray Data. Science 298(5593):539–539. https://doi.org/10.1126/science.298.5593.539b

Benjamini Y, Hochberg Y (1995) Controlling the false discovery rate: a practical and powerful approach to multiple testing. J Royal Statistic Soc Series B (Methodological) 57(1):289–300. https://doi.org/10.1111/j.2517-6161.1995.tb02031.x

Bibikova M, Le J, Barnes B, Saedinia-Melnyk S, Zhou L, Shen R, Gunderson KL (2009) Genome-wide DNA methylation profiling using Infinium assay. Epigenomics 1(1):177–200. https://doi.org/10.2217/epi.09.14

Blanchard AP, Friend SH (1999) Cheap DNA arrays—it's not all smoke and mirrors. Nat Biotechnol 17(10):953–953. https://doi.org/10.1038/13644

Blanchard AP, Kaiser RJ, Hood LE (1996) High-density oligonucleotide arrays. Biosens Bioelectron 11(6–7) (687):–690. https://doi.org/10.1016/0956-5663(96)83302-1

Boguski MS (1995) The turning point in genome research. Trends Biochem Sci 20(8):295–296. https://doi.org/10.1016/s0968-0004(00)89051-9

Boguski MS, Lowe TMJ, Tolstoshev CM (1993) dbEST—database for "expressed sequence tags". Nat Genet 4(4):332–333. https://doi.org/10.1038/ng0893-332

Bolstad BM, Irizarry RA, Astrand M, Speed TP (2003) A comparison of normalization methods for high density oligonucleotide array data based on variance and bias. Bioinformatics 19(2):185–193. https://doi.org/10.1093/bioinformatics/19.2.185

Bono H, Kasukawa T, Hayashizaki Y, Okazaki Y (2002) READ: RIKEN expression array database. Nucleic Acids Res 30(1):211–213. https://doi.org/10.1093/nar/30.1.211

Bowtell DDL (1999) Options available—from start to finish—for obtaining expression data by microarray. Nat Genet 21(Suppl 1):25–32. https://doi.org/10.1038/4455

Brazma A, Robinson A, Cameron G, Ashburner M (2000) One-stop shop for microarray data. Nature 403(6771):699–700. https://doi.org/10.1038/35001676

Brazma A, Hingamp P, Quackenbush J, Sherlock G, Spellman P, Stoeckert C, Aach J, Ansorge W, Ball CA, Causton HC, Gaasterland T, Glenisson P, Holstege FCP, Kim IF, Markowitz V, Matese JC, Parkinson H, Robinson A, Sarkans U, Schulze-Kremer S, Stewart J, Taylor R, Vilo J, Vingron M (2001) Minimum information about a microarray experiment (MIAME)—toward standards for microarray data. Nat Genet 29(4):365–371. https://doi.org/10.1038/ng1201-365

Brazma A, Parkinson H, Sarkans U, Shojatalab M, Vilo J, Abeygunawardena N, Holloway E, Kapushesky M, Kemmeren P, Lara GG, Oezcimen A, Rocca-Serra P, Sansone S-A (2003) ArrayExpress—a public repository for microarray gene expression data at the EBI. Nucleic Acids Res 31(1):68–71. https://doi.org/10.1093/nar/gkg091

Brett D, Hanke J, Lehmann G, Haase S, Delbrück S, Krueger S, Reich J, Bork P (2000) EST comparison indicates 38% of human mRNAs contain possible alternative splice forms. FEBS Lett 474(1):83–86. https://doi.org/10.1016/s0014-5793(00)01581-7

Brown PO, Botstein D (1999) Exploring the new world of the genome with DNA microarrays. Nat Genet 21 (Suppl 1):33–37. https://doi.org/10.1038/4462

Buck MJ, Lieb JD (2004) ChIP-chip: considerations for the design, analysis, and application of genome-wide chromatin immunoprecipitation experiments. Genomics 83(3):349–360. https://doi.org/10.1016/j.ygeno.2003.11.004

Castle J, Garrett-Engele P, Armour CD, Duenwald SJ, Loerch PM, Meyer MR, Schadt EE, Stoughton R, Parrish ML, Shoemaker DD, Johnson JM (2003) Optimization of oligonucleotide arrays and RNA amplification protocols for analysis of transcript structure and alternative splicing. Genome Biol 4(10):R66. https://doi.org/10.1186/gb-2003-4-10-r66

Celniker SE, Dillon LAL, Gerstein MB, Gunsalus KC, Henikoff S, Karpen GH, Kellis M, Lai EC, Lieb JD, MacAlpine DM, Micklem G, Piano F, Snyder M, Stein L, White KP, Waterston RH, Consortium modENCODE (2009) Unlocking the secrets of the genome. Nature 459(7249):927–930. https://doi.org/10.1038/459927a

Chee M, Yang R, Hubbell E, Berno A, Huang XC, Stern D, Winkler J, Lockhart DJ, Morris MS, Fodor SPA (1996) Accessing genetic information with high-density DNA arrays. Science 274(5287):610–614. https://doi.org/10.1126/science.274.5287.610

Cheung VG, Morley M, Aguilar F, Massimi A, Kucherlapati R, Childs G (1999) Making and reading microarrays. Nat Genet 21(Suppl 1):15–19. https://doi.org/10.1038/4439

Churchill GA (2002) Fundamentals of experimental design for cDNA microarrays. Nat Genet 32:490–495. https://doi.org/10.1038/ng1031

Churchill GA, Oliver B (2001) Sex, flies and microarrays. Nat Genet 29(4):355–356. https://doi.org/10.1038/ng1201-355

Clark TA, Sugnet CW, Ares M (2002) Genomewide analysis of mRNA processing in yeast using splicing-specific microarrays. Science 296(5569):907–910. https://doi.org/10.1126/science.1069415

Clark TA, Schweitzer AC, Chen TX, Staples MK, Lu G, Wang H, Williams A, Blume JE (2007) Discovery of tissue-specific exons using comprehensive human exon microarrays. Genome Biol 8(4):R64. https://doi.org/10.1186/gb-2007-8-4-r64

Collins RL, Brand H, Karczewski KJ, Zhao X, Alföldi J, Francioli LC, Khera AV, Lowther C, Gauthier LD, Wang H, Watts NA, Solomonson M, O'Donnell-Luria A, Baumann A, Munshi R, Walker M, Whelan CW, Huang Y, Brookings T, Sharpe T, Stone MR, Valkanas E, Fu J, Tiao G, Laricchia KM, Ruano-Rubio V, Stevens C, Gupta N, Cusick C, Margolin L, Alföldi J, Armean IM, Banks E, Bergelson L, Cibulskis K, Collins RL, Connolly KM, Covarrubias M, Cummings B, Daly MJ, Donnelly S, Farjoun Y, Ferriera S, Francioli L, Gabriel S, Gauthier LD, Gentry J, Gupta N, Jeandet T, Kaplan D, Karczewski KJ, Laricchia KM, Llanwarne C, Minikel EV, Munshi R, Neale BM, Novod S, O'Donnell-Luria AH, Petrillo N, Poterba T, Roazen D, Ruano-Rubio V, Saltzman A, Samocha KE, Schleicher M, Seed C, Solomonson M, Soto J, Tiao G, Tibbetts K, Tolonen C, Vittal C, Wade G, Wang A, Wang Q, Ware JS, Watts NA, Weisburd B, Whiffin N, Salinas CAA, Ahmad T, Albert CM, Ardissino D, Atzmon G, Barnard J, Beaugerie L, Benjamin EJ, Boehnke M, Bonnycastle LL, Bottinger EP, Bowden DW, Bown MJ, Chambers JC, Chan JC, Chasman D, Cho J, Chung MK, Cohen B, Correa A, Dabelea D, Daly MJ, Darbar D, Duggirala R, Dupuis J, Ellinor PT, Elosua R, Erdmann J, Esko T, Färkkilä M, Florez J, Franke A, Getz G, Glaser B, Glatt SJ, Goldstein D, Gonzalez C, Groop L, Haiman C, Hanis C, Harms M, Hiltunen M, Holi MM, Hultman CM, Kallela M, Kaprio J, Kathiresan S, Kim B-J, Kim YJ, Kirov G, Kooner J, Koskinen S, Krumholz HM, Kugathasan S, Kwak SH, Laakso M, Lehtimäki T, Loos RJF, Lubitz SA, Ma RCW, MacArthur DG, Marrugat J, Mattila KM, McCarroll S, McCarthy MI, McGovern D, McPherson R, Meigs JB, Melander O, Metspalu A, Neale BM, Nilsson PM, O'Donovan MC, Ongur D, Orozco L, Owen MJ, Palmer CNA, Palotie A, Park KS, Pato C, Pulver AE, Rahman N, Remes AM, Rioux JD, Ripatti S, Roden DM, Saleheen D, Salomaa V, Samani NJ, Scharf J, Schunkert H, Shoemaker MB, Sklar P, Soininen H, Sokol H, Spector T, Sullivan PF, Suvisaari J, Tai ES, Teo YY, Tiinamaija T, Tsuang M, Turner D, Tusie-Luna T, Vartiainen E, Vawter MP, Ware JS, Watkins H, Weersma RK, Wessman M, Wilson JG, Xavier RJ, Taylor KD, Lin HJ, Rich SS, Post WS, Chen Y-DI, Rotter JI, Nusbaum C, Philippakis A, Lander E, Gabriel S, Neale BM, Kathiresan S, Daly MJ, Banks E, MacArthur DG, Talkowski ME (2020) A structural variation reference for medical and population genetics. Nature 581(7809):444–451. https://doi.org/10.1038/s41586-020-2287-8

Comaills V, Castellano-Pozo M (2023) Chromosomal instability in genome evolution: from cancer to macroevolution. Biology 12(5):671. https://doi.org/10.3390/biology12050671

Consortium TEP (2011) A user's guide to the encyclopedia of DNA elements (ENCODE). PLoS Biol 9(4): e1001046. https://doi.org/10.1371/journal.pbio.1001046

Council NR (1988) Mapping and sequencing the human genome. The National Academies Press, Washington, DC

Croft L, Schandorff S, Clark F, Burrage K, Arctander P, Mattick JS (2000) ISIS, the intron information system, reveals the high frequency of alternative splicing in the human genome. Nat Genet 24(4):340–341. https://doi.org/10.1038/74153

DeRisi J, Penland L, Brown PO, Bittner ML, Meltzer PS, Ray M, Chen Y, Su YA, Trent JM (1996) Use of a cDNA microarray to analyse gene expression patterns in human cancer. Nat Genet 14(4):457–460. https://doi.org/10.1038/ng1296-457

DeRisi JL, Iyer VR, Brown PO (1997) Exploring the metabolic and genetic control of gene expression on a genomic scale. Science 278(5338):680–686. https://doi.org/10.1126/science.278.5338.680

Dhami P, Coffey AJ, Abbs S, Vermeesch JR, Dumanski JP, Woodward KJ, Andrews RM, Langford C, Vetrie D (2005) Exon array CGH: detection of copy-number changes at the resolution of individual exons in the human genome. Am J Hum Genet 76(5):750–762. https://doi.org/10.1086/429588

Dölken L, Ruzsics Z, Rädle B, Friedel CC, Zimmer R, Mages J, Hoffmann R, Dickinson P, Forster T, Ghazal P, Koszinowski UH (2008) High-resolution gene expression profiling for simultaneous kinetic parameter analysis of RNA synthesis and decay. RNA 14(9):1959–1972. https://doi.org/10.1261/rna.1136108

Dramanac R, Labat I, Brukner I, Crkvenjakov R (1989) Sequencing of megabase plus DNA by hybridization: theory of the method. Genomics 4(2):114–128. https://doi.org/10.1016/0888-7543(89)90290-5

Duggan DJ, Bittner M, Chen Y, Meltzer P, Trent JM (1999) Expression profiling using cDNA microarrays. Nat Genet 21(Suppl 1):10–14. https://doi.org/10.1038/4434

Edgar R, Domrachev M, Lash AE (2002) Gene expression omnibus: NCBI gene expression and hybridization array data repository. Nucleic Acids Res 30(1): 207–210. https://doi.org/10.1093/nar/30.1.207

Eisen MB, Spellman PT, Brown PO, Botstein D (1998) Cluster analysis and display of genome-wide expression patterns. Proc Natl Acad Sci 95(25): 14863–14868. https://doi.org/10.1073/pnas.95.25.14863

Fan J-B, Oliphant A, Shen R, Kermani BG, Garcia F, Gunderson KL, Hansen M, Steemers F, Butler SL, Deloukas P, Galver L, Hunt S, Mcbride C, Bibikova M, Rubano T, Chen J, Wickham E, Doucet D, Chang W, Campbell D, Zhang B, Kruglyak S, Bentley D, Haas J, Rigault P, Zhou L, Stuelpnagel J, Chee MS (2003) Highly parallel SNP genotyping. Cold Spring Harb Symp Quant Biol 68(1): 69–78. https://doi.org/10.1101/sqb.2003.68.69

Fan J-B, Yeakley JM, Bibikova M, Chudin E, Wickham E, Chen J, Doucet D, Rigault P, Zhang B, Shen R, McBride C, Li H-R, Fu X-D, Oliphant A, Barker DL, Chee MS (2004) A versatile assay for high-throughput gene expression profiling on universal array matrices. Genome Res 14(5):878–885. https://doi.org/10.1101/gr.2167504

Fodor SPA, Read JL, Pirrung MC, Stryer L, Lu AT, Solas D (1991) Light-directed, spatially addressable parallel chemical synthesis. Science 251(4995):767–773. https://doi.org/10.1126/science.1990438

Fodor SPA, Rava RP, Huang XC, Pease AC, Holmes CP, Adams CL (1993) Multiplexed biochemical assays with biological chips. Nature 364(6437):555–556. https://doi.org/10.1038/364555a0

Friedel CC, Dölken L, Ruzsics Z, Koszinowski UH, Zimmer R (2009) Conserved principles of mammalian transcriptional regulation revealed by RNA half-life. Nucleic Acids Res 37(17):e115–e115. https://doi.org/10.1093/nar/gkp542

Frommer M, McDonald LE, Millar DS, Collis CM, Watt F, Grigg GW, Molloy PL, Paul CL (1992) A genomic sequencing protocol that yields a positive display of 5-methylcytosine residues in individual DNA strands. Proc Natl Acad Sci 89(5):1827–1831. https://doi.org/10.1073/pnas.89.5.1827

Gama-Carvalho M, Barbosa-Morais NL, Brodsky AS, Silver PA, Carmo-Fonseca M (2006) Genome-wide identification of functionally distinct subsets of cellular mRNAs associated with two nucleocytoplasmic-shuttling mammalian splicing factors. Genome Biol 7(11):R113. https://doi.org/10.1186/gb-2006-7-11-r113

Gentleman RC, Carey VJ, Bates DM, Bolstad BM, Dettling M, Dudoit S, Ellis B, Gautier L, Ge Y, Gentry J, Hornik K, Hothorn T, Huber W, Iacus S, Irizarry R, Leisch F, Li C, Maechler M, Rossini AJ, Sawitzki G, Smith C, Smyth G, Tierney L, Yang YH, Zhang J (2004) Bioconductor: open software development for computational biology and bioinformatics. Genome Biol 5(10):R80. https://doi.org/10.1186/gb-2004-5-10-r80

Gitan RS, Shi H, Chen C-M, Yan PS, Huang TH-M (2002) Methylation-specific oligonucleotide microarray: a new potential for high-throughput methylation analysis. Genome Res 12(1):158–164. https://doi.org/10.1101/gr.202801

Golub TR, Slonim DK, Tamayo P, Huard C, Gaasenbeek M, Mesirov JP, Coller H, Loh ML, Downing JR, Caligiuri MA, Bloomfield CD, Lander ES (1999) Molecular classification of cancer: class discovery and class prediction by gene expression monitoring. Science 286(5439):531–537. https://doi.org/10.1126/science.286.5439.531

Grant SFA, Hakonarson H (2008) Microarray technology and applications in the arena of genome-wide association. Clin Chem 54(7):1116–1124. https://doi.org/10.1373/clinchem.2008.105395

Grayton HM, Fernandes C, Rujescu D, Collier DA (2012) Copy number variations in neurodevelopmental disorders. Prog Neurobiol 99(1):81–91. https://doi.org/10.1016/j.pneurobio.2012.07.005

Gunderson KL, Kruglyak S, Graige MS, Garcia F, Kermani BG, Zhao C, Che D, Dickinson T, Wickham E, Bierle J, Doucet D, Milewski M, Yang R, Siegmund C, Haas J, Zhou L, Oliphant A, Fan J-B, Barnard S, Chee MS (2004) Decoding randomly ordered DNA arrays. Genome Res 14(5): 870–877. https://doi.org/10.1101/gr.2255804

Gunderson KL, Steemers FJ, Lee G, Mendoza LG, Chee MS (2005) A genome-wide scalable SNP genotyping assay using microarray technology. Nat Genet 37(5): 549–554. https://doi.org/10.1038/ng1547

Gunderson KL, Steemers FJ, Ren H, Ng P, Zhou L, Tsan C, Chang W, Bullis D, Musmacker J, King C, Lebruska LL, Barker D, Oliphant A, Kuhn KM, Shen R (2006) Whole-genome genotyping Methods. Enzym 410:359–376. https://doi.org/10.1016/s0076-6879(06)10017-8

Guo Z, Guilfoyle RA, Thiel AJ, Wang R, Smith LM (1994) Direct fluorescence analysis of genetic polymorphisms by hybridization with oligonucleotide arrays on glass supports. Nucleic Acids Res 22(24): 5456–5465. https://doi.org/10.1093/nar/22.24.5456

Hacia JG, Brody LC, Chee MS, Fodor SPA, Collins FS (1996) Detection of heterozygous mutations in BRCA1 using high density oligonucleotide arrays and two–colour fluorescence analysis. Nat Genet 14(4):441–447. https://doi.org/10.1038/ng1296-441

Hillier LD, Lennon G, Becker M, Bonaldo MF, Chiapelli B, Chissoe S, Dietrich N, DuBuque T, Favello A, Gish W, Hawkins M, Hultman M, Kucaba T, Lacy M, Le M, Le N, Mardis E, Moore B, Morris M, Parsons J, Prange C, Rifkin L, Rohlfing T, Schellenberg K, Soares MB, Tan F, Thierry-Meg J, Trevaskis E, Underwood K, Wohldman P, Waterston R, Wilson R, Marra M (1996) Generation and analysis of 280,000 human expressed sequence tags. Genome Res 6(9):807–828. https://doi.org/10.1101/gr.6.9.807

Hong F, Breitling R, Mc Entee CW, Wittner BS, Nemhauser JL, Chory J (2006) RankProd: a bioconductor package for detecting differentially expressed genes in meta-analysis. Bioinformatics 22(22):2825–2402. https://doi.org/10.1093/bioinformatics/btl476

Hu GK, Madore SJ, Moldover B, Jatkoe T, Balaban D, Thomas J, Wang Y (2001) Predicting splice variant from DNA Chip expression Data. Genome Res 11(7): 1237–1245. https://doi.org/10.1101/gr.165501

Hughes TR, Mao M, Jones AR, Burchard J, Marton MJ, Shannon KW, Lefkowitz SM, Ziman M, Schelter JM, Meyer MR, Kobayashi S, Davis C, Dai H, He YD, Stephaniants SB, Cavet G, Walker WL, West A,

Coffey E, Shoemaker DD, Stoughton R, Blanchard AP, Friend SH, Linsley PS (2001) Expression profiling using microarrays fabricated by an ink-jet oligonucleotide synthesizer. Nat Biotechnol 19(4):342–347. https://doi.org/10.1038/86730

Irizarry RA, Hobbs B, Collin F, Beazer-Barclay YD, Antonellis KJ, Scherf U, Speed TP (2003) Exploration, normalization, and summaries of high density oligonucleotide array probe level data. Biostatistics 4(2): 249–264. https://doi.org/10.1093/biostatistics/4.2.249

Ishkanian AS, Malloff CA, Watson SK, de Leeuw RJ, Chi B, Coe BP, Snijders A, Albertson DG, Pinkel D, Marra MA, Ling V, MacAulay C, Lam WL (2004) A tiling resolution DNA microarray with complete coverage of the human genome. Nat Genet 36(3):299–303. https://doi.org/10.1038/ng1307

Iyer VR, Eisen MB, Ross DT, Schuler G, Moore T, Lee JCF, Trent JM, Staudt LM, Hudson J Jr, Boguski MS, Lashkari D, Shalon D, Botstein D, Brown PO (1999) The transcriptional program in the response of human fibroblasts to serum. Science 283(5398):83–87. https://doi.org/10.1126/science.283.5398.83

Iyer VR, Horak CE, Scafe CS, Botstein D, Snyder M, Brown PO (2001) Genomic binding sites of the yeast cell-cycle transcription factors SBF and MBF. Nature 409(6819):533–538. https://doi.org/10.1038/35054095

Jiang L, Schlesinger F, Davis CA, Zhang Y, Li R, Salit M, Gingeras TR, Oliver B (2011) Synthetic spike-in standards for RNA-seq experiments. Genome Res 39(18):e129. https://doi.org/10.1101/gr.121095.111

Jin W, Riley RM, Wolfinger RD, White KP, Passador-Gurgel G, Gibson G (2001) The contributions of sex, genotype and age to transcriptional variance in Drosophila melanogaster. Nat Genet 29(4):389–395. https://doi.org/10.1038/ng766

Johannes G, Carter MS, Eisen MB, Brown PO, Sarnow P (1999) Identification of eukaryotic mRNAs that are translated at reduced cap binding complex eIF4F concentrations using a cDNA microarray. Proc Natl Acad Sci 96(23):13118–13123. https://doi.org/10.1073/pnas.96.23.13118

Johnson JM, Castle J, Garrett-Engele P, Kan Z, Loerch PM, Armour CD, Santos R, Schadt EE, Stoughton R, Shoemaker DD (2003) Genome-wide survey of human alternative pre-mRNA splicing with exon junction microarrays. Science 302(5653):2141–2144. https://doi.org/10.1126/science.1090100

Johnson JM, Edwards S, Shoemaker D, Schadt EE (2005) Dark matter in the genome: evidence of widespread transcription detected by microarray tiling experiments. Trends Genet 21(2):93–102. https://doi.org/10.1016/j.tig.2004.12.009

Juneau K, Palm C, Miranda M, Davis RW (2007) High-density yeast-tiling array reveals previously undiscovered introns and extensive regulation of meiotic splicing. Proc Natl Acad Sci 104(5):1522–1527. https://doi.org/10.1073/pnas.0610354104

Kafatos FC, Jones CW, Efstratiadis A (1979) Determination of nucleic acid sequence homologies and relative concentrations by a dot hybridization procedure. Nucleic Acids Res 7(6):1541–1552. https://doi.org/10.1093/nar/7.6.1541

Kallioniemi A, Kallioniemi O-P, Sudar D, Rutovitz D, Gray JW, Waldman F, Pinkel D (1992) Comparative genomic hybridization for molecular cytogenetic analysis of solid tumors. Science 258(5083):818–821. https://doi.org/10.1126/science.1359641

Kellam P (2001) Microarray gene expression database: progress towards an international repository of gene expression data. Genome Biol 2(5):reports4011.1. https://doi.org/10.1186/gb-2001-2-5-reports4011

Kenzelmann M, Maertens S, Hergenhahn M, Kueffer S, Hotz-Wagenblatt A, Li L, Wang S, Ittrich C, Lemberger T, Arribas R, Jonnakuty S, Hollstein MC, Schmid W, Gretz N, Gröne HJ, Schütz G (2007) Microarray analysis of newly synthesized RNA in cells and animals. Proc Natl Acad Sci 104(15): 6164–6169. https://doi.org/10.1073/pnas.0610439104

Keshet I, Schlesinger Y, Farkash S, Rand E, Hecht M, Segal E, Pikarski E, Young RA, Niveleau A, Cedar H, Simon I (2006) Evidence for an instructive mechanism of de novo methylation in cancer cells. Nat Genet 38(2):149–153. https://doi.org/10.1038/ng1719

Khrapko KR, Lysov YP, Khorlyn AA, Shick VV, Florentiev VL, Mirzabekov AD (1989) An oligonucleotide hybridization approach to DNA sequencing. FEBS Lett 256(1–2)(122):118. https://doi.org/10.1016/0014-5793(89)81730-2

Kruglyak L, Nickerson DA (2001) Variation is the spice of life. Nat Genet 27(3):234–236. https://doi.org/10.1038/85776

Kuhn K, Baker SC, Chudin E, Lieu M-H, Oeser S, Bennett H, Rigault P, Barker D, McDaniel TK, Chee MS (2004) A novel, high-performance random array platform for quantitative gene expression profiling. Genome Res 14(11):2347–2356. https://doi.org/10.1101/gr.2739104

Laird PW (2010) Principles and challenges of genome-wide DNA methylation analysis. Nat Rev Genet 11(3): 191–203. https://doi.org/10.1038/nrg2732

Lennon G, Auffray C, Polymeropoulos M, Soares MB (1996) The I.M.A.G.E. Consortium: an integrated molecular analysis of genomes and their expression. Genomics 33(1):151–152. https://doi.org/10.1006/geno.1996.0177

Lenoir T, Giannella E (2006) The emergence and diffusion of DNA microarray technology. J Biomed Discov Collab 1(1):11. https://doi.org/10.1186/1747-5333-1-11

Lipshutz RJ, Fodor SPA, Gingeras TR, Lockhart DJ (1999) High density synthetic oligonucleotide arrays. Nat Genet 21(Suppl 1):20–24. https://doi.org/10.1038/4447

Manolio TA, Collins FS (2009) The HapMap and genome-wide association studies in diagnosis and therapy*.

Annu Rev Med 60(1):443–456. https://doi.org/10.1146/annurev.med.60.061907.093117

Maskos U, Southern EM (1993) A novel method for the parallel analysis of multiple mutations in multiple samples. Nucleic Acids Res 21(9):2269–2270. https://doi.org/10.1093/nar/21.9.2269

McGeachie AB, Koishi K, Andrews ZB, McLennan IS (2005) Analysis of mRNAs that are enriched in the post-synaptic domain of the neuromuscular junction. Mol Cell Neurosci 30(2):173–185. https://doi.org/10.1016/j.mcn.2005.07.002

Meijer HA, Bushell M, Hill K, Gant TW, Willis AE, Jones P, de Moor CH (2007) A novel method for poly(A) fractionation reveals a large population of mRNAs with a short poly(A) tail in mammalian cells. Nucleic Acids Res 35(19):e132. https://doi.org/10.1093/nar/gkm830

Michael KL, Taylor LC, Schultz SL, Walt DR (1998) Randomly ordered addressable high-density optical sensor arrays. Anal Chem 70(7):1242–1248. https://doi.org/10.1021/ac971343r

Miller DT, Adam MP, Aradhya S, Biesecker LG, Brothman AR, Carter NP, Church DM, Crolla JA, Eichler EE, Epstein CJ, Faucett WA, Feuk L, Friedman JM, Hamosh A, Jackson L, Kaminsky EB, Kok K, Krantz ID, Kuhn RM, Lee C, Ostell JM, Rosenberg C, Scherer SW, Spinner NB, Stavropoulos DJ, Tepperberg JH, Thorland EC, Vermeesch JR, Waggoner DJ, Watson MS, Martin CL, Ledbetter DH (2010) Consensus statement: chromosomal microarray is a first-tier clinical diagnostic test for individuals with developmental disabilities or congenital anomalies. Am J Hum Genet 86(5):749–764. https://doi.org/10.1016/j.ajhg.2010.04.006

Mironov AA, Fickett JW, Gelfand MS (1999) Frequent alternative splicing of human genes. Genome Res 9(12):1288–1293. https://doi.org/10.1101/gr.9.12.1288

Monti S, Tamayo P, Mesirov J, Golub T (2003) Consensus clustering: a resampling-based method for class discovery and visualization of gene expression microarray data. Mach Learn 52:91–118. https://doi.org/10.1023/A:1023949509487

Moore MJ, Silver PA (2008) Global analysis of mRNA splicing. RNA 14(2):197–203. https://doi.org/10.1261/rna.868008

Nagaoka SI, Hassold TJ, Hunt PA (2012) Human aneuploidy: mechanisms and new insights into an age-old problem. Nat Rev Genet 13(7):493–504. https://doi.org/10.1038/nrg3245

Nizetić D, Zehetner G, Monaco AP, Gellen L, Young BD, Lehrach H (1991) Construction, arraying, and high-density screening of large insert libraries of human chromosomes X and 21: their potential use as reference libraries. Proc Natl Acad Sci 88(8):3233–3237. https://doi.org/10.1073/pnas.88.8.3233

Ohtsu M, Kawate M, Fukuoka M, Gunji W, Hanaoka F, Utsugi T, Onoda F, Murakami Y (2008) Novel DNA microarray system for analysis of nascent mRNAs. DNA Res 15(4):241–251. https://doi.org/10.1093/dnares/dsn015

Okoniewski MJ, Hey Y, Pepper SD, Miller CJ (2007) High correspondence between Affymetrix exon and standard expression arrays. BioTechniques 42(2):181–185. https://doi.org/10.2144/000112315

Pease AC, Solas D, Sullivan EJ, Cronin MT, Holmes CP, Fodor SP (1994) Light-generated oligonucleotide arrays for rapid DNA sequence analysis. Proc Natl Acad Sci 91(11):5022–5026. https://doi.org/10.1073/pnas.91.11.5022

Perou CM, Sørlie T, Eisen MB, van de Rijn M, Jeffrey SS, Rees CA, Pollack JR, Ross DT, Johnsen H, Akslen LA, Fluge Ø, Pergamenschikov A, Williams C, Zhu SX, Lønning PE, Børresen-Dale A-L, Brown PO, Botstein D (2000) Molecular portraits of human breast tumours. Nature 406(6797):747–752. https://doi.org/10.1038/35021093

Phimister B (1999) Going global. Nat Genet 21(Suppl 1):1–1. https://doi.org/10.1038/4423

Piétu G, Mariage-Samson R, Fayein N-A, Matingou C, Eveno E, Houlgatte R, Decraene C, Vandenbrouck Y, Tahi F, Devignes M-D, Wirkner U, Ansorge W, Cox D, Nagase T, Nomura N, Auffray C (1999) The Genexpress IMAGE knowledge base of the human brain transcriptome: a prototype integrated resource for functional and computational genomics. Genome Res 9(2):195–209. https://doi.org/10.1101/gr.9.2.195

Pinkel D, Segraves R, Sudar D, Clark S, Poole I, Kowbel D, Collins C, Kuo W-L, Chen C, Zhai Y, Dairkee SH, Ljung B, Gray JW, Albertson DG (1998) High resolution analysis of DNA copy number variation using comparative genomic hybridization to microarrays. Nat Genet 20(2):207–211. https://doi.org/10.1038/2524

Pollack JR, Perou CM, Alizadeh AA, Eisen MB, Pergamenschikov A, Williams CF, Jeffrey SS, Botstein D, Brown PO (1999) Genome-wide analysis of DNA copy-number changes using cDNA microarrays. Nat Genet 23(1):41–46. https://doi.org/10.1038/12640

Raghavan A, Ogilvie RL, Reilly C, Abelson ML, Raghavan S, Vasdewani J, Krathwohl M, Bohjanen PR (2002) Genome-wide analysis of mRNA decay in resting and activated primary human T lymphocytes. Nucleic Acids Res 30(24):5529–5538. https://doi.org/10.1093/nar/gkf682

Ren B, Robert F, Wyrick JJ, Aparicio O, Jennings EG, Simon I, Zeitlinger J, Schreiber J, Hannett N, Kanin E, Volkert TL, Wilson CJ, Bell SP, Young RA (2000) Genome-wide location and function of DNA binding proteins. Science 290(5500):2306–2309. https://doi.org/10.1126/science.290.5500.2306

Rhodes DR, Yu J, Shanker K, Deshpande N, Varambally R, Ghosh D, Barrette T, Pandey A, Chinnaiyan AM (2004) Large-scale meta-analysis of cancer microarray data identifies common transcriptional profiles of neoplastic transformation and progression. Proc Natl Acad Sci 101(25): 9309–9314. https://doi.org/10.1073/pnas.0401994101

Sarkans U, Füllgrabe A, Ali A, Athar A, Behrangi E, Diaz N, Fexova S, George N, Iqbal H, Kurri S, Munoz J, Rada J, Papatheodorou I, Brazma A (2020) From ArrayExpress to BioStudies. Nucleic Acids Res 49(D1):D1502–D1506. https://doi.org/10.1093/nar/gkaa1062

Schadt EE, Li C, Ellis B, Wong WH (2002) Feature extraction and normalization algorithms for high-density oligonucleotide gene expression array data. J Cell Biochem 84(S37):120–125. https://doi.org/10.1002/jcb.10073

Schena M, Shalon D, Davis RW, Brown PO (1995) Quantitative monitoring of gene expression patterns with a complementary DNA microarray. Science 270(5235): 467–470. https://doi.org/10.1126/science.270.5235.467

Schena M, Shalon D, Heller R, Chai A, Brown PO, Davis RW (1996) Parallel human genome analysis: microarray-based expression monitoring of 1000 genes. Proc Natl Acad Sci 93(20):10614–10619. https://doi.org/10.1073/pnas.93.20.10614

Shalon D, Smith SJ, Brown PO (1996) A DNA microarray system for analyzing complex DNA samples using two-color fluorescent probe hybridization. Genome Res 6(7):639–645. https://doi.org/10.1101/gr.6.7.639

Shao L, Akkari Y, Cooley LD, Miller DT, Seifert BA, Wolff DJ, Mikhail FM, Committee ALQA (2021) Chromosomal microarray analysis, including constitutional and neoplastic disease applications, 2021 revision: a technical standard of the American College of Medical Genetics and Genomics (ACMG). Genet Med 23(10):1818–1829. https://doi.org/10.1038/s41436-021-01214-w

Sherlock G, Hernandez-Boussard T, Kasarskis A, Binkley G, Matese JC, Dwight SS, Kaloper M, Weng S, Jin H, Ball CA, Eisen MB, Spellman PT, Brown PO, Botstein D, Cherry JM (2001) The Stanford microarray database. Nucleic Acids Res 29(1):152–155. https://doi.org/10.1093/nar/29.1.152

Shoemaker DD, Schadt EE, Armour CD, He YD, Garrett-Engele P, McDonagh PD, Loerch PM, Leonardson A, Lum PY, Cavet G, Wu LF, Altschuler SJ, Edwards S, King J, Tsang JS, Schimmack G, Schelter JM, Koch J, Ziman M, Marton MJ, Li B, Cundiff P, Ward T, Castle J, Krolewski M, Meyer MR, Mao M, Burchard J, Kidd MJ, Dai H, Phillips JW, Linsley PS, Stoughton R, Scherer S, Boguski MS (2001) Experimental annotation of the human genome using microarray technology. Nature 409(6822):922–927. https://doi.org/10.1038/35057141

Simon R, Radmacher MD, Dobbin K, McShane LM (2003) Pitfalls in the use of DNA microarray data for diagnostic and prognostic classification. J Natl Cancer Inst 95(1):14–18. https://doi.org/10.1093/jnci/95.1.14

Singh-Gasson S, Green RD, Yue Y, Nelson C, Blattner F, Sussman MR, Cerrina F (1999) Maskless fabrication of light-directed oligonucleotide microarrays using a digital micromirror array. Nat Biotechnol 17(10): 974–978. https://doi.org/10.1038/13664

Smyth GK (2004) Linear models and empirical Bayes methods for assessing differential expression in microarray experiments. Stat Appl Genet Mol Biol 3(1):Article 3. https://doi.org/10.2202/1544-6115.1027

Solinas-Toldo S, Lampel S, Stilgenbauer S, Nickolenko J, Benner A, Döhner H, Cremer T, Lichter P (1997) Matrix-based comparative genomic hybridization: biochips to screen for genomic imbalances. Genes Chromosom Cancer 20(4):399–407

Solomon MJ, Larsen PL, Varshavsky A (1988) Mapping proteinDNA interactions in vivo with formaldehyde: evidence that histone H4 is retained on a highly transcribed gene. Cell 53(6):937–947. https://doi.org/10.1016/s0092-8674(88)90469-2

Southern EM (1975) Detection of specific sequences among DNA fragments separated by gel electrophoresis. J Mol Biol 98(3):503–517. https://doi.org/10.1016/s0022-2836(75)80083-0

Southern EM (1996) DNA chips: analysing sequence by hybridization to oligonucleotides on a large scale. Trends Genet 12(3):110–115. https://doi.org/10.1016/0168-9525(96)81422-3

Southern EM, Maskos U, Elder JK (1992) Analyzing and comparing nucleic acid sequences by hybridization to arrays of oligonucleotides: evaluation using experimental models. Genomics 13(4):1008–1017. https://doi.org/10.1016/0888-7543(92)90014-j

Southern EM, Case-Green SC, Elder JK, Johnson M, Mir KU, Wang L, Williams JC (1994) Arrays of complementary oligonucleotides for analysing the hybridisation behaviour of nucleic acids. Nucleic Acids Res 22(8):1368–1373. https://doi.org/10.1093/nar/22.8.1368

Subramanian A, Tamayo P, Mootha VK, Mukherjee S, Ebert BL, Gillette MA, Paulovich A, Pomeroy SL, Golub TR, Lander ES, Mesirov JP (2005) Gene set enrichment analysis: a knowledge-based approach to interpreting genome-wide expression profiles. Proc Natl Acad Sci 102(43):15545–15550. https://doi.org/10.1073/pnas.0506580102

Tenenbaum SA, Carson CC, Lager PJ, Keene JD (2000) Identifying mRNA subsets in messenger ribonucleoprotein complexes by using cDNA arrays. Proc Natl Acad Sci 97(26):14085–14090. https://doi.org/10.1073/pnas.97.26.14085

Tusher VG, Tibshirani R, Chu G (2001) Significance analysis of microarrays applied to the ionizing radiation response. Proc Natl Acad Sci 98(9):5116–5121. https://doi.org/10.1073/pnas.091062498

Uffelmann E, Huang QQ, Munung NS, de Vries J, Okada Y, Martin AR, Martin HC, Lappalainen T,

Posthuma D (2021) Genome-wide association studies. Nat Rev Methods Prim 1(1):59. https://doi.org/10.1038/s43586-021-00056-9

van den IJssel P, Tijssen M, Chin S-F, Eijk P, Carvalho B, Hopmans E, Holstege H, Bangarusamy DK, Jonkers J, Meijer GA, Caldas C, Ylstra B (2005) Human and mouse oligonucleotide-based array CGH. Nucleic Acids Res 33(22):e192–e192. https://doi.org/10.1093/nar/gni191

Velculescu VE, Zhang L, Vogelstein B, Kinzler KW (1995) Serial analysis of gene expression. Science 270(5235):484–487

Velculescu VE, Zhang L, Zhou W, Vogelstein J, Basrai MA, Bassett DE, Hieter P, Vogelstein B, Kinzler KW (1997) Characterization of the yeast transcriptome. Cell 88(2):243–251. https://doi.org/10.1016/s0092-8674(00)81845-0

Walt DR (2000) Bead-based fiber-optic arrays. Science 287(5452):451–452. https://doi.org/10.1126/science.287.5452.451

Wang DG, Fan J-B, Siao C-J, Berno A, Young P, Sapolsky R, Ghandour G, Perkins N, Winchester E, Spencer J, Kruglyak L, Stein L, Hsie L, Topaloglou T, Hubbell E, Robinson E, Mittmann M, Morris MS, Shen N, Kilburn D, Rioux J, Nusbaum C, Rozen S, Hudson TJ, Lipshutz R, Chee M, Lander ES (1998) Large-scale identification, mapping, and genotyping of single-nucleotide polymorphisms in the human genome. Science 280(5366):1077–1082. https://doi.org/10.1126/science.280.5366.1077

Wang W-X, Wilfred BR, Hu Y, Stromberg AJ, Nelson PT (2010) Anti-Argonaute RIP-Chip shows that miRNA transfections alter global patterns of mRNA recruitment to microribonucleoprotein complexes. RNA (New York) 16(2):394–404. https://doi.org/10.1261/rna.1905910

Weber M, Davies JJ, Wittig D, Oakeley EJ, Haase M, Lam WL, Schübeler D (2005) Chromosome-wide and promoter-specific analyses identify sites of differential DNA methylation in normal and transformed human cells. Nat Genet 37(8):853–862. https://doi.org/10.1038/ng1598

Yamada K, Lim J, Dale JM, Chen H, Shinn P, Palm CJ, Southwick AM, Wu HC, Kim C, Nguyen M, Pham P, Cheuk R, Karlin-Newmann G, Liu SX, Lam B, Sakano H, Wu T, Yu G, Miranda M, Quach HL, Tripp M, Chang CH, Lee JM, Toriumi M, Chan MMH, Tang CC, Onodera CS, Deng JM, Akiyama K, Ansari Y, Arakawa T, Banh J, Banno F, Bowser L, Brooks S, Carninci P, Chao Q, Choy N, Enju A, Goldsmith AD, Gurjal M, Hansen NF, Hayashizaki Y, Johnson-Hopson C, Hsuan VW, Iida K, Karnes M, Khan S, Koesema E, Ishida J, Jiang PX, Jones T, Kawai J, Kamiya A, Meyers C, Nakajima M, Narusaka M, Seki M, Sakurai T, Satou M, Tamse R, Vaysberg M, Wallender EK, Wong C, Yamamura Y, Yuan S, Shinozaki K, Davis RW, Theologis A, Ecker JR (2003) Empirical analysis of transcriptional activity in the Arabidopsis genome. Science 302(5646):842–846. https://doi.org/10.1126/science.1088305

Yang YH, Speed T (2002) Design issues for cDNA microarray experiments. Nat Rev Genet 3(8):579–588. https://doi.org/10.1038/nrg863

Yang YH, Dudoit S, Luu P, Lin DM, Peng V, Ngai J, Speed TP (2002) Normalization for cDNA microarray data: a robust composite method addressing single and multiple slide systematic variation. Nucleic Acids Res 30(4):e15. https://doi.org/10.1093/nar/30.4.e15

Zhang L, Zhou W, Velculescu VE, Kern SE, Hruban RH, Hamilton SR, Vogelstein B, Kinzler KW (1997) Gene expression profiles in Normal and cancer cells. Science 276(5316):1268–1272. https://doi.org/10.1126/science.276.5316.1268

Advances in Massive Parallel Sequencing: From Genomics to Spatial Transcriptomics

3

Tomás Araújo and Margarida Gama-Carvalho

Abstract

The ability to decode nucleic acids has reshaped biological research and biotechnology, enabling systematic analysis of genome structure, gene regulation, and cellular heterogeneity. This chapter reviews the major technological advances that have driven this transformation. We begin with the historical development of Sanger sequencing and its role in establishing the first genome-scale analyses. We then examine the emergence of next-generation sequencing, highlighting the conceptual innovations—massive parallelization, clonal amplification, and cyclic detection chemistries—that enabled high-throughput, cost-effective sequencing. Building on these foundations, we discuss the shift toward single-cell and spatial transcriptomics, which extend sequencing from bulk measurements to the resolution of individual cells and their tissue contexts. Together, these developments illustrate how sequencing technologies have progressed from early linear workflows to multimodal, high-resolution platforms that now support comprehensive interrogation of biological systems.

Keywords

Massive parallel sequencing · Next-generation sequencing (NGS) · Library preparation · Single-cell RNA-seq (scRNA-seq) · Spatial transcriptomics

3.1 Introduction

Few technologies have influenced modern biology as profoundly as nucleic acid sequencing. It provides means to read the genetic information and unravel gene expression processes that underlie cellular function, development, and disease, linking genotype to phenotype and enabling reconstruction of evolutionary relationships. The development of high-throughput methods for sequencing in the early 2000s has dramatically expanded the scale and scope of questions that can be addressed. Together with other omics approaches, sequencing transformed the field from predominantly descriptive research into one capable of systematic, data-driven investigation.

Sequencing is now central to multiple levels of biological inquiry. In genomics, it enables genome assembly, detection of structural and copy-number variation, identification of disease-associated polymorphisms, and profiling of

T. Araújo · M. Gama-Carvalho (✉)
Faculdade de Ciências, Universidade de Lisboa, Lisbon, Portugal

BioISI – Biosystems & Integrative Sciences Institute, Faculty of Sciences, Universidade de Lisboa, Lisbon, Portugal
e-mail: mhcarvalho@fc.ul.pt

M. Gama-Carvalho et al. (eds.), *Omics Approaches in Biomedicine and Biotechnology*, Advances in Experimental Medicine and Biology 1504, https://doi.org/10.1007/978-3-032-18966-0_3

biodiversity across diverse sample types. In transcriptomics, it supports quantification of gene expression, characterization of isoform diversity, and assessment of cellular heterogeneity. The integration of sequencing datasets underpins systems-level studies of regulatory networks in physiological and pathological contexts, driving progress in predictive modeling and therapeutic development.

This chapter reviews the evolution of sequencing technologies from the Sanger method to contemporary massively parallel methods, commonly referred to as next-generation sequencing (NGS). Emphasis is placed on currently used platforms, with only brief reference to intermediate methods that played a relevant role in the development of the field. The chapter discusses the principles, performance, and limitations of each method, highlighting innovations that contributed to increased throughput, resolution, and analytical depth. The final sections focus on transcriptomic applications and recent advances that integrate sequence information with cellular and spatial context. Through this perspective, the chapter illustrates how sequencing continues to expand the resolution at which biological systems can be interrogated and outlines future directions in omics-driven research.

3.2 Historical Perspective: The Era of Sanger Sequencing

3.2.1 Frederick Sanger's Pioneer Work on Sequence-Defined Biopolymers

Despite prior attempts to determine DNA sequences in the early 1970s (Heather and Chain 2016), the first widely applicable and efficient sequencing method was developed by Frederick (Fred) Sanger in 1977, which remained the field standard for approximately 25 years (Sanger et al. 1977b). The significance of his contributions is underscored by the rare distinction of having received two Nobel Prizes (Brownlee 2013).

A biochemist by training, Sanger was the first to determine the complete amino acid sequence of a protein, elucidating the A and B chains of insulin and demonstrating that proteins are sequence-defined biopolymers with a specific, ordered arrangement of residues (Sanger and Tuppy 1951; Sanger and Thompson 1953). He was awarded the 1958 Nobel Prize in Chemistry for "his work on the structure of proteins" and for developing methods to determine their sequences.

Sanger subsequently directed his attention to nucleic acids. In 1967, he published the first sequence of a ribosomal RNA (rRNA) molecule, *E. coli* 5S rRNA (Brownlee et al. 1967). A decade later, he reported the first complete DNA genome sequence, that of bacteriophage φX174, using an early "plus-minus" sequencing method (Sanger et al. 1977a). In that same year, Maxam and Gilbert introduced a chemical sequencing method based on base-specific cleavage of terminally labeled DNA followed by electrophoretic analysis of the fragments (Maxam and Gilbert 1977), while Sanger and colleagues developed the dideoxynucleotide chain-termination method ("Sanger sequencing"), which greatly improved the efficiency, accuracy, and general applicability of DNA sequencing (Sanger et al. 1977b). This method underpinned most sequencing technologies for the next several decades. For these contributions, Sanger was awarded the 1980 Nobel Prize in Chemistry for "his work concerning the determination of base sequences in nucleic acids."

3.2.2 Principles of Sanger Sequencing

Fred Sanger introduced two major conceptual advances that enabled efficient DNA sequencing. The first one was to base the method on de novo synthesis of DNA fragments by polymerase extension from a defined primer. The second one was to exploit 2′,3′-dideoxynucleotides (ddNTPs) to induce precise chain termination. ddNTPs—initially synthesized in Sanger's laboratory—are modified nucleotides that lack the 3-′-hydroxyl required for phosphodiester bond formation. When incorporated into an elongating chain by DNA polymerase, they irreversibly terminate strand extension. In the original protocol, four parallel reactions were prepared, each containing all four dNTPs and one ddNTP, together with DNA polymerase I (Klenow fragment). Primer extension generated nested sets of termination fragments corresponding to every position at which the specific ddNTP could be incorporated. Electrophoretic separation of the four reactions enabled inference of the template sequence (Sanger et al. 1977b). Early implementations used radioactive detection through incorporation of radiolabeled nucleotides (typically α-32P-dATP) or radiolabeled primers (McGraw 1984), allowing autoradiographic visualization of terminated fragments.

A major practical limitation of the original method was the requirement for a single-stranded DNA (ssDNA) template. The φX174 and G4 bacteriophage genomes used by Sanger to set up the method naturally provide ssDNA, but most biological DNA is double-stranded (dsDNA), and strand separation was technically difficult. This limitation could be overcome by exonuclease III treatment of the sample to generate predominantly ssDNA structures from double-stranded molecules (Smith 1979), but cloning of the target sequence into the single-stranded M13 bacteriophage vector soon became the preferred strategy (Sanger et al. 1980).

Later developments included thermal-cycle sequencing enabled by thermostable polymerases, which allowed direct sequencing from double-stranded templates, improved signal intensity, and eliminated the requirement for ssDNA preparation (Murray 1989). The use of thermal cycling allowed for the direct sequencing of dsDNA, although it still requires a cloned template. This was followed by the transition to fluorescence-based detection. The first demonstration of fluorescence-based Sanger sequencing used fluorescently labeled primers in which each base-specific reaction carried a distinct fluorophore; the four reactions were combined and resolved in a single electrophoretic run in an acrylamide gel tube, enabling direct computer-assisted sequence acquisition (Smith et al. 1986). Shortly thereafter, fluorescent dye-terminator chemistry was introduced, where each ddNTP carried a spectrally distinct fluorophore, enabling the entire sequencing reaction to proceed in one tube (Prober et al. 1987). These advances laid the foundation for slab gel-based automated Sanger sequencers and later capillary-electrophoresis instruments in the 1990s.

3.2.3 The Emergence of Automated Sequencers

The development of automated sequencing systems took place in parallel with the early discussions proposing the detailed mapping and sequence analysis of the human genome (Hood et al. 1987). Building on the work developed at the California Institute of Technology by Leroy Hood and Loyd Smith, the company Applied Biosystems Inc. (ABI) played a major role in the process, releasing the first automated sequencer on the market, the ABI 370A in 1986, which was able to run 16 sequencing reactions in parallel using slab gel electrophoresis (Hood et al. 1987). Other early commercial implementations were based on work developed at the European Molecular Biology Laboratory by Ansorge and colleagues (Ansorge et al. 1987; Freeman et al. 1990) and by Hitachi in Japan (Kambara et al. 1988).

Craig Venter's laboratory at the NIH reported the first application of automated sequencing to gene-scale projects (Gocayne et al. 1987). Between 1987 and 1991, this group demonstrated the feasibility of increasing sequencing throughput at a single site by parallelizing the operation of six automated sequencers with two robots for automated sample preparation. Over a 3-year period, this facility produced significant datasets, including sequencing of the entire small-pox genome, 6 human cosmid clones, and 11,000 human and *C. elegans* expressed sequence tags (Adams et al. 1994). With the founding of TIGR in 1992, Venter pioneered integrated, production-scale sequencing and analysis workflows that combined dozens of automated instruments with dedicated robotics and local computational infrastructure (Adams et al. 1994). TIGR's model illustrates how operational scale required not only instrument count, which included 30 ABI 373A sequencers, but also sample-preparation robotics and networked informatics to produce a continuous sequencing-to-data pipeline. In the early 1990s, the Hood lab reported being able to consistently resolve 450–500 bp of sequence from the priming site in 24 samples run in parallel per 12-14 h run in the ABI 737A sequencers, generating ~3.2 Mbp of sequence per working year (Hunkapiller et al. 1991), which is three orders of magnitude below the size of the haploid human genome. At the time, the cost of sequencing was around $3–5 (US) per base pair (Cathcart 1990), with estimates for the completion of the human genome sequence ranging from 10 to 30 billion dollars and a timeframe of 10–15 years (Hood et al. 1987). It was however widely anticipated that developments in sequencing would lead to increased throughputs and decreased costs, making these numbers obsolete.

In 1990, three groups independently reported what would be a major advance in sequencing technology—the development of capillary electrophoresis (Luckey et al. 1990; Swerdlow et al. 1990; Ruiz-Martinez et al. 1993). The commercial adaptation of this technology by ABI resulted in the introduction of the ABI Prism 310 capillary sequencer in 1996, with further developments supporting sequencing of 1000 bp in 1 h running 96 samples in parallel (Hutchison III 2007).

3.2.4 Sequencing the Human Genome: Technological Achievements and Limitations

3.2.4.1 High-Throughput Sequencing Centers and the Scaling of Sanger Technology

The advent of automated sequencers and their rapid evolution was essential for the undertaking of the first "big science" project in biology: sequencing of the human genome. The International Human Genome Sequencing Consortium brought together more than 20 sequencing centers across the United States, the United Kingdom, continental Europe, China, and Japan. Data generation, however, was concentrated in five major institutions: the Wellcome Trust Sanger Institute (UK), the Whitehead Institute/MIT Center for Genome Research (USA), Washington University Genome Sequencing Center (USA), the DOE Joint Genome Institute (USA), and the Baylor College of Medicine Human Genome Sequencing Center (USA) (Lander et al. 2001).

To achieve the throughput required for producing the human genome sequence, these institutions established what were literal "sequencing factories." At the Sanger Centre in Cambridge, opened in 1993, eight sequencing teams were formed within 2 years, supported by centralized groups for subcloning, gel and media preparation, oligonucleotide synthesis, robotics, and software development (Rogers 2025). As the efforts of the project moved forward, the facility later expanded to a new building operating 140 slab-gel sequencers on a continuous schedule, consuming over 2000 gels per week and maintaining a sequence producing rate of ~400,000 reads per month, becoming the largest individual contributor of the final genome sequence. The expansion of sequencing capacity continued at an unprecedented pace, driven by coordinated efforts across the multiple high-throughput centers (Lander et al. 2001). Automation enabled factory-scale workflows capable of processing more than 100,000 reactions in 12 h, while the introduction of new detectors in mid-1999 produced an eightfold increase in output within 8 months, reaching nearly seven million samples per month without loss of read quality. By the end of the project, the collective network of centers was generating raw sequence at a rate equivalent to onefold coverage of the human genome in under 6 weeks, corresponding to more than 1000 nucleotides per second, operating continuously. These advances allowed the consortium to finish the first production phase, producing a first draft sequence covering 90% of the human genome, in June 2000, 10 years after the start of the project. Technological optimization supported a sustained reduction of the per base cost, which decreased roughly a 100-fold over that decade (Lander et al. 2001).

3.2.4.2 Methodological Foundations and Limitations of Sanger Sequencing

A key limitation of the Sanger sequencing method is the need to use cloned DNA segments as templates for each sequencing reaction. With read lengths limited to 500–1000 bp, tackling large sequences required a significant effort in cloning and assembly. Two main strategies were developed early on to tackle this challenge: random or "shotgun" strategies and directed sequencing (Hood et al. 1987).

In the shotgun approach, genomic DNA is randomly fragmented (e.g., by sonication) into 500–1000 bp pieces, which are cloned into a suitable vector. Individual clones are then randomly picked and sequenced, and the resulting reads are assembled. Assembly algorithms capable of handling shotgun data were developed as early as 1980 (Staden 1980). This approach was used by Sanger's lab in 1982 to perform the first complete sequencing of a complex DNA virus, bacteriophage λ, with a 48,502 bp genome. This work was the first to demonstrate that systematic, high-coverage sequencing and computational assembly could scale to tens of kilobases, paving the way for strategies used in later projects. The approach was gradually improved, namely, through better protocols for fragmenting and cloning, resulting in more representative libraries, and the use of paired-end sequencing—introduced by the Ansorge laboratory—which provided linking information between sequence fragments (Edwards et al. 1990).

Directed sequencing methods use ordered exonuclease fragmentation, detailed restriction map-based clone identification, and primer-directed approaches. Compared to shotgun sequencing, they require less sequencing but demand more time and are more labor-intensive because of complex mapping and cloning operations, which reduce the overall efficiency of the process. Consequently, shotgun sequencing became the fundamental method for sequencing large genomic fragments.

However, several limitations to the approach were recognized early on: increased redundancy of the sequencing output as the number of sequenced clones grows leads to the generation of significant amounts of duplicated data, while the difficulty of capturing specific missing regions makes it difficult to close final assembly gaps (Sanger et al. 1982; Hood et al. 1987). Furthermore, for highly repetitive genomes, the assembly of a correct and finished genome sequence can become very challenging with a shotgun approach. In an outbred species like humans, it is further complicated by single-nucleotide polymorphisms (SNPs) and larger-scale structural heterozygosity.

3.2.4.3 Hierarchical Shotgun Strategy, the Celera Challenge, and Comparative Assemblies

Given these challenges, and after a lively scientific debate that also factored in the multicentered approach underlying the Human Genome Project (HGP), the consortium decided to pursue a hierarchical shotgun approach: start by building a high-resolution physical map of overlapping large-insert clones (typically BACs of 100–200 kb) spanning the genome, followed by shotgun sequencing within each selected clone (Lander et al. 2001). As a consequence, the first 9 years of the project were mostly devoted to cloning, mapping, and optimization of sequencing methods through pilot projects (including sequencing of model organism and cDNA), with the full-scale production phase to generate the draft sequence taking place between March 1999 and June 2000 (Lander et al. 2001). Work was divided by allocating specific chromosomes or chromosomal regions to individual centers, ensuring parallelization of sequencing and assembly while maintaining standardized protocols for map-based clone selection and quality assessment.

In 1998, Craig Venter announced the creation of a private company, Celera Genomics, that aimed to implement a whole-genome shotgun sequencing and assembly strategy to complete the human genome in just 3 years, a full 4 years ahead of the HGP consortium plan (Venter et al. 1998). Funded by Perkin Elmer, the parent company of Applied Biosystems, Celera would take advantage of the newly available high-throughput, automated capillary sequencers (ABI PRISM 3700), and associated automation platforms to perform the tens of millions of reactions needed to sequence the human genome more rapidly and at lower cost than previously anticipated, building a commercial business around the sequencing effort. Taking advantage of the ability of the ABI PRISM 3700 capillary sequencer to process 1000 samples per day with about 15 min of hands-on operator time, compared with roughly 8 h of operator time on the earlier instruments, the project cost was estimated at $200–250 million (US). This estimate included the complete laboratory and computational infrastructure needed to develop the finished sequence and the informatics tools to do it, substantially below the $3 billion budget of the HGP (Venter et al. 1998). The laboratory infrastructure that was set up was equipped with 230 sequencers, able to produce a combined daily capacity of 100 Mbp of raw sequence, plus the infrastructure required to produce 100,000 template preps and 200,000 sequencing reactions daily.

Celera's announcement pushed the public HGP consortium to speed up the program. As a consequence, both entities announced the completion of the draft sequence in June 2000 and published the results the following year (Lander et al. 2001; Venter et al. 2001). Of note, Celera's assembly integrated its own shotgun reads with publicly available hierarchical data from the public consortium, leveraging the map-based reads for scaffolding and verification. In the following 2 years, the public consortium was dedicated to "finishing" the draft sequence: resolving problem regions, eliminating gaps and ensuring an error rate of ~1 per 100,000 bases in the assembled near-complete genome (International Human Genome Sequencing Consortium 2004).

In 2004, the team behind the Celera sequencing efforts used the NCBI Build 34 of the human genome, released in 2003, to compare the 2001 draft sequences, assessing the strengths and weaknesses of whole-genome shotgun versus hierarchical shotgun (Istrail et al. 2004). They concluded that Celera's 2001 whole-genome shotgun assembly provided superior long-range order and orientation of human genomic sequence, whereas the public consortium's BAC-based assembly offered better coverage of exact and nearly exact repeat regions, with both approaches covering comparable overall portions of the genome but in clearly different patterns. By performing a new assembly of Celera's proprietary data and comparing it to the NCBI Build 34 reference, they further demonstrated that whole-genome shotgun sequencing, at modest coverage, could produce a high-quality draft of a large, complex eukaryotic genome, independent of clone-by-clone data (Istrail et al. 2004). These results confirmed the feasibility of assembling large, complex genomes from fragmented sequence data alone, a prerequisite that directly paved the way for the development and adoption of the next generation of sequencing technologies.

3.3 Next-Generation Sequencing: The Advent of Massive Parallelization

With a high-quality reference map of the human genome in place, attention shifted toward characterizing human genomic variation. It was, however, evident that existing technology could not support this objective and that this would require a significant technological breakthrough. Although by the early 2000s sequencing throughput had increased by several orders of magnitude, the factory-scale Sanger workflows remained constrained by read length, cloning requirements, reaction-limited scaling, and a cost per base that was still prohibitive. Microarray platforms provided an alternative for SNP profiling, but hybridization-based assays could not substitute for comprehensive sequencing. Addressing genomic variation at population scale and reasonable costs required a technological breakthrough that eliminated the need to isolate individual genome fragments and enabled their parallel sequencing within a single assay. This then could be followed by efficient assembly using and extending the computational strategies previously developed for shotgun sequencing.

3.3.1 Foundational Concepts Underlying Massive Parallel Sequencing

Breaking with the paradigm of automated capillary sequencers, which relied on cloned or highly purified DNA fragments as sequencing templates, a new generation of technologies emerged that fundamentally redefined how nucleic acids could be analyzed. These approaches, collectively termed next-generation sequencing (NGS), share three defining principles (Metzker 2010):

1. Library preparation through random fragmentation and adapter ligation—a single library is generated per genome sample via random fragmentation of nucleic acids, followed by adapter ligation in a unified reaction.
2. Massive parallelization through spatial segregation—millions of individual DNA fragments are physically separated on a solid support, enabling their simultaneous sequencing.
3. Signal generation and base detection—clonal amplification produces detectable signals, while cyclic chemistries allow sequential identification of each base. Later developments in single-molecule sequencing eliminated the need for amplification but required real-time, high-sensitivity detection systems.

At the core of all NGS technologies lies the concept of random fragmentation and adapter ligation as the central strategy for library preparation. Random fragmentation minimizes biases inherent to cloning, ensuring more uniform genomic coverage. The ligation of synthetic adapters provides defined priming sites for both amplification and sequencing reactions. This design also enables molecular barcoding, facilitating multiplexing of multiple samples within a single sequencing run. Moreover, adapters can incorporate unique molecular identifiers (UMIs), which enhance quantification accuracy and error correction (see Sect. 3.3.3.2).

The ability to perform massive parallel sequencing depends critically on spatially segregating individual DNA molecules so that each fragment's sequence can be distinguished. Different platforms have implemented this principle through various physical strategies, often centered around distinct flow cell architectures. Some systems employ emulsion PCR to form bead-based "microreactors," with each bead capturing and amplifying copies of a single DNA fragment. Others immobilize DNA fragments directly onto surfaces at high density, where they undergo in situ clonal amplification. Despite their technical differences, all designs share the common goal of resolving signals from individual fragments during sequencing.

The final defining component of NGS is the detection chemistry, which identifies nucleotide incorporation events through iterative biochemical cycles. Platforms differ in their modes of signal acquisition—using luminescent, fluorescent, or semiconductor-based detection of ion currents—to record the identity of each incorporated base.

Together, these innovations enabled the transition from capillary-based, single-fragment sequencing to massively parallel systems capable of generating millions to billions of reads in a single run, revolutionizing genomic analysis and paving the way for modern high-throughput biology. The following section presents major commercial implementations of NGS technologies that were historically significant or remain in widespread use.

3.3.2 Commercial Implementations of Next-Generation Sequencing

The mid-2000s marked the transition from conceptual frameworks to deployable platforms that operationalized massive parallelization. These systems shared the key principles described above but differed in how they implemented clonal amplification, immobilization of DNA

Table 3.1 Current NGS platforms comparison highlighting emerging challengers to Illumina dominance

Platform	Chemistry	Read length	Accuracy	Output/flow cell	Cost/Gb	Key strength
Illumina NovaSeq X	SBS (reversible terminators)	2 × 150 bp	Q30–40 (85–90%)	16-26B reads (4–8 Tb)	\$10–20	High-throughput market leader
Element AVITI	Avidity SBS (multivalent binding)	2 × 150 bp	Q40+ (>90% Q30)	800 M-1.6B reads (300–480 Gb)	\$7	Accuracy + cost for mid-scale
MGI DNBSEQ-T20	cPAL (rolling circle ligation)	2 × 150 bp	Q38 (>99.9%)	6-12B reads (1.8–3.6 Tb)	\$8–12	Cost-effective Illumina alternative
Ultima UG 100	Wafer-scale continuous flow (ppmSeq)	2 × 100 bp	Q60 (1 ppm SNV)	10-26B reads/ wafer(~\$100G)	\$1–2	Ultra-low cost, high accuracy
PacBio Revio2	SMRT (HiFi long-read)	15–25 kb	99.8% (HiFi)	7.3 M reads (730 Gb)	\$25–40	Isoforms + structural variants
ONT PromethION 2	Nanopore (electrical)	100 kb–4 Mb	99.8% (Q20 duplex)	290 Gb/flow cell	\$50–80	Native RNA + modifications

Current next-generation sequencing (NGS) platforms benchmarked by chemistry, performance metrics, and application strengths. Short-read platforms dominate bulk/single-cell/spatial transcriptomics applications; long-read technologies excel in isoform detection and structural variant calling. Ultima UG 100: Emerging wafer-scale platform (2025 commercial launch); no methods publication available. SBS, sequencing-by-synthesis. Data compiled from manufacturer specifications and independent benchmarks (Dec 2025)

fragments, and signal detection. Current commercial platforms are summarized in Table 3.1.

3.3.2.1 454 Life Sciences (Roche)

454 Life Sciences developed the first NGS method to demonstrate large-scale parallel sequencing, offering roughly a 100-fold increase in throughput compared with conventional Sanger sequencing. The technology was based on pyrosequencing, an approach initially developed in the 1990s as a low-throughput method in which each deoxynucleoside triphosphate (dNTP) was added cyclically. When a nucleotide complementary to the template was incorporated, pyrophosphate (PPi) was released and converted to ATP, which in turn fueled a luciferase reaction producing light. The emitted light intensity, correlated with nucleotide incorporation, allowed sequence inference (Ronaghi et al. 1996; Ronaghi et al. 1998).

454 Life Sciences adapted this chemistry to high-throughput sequencing by miniaturizing the reaction into picoliter-scale emulsion PCR microreactors, each containing a single bead carrying clonally amplified copies of a DNA fragment (Margulies et al. 2005). Despite its early success, Life Sciences pyrosequencing was later supplanted by platforms offering higher throughput and lower cost per base.

3.3.2.2 Solexa/Illumina Sequencing-by-Synthesis

The company Solexa developed a sequencing-by-synthesis (SBS) approach that uses reversible chain-terminating nucleotides linked to a fluorophore, a principle similar to Sanger sequencing but optimized for parallelization. The main innovation of this platform, later commercialized by Illumina, was the clonal bridge amplification of DNA fragments immobilized on a flow cell surface to generate dense single-molecule clusters (Bentley et al. 2008).

During sequencing, all four fluorescently labeled, reversibly terminated dNTPs are added simultaneously. Incorporation of one nucleotide into a growing strand is detected by fluorescence imaging, after which the terminator and fluorophore are cleaved (commonly using tris (2-carboxyethyl)phosphine), allowing continuation to the next cycle. The raw output of this sequencing method is a succession of images representing the nucleotides incorporated on each nucleotide cluster, which are then analyzed for conversion to sequence information. Illumina's system quickly became—and still is—the dominant NGS platform due to its high accuracy, scalability, and cost efficiency, in spite of the relatively short sequencing read lengths it produces.

3.3.2.3 Applied Biosystems SOLiD

The SOLiD (Sequencing by Oligonucleotide Ligation and Detection) platform, initially developed by Agencourt and later commercialized by Applied Biosystems, introduced sequencing by ligation (McKernan et al. 2009). This method employs octamer probes in which the two 3′ nucleotides are specific, while the remaining positions are degenerate. Each probe carries a fluorophore attached via a cleavable linker.

After adapter ligation to DNA fragments, sequencing primers hybridize to the adapters in offset positions. During each cycle, probes are ligated if their variable bases are complementary to the target, and the incorporated dinucleotide pair is identified through fluorescence emission. Following cleavage of the probe to remove the fluorophore, additional cycles proceed to interrogate the next positions. Five primers offset by one base each ensure overlapping coverage, allowing each base to be read twice—a two-base encoding strategy that confers high accuracy. Despite this accuracy, the method's complexity and read length limitations restricted its long-term adoption.

3.3.2.4 Ion Torrent (Life Technologies)

Ion Torrent technology, developed by Life Technologies, replaced optical detection with semiconductor-based sensing to detect the hydrogen ions released by DNA polymerase during SBS (Rothberg et al. 2011). Using complementary metal–oxide–semiconductor (CMOS) technology, the Ion Torrent sequencer integrates ion-sensitive field-effect transistors (ISFETs) associated with reaction microwells to monitor pH changes in real time. During the sequencing process, individual dNTPs are flowed sequentially into the wells containing the DNA template and DNA polymerase. Each incorporation event releases a proton, causing a measurable drop in pH inside the well. These changes are detected electronically, translating the chemical reaction directly into a digital signal which, in turn, is converted into a sequence. Ion Torrent offered a cost-effective, compact alternative to optical systems, though with lower accuracy for homopolymeric regions.

3.3.2.5 Helicos Heliscope

Launched in 2008 by Helicos BioSciences, the HeliScope platform shares some similarities with Illumina sequencing, such as SBS and detection via reversibly terminable, fluorescently labeled nucleotides. However, building upon earlier work demonstrating that DNA polymerase activity could be monitored at the single-molecule level (Braslavsky et al. 2003), it became the first commercially available single-molecule sequencing platform. HeliScope immobilized single-stranded DNA fragments on a flow cell via poly(A)-tailing and poly(T)-coated surfaces and used fluorescently labeled, reversibly terminable nucleotides for SBS. Detection of nucleotide incorporation was achieved in real time through total internal reflection fluorescence microscopy (Harris et al. 2008).

Although innovative, the platform suffered from high indel error rates, short read lengths, and low throughput, leading to limited adoption and the company's eventual bankruptcy (Carroll 2012). Nonetheless, it paved the way for subsequent technologies by demonstrating the feasibility of direct single-molecule sequencing.

3.3.2.6 Pacific Biosciences SMRT Sequencing

Pacific Biosciences (PacBio) introduced single-molecule real-time (SMRT) sequencing, first demonstrated in 2009 (Eid et al. 2009) and commercialized in 2011. This platform integrated several technological innovations including precision nanofabrication and optical confinement to enable long-read sequencing—up to 4000 bp in the original implementation—taking a distinctive approach compared to other NGS platforms.

SMRT technology uses zero-mode waveguides (ZMWs)—nanometer-scale wells in which a single DNA polymerase is fixed at the base. Fluorescently labeled nucleotides diffuse into the well and are incorporated by the enzyme. A laser excites fluorophores within the limited depth of the evanescent field, allowing real-time detection with minimal background noise (Eid et al. 2009). To improve the accuracy of SMRT sequencing, library preparation methods were adapted to produce SMRTbell™ templates: circularized structures enabling multiple passes of the polymerase and generation of circular consensus sequences (CCS) (Travers et al. 2010). PacBio technology has been continuously improved, with optimized CCS on the Sequel system yielding long HiFi reads with 99.8% average accuracy and 13.5 kb average length; this work turns PacBio from a primarily long-but-noisy technology into one that delivers long, highly accurate reads, improving variant detection, assembly accuracy, difficult-region coverage, and high computational efficiency compared to previous approaches (Wenger et al. 2019).

In addition to the unparallelled capacity to deliver highly accurate long reads, with significant impact on the characterization of genome structural variants and transcript isoforms, SMRT sequencing on the PacBio platform can be used to detect DNA modifications directly by monitoring how they alter DNA polymerase kinetics during sequencing (Flusberg et al. 2010). The unique qualities of the progressive implementations of the PacBio technology are currently having a profound impact on genomics and, in particular, in medical diagnostics (Wang (王博) et al. 2025).

3.3.2.7 Oxford Nanopore Technologies

Following the initial studies reporting a sequencing approach using nanopores in 2009 (Clarke et al. 2009), Oxford Nanopore Technologies (ONT) introduced its commercial platform in 2014, marking another major innovation in sequencing technologies (Pugh 2023). Unlike optical or semiconductor-based detection, this platform measures changes in ionic current as single nucleic acid molecules pass through nanoscale pores embedded in membranes. The underlying method is based on the flow of ions through nanopores when a high voltage is applied across the membrane, establishing an ionic flux and an associated baseline current that can easily be measured. As nucleic acid molecules pass through the pore, distinct ionic flux disruptions occur, reflecting the nucleotides passing through the channel at each moment. Machine learning models interpret these signals into base calls in real time. A helicase or similar motor protein controls translocation speed, allowing base-level resolution. Because of its unique detection mechanism, nanopore sequencing is the only technology so far that supports direct sequencing of RNA molecules, in addition to DNA (Garalde et al. 2018). It can also directly identify base modifications (Hewel et al. 2025).

Modern devices—from portable MinION sequencers to high-throughput PromethION platforms—produce ultra-long reads, often >100 kb, and up to the megabase scale and currently with high consensus accuracy (>99% in duplex mode). ONT was also the first company to launch a portable device, with the size of a large USB pen drive, to be used onsite, along with a range of benchtop instruments that can generate up to 14 Tb of data (Pugh 2023). ONT's continuous improvements in pore design, chemistry, and base-calling algorithms have established it as one of the most flexible and rapidly evolving sequencing technologies in both research and applied genomics.

3.3.2.8 MGI/BGI Platforms

Complete Genomics, later acquired by the BGI Group as its MGI Tech division, developed a sequencing-by-ligation platform based on earlier work with combinatorial probe anchor ligation (cPAL) chemistry (Drmanac et al. 1993). The new method coupled circularized DNA fragments with rolling circle amplification (RCA) to produce dense arrays of DNA nanoballs (DNBs) immobilized on patterned silicon substrates, supporting increased parallelization, longer reads, and accuracy (Drmanac et al. 2010). The linear nature of rolling circle amplification can reduce biases introduced by PCR during library preparation, as discussed in the next section. Following immobilization, each DNB serves as a sequencing unit to which a primer anneals. Sets of probes containing defined "interrogating" bases, degenerate regions, cleavable linkers, and fluorescent or hapten labels are sequentially introduced. Fluorescence signals from successful ligations identify the incorporated nucleotides and subsequent cleavage steps remove probes for the next cycle. This strategy yields high accuracy and cost-efficiency, enabling high-throughput sequencing at a reduced reagent cost. Accuracy typically reaches one false variant per 100 kilobases, positioning MGI's DNBSEQ series among the leading alternatives to Illumina's SBS platforms (Kim et al. 2021).

More recently, MGI has developed a new sequencing chemistry called CoolMPS compatible with its DNB technology. This is a novel SBS approach that relies on natural nucleotide labeling with reusable antibodies (Drmanac et al. 2020). CoolMPS has shown higher per-base sequencing quality with a more favorable and stable quality profile across the read length, and an increased fraction of mapped reads at a lower cost compared with standard SBS on the same platform (Li et al. 2021).

3.3.2.9 Element Biosciences

Element Biosciences launched its benchtop NGS platform in 2022 using proprietary Avidity Sequencing, a short-read DNA sequencing chemistry that decouples the processes of moving along the DNA template from identifying each incorporated nucleotide (Arslan et al. 2024). Instead of using single modified nucleotides to both extend and label DNA, it employs "avidites": multivalent nucleotide ligands on dye-labeled polymer cores that form stable complexes with polymerase and clonal DNA targets. These avidites act as polymer–nucleotide substrates that reduce the required concentration of reporting nucleotides from micromolar to nanomolar and exhibit negligible dissociation, lowering reagent consumption while maintaining a strong, persistent signal. Template stepping itself is achieved with reversibly terminated nucleotides carrying a 3′ blocking group and dye label, in cycles where the blocking group and label are removed after imaging, similar in structure to standard SBS but optimized separately from base identification. During each sequencing cycle, four images are captured per field of view, corresponding to the dyes used to label each avidite, and intensities are extracted for each clonal cluster (polony) in each color channel. This method achieves high base-calling accuracy and maintains a stable average error rate even after long homopolymers, supporting applications such as whole-human-genome and single-cell RNA sequencing.

3.3.3 Library Preparation Strategies

Library preparation is the central enabling step of all next-generation sequencing workflows, translating native nucleic acids into platform-compatible molecules that can be clonally amplified or sequenced as single molecules. Despite the diversity of commercial platforms, most protocols share a common backbone of fragmentation, end repair, adapter ligation, optional library amplification, and size selection, with each stage offering opportunities to tune performance for specific applications. Modifications to this basic scheme, such as incorporating barcodes and UMIs, altering adapter architecture, or adding capture or tagmentation steps, underpin the broad spectrum of sequencing strategies, from whole-genome and whole-exome sequencing to targeted panels, RNA-seq, single-cell assays, and epigenomic protocols (Head et al. 2014).

3.3.3.1 Fragmentation and Size Selection

With the exception of long-read sequencing platforms, library preparation methods typically begin by fragmenting nucleic acids into shorter molecules within a platform-dependent size range. Early approaches relied on physical methods. Sonication was widely used in early shotgun strategies (Deininger 1983) and was rapidly adopted for NGS because it produces near-random breaks and allows fragment size to be tuned by adjusting sonication time and energy, enabling generation of libraries with desired size distributions. Alternatively, nebulization fragments DNA by forcing solutions through a narrow orifice to create aerosols; varying pressure, duration, temperature, orifice size, and solution properties allows limited control of fragment length distributions, with comparable results to sonication (Knierim et al. 2011).

Enzymatic fragmentation methods provide alternatives to physical shearing, with distinct trade-offs in fragment size control, bias, and integration with library construction. DNase I is a nonspecific endonuclease that introduces single-strand nicks and, under appropriate conditions, double-strand breaks. It can be used when input amounts are low and specialized equipment is not available, but it generates poorly controlled fragment size distributions and can introduce sequence bias, which limits its use in routine library preparation. Instead, DNase I is particularly important in DNase-seq, where its requirement for accessible chromatin is exploited to map regulatory regions genome-wide (Head et al. 2014). Commercial "fragmentase" systems, such as NEBNext dsDNA Fragmentase, combine a random nicking activity with a structure-dependent nuclease that cleaves opposite the nicks, producing double-stranded fragments in a time-dependent manner. By adjusting incubation time, fragment sizes can be tuned over an approximate 50–1000 bp range, providing more precise, efficient, and reproducible fragmentation than DNase I, including in regions that are relatively refractory to DNase-based methods (Knierim et al. 2011). Finally, transposase-mediated tagmentation uses pre-assembled transposomes (e.g., Tn5 loaded with adapter sequences) to couple DNA cleavage with simultaneous adaptor insertion (Adey et al. 2010). This approach yields near-random fragmentation, reduces hands-on time and protocol complexity, and substantially lowers the required input amount, which has led to its widespread adoption as a core step in many modern NGS library preparation kits.

In addition to these random fragmentation strategies, several approaches generate NGS-compatible fragments through amplification or targeted cleavage rather than bulk shearing. PCR can serve as a functional substitute for fragmentation in NGS workflows by generating size-defined amplicons from specific loci, a strategy widely used in targeted gene panels, microbial marker sequencing, or viral tiling approaches. PCR-based methods such as double-priming and sequence-independent single primer amplification (SISPA) stochastically amplify short regions across the template using random primers carrying adapter sequences, thereby producing small amplicons suitable for sequencing even from very low input (Djikeng et al. 2008). Although fragment size is less tightly controlled and typically requires downstream size selection, such strategies offer advantages in terms of minimal input requirements, simple equipment, and direct adapter incorporation (Parras-Moltó et al. 2018). By contrast, CRISPR/Cas9-based fragmentation is inherently targeted: guide RNAs direct Cas9 to specific sites, enabling excision and enrichment of predefined regions before library construction. Protocols such as CRISPR-DS and CRISPR-Cap combine Cas9 cleavage with molecular barcoding or biotin–streptavidin capture to generate highly enriched libraries of regions of interest at reduced sequencing cost, which is particularly advantageous in clinical applications focused on limited gene panels (Nachmanson et al. 2018; Lee et al. 2019).

Because most NGS platforms sequence a restricted fragment length range, fragmentation is typically followed by a size-selection step to enrich for molecules within the desired insert size window. Historically, this was achieved by agarose or polyacrylamide gel electrophoresis with excision of the appropriate band, but this approach is labor-intensive and low throughput. The introduction of solid-phase reversible immobilization (SPRI) bead-based methods, which exploit differential binding of DNA to PEG/NaCl-treated magnetic beads as a function of fragment length and bead-to-sample ratio, enabled tunable, double-sided size selection with markedly improved scalability and automation. As a result, SPRI-based workflows have become the *de facto* standard for NGS library size selection in both research and clinical laboratories (Head et al. 2014).

3.3.3.2 End Repair, A-Tailing, and Adapter Ligation

NGS protocols require that DNA fragments are flanked by adapeors to support downstream enzymatic reactions, including sequencing, as well as interaction with the sequencing platform. Addition of these adapter sequences is often performed by ligation, requiring prior enzymatic end repair and A-tailing of the library DNA, as the fragmentation process often leads to unphosphorylated ends and/or overhangs (Head et al. 2014). End repair enzymatically fills in 5′ overhangs via polymerase activity, removes 3′ overhangs through exonuclease action, and adds 5′ phosphate groups with kinase activity to prepare fragments for ligation. A-tailing adds a single adenine nucleotide to 3′ ends of double-stranded fragments, facilitating ligation to adapters with complementary 3′ thymine overhangs and preserving strand directionality. Library preparation protocols based on tagmentation or PCR integrate adapter sequences directly, thereby not requiring these steps. Additionally, platforms using blunt-end adapter ligation, such as some Ion Torrent workflows, may bypass A-tailing altogether (Forth and Höper 2019). Ensuring effective end repair is a crucial step to avoid bias in the sequence coverage of the library (Zhang et al. 2018).

Adapter ligation attaches platform-specific oligonucleotides to both ends of each fragment. Adapters typically include (1) sequences ensuring binding to the platform (e.g., flow cell or beads), (2) optional sample indexes for multiplexing, (3) optional PCR primer binding sites, (4) unique molecular identifiers (UMIs) to reduce amplification bias, (5) sequencing primer binding sites, (6) thymine overhangs to facilitate ligation when A-tailing is used, and (7) optional strand directionality tags to enable bioinformatic distinction of strands (Head et al. 2014). Strand directionality tags can involve asymmetric ligation using chemical blocking or phosphate groups to ensure distinct adapters ligate to fragment ends, important for resolving overlapping or antisense transcripts. Some adapters may contain structural elements, such as complementary sequences in DNA nanoball sequencing that promote circularization (Xu et al. 2019).

Unique molecular identifiers (UMIs) are short random nucleotide sequences incorporated into sequencing adapters during library preparation to uniquely tag individual DNA or RNA molecules before any amplification step. By assigning a distinct identifier to each original template, UMIs allow computational collapsing of PCR duplicates and thus enable accurate molecule counting, reducing amplification bias and improving quantitative precision (Kivioja et al. 2011; Fu et al. 2018). This strategy became particularly relevant for single-cell RNA sequencing (discussed in Sect. 3.4.1), where starting material is extremely limited and amplification introduces substantial distortions; UMIs allow transcript abundance to be measured at the molecule level, supporting reliable comparisons across cells and conditions.

Adapter ligation can be achieved either in a single-step (common in PCR-free workflows) or in two steps: a stub adapter ligated directly to fragments after end repair and A-tailing, followed by PCR primers introducing the platform recognition sequences to ensure their positioning at fragment termini (Head et al. 2014). As mentioned, adapters can be directly introduced during tagmentation or as PCR overhangs. Additionally, for RNA sequencing, adapters may be introduced by reverse-transcription primers (see Sect. 3.3.3.4).

3.3.3.3 Library Amplification and Quality Control

Library amplification via PCR is commonly required across most NGS platforms—except in PCR-free workflows—to generate sufficient DNA for sequencing coverage, especially when input material is limited (Head et al. 2014). Primers hybridize to sequences included in library adapters and may simultaneously complete the adapter construct during amplification. PCR amplification remains a major source of sequence composition bias, notably underrepresenting regions with extreme GC content. While advances in amplification protocols have reduced these biases, they cannot be fully eliminated (Aird et al. 2011). Therefore, PCR-free workflows, which omit amplification, offer advantages in reducing bias but require greater input amounts and/or accept lower coverage thresholds (Kozarewa and Turner 2011).

Quality control (QC) is integral throughout library preparation, with the critical assessment typically occurring post-library construction and before sequencing to ensure sample suitability. Key QC parameters include library concentration and fragment size distribution since fragments must fall within the sequencing platform's supported size range, and library concentrations outside optimal ranges can lead to insufficient data or overloaded runs. Concentration is most commonly measured by UV spectrophotometry (e.g., NanoDrop), fluorescence spectroscopy (e.g., Qubit), and quantitative PCR (e.g., SYBR Green or TaqMan assays). Size distribution is assessed by electrophoretic methods, including microfluidic chip- or capillary-based systems (e.g., Bioanalyzer, TapeStation, Fragment Analyzer) or agarose gel electrophoresis. Selection of specific QC methods and thresholds can vary depending on workflow requirements and expected library yields (Robin et al. 2016; Hussing et al. 2018).

3.3.3.4 Assay-Specific Library Preparation Strategies

NGS can be tailored to answer diverse biological questions by adapting library preparation workflows and overall study design. The most widely used approaches include whole-genome sequencing (WGS), whole-exome sequencing (WES), targeted sequencing, and RNA sequencing (RNA-seq), each differing in coverage, cost, and data type.

Two fundamental concepts guide the design of NGS-based studies: sequencing depth and coverage (Sims et al. 2014). Sequencing depth refers to the average number of independent sequencing reads that cover a specific nucleotide position or genomic region, indicating how many times that target has been observed. Coverage describes the fraction or percentage of the target sequence (e.g., genome or exome) represented in the sequencing data. Adequate depth and broad coverage improve variant detection sensitivity, expression quantification, and overall data reliability, but both increase sequencing cost and data volume. Library preparation methods can be adapted to enrich desired genomic or transcriptomic regions, enabling higher effective coverage of targets while controlling cost.

Whole-genome sequencing (WGS) provides comprehensive coverage of the entire genome, making it the method of choice for de novo genome assembly, population genomics, and the characterization of novel organisms. Depending on the final application, WGS may be preferable to targeted approaches; however, the required sequencing depth/coverage and higher cost make careful budgeting and project scope planning essential.

Whole-exome sequencing (WES) focuses on the coding regions of genes, capturing the exons where most known disease-associated variants reside (Hodges et al. 2007). It serves as a cost-effective alternative to WGS for variant discovery in clinical and research settings when noncoding regions are of less interest. WES libraries incorporate targeted capture steps after fragmentation and adapter ligation to enrich exonic regions. Targeted sequencing panels are designed for highly focused analysis of known genes or loci related to specific diseases or traits. These approaches further reduce sequencing costs and increase sensitivity for low-frequency variants through deep coverage of selected regions. Target enrichment before sequencing may use hybrid capture or amplicon-based methods such as PCR or CRISPR/Cas9 cleavage, depending on study requirements, with commercial panels for clinical diagnosis being widely used.

In addition to sequencing of genomic DNA, NGS library preparation can be adapted to support the characterization of transcriptomes. RNA sequencing (RNA-seq) enables quantification of gene expression and the discovery of novel transcripts or isoforms (Stark et al. 2019). Protocols vary from full-length transcript sequencing (capturing entire RNA molecules) to 3′ or 5′ end counting approaches, which reduce costs and sequencing bias. RNA-seq library preparation generally begins with reverse transcription of RNA into cDNA, often including depletion of abundant ribosomal RNA (>90% of total RNA) or enrichment of target RNA populations such as polyadenylated transcripts or small RNAs (small RNA-seq) (Zhao et al. 2014).

Specialized NGS assays adapt library workflows to study functional features of the genome, such as nucleotide modifications, chromatin accessibility, molecular interactions, or translation efficiency. The adaptations are countless and new methods to answer ever-more ingenious questions are being constantly developed. For example, ATAC-seq maps open chromatin regions by tagmentation-mediated fragmentation of accessible DNA (Buenrostro et al. 2013), while ChIP-seq profiles protein-DNA binding through chromatin immunoprecipitation followed by sequencing (Park 2009). The Ribo-seq workflow adapts NGS to characterize mRNA translation by focusing sequencing on ribosome-protected mRNA fragments (Ingolia et al. 2009). In common, these protocols incorporate tailored enrichment steps that modify the standard library construction scheme to selectively capture target sequences of interest, providing insights into dynamic aspects of genome function.

Each of these assay-specific strategies reflects a compromise between biological scope, sequencing cost, data resolution, and complexity. While the above methods underpin bulk sequencing applications, recent innovations have enabled massively parallel analyses at the single-cell level and within spatial tissue contexts. These approaches incorporate cellular barcoding, droplet microfluidics, and imaging modalities to revolutionize transcriptomics and are described in the following section.

3.4 Single-Cell and Spatial Transcriptomics

The ability to characterize the genome and transcriptome of individual cells is of central relevance for both clinical applications and research-driven questions. PCR-based screening of single-cell biopsies of six to eight cell-stage embryos was developed in 1990 for preimplantation diagnosis of severe genetic disorders (Handyside et al. 1990). Shortly thereafter, the first profiling of gene expression in individual neurons through RNA captured via micromanipulation methods like patch clamp was reported (Eberwine et al. 1992; Lambolez et al. 1992). The ability to analyze the genome and transcriptome of single cells opens the door to many relevant biological questions that cannot be addressed by profiling bulk tissue samples. In the last decade, the development of high-throughput cell isolation and barcoding methods has opened the world of single cells to NGS technologies. Given the central importance of transcriptome diversity to understanding cellular function and development, single-cell RNA sequencing (scRNA-seq) has been at the core of this methodological revolution (Aldridge and Teichmann 2020). More recent developments have allowed the preservation of spatial information, providing insights into tissue organization (Liu et al. 2024). This section provides an overview of the core methodologies underlying these approaches and the emergence of novel, multimodal methods that are currently evolving.

3.4.1 Single-Cell RNA Sequencing (scRNA-Seq)

scRNA-seq enables transcriptome profiling at single-cell resolution by isolating individual cells, converting their RNA into barcoded cDNA, and generating sequencing libraries compatible with standard NGS platforms. The first demonstration of NGS-based whole-transcriptome analysis from a single mammalian cell was reported in 2009 by Tang and colleagues using manual micromanipulation to isolate cells under a microscope (Tang et al. 2009). Subsequent methodological developments introduced high-throughput systems capable of profiling thousands to tens of thousands of cells per experiment, facilitating analyses of cellular heterogeneity, lineage relationships, and rare subpopulations in complex tissues.

A central feature of scRNA-seq is cellular barcoding, in which each cell's transcripts receive a unique sequence tag during reverse transcription or early library preparation (Kolodziejczyk et al. 2015). This enables pooling of all cDNA into a single library that can be sequenced together and computationally demultiplexed into individual transcriptomes. In parallel, UMIs allow recovery of molecule-level counts after amplification, substantially reducing biases introduced by PCR. UMIs are therefore essential for accurate quantification in most high-throughput scRNA-seq platforms.

Two major approaches currently support the generation of scRNA-seq libraries, differing in throughput, read coverage, and protocol complexity. Droplet-based systems (currently the most widely used approach, with 10x Genomics platforms dominating the scRNA-seq market) rely on microfluidics to encapsulate individual cells with barcoded beads in aqueous droplets (Macosko et al. 2015; Zheng et al. 2017). The Drop-seq framework was the first to be described, with the authors providing detailed protocols for quick and inexpensive construction in a standard biology lab using readily available equipment. The approach by Zheng et al. (2017) underpins the 10x Genomics Chromium system, which generates gel beads-in-emulsion (GEMs) carrying cell barcodes, UMIs, and oligo-dT sequences (Zheng et al. 2017). The system achieves high cell throughput (typically 3000–10,000 cells per channel), with ~50% loading efficiency, and processes up to eight samples in parallel on an eight-lane chip. Droplet-based methods provide high cell numbers with 3′-biased transcript coverage.

Plate-based protocols (e.g., Smart-seq2) are compatible with single-cell isolation methods based on FACS sorting, micromanipulation, or limiting dilution (Ramsköld et al. 2012; Picelli et al. 2013). Reverse transcription uses oligo-dT priming and template switching, producing full-length cDNA followed by PCR amplification, fragmentation/tagmentation, and indexing. Although lower throughput (up to 384 cells per plate) and more labor-intensive than droplet methods, plate-based approaches provide superior transcript coverage, enabling isoform detection and improved analysis of splice variants (Picelli et al. 2014).

Library preparation proceeds through reverse transcription with barcoded primers (introducing cell barcodes and UMIs when applicable), cDNA amplification, fragmentation or tagmentation, adapter ligation, and final indexing. Quality control of cell integrity, cDNA yield, and library complexity is required at multiple steps. The point at which barcodes and UMIs are introduced varies by platform and influences downstream computational workflows.

scRNA-seq data has several recurrent limitations. First, joint capture of two or more cells, or apoptotic cells, leads to artifacts; computational tools identify and remove these based on gene expression anomalies and barcode co-occurrence (Wolock et al. 2019). Second, data is characterized by high sparsity and dropout: low mRNA content per cell causes many genes to be undetected, complicating normalization and downstream analyses (Andrews and Hemberg 2019; Hafemeister and Satija 2019). Finally, the method is prone to significant batch effects: variations due to technical factors (different runs, library preps, or instruments) can confound biological signal; batch correction and integration methods help harmonize multi-batch datasets. Despite these challenges, advances in experimental design, sequencing chemistry, and bioinformatics have made scRNA-seq an indispensable tool for dissecting cellular complexity in health and disease.

3.4.2 Spatial Transcriptomics

Spatial transcriptomics integrates gene expression profiling with spatial localization within tissue sections, preserving architectural context and enabling the study of cellular neighborhoods, tissue heterogeneity, and microenvironmental organization. It bridges traditional histology with high-throughput genomics and encompasses diverse experimental strategies that differ in spatial resolution, throughput, and molecular complexity (Liu et al. 2024).

Broadly, sequencing-based characterization of localized gene expression can be performed either by direct, in situ sequencing or by advanced barcoding methods that preserve spatial information regarding the origin of sequencing reads.

In situ sequencing was first described by Ke and colleagues (Ke et al. 2013), as a targeted method applied directly in fixed cells and tissues while maintaining morphology. It relied on padlock probing, rolling-circle amplification, and in situ sequencing-by-ligation but had limited multiplexing capacity (<50 genes). Subsequent advancements, such as Fluorescent In Situ Sequencing (FISSEQ) (Lee et al. 2014) and STARmap (Wang et al. 2018), improved efficiency and multiplexing. STARmap achieved >1000-gene detection by combining hydrogel-tissue chemistry, DNA nanospheres, and error-corrected two-base encoding. These methods still face enzymatic and diffusion limitations imposed by complex intracellular environments. Expansion sequencing (ExSeq) addressed some of these constraints by integrating expansion microscopy with FISSEQ principles, physically expanding tissues ~4× and enabling nanoscale RNA localization (Alon et al. 2021).

A second class of methods captures RNA molecules together with spatial barcodes and subsequently reads them out by NGS. These approaches are commonly referred to as spatial transcriptomics. Numerous strategies have been developed and can be grouped by barcoding modality (Liu et al. 2024). In physical-region barcoding, tissue is sectioned or microdissected, and each region is separately barcoded before sequencing. In in situ capture on barcoded substrates, spatial barcodes are immobilized on glass slides, beads, chips, or microfluidic grids and capture mRNA directly from the overlaid tissue, enabling much finer spatial resolution. Region-based approaches operate at the tissue or cluster level, whereas in situ capture technologies reach micro- to nanoscale resolution.

Microscale methods include the commercial 10× Genomics Visium system, based on the pioneer arrayed spatial transcriptomics method by Ståhl et al. (Ståhl et al. 2016), as well as Slide-seq and Slide-seqV2 (Rodriques et al. 2019; Stickels et al. 2021), with resolutions ranging from ~100 μm down to 2–10 μm. Nanoscale approaches such as SeqScope (Cho et al. 2021), Stereo-seq (Chen et al. 2022), and Open-ST (Schott et al. 2024) use ultra-dense nano- or microscale arrays to achieve resolutions near ~500 nm and enable high-resolution 2D and 3D transcriptomic mapping. High-resolution tissue imaging is aligned with barcode-derived maps, enabling precise integration of transcriptomic and anatomical information.

The most widely used platform is 10× Genomics Visium (including Visium HD), largely due to its turnkey, commercially supported workflow, compatibility with fresh-frozen and FFPE samples, and straightforward integration with standard Illumina sequencing. Other technologies such as Slide-seq remain influential but are less widely adopted due to technical complexity and infrastructure requirements. Visium uses glass slides spotted with barcoded oligonucleotide capture probes arranged in defined ~55 μm spots, allowing spatially resolved transcript capture from overlaid tissue sections. It provides transcriptome-wide data at near-multicellular resolution. Slide-seq uses beads with unique barcodes arranged as a dense monolayer; transcripts released during permeabilization hybridize to these bead-bound probes. Slide-seq achieves ~10 μm resolution, approaching single-cell profiling while retaining whole-transcriptome coverage.

Each platform represents a balance among spatial resolution, molecular throughput, and practical implementation, allowing selection according to specific research needs. Library preparation workflows vary across technologies but generally include (i) tissue sectioning and fixation/preservation to maintain spatial structure and RNA integrity; (ii) transcript capture through spatially barcoded oligonucleotides; (iii) reverse transcription incorporating spatial barcodes and UMIs to map reads back to physical coordinates; and (iv) PCR amplification and sequencing adapter addition compatible with standard NGS systems. Protocols require optimization for tissue type, section thickness, and RNA quality, with quality control steps to assess spatial barcode distribution and library complexity.

The rapid expansion of spatial transcriptomics is enabling unprecedented insights into organ organization, developmental processes, species-specific tissue architecture, and disease pathogenesis.

3.5 Conclusion

The past two decades have seen a major transformation in sequencing technologies, reshaping research in the life sciences and expanding clinical and biotechnological applications. Current platforms provide an extensive set of tools for nucleic acid profiling, spanning detection of sequence variants and base modifications, bulk and single-cell analyses, short- and long-read approaches, and population-scale to spatially resolved measurements. As throughput and accuracy continue to increase, library construction and downstream computational analysis have become the principal rate-limiting steps in many workflows. Emerging technologies are now extending sequencing beyond traditional genomic and transcriptomic measurements. Multimodal profiling, integrating RNA, chromatin accessibility, protein expression, and epigenetic features within the same cells or tissue contexts represents the next major advance. These developments will support more comprehensive characterization of cellular states, regulatory programs, and tissue organization, with deep implications in both basic and clinical research.

References

Adams MD, Kerlavage AR, Kelley JM, Gocayne JD, Fields C, Fraser CM, Venter JC (1994) A model for high-throughput automated DNA sequencing and analysis core facilities. Nature 368(6470):474–475. https://doi.org/10.1038/368474a0

Adey A, Morrison HG, Asan XX, Kitzman JO, Turner EH, Stackhouse B, MacKenzie AP, Caruccio NC, Zhang X, Shendure J (2010) Rapid, low-input, low-bias construction of shotgun fragment libraries by high-density in vitro transposition. Genome Biol 11(12):R119. https://doi.org/10.1186/gb-2010-11-12-r119

Aird D, Ross MG, Chen W-S, Danielsson M, Fennell T, Russ C, Jaffe DB, Nusbaum C, Gnirke A (2011) Analyzing and minimizing PCR amplification bias in Illumina sequencing libraries. Genome Biol 12(2): R18. https://doi.org/10.1186/gb-2011-12-2-r18

Aldridge S, Teichmann SA (2020) Single cell transcriptomics comes of age. Nat Commun 11(1): 4307. https://doi.org/10.1038/s41467-020-18158-5

Alon S, Goodwin DR, Sinha A, Wassie AT, Chen F, Daugharthy ER, Bando Y, Kajita A, Xue AG, Marrett K, Prior R, Cui Y, Payne AC, Yao C-C, Suk H-J, Wang R, Yu C-C(J), Tillberg P, Reginato P, Pak N, Liu S, Punthambaker S, Iyer EPR, Kohman RE, Miller JA, Lein ES, Lako A, Cullen N, Rodig S, Helvie K, Abravanel DL, Wagle N, Johnson BE, Klughammer J, Slyper M, Waldman J, Jané-Valbuena J, Rozenblatt-Rosen O, Regev A, IMAXT CONSORTIUM, Church GM, Marblestone AH, Boyden ES (2021) Expansion sequencing: spatially precise in situ transcriptomics in intact biological systems. Science 371(6528):eaax2656. https://doi.org/10.1126/science.aax2656

Andrews TS, Hemberg M (2019) M3Drop: dropout-based feature selection for scRNASeq. Bioinformatics 35(16):2865–2867. https://doi.org/10.1093/bioinformatics/bty1044

Ansorge W, Sproat B, Stegemann J, Schwager C, Zenke M (1987) Automated DNA sequencing: ultrasensitive detection of fluorescent bands during electrophoresis. Nucleic Acids Res 15(11):4593–4602. https://doi.org/10.1093/nar/15.11.4593

Arslan S, Garcia FJ, Guo M, Kellinger MW, Kruglyak S, LeVieux JA, Mah AH, Wang H, Zhao J, Zhou C, Altomare A, Bailey J, Byrne MB, Chang C, Chen SX, Cho B, Dennler CN, Dien VT, Fuller D, Kelley R, Khandan O, Klein MG, Kim M, Lajoie BR, Lin B, Liu Y, Lopez T, Mains PT, Price AD, Robertson SR, Taylor-Weiner H, Tippana R, Tomaney AB, Zhang S, Abtahi M, Ambroso MR, Bajari R, Bellizzi AM, Benitez CB, Berard DR, Berti L, Blease KN, Blum AP, Boddicker AM, Bondar L, Brown C, Bui CA, Calleja-Aguirre J, Cappa K, Chan J, Chang VW, Charov K, Chen X, Constandse RM, Damron W, Dawood M, DeBuono N, Dimalanta JD, Edoli L, Elango K, Faustino N, Feng C, Ferrari M, Frankie K, Fries A, Galloway A, Gavrila V, Gemmen GJ, Ghadiali J, Ghorbani A, Goddard LA, Guetter AR, Hendricks GL, Hentschel J, Honigfort DJ, Hsieh Y-T, Hwang Fu Y-H, Im SK, Jin C, Kabu S, Kincade DE, Levy S, Li Y, Liang VK, Light WH, Lipsher JB, Liu T, Long G, Ma R, Mailloux JM, Mandla KA, Martinez AR, Mass M, McKean DT, Meron M, Miller EA, Moh CS, Moore RK, Moreno J, Neysmith JM, Niman CS, Nunez JM, Ojeda MT, Ortiz SE, Owens J, Piland G, Proctor DJ, Purba JB, Ray M, Rong D, Saade VM, Saha S, Tomas GS, Scheidler N, Sirajudeen LH, Snow S, Stengel G, Stinson R, Stone MJ, Sundseth KJ, Thai E, Thompson CJ, Tjioe M, Trejo CL, Trieger G, Truong DN, Tse B, Voiles B, Vuong H, Wong JC, Wu C-T, Yu H, Yu Y, Yu M, Zhang X, Zhao D, Zheng G, He M, Previte M (2024) Sequencing by avidity enables high accuracy with low reagent consumption. Nat Biotechnol 42(1):132–138. https://doi.org/10.1038/s41587-023-01750-7

Bentley DR, Balasubramanian S, Swerdlow HP, Smith GP, Milton J, Brown CG, Hall KP, Evers DJ, Barnes CL, Bignell HR, Boutell JM, Bryant J, Carter RJ, Keira Cheetham R, Cox AJ, Ellis DJ, Flatbush MR, Gormley NA, Humphray SJ, Irving LJ, Karbelashvili MS, Kirk SM, Li H, Liu X, Maisinger KS, Murray LJ, Obradovic B, Ost T, Parkinson ML, Pratt MR, Rasolonjatovo IMJ, Reed MT, Rigatti R, Rodighiero C, Ross MT, Sabot A, Sankar SV, Scally A, Schroth GP, Smith ME, Smith VP, Spiridou A, Torrance PE, Tzonev SS, Vermaas EH, Walter K, Wu X, Zhang L, Alam MD, Anastasi C, Aniebo IC, Bailey DMD, Bancarz IR, Banerjee S, Barbour SG, Baybayan PA, Benoit VA, Benson KF, Bevis C, Black PJ, Boodhun A, Brennan JS, Bridgham JA, Brown RC, Brown AA, Buermann DH, Bundu AA, Burrows JC, Carter NP, Castillo N, Chiara E, Catenazzi M, Chang S, Neil Cooley R, Crake NR, Dada OO, Diakoumakos KD, Dominguez-Fernandez-B, Earnshaw DJ, Egbujor UC, Elmore DW, Etchin SS, Ewan MR, Fedurco M, Fraser LJ, Fuentes Fajardo KV, Scott Furey W, George D, Gietzen KJ, Goddard CP, Golda GS, Granieri PA, Green DE, Gustafson DL, Hansen NF, Harnish K, Haudenschild CD, Heyer NI, Hims MM, Ho JT, Horgan AM, Hoschler K, Hurwitz S, Ivanov DV, Johnson MQ, James T, Huw Jones TA, Kang G-D, Kerelska TH, Kersey AD, Khrebtukova I, Kindwall AP, Kingsbury Z, Kokko-Gonzales PI, Kumar A, Laurent MA, Lawley CT, Lee SE, Lee X, Liao AK, Loch JA, Lok M, Luo S, Mammen RM, Martin JW, McCauley PG, McNitt P, Mehta P, Moon KW, Mullens JW, Newington T, Ning Z, Ling Ng B, Novo SM, O'Neill MJ, Osborne MA, Osnowski A, Ostadan O, Paraschos LL, Pickering L, Pike AC, Pike AC, Chris Pinkard D, Pliskin DP, Podhasky J, Quijano VJ, Raczy C, Rae VH, Rawlings SR, Chiva Rodriguez A, Roe PM, Rogers J, Rogert Bacigalupo MC, Romanov N, Romieu A, Roth RK, Rourke NJ, Ruediger ST, Rusman E, Sanches-Kuiper RM, Schenker MR, Seoane JM, Shaw RJ, Shiver MK, Short SW, Sizto NL, Sluis JP, Smith MA, Ernest Sohna Sohna J, Spence EJ, Stevens K, Sutton N, Szajkowski L, Tregidgo CL, Turcatti G, vandeVondele S, Verhovsky Y, Virk SM, Wakelin S, Walcott GC, Wang J, Worsley GJ, Yan J, Yau L, Zuerlein M, Rogers J, Mullikin JC, Hurles ME, McCooke NJ, West JS, Oaks FL, Lundberg PL, Klenerman D, Durbin R, Smith AJ (2008) Accurate whole human genome sequencing using reversible terminator chemistry. Nature 456(7218):53–59. https://doi.org/10.1038/nature07517

Braslavsky I, Hebert B, Kartalov E, Quake SR (2003) Sequence information can be obtained from single DNA molecules. Proc Natl Acad Sci USA 100 (7):3960–3964. https://doi.org/10.1073/pnas.0230489100

Brownlee GG (2013) Frederick Sanger (1918–2013). Curr Biol 23(24):R1074–R1076. https://doi.org/10.1016/j.cub.2013.11.037

Brownlee GG, Sanger F, Barrell BG (1967) Nucleotide sequence of 5S-ribosomal RNA from Escherichia coli. Nature 215(5102):735–736. https://doi.org/10.1038/215735a0

Buenrostro JD, Giresi PG, Zaba LC, Chang HY, Greenleaf WJ (2013) Transposition of native chromatin for fast and sensitive epigenomic profiling of open chromatin, DNA-binding proteins and nucleosome position. Nat Methods 10(12):1213–1218. https://doi.org/10.1038/nmeth.2688

Carroll J (2012) Troubled sequencing pioneer Helicos shelters in bankruptcy court | Fierce Biotech. https://www.fiercebiotech.com/r-d/troubled-sequencing-pioneer-helicos-shelters-bankruptcy-court. Accessed 3 Aug 2025

Cathcart R (1990) Advances in automated DNA sequencing. Nature 347(6290):310–310. https://doi.org/10.1038/347310a0

Chen A, Liao S, Cheng M, Ma K, Wu L, Lai Y, Qiu X, Yang J, Xu J, Hao S, Wang X, Lu H, Chen X, Liu X, Huang X, Li Z, Hong Y, Jiang Y, Peng J, Liu S, Shen M, Liu C, Li Q, Yuan Y, Wei X, Zheng H, Feng W, Wang Z, Liu Y, Wang Z, Yang Y, Xiang H, Han L, Qin B, Guo P, Lai G, Muñoz-Cánoves P, Maxwell PH, Thiery JP, Wu Q-F, Zhao F, Chen B, Li M, Dai X, Wang S, Kuang H, Hui J, Wang L, Fei J-F, Wang O, Wei X, Lu H, Wang B, Liu S, Gu Y, Ni M, Zhang W, Mu F, Yin Y, Yang H, Lisby M, Cornall RJ, Mulder J, Uhlén M, Esteban MA, Li Y, Liu L, Xu X, Wang J (2022) Spatiotemporal transcriptomic atlas of mouse organogenesis using DNA nanoball-patterned arrays. Cell 185(10):1777–1792.e21. https://doi.org/10.1016/j.cell.2022.04.003

Cho C-S, Xi J, Si Y, Park S-R, Hsu J-E, Kim M, Jun G, Kang HM, Lee JH (2021) Microscopic examination of spatial transcriptome using Seq-scope. Cell 184(13):3559–3572.e22. https://doi.org/10.1016/j.cell.2021.05.010

Clarke J, Wu H-C, Jayasinghe L, Patel A, Reid S, Bayley H (2009) Continuous base identification for single-molecule nanopore DNA sequencing. Nat Nanotechnol 4(4):265–270. https://doi.org/10.1038/nnano.2009.12

Deininger PL (1983) Random subcloning of sonicated DNA: application to shotgun DNA sequence analysis. Anal Biochem 129(1):216–223. https://doi.org/10.1016/0003-2697(83)90072-6

Djikeng A, Halpin R, Kuzmickas R, DePasse J, Feldblyum J, Sengamalay N, Afonso C, Zhang X, Anderson NG, Ghedin E, Spiro DJ (2008) Viral genome sequencing by random priming methods. BMC Genomics 9(1):5. https://doi.org/10.1186/1471-2164-9-5

Drmanac R, Drmanac S, Strezoska Z, Paunesku T, Labat I, Zeremski M, Snoddy J, Funkhouser WK, Koop B,

Hood L, Crkvenjakov R (1993) DNA sequence determination by hybridization: a strategy for efficient large-scale sequencing. Science 260(5114): 1649–1652. https://doi.org/10.1126/science.8503011

Drmanac R, Sparks AB, Callow MJ, Halpern AL, Burns NL, Kermani BG, Carnevali P, Nazarenko I, Nilsen GB, Yeung G, Dahl F, Fernandez A, Staker B, Pant KP, Baccash J, Borcherding AP, Brownley A, Cedeno R, Chen L, Chernikoff D, Cheung A, Chirita R, Curson B, Ebert JC, Hacker CR, Hartlage R, Hauser B, Huang S, Jiang Y, Karpinchyk V, Koenig M, Kong C, Landers T, Le C, Liu J, McBride CE, Morenzoni M, Morey RE, Mutch K, Perazich H, Perry K, Peters BA, Peterson J, Pethiyagoda CL, Pothuraju K, Richter C, Rosenbaum AM, Roy S, Shafto J, Sharanhovich U, Shannon KW, Sheppy CG, Sun M, Thakuria JV, Tran A, Vu D, Zaranek AW, Wu X, Drmanac S, Oliphant AR, Banyai WC, Martin B, Ballinger DG, Church GM, Reid CA (2010) Human genome sequencing using unchained base reads on self-assembling DNA nanoarrays. Science 327(5961):78–81. https://doi.org/10.1126/science.1181498

Drmanac S, Callow M, Chen L, Zhou P, Eckhardt L, Xu C, Gong M, Gablenz S, Rajagopal J, Yang Q, Villarosa C, Au A, Davis K, Jorjorian A, Wang J, Chen A, Zhang X, Borcherding A, Wei X, Zhang M, Xie Y, Barua N, Shafto J, Dong Y, Zheng Y, Wang L, Zhai L, Li J, Liao S, Zhang W, Liu J, Jiang H, Wang J, Li H, Xu X, Drmanac R (2020) CoolMPS™: advanced massively parallel sequencing using antibodies specific to each natural nucleobase. https://doi.org/10.1101/2020.02.19.953307

Eberwine J, Yeh H, Miyashiro K, Cao Y, Nair S, Finnell R, Zettel M, Coleman P (1992) Analysis of gene expression in single live neurons. Proc Natl Acad Sci 89(7):3010–3014. https://doi.org/10.1073/pnas.89.7.3010

Edwards A, Voss H, Rice P, Civitello A, Stegemann J, Schwager C, Zimmermann J, Erfle H, Caskey CT, Ansorge W (1990) Automated DNA sequencing of the human HPRT locus. Genomics 6(4):593–608. https://doi.org/10.1016/0888-7543(90)90493-e

Eid J, Fehr A, Gray J, Luong K, Lyle J, Otto G, Peluso P, Rank D, Baybayan P, Bettman B, Bibillo A, Bjornson K, Chaudhuri B, Christians F, Cicero R, Clark S, Dalal R, deWinter A, Dixon J, Foquet M, Gaertner A, Hardenbol P, Heiner C, Hester K, Holden D, Kearns G, Kong X, Kuse R, Lacroix Y, Lin S, Lundquist P, Ma C, Marks P, Maxham M, Murphy D, Park I, Pham T, Phillips M, Roy J, Sebra R, Shen G, Sorenson J, Tomaney A, Travers K, Trulson M, Vieceli J, Wegener J, Wu D, Yang A, Zaccarin D, Zhao P, Zhong F, Korlach J, Turner S (2009) Real-time DNA sequencing from single polymerase molecules. Science 323(5910):133–138. https://doi.org/10.1126/science.1162986

Flusberg BA, Webster DR, Lee JH, Travers KJ, Olivares EC, Clark TA, Korlach J, Turner SW (2010) Direct detection of DNA methylation during single-molecule, real-time sequencing. Nat Methods 7(6):461–465. https://doi.org/10.1038/nmeth.1459

Forth LF, Höper D (2019) Highly efficient library preparation for ion torrent sequencing using Y-adapters. BioTechniques 67(5):229–237. https://doi.org/10.2144/btn-2019-0035

Freeman M, Baehler C, Spotts S (1990) Automated laser-fluorescence sequencing. Bio/Technology 8(2): 147–148. https://doi.org/10.1038/nbt0290-147

Fu Y, Wu P-H, Beane T, Zamore PD, Weng Z (2018) Elimination of PCR duplicates in RNA-seq and small RNA-seq using unique molecular identifiers. BMC Genomics 19(1):531. https://doi.org/10.1186/s12864-018-4933-1

Garalde DR, Snell EA, Jachimowicz D, Sipos B, Lloyd JH, Bruce M, Pantic N, Admassu T, James P, Warland A, Jordan M, Ciccone J, Serra S, Keenan J, Martin S, McNeill L, Wallace EJ, Jayasinghe L, Wright C, Blasco J, Young S, Brocklebank D, Juul S, Clarke J, Heron AJ, Turner DJ (2018) Highly parallel direct RNA sequencing on an array of nanopores. Nat Methods 15(3):201–206. https://doi.org/10.1038/nmeth.4577

Gocayne J, Robinson DA, FitzGerald MG, Chung FZ, Kerlavage AR, Lentes KU, Lai J, Wang CD, Fraser CM, Venter JC (1987) Primary structure of rat cardiac beta-adrenergic and muscarinic cholinergic receptors obtained by automated DNA sequence analysis: further evidence for a multigene family. Proc Natl Acad Sci USA 84(23):8296–8300. https://doi.org/10.1073/pnas.84.23.8296

Hafemeister C, Satija R (2019) Normalization and variance stabilization of single-cell RNA-seq data using regularized negative binomial regression. Genome Biol 20(1):296. https://doi.org/10.1186/s13059-019-1874-1

Handyside AH, Kontogianni EH, Hardy K, Winston RM (1990) Pregnancies from biopsied human preimplantation embryos sexed by Y-specific DNA amplification. Nature 344(6268):768–770. https://doi.org/10.1038/344768a0

Harris TD, Buzby PR, Babcock H, Beer E, Bowers J, Braslavsky I, Causey M, Colonell J, DiMeo J, Efcavitch JW, Giladi E, Gill J, Healy J, Jarosz M, Lapen D, Moulton K, Quake SR, Steinmann K, Thayer E, Tyurina A, Ward R, Weiss H, Xie Z (2008) Single-molecule DNA sequencing of a viral genome. Science 320(5872):106–109. https://doi.org/10.1126/science.1150427

Head SR, Komori HK, LaMere SA, Whisenant T, Van Nieuwerburgh F, Salomon DR, Ordoukhanian P (2014) Library construction for next-generation sequencing: overviews and challenges. BioTechniques 56(2):61.-passim. https://doi.org/10.2144/000114133

Heather JM, Chain B (2016) The sequence of sequencers: the history of sequencing DNA. Genomics 107(1):1–8. https://doi.org/10.1016/j.ygeno.2015.11.003

Hewel C, Wierczeiko A, Miedema J, Friedrich J, Hofmann F, Weißbach S, Dietrich V, Holthöfer L, Haug V, Mündnich S, Schartel L, Lehmann L, Jensen KL, Diederich S, Sys S, Butto T, Paul NW, Koch J, Lyko F, Kraft F, Russo A, Schweiger S, Lemke EA, Helm M, Linke M, Gerber S (2025) Direct RNA sequencing enables improved transcriptome assessment and tracking of RNA modifications for medical applications. Nucleic Acids Res 53(22):gkaf1314. https://doi.org/10.1093/nar/gkaf1314

Hodges E, Xuan Z, Balija V, Kramer M, Molla MN, Smith SW, Middle CM, Rodesch MJ, Albert TJ, Hannon GJ, McCombie WR (2007) Genome-wide in situ exon capture for selective resequencing. Nat Genet 39(12): 1522–1527. https://doi.org/10.1038/ng.2007.42

Hood LE, Hunkapiller MW, Smith LM (1987) Automated DNA sequencing and analysis of the human genome. Genomics 1(3):201–212. https://doi.org/10.1016/0888-7543(87)90046-2

Hunkapiller T, Kaiser RJ, Koop BF, Hood L (1991) Large-scale and automated DNA sequence determination. Science 254(5028):59–67. https://doi.org/10.1126/science.1925562

Hussing C, Kampmann M-L, Mogensen HS, Børsting C, Morling N (2018) Quantification of massively parallel sequencing libraries—a comparative study of eight methods. Sci Rep 8:1110. https://doi.org/10.1038/s41598-018-19574-w

Hutchison CA III (2007) DNA sequencing: bench to bedside and beyond †. Nucleic Acids Res 35(18): 6227–6237. https://doi.org/10.1093/nar/gkm688

Ingolia NT, Ghaemmaghami S, Newman JRS, Weissman JS (2009) Genome-wide analysis in vivo of translation with nucleotide resolution using ribosome profiling. Science 324(5924):218–223. https://doi.org/10.1126/science.1168978

International Human Genome Sequencing Consortium (2004) Finishing the euchromatic sequence of the human genome. Nature 431(7011):931–945. https://doi.org/10.1038/nature03001

Istrail S, Sutton GG, Florea L, Halpern AL, Mobarry CM, Lippert R, Walenz B, Shatkay H, Dew I, Miller JR, Flanigan MJ, Edwards NJ, Bolanos R, Fasulo D, Halldorsson BV, Hannenhalli S, Turner R, Yooseph S, Lu F, Nusskern DR, Shue BC, Zheng XH, Zhong F, Delcher AL, Huson DH, Kravitz SA, Mouchard L, Reinert K, Remington KA, Clark AG, Waterman MS, Eichler EE, Adams MD, Hunkapiller MW, Myers EW, Venter JC (2004) Whole-genome shotgun assembly and comparison of human genome assemblies. Proc Natl Acad Sci 101(7):1916–1921. https://doi.org/10.1073/pnas.0307971100

Kambara H, Nishikawa T, Katayama Y, Yamaguchi T (1988) Optimization of parameters in a DNA Sequenator using fluorescence detection. Bio/Technology 6(7):816–821. https://doi.org/10.1038/nbt0788-816

Ke R, Mignardi M, Pacureanu A, Svedlund J, Botling J, Wählby C, Nilsson M (2013) In situ sequencing for RNA analysis in preserved tissue and cells. Nat Methods 10(9):857–860. https://doi.org/10.1038/nmeth.2563

Kim H-M, Jeon S, Chung O, Jun JH, Kim H-S, Blazyte A, Lee H-Y, Yu Y, Cho YS, Bolser DM, Bhak J (2021) Comparative analysis of 7 short-read sequencing platforms using the Korean reference genome: MGI and Illumina sequencing benchmark for whole-genome sequencing. Gigascience 10(3): giab014. https://doi.org/10.1093/gigascience/giab014

Kivioja T, Vähärautio A, Karlsson K, Bonke M, Enge M, Linnarsson S, Taipale J (2011) Counting absolute numbers of molecules using unique molecular identifiers. Nat Methods 9(1):72–74. https://doi.org/10.1038/nmeth.1778

Knierim E, Lucke B, Schwarz JM, Schuelke M, Seelow D (2011) Systematic comparison of three methods for fragmentation of long-range PCR products for next generation sequencing. PLoS One 6(11):e28240. https://doi.org/10.1371/journal.pone.0028240

Kolodziejczyk AA, Kim JK, Svensson V, Marioni JC, Teichmann SA (2015) The technology and biology of single-cell RNA sequencing. Mol Cell 58(4):610–620. https://doi.org/10.1016/j.molcel.2015.04.005

Kozarewa I, Turner DJ (2011) Amplification-free library preparation for paired-end Illumina sequencing. Methods Mol Biol (Clifton NJ) 733:257–266. https://doi.org/10.1007/978-1-61779-089-8_18

Lambolez B, Audinat E, Bochet P, Crépel F, Rossier J (1992) AMPA receptor subunits expressed by single Purkinje cells. Neuron 9(2):247–258. https://doi.org/10.1016/0896-6273(92)90164-9

Lander ES, Linton LM, Birren B, Nusbaum C, Zody MC, Baldwin J, Devon K, Dewar K, Doyle M, FitzHugh W, Funke R, Gage D, Harris K, Heaford A, Howland J, Kann L, Lehoczky J, LeVine R, McEwan P, McKernan K, Meldrim J, Mesirov JP, Miranda C, Morris W, Naylor J, Raymond C, Rosetti M, Santos R, Sheridan A, Sougnez C, Stange-Thomann N, Stojanovic N, Subramanian A, Wyman D, Rogers J, Sulston J, Ainscough R, Beck S, Bentley D, Burton J, Clee C, Carter N, Coulson A, Deadman R, Deloukas P, Dunham A, Dunham I, Durbin R, French L, Grafham D, Gregory S, Hubbard T, Humphray S, Hunt A, Jones M, Lloyd C, McMurray A, Matthews L, Mercer S, Milne S, Mullikin JC, Mungall A, Plumb R, Ross M, Shownkeen R, Sims S, Waterston RH, Wilson RK, Hillier LW, McPherson JD, Marra MA, Mardis ER, Fulton LA, Chinwalla AT, Pepin KH, Gish WR, Chissoe SL, Wendl MC, Delehaunty KD, Miner TL, Delehaunty A, Kramer JB, Cook LL, Fulton RS, Johnson DL, Minx PJ, Clifton SW, Hawkins T, Branscomb E, Predki P, Richardson P, Wenning S, Slezak T, Doggett N, Cheng J-F, Olsen A, Lucas S, Elkin C, Uberbacher E, Frazier M, Gibbs RA, Muzny DM, Scherer SE, Bouck JB, Sodergren EJ, Worley KC, Rives CM, Gorrell JH, Metzker ML, Naylor SL, Kucherlapati RS, Nelson DL, Weinstock GM,

Sakaki Y, Fujiyama A, Hattori M, Yada T, Toyoda A, Itoh T, Kawagoe C, Watanabe H, Totoki Y, Taylor T, Weissenbach J, Heilig R, Saurin W, Artiguenave F, Brottier P, Bruls T, Pelletier E, Robert C, Wincker P, Rosenthal A, Platzer M, Nyakatura G, Taudien S, Rump A, Smith DR, Doucette-Stamm L, Rubenfield M, Weinstock K, Lee HM, Dubois J, Yang H, Yu J, Wang J, Huang G, Gu J, Hood L, Rowen L, Madan A, Qin S, Davis RW, Federspiel NA, Abola AP, Proctor MJ, Roe BA, Chen F, Pan H, Ramser J, Lehrach H, Reinhardt R, McCombie WR, de la Bastide M, Dedhia N, Blöcker H, Hornischer K, Nordsiek G, Agarwala R, Aravind L, Bailey JA, Bateman A, Batzoglou S, Birney E, Bork P, Brown DG, Burge CB, Cerutti L, Chen H-C, Church D, Clamp M, Copley RR, Doerks T, Eddy SR, Eichler EE, Furey TS, Galagan J, Gilbert JGR, Harmon C, Hayashizaki Y, Haussler D, Hermjakob H, Hokamp K, Jang W, Johnson LS, Jones TA, Kasif S, Kaspryzk A, Kennedy S, Kent WJ, Kitts P, Koonin EV, Korf I, Kulp D, Lancet D, Lowe TM, McLysaght A, Mikkelsen T, Moran JV, Mulder N, Pollara VJ, Ponting CP, Schuler G, Schultz J, Slater G, Smit AFA, Stupka E, Szustakowki J, Thierry-Mieg D, Thierry-Mieg J, Wagner L, Wallis J, Wheeler R, Williams A, Wolf YI, Wolfe KH, Yang S-P, Yeh R-F, Collins F, Guyer MS, Peterson J, Felsenfeld A, Wetterstrand KA, Myers RM, Schmutz J, Dickson M, Grimwood J, Cox DR, Olson MV, Kaul R, Raymond C, Shimizu N, Kawasaki K, Minoshima S, Evans GA, Athanasiou M, Schultz R, Patrinos A, Morgan MJ, International Human Genome Sequencing Consortium, Whitehead Institute for Biomedical Research C for GR, The Sanger Centre:, Washington University Genome Sequencing Center, US DOE Joint Genome Institute:, Baylor College of Medicine Human Genome Sequencing Center:, RIKEN Genomic Sciences Center:, Genoscope and CNRS UMR-8030:, Department of Genome Analysis I of MB, GTC Sequencing Center:, Beijing Genomics Institute/Human Genome Center:, Multimegabase Sequencing Center TI for SB, Stanford Genome Technology Center:, University of Oklahoma's Advanced Center for Genome Technology:, Max Planck Institute for Molecular Genetics:, Cold Spring Harbor Laboratory LAHGC, GBF—German Research Centre for Biotechnology:, *Genome Analysis Group (listed in alphabetical order also includes individuals listed under other headings):, Scientific management: National Human Genome Research Institute UNI of H, Stanford Human Genome Center:, University of Washington Genome Center:, Department of Molecular Biology KUS of M, University of Texas Southwestern Medical Center at Dallas:, Office of Science UD of E, The Wellcome Trust (2001) Initial sequencing and analysis of the human genome. Nature 409(6822): 860–921. https://doi.org/10.1038/35057062

Lee JH, Daugharthy ER, Scheiman J, Kalhor R, Yang JL, Ferrante TC, Terry R, Jeanty SSF, Li C, Amamoto R, Peters DT, Turczyk BM, Marblestone AH, Inverso SA, Bernard A, Mali P, Rios X, Aach J, Church GM (2014) Highly multiplexed subcellular RNA sequencing in situ. Science 343(6177):1360–1363. https://doi.org/10.1126/science.1250212

Lee J, Lim H, Jang H, Hwang B, Lee JH, Cho J, Lee JH, Bang D (2019) CRISPR-Cap: multiplexed double-stranded DNA enrichment based on the CRISPR system. Nucleic Acids Res 47(1):e1. https://doi.org/10.1093/nar/gky820

Li Y, Fehlmann T, Borcherding A, Drmanac S, Liu S, Groeger L, Xu C, Callow M, Villarosa C, Jorjorian A, Kern F, Grammes N, Meese E, Jiang H, Drmanac R, Ludwig N, Keller A (2021) CoolMPS: evaluation of antibody labeling based massively parallel non-coding RNA sequencing. Nucleic Acids Res 49(2):e10. https://doi.org/10.1093/nar/gkaa1122

Liu L, Chen A, Li Y, Mulder J, Heyn H, Xu X (2024) Spatiotemporal omics for biology and medicine. Cell 187(17):4488–4519. https://doi.org/10.1016/j.cell.2024.07.040

Luckey JA, Drossman H, Kostichka AJ, Mead DA, D'Cunha J, Norris TB, Smith LM (1990) High speed DNA sequencing by capillary electrophoresis. Nucleic Acids Res 18(15):4417–4421. https://doi.org/10.1093/nar/18.15.4417

Macosko EZ, Basu A, Satija R, Nemesh J, Shekhar K, Goldman M, Tirosh I, Bialas AR, Kamitaki N, Martersteck EM, Trombetta JJ, Weitz DA, Sanes JR, Shalek AK, Regev A, McCarroll SA (2015) Highly parallel genome-wide expression profiling of individual cells using Nanoliter droplets. Cell 161(5): 1202–1214. https://doi.org/10.1016/j.cell.2015.05.002

Margulies M, Egholm M, Altman WE, Attiya S, Bader JS, Bemben LA, Berka J, Braverman MS, Chen Y-J, Chen Z, Dewell SB, Du L, Fierro JM, Gomes XV, Godwin BC, He W, Helgesen S, Ho CH, Irzyk GP, Jando SC, Alenquer MLI, Jarvie TP, Jirage KB, Kim J-B, Knight JR, Lanza JR, Leamon JH, Lefkowitz SM, Lei M, Li J, Lohman KL, Lu H, Makhijani VB, McDade KE, McKenna MP, Myers EW, Nickerson E, Nobile JR, Plant R, Puc BP, Ronan MT, Roth GT, Sarkis GJ, Simons JF, Simpson JW, Srinivasan M, Tartaro KR, Tomasz A, Vogt KA, Volkmer GA, Wang SH, Wang Y, Weiner MP, Yu P, Begley RF, Rothberg JM (2005) Genome sequencing in microfabricated high-density picolitre reactors. Nature 437(7057):376–380. https://doi.org/10.1038/nature03959

Maxam AM, Gilbert W (1977) A new method for sequencing DNA. Proc Natl Acad Sci USA 74(2):560–564. https://doi.org/10.1073/pnas.74.2.560

McGraw RA (1984) Dideoxy DNA sequencing with end-labeled oligonucleotide primers. Anal Biochem 143(2):298–303. https://doi.org/10.1016/0003-2697(84)90666-3

McKernan KJ, Peckham HE, Costa GL, McLaughlin SF, Fu Y, Tsung EF, Clouser CR, Duncan C, Ichikawa JK, Lee CC, Zhang Z, Ranade SS, Dimalanta ET, Hyland

FC, Sokolsky TD, Zhang L, Sheridan A, Fu H, Hendrickson CL, Li B, Kotler L, Stuart JR, Malek JA, Manning JM, Antipova AA, Perez DS, Moore MP, Hayashibara KC, Lyons MR, Beaudoin RE, Coleman BE, Laptewicz MW, Sannicandro AE, Rhodes MD, Gottimukkala RK, Yang S, Bafna V, Bashir A, MacBride A, Alkan C, Kidd JM, Eichler EE, Reese MG, De La Vega FM, Blanchard AP (2009) Sequence and structural variation in a human genome uncovered by short-read, massively parallel ligation sequencing using two-base encoding. Genome Res 19(9):1527–1541. https://doi.org/10.1101/gr.091868.109

Metzker ML (2010) Sequencing technologies—the next generation. Nat Rev Genet 11(1):31–46. https://doi.org/10.1038/nrg2626

Murray V (1989) Improved double-stranded DNA sequencing using the linear polymerase chain reaction. Nucleic Acids Res 17(21):8889. https://doi.org/10.1093/nar/17.21.8889

Nachmanson D, Lian S, Schmidt EK, Hipp MJ, Baker KT, Zhang Y, Tretiakova M, Loubet-Senear K, Kohrn BF, Salk JJ, Kennedy SR, Risques RA (2018) Targeted genome fragmentation with CRISPR/Cas9 enables fast and efficient enrichment of small genomic regions and ultra-accurate sequencing with low DNA input (CRISPR-DS). Genome Res 28(10):1589–1599. https://doi.org/10.1101/gr.235291.118

Park PJ (2009) ChIP-seq: advantages and challenges of a maturing technology. Nat Rev Genet 10(10):669–680. https://doi.org/10.1038/nrg2641

Parras-Moltó M, Rodríguez-Galet A, Suárez-Rodríguez P, López-Bueno A (2018) Evaluation of bias induced by viral enrichment and random amplification protocols in metagenomic surveys of saliva DNA viruses. Microbiome 6(1):119. https://doi.org/10.1186/s40168-018-0507-3

Picelli S, Björklund ÅK, Faridani OR, Sagasser S, Winberg G, Sandberg R (2013) Smart-seq2 for sensitive full-length transcriptome profiling in single cells. Nat Methods 10(11):1096–1098. https://doi.org/10.1038/nmeth.2639

Picelli S, Faridani OR, Björklund AK, Winberg G, Sagasser S, Sandberg R (2014) Full-length RNA-seq from single cells using Smart-seq2. Nat Protoc 9(1):171–181. https://doi.org/10.1038/nprot.2014.006

Prober JM, Trainor GL, Dam RJ, Hobbs FW, Robertson CW, Zagursky RJ, Cocuzza AJ, Jensen MA, Baumeister K (1987) A system for rapid DNA sequencing with fluorescent Chain-terminating Dideoxynucleotides. Science 238(4825):336–341. https://doi.org/10.1126/science.2443975

Pugh J (2023) The current state of Nanopore sequencing. Methods Mol Biol (Clifton NJ) 2632:3–14. https://doi.org/10.1007/978-1-0716-2996-3_1

Ramsköld D, Luo S, Wang Y-C, Li R, Deng Q, Faridani OR, Daniels GA, Khrebtukova I, Loring JF, Laurent LC, Schroth GP, Sandberg R (2012) Full-Length mRNA-Seq from single cell levels of RNA and individual circulating tumor cells. Nat Biotechnol 30(8):777–782. https://doi.org/10.1038/nbt.2282

Robin JD, Ludlow AT, LaRanger R, Wright WE, Shay JW (2016) Comparison of DNA quantification methods for next generation sequencing. Sci Rep 6:24067. https://doi.org/10.1038/srep24067

Rodriques SG, Stickels RR, Goeva A, Martin CA, Murray E, Vanderburg CR, Welch J, Chen LM, Chen F, Macosko EZ (2019) Slide-seq: a scalable technology for measuring genome-wide expression at high spatial resolution. Science 363(6434):1463–1467. https://doi.org/10.1126/science.aaw1219

Rogers J (2025) Memories of the human genome project at the Sanger Centre. Annu Rev Genomics Hum Genet 26:25–43. https://doi.org/10.1146/annurev-genom-111422-103029

Ronaghi M, Karamohamed S, Pettersson B, Uhlén M, Nyrén P (1996) Real-time DNA sequencing using detection of pyrophosphate release. Anal Biochem 242(1):84–89. https://doi.org/10.1006/abio.1996.0432

Ronaghi M, Uhlén M, Nyrén P (1998) A sequencing method based on real-time pyrophosphate. Science 281(5375):363–365. https://doi.org/10.1126/science.281.5375.363

Rothberg JM, Hinz W, Rearick TM, Schultz J, Mileski W, Davey M, Leamon JH, Johnson K, Milgrew MJ, Edwards M, Hoon J, Simons JF, Marran D, Myers JW, Davidson JF, Branting A, Nobile JR, Puc BP, Light D, Clark TA, Huber M, Branciforte JT, Stoner IB, Cawley SE, Lyons M, Fu Y, Homer N, Sedova M, Miao X, Reed B, Sabina J, Feierstein E, Schorn M, Alanjary M, Dimalanta E, Dressman D, Kasinskas R, Sokolsky T, Fidanza JA, Namsaraev E, McKernan KJ, Williams A, Roth GT, Bustillo J (2011) An integrated semiconductor device enabling non-optical genome sequencing. Nature 475(7356):348–352. https://doi.org/10.1038/nature10242

Ruiz-Martinez MC, Jan B, Alexei B, Frantisek F, Miller AW, Karger BL (1993) DNA sequencing by capillary electrophoresis with replaceable linear polyacrylamide and laser-induced fluorescence detection. Anal Chem 65(20):2851–2858. https://doi.org/10.1021/ac00068a023

Sanger F, Thompson EOP (1953) The amino-acid sequence in the glycyl chain of insulin. 1. The identification of lower peptides from partial hydrolysates. Biochem J 53(3):353–366

Sanger F, Tuppy H (1951) The amino-acid sequence in the phenylalanyl chain of insulin. 2. The investigation of peptides from enzymic hydrolysates. Biochem J 49(4):481–490. https://doi.org/10.1042/bj0490481

Sanger F, Air GM, Barrell BG, Brown NL, Coulson AR, Fiddes JC, Hutchison CA, Slocombe PM, Smith M (1977a) Nucleotide sequence of bacteriophage φX174 DNA. Nature 265(5596):687–695. https://doi.org/10.1038/265687a0

Sanger F, Nicklen S, Coulson AR (1977b) DNA sequencing with chain-terminating inhibitors. Proc Natl Acad Sci USA 74(12):5463–5467

Sanger F, Coulson AR, Barrell BG, Roe BA (1980) Cloning in single-stranded bacteriophage as an aid to rapid DNA sequencing. J Mol Biol 143(2):161–178. https://doi.org/10.1016/0022-2836(80)90196-5

Sanger F, Coulson AR, Hong GF, Hill DF, Petersen GB (1982) Nucleotide sequence of bacteriophage λ DNA. J Mol Biol 162(4):729–773. https://doi.org/10.1016/0022-2836(82)90546-0

Schott M, León-Periñán D, Splendiani E, Strenger L, Licha JR, Pentimalli TM, Schallenberg S, Alles J, Tagliaferro SS, Boltengagen A, Ehrig S, Abbiati S, Dommerich S, Pagani M, Ferretti E, Macino G, Karaiskos N, Rajewsky N (2024) Open-ST: high-resolution spatial transcriptomics in 3D. Cell 187(15): 3953–3972.e26. https://doi.org/10.1016/j.cell.2024.05.055

Sims D, Sudbery I, Ilott NE, Heger A, Ponting CP (2014) Sequencing depth and coverage: key considerations in genomic analyses. Nat Rev Genet 15(2):121–132. https://doi.org/10.1038/nrg3642

Smith AJ (1979) The use of exonuclease III for preparing single stranded DNA for use as a template in the chain terminator sequencing method. Nucleic Acids Res 6(3):831–848. https://doi.org/10.1093/nar/6.3.831

Smith LM, Sanders JZ, Kaiser RJ, Hughes P, Dodd C, Connell CR, Heiner C, Kent SBH, Hood LE (1986) Fluorescence detection in automated DNA sequence analysis. Nature 321(6071):674–679. https://doi.org/10.1038/321674a0

Staden R (1980) A new computer method for the storage and manipulation of DNA gel reading data. Nucleic Acids Res 8(16):3673–3694. https://doi.org/10.1093/nar/8.16.3673

Ståhl PL, Salmén F, Vickovic S, Lundmark A, Navarro JF, Magnusson J, Giacomello S, Asp M, Westholm JO, Huss M, Mollbrink A, Linnarsson S, Codeluppi S, Borg Å, Pontén F, Costea PI, Sahlén P, Mulder J, Bergmann O, Lundeberg J, Frisén J (2016) Visualization and analysis of gene expression in tissue sections by spatial transcriptomics. Science 353(6294):78–82. https://doi.org/10.1126/science.aaf2403

Stark R, Grzelak M, Hadfield J (2019) RNA sequencing: the teenage years. Nat Rev Genet 20(11):631–656. https://doi.org/10.1038/s41576-019-0150-2

Stickels RR, Murray E, Kumar P, Li J, Marshall JL, Di Bella DJ, Arlotta P, Macosko EZ, Chen F (2021) Highly sensitive spatial transcriptomics at near-cellular resolution with Slide-seqV2. Nat Biotechnol 39(3): 313–319. https://doi.org/10.1038/s41587-020-0739-1

Swerdlow H, Wu SL, Harke H, Dovichi NJ (1990) Capillary gel electrophoresis for DNA sequencing. Laser-induced fluorescence detection with the sheath flow cuvette. J Chromatogr 516(1):61–67. https://doi.org/10.1016/s0021-9673(01)90204-3

Tang F, Barbacioru C, Wang Y, Nordman E, Lee C, Xu N, Wang X, Bodeau J, Tuch BB, Siddiqui A, Lao K, Surani MA (2009) mRNA-Seq whole-transcriptome analysis of a single cell. Nat Methods 6(5):377–382. https://doi.org/10.1038/nmeth.1315

Travers KJ, Chin C-S, Rank DR, Eid JS, Turner SW (2010) A flexible and efficient template format for circular consensus sequencing and SNP detection. Nucleic Acids Res 38(15):e159. https://doi.org/10.1093/nar/gkq543

Venter JC, Adams MD, Sutton GG, Kerlavage AR, Smith HO, Hunkapiller M (1998) Shotgun sequencing of the human genome. Science 280(5369):1540–1542. https://doi.org/10.1126/science.280.5369.1540

Venter JC, Adams MD, Myers EW, Li PW, Mural RJ, Sutton GG, Smith HO, Yandell M, Evans CA, Holt RA, Gocayne JD, Amanatides P, Ballew RM, Huson DH, Wortman JR, Zhang Q, Kodira CD, Zheng XH, Chen L, Skupski M, Subramanian G, Thomas PD, Zhang J, Gabor Miklos GL, Nelson C, Broder S, Clark AG, Nadeau J, McKusick VA, Zinder N, Levine AJ, Roberts RJ, Simon M, Slayman C, Hunkapiller M, Bolanos R, Delcher A, Dew I, Fasulo D, Flanigan M, Florea L, Halpern A, Hannenhalli S, Kravitz S, Levy S, Mobarry C, Reinert K, Remington K, Abu-Threideh J, Beasley E, Biddick K, Bonazzi V, Brandon R, Cargill M, Chandramouliswaran I, Charlab R, Chaturvedi K, Deng Z, Francesco VD, Dunn P, Eilbeck K, Evangelista C, Gabrielian AE, Gan W, Ge W, Gong F, Gu Z, Guan P, Heiman TJ, Higgins ME, Ji R-R, Ke Z, Ketchum KA, Lai Z, Lei Y, Li Z, Li J, Liang Y, Lin X, Lu F, Merkulov GV, Milshina N, Moore HM, Naik AK, Narayan VA, Neelam B, Nusskern D, Rusch DB, Salzberg S, Shao W, Shue B, Sun J, Wang ZY, Wang A, Wang X, Wang J, Wei M-H, Wides R, Xiao C, Yan C, Yao A, Ye J, Zhan M, Zhang W, Zhang H, Zhao Q, Zheng L, Zhong F, Zhong W, Zhu SC, Zhao S, Gilbert D, Baumhueter S, Spier G, Carter C, Cravchik A, Woodage T, Ali F, An H, Awe A, Baldwin D, Baden H, Barnstead M, Barrow I, Beeson K, Busam D, Carver A, Center A, Cheng ML, Curry L, Danaher S, Davenport L, Desilets R, Dietz S, Dodson K, Doup L, Ferriera S, Garg N, Gluecksmann A, Hart B, Haynes J, Haynes C, Heiner C, Hladun S, Hostin D, Houck J, Howland T, Ibegwam C, Johnson J, Kalush F, Kline L, Koduru S, Love A, Mann F, May D, McCawley S, McIntosh T, McMullen I, Moy M, Moy L, Murphy B, Nelson K, Pfannkoch C, Pratts E, Puri V, Qureshi H, Reardon M, Rodriguez R, Rogers Y-H, Romblad D, Ruhfel B, Scott R, Sitter C, Smallwood M, Stewart E, Strong R, Suh E, Thomas R, Tint NN, Tse S, Vech C, Wang G, Wetter J, Williams S, Williams M, Windsor S, Winn-Deen E, Wolfe K, Zaveri J, Zaveri K, Abril JF, Guigó R, Campbell MJ, Sjolander KV, Karlak B, Kejariwal A, Mi H, Lazareva B, Hatton T, Narechania A, Diemer K, Muruganujan A, Guo N, Sato S, Bafna V, Istrail S, Lippert R, Schwartz R, Walenz B, Yooseph S, Allen D, Basu A, Baxendale J, Blick L, Caminha M, Carnes-Stine J, Caulk P, Chiang Y-H, Coyne M, Dahlke C, Mays AD, Dombroski M, Donnelly M, Ely D, Esparham S, Fosler C, Gire H, Glanowski S, Glasser K, Glodek A,

Gorokhov M, Graham K, Gropman B, Harris M, Heil J, Henderson S, Hoover J, Jennings D, Jordan C, Jordan J, Kasha J, Kagan L, Kraft C, Levitsky A, Lewis M, Liu X, Lopez J, Ma D, Majoros W, McDaniel J, Murphy S, Newman M, Nguyen T, Nguyen N, Nodell M, Pan S, Peck J, Peterson M, Rowe W, Sanders R, Scott J, Simpson M, Smith T, Sprague A, Stockwell T, Turner R, Venter E, Wang M, Wen M, Wu D, Wu M, Xia A, Zandieh A, Zhu X (2001) The sequence of the human genome. Science 291(5507):1304–1351. https://doi.org/10.1126/science.1058040

Wang (王博) B, Jia (贾鹏) P, Gao (高胜寒) S, Zhao (赵焕焕) H, Zheng (郑高洋) G, Xu (许林峰) L, Ye (叶凯) K (2025) Long and accurate: how HiFi sequencing is transforming genomics. Genomics Proteomics Bioinformatics 23(1):qzaf003. https://doi.org/10.1093/gpbjnl/qzaf003

Wang X, Allen WE, Wright MA, Sylwestrak EL, Samusik N, Vesuna S, Evans K, Liu C, Ramakrishnan C, Liu J, Nolan GP, Bava F-A, Deisseroth K (2018) Three-dimensional intact-tissue sequencing of single-cell transcriptional states. Science 361(6400):eaat5691. https://doi.org/10.1126/science.aat5691

Wenger AM, Peluso P, Rowell WJ, Chang P-C, Hall RJ, Concepcion GT, Ebler J, Fungtammasan A, Kolesnikov A, Olson ND, Töpfer A, Alonge M, Mahmoud M, Qian Y, Chin C-S, Phillippy AM, Schatz MC, Myers G, DePristo MA, Ruan J, Marschall T, Sedlazeck FJ, Zook JM, Li H, Koren S, Carroll A, Rank DR, Hunkapiller MW (2019) Accurate circular consensus long-read sequencing improves variant detection and assembly of a human genome. Nat Biotechnol 37(10):1155–1162. https://doi.org/10.1038/s41587-019-0217-9

Wolock SL, Lopez R, Klein AM (2019) Scrublet: computational identification of cell doublets in single-cell transcriptomic data. Cell Syst 8(4):281–291.e9. https://doi.org/10.1016/j.cels.2018.11.005

Xu Y, Lin Z, Tang C, Tang Y, Cai Y, Zhong H, Wang X, Zhang W, Xu C, Wang J, Wang J, Yang H, Yang L, Gao Q (2019) A new massively parallel nanoball sequencing platform for whole exome research. BMC Bioinformatics 20:153. https://doi.org/10.1186/s12859-019-2751-3

Zhang A, Li S, Apone L, Sun X, Chen L, Ettwiller LM, Langhorst BW, Noren CJ, Xu M-Q (2018) Solid-phase enzyme catalysis of DNA end repair and 3′ A-tailing reduces GC-bias in next-generation sequencing of human genomic DNA. Sci Rep 8(1):15887. https://doi.org/10.1038/s41598-018-34079-2

Zhao W, He X, Hoadley KA, Parker JS, Hayes DN, Perou CM (2014) Comparison of RNA-Seq by poly (A) capture, ribosomal RNA depletion, and DNA microarray for expression profiling. BMC Genomics 15(1):419. https://doi.org/10.1186/1471-2164-15-419

Zheng GXY, Terry JM, Belgrader P, Ryvkin P, Bent ZW, Wilson R, Ziraldo SB, Wheeler TD, McDermott GP, Zhu J, Gregory MT, Shuga J, Montesclaros L, Underwood JG, Masquelier DA, Nishimura SY, Schnall-Levin M, Wyatt PW, Hindson CM, Bharadwaj R, Wong A, Ness KD, Beppu LW, Deeg HJ, McFarland C, Loeb KR, Valente WJ, Ericson NG, Stevens EA, Radich JP, Mikkelsen TS, Hindson BJ, Bielas JH (2017) Massively parallel digital transcriptional profiling of single cells. Nat Commun 8:14049. https://doi.org/10.1038/ncomms14049

Latest Developments in Mass Spectrometry-Based Techniques for Metabolomics Analysis

4

Marisa Maia, Jasmine Hertzog, and Vincent Carré

Abstract

Metabolomics is one of the most recent OMICs sciences and considered one of the most complex. It aims to provide a comprehensive study of small molecules (metabolites) in a biological system to obtain insights of cellular processes and metabolic networks. For such purpose, several advanced analytical techniques such as mass spectrometry (MS) are employed to detect, identify, and even quantify metabolites.

In this book chapter, we present the two main approaches (targeted and untargeted) used in metabolomics, as well as their individual advantages and limitations and their complementary strengths. As far as MS analytical techniques are concerned, both the various mass spectrometers and ionization sources, currently used for studying metabolites, are discussed, as well as the field of application linked to their performance.

Lastly, recent advances made in exploring the different organisms through single-cell metabolomics as well as metabolomics data integration with other OMIC sciences are also discussed.

Keywords

Metabolomics · Targeted and untargeted analysis · Mass analyzers · Ionization sources · Mass spectrometry imaging

4.1 Introduction

Metabolites are small molecular weight molecules produced during metabolic processes in living organisms from all different kingdoms. These biomolecules are extremely reactive to any type of changes, which allows researchers to better understand the intricate web of biochemical reactions and pathways within cells and tissues (Bino et al. 2004; Klassen et al. 2017). Thus, metabolites provide a real picture of the physiological state of an organism at a given time (Fiehn 2002). The study of metabolites, in a large-scale, denominated "metabolomics," is an OMICs science focused on the comprehensive analysis, through detection, identification, and quantification of metabolites, composing the metabolome, with the main goal of understanding the

Marisa Maia, Jasmine Hertzog and Vincent Carré contributed equally to this book chapter.

M. Maia
Grapevine Pathogen Systems Lab (GPS Lab), Biosystems & Integrative Sciences Institute (BioISI), Faculdade de Ciências, Universidade de Lisboa, Lisbon, Portugal

Departamento de Biologia, Faculdade de Ciências, Universidade de Lisboa, Lisbon, Portugal
e-mail: mrmaia@ciencias.ulisboa.pt

J. Hertzog (✉) · V. Carré (✉)
Université de Lorraine, Metz, France
e-mail: jasmine.hertzog@univ-lorraine.fr; vincent.carre@univ-lorraine.fr

M. Gama-Carvalho et al. (eds.), *Omics Approaches in Biomedicine and Biotechnology*, Advances in Experimental Medicine and Biology 1504, https://doi.org/10.1007/978-3-032-18966-0_4

molecular footprints and biochemical pathways behind cellular regulatory processes in living organisms (Klassen et al. 2017; Giera et al. 2022).

Compared to other OMICs sciences, such as genomics, transcriptomics, and proteomics, metabolomics is considered quite recent with the concepts of "metabolome" and "metabolomics" only appearing in the late 1990s/beginning of 2000s (Klassen et al. 2017; Maia et al. 2023; Giera et al. 2022; Oliver et al. 1998). Nevertheless, it has been rapidly expanding and considered as one of the most promising approaches, being nowadays essential in a variety of fields such as biological processes, metabolic networks, biomedical and clinical applications, health and nutrition studies, environmental and agricultural applications, drug development, toxicology, food security, and biotechnology (Reaves and Rabinowitz 2011; Bouslimani et al. 2014; Yang et al. 2019; Razzaq et al. 2019; Gonzalez-Covarrubias et al. 2022; Bauermeister et al. 2022; Katam et al. 2022; Pang and Hu 2023).

The compelling evidence of the importance of these compounds has stimulated researchers to focus on this particular OMIC. Figure 4.1a clearly shows the increase of the publication number focused on metabolomics, reaching almost the same number of publications dedicated to proteomics last year (2024).

Having this in mind, researchers have been applying and developing protocols and analytical techniques to identify, characterize, and quantify metabolites in different living systems. However, the metabolomic workflow is quite challenging, when considering their number, their wide range of structures and chemical properties, their vast differences in level, their broad biological variance, their overlap between species, and the presence of other biomolecules. Several extensive reviews, protocols, and original studies have been published in the last decade focusing specifically on this topic (Hall et al. 2002; Dettmer et al. 2007; Lei et al. 2011; Gowda and Djukovic 2014; Wang et al. 2015; Ren et al. 2018; Pereira Braga and Adamec 2019; Santamaria and Pinto 2024). Shortly, from a sample preparation point of view, since metabolites are generally the first to be affected by changing conditions, sample collection, storage, and preparation steps must be quickly and precisely done to avoid any bias (Vuckovic 2012; Mushtaq et al. 2014; Dietz 2017; Salem et al. 2020; de Souza et al. 2024). The sample itself needs to be treated by removing other biomolecules, such as proteins and salts, which can hamper the accuracy and detection of the metabolites (Causon and Hann 2016; de Souza et al. 2024). Also, although many protocols are already established, to get optimal results and comparable datasets and to perform large-scale studies, it is fundamental to limit the variability factors such as the number of operators or changes of consumable suppliers, throughout the sample preparation steps. Lastly and most importantly, it is imperative that all these steps are compatible with the analytical technique that is going to be used.

Nuclear magnetic resonance (NMR) and mass spectrometry (MS) are currently the two main analytical techniques used to study, identify, and quantify metabolites (Fig. 4.1b) (Gowda and Djukovic 2014; Klassen et al. 2017; Ren et al. 2018; Muthubharathi et al. 2021). NMR is a nondestructive and highly reproducible technique that exploits the magnetic properties of certain atomic nuclei. This analytical technique has the capacity to analyze and quantify metabolites individually, allowing the determination of their molecular structure and concentration within a biological sample (Wishart 2008; Gowda and Raftery 2021; Moco 2022). Nevertheless, NMR has several limitations that can make it less suitable for certain metabolomic applications (Emwas 2015). NMR has relatively lower sensitivity, requires large sample volumes, and may take longer data acquisition time. Also, complex samples are difficult to analyze since many metabolites have similar chemical structures, leading to overlapping NMR signals, which are difficult to resolve and identify with certainty individual metabolites (Emwas 2015).

In contrast, MS offers a good combination of accuracy, high throughput, and resolution (Lei et al. 2011; Wang et al. 2015; Ren et al. 2018). In this technique, the metabolites present in the sample are ionized, giving at least positively or

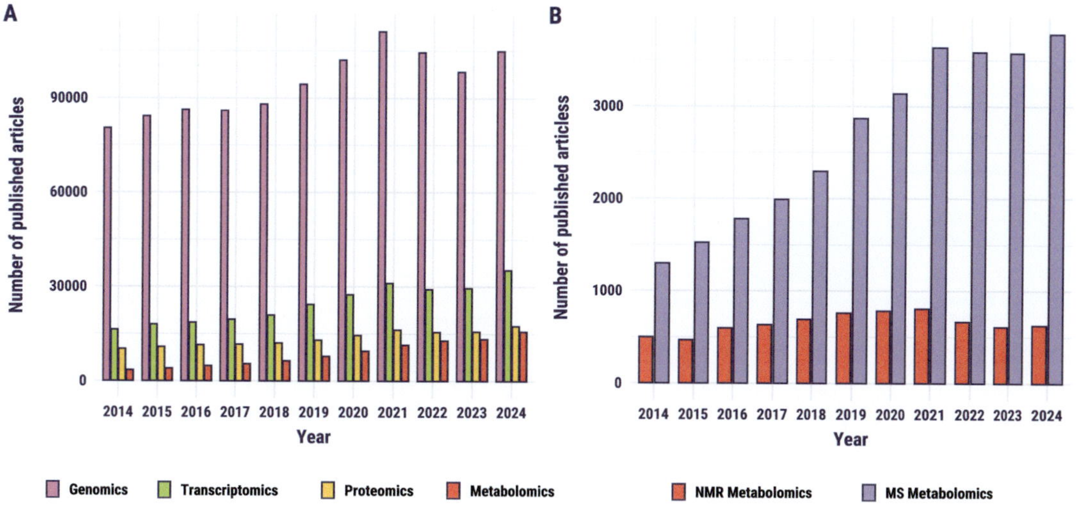

Fig. 4.1 Representation of the number of research manuscripts published during the past 10 years (2014–2024). Numbers were obtained from the PubMed search (https://pubmed.ncbi.nlm.nih.gov/) using the following keywords: (**a**) "Genomics," "Transcriptomics," "Proteomics," and "Metabolomics" and (**b**) "Nuclear Magnetic Resonance Metabolomics" and "Mass Spectrometry Metabolomics"

negatively charged ions, and are separated based on their mass-to-charge ratio (*m/z*). The output is a mass spectrum representing the intensity of each ion as a function of their *m/z* value. The MS analysis provides information on the molecular weight of each metabolite and may also give access to their structures according to the MS operating mode. This analysis method is highly sensitive allowing the detection of metabolites in the picomole to femtomole concentrations within highly complex biological samples (Tomita and Nishioka 2005; Maia et al. 2023). Moreover, MS is a highly flexible and versatile technique. It can be coupled with various separation techniques such as gas chromatography (GC) and liquid chromatography (LC), to improve metabolite detection and identification (Gika et al. 2014; Begou et al. 2017; Ren et al. 2018; Lacalle-Bergeron et al. 2021). Moreover, different ionization sources can be used to optimize the detection of metabolites based on their chemical properties (Wang et al. 2015).

Over the last decade, NMR was shown to offer advantages for studying the metabolome. However, the number of publications using NMR does not come close to those involving MS (Fig. 4.1b). This is mainly due to the analytical improvements and developments of analytical methods by mass spectrometry for the detection, identification, and quantification of a wide range of metabolites in different samples independently of its complexity (Wang et al. 2015; Ren et al. 2018).

4.2 Targeted and Untargeted MS-Based Metabolomics

In MS-based metabolomics, there are two distinct approaches to analyze the metabolome: targeted and untargeted (also known as nontargeted) (Ribbenstedt et al. 2018; Pereira Braga and Adamec 2019; Lelli et al. 2021). In both approaches, the main aim is to analyze the metabolites present in the sample. However, the choice between targeted and untargeted methods depends on the research objectives (Klassen et al. 2017; Lelli et al. 2021) (Fig. 4.2). The analytical methodologies and MS techniques applied for both approaches can be the same; however, some adaptations need to be made for sample preparation and/or analysis and data treatment.

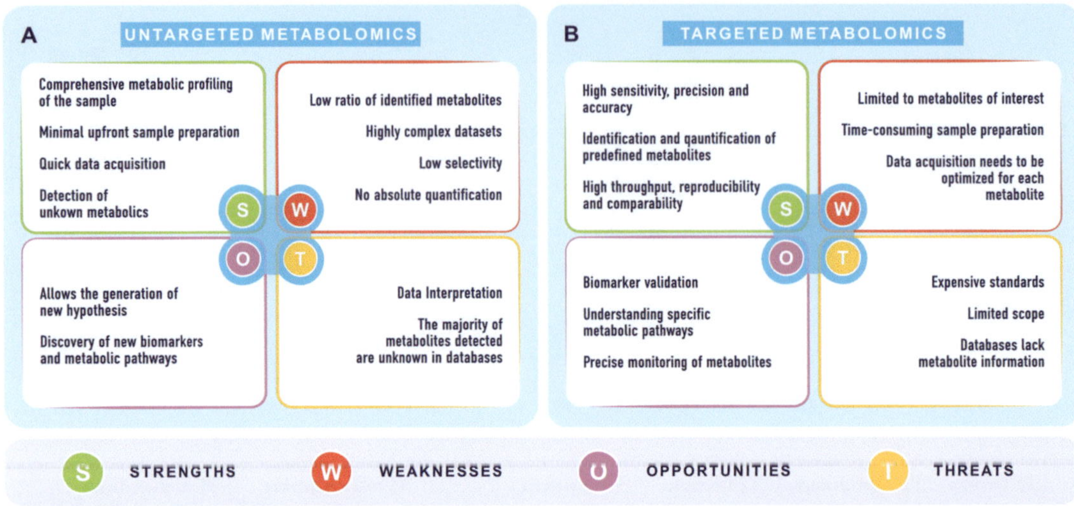

Fig. 4.2 SWOT (strengths, weaknesses, opportunities, and threats) analysis of (**a**) untargeted MS-based metabolomics and (**b**) targeted MS-based metabolomics

Both targeted and untargeted approaches are essential in everyday life due to their applications in numerous fields such as disease diagnostics, pharmaceutical, quality and food control, environment, ecology, toxicology, and archaeology and heritage (Clish 2015; Razzaq et al. 2019; Li et al. 2021b; Elshafie et al. 2023; Qiu et al. 2023; Badillo-Sanchez et al. 2023; Caldwell and Leo 2017). In this section, advantages and limitations of each approach are going to be explored.

4.2.1 Untargeted Metabolomics

The main goal of the untargeted approach is to have a holistic overview of the sample metabolome, with minimal sample preparation and a high-throughput data acquisition. With this method, an exploratory methodology is used where all detectable metabolites are profiled without prior knowledge or guidance. The goal is to detect as many metabolites as possible, covering a wide range of chemical classes, to obtain a molecular fingerprint that can be compared with other samples. Thus, it allows capturing unexpected metabolic changes, discovering new metabolites, and understanding metabolic pathways (Fig. 4.2a) (Sévin et al. 2015). Ultimately, this approach helps to better understand biological changes, as well as generate new biological hypotheses.

In the last decade, untargeted metabolomics had a significant importance in different research fields (Sévin et al. 2015; Clish 2015; Qiu et al. 2023; Caldwell and Leo 2017). For example, untargeted MS-based metabolomics has been aiding in the medicine research field by helping on the diagnostic and screening of diseases (Gonzalez-Covarrubias et al. 2022). LC-MS approach was used to identify novel biomarkers for incident coronary heart disease (Ganna et al. 2014). Ganna and colleagues investigated the metabolic profiles of more than 3600 individuals in positive-ion mode and discovered that certain metabolites, i.e., lysophosphatidylcholines 18:1, 18:2, monoglyceride 18:2, and sphingomyelin 28:1, are risk factors of coronary heart disease. More recently, in a similar topic, Willems and co-workers (Willems et al. 2023) showed how untargeted metabolomic approach, using high-resolution mass spectrometry (Orbitrap) with direct-infusion (DI), can help to diagnose and screen inherited metabolic diseases. The authors gathered, over 1 year, plasma samples of patients suspected to have an inherited metabolic disease and focused their analysis on polar metabolites, a class of compounds known to be suitable for diagnosing inherited metabolic diseases. The results showed that untargeted metabolomic approach can be used as a first-tier approach replacing targeted assays (Willems et al. 2023). Untargeted MS-based techniques have also emerged as a valuable tool in drug discovery and development, offering insights into novel therapeutic targets, understanding drug efficiency and interaction with biological systems, preclinical safety assessment, and understanding drug-induced toxicity (Sindelar and Patti 2020; Caldwell and Leo 2017).

The use of MS untargeted approaches has also been key on exposome studies, helping associating exogenous exposures to complex environmental mixtures, such as air pollution, with endogenous toxic doses and biological metabolic changes (Lankadurai et al. 2013; Vlaanderen et al. 2017; Liang et al. 2018; van Veldhoven et al. 2019; Liang et al. 2023). A very interesting example of this application was the study of the human metabolome before, during, and after the 2008 Beijing Olympics Games, when air pollution was high, low, and high, respectively (Mu et al. 2019). In this work, Mu and colleagues collected serum samples from 26 participants and used GC-MS and LC-MS. The results showed that upon a drastic change of the air pollution, two major metabolic signatures could be identified in lipid and peptide metabolisms (Mu et al. 2019).

Untargeted metabolomics has also been used to explore the microbial metabolome (Bauermeister et al. 2022). The employment of this approach has the capability to illustrate gut microbial metabolic profile but also help to differentiate between closely related pathogenic and nonpathogenic bacteria species. By LC-MS-based untargeted metabolomics, it was possible to explore the microbial metabolome of bacteria single cultures and co-culture systems and to create a workflow to differentiate gut bacterial species and detecting their characteristic metabolites proportionally to the microbial population in co-culture systems (Zhang and Zhu 2022).

As a last example, MS-based untargeted metabolomic approaches have also been used in agricultural and sustainable research (Gemperline et al. 2016; Razzaq et al. 2019; Castro-Moretti et al. 2020; Maia et al. 2023; Liu et al. 2022). It can be applied to discriminate disease resistant and susceptible plants in a quick and simple way, not only upon infection but also at the constitutive level. As grapevine (*Vitis vinifera* L.) is one of the most economically important crops in the world, an untargeted metabolomic analysis by Fourier transform ion cyclotron resonance mass spectrometry (FT-ICR MS) was used to discriminate susceptible and resistant *Vitis* genotypes (Maia et al. 2020). The results showed that the metabolome profile, obtained for the different *Vitis* leaves, appears to be sufficiently sensitive to distinguish species and cultivars but also to susceptible and resistant genotypes. Also, using a nontargeted GC-MS metabolomic approach, it was possible to identify volatile organic compounds associated with different immune responses of tomato leaves infected with either virulent or avirulent strains of *Pseudomonas syringae* (López-Gresa et al. 2017).

The studies mentioned above serve as a representation of the wide range of research fields where untargeted MS-approaches are essential. In fact, innumerable studies are published every year pushing the boundaries and demonstrating the various capabilities of this approach.

However, an untargeted MS approach has also several limitations (Fig. 4.2a). Focusing on the full metabolome coverage of a sample may result in lower sensitivity for specific compounds, and due to the high complexity of generated datasets, specific bioinformatic tools are needed to process and interpret the data. Also, it is important to have in mind that despite all the technological advances in detection and resolving in MS equipment's, the metabolites complexity, similarity, and chemical properties surpass these improvements. With untargeted MS studies, a full profile of the sample is provided but neither a full understanding of each metabolite function is possible nor the identification of all detected metabolites, especially for unknown ones.

4.2.2 Targeted Metabolomics

Targeted approaches are focused on the detection and identification of specific and known metabolites (Ribbenstedt et al. 2018; Pereira Braga and Adamec 2019; Lelli et al. 2021). Opposed to untargeted approaches, targeted methods offer better selectivity and precision, higher reproducibility, and the possibility of identification of predefined metabolites (Fig. 4.2b) (Griffiths et al. 2010).

In targeted studies, prior data acquisition, sample preparation needs to be optimized (Klassen et al. 2017). Due to biological sample complexity, other biomolecules and metabolites, of different chemical properties, need to be eliminated to retain only the metabolites of interest. Also, the method for data acquisition needs to be optimized to ensure the detection and proper identification of the selected metabolites.

Unlike untargeted methods, targeted analysis is defined as a quantitative or semiquantitative analysis providing, respectively, precise amount in a sample or the registration of relative intensities of specific metabolites with high accuracy (Pereira Braga and Adamec 2019; Lelli et al. 2021). This ensures an easier data interpretation, providing detailed information not only on the metabolites of interest but also on metabolites interactions and metabolic pathways. Quantitative MS-based metabolomics generally involves chromatographic separation and tandem MS (MS/MS) approaches (Roberts et al. 2012; Bouslimani et al. 2014; Crotti et al. 2023). For the chromatographic separation, during the analytical process, a chromatogram of the MS signal over a defined time is generated, from which it is possible to retrieve the peak area and the retention time of each metabolites of interest (Patel et al. 2010; Begou et al. 2017). For quantitation approach, calibration curves are needed not only to control accuracy and reproducibility of the system but also to determine the concentration of the specific metabolites in complex samples. For MS/MS, ion species are selected and isolated from the mixture and fragmented into smaller ions. The output is a mass spectrum of all the fragments generated. Both in chromatography separation and tandem MS, standards are essential (Bergeron et al. 2009; Rocca-Serra et al. 2015; Ghafari and Sleno 2024). After sample analysis, both the known MS fragmentation patterns and retention times of standards (obtained in-house or through online spectral libraries) can be compared with the experimental data obtained for the metabolites of interest. This strategy enables the user to obtain accurate metabolic identifications and quantifications, with high reproducibility and data quality (Rocca-Serra et al. 2015).

Targeted metabolomics also has a wide range of applications. However, it is particularly used in clinical diagnostics, biomarker validation, drug quantification, and understanding metabolic changes (Zhou and Zhong 2022; Lin et al. 2024).

The study of Lee et al. (2019) can be used as an example of the advantages of using a targeted MS approach for the study and comparison of different human experimental samples (Lee et al. 2019). The authors developed a tandem MS-based label-free targeted metabolomic analysis framework coupled to two distinct chromatographic methods: reversed-phase liquid chromatography (RPLC) and hydrophilic interaction liquid chromatography (HILIC). This method allowed for the simultaneous detection of over 200 metabolites enabling the authors to study metabolic dynamics. This methodology was tested in several heterogeneous biological samples including cancer cell lines, extracellular media, immune cells, organs, and tissues. The global characterization of the samples revealed that there are some specific metabolic features and metabolic changes, which are common across all samples. These results led the authors to suggest that some metabolic pathways could systematically play a more significant role than others.

In the biomarker discovery field applied to disease diagnostics, Ward-Caviness and colleagues used MS targeted metabolomics to identify biomarkers for myocardial infarction risk prediction and evaluate their involvement in the pathogenesis on this condition (Ward-Caviness et al. 2017). The authors used LC and flow injection analysis–mass spectrometry to identify and quantify simultaneous 188 metabolites. The combination of the obtained MS data with bioinformatic and statistical tools allowed to identify three metabolites (arginine, lysophosphatidylcholines (LPC 17:0 and LPC 18:2)), which, in combination, are associated with myocardial infarction and increase the predictive value of the estimation of a risk of a heart attack in 10 years. In the same frame, MS-based targeted metabolomics was also used to predict stroke recurrence after a transient ischemic attack (Jové et al. 2015). LC-MS was also used, and due to the sensibility of this technique, low concentration of a specific metabolite lysophosphatidylcholine (LysoPC[16:0]) was detected and significantly associated with stroke recurrence. Moreover, LysoPC(20:4) and LysoPC(22:6) were also pointed out as potential biomarkers for, respectively, stroke recurrence and large-artery atherosclerosis.

Also, the study of natural based products is highly interesting for the search of new compounds with several biological activities. Since these compounds tend to be present in low concentrations and are highly diverse in chemical structures, targeted MS-based methods are the approach of choice. One example are phenolic compounds, known for their health benefits, plant protection, and their impact on food quality (Cao-Ngoc et al. 2019; Cao-Ngoc et al. 2020; Maia et al. 2021; Nardini 2022). Assessing the identification and quantification of these compounds in different natural products is a valuable tool. Vrhovsek et al. (2012) developed a targeted metabolomic method for the quantification of phenolic compounds, in fruits, such as grapes and berries, and green tea extracts and red wine using UPLC-MS/MS (Vrhovsek et al. 2012).

Regardless of the essential role in studying the metabolome, targeted MS approaches also present some limitations (Fig. 4.2b). Targeted studies can only be performed if genuine chemical standards of the metabolites in study are available (Bergeron et al. 2009), which is difficult to obtain since there are no pure standards available for all metabolites of interest. Moreover, targeted studies also tend to be limited to individual biological processes due to their limited coverage of the metabolome. Although this helps to focus on a specific biological question, this approach represents a reductionist view of the functions of the global system, leading to the creation of limited hypotheses.

4.3 MS-Based Techniques in Metabolomics

Targeted and untargeted MS-based metabolomic approaches play essential roles in studying the metabolome (Ribbenstedt et al. 2018). Although each one has its own advantages and limitations, and having in mind all the research studies mentioned above, there is no optimal choice for a specific approach or methodology of sample analysis that is considered more correct for targeted or untargeted metabolomic studies. Together, these approaches can provide an overall view of a biological system, highlighting the importance of specific metabolites, metabolites similarities and differences between samples, and metabolites relative and absolute abundances in samples. Indeed, quite often, a combination of both approaches is used. First, an untargeted approach is performed to explore the overall metabolome and understand the changes in metabolite abundancies, and then a targeted approach is employed to identify and quantify the metabolites of interest (Cajka and Fiehn 2016; Klassen et al. 2017; Ribbenstedt et al. 2018; Pereira Braga and Adamec 2019).

Considering the different MS equipment's currently available, no single MS technique has the capacity of detecting all metabolites present in a biological sample. It is therefore necessary to combine several analytical methods, such as different ion sources, separative techniques, and mass analyzers to improve the metabolome coverage.

For untargeted approaches, direct injection (DI) is commonly used (Zahn et al. 2001; Draper et al. 2013; Kirwan et al. 2014; González-Domínguez et al. 2017; Gutbrod et al. 2021; Romano et al. 2023; Meijer et al. 2024). It is the most straightforward method for sample analysis.

The sample is injected directly into the mass spectrometer, without prior chromatographic separation, typically after minimal sample preparation such as dilution or filtration. This allows for a rapid and high-throughput profiling with high sensitivity, generating a mass spectrum that reflects the overall metabolite composition of the sample. Nevertheless, it also offers significant disadvantages (Kirwan et al. 2014). Ion suppression may be encountered, especially with electrospray ionization source (Draper et al. 2013; Furey et al. 2013), making sample comparison difficult, in that case. Moreover, isomers cannot be differentiated, which can be relevant for the accurate identification of metabolites of interest (Gertsman and Barshop 2018; Alseekh et al. 2021). Lastly, the data obtained are highly complex, with thousands of *m/z* features, being difficult to directly analyze and interpret. A way to overcome these issues is to couple separative techniques to MS, such as chromatography, capillary electrophoresis, and ion mobility (Gika et al. 2014; Begou et al. 2017; Crotti et al. 2023; Soga 2023; Christofi and Barran 2023). Depending on the methods that can be applied on these combined systems, separation of a range of metabolites and an overall resolution of their complexity can be achieved compared to direct injection methods (this topic is detailed in the Chapter "Separative Techniques Coupled to Mass Spectrometry for Metabolomic Analysis" of this book). These coupled techniques can be used for nontargeted and targeted analyses. For nontargeted metabolomic, tandem mass spectrometry (MS/MS) is particularly used combined with liquid chromatography (LC). Two main operating modes can be implemented: data-dependent analysis (DDA) and data-independent analysis (DIA). The first one works by repeating the sequence: a first MS scan is acquired; the top N ions are selected from this mass spectrum and successively fragmented by MS/MS experiments (usually by collision-induced dissociation, CID). Additionally, an active list of ions can be defined to avoid the fragmentation of the same ions during a defined period of time. This mode links one precursor with a list of fragment ions. For DIA, multi-cycles of MS/MS are generated. For one cycle, a first large width of *m/z* range is selected, and all ions are fragmented at a specific collision energy (in CID). Then, the experiment is repeated on the next *m/z* ranges until the full *m/z* range is analyzed. As the links between precursors and their fragment are not obvious, the deconvolution of the MS/MS spectra is required. Both operating modes have their strengths and weaknesses (Guo and Huan 2020), but for metabolomics, these MS/MS data combined with the chromatography data can considerably help the annotation of metabolites asking the MS/MS databases.

For targeted approaches, the coupling of chromatographic separation techniques to mass spectrometry is the most suitable analysis system. This improves both the selectivity and identification and can lead to the quantification of metabolites. Gas chromatography (GC) is commonly used for analyzing volatile, thermally stable, and small metabolites. This technique is highly efficient for metabolomics, particularly for compounds such as organic acids, amino acids, fatty acids, sugars, and lipophilic compounds. LC-MS analysis is considered more versatile in terms of sample handling (Gika et al. 2014; Begou et al. 2017). Also, LC is a more versatile method used for analyzing a broader range of metabolites, from highly polar to non-polar, without the need for derivatization.

Compared to other approaches, the field of MS-based metabolomics is recently young, and considering the possible approaches, it is a trendy research area with new methods, software, and platforms being regularly published or updated (Damiani et al. 2020; Misra 2021; Chen et al. 2022).

In this next section, the recent advances made in different MS instruments and ionization sources, for metabolomic studies are going to be described. Input of the different separation techniques coupled to mass spectrometry, as part of metabolomics, is discussed in the next chapter.

4.3.1 Mass Spectrometers

A mass spectrometer is an analytical instrument comprising at minimum an ion source, a mass

Table 4.1 Performances of common mass spectrometers used for MS-based metabolomics

Mass spectrometer	Ion source	Mass resolution	Mass accuracy[a]	Dynamic range (one scan)	Sensitivity (LOD)	Type of study
Ion trap MS	ESI, APCI	1000–4000	0.1 Da	Four orders of magnitude	Femtomole	Structural elucidation and targeted metabolomics
Triple quadrupole (QqQ)	ESI, APCI, EI, CI	1000–4000	0.1 Da	Six orders of magnitude	Femtomole	Targeted metabolomics and quantification
Quadrupole-TOF (Q-TOF)	ESI, APCI (MALDI) (DESI)	40,000–200,000	<3 ppm	Four to five orders of magnitude	Femtomole	Untargeted metabolomics, metabolite identification, and targeted metabolomics
Orbitrap (Fusion)	ESI, APCI	Up to 500,000 at *m/z* 200	<3 ppm	Four orders of magnitude	Femtomole	Untargeted metabolomics, metabolite identification, and targeted metabolomics (low resolution)
FT-ICR MS (7 T 2xR)	ESI, APCI, APPI, MALDI	>1000,000 at *m/z* 200	<1 ppm	Four orders of magnitude	Femtomole	Untargeted metabolomics, metabolite identification, and high complex metabolomics

Information adapted from technical specifications of companies and Peters-Clarke et al. (2024)
[a]Mass accuracy with external calibration

analyzer, and an ion detector (Table 4.1) (Gross 2017). Additional components, such as ion optics (Covey et al. 2009), may be positioned between the ion source and the mass analyzer to focus, select, or induce fragmentation of ions. While the ion source may operate at atmospheric pressure, all mass analyzers function under high vacuum, necessitating a multi-stage pumping system to achieve pressures between 10^{-6} and 10^{-10} mbar (Nakorchevsky and Yates 2012; Li et al. 2021a). The instrument is governed by software that applies a predefined method, incorporating all operational parameters and facilitating data acquisition. In metabolomics, the selection of a mass spectrometer is influenced by the chemical nature of the metabolites under investigation and the desired analytical approach, which may range from targeted to untargeted, or quantitative, and can be performed with or without prior separation technique.

Mass spectrometers commonly used for metabolomics are hybrids, i.e., combining different mass analyzers connected with ion guides. At least, they have a first quadrupole MS system allowing to select from a tight (0.4) to a wide (1200) *m/z* range. Quadrupole is constituted by four parallel rods on which continuous and alternative electric fields are applied (Gross 2017). Depending on the voltage values, they act as an ion filter to allow a specific *m/z* range to travel from the entrance to the exit. After the first quadrupole, a multipole device (quadrupole or hexapole) is used to cool and store ions and/or to fragment them by adding a collision gas (for collision-induced dissociation, CID) (Murrell et al. 2003). Then, ions are pushed to a second mass analyzer. Different configurations exist affecting the performance of the mass measurement (speed, *m/z* range, mass resolution, mass accuracy, dynamic range, sensitivity). Triple quadrupole (QqQ) is well recognized for targeted quantification due to its high sensitivity and dynamic range, but it is limited to a low mass resolution and mass accuracy (Hopfgartner et al. 2004; Yost 2022). Typically, it is used by selecting one *m/z* with the first Q (precursor ion), fragmenting them by CID in the second q, and selecting one specific *m/z* of ion fragment with the last Q (daughter ion). As its experiment time lasts maximum tens of milliseconds, numerous couples of precursor-daughter ions can be probed very quickly (Cheng et al. 2020).

An alternative system to the triple quadruple is the ion trap mass spectrometer. For this system, ions are captured in a three-dimensional electric field, and a modification of alternative electric fields allows for ion selection, activation, and their mass analysis by ejection in the direction of detectors. It allows multiple rounds of mass selection and fragmentation (MS^n), making it useful for structural elucidation (Chen et al. 2023).

Q-TOF MS combines moderate-to-high mass resolution and mass accuracy, making it versatile for both profiling and quantifying metabolites. After selection of a *m/z* range by the first quadrupole, ions (precursor and/or fragment ions) are transferred and accelerated to a time of flight (TOF) mass analyzer that measures the time that ions take to travel a fixed distance to a detector (Chernushevich et al. 2001). The time needed for a full mass spectrum is around 100 μs giving this system the highest ability to analyze the mass of many fragment ions in a minimal time, in high-resolution mode. As one or more ion reflectors are used, Q-TOF MS are relatively compact, and mass resolution can reach from 40,000 to 200,000 with multi-reflector system (Waters 2022). As a consequence of the high ability of the Q-TOF MS to fast switch from one scan mode to another (e.g., from MS to MS/MS of different ion precursors) with high resolution, this MS system is preferred for DDA and DIA analyses.

Last but not least, Orbitrap-MS and Fourier transform ion cyclotron resonance mass spectrometer (FT-ICR MS) allow to reach very high mass resolution and mass accuracy measurements, typically 120,000 to 500,000 for Orbitrap and more than 1,000,000 for FT-ICR MS at *m/z* 200 (Schuhmann et al. 2011; Schuhmann et al. 2012; Ghaste et al. 2016). To trap ions in an orbital motion, Orbitrap uses an electrostatic field and FT-ICR MS a strong magnetic field (7–21 T) (Hu et al. 2005; Scigelova et al. 2011; Hendrickson et al. 2015). Both systems detect oscillating motions of ions during a fixed period of time, and the signal is Fourier-transformed to get the ion frequency that is specific to each ion mass (Easterling 2019; Makarov et al. 2019). Mass measurement accuracy is usually better than 1 ppm for Orbitrap and better than 0.2 ppm for FT-ICR MS (Huang et al. 2021). These systems are the most straightforward approach for measuring large numbers of metabolites in full-scan mode, but they can also work in MS^2 mode thanks to the front-end quadrupole and collision trap. However, due to their dependence on the time required for signal measurement, their performances decrease as acquisition frequency increases, which may lead to less applications for more complex ion experiments, especially during a chromatographic run (Gosset-Erard et al. 2023).

These recent instruments are selected based on their performance across diverse metabolomic applications, demonstrating their capabilities in both discovery and quantification workflows.

4.3.2 Ionization Sources

For volatile metabolites, their introduction to the mass spectrometer can be done in gas phase, usually using a gas chromatograph. In many cases, ions are produced at low pressure by electron ionization (EI) (Busch 2003; Famiglini et al. 2021; Maciel et al. 2022; Dantus 2024). As electron impact is preferably done at 70 eV, EI is considered a "hard" ionization method that leads to generate reproducible mass spectra including specific ion fragments from the molecular radical ions (Margolin Eren et al. 2020; Famiglini et al. 2021). This characteristic and standardization of EI source allow an easy use of mass spectral libraries for searching of compounds by scoring their mass spectra similarities (NIST Research Library 2009; Hamow et al. 2021). While chemical ionization with a regent gas such as methane is an alternative to limit the ion fragmentation within an EI source, the atmospheric pressure chemical ionization (APCI) is becoming popular as a soft ionization source in gas phase (Rebane et al. 2016; Wentrup et al. 2024a; Wentrup et al. 2024b). APCI involves a current corona discharge on a needle placed between the gas sample flow and the entrance to the first low pressure part of the mass spectrometer (Covey et al. 2009; Awad et al. 2015; Peacock et al. 2017). Reactive species from the background atmospheric gas allows to generate radical or (de)protonated ions from gas analytes by charge transfer. The simplicity of the system means that it can be easily coupled to numerous mass spectrometers equipped with an atmospheric pressure interface.

Metabolites can also be analyzed in liquid phase. For this, different steps of sample preparation and purification can be performed in order to extract the metabolites of interest from a biological sample (Causon and Hann 2016; de Souza et al. 2024). At the end, this fraction is reconstituted in a solvent adapted to the used analyzer. In numerous metabolomic studies by mass spectrometry, a mix of water with an organic solvent (typically methanol or acetonitrile) is preferred to solubilize the metabolites and to allow their direct injection to the mass spectrometer or their previous separation by liquid chromatography (Mushtaq et al. 2014; Lepoittevin et al. 2023). An ionization source working with solubilized analytes is then required. The most common to all instrument is the electrospray ion source (ESI), a soft-ionization source working at atmospheric pressure (Hiraoka 2013; Awad et al. 2015; Gross 2017). Briefly, the liquid sample is pushed through a thin conductive capillary, whose end is placed at a few centimeters from the entrance of the mass spectrometer. An electric potential of thousands of volts is applied continuously between the capillary and the interface of the mass spectrometer. At the end of the capillary, charged droplets are produced, and different processes may lead to gas phase ions from analytes (Fenn et al. 1989; Konermann et al. 2013; Awad et al. 2015). While multicharged ions can be produced by this soft ion source, main ions generated from metabolites correspond to (de)-protonated molecules or ion adducts (alkaline, halogen, or little acids or bases ions from the spray solution). This ion source works ideally for a vast range of polar metabolites (Wang et al. 2015). One of the main drawbacks of this ion source is the risk of compounds competing to ionize, which can lead to ion suppression (Müller et al. 2002; Tang et al. 2004). For less polar compounds, other atmospheric ion sources such as APCI and atmospheric pressure photoionization (APPI) are preferred for their higher sensitivity (Wang et al. 2015). For APCI, the solution containing analytes is sprayed in a block cylinder heated between 250 and 450 °C to ensure its total transition to gas phase before arriving to the corona discharge area (Awad et al. 2015; Gates 2021). Compared with gas phase analytes, ionization processes may also involve the solvent gas that is continuously present in the source. For APPI, the same configuration as for APCI is used, but a far-UV lamp is arranged in place of the corona needle (Raffaelli and Saba 2003; Awad et al. 2015). Krypton lamp is commonly used as it generates photons at 10.6 eV that are upper to the ionization potential of numerous molecules and just below that of usual solvents. For such ion source, the addition of a dopant such as toluene or acetone in the spray solvent may promote the analyte ionization (Smith et al. 2009; Zhang et al. 2014; Kauppila et al. 2014).

Matrix-assisted laser desorption ionization (MALDI) is also an ion source implemented for metabolomic studies (Sun et al. 2007; Ren et al. 2018). It is suitable for solid samples: dried sample, tissue sections, or complex biological matrices. Commonly, sample preparation can be performed by the deposition of a large amount of a matrix (a specific chemical compound that helps to ionize and desorb the analyte molecules from the sample) on a flat sample or by mixing the matrix with a solubilized sample. Then, a laser shot of several nanoseconds of time duration irradiates at one position of this deposit with a laser fluence around 10 mJ.cm^{-2}. As numerous matrix molecules absorbe the laser energy, desorption and ionization of matrix clusters lead to the desorption of sample compound and their ionization by charge transfer (Jaskolla and Karas 2011). Intact metabolite profiling is then possible, with minimal fragmentation. As the released ions come from a specific location of the sample, this ion source is ideally for mapping metabolites on a sample surface, by mass spectrometry imaging experiment (McDonnell and Heeren 2007; Bodzon-Kulakowska and Suder 2016).

Despite its advantages, metabolomic data acquisition with a MALDI source can be complex. It is imperative to choose a good matrix for the specific class of compounds of interest and at the same time to avoid possible interferences of matrix ions in the low *m/z* region. Also, since it is not possible to couple separative techniques, very high-resolution MS facilitates the exploration of the high diversity of MALDI ions.

For metabolomic applications, numerous other ion sources were developed for direct solid sample analyses at ambient pressure (Shrestha and Cha 2021). Most of them are based on the previous ion sources described above and may combine several of them such as MALDI and ESI (MALDESI). Desorption electrospray (DESI) is a specific one for which a solid sample, such as a biological tissue, is placed between the ion source and the mass spectrometer. The ESI spray, containing primary ions, impacts on a specific location of the sample, allowing both solvent extraction and charge transfer. This generates secondary ions from the sample surface, which are then transferred to the mass spectrometer (Takáts et al. 2005). A version of this ion source, called nanoDESI, constrains by using capillaries both the primary and secondary sprays to be focused on a very little area of the analyzed sample.

4.3.3 Mass Spectrometry Imaging

Investigation of metabolites involved in a specific state of different living systems may require a holistic exploration, especially to look at their spatial distribution within a sample. In this frame, mass spectrometry imaging (MSI) has been currently pointed out as one of the most promising techniques for metabolomics studies.

The idea of MSI was introduced in 1997 by Caprioli group (Caprioli et al. 1997). Despite its first application to peptides and proteins localizations within tissue, the interest for the application of this molecular imaging technology to a lot of different biomolecules grew exponentially. In the last 10 years, MSI has emerged as a leading technique for spatially revealing the metabolome and lipidome, across various tissue and cell samples. Numerous reviews on MSI were recently published for general developments (Buchberger et al. 2018) and different applications (Sgobba et al. 2021), such as drug development and clinical research (Vaysse et al. 2017; Schulz et al. 2019; Swales et al. 2019; Granborg et al. 2022; Song et al. 2022; Krestensen et al. 2023), food security and quality (Li et al. 2016; Kokesch-Himmelreich et al. 2022), environmental sciences (Maloof et al. 2020), plants and microbial interaction (Dong and Aharoni 2022; Parker et al. 2023; Horn and Chapman 2023; Chen et al. 2024; Maia et al. 2024; Vats et al. 2024), their integration to multi-omics data (Zhang et al. 2024), or other imaging technologies (Tuck et al. 2022; Guo et al. 2023; Bischof et al. 2024).

Essentially, MSI takes advantage of the ionization sources able to generate ions from a specific location of a solid sample. Indeed, laser (MALDI) or spray solvent (DESI) can be focused on very small surface of a sample, until a spot with a diameter next to 5 μm. The mass of expelled ions from this area are then measured, generating a mass spectrum. Registering the X,Y coordinates from this position, next MS analyses are done on other sample locations following a predefined raster. Finally, as for each coordinate, a mass spectrum is stored, the selection of one specific *m/z* value in all mass spectra of the total area investigated will give an image with pixels of varying brightness depending on the signal intensity at these ions.

Although it seems pretty straightforward, it is quite a complex technique from the sample preparation, analytical, and data interpretation point of views pushing discussion about guidelines covering the entire MSI workflow (McDonnell et al. 2015).

In fact, it is almost safe to say that sample preparation is the most challenging part when one is using this technique (Fig. 4.3). In MSI, the spatial resolution is directly related to the sample preparation and more specifically with the sectioning and mounting of the sample. To avoid any artifacts and to determine the precise location of the metabolites, it is important to pay attention to the sectioning. Moreover, the sample needs to be mounted in the analysis plate as flat as

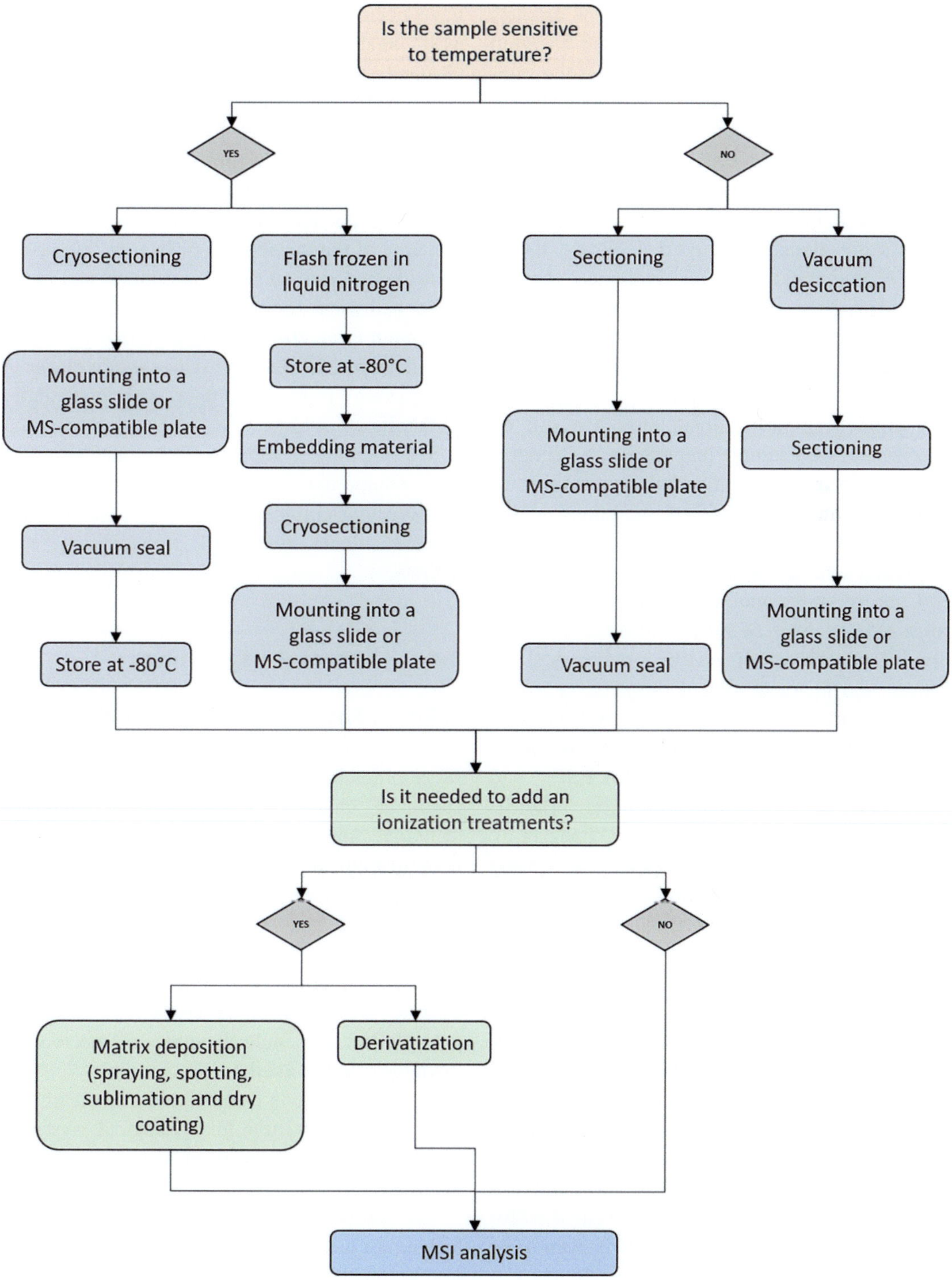

Fig. 4.3 Schematics of the most common steps in sample preparation for MSI analysis

possible so that the beams or solvent droplets reach all locations in the same manner ensuring equal ion signals. Any mis-handling of the sample can cause the delocation of molecules, lateral displacement of the ions, bad signals, incorrect ionization, and wrong assessment of the spatial distribution (Goodwin 2012; Bodzon-Kulakowska and Suder 2016). Several extensive reviews are already published dedicated to the sample preparation of all biological tissues (Goodwin 2012; Shimma and Sugiura 2014; Dong et al. 2016; Edwards et al. 2017; Zhou et al. 2021; Cillero-Pastor and Cuypers 2021). Despite the advances made on sample preparation and protocols optimizations, there are still no specific steps for sample preparation. It is important to keep in mind that sample preparation for MSI experiments depends on the nature of the samples, the metabolites to be analyzed, and the ionization source/MS instrument to be used.

Concerning the analytical equipment's to perform MSI analysis of metabolites, these can be divided into three categories, ion beams, laser beams, and charged solvent droplets, including secondary ion mass spectrometry (SIMS), laser desorption/ionization (LDI), MALDI, laser ablation electrospray ionization (LA-ESI), and desorption electrospray ionization (DESI) (Ma et al. 2023; Guo et al. 2023; Zhang et al. 2024). However, the main ionization sources mainly used for metabolomics MSI experiments are currently MALDI and DESI (Maciel et al. 2023).

The two main areas mentioned above, sample preparation and MSI instruments, indeed have evolved tremendously. However, there are still some analytical points that could be improved (Buchberger et al. 2018) such as spatial resolution and analysis time. Improvements on data acquisition time should be done. When one intends to use an equipment with high resolution and at a low spatial resolution ($\approx$5 μm), the time used for the analysis can be very long. This is even more problematic when using limited sample material and/or performing replicates.

Concerning data visualization, treatment, and comparison, this has been an area that has been advancing at a very steady pace. Several software have been developed to improve and facilitate MSI analysis, statistics, and quantification (Bruand et al. 2011; He et al. 2018; Baquer et al. 2023; Abu Sammour et al. 2023; Bemis et al. 2023; Morosi et al. 2024). Also, the integration of data originating from different analytical techniques with data obtained from a MSI analysis allows not only a deeper understanding and certainty of the metabolites of interest, but also a better knowledge on metabolite fluxes and metabolic pathways (Schwaiger-Haber et al. 2023; Levasseur et al. 2024; Zhang et al. 2024).

Altogether, MSI data provides another layer of information from a sample and particularly metabolites.

By mapping a wide range of metabolites, a more comprehensive and detailed understanding of molecular complexity in living systems may unravel their functional interactions and pathways.

4.4 Future Perspectives

4.4.1 Single-Cell Metabolomics

Metabolites are continuously being submitted to chemical changes. The different analytical techniques mentioned above allow the detection, identification, and quantification of metabolites of interest or new ones. Nevertheless, as demonstrated, majority of the studies are focused on the analysis of specific standards or a pool of samples of interest. With this method, a global overview of the metabolome of the samples is possible but the single dynamics of metabolites is impossible to unveil.

With the growing understanding of the importance of metabolites for biological systems, researchers realized the need to reach cell level analysis.

To do so, in the last decade, MS-based techniques have undergone massive instrumental developments to allow cell analysis (Zhang and Vertes 2018; Duncan et al. 2019; Gilmore et al. 2019; Shrestha 2020; Lanekoff et al. 2022). However, the main challenge is the small sample size. Cell size range is very wide as it can go as small as 1 μm. To ensure such analyses, spatial and analytical resolutions have to be increased in MS equipments, allowing not only the detection of a cell but also the ability to analyze it and distinguish closely related metabolites. The sensitivity was also improved, permitting the detection of low concentrations of metabolites (Emara et al. 2017; Duncan et al. 2019; Shrestha 2020; Lanekoff et al. 2022).

As an example, one research field where single-cell metabolomics can have a major impact is in cancer (Wei et al. 2022; Danzi et al. 2023). It is known that cancer normally starts from one abnormal cell that divides afterward. In this particular case, the study of individual cells could lead to early detections of the disease, identification of the type and aggressiveness of cancer, and a better understanding of drug action and resistances (Wei et al. 2022; Zhu et al. 2023).

In this thematic, Liu and co-authors combined single-cell metabolomic analysis with machine learning models to monitor and predict at the singe-cell level chemotherapy-induced drug resistance (Liu et al. 2019). In this study the authors analyzed individual live cancer cells, with different levels of drug resistance, using the single-probe single cell mass spectrometry (SCMS) technique and constructed three algorithms including random forest, artificial neural network, and penalized logistic regression, to predict the degrees of drug resistance of individual cells. The results obtained by the group indicated that the model constructed based on the SCMS datasets could provide an accurate and rapid prediction of the level of resistance of individual cells and that this model can be used by other labs.

More recently, Zhu and colleagues investigated the mechanism of drug action at the single-cell level (Zhu et al. 2023). The authors developed a label-free platform by combining intact living-cell electro-launching ionization mass spectrometry with metabolomic analysis. With this approach, the authors revealed that non-small-cell lung cancer cells, treated with gefitinib (commonly used drug for the treatment of this type of cancer), can be clustered into two cell subpopulations, with different metabolic responses. One of the subpopulations was drug-resistant, demonstrating the heterogeneity in the response of single cells to gefitinib treatment. These findings open new doors to a better evaluation of the therapeutic effects of drugs and to overcome drug resistance.

Hence, to fully understand the different functions associated with cells, to characterize cell-to-cell heterogenicity, to differentiate cellular subpopulations, and to unveil the signaling responses and cell differentiation, single-cell studies are required (Emara et al. 2017; Duncan et al. 2019).

4.4.2 Multi-OMICs

OMICs technologies aim at detecting and identifying an entire set of biomolecules such as genes, transcripts, proteins, and metabolites, which are respectively so-called genomics, transcriptomics, proteomics, and metabolomics (Pereira Braga and Adamec 2019; Gutierrez Reyes et al. 2024). However, a stand-alone OMIC does not have the capacity to offer a full insight of the whole system and lacks having a comprehensive understanding of the dynamics of a biological process.

Regarding metabolomics, as mentioned above, NMR and MS techniques are the two main instruments for the analysis of metabolites. Their mode of sample analysis, detection, and identification capacities are different, and the datasets obtained although complementary are entirely different. Moreover, in the group of MS techniques, the list of instruments currently available is quite long. Hence, due to the plethora of possible analytical approaches, the combination of datasets from different MS techniques, by itself, is already challenging since not all have the same sensitivity, accuracy, and resolution. In this frame, there is a need for algorithms, which could allow the simultaneous integration, calibration, processing, and alignment of MS data from different equipments. In the past years, remarkable advances in software for MS data analysis have been made. Open software, such as MZmine (Schmid et al. 2023) and Global Natural Products

Social Molecular Networking (GNPS) (Wang et al. 2016), were created, allowing MS data processing and analysis. Both tools are compatible with data from various MS instrumental setups, including hyphenated techniques, ion mobility spectrometry, and MS imaging. Moreover, they allow the annotation of fragmentation mass spectrometry data (MS/MS), aiding on the discovery and identification of novel compounds.

Also, a higher effort should be made to improve data accessibility and databases, since the datasets generated are highly complex and most of the detected metabolites are still unknown, adding another layer of difficulty to data integration.

Nevertheless, the integration of data from different OMICs, also called multi-omics, is crucial to unveil the interactions between different molecules and interconnect networks, the changes happening in the organism, and to have a holistic view of the biological system.

Although in theory the idea seems straightforward, the integration of OMICs data is quite challenging, and it is still considered the major bottleneck due to the data complexity (Pinu et al. 2019).

Based on the research community efforts, tremendous progresses have been done in this topic, and by using advanced statistical methods and computational approaches, various web applications, algorithms, and software have emerged (Karahalil 2016; Misra et al. 2019; Pinu et al. 2019; Gutierrez Reyes et al. 2024). Some of the tools and web applications that are already available, which can aid on the integration of OMICs data are MiBiOmics (Zoppi et al. 2021), TiMEG (Das and Mukhopadhyay 2021), OmicsTIDE (Harbig et al. 2023), Asterics (Maigné et al. 2023), and others that were extensively reviewed in Misra et al. (2019) and Pinu et al. (2019). Also, some commercial software were also developed, such as Ingenuity Pathway Analysis from QIAGEN Co. (Krämer et al. 2014) and iPathwayGuide developed from ADVAITA Co. (Ahsan and Drăghici 2017).

4.5 Conclusion

Considering what has been explored in this chapter, the selection of the metabolomics approach (targeted *vs.* untargeted), as well as the mass spectrometer and the ionization source to use, are key points to fully comprehend the metabolome of an organism at the different levels. Moreover, since MS allows the combination of various ionization sources and separative techniques, it can be considered as a very versatile technique to explore the metabolome.

Also, the application of mass spectrometry techniques to single-cell scale, as well as data integration have allowed to deepen our understanding of the metabolome and underlining its importance within the OMICs sciences.

Acknowledgments The authors acknowledge the financial support from Fundação para a Ciência e a Tecnologia (FCT-Portugal) through Marisa Maia research contract 2022.07433.CEECIND (https://doi.org/10.54499/2022.07433.CEECIND/CP1715/CT0009), the DEVINE-ME exploratory project 2023.14736.PEX (https://doi.org/10.54499/2023.14736.PEX), and center grant to BioISI UID/04046/2025 (https://doi.org/10.54499/UID/04046/2025)

The authors also acknowledge the financial support obtained from FCT and Campus France through the bilateral cooperation program—Franco-Portuguese Hubert Curien Partnership (Pessoa Program 2024) (Number 50898WC). Support from the MassLor research infrastructure at the University of Lorraine is also acknowledged.

References

Abu Sammour D, Cairns JL, Boskamp T, Marsching C, Kessler T, Ramallo Guevara C, Panitz V, Sadik A, Cordes J, Schmidt S, Mohammed SA, Rittel MF, Friedrich M, Platten M, Wolf I, von Deimling A, Opitz CA, Wick W, Hopf C (2023) Spatial probabilistic mapping of metabolite ensembles in mass spectrometry imaging. Nat Commun 14(1):1823. https://doi.org/10.1038/s41467-023-37394-z

Ahsan S, Drăghici S (2017) Identifying significantly impacted pathways and putative mechanisms with iPathwayGuide. Curr Protoc Bioinformatics 57(1): 7.15.1–7.15.30. https://doi.org/10.1002/cpbi.24

Alseekh S, Aharoni A, Brotman Y, Contrepois K, D'Auria J, Ewald J, Ewald JC, Fraser PD, Giavalisco P, Hall RD, Heinemann M, Link H, Luo J, Neumann S, Nielsen J, Perez de Souza L, Saito K, Sauer U, Schroeder FC, Schuster S, Siuzdak G, Skirycz A, Sumner LW, Snyder MP, Tang H, Tohge T, Wang Y, Wen W, Wu S, Xu G, Zamboni N, Fernie AR (2021) Mass spectrometry-based metabolomics: a guide for annotation, quantification and best reporting practices. Nat Methods 18(7): 747–756. https://doi.org/10.1038/s41592-021-01197-1

Awad H, Khamis MM, El-Aneed A (2015) Mass spectrometry, review of the basics: ionization. Appl Spectrosc Rev 50(2):158–175. https://doi.org/10.1080/05704928.2014.954046

Badillo-Sanchez D, Serrano Ruber M, Davies-Barrett A, Jones DJL, Hansen M, Inskip S (2023) Metabolomics in archaeological science: a review of their advances and present requirements. Sci Adv 9(32):eadh0485. https://doi.org/10.1126/sciadv.adh0485

Baquer G, Sementé L, Mahamdi T, Correig X, Ràfols P, García-Altares M (2023) What are we imaging? Software tools and experimental strategies for annotation and identification of small molecules in mass spectrometry imaging. Mass Spectrom Rev 42(5): 1927–1964. https://doi.org/10.1002/mas.21794

Bauermeister A, Mannochio-Russo H, Costa-Lotufo LV, Jarmusch AK, Dorrestein PC (2022) Mass spectrometry-based metabolomics in microbiome investigations. Nat Rev Microbiol 20(3):143–160. https://doi.org/10.1038/s41579-021-00621-9

Begou O, Gika HG, Wilson ID, Theodoridis G (2017) Hyphenated MS-based targeted approaches in metabolomics. Analyst 142(17):3079–3100. https://doi.org/10.1039/C7AN00812K

Bemis KA, Föll MC, Guo D, Lakkimsetty SS, Vitek O (2023) Cardinal v.3: a versatile open-source software for mass spectrometry imaging analysis. Nat Methods 20(12):1883–1886. https://doi.org/10.1038/s41592-023-02070-z

Bergeron A, Furtado M, Garofolo F (2009) Importance of using highly pure internal standards for successful liquid chromatography/tandem mass spectrometric bioanalytical assays. Rapid Commun Mass Spectrom 23(9):1287–1297. https://doi.org/10.1002/rcm.4001

Bino RJ, Hall RD, Fiehn O, Kopka J, Saito K, Draper J, Nikolau BJ, Mendes P, Roessner-Tunali U, Beale MH, Trethewey RN, Lange BM, Wurtele ES, Sumner LW (2004) Potential of metabolomics as a functional genomics tool. Trends Plant Sci 9(9):418–425. https://doi.org/10.1016/j.tplants.2004.07.004

Bischof J, Fletcher G, Verkade P, Kuntner C, Fernandez-Rodriguez J, Chaabane L, Rose LA, Walter A, Vandenbosch M, van Zandvoort MAMJ, Zaritsky A, Keppler A, Parsons M (2024) Multimodal bioimaging across disciplines and scales: challenges, opportunities and breaking down barriers. npj Imaging 2(1):1–6. https://doi.org/10.1038/s44303-024-00010-w

Bodzon-Kulakowska A, Suder P (2016) Imaging mass spectrometry: instrumentation, applications, and combination with other visualization techniques. Mass Spectrom Rev 35(1):147–169. https://doi.org/10.1002/mas.21468

Bouslimani A, Sanchez LM, Garg N, Dorrestein PC (2014) Mass spectrometry of natural products: current, emerging and future technologies. Nat Prod Rep 31(6): 718–729. https://doi.org/10.1039/C4NP00044G

Bruand J, Alexandrov T, Sistla S, Wisztorski M, Meriaux C, Becker M, Salzet M, Fournier I, Macagno E, Bafna V (2011) AMASS: algorithm for MSI analysis by semi-supervised segmentation. J Proteome Res 10(10):4734. https://doi.org/10.1021/pr2005378

Buchberger AR, DeLaney K, Johnson J, Li L (2018) Mass spectrometry imaging: a review of emerging advancements and future insights. Anal Chem 90(1): 240–265. https://doi.org/10.1021/acs.analchem.7b04733

Busch KL (2003) Mass spectrometry. In: Meyers RA (ed) Encyclopedia of physical science and technology, 3rd edn. Academic Press, New York, pp 145–158

Cajka T, Fiehn O (2016) Toward merging untargeted and targeted methods in mass spectrometry-based metabolomics and lipidomics. Anal Chem 88(1): 524–545. https://doi.org/10.1021/acs.analchem.5b04491

Caldwell GW, Leo GC (2017) Can untargeted metabolomics be utilized in drug discovery/development? Curr Top Med Chem 17(24):2716–2739. https://doi.org/10.2174/1568026617666170707130032

Cao-Ngoc P, Leclercq L, Rossi J-C, Desvignes I, Hertzog J, Fabiano-Tixier A-S, Chemat F, Schmitt-Kopplin P, Cottet H (2019) Optimizing water-based extraction of bioactive principles of hawthorn: from experimental laboratory research to homemade preparations. Molecules 24(23):4420. https://doi.org/10.3390/molecules24234420

Cao-Ngoc P, Leclercq L, Rossi J-C, Hertzog J, Tixier A-S, Chemat F, Nasreddine R, Al Hamoui Dit Banni G, Nehmé R, Schmitt-Kopplin P, Cottet H (2020) Water-based extraction of bioactive principles from blackcurrant leaves and Chrysanthellum americanum: a comparative study. Foods 9(10):1478. https://doi.org/10.3390/foods9101478

Caprioli RM, Farmer TB, Gile J (1997) Molecular imaging of biological samples: localization of peptides and proteins using MALDI-TOF MS. Anal Chem 69(23): 4751–4760. https://doi.org/10.1021/ac970888i

Castro-Moretti FR, Gentzel IN, Mackey D, Alonso AP (2020) Metabolomics as an emerging tool for the study of plant-pathogen interactions. Meta 10(2). https://doi.org/10.3390/metabo10020052

Causon TJ, Hann S (2016) Review of sample preparation strategies for MS-based metabolomic studies in industrial biotechnology. Anal Chim Acta 938:18–32. https://doi.org/10.1016/j.aca.2016.07.033

Chen C-L, Chen Y-T, Liao W-Y, Chang Y-S, Yu J-S, Juo B-R (2023) Urinary metabolomic analysis of prostate cancer by UPLC-FTMS and UPLC-ion trap MSn. Diagnostics 13(13):2270. https://doi.org/10.3390/diagnostics13132270

Chen Y, Li E-M, Xu L-Y (2022) Guide to metabolomics analysis: a bioinformatics workflow. Meta 12(4):357. https://doi.org/10.3390/metabo12040357

Chen Y-J, Zeng H-S, Jin H, Wang H (2024) Applications of mass spectrometry imaging in botanical research. Adv Biotechnol 2(1):6. https://doi.org/10.1007/s44307-024-00014-y

Cheng S-C, Lee R-H, Jeng J-Y, Lee C-W, Shiea J (2020) Fast screening of trace multiresidue pesticides on fruit and vegetable surfaces using ambient ionization tandem mass spectrometry. Anal Chim Acta 1102:63–71. https://doi.org/10.1016/j.aca.2019.12.038

Chernushevich IV, Loboda AV, Thomson BA (2001) An introduction to quadrupole–time-of-flight mass spectrometry. J Mass Spectrom 36(8):849–865. https://doi.org/10.1002/jms.207

Christofi E, Barran P (2023) Ion mobility mass spectrometry (IM-MS) for structural biology: insights gained by measuring mass, charge, and collision cross section. Chem Rev 123(6):2902–2949. https://doi.org/10.1021/acs.chemrev.2c00600

Cillero-Pastor B, Cuypers E (2021) Sample preparation of biological tissues for MALDI-MSI. https://doi.org/10.1039/9781839165191-00087

Clish CB (2015) Metabolomics: an emerging but powerful tool for precision medicine. Cold Spring Harb Mol Case Stud 1(1):a000588. https://doi.org/10.1101/mcs.a000588

Covey TR, Thomson BA, Schneider BB (2009) Atmospheric pressure ion sources. Mass Spectrom Rev 28(6):870–897. https://doi.org/10.1002/mas.20246

Crotti S, Menicatti M, Pallecchi M, Bartolucci G (2023) Tandem mass spectrometry approaches for recognition of isomeric compounds mixtures. Mass Spectrom Rev 42(4):1244–1260. https://doi.org/10.1002/mas.21757

Damiani C, Gaglio D, Sacco E, Alberghina L, Vanoni M (2020) Systems metabolomics: from metabolomic snapshots to design principles. Curr Opin Biotechnol 63:190–199. https://doi.org/10.1016/j.copbio.2020.02.013

Dantus M (2024) Tracking molecular fragmentation in electron–ionization mass spectrometry with ultrafast time resolution. Acc Chem Res 57(6):845. https://doi.org/10.1021/acs.accounts.3c00713

Danzi F, Pacchiana R, Mafficini A, Scupoli MT, Scarpa A, Donadelli M, Fiore A (2023) To metabolomics and beyond: a technological portfolio to investigate cancer metabolism. Sig Transduct Target Ther 8(1):1–22. https://doi.org/10.1038/s41392-023-01380-0

Das S, Mukhopadhyay I (2021) TiMEG: an integrative statistical method for partially missing multi-omics data. Sci Rep 11(1):24077. https://doi.org/10.1038/s41598-021-03034-z

de Souza HMR, Pereira TTP, de Sá HC, Alves MA, Garrett R, Canuto GAB (2024) Critical factors in sample collection and preparation for clinical metabolomics of underexplored biological specimens. Meta 14(1):36. https://doi.org/10.3390/metabo14010036

Dettmer K, Aronov PA, Hammock BD (2007) Mass spectrometry-based metabolomics. Mass Spectrom Rev 26(1):51–78. https://doi.org/10.1002/mas.20108

Dietz K-J (2017) Subcellular metabolomics: the choice of method depends on the aim of the study. J Exp Bot 68(21–22):5695–5698. https://doi.org/10.1093/jxb/erx406

Dong Y, Aharoni A (2022) Image to insight: exploring natural products through mass spectrometry imaging. Nat Prod Rep 39(7):1510–1530. https://doi.org/10.1039/d2np00011c

Dong Y, Li B, Malitsky S, Rogachev I, Aharoni A, Kaftan F, Svatoš A, Franceschi P (2016) Sample preparation for mass spectrometry imaging of plant tissues: a review. Front Plant Sci 7. https://doi.org/10.3389/fpls.2016.00060

Draper J, Lloyd AJ, Goodacre R, Beckmann M (2013) Flow infusion electrospray ionisation mass spectrometry for high throughput, non-targeted metabolite fingerprinting: a review. Metabolomics 9(1):4–29. https://doi.org/10.1007/s11306-012-0449-x

Duncan KD, Fyrestam J, Lanekoff I (2019) Advances in mass spectrometry based single-cell metabolomics. Analyst 144(3):782–793. https://doi.org/10.1039/C8AN01581C

Edwards G, Mesa A, Vazquez-Padron RI, Kowalski P, Bhattacharya SK (2017) Sample preparation and analysis for imaging mass spectrometry. Methods Mol Biol (Clifton, NJ) 1609:43. https://doi.org/10.1007/978-1-4939-6996-8_5

Elshafie HS, Camele I, Mohamed AA (2023) A comprehensive review on the biological, agricultural and pharmaceutical properties of secondary metabolites based-plant origin. Int J Mol Sci 24(4):3266. https://doi.org/10.3390/ijms24043266

Emara S, Amer S, Ali A, Abouleila Y, Oga A, Masujima T (2017) Single-cell metabolomics. Adv Exp Med Biol 965:323–343. https://doi.org/10.1007/978-3-319-47656-8_13

Emwas A-HM (2015) The strengths and weaknesses of NMR spectroscopy and mass spectrometry with particular focus on metabolomics research. Methods Mol Biol 1277:161–193. https://doi.org/10.1007/978-1-4939-2377-9_13

Easterling ML, Agar JN (2019) Fundamentals, strengths, and future directions for Fourier transform ion cyclotron resonance mass spectrometry. In: Fundamentals and Applications of Fourier Transform Mass Spectrometry. Elsevier, pp 63–88

Famiglini G, Palma P, Termopoli V, Cappiello A (2021) The history of electron ionization in LC-MS, from the early days to modern technologies: a review. Anal

Chim Acta 1167:338350. https://doi.org/10.1016/j.aca.2021.338350

Fenn JB, Mann M, Meng CK, Wong SF, Whitehouse CM (1989) Electrospray ionization for mass spectrometry of large biomolecules. Science 246(4926):64–71. https://doi.org/10.1126/science.2675315

Fiehn O (2002) Metabolomics—the link between genotypes and phenotypes. Plant Mol Biol 48(1–2): 155–171. https://doi.org/10.1023/A:1013713905833

Furey A, Moriarty M, Bane V, Kinsella B, Lehane M (2013) Ion suppression; a critical review on causes, evaluation, prevention and applications. Talanta 115: 104–122. https://doi.org/10.1016/j.talanta.2013.03.048

Ganna A, Salihovic S, Sundström J, Broeckling CD, Hedman ÅK, Magnusson PKE, Pedersen NL, Larsson A, Siegbahn A, Zilmer M, Prenni J, Ärnlöv J, Lind L, Fall T, Ingelsson E (2014) Large-scale metabolomic profiling identifies novel biomarkers for incident coronary heart disease. PLoS Genet 10(12):e1004801. https://doi.org/10.1371/journal.pgen.1004801

Gates PJ (2021) Atmospheric pressure chemical ionisation mass spectrometry for the routine analysis of low molecular weight analytes. Eur J Mass Spectrom (Chichester, Eng) 27(1):13. https://doi.org/10.1177/14690667211005055

Gemperline E, Keller C, Li L (2016) Mass spectrometry in plant-omics. Anal Chem 88(7):3422–3434. https://doi.org/10.1021/acs.analchem.5b02938

Gertsman I, Barshop BA (2018) Promises and pitfalls of untargeted metabolomics. J Inherit Metab Dis 41(3): 355–366. https://doi.org/10.1007/s10545-017-0130-7

Ghafari N, Sleno L (2024) Challenges and recent advances in quantitative mass spectrometry-based metabolomics. Anal Sci Adv 5(5–6):e2400007. https://doi.org/10.1002/ansa.202400007

Ghaste M, Mistrik R, Shulaev V (2016) Applications of Fourier transform ion cyclotron resonance (FT-ICR) and orbitrap based high resolution mass spectrometry in metabolomics and lipidomics. Int J Mol Sci 17(6). https://doi.org/10.3390/ijms17060816

Giera M, Yanes O, Siuzdak G (2022) Metabolite discovery: biochemistry's scientific driver. Cell Metab 34(1): 21–34. https://doi.org/10.1016/j.cmet.2021.11.005

Gika HG, Theodoridis GA, Plumb RS, Wilson ID (2014) Current practice of liquid chromatography–mass spectrometry in metabolomics and metabonomics. J Pharm Biomed Anal 87:12–25. https://doi.org/10.1016/j.jpba.2013.06.032

Gilmore IS, Heiles S, Pieterse CL (2019) Metabolic imaging at the single-cell scale: recent advances in mass spectrometry imaging. Ann Rev Anal Chem 12(2019): 201–224. https://doi.org/10.1146/annurev-anchem-061318-115516

Gonzalez-Covarrubias V, Martínez-Martínez E, del Bosque-Plata L (2022) The potential of metabolomics in biomedical applications. Meta 12(2):194. https://doi.org/10.3390/metabo12020194

González-Domínguez R, Sayago A, Fernández-Recamales Á (2017) Direct infusion mass spectrometry for metabolomic phenotyping of diseases. Bioanalysis 9(1):131–148. https://doi.org/10.4155/bio-2016-0202

Goodwin RJA (2012) Sample preparation for mass spectrometry imaging: small mistakes can lead to big consequences. J Proteome 75(16):4893–4911. https://doi.org/10.1016/j.jprot.2012.04.012

Gosset-Erard C, Aubriet F, Leize-Wagner E, François Y-N, Chaimbault P (2023) Hyphenation of Fourier transform ion cyclotron resonance mass spectrometry (FT-ICR MS) with separation methods: the art of compromises and the possible - a review. Talanta 257:124324. https://doi.org/10.1016/j.talanta.2023.124324

Gowda GAN, Djukovic D (2014) Overview of mass spectrometry-based metabolomics: opportunities and challenges. Methods Mol Biol 1198:3–12. https://doi.org/10.1007/978-1-4939-1258-2_1

Gowda GAN, Raftery D (2021) NMR based metabolomics. Adv Exp Med Biol 1280:19–37. https://doi.org/10.1007/978-3-030-51652-9_2

Granborg JR, Handler AM, Janfelt C (2022) Mass spectrometry imaging in drug distribution and drug metabolism studies—principles, applications and perspectives. TrAC Trends Anal Chem 146:116482. https://doi.org/10.1016/j.trac.2021.116482

Griffiths WJ, Koal T, Wang Y, Kohl M, Enot DP, Deigner H-P (2010) Targeted metabolomics for biomarker discovery. Angew Chem Int Ed 49(32):5426–5445. https://doi.org/10.1002/anie.200905579

Gross JH (2017) Mass spectrometry: a textbook, 3rd edn. Springer

Guo J, Huan T (2020) Comparison of full-scan, data-dependent, and data-independent acquisition modes in liquid chromatography–mass spectrometry based untargeted metabolomics. Anal Chem 92(12): 8072–8080. https://doi.org/10.1021/acs.analchem.9b05135

Guo X, Wang X, Tian C, Dai J, Zhao Z, Duan Y (2023) Development of mass spectrometry imaging techniques and its latest applications. Talanta 264: 124721. https://doi.org/10.1016/j.talanta.2023.124721

Gutbrod K, Peisker H, Dörmann P (2021) Direct infusion mass spectrometry for complex lipid analysis. Methods Mol Biol 2295:101–115. https://doi.org/10.1007/978-1-0716-1362-7_7

Gutierrez Reyes CD, Alejo-Jacuinde G, Perez Sanchez B, Chavez Reyes J, Onigbinde S, Mogut D, Hernández-Jasso I, Calderón-Vallejo D, Quintanar JL, Mechref Y (2024) Multi omics applications in biological systems. Curr Issues Mol Biol 46(6):5777–5793. https://doi.org/10.3390/cimb46060345

Hall R, Beale M, Fiehn O, Hardy N, Sumner L, Bino R (2002) Plant metabolomics: the missing link in functional genomics strategies. Plant Cell 14(7): 1437–1440. https://doi.org/10.1105/tpc.140720

Hamow KÁ, Ambrózy Z, Puskás K, Majláth I, Cséplő M, Mátyus R, Posta K, Lukács P, Sági L (2021) Emission of novel volatile biomarkers for wheat powdery mildew. Sci Total Environ 781:146767. https://doi.org/10.1016/j.scitotenv.2021.146767

Harbig TA, Fratte J, Krone M, Nieselt K (2023) OmicsTIDE: interactive exploration of trends in multi-omics data. Bioinform Adv 3(1):vbac093. https://doi.org/10.1093/bioadv/vbac093

He J, Huang L, Tian R, Li T, Sun C, Song X, Lv Y, Luo Z, Li X, Abliz Z (2018) MassImager: a software for interactive and in-depth analysis of mass spectrometry imaging data. Anal Chim Acta 1015:50–57. https://doi.org/10.1016/j.aca.2018.02.030

Hendrickson CL, Quinn JP, Kaiser NK, Smith DF, Blakney GT, Chen T, Marshall AG, Weisbrod CR, Beu SC (2015) 21 Tesla Fourier transform ion cyclotron resonance mass spectrometer: a national resource for ultrahigh resolution mass analysis. J Am Soc Mass Spectrom 26(9):1626–1632. https://doi.org/10.1007/s13361-015-1182-2

Hiraoka K (2013) Fundamentals of mass spectrometry. Springer

Hopfgartner G, Varesio E, Tschäppät V, Grivet C, Bourgogne E, Leuthold LA (2004) Triple quadrupole linear ion trap mass spectrometer for the analysis of small molecules and macromolecules. J Mass Spectrom 39(8):845–855. https://doi.org/10.1002/jms.659

Horn PJ, Chapman KD (2023) Imaging plant metabolism in situ. J Exp Bot:erad423. https://doi.org/10.1093/jxb/erad423

Hu Q, Noll RJ, Li H, Makarov A, Hardman M, Graham Cooks R (2005) The Orbitrap: a new mass spectrometer. J Mass Spectrom 40(4):430–443. https://doi.org/10.1002/jms.856

Huang D, Bouza M, Gaul DA, Leach III FE, Amster IJ, Schroeder FC, Edison AS, Fernández FM (2021) Comparison of high-resolution Fourier transform mass spectrometry platforms for putative metabolite annotation. Anal Chem 93(36):12374. https://doi.org/10.1021/acs.analchem.1c02224

Jaskolla TW, Karas M (2011) Compelling evidence for lucky survivor and gas phase protonation: the unified MALDI analyte protonation mechanism. J Am Soc Mass Spectrom 22(6):976–988. https://doi.org/10.1007/s13361-011-0093-0

Jové M, Mauri-Capdevila G, Suárez I, Cambray S, Sanahuja J, Quílez A, Farré J, Benabdelhak I, Pamplona R, Portero-Otín M, Purroy F (2015) Metabolomics predicts stroke recurrence after transient ischemic attack. Neurology 84(1):36–45. https://doi.org/10.1212/WNL.0000000000001093

Karahalil B (2016) Overview of systems biology and omics technologies. Curr Med Chem 23(37): 4221–4230. https://doi.org/10.2174/0929867323666160926150617

Katam R, Lin C, Grant K, Katam CS, Chen S (2022) Advances in plant metabolomics and its applications in stress and single-cell biology. Int J Mol Sci 23(13): 6985. https://doi.org/10.3390/ijms23136985

Kauppila TJ, Kersten H, Benter T (2014) The ionization mechanisms in direct and dopant-assisted atmospheric pressure photoionization and atmospheric pressure laser ionization. J Am Soc Mass Spectrom 25(11): 1870–1881. https://doi.org/10.1007/s13361-014-0988-7

Kirwan JA, Weber RJM, Broadhurst DI, Viant MR (2014) Direct infusion mass spectrometry metabolomics dataset: a benchmark for data processing and quality control. Sci Data 1. https://doi.org/10.1038/sdata.2014.12

Klassen A, Faccio AT, Canuto GAB, da Cruz PLR, Ribeiro HC, Tavares MFM, Sussulini A (2017) Metabolomics: definitions and significance in systems biology. In: Sussulini A (ed) Metabolomics: from fundamentals to clinical applications. Springer, Cham, pp 3–17

Kokesch-Himmelreich J, Wittek O, Race AM, Rakete S, Schlicht C, Busch U, Römpp A (2022) MALDI mass spectrometry imaging: from constituents in fresh food to ingredients, contaminants and additives in processed food. Food Chem 385:132529. https://doi.org/10.1016/j.foodchem.2022.132529

Konermann L, Ahadi E, Rodriguez AD, Vahidi S (2013) Unraveling the mechanism of electrospray ionization. Anal Chem 85(1):2–9. https://doi.org/10.1021/ac302789c

Krämer A, Green J, Pollard J Jr, Tugendreich S (2014) Causal analysis approaches in ingenuity pathway analysis. Bioinformatics 30(4):523–530. https://doi.org/10.1093/bioinformatics/btt703

Krestensen KK, Heeren RMA, Balluff B (2023) State-of-the-art mass spectrometry imaging applications in biomedical research. Analyst 148(24):6161–6187. https://doi.org/10.1039/d3an01495a

Lacalle-Bergeron L, Izquierdo-Sandoval D, Sancho JV, López FJ, Hernández F, Portolés T (2021) Chromatography hyphenated to high resolution mass spectrometry in untargeted metabolomics for investigation of food (bio)markers. TrAC Trends Anal Chem 135:116161. https://doi.org/10.1016/j.trac.2020.116161

Lanekoff I, Sharma VV, Marques C (2022) Single-cell metabolomics: where are we and where are we going? Curr Opin Biotechnol 75:102693. https://doi.org/10.1016/j.copbio.2022.102693

Lankadurai BP, Nagato EG, Simpson MJ (2013) Environmental metabolomics: an emerging approach to study organism responses to environmental stressors. Environ Rev 21(3):180–205. https://doi.org/10.1139/er-2013-0011

Lee H-J, Kremer DM, Sajjakulnukit P, Zhang L, Lyssiotis CA (2019) A large-scale analysis of targeted metabolomics data from heterogeneous biological samples provides insights into metabolite dynamics. Metabolomics 15(7):103. https://doi.org/10.1007/s11306-019-1564-8

Lei Z, Huhman DV, Sumner LW (2011) Mass spectrometry strategies in metabolomics. J Biol Chem 286(29): 25435–25442. https://doi.org/10.1074/jbc.R111.238691

Lelli V, Belardo A, Timperio AM, Lelli V, Belardo A, Timperio AM (2021) From targeted quantification to untargeted metabolomics. In: Metabolomics - methodology and applications in medical sciences and life sciences. IntechOpen

Lepoittevin M, Blancart-Remaury Q, Kerforne T, Pellerin L, Hauet T, Thuillier R (2023) Comparison between 5 extractions methods in either plasma or serum to determine the optimal extraction and matrix combination for human metabolomics. Cell Mol Biol Lett 28(1):43. https://doi.org/10.1186/s11658-023-00452-x

Levasseur M, Nicol E, Elie N, Houël E, Eparvier V, Touboul D (2024) Spatialized metabolomic annotation combining MALDI imaging and molecular networks. Anal Chem 96(1):18–22. https://doi.org/10.1021/acs.analchem.3c03482

Li B, Dunham SJB, Dong Y, Yoon S, Zeng M, Sweedler JV (2016) Analytical capabilities of mass spectrometry imaging and its potential applications in food science. Trends Food Sci Technol 47:50–63. https://doi.org/10.1016/j.tifs.2015.10.018

Li C, Chu S, Tan S, Yin X, Jiang Y, Dai X, Gong X, Fang X, Tian D (2021a) Towards higher sensitivity of mass spectrometry: a perspective from the mass analyzers. Front Chem 9. https://doi.org/10.3389/fchem.2021.813359

Li S, Tian Y, Jiang P, Lin Y, Liu X, Yang H (2021b) Recent advances in the application of metabolomics for food safety control and food quality analyses. Crit Rev Food Sci Nutr 61(9):1448–1469. https://doi.org/10.1080/10408398.2020.1761287

Liang D, Li Z, Vlaanderen J, Tang Z, Jones DP, Vermeulen R, Sarnat JA (2023) A state-of-the-science review on high-resolution metabolomics application in air pollution health research: current progress, analytical challenges, and recommendations for future direction. Environ Health Perspect 131(5):056002. https://doi.org/10.1289/EHP11851

Liang D, Moutinho JL, Golan R, Yu T, Ladva CN, Niedzwiecki M, Walker DI, Sarnat SE, Chang HH, Greenwald R, Jones DP, Russell AG, Sarnat JA (2018) Use of high-resolution metabolomics for the identification of metabolic signals associated with traffic-related air pollution. Environ Int 120:145–154. https://doi.org/10.1016/j.envint.2018.07.044

Lin C, Tian Q, Guo S, Xie D, Cai Y, Wang Z, Chu H, Qiu S, Tang S, Zhang A (2024) Metabolomics for clinical biomarker discovery and therapeutic target identification. Molecules 29(10):2198. https://doi.org/10.3390/molecules29102198

Liu R, Sun M, Zhang G, Lan Y, Yang Z (2019) Towards early monitoring of chemotherapy-induced drug resistance based on single cell metabolomics: combining single-probe mass spectrometry with machine learning. Anal Chim Acta 1092:42–48. https://doi.org/10.1016/j.aca.2019.09.065

Liu Z, Zhang M, Chen P, Harnly JM, Sun J (2022) Mass spectrometry-based nontargeted and targeted analytical approaches in fingerprinting and metabolomics of food and agricultural research. J Agric Food Chem 70(36): 11138–11153. https://doi.org/10.1021/acs.jafc.2c01878

López-Gresa MP, Lisón P, Campos L, Rodrigo I, Rambla JL, Granell A, Conejero V, Bellés JM (2017) A non-targeted metabolomics approach unravels the VOCs associated with the tomato immune response against pseudomonas syringae. Front Plant Sci 8. https://doi.org/10.3389/fpls.2017.01188

Ma S, Leng Y, Li X, Meng Y, Yin Z, Hang W (2023) High spatial resolution mass spectrometry imaging for spatial metabolomics: advances, challenges, and future perspectives. TrAC Trends Anal Chem 159:116902. https://doi.org/10.1016/j.trac.2022.116902

Maciel EVS, Pereira dos Santos NG, Vargas Medina DA, Lanças FM (2022) Electron ionization mass spectrometry: quo vadis? Electrophoresis 43(15):1587–1600. https://doi.org/10.1002/elps.202100392

Maciel LÍL, Bernardo RA, Martins RO, Batista Junior AC, Oliveira JVA, Chaves AR, Vaz BG (2023) Desorption electrospray ionization and matrix-assisted laser desorption/ionization as imaging approaches for biological samples analysis. Anal Bioanal Chem 415(18):4125–4145. https://doi.org/10.1007/s00216-023-04783-8

Maia M, Aziz A, Jeandet P, Figueiredo A, Carré V (2024) Mass spectrometry imaging to map metabolites in plant-microbe interactions: grapevine as a case study: this article is published in cooperation with 2ICWGS (II International Conference of Grapevine and Wine Science), 8-10 November 2023, Logroño, La Rioja, Spain. Invited associate editors: José Miguel Martínez Zapater, Javier Tello Moro and María Pilar Sáenz Navajas. OENO One 58(3). https://doi.org/10.20870/oeno-one.2024.58.3.8059

Maia M, Cavaco AR, Laureano G, Cunha J, Eiras-Dias J, Matos AR, Duarte B, Figueiredo A (2021) More than just wine: the nutritional benefits of grapevine leaves. Foods 10(10):2251. https://doi.org/10.3390/foods10102251

Maia M, Ferreira AEN, Nascimento R, Monteiro F, Traquete F, Marques AP, Cunha J, Eiras-Dias JE, Cordeiro C, Figueiredo A, Sousa Silva M (2020) Integrating metabolomics and targeted gene expression to uncover potential biomarkers of fungal/oomycetes-associated disease susceptibility in grapevine. Sci Rep 10(1):15688. https://doi.org/10.1038/s41598-020-72781-2

Maia M, Figueiredo A, Cordeiro C, Sousa Silva M (2023) FT-ICR-MS-based metabolomics: a deep dive into plant metabolism. Mass Spectrom Rev 42(5): 1535–1556. https://doi.org/10.1002/mas.21731

Maigné É, Noirot C, Henry J, Adu Kesewaah Y, Badin L, Déjean S, Guilmineau C, Krebs A, Mathevet F, Segalini A, Thomassin L, Colongo D, Gaspin C, Liaubet L, Vialaneix N (2023) Asterics: a simple tool for the ExploRation and integration of omiCS data. BMC Bioinformatics 24(1):391. https://doi.org/10.1186/s12859-023-05504-9

Maloof KA, Reinders AN, Tucker KR (2020) Applications of mass spectrometry imaging in the environmental sciences. Curr Opin Environ Sci Health 18:54–62. https://doi.org/10.1016/j.coesh.2020.07.005

Margolin Eren KJ, Elkabets O, Amirav A (2020) A comparison of electron ionization mass spectra obtained at 70 eV, low electron energies, and with cold EI and their NIST library identification probabilities. J Mass Spectrom 55(12):e4646. https://doi.org/10.1002/jms.4646

Makarov A, Grindfeld D, Ayzikov K (2019) Fundamentals of Orbitrap analyzer. In: Fundamentals and Applications of Fourier Transform Mass Spectrometry. Elsevier, pp 37–61

Marshall AG, Hendrickson CL, Jackson GS (1998) Fourier transform ion cyclotron resonance mass spectrometry: a primer. Mass Spectrom Rev 17(1):1–35. https://doi.org/10.1002/(SICI)1098-2787(1998)17:1<1::AID-MAS1>3.0.CO;2-K

McDonnell LA, Heeren RMA (2007) Imaging mass spectrometry. Mass Spectrom Rev 26(4):606–643. https://doi.org/10.1002/mas.20124

McDonnell LA, Römpp A, Balluff B, Heeren RMA, Albar JP, Andrén PE, Corthals GL, Walch A, Stoeckli M (2015) Discussion point: reporting guidelines for mass spectrometry imaging. Anal Bioanal Chem 407(8):2035–2045. https://doi.org/10.1007/s00216-014-8322-6

Meijer NWF, Zwakenberg S, Gerrits J, Westland D, Ardisasmita AI, Fuchs SA, Verhoeven-Duif NM, Jans JJM, Zwartkruis FJT (2024) Direct infusion mass spectrometry to rapidly map metabolic flux of substrates labeled with stable isotopes. Meta 14(5):246. https://doi.org/10.3390/metabo14050246

Misra BB (2021) New software tools, databases, and resources in metabolomics: updates from 2020. Metabolomics 17(5):49. https://doi.org/10.1007/s11306-021-01796-1

Misra BB, Langefeld CD, Olivier M, Cox LA (2019) Integrated omics: tools, advances, and future approaches. J Mol Endocrinol 62(1):R21–R45. https://doi.org/10.1530/JME-18-0055

Moco S (2022) Studying metabolism by NMR-based metabolomics. Front Mol Biosci 9

Morosi L, Miotto M, Timo S, Carloni S, Bruno E, Meroni M, Menna E, Lodato S, Rescigno M, Martano G (2024) MSIpixel: a fully automated pipeline for compound annotation and quantitation in mass spectrometry imaging experiments. Brief Bioinform 25(1):bbad463. https://doi.org/10.1093/bib/bbad463

Mu L, Niu Z, Blair RH, Yu H, Browne RW, Bonner MR, Fanter T, Deng F, Swanson M (2019) Metabolomics profiling before, during, and after the Beijing Olympics: a panel study of within-individual differences during periods of high and low air pollution. Environ Health Perspect 127(5):057010. https://doi.org/10.1289/EHP3705

Müller C, Schäfer P, Störtzel M, Vogt S, Weinmann W (2002) Ion suppression effects in liquid chromatography–electrospray-ionisation transport-region collision induced dissociation mass spectrometry with different serum extraction methods for systematic toxicological analysis with mass spectra libraries. J Chromatogr B 773(1):47–52. https://doi.org/10.1016/S1570-0232(02)00142-3

Murrell J, Despeyroux D, Lammert SA, Stephenson JL, Goeringer DE (2003) "Fast excitation" cid in a quadrupole ion trap mass spectrometer. J Am Soc Mass Spectrom 14(7):785–789. https://doi.org/10.1016/S1044-0305(03)00326-X

Mushtaq MY, Choi YH, Verpoorte R, Wilson EG (2014) Extraction for metabolomics: access to the metabolome. Phytochem Anal 25(4):291–306. https://doi.org/10.1002/pca.2505

Muthubharathi BC, Gowripriya T, Balamurugan K (2021) Metabolomics: small molecules that matter more. Mol Omics 17(2):210–229. https://doi.org/10.1039/D0MO00176G

Nakorchevsky A, Yates JR (2012) 1.17 mass spectrometry. In: Egelman EH (ed) Comprehensive biophysics. Elsevier, Amsterdam, pp 341–375

Nardini M (2022) Phenolic compounds in food: characterization and health benefits. Molecules 27(3):783. https://doi.org/10.3390/molecules27030783

NIST Research Library (2009) NIST Research Library. In: NIST. https://www.nist.gov/nist-research-library. Accessed 22 Oct 2024

Oliver SG, Winson MK, Kell DB, Baganz F (1998) Systematic functionalbr/analysis of the yeast genome. Trends Biotechnol 16(9):373–378

Pang H, Hu Z (2023) Metabolomics in drug research and development: the recent advances in technologies and applications. Acta Pharm Sin B 13(8):3238–3251. https://doi.org/10.1016/j.apsb.2023.05.021

Parker GD, Hanley L, Yu X-Y (2023) Mass spectral imaging to map plant–microbe interactions. Microorganisms 11(8):2045. https://doi.org/10.3390/microorganisms11082045

Patel KN, Patel JK, Patel MP, Rajput GC, Patel HA (2010) Introduction to hyphenated techniques and their applications in pharmacy. Pharm Methods 1(1):2–13. https://doi.org/10.4103/2229-4708.72222

Peacock PM, Zhang W-J, Trimpin S (2017) Advances in ionization for mass spectrometry. Anal Chem 89(1):372–388. https://doi.org/10.1021/acs.analchem.6b04348

Pereira Braga C, Adamec J (2019) Metabolome analysis. In: Ranganathan S, Gribskov M, Nakai K, Schönbach C (eds) Encyclopedia of bioinformatics and computational biology. Academic Press, Oxford, pp 463–475

Peters-Clarke TM, Coon JJ, Riley NM (2024) Instrumentation at the leading edge of proteomics. Anal Chem 96(20):7976–8010. https://doi.org/10.1021/acs.analchem.3c04497

Pinu FR, Beale DJ, Paten AM, Kouremenos K, Swarup S, Schirra HJ, Wishart D (2019) Systems biology and multi-omics integration: viewpoints from the metabolomics research community. Meta 9(4):76. https://doi.org/10.3390/metabo9040076

Qiu S, Cai Y, Yao H, Lin C, Xie Y, Tang S, Zhang A (2023) Small molecule metabolites: discovery of biomarkers and therapeutic targets. Sig Transduct Target Ther 8(1):1–37. https://doi.org/10.1038/s41392-023-01399-3

Raffaelli A, Saba A (2003) Atmospheric pressure photoionization mass spectrometry. Mass Spectrom Rev 22(5):318–331. https://doi.org/10.1002/mas.10060

Razzaq A, Sadia B, Raza A, Khalid Hameed M, Saleem F (2019) Metabolomics: a way forward for crop improvement. Meta 9(12). https://doi.org/10.3390/metabo9120303

Reaves ML, Rabinowitz JD (2011) Metabolomics in systems microbiology. Curr Opin Biotechnol 22(1): 17–25. https://doi.org/10.1016/j.copbio.2010.10.001

Rebane R, Kruve A, Liigand P, Liigand J, Herodes K, Leito I (2016) Establishing atmospheric pressure chemical ionization efficiency scale. Anal Chem 88(7):3435–3439. https://doi.org/10.1021/acs.analchem.5b04852

Ren J-L, Zhang A-H, Kong L, Wang X-J (2018) Advances in mass spectrometry-based metabolomics for investigation of metabolites. RSC Adv 8(40):22335–22350. https://doi.org/10.1039/C8RA01574K

Ribbenstedt A, Ziarrusta H, Benskin JP (2018) Development, characterization and comparisons of targeted and non-targeted metabolomics methods. PLoS One 13(11):e0207082. https://doi.org/10.1371/journal.pone.0207082

Roberts LD, Souza AL, Gerszten RE, Clish CB (2012) Targeted metabolomics. Curr Protoc Mol Biol 98(1). https://doi.org/10.1002/0471142727.mb3002s98

Rocca-Serra P, Salek RM, Arita M, Correa E, Dayalan S, Gonzalez-Beltran A, Ebbels T, Goodacre R, Hastings J, Haug K, Koulman A, Nikolski M, Oresic M, Sansone S-A, Schober D, Smith J, Steinbeck C, Viant MR, Neumann S (2015) Data standards can boost metabolomics research, and if there is a will, there is a way. Metabolomics 12(1):14. https://doi.org/10.1007/s11306-015-0879-3

Romano A, Capozzi V, Khomenko I, Biasioli F (2023) Advances in the application of direct injection mass spectrometry techniques to the analysis of grape, wine and other alcoholic beverages. Molecules 28(22):7642. https://doi.org/10.3390/molecules28227642

Salem MA, Perez de Souza L, Serag A, Fernie AR, Farag MA, Ezzat SM, Alseekh S (2020) Metabolomics in the context of plant natural products research: from sample preparation to metabolite analysis. Meta 10(1):37. https://doi.org/10.3390/metabo10010037

Santamaria G, Pinto FR (2024) Bioinformatic analysis of metabolomic data: from raw spectra to biological insight. Biochemist 4(2):90–114. https://doi.org/10.3390/biochem4020005

Schmid R, Heuckeroth S, Korf A, Smirnov A, Myers O, Dyrlund TS, Bushuiev R, Murray KJ, Hoffmann N, Lu M, Sarvepalli A, Zhang Z, Fleischauer M, Dührkop K, Wesner M, Hoogstra SJ, Rudt E, Mokshyna O, Brungs C, Ponomarov K, Mutabdžija L, Damiani T, Pudney CJ, Earll M, Helmer PO, Fallon TR, Schulze T, Rivas-Ubach A, Bilbao A, Richter H, Nothias L-F, Wang M, Orešič M, Weng J-K, Böcker S, Jeibmann A, Hayen H, Karst U, Dorrestein PC, Petras D, Du X, Pluskal T (2023) Integrative analysis of multimodal mass spectrometry data in MZmine 3. Nat Biotechnol 41(4):447–449. https://doi.org/10.1038/s41587-023-01690-2

Schuhmann K, Almeida R, Baumert M, Herzog R, Bornstein SR, Shevchenko A (2012) Shotgun lipidomics on a LTQ Orbitrap mass spectrometer by successive switching between acquisition polarity modes. J Mass Spectrom 47(1):96–104. https://doi.org/10.1002/jms.2031

Schuhmann K, Herzog R, Schwudke D, Metelmann-Strupat W, Bornstein SR, Shevchenko A (2011) Bottom-up shotgun lipidomics by higher energy collisional dissociation on LTQ Orbitrap mass spectrometers. Anal Chem 83(14):5480–5487. https://doi.org/10.1021/ac102505f

Schulz S, Becker M, Groseclose MR, Schadt S, Hopf C (2019) Advanced MALDI mass spectrometry imaging in pharmaceutical research and drug development. Curr Opin Biotechnol 55:51–59. https://doi.org/10.1016/j.copbio.2018.08.003

Schwaiger-Haber M, Stancliffe E, Anbukumar DS, Sells B, Yi J, Cho K, Adkins-Travis K, Chheda MG, Shriver LP, Patti GJ (2023) Using mass spectrometry imaging to map fluxes quantitatively in the tumor ecosystem. Nat Commun 14(1):2876. https://doi.org/10.1038/s41467-023-38403-x

Scigelova M, Hornshaw M, Giannakopulos A, Makarov A (2011) Fourier transform mass spectrometry. Mol Cell Proteomics 10(7):M111–009431

Sévin DC, Kuehne A, Zamboni N, Sauer U (2015) Biological insights through nontargeted metabolomics. Curr Opin Biotechnol 34:1–8. https://doi.org/10.1016/j.copbio.2014.10.001

Sgobba E, Daguerre Y, Giampà M (2021) Unravel the local complexity of biological environments by MALDI mass spectrometry imaging. Int J Mol Sci 22(22):12393. https://doi.org/10.3390/ijms222212393

Shimma S, Sugiura Y (2014) Effective sample preparations in imaging mass spectrometry. Mass

Spectrom 3(Spec Issue):S0029. https://doi.org/10.5702/massspectrometry.S0029

Shrestha B (2020) Single-cell metabolomics by mass spectrometry. In: Shrestha B (ed) Single cell metabolism: methods and protocols. Springer, New York, NY, pp 1–8

Shrestha SA, Cha S (2021) Ambient desorption/ionization mass spectrometry for direct solid material analysis. TrAC Trends Anal Chem 144:116420. https://doi.org/10.1016/j.trac.2021.116420

Sindelar M, Patti GJ (2020) Chemical discovery in the era of metabolomics. J Am Chem Soc 142(20):9097–9105. https://doi.org/10.1021/jacs.9b13198

Smith DR, Robb DB, Blades MW (2009) Comparison of dopants for charge exchange ionization of nonpolar polycyclic aromatic hydrocarbons with reversed-phase LC-APPI-MS. J Am Soc Mass Spectrom 20(1):73–79. https://doi.org/10.1016/j.jasms.2008.09.012

Soga T (2023) Advances in capillary electrophoresis mass spectrometry for metabolomics. TrAC Trends Anal Chem 158:116883. https://doi.org/10.1016/j.trac.2022.116883

Song X, Li C, Meng Y (2022) Mass spectrometry imaging advances and application in pharmaceutical research. Acta Mater Med 1:507–533. https://doi.org/10.15212/AMM-2022-0046

Sun G, Yang K, Zhao Z, Guan S, Han X, Gross RW (2007) Shotgun metabolomics approach for the analysis of negatively charged water-soluble cellular metabolites from mouse heart tissue. Anal Chem 79(17):6629. https://doi.org/10.1021/ac070843+

Swales JG, Hamm G, Clench MR, Goodwin RJA (2019) Mass spectrometry imaging and its application in pharmaceutical research and development: a concise review. Int J Mass Spectrom 437:99–112. https://doi.org/10.1016/j.ijms.2018.02.007

Takáts Z, Wiseman JM, Cooks RG (2005) Ambient mass spectrometry using desorption electrospray ionization (DESI): instrumentation, mechanisms and applications in forensics, chemistry, and biology. J Mass Spectrom 40(10):1261–1275. https://doi.org/10.1002/jms.922

Tang K, Page JS, Smith RD (2004) Charge competition and the linear dynamic range of detection in electrospray ionization mass spectrometry. J Am Soc Mass Spectrom 15(10):1416. https://doi.org/10.1016/j.jasms.2004.04.034

Tomita M, Nishioka T (2005) Metabolomics: the frontier of systems biology. Springer, Tokyo/New York

Tuck M, Grélard F, Blanc L, Desbenoit N (2022) MALDI-MSI towards multimodal imaging: challenges and perspectives. Front Chem 10. https://doi.org/10.3389/fchem.2022.904688

van Veldhoven K, Kiss A, Keski-Rahkonen P, Robinot N, Scalbert A, Cullinan P, Chung KF, Collins P, Sinharay R, Barratt BM, Nieuwenhuijsen M, Rodoreda AA, Carrasco-Turigas G, Vlaanderen J, Vermeulen R, Portengen L, Kyrtopoulos SA, Ponzi E, Chadeau-Hyam M, Vineis P (2019) Impact of short-term traffic-related air pollution on the metabolome—results from two metabolome-wide experimental studies. Environ Int 123:124–131. https://doi.org/10.1016/j.envint.2018.11.034

Vats M, Cillero-Pastor B, Cuypers E, Heeren RMA (2024) Mass spectrometry imaging in plants, microbes, and food: a review. Analyst 149(18):4553–4582. https://doi.org/10.1039/D4AN00644E

Vaysse P-M, Heeren RMA, Porta T, Balluff B (2017) Mass spectrometry imaging for clinical research—latest developments, applications, and current limitations. Analyst 142(15):2690–2712. https://doi.org/10.1039/C7AN00565B

Vlaanderen JJ, Janssen NA, Hoek G, Keski-Rahkonen P, Barupal DK, Cassee FR, Gosens I, Strak M, Steenhof M, Lan Q, Brunekreef B, Scalbert A, Vermeulen RCH (2017) The impact of ambient air pollution on the human blood metabolome. Environ Res 156:341–348. https://doi.org/10.1016/j.envres.2017.03.042

Vrhovsek U, Masuero D, Gasperotti M, Franceschi P, Caputi L, Viola R, Mattivi F (2012) A versatile targeted metabolomics method for the rapid quantification of multiple classes of phenolics in fruits and beverages. J Agric Food Chem 60(36):8831–8840. https://doi.org/10.1021/jf2051569

Vuckovic D (2012) Current trends and challenges in sample preparation for global metabolomics using liquid chromatography-mass spectrometry. Anal Bioanal Chem 403(6):1523–1548. https://doi.org/10.1007/s00216-012-6039-y

Wang M, Carver JJ, Phelan VV, Sanchez LM, Garg N, Peng Y, Nguyen DD, Watrous J, Kapono CA, Luzzatto-Knaan T, Porto C, Bouslimani A, Melnik AV, Meehan MJ, Liu W-T, Crüsemann M, Boudreau PD, Esquenazi E, Sandoval-Calderón M, Kersten RD, Pace LA, Quinn RA, Duncan KR, Hsu C-C, Floros DJ, Gavilan RG, Kleigrewe K, Northen T, Dutton RJ, Parrot D, Carlson EE, Aigle B, Michelsen CF, Jelsbak L, Sohlenkamp C, Pevzner P, Edlund A, McLean J, Piel J, Murphy BT, Gerwick L, Liaw C-C, Yang Y-L, Humpf H-U, Maansson M, Keyzers RA, Sims AC, Johnson AR, Sidebottom AM, Sedio BE, Klitgaard A, Larson CB, Boya P CA, Torres-Mendoza-D, Gonzalez DJ, Silva DB, Marques LM, Demarque DP, Pociute E, O'Neill EC, Briand E, Helfrich EJN, Granatosky EA, Glukhov E, Ryffel F, Houson H, Mohimani H, Kharbush JJ, Zeng Y, Vorholt JA, Kurita KL, Charusanti P, McPhail KL, Nielsen KF, Vuong L, Elfeki M, Traxler MF, Engene N, Koyama N, Vining OB, Baric R, Silva RR, Mascuch SJ, Tomasi S, Jenkins S, Macherla V, Hoffman T, Agarwal V, Williams PG, Dai J, Neupane R, Gurr J, Rodríguez AMC, Lamsa A, Zhang C, Dorrestein K, Duggan BM, Almaliti J, Allard P-M, Phapale P, Nothias L-F, Alexandrov T, Litaudon M, Wolfender J-L, Kyle JE, Metz TO, Peryea T, Nguyen D-T, VanLeer D, Shinn P, Jadhav A, Müller R, Waters KM, Shi W, Liu X, Zhang L, Knight R, Jensen PR, Palsson BO, Pogliano K, Linington RG, Gutiérrez M, Lopes NP,

Gerwick WH, Moore BS, Dorrestein PC, Bandeira N (2016) Sharing and community curation of mass spectrometry data with global natural products social molecular networking. Nat Biotechnol 34(8): 828–837. https://doi.org/10.1038/nbt.3597

Wang Y, Liu S, Hu Y, Li P, Wan J-B (2015) Current state of the art of mass spectrometry-based metabolomics studies—a review focusing on wide coverage, high throughput and easy identification. RSC Adv 5(96): 78728–78737. https://doi.org/10.1039/C5RA14058G

Ward-Caviness CK, Xu T, Aspelund T, Thorand B, Montrone C, Meisinger C, Dunger-Kaltenbach I, Zierer A, Yu Z, Helgadottir IR, Harris TB, Launer LJ, Ganna A, Lind L, Eiriksdottir G, Waldenberger M, Prehn C, Suhre K, Illig T, Adamski J, Ruepp A, Koenig W, Gudnason V, Emilsson V, Wang-Sattler R, Peters A (2017) Improvement of myocardial infarction risk prediction via inflammation-associated metabolite biomarkers. Heart 103(16):1278–1285. https://doi.org/10.1136/heartjnl-2016-310789

Waters (2022) UPLC with the SELECT SERIES MRT for Metabolite Identification—Product Solution | Waters. https://www.waters.com/content/dam/waters/pt_br/library/brochures/2022/waters-brochure-UPLCwiththeSELECTSERIESMRTforMetaboliteIdentificationProductSolution-720007657.pdf. Accessed 22 Oct 2024

Wei D, Xu M, Wang Z, Tong J (2022) The development of single-cell metabolism and its role in studying cancer emergent properties. Front Oncol 11. https://doi.org/10.3389/fonc.2021.814085

Wentrup J, Bösing I, Dülcks T, Thöming J (2024a) Rapid online analysis of n-alkanes in gaseous streams via APCI mass spectrometry. Anal Bioanal Chem 416(8): 1843. https://doi.org/10.1007/s00216-024-05182-3

Wentrup J, Dülcks T, Thöming J (2024b) The role of water in APCI-MS online monitoring of gaseous n-alkanes. Anal Bioanal Chem 416(22):4961–4971. https://doi.org/10.1007/s00216-024-05431-5

Willems AP, van der Ham M, Schiebergen-Bronkhorst BGM, van Aalderen M, de Barse MMJ, De Gruyter FE, van Hoek IN, Pras-Raves ML, de Sain-van der Velden MGM, Prinsen HCMT, Verhoeven-Duif NM, Jans JJM (2023) A one-year pilot study comparing direct-infusion high resolution mass spectrometry based untargeted metabolomics to targeted diagnostic screening for inherited metabolic diseases. Front Mol Biosci 10. https://doi.org/10.3389/fmolb.2023.1283083

Wishart DS (2008) Quantitative metabolomics using NMR. TrAC Trends Anal Chem 27(3):228–237. https://doi.org/10.1016/j.trac.2007.12.001

Yang Q, Zhang A, Miao J, Sun H, Han Y, Yan G, Wu F, Wang X (2019) Metabolomics biotechnology, applications, and future trends: a systematic review. RSC Adv 9(64):37245–37257. https://doi.org/10.1039/C9RA06697G

Yost RA (2022) The triple quadrupole: innovation, serendipity and persistence. J Mass Spectrom Adv Clin Lab 24:90–99. https://doi.org/10.1016/j.jmsacl.2022.05.001

Zahn JA, Higgs RE, Hilton MD (2001) Use of direct-infusion electrospray mass spectrometry to guide empirical development of improved conditions for expression of secondary metabolites from actinomycetes. Appl Environ Microbiol 67(1): 377–386. https://doi.org/10.1128/AEM.67.1.377-386.2001

Zhang H, Lu KH, Ebbini M, Huang P, Lu H, Li L (2024) Mass spectrometry imaging for spatially resolved multi-omics molecular mapping. npj Imaging 2(1): 1–15. https://doi.org/10.1038/s44303-024-00025-3

Zhang K, Wong JW, Mai H, Trucksess MW (2014) Dopant-assisted atmospheric pressure photoionization of patulin in apple juice and apple-based food with liquid chromatography-tandem mass spectrometry. J Agric Food Chem 62(18):4112–4118. https://doi.org/10.1021/jf5006726

Zhang L, Vertes A (2018) Single-cell mass spectrometry approaches to explore cellular heterogeneity. Angew Chem Int Ed 57(17):4466–4477. https://doi.org/10.1002/anie.201709719

Zhang S, Zhu J (2022) Untargeted metabolomics sensitively differentiates gut bacterial species in single culture and co-culture systems. ACS Omega 7(17): 14643–14652. https://doi.org/10.1021/acsomega.1c07114

Zhou J, Zhong L (2022) Applications of liquid chromatography-mass spectrometry based metabolomics in predictive and personalized medicine. Front Mol Biosci 9. https://doi.org/10.3389/fmolb.2022.1049016

Zhou Q, Fülöp A, Hopf C (2021) Recent developments of novel matrices and on-tissue chemical derivatization reagents for MALDI-MSI. Anal Bioanal Chem 413(10):2599–2617. https://doi.org/10.1007/s00216-020-03023-7

Zhu G, Zhang W, Zhao Y, Chen T, Yuan H, Liu Y, Guo G, Liu Z, Wang X (2023) Single-cell metabolomics-based strategy for studying the mechanisms of drug action. Anal Chem 95(10):4712–4720. https://doi.org/10.1021/acs.analchem.2c05351

Zoppi J, Guillaume J-F, Neunlist M, Chaffron S (2021) MiBiOmics: an interactive web application for multi-omics data exploration and integration. BMC Bioinformatics 22(1):6. https://doi.org/10.1186/s12859-020-03921-8

5 Separative Techniques Coupled to Mass Spectrometry for Metabolomic Analysis

Jasmine Hertzog, Marisa Maia, and Vincent Carré

Abstract

Advanced analytical techniques are required to decipher metabolome complexity. The previous chapter ("Latest Developments in Mass Spectrometry-Based Techniques for Metabolomics Analysis") shows general aspect of mass spectrometry techniques for the detection, identification, and quantification of metabolites by target and non-targeted approaches. It also points out that, in some instances, the sole direct analysis by mass spectrometry is a limitation for deciphering the high molecular complexity of the metabolome. Therefore, upstream separative methods such as chromatography, capillary electrophoresis, or ion mobility can be an effective solution. Indeed, these techniques reduce the ion suppression effects, separate some isomers, and increase the capacity measurement for metabolite identification and quantification. This chapter focuses on the separative methods most frequently used in metabolomic studies. Their principles and some examples are addressed along with their coupling with MS.

Lastly, analytical improvements and recent advances in MS hyphenated approaches for metabolomic study are also discussed with multidimensional analysis.

Keywords

Metabolomics · Separative techniques · Chromatography · Capillary electrophoresis · Ion mobility

Jasmine Hertzog, Marisa Maia and Vincent Carré contributed equally with all other contributors.

J. Hertzog (✉) · V. Carré
Université de Lorraine, Metz, France
e-mail: jasmine.hertzog@univ-lorraine.fr; vincent.carre@univ-lorraine.fr

M. Maia
Grapevine Pathogen Systems Lab (GPS Lab), Biosystems and Integrative Sciences Institute (BioISI), Faculdade de Ciências, Universidade de Lisboa, Lisbon, Portugal

Departamento de Biologia, Faculdade de Ciências, Universidade de Lisboa, Lisbon, Portugal
e-mail: mrmaia@ciencias.ulisboa.pt

5.1 Introduction

Different separative techniques can be coupled to mass spectrometry (MS) providing advantages and complementary information to direct MS (Begou et al. 2017). Indeed, ion suppression and matrix effects can be reduced, and isomers can be differentiated based on their physicochemical parameters such as volatility, polarity, and shape (Fig. 5.1). Thus, with chromatographic techniques, retention time is obtained, while in capillary electrophoresis, it is migration time. In ion mobility, the mobility of the ions is measured and can be then used to calculate their collision cross sections (CCS). Each of these parameters are complementary to mass-to-charge ratio (*m/z*)

M. Gama-Carvalho et al. (eds.), *Omics Approaches in Biomedicine and Biotechnology*, Advances in Experimental Medicine and Biology 1504, https://doi.org/10.1007/978-3-032-18966-0_5

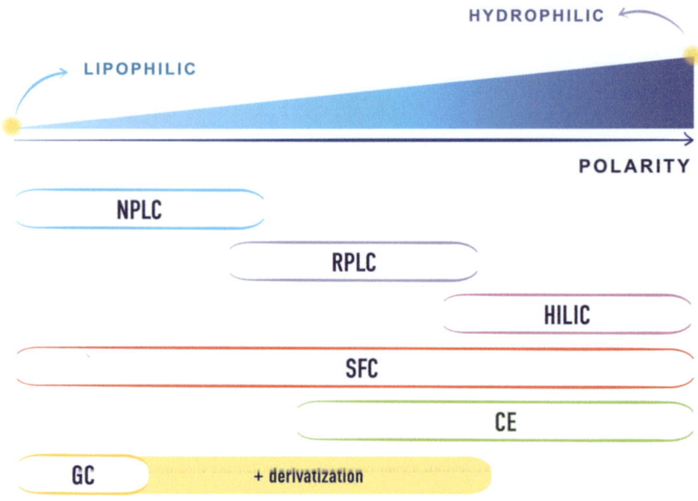

Fig. 5.1 Diagram illustrating the application range of various separation techniques according to the metabolite polarity. Techniques are positioned along this gradient: NPLC (Normal Phase Liquid Chromatography), RPLC (Reversed Phase Liquid Chromatography), HILIC (Hydrophilic Interaction Liquid Chromatography), SFC (Supercritical Fluid Chromatography), CE (Capillary Electrophoresis), and GC (Gas Chromatography) with derivatization

data and provide higher confidence in metabolite identification. It eventually enables a more precise quantification of single metabolite compounds.

Coupled MS techniques can also be of interest for the characterization of some complex metabolomic samples. It not only decreases their complexity but also, by using different and complementary separation techniques, helps to decipher most of the metabolome.

In this section, the different separative techniques, which can be coupled to MS, will be described, mainly in the frame of metabolomic analyses, with their advantages and limitations and for analytical and non-preparative purposes. Operating principles of both mass spectrometers and ionization sources are described in the Chapter 4. In this chapter, their use will be presented with their specificities, depending on the coupled separative technique and on the targeted metabolites. Some applications will be also discussed to illustrate each technique.

5.2 Gas Chromatography

Gas chromatography (GC) is employed for the analysis of volatile or semi-volatile metabolites from food (Kataoka et al. 2000), plants (Hazrati et al. 2022), humans (Vernocchi et al. 2012), and animals (Ma et al. 2015; Xiao et al. 2020). These metabolites can be directly analyzed by GC-MS, or they can be first chemically modified and/or extracted from sample using solvent or sorbent, as, for example, by solid-phase microextraction (SPME). All these steps are intended to prepare the sample to be introduced in a GC system.

5.2.1 GC Technique

A gas chromatograph is composed of a heated injector connected to a capillary column placed in an oven, ended by a heated transfer line linked to a detector such as a mass spectrometer. Injectors can have different configurations such as split/splitless, programmable in temperature (PTV) in the objective to vaporize a small volume of solubilized sample, typically 1 μL, or to desorb compounds from solid support such as SPME fiber or directly transfer gas phase compounds to the column. Compound separation takes place into the capillary column, whose inner wall is covered with a stationary phase and continuously swept by a carrier gas, such as helium, at a flow rate close to 1 mL.min^{-1}. Increasing the column temperature during a chromatographic run shifts the balance of compounds from the stationary phase to the gas phase, which is then pushed toward the detector by the carrier gas. Typically, with a capillary column of 30 m long, 0.25 mm of inner diameter, and a stationary phase ranging from 0.1 to 0.53 μm, GC is adapted to the low-pressure ion sources of MS and offers the most efficient separation of volatile and semi-volatiles compounds (Beale et al. 2018). As the nonpolar stationary phase, constituted by 5% phenyl-95% dimethyl-siloxane, is very stable until temperature of 330–350 °C, it is often used for the separation of numerous compounds in gas phase; other stationary phases are also usefull depending on the applications (Beale et al. 2018).

5.2.2 Sample Preparation

5.2.2.1 Solid-Phase Microextraction (SPME)

SPME fiber can be either directly immersed in the sample in liquid phase (DI SPME) or can be used to adsorb the volatile metabolites present in the headspace of the vial comprising the solid or liquid sample (HS SPME) or can be used for in vivo SPME (Jalili et al. 2020). SPME has the advantage of concentrating metabolites present in trace amount and, depending on the coating sorbent of the fiber, to extract compounds of various polarities (Kataoka et al. 2000). This extraction technique can also be used prior to liquid chromatography MS analysis (Kanu 2021; Hu and Ouyang 2021). Some recent studies were reported using SPME combined with GC-MS-based techniques for the characterization of wine volatilome, responsible for the wine flavor (Muñoz-Redondo et al. 2020; Pérez-Jiménez et al. 2021). This approach allows for wine volatile profiling but was also showed to be informative on the chemical changes throughout winemaking and on the influence of the region on the wine volatilome (Šuklje et al. 2019; Furdíková et al. 2020; Arcena et al. 2020).

This approach is also applied to study plant and food flavors (Pavagadhi and Swarup 2020; Rodeo and Mitcham 2023; Deng et al. 2023; Wan et al. 2024; Wang et al. 2024). Thus, flavor profiles of strawberries grown with different fertilizers were studied by HS-SPME-GC-MS, with 92 identified compounds (Liu et al. 2024). It was notably evidenced that the use of fertilizers promoted the esters responsible for the strawberry aroma. HS-SPME-GC-MS is also used to investigate volatile organic compounds related to plants that can be involved in plant defense mechanisms or growth (Farag et al. 2006; Chen et al. 2020, 2021; Gamboa-Becerra et al. 2022; Mhlongo et al. 2022; Poria et al. 2024). For instance, five SPME fibers of four different coating phases were tested for the optimal extraction of aldehydes secreted by insects (Bojke et al. 2020). Among the Divinylbenzene/Carboxen/Polydimethylsiloxane (DVB/CAR/PDMS), PDMS, PDMS/DVB, and polyacrylate PA, the DVB/CAR/PDMS fiber showed the best recoveries for 17 aldehydes. Aldehydes such as heptanal, 2-decanal, and undecanal were demonstrated to be effective against entomopathogenic fungi.

Eventually, human metabolome can also be studied by means of *in situ* SPME-MS to detect analytes from skin surface, body fluids (Khodadadi and Pourfarzam 2020), and human exhaled breath (Fig. 5.2) (Hu and Ouyang 2021). This was also extended to other species (such as mammals, frog, and fishes) and other organs (muscles and brain) for doping control, bioaccumulation, and pharmacokinetics studies (Queiroz et al. 2022).

5.2.2.2 Derivatization

Furthermore, GC can also be used for the study of less volatile and polar metabolites, such as carbohydrates and fatty and amino acids (Fiehn 2016). However, it requires an upstream derivatization step to increase their volatility and stability. Several derivatization techniques exist that are aimed at protecting some functional groups such as carboxylic acids, alcohols, amines, amines, and thiols. The most commonly used are silylation, alkylation, and acylation (Koek et al. 2011; Khodadadi and Pourfarzam 2020). Derivatization step is necessary for the investigation of such metabolites. However, reaction can be incomplete leading to artifact compounds that bias the analyses; therefore, derivatization reaction must be well-controlled (Kanani and Klapa 2007; Rodríguez-Sánchez et al. 2011; Jemmali et al. 2016). This step is often observed as part

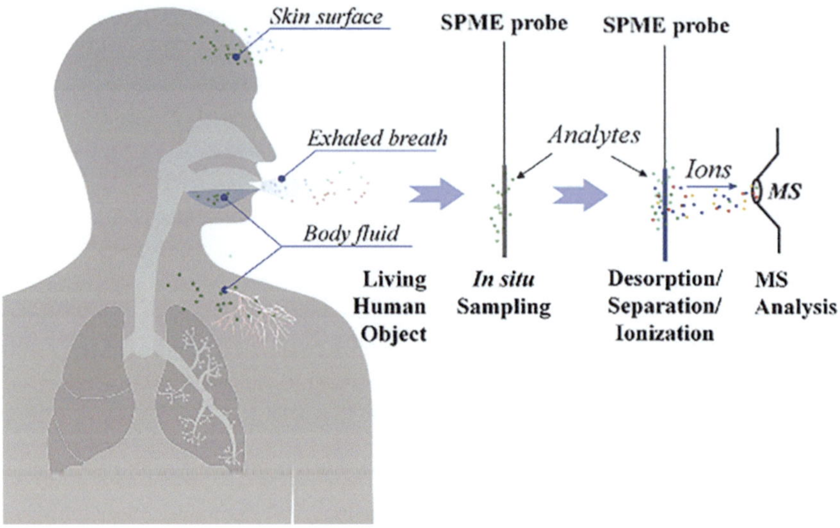

Fig. 5.2 Diagram illustrating *in situ* SPME sampling applied to the study of human metabolome from different body parts and analysis by mass spectrometry-based techniques (Image retrieved from Hu and Ouyang (2021))

of studies related to human body fluids, such as urine (Khodadadi and Pourfarzam 2020) and plasma (Srivastava et al. 2023), but also in plants (Marinaki et al. 2023).

5.2.3 Ionization Sources and MS Detection

Regarding the ionization sources, GC-MS metabolomic analyses are frequently performed with standard electron ionization (EI) at 70 eV energy (Lebanov et al. 2021; Kiseleva et al. 2022; Shahid et al. 2023; Weiller et al. 2024). The high reproducibility of EI mass spectra allows easy compound identification based on their fragmentation patterns using EI mass spectra databases. Additionally, the high separation efficiency of the capillary columns offers the possibility to use the retention time for identifying and quantifying isomers non-discriminated by their unique EI mass spectra.

Alternative ionization techniques, such as chemical ionization (CI), are also used as part of GC-MS analysis. Recently, EI and CI were compared for the targeted analyses of biological matrices (Misra and Olivier 2020). Positive CI (PCI), based on proton-transfer from ion-molecule collisions, and negative CI (NCI), with an electron capture process from electrons transferred from the reagent gas (Misra and Olivier 2020), were performed. These three analytical conditions (EI, PCI, and NCI) were showed to provide complementary information on human plasma analyses by GC-Orbitrap MS with the ionization of different subsets of metabolites and the global identification of 270 metabolites. However, CI-MS analyses notably suffer from less exhaustive metabolite databases compared to EI-MS requiring the use of in-house databases. As CI can be performed with different reagent gases, influence of its nature during GC-CI-MS analysis of stable isotope-labeled metabolites was assessed (Capellades et al. 2021). Isobutane demonstrated the only detection of $[M + H]^+$ cell metabolites and therefore to achieve the isotopic enrichment, while methane was observed to fail by inducing more unwanted fragmentation.

The last 15 years have seen the increase of GC-MS studies with atmospheric pressure ionization techniques based on laser, plasma, electrospray, and photo- and chemical ionizations (Ayala-Cabrera et al. 2023a). Those techniques present some advantages over vacuum EI and CI ones, such as lower fragmentation and higher sensitivity. Most of the published works are performed with atmospheric pressure chemical ionization (APCI), considered as a soft ionization source (Ayala-Cabrera et al. 2023a). GC-APCI-MS was compared to GC-EI-MS for the identification of avocado metabolites. Advantages and limitations of each technique were discussed, as well as corresponding databases. It was notably observed that APCI-based analysis allowed the identification of hitherto unknown metabolites (Hurtado-Fernández et al. 2013). Similar conclusions were achieved after analysis of root extract of *Arabidopsis thaliana* by GC-EI-MS and GC-APCI-MS (Strehmel et al. 2014).

MS detectors coupled to GC are often quadrupoles due to their good dynamic range, sensitivity, and robustness and are affordable (Lei et al. 2011; Beale et al. 2018). As ion trap, these detectors can be used as a single or triple quadrupole. The last configuration ensured to perform MS^2 analyses and multiple reaction monitoring (MRM). With MRM, it is possible to identify and quantify specific analytes, by avoiding co-eluting issues, thanks to their MS^2 ions. Single and triple quadrupole mass spectrometers are mostly used for targeted analyses.

Untargeted GC-MS analyses are mainly performed with time of flight (TOF) MS (Beale et al. 2018). Compared to quadrupole MS, TOF-MS have higher mass accuracy and resolution. It also offers a higher scan rate allowing to have a better deconvolution of GC peaks.

As previously said, with the implementation of atmospheric pressure ionization techniques, it is easier to couple more advanced mass spectrometers such as Orbitrap MS, used mainly for metabolomic profiling analyses thanks to its high-mass accuracy and high-mass resolution (Escolà Casas and Matamoros 2022; He et al. 2023).

5.2.4 Two-Dimensional GC

Due to the complexity of some samples, there is the need to increase metabolite separation. Therefore, two-dimensional GC or GC x GC was developed increasing both the peak capacity and the sensitivity (Wolfender et al. 2015). In this approach, two separations are performed. Compounds are firstly separated on a first column, usually a nonpolar column, and a second separation is done on a second column, which is commonly polar. However, the low-temperature stability of the polar stationary phase could be a limitation to the analysis of high-boiling point compounds. Transfer onto the second column is done by a modulator that successively traps, focuses, and transfers the eluent segment from the first column onto the second column. The separation in the second dimension must be fast, a few seconds, since it needs to be done before the next eluents are transferred on the second column from the modulator. Therefore, the second column is much shorter than the first one. For GC x GC analyses, the scan rate of the MS detector is particularly paramount; therefore TOF and Orbitrap MS are preferred.

In several studies, performances of GC x GC-MS over GC-MS were demonstrated for new biomarker discovery (Winnike et al. 2015; Yu et al. 2017) or metabolic profiling (Wong et al. 2017). In a recent study, GC x GC and GC, both coupled to TOF-MS, were compared as part of metabolite identification from plasma samples originating from healthy and diabetes patients with kidney failure (Duangkumpha et al. 2022). GC x GC-MS analyses showed to both detect and identify more features than GC-MS, irrespective of the patient type, which can be relevant for the better understanding of metabolic diseases.

GC-MS-based techniques were demonstrated to be powerful for the targeted and nontargeted analyses of volatile compounds from plants, animals, and humans. Especially, quantitative metabolomics, with hyphenated MS-based methods, must follow validation steps to ascertain quality and accuracy of the results (Koek et al. 2011; Ghafari and Sleno 2024).

However, it represents a limited fraction of the metabolome, as it mainly refers to thermally stable, high volatility, and low-molecular-weight compounds. Therefore, complementary techniques must be applied to access the more polar and higher-mass part of the metabolome, which are developed in the following sections.

5.3 Liquid Chromatography

Liquid chromatography (LC) coupled with MS is largely used as part of targeted and nontargeted metabolomics (Kuehnbaum and Britz-McKibbin 2013; Zhou and Yin 2016; Zhou and Zhong 2022; Queiroz et al. 2024; Ovbude et al. 2024; Han et al. 2024; Cajka 2024). Indeed, despite the lower efficiency of the filled columns used in LC compared to the GC capillary ones, this technique offers good reproducibility and allows wide metabolome coverage. In addition, it not only decreases sample complexity and ionization suppression, but it also enables to have higher confidence in metabolite annotation thanks to retention time, as a complementary experimental data to *m/z*. Strategies developed for LC-MS metabolomics notably comprise the use of a sample preparation step and the choice of the LC technique.

5.3.1 Sample Preparation

Sample preparation prior to LC-MS analysis can be relevant for extracting metabolites of interest and decrease sample complexity. There are different extraction methods with liquid-liquid (LLE), solid-liquid (SLE), solid phase (SPE), and SPME being the most employed (Zhou et al. 2012). More details and discussions on the sample preparations and extractions for LC-MS-based metabolomic studies can be found in dedicated reviews and research papers (Mushtaq et al. 2014; Verpoorte et al. 2022; Cuevas-Delgado et al. 2024; Li et al. 2024a). As well as GC, derivatization can also be performed on some chemical functions of metabolites prior LC-MS (Chen et al. 2023).

LLE consists in extracting metabolites from the sample matrix based on their affinity with two immiscible liquids. For example, LLE at room temperature was used for leaf extraction and compared with SPE performed with two different hydrophilic polymers as sorbent (Lang et al. 2024). Leaf tissue was first ground to be then extracted with a MeOH/H_2O solution. Dichloromethane was then used to perform LLE. Compared to SPE, although a lower selectivity was achieved, more compounds were recovered by LLE belonging to different biochemical classes such as flavonoids, carboxylic acids, and coumarins. Even if this technique is easy to achieve, it presents some limitations such as its solvent needs, poor analyte enrichment, and potential matrix interferences (Młynarczyk et al. 2024).

SP(M)E is regarded as a flexible extraction technique for various metabolomic samples due to the multiple types of sorbents covering a large range of polarity and functionality (Lang et al. 2024). Several reviews have been recently published, which highlighted its advantages and limitations (Owczarzy et al. 2024; Młynarczyk et al. 2024).

The microextraction efficiency of 42 different SPME coatings was evaluated by means of 36 metabolites belonging to different classes: amino and organics acids, sugars, hormones, and peptides (Vuckovic and Pawliszyn 2011). Authors observed that mixed-mode—phenylboronic acid and polystyrene-divinylbenzene coatings—allowed the best simultaneous extraction of both hydrophilic and hydrophobic metabolites. In the same study, they also compared SPME method with two other extraction techniques, namely, ultrafiltration and solvent precipitation. SPME was showed to have higher precision based on the relative standard deviation (RSD) values compared to the two other methods. They also mentioned the reduction of ionization suppression with the SPME methodology.

5.3.2 LC Technique

Various LC techniques exist that are capable of separating metabolites over a wide polarity range thanks to the different stationary phases. Briefly, liquid chromatography system is set up by different parts. First, the mobile phase consists in one or a mix of solvents coming from stock solvent bottles through different lines, and their proportion can be changed along a chromatographic run. In addition, binary or quaternary pumps are used to deliver the mobile phase at a constant and precise flowrate to and through the column before reaching one or more detectors (MS, UV…). As each column is filled with a solid stationary phase (porous polymers or silica particles surrounded by functional grafting), the pumping system is designed to operate at high pressure (up to over 1000 bar for UPLC columns). A liquid sample is then taken from a sample vial and loaded (automatically) in a loop of a specific volume (partially or fully filled), which is connected to a six-way valve. As this valve is switched in injection mode, the flow of the mobile phase flushes the loop and pushes the sample onto the LC column. The component separation takes place into the column, at a specific controlled temperature, according to their different affinities between the mobile and the stationary phases.

There is the reversed phase LC (RPLC), hydrophilic interaction LC (HILIC), normal-phase liquid chromatography (NPLC), ion-exchange chromatography, size-exclusion chromatography (SEC), adsorption chromatography, and affinity chromatography (AC) (Ovbude et al. 2024). More details are given below on the three first LC techniques, since they are the most used for LC-MS metabolomic studies.

RPLC is the most employed in LC-MS metabolomics (Kuehnbaum and Britz-McKibbin 2013). While an apolar stationary phase is used, the elution is usually performed in gradient mode where the proportion of the aqueous solvent decreases during the chromatographic run in favor of the organic solvent such as acetonitrile. It can be used for the analysis of a wide variety of samples ranging from microorganisms and plant extracts (Ryparova Kvirencova et al. 2023) to human and animal biological fluids (Saigusa et al. 2016; Uchida et al. 2024; Dorakumbura et al. 2024). The apolar stationary phases, such as C_8, C_{18}, or C_{30}, are used to separate semi-polar compounds such as flavonoids, alkaloids, phenolic acids, and glycosylated species (Zhou et al. 2012; Wolfender et al. 2015). An example of the diversity of metabolites that can be separated and identified by RPLC-MS of tomato extract is illustrated in Fig. 5.3 (Gómez-Romero et al. 2010).

Ion-paring reagents can also be added to the mobile phase to expand separation selectivity by RPLC (Kuehnbaum and Britz-McKibbin 2013).

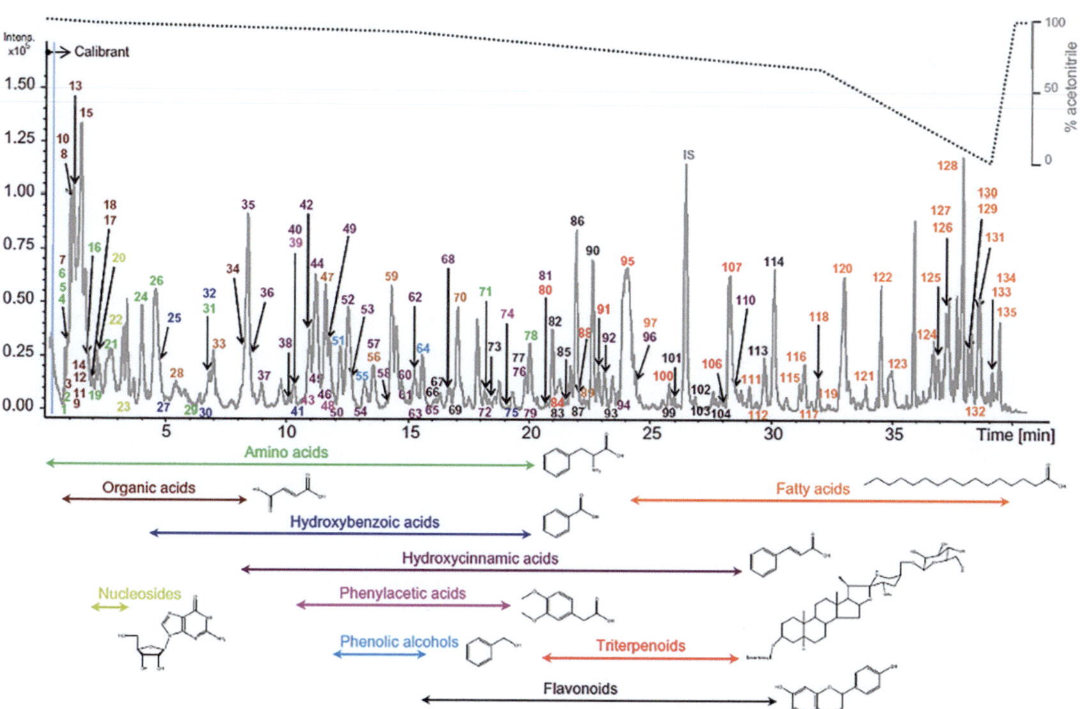

Fig. 5.3 Base peak chromatogram of tomato extract obtained by RPLC-MS with 135 identified compounds of different chemical classes. (Image retrieved from Gómez-Romero et al. (2010))

Thus, basic metabolites are separated with perchlorate or alkane-sulfonic acids, whereas separation of anions is done with tetrabutylammonium hydroxide (Zhou et al. 2012; Ovbude et al. 2024). However, the use of such nonvolatile agents can lead to signal suppression and/or ion source contamination.

For the separation of more polar and ionic metabolites, such as sugars, vitamins, and amino and carboxylic acids, HILIC is a good alternative to RPLC (Zhou et al. 2012; Tang et al. 2016). HILIC columns can have different polar stationary phases such as non-ionic (i.e., unmodified silica), ionic (i.e., amine or cyano groups), or zwitterionic phases (i.e., sulfoalkylbetaine or ammonium-sulfonic acid) columns (Kuehnbaum and Britz-McKibbin 2013; Ovbude et al. 2024). The HILIC mobile phase usually requires a high proportion of organic solvent, to maintain at least 3% of water, and a well-controlled pH. Complementarity of HILIC and RPLC was evidenced as part of human cerebrospinal fluid analysis with HILIC-MS/MS capable of detecting exclusive 273 metabolites against 218 unique metabolites by RPLC-MS/MS (Hooshmand et al. 2024). Among the features annotated in HILIC, there are amino acids and derivatives, carbohydrates, organic acids, and nucleosides. In RPLC, triglycerides, glycerophosphocholines, and sphingomyelin were mainly detected. In another study, different HILIC and RP stationary phases were investigated for the untargeted analysis of human liver microsomes, and rat urine and plasma (Hemmer et al. 2024). The study evidenced that combination of ZicHILIC (sulfobetaine) and RP BEH C_{18} columns was suitable for plasma sample analysis. But overall, ZicHILIC and RP-Phenyl-Hexyl columns showed the best complementarity.

Another technique capable of separating polar compounds, such as polar lipids and sugars, is the so-called normal phase liquid chromatography (NPLC). Stationary phases are polar, with diol, aminopropyl, or cyano groups. However, it requires nonpolar organic mobile phases for analyte elution, which makes it less compatible with electrospray ionization source (ESI)-MS (Zhou et al. 2012; Ovbude et al. 2024).

Among commercial instruments for LC, it is possible to distinguish between high-performance liquid chromatography (HPLC) and ultra HPLC (UHPLC). The second one differs from HPLC by its higher back pressure (up to 15,000 psi vs. 6000 psi in HPLC) and lower packing material (< 2 μm). These properties allow to reduce time analysis and signal suppression and to increase resolution, peak capacity, and sensitivity (Nordström et al. 2006; Anspach et al. 2007; Evans et al. 2009; Guillarme et al. 2010).

5.3.3 Ionization Sources and MS Detection

The most used ionization source, in metabolomic LC-MS, is ESI. Indeed, most of the metabolites are mid-polar to polar or ionic and are already in liquid phase (Theodoridis et al. 2012; Forcisi et al. 2013). Other sources can be employed in metabolomic LC-MS studies, such as atmospheric pressure chemical ionization (APCI) and atmospheric pressure photoionization (APPI), which allows to extend metabolome characterization to less polar and thermally stable metabolites. As an example, Cirigliano and Cabrera (2020) used ESI, APCI, and APPI for LC-MS analysis of plant extract products (Cirigliano and Cabrera 2020).

LC-MS with ESI and APCI sources, both in positive and negative-modes, were employed for the characterization of erythrocyte metabolome (Sana et al. 2008). Combination of both techniques ensured to have a more extensive coverage of the metabolome with 1596 features exclusive to ESI (+/−) and 600 features only observed by APCI (+/−) and only 174 common features. A similar approach was done for the untargeted and targeted LC-MS analysis of grape berry (Commisso et al. 2017). There are also few studies involving APPI with LC-MS-based metabolomics (Tian et al. 2013; Popov et al. 2015). For instance, plasma metabolome was studied with different ionization sources (Tian et al. 2013). ESI was more adapted to the detection of bile acids, glycerophosphocholines, glycerophosphoethanolamines, and sulfate; APCI to cyclic alcohols, fatty acids, and linoleic acids; and APPI to steroids, sphingolipids, some amino acids, nucleosides, and purines.

Reviews done on metabolomic LC-MS show different MS detectors, such as triple quadrupole, ion trap, Q-TOF, TOF, Orbitrap, and LTQ-Orbitrap, for targeted and untargeted analyses (Tang et al. 2016; Aszyk et al. 2018; Chen et al. 2023). Some LC-based metabolomic studies were also performed with Fourier transform ion cyclotron resonance mass spectrometry (FT-ICR MS) despite its lower scan rate (Scherling et al. 2010; Park et al. 2013; Forcisi et al. 2013; Peng et al. 2014). In addition, given the higher chromatographic resolution and peak capacity of UPLC, the scan speed must be high enough for targeted analyses, which tends to disqualify very high-resolution mass analyzers.

5.3.4 Two-Dimensional LC

Some samples are too complex to have their metabolites separated in one chromatographic dimension, which prevents their accurate detection and quantification. For instance, the more polar compounds will not be retained on a RPLC column. Therefore, two-dimensional LC (2DLC) must be used, requiring different chromatographic techniques, in order to obtain the highest metabolome coverage (Collins et al. 2021). However, the 2DLC setup is not easy, and loss of sensitivity can happen due to dilution effect in the second dimension (Mondello et al. 2008).

Among the 2DLC modes, there are the heart-cutting 2DLC and the comprehensive 2DLC (Lv et al. 2019). The first one is more adapted to targeted analyses by cutting specific peaks within the first dimension to be sent on the second LC column (Venkatramani et al. 2014). In comprehensive 2DLC, all fractions of the first LC dimension are transferred onto the second LC column, which is appropriate for nontargeted metabolomics (Murphy et al. 1998).

There are different existing LC combinations, but RP-HILIC is the most common due to its high orthogonality. It was used for characterization of plants (Fu et al. 2020), human and animal biological (Tang et al. 2016; Lv et al. 2019; Guo et al. 2023), biomedical samples (Basov et al. 2024; Xin et al. 2024), and foods (Hemmler et al. 2018; Aly and Górecki 2024). Interestingly, degree of orthogonality of 23 chromatographic setups, with notably RP and HILIC phases, was determined based the separation of human urine metabolites (Gruebner et al. 2021). Principal component analysis (PCA) was achieved from the retention data of 355 analytes obtained from the 23 chromatographic conditions. According to the PCA, RP columns present less retention time variations compared to the HILIC columns. In fact, the latter columns gather various stationary phases, which explains the higher variability. PCA also shows the orthogonality of the different columns.

Other combinations are possible, such as RPLC x RPLC that is applied to various metabolomic samples (Iguiniz et al. 2019; Liu et al. 2022; Aly and Górecki 2024). Chiral columns can also be combined with RPLC for stereoisomer separation (Aturki et al. 2004). NPLC is also combined with RPLC for a higher metabolome coverage (François et al. 2008). However, the use of organic solvent for NPLC makes difficult the interface either with the RPLC column or with the mass detector. Therefore, an alternative to NPLC is the supercritical fluid chromatography (SFC) discussed in the next section.

5.4 Supercritical Fluid Chromatography

Supercritical fluid chromatography (SFC) technique uses CO_2 in supercritical phase ($SCCO_2$) (from 74 bar and 31 °C) as the main mobile phase, the stationary phase corresponding to what used in conventional LC (Laboureur et al. 2015). Low viscosity of $SCCO_2$ enables faster compound separation compared to HPLC resulting to shorter analysis times and higher chromatographic resolution (Gordillo 2021).

5.4.1 Modifiers and Stationary Phases

When used alone, $SCCO_2$ is nonpolar, similar as hexane (Cajka and Fiehn 2016). But modifiers can be added to the mobile phase to increase the mobile phase polarity. Thus, polar organic solvents, such as methanol, ethanol, and acetonitrile, can be combined to $SCCO_2$ at given percentage (Gordillo 2021). These cosolvents are less harmful and are used in lower quantities than what is used in NPLC and HPLC, respectively, making SFC a more environmentally friendly technique. Water can also be added, with methanol, which was notably showed to improve peak shape in polar compound analyses and repeatability of retention times (Losacco et al. 2020; Gordillo 2021). For nonpolar metabolites, nonconventional modifiers can be added such as dichloromethane, acetone, or toluene (Gordillo 2021).

Additives such as acids, bases, or buffers can be used in small quantity in the mobile phase, often with a modifier, to improve chromatographic performances (van de Velde et al. 2020; Gordillo 2021). They can act as ion-pairing agents and cover active sites of the stationary phase resulting in better peak shape and separation efficiency of some polar compounds. However, use of additives is dependent on the nature of both the targeted metabolites and the stationary phase (Chen et al. 2022).

Recently, effect of four different additives—trifluoroacetic acid, diethylamine, ammonium acetate, and water—with six different stationary phase SFC columns, bare silica, cyano, 2-ethylpyridine, zwitterionic sulfobetaine, octadecylsilane, and pentafluorophenyl, was investigated (Ovchinnikov et al. 2022). The effects of the additives on the SFC retention appeared to be complex and result from a combination of different parameters: type of stationary phase, type and concentration of additive, and amount of cosolvent.

The nature of the stationary phase is indeed relevant for the separation efficiency of compounds by SFC. SFC columns of various stationary phases exist, which can make the column selection complicated. A classification of SFC columns, based on linear solvation energy relationships of 109 analytes on 31 different stationary phases, was proposed (West et al. 2016). There are recent reviews reporting metabolomic studies performed in SFC-MS with stationary phase nature reported (Shulaev and Isaac 2018; van de Velde et al. 2020; Gordillo 2021; Chen et al. 2022). Thus, depending on the stationary and mobile phases, SFC has the capability to separate compounds over a large polarity range compared to HPLC and, combined with MS, to characterize metabolites covering a broad polarity range, from lipids to carbohydrates and amino acids (Cajka and Fiehn 2016; Gordillo 2021).

It is also noteworthy to mention that chiral SFC columns ensure to differentiate and quantify enantiomers, which is relevant as part of pharmaceutical and drugs analysis, insofar as two enantiomers can have different biological activities (Hofstetter et al. 2018; Quaranta et al. 2022). It was also recently used for amino acids differentiation in black vinegar spiked with different L and D amino acids (Konya et al. 2022).

Furthermore, new-generation instruments led to the emergence of ultrahigh-performance SFC-MS (UHPSFC-MS). This technique was recently used for the qualitative analysis of 78 different metabolites classified as nucleosides, biogenic amines, amino acids, sugars, and polar and nonpolar lipids (Antonelli et al. 2022). Eight different stationary phases, influence of additives, and two ionization sources were investigated. Diol column provided the highest number of detected species (67), and higher sensitivity was obtained with ammonium acetate than with ammonium formate. Eventually, ESI allowed observing more metabolites than in APCI.

In a second study, ultrahigh-performance SFC-MS was used for the quantitation of nonpolar and polar lipids from pooled human plasma and serum (Wolrab et al. 2020). The same samples were also subject to HILIC-UHPLC-MS analysis. Results demonstrated that UHPSFC-MS was able to separate and therefore quantify both classes of lipids contrary to HILIC-UHPLC-MS.

5.4.2 Ionization Sources and MS Detection

In SFC, the post-column introduction of a make-up solvent makes hyphenation with MS detectors possible, notably with various ambient and soft ionization sources such as APCI or APPI (Gazárková et al. 2022). ESI remains the most used ionization source as part of SFC-MS analysis and more specifically to metabolomic analysis (van de Velde et al. 2020; Gazárková et al. 2022). Depending on the ionization source, make-up solvent is aimed at increasing metabolite ionization.

5.4.3 Hyphenation with Other Chromatographic Techniques

Due to the versatility of the SFC stationary phases and multiple mobile phase compositions, SFC analyses can be conducted in reversed or normal phase modes. Therefore, SFC can increase peak capacity by being hyphenated with other chromatographic separations, mainly LC, as first or second dimension.

SFC x RPLC is more commonly observed, as an alternative to NPLC x RPLC. Indeed, the pairing between the last two techniques is difficult because of mobile phase immiscibility (Kaplitz et al. 2021). Different pairing combinations were found where SFC was placed before or after LC, online or offline, with MS detection, with applications to various biological matrices (Yang et al. 2020; Kaplitz et al. 2021; Burlet-Parendel and Faure 2021; Chen et al. 2022).

5.5 Capillary Electrophoresis

Another separation technique used for targeted and nontargeted metabolomics is the capillary electrophoresis (CE), also often referred to as capillary zone electrophoresis (CZE) (Dettmer et al. 2007). In CE, separation occurs in a liquid phase, in presence of background electrolyte (BGE), by applying an electric field on the electrophoresis capillary. Mobility of an analyte in CE is determined by both its electrophoretic mobility and electroosmotic flow (Tagliaro et al. 1998). Thus, analytes are separated based on their charge and size, making CE a powerful technique for the analysis of highly polar and ionic metabolites (Lei et al. 2011; Soga 2023; Li et al. 2024b). Therefore, CE applications are covering various fields such as living-system metabolites (Harada and Fukusaki 2009; Wang et al. 2023), food (Castro-Puyana and Marina 2023), and pharmaceuticals (Lei et al. 2011; Maier and Schmitt-Kopplin 2016; Seyfinejad and Jouyban 2022; Soga 2023; Li et al. 2024b).

One of the paramount parameters to consider for CE-MS analysis is the composition of the separation medium. Indeed, its composition depends on the physicochemical properties of the analytes to be separated and the used detection mode (Li et al. 2024b). Commonly, they are composed of salt and buffering agents such as ammonium acetate or sodium phosphate (Seyfinejad and Jouyban 2022). For instance, in positive-mode CE-MS, low-pH BGE are employed, such as formic acid, giving positive charge to cationic metabolites. These conditions are specifically used for the analysis of amino acids, small peptides, nucleobases, and nucleosides (Gosset-Erard et al. 2022; Kasicka 2024; Stefanik et al. 2024; Purohit et al. 2024). On the other hand, the use of neutral or alkaline pH conditions can be of interest to the analysis of anionic metabolites such as phosphorylated compounds or carboxylic acids (Soga 2023). In a recent study, the cationic metabolic profile of plasma samples from both cancer and control patients was studied (López-López et al. 2022). They notably tested two types of CE capillaries, namely, a fused silica and a polyvinyl alcohol (PVA)-coated capillary. The influence of two BGE was also investigated in normal and reversed polarity, which are formic acid and ammonium acetate. Relative standard deviation (RSD) of the relative migration times (RMT) obtained on 37 different cationic standard compounds were calculated for each of the 4 analytical conditions. With both capillaries, ammonium acetate BGE led to the highest RDS values than with formic acid. Regarding the capillary type, PVA one resulted in lower RSD values compared to silica. Results achieved from the optimized CE-MS method enabled to differentiate cancer patients from control ones based on their anionic metabolite profile. They notably found novel metabolites such as N-formyl-L-aspartate, specifically observed in cancer patients.

It is also possible to perform chiral CE for enantiomer separation with chiral selectors such as cyclodextrins (Chankvetadze and Scriba 2023). It can be used for separation of oligonucleotides (Ghassemi et al. 2022), drugs (Tang et al. 2022; Alawadi et al. 2023), imidazole ring, and amino group species (Hu et al. 2022).

Ionic liquids, deep eutectic, and organic solvents can also be used in CE. Each of them has the capability to increase the application of CE (Kumar et al. 2021; Li et al. 2024b).

In addition to ionic metabolite separation, CE has also the capacity to separate neutral compounds. This is performed by adding a surfactant to the BGE, such as sodium dodecyl sulfate (SDS) that forms charged micelles containing the neutral compound. This technique is known as micellar electrokinetic chromatography (MEKC) (Tagliaro et al. 1998). MEKC was recently reported to be used for plant alkaloid analysis (Kumar et al. 2021). However, they may not be compatible with ESI-MS (Seyfinejad and Jouyban 2022). Other types of CE techniques exist that are not discussed here but can be found in reviews (Altria 1999; Jankech et al. 2024).

In some studies, CE-MS was performed and compared with HILIC-MS for metabolome analysis. Regarding method development, HILIC appears to be more tedious due to the wide variety of columns, re-equilibration of the system, or dissolution of sample in specific solvent (Mamani-Huanca et al. 2023). Recently, *C. elegans* metabolomics was carried out by CE- and HILIC-MS. CE-MS was notably showed to be complementary to HILIC-MS, with 41 metabolites observed exclusively in CE-MS (Salzer et al. 2023).

Overall, CE has the potential to rapidly separate metabolites and with a high chromatographic resolution. Also, it requires only a small amount of sample, typically in the range of nanoliters, which is convenient for biomedicals studies (Kartsova and Makeeva 2023) and consumes less solvent than conventional LC techniques (Jankech et al. 2024). However, CE-MS presents some limitations such as lower sensitivity and robustness compared to other chromatographic techniques, as well as some electrochemical reactions of metabolites (Büscher et al. 2009; Soga 2023). Nevertheless, tools for the treatment of CE-MS are being developed to overcome the migration time shifting and metabolite annotation (Mamani-Huanca et al. 2023).

5.5.1 Ionization Sources and MS Detection

CE can be coupled with different types of MS detectors. Most of the CE-MS interfaces, sheath or sheathless, comprise electrospray ionization (ESI) as the analytes are already in liquid phase in the capillary (Schmitt-Kopplin 2008; Soga 2023). In addition, most of the metabolites studied in CE are polar and ionic, which can be easily ionized by ESI. However, to obtain the optimal sampling of the CE peaks, fast acquisition rate MS is required. Therefore, TOF-MS are often employed (Kumar et al. 2021). Also based on the recent reviews, other mass spectrometers are coupled to CE, to perform targeted and untargeted analyses: triple quadrupole, Q-TOF MS, quadrupole ion trap, and ion trap (Seyfinejad and Jouyban 2022; Mamani-Huanca et al. 2023; Castro-Puyana and Marina 2023). Recently, some metabolomic studies were published using Fourier transform mass spectrometry with CE (Sauer et al. 2020; Maruyama et al. 2021; Ly et al. 2022; Gosset-Erard et al. 2023). Apart from ESI, offline coupling of CE with matrix assisted laser desorption ionization (MALDI)-MS were also (Zuberovic et al. 2009; Gasilova et al. 2012; Zhong et al. 2014).

5.6 Ion Mobility Spectrometry

Chromatographic and CE techniques, previously discussed, are capable of distinguishing isomers, but to a certain extent (Zhang et al. 2017). To overcome this limitation and increase metabolomic coverage, ion mobility spectrometry (IMS) is a promising solution. In IMS, gas-phase ions are separated in an electric field in a chamber filled with an inert gas also called collision gas (Gabelica 2022). In fact, there is a proportionality factor between the ion's drift velocity and the strength of the electric field (Kirk et al. 2019).

The last decades have seen the development of hybrid instruments combining ion mobility spectrometry with mass spectrometry (IM-MS). It is therefore possible to separate and detect ions according to their size, shape, and mass, and by extension to distinguish some isomers. There are different IM-MS technologies whose principles are reported in different reviews (Cajka and Fiehn 2016; Delvaux et al. 2022; Gabelica 2022). However, not all of them enable to determine the so-called collision cross section (CCS) (Ieritano et al. 2021; Gabelica 2022). CCS is a parameter that is directly related to the physicochemical properties of the compound crossing the gas chamber. Therefore, IMS is regarded as an orthogonal separation technique to chromatography and allows higher confidence in compound identification, increasing signal-to-noise ratio and peak capacity (Wolfender et al. 2015; Cajka and Fiehn 2016; Zhang et al. 2017).

IM-MS is now used for targeted and nontargeted metabolomic analyses of various samples (Masike et al. 2021; Paglia et al. 2022). It can be used in direct infusion mode or with an upstream chromatographic step (Levy et al. 2019).

Direct IM-MS was used for the metabolic profiling of human blood (Dwivedi et al. 2010). Over 1100 metabolites were detected and 300 isomers separated that were identified as amines, organic acids, carbohydrates, and lipids. In another study, lipids from human plasma were studied by LC x LC-MS and LC-IM-MS (Baglai et al. 2017). Although more features were obtained within the first method (1100 vs. 800), higher peak capacity was obtained with the LC-IM-MS approach (Fig. 5.4). A similar approach was carried out with the use of direct infusion IM-MS, LC-MS, and LC-IM-MS for the profiling of biological matrices (Zhang et al. 2017). In this work, the advantages and limitations of each technique were discussed. For instance, direct infusion led to ionization suppression, while LC-based methods ensured the detection of more metabolites.

Ion separation in IMS occurs in millisecond timescale, making the upstream coupling of a chromatographic device possible, but also requires a fast acquisition rate MS detector. Therefore TOF-MS are often used as part of IM-MS instruments (Delvaux et al. 2022). Hybridization of IMS with MS also opens the possibility to use different ionization sources (Delvaux et al. 2022). However, as previously observed, ESI remains the most used (Kohoutek and Harrington 2023).

Furthermore, if most of the hyphenated IM-MS techniques involve LC, GC can also be hyphenated with IM-MS by means of atmospheric pressure ionization (Ayala-Cabrera et al. 2023a; Ayala-Cabrera et al. 2023b).

5.7 Analytical Improvements in Ion Mobility and Hyphenated Techniques

As previously seen, the implementation of IM-MS with chromatographic technique has multiple advantages, such as the reduction of both ion suppression and co-elution issues and increase of peak capacity, resulting in a higher coverage of the metabolome description. In addition, by acquiring four-dimensional data, namely, retention time, CCS, *m/z*, and MS/MS spectra, higher confidence in metabolite identification and better quantification are obtained. This was notably used for lipidomics in food and biological human matrices (Lerner et al. 2023; Kurilung et al. 2024; Martakos et al. 2024). For instance, olive oil was analyzed by LC-IM-MS, which evidenced new species, notably isomers hitherto unknown that can be used as potential authenticity markers (Drakopoulou et al. 2021; Drakopoulou et al. 2024).

Other studies were performed where 2DLC was coupled with IM-MS that is regarded as four-dimensional separation analysis. Thus, SFC x RPLC-IM-MS/MS was employed for the separation and identification of carotenoids in red

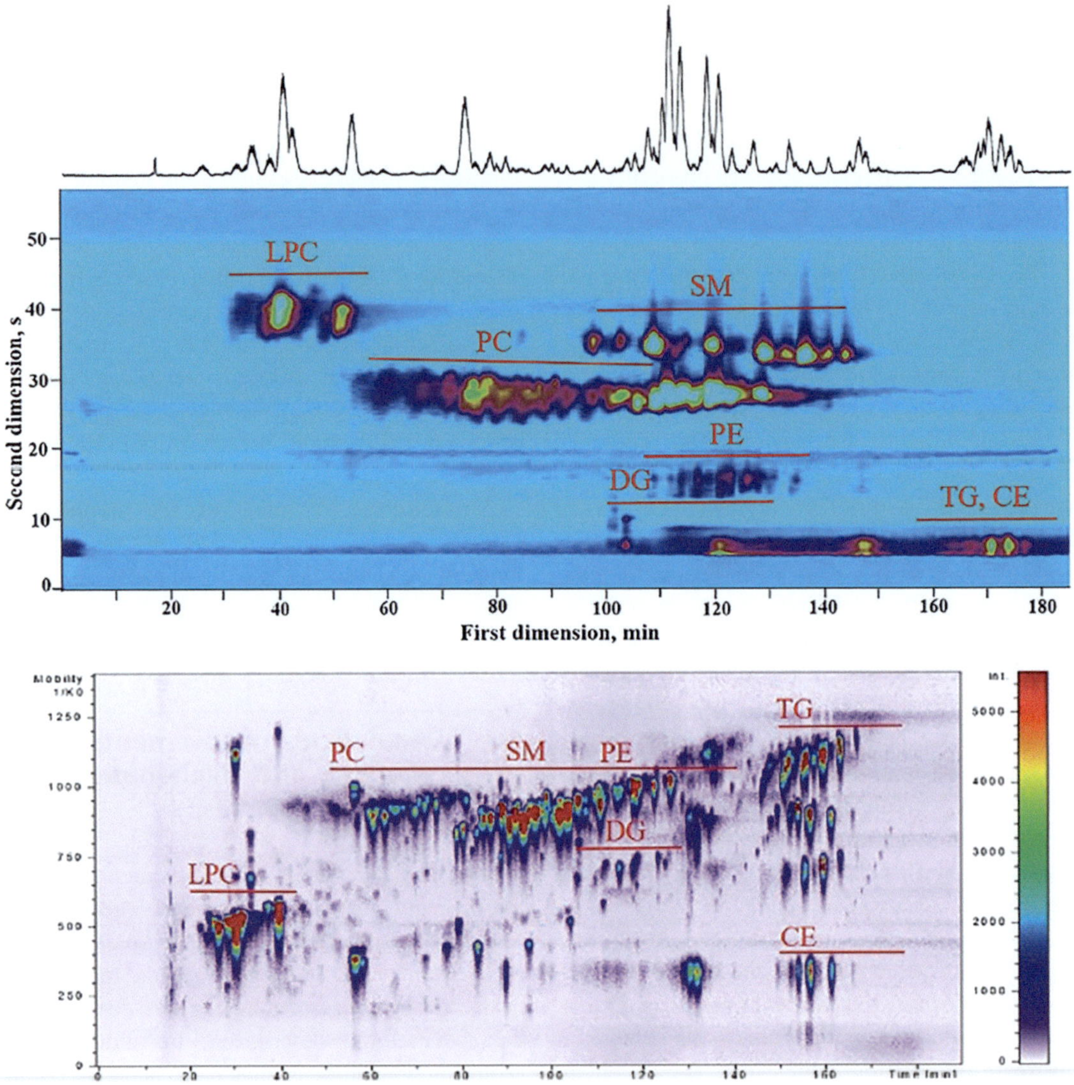

Fig. 5.4 Heatmap of (top) reconstructed chromatogram from RPLC × HILIC-ESI (+)-MS analysis and of (bottom) RPLC-TIMS-MS separation of human plasma. *LPC* lysophosphatidylcholine, *PC* phosphatidylcholine, *SM* sphingomyelin, *PE* phosphatidylethanolamine, *DG* diacylglycerol, *TG* triacylglycerol, *CE* ceramide. (Image retrieved from Baglai et al. (2017))

chili pepper (Donato et al. 2018). Fifty carotenoids were identified, and *cis/trans* isomers were differentiated thanks to the ion mobility separation. *Ginkgo biloba* extract was also subject to 2DLC-IM-MS analysis, with an achieved peak capacity of 8700, notably due to the separation of coeluting isobaric compounds (Stephan et al. 2016).

If the implantation of 1D or 2D chromatography with IM-MS leads to a higher coverage of the metabolome and its identification (Fig. 5.5), the data treatment can be complicated with the handling of data from the various dimensions. Therefore, several tools, in-house or commercial, are developed aimed at assembling these data and detecting peaks, for quantification and

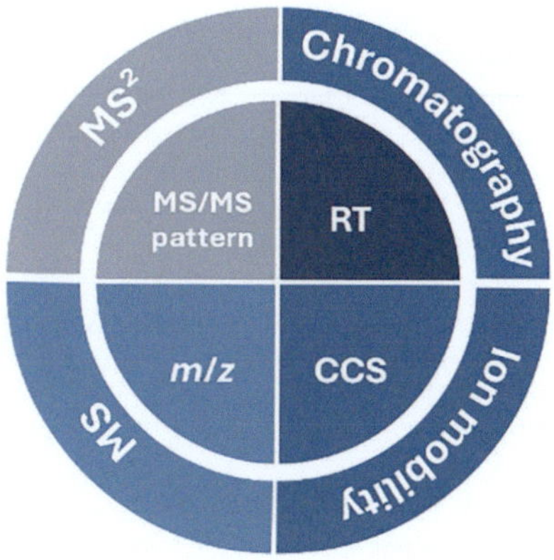

Fig. 5.5 Data dimensions achieved for metabolite identification

identification of metabolites collectively (Zhou et al. 2019; Guerrero Esperanza et al. 2020; Luo et al. 2023; Lou et al. 2024; Kurilung et al. 2024).

5.8 Conclusion

Mass spectrometry is a key technique to fully comprehend the metabolome due to its capacity to analyze a vast number of compounds with high sensitivity, specificity, and accuracy, providing both qualitative and quantitative data. Moreover, through the coupling with different separative techniques, MS offers precise insights into the complex biochemical processes in living organisms.

The possibility of different instrumental combinations, in connection with instrumental development, allows to push boundaries in analytical methodologies enabling a deeper understanding of the different organism's metabolome.

Acknowledgments The present chapter was financially supported by Fundação para a Ciência e a Tecnologia (Portugal) through the exploratory project 2023.14736.PEX (https://doi.org/10.54499/2023.14736.PEX), Marisa Maia CEEC program (2022.07433.CEECIND; https://doi.org/10.54499/2022.07433.CEECIND/CP1715/CT0009) and the Centre grant to BioISI UID/04046/2025 (https://doi.org/10.54499/UID/04046/2025).

We also acknowledge the bilateral cooperation program—Franco-Portuguese Hubert Curien Partnership (Pessoa Program 2024) (Number 50898WC) funded by FCT and CAMPUS FRANCE. Support from the MassLor research infrastructure at the University of Lorraine is also acknowledged.

References

Alawadi M, Fakhari AR, Bayatloo MR, Nojavan S (2023) Carboxymethylated maltodextrin as a chiral selector for the separation of some basic drug enantiomers using capillary electrophoresis. J Chromatogr A 1708:464335. https://doi.org/10.1016/j.chroma.2023.464335

Altria KD (1999) Overview of capillary electrophoresis and capillary electrochromatography. J Chromatogr A 856(1):443–463. https://doi.org/10.1016/S0021-9673(99)00830-4

Aly AA, Górecki T (2024) Two-dimensional liquid chromatography with reversed phase in both dimensions: a review. J Chromatogr A 1721:464824. https://doi.org/10.1016/j.chroma.2024.464824

Anspach JA, Maloney TD, Colón LA (2007) Ultrahigh-pressure liquid chromatography using a 1-mm id column packed with 1.5-μm porous particles. J Sep Sci 30(8):1207–1213. https://doi.org/10.1002/jssc.200600535

Antonelli M, Holčapek M, Wolrab D (2022) Ultrahigh-performance supercritical fluid chromatography—mass spectrometry for the qualitative analysis of metabolites covering a large polarity range. J Chromatogr A 1665:462832. https://doi.org/10.1016/j.chroma.2022.462832

Arcena MR, Kebede B, Leong SY, Silcock P, Oey I (2020) Feasibility of using integrated fingerprinting, profiling and chemometrics approach to understand (bio) chemical changes throughout commercial red winemaking: a case study on merlot. Food Res Int 127:108767. https://doi.org/10.1016/j.foodres.2019.108767

Aszyk J, Byliński H, Namieśnik J, Kot-Wasik A (2018) Main strategies, analytical trends and challenges in LC-MS and ambient mass spectrometry–based metabolomics. TrAC Trends Anal Chem 108:278–295. https://doi.org/10.1016/j.trac.2018.09.010

Aturki Z, Brandi V, Sinibaldi M (2004) Separation of Flavanone-7-O-glycoside Diastereomers and analysis in citrus juices by multidimensional liquid chromatography coupled with mass spectrometry. J Agric Food Chem 52(17):5303–5308. https://doi.org/10.1021/jf0400967

Ayala-Cabrera JF, Montero L, Meckelmann SW, Uteschil F, Schmitz OJ (2023a) Review on atmospheric pressure ionization sources for gas chromatography-mass spectrometry. Part I: current ion source developments and improvements in ionization strategies. Anal Chim Acta 1238:340353. https://doi.org/10.1016/j.aca.2022.340353

Ayala-Cabrera JF, Montero L, Meckelmann SW, Uteschil F, Schmitz OJ (2023b) Review on atmospheric pressure ionization sources for gas chromatography-mass spectrometry. Part II: current applications. Anal Chim Acta 1238:340379. https://doi.org/10.1016/j.aca.2022.340379

Baglai A, Gargano AFG, Jordens J, Mengerink Y, Honing M, van der Wal S, Schoenmakers PJ (2017) Comprehensive lipidomic analysis of human plasma using multidimensional liquid- and gas-phase separations: two-dimensional liquid chromatography–mass spectrometry *vs.* liquid chromatography–trapped-ion-mobility–mass spectrometry. J Chromatogr A 1530:90–103. https://doi.org/10.1016/j.chroma.2017.11.014

Basov NV, Rogachev AD, Aleshkova MA, Gaisler EV, Sotnikova YS, Patrushev YV, Tolstikova TG, Yarovaya OI, Pokrovsky AG, Salakhutdinov NF (2024) Global LC-MS/MS targeted metabolomics using a combination of HILIC and RP LC separation modes on an organic monolithic column based on 1-vinyl-1,2,4-triazole. Talanta 267:125168. https://doi.org/10.1016/j.talanta.2023.125168

Beale DJ, Pinu FR, Kouremenos KA, Poojary MM, Narayana VK, Boughton BA, Kanojia K, Dayalan S, Jones OAH, Dias DA (2018) Review of recent developments in GC–MS approaches to metabolomics-based research. Metabolomics 14(11):152. https://doi.org/10.1007/s11306-018-1449-2

Begou O, Gika HG, Wilson ID, Theodoridis G (2017) Hyphenated MS-based targeted approaches in metabolomics. Analyst 142(17):3079–3100. https://doi.org/10.1039/C7AN00812K

Bojke A, Tkaczuk C, Bauer M, Kamysz W, Gołębiowski M (2020) Application of HS-SPME-GC-MS for the analysis of aldehydes produced by different insect species and their antifungal activity. J Microbiol Methods 169:105835. https://doi.org/10.1016/j.mimet.2020.105835

Burlet-Parendel M, Faure K (2021) Opportunities and challenges of liquid chromatography coupled to supercritical fluid chromatography. TrAC Trends Anal Chem 144:116422. https://doi.org/10.1016/j.trac.2021.116422

Büscher JM, Czernik D, Ewald JC, Sauer U, Zamboni N (2009) Cross-platform comparison of methods for quantitative metabolomics of primary metabolism. Anal Chem 81(6):2135–2143. https://doi.org/10.1021/ac8022857

Cajka T (2024) Liquid chromatography–mass spectrometry–based metabolomics approaches for foodomics research. Curr Opin Food Sci 58:101201. https://doi.org/10.1016/j.cofs.2024.101201

Cajka T, Fiehn O (2016) Toward merging untargeted and targeted methods in mass spectrometry-based metabolomics and Lipidomics. Anal Chem 88(1):524–545. https://doi.org/10.1021/acs.analchem.5b04491

Capellades J, Junza A, Samino S, Brunner JS, Schabbauer G, Vinaixa M, Yanes O (2021) Exploring the use of gas chromatography coupled to chemical ionization mass spectrometry (GC-CI-MS) for stable isotope labeling in metabolomics. Anal Chem 93(3):1242–1248. https://doi.org/10.1021/acs.analchem.0c02998

Castro-Puyana M, Marina ML (2023) Capillary electrophoresis–mass spectrometry-based approaches for food analysis and food metabolomics. Curr Opin Food Sci 54:101088. https://doi.org/10.1016/j.cofs.2023.101088

Chankvetadze B, Scriba GKE (2023) Cyclodextrins as chiral selectors in capillary electrophoresis: recent trends in mechanistic studies. TrAC Trends Anal Chem 160:116987. https://doi.org/10.1016/j.trac.2023.116987

Chen J-H, Xiang W, Cao K-X, Lu X, Yao S-C, Hung D, Huang R-S, Li L-B (2020) Characterization of volatile organic compounds emitted from Endophytic *Burkholderia cenocepacia* ETR-B22 by SPME-GC-MS and their inhibitory activity against various plant fungal pathogens. Molecules 25(17):3765. https://doi.org/10.3390/molecules25173765

Chen L, Li X, Zhang J, He T, Huang J, Zhang Z, Wang Y, Hafeez M, Zhou S, Ren X, Hou Y, Lu Y (2021) Comprehensive metabolome and Volatilome analyses in eggplant and tomato reveal their differential responses to *Tuta absoluta* infestation. Front Plant Sci 12:757230. https://doi.org/10.3389/fpls.2021.757230

Chen M, Wen S-S, Wang R, Ren Q-X, Guo C-W, Li P, Gao W (2022) Advanced development of supercritical fluid chromatography in herbal medicine analysis. Molecules 27(13):4159. https://doi.org/10.3390/molecules27134159

Chen C-J, Lee D-Y, Yu J, Lin Y-N, Lin T-M (2023) Recent advances in LC-MS-based metabolomics for clinical biomarker discovery. Mass Spectrom Rev 42(6):2349–2378. https://doi.org/10.1002/mas.21785

Cirigliano AM, Cabrera GM (2020) Post-column in-source derivatisation in LC-MS: a tool for natural products characterisation and metabolomics. Phytochem Anal 31(5):606–615. https://doi.org/10.1002/pca.2926

Collins SL, Koo I, Peters JM, Smith PB, Patterson AD (2021) Current challenges and recent developments in mass spectrometry–based metabolomics. Annu Rev Anal Chem 14:467–487. https://doi.org/10.1146/annurev-anchem-091620-015205

Commisso M, Anesi A, Dal Santo S, Guzzo F (2017) Performance comparison of electrospray ionization and atmospheric pressure chemical ionization in untargeted and targeted liquid chromatography/mass spectrometry based metabolomics analysis of grapeberry metabolites. Rapid Commun Mass Spectrom 31(3):292–300. https://doi.org/10.1002/rcm.7789

Cuevas-Delgado P, Warmuzinska N, Luczykowski K, Bojko B, Barbas C (2024) Exploring sample treatment strategies for untargeted metabolomics: a comparative study of solid phase microextraction (SPME) and homogenization with solid-liquid extraction (SLE) in renal tissue. Anal Chim Acta 1312:342758. https://doi.org/10.1016/j.aca.2024.342758

Delvaux A, Rathahao-Paris E, Alves S (2022) Different ion mobility-mass spectrometry coupling techniques to promote metabolomics. Mass Spectrom Rev 41(5):695–721. https://doi.org/10.1002/mas.21685

Deng Y, Kan H, Li Y, Liu Y, Qiu X (2023) Analysis of volatile components in *Rosa roxburghii* Tratt. And *Rosa sterilis* using headspace-solid-phase microextraction-gas chromatography-mass spectrometry. Molecules 28(23):7879. https://doi.org/10.3390/molecules28237879

Dettmer K, Aronov PA, Hammock BD (2007) Mass spectrometry-based metabolomics. Mass Spectrom Rev 26(1):51–78. https://doi.org/10.1002/mas.20108

Donato P, Giuffrida D, Oteri M, Inferrera V, Dugo P, Mondello L (2018) Supercritical fluid chromatography × ultra-high pressure liquid chromatography for red Chilli pepper fingerprinting by photodiode Array, quadrupole-time-of-flight and ion mobility mass spectrometry (SFC × RP-UHPLC-PDA-Q-ToF MS-IMS). Food Anal Methods 11(12):3331–3341. https://doi.org/10.1007/s12161-018-1307-x

Dorakumbura BN, Beckett NM, Cook EC, Nolan A-N, Brown D, Douglas B (2024) Emerging human medications in racing animals: analytical and regulatory challenges. Drug Test Anal. https://doi.org/10.1002/dta.3805

Drakopoulou SK, Damalas DE, Baessmann C, Thomaidis NS (2021) Trapped ion mobility incorporated in LC–HRMS workflows as an integral analytical platform of high sensitivity: targeted and untargeted 4D-metabolomics in extra virgin olive oil. J Agric Food Chem 69(51):15728–15737. https://doi.org/10.1021/acs.jafc.1c04789

Drakopoulou SK, Kritikou AS, Baessmann C, Thomaidis NS (2024) Untargeted 4D-metabolomics using trapped ion mobility combined with LC-HRMS in extra virgin olive oil adulteration study with lower-quality olive oils. Food Chem 434:137410. https://doi.org/10.1016/j.foodchem.2023.137410

Duangkumpha K, Jariyasopit N, Wanichthanarak K, Dhakal E, Wisanpitayakorn P, Thotsiri S, Sirivatanauksorn Y, Kitiyakara C, Sathirapongsasuti N, Khoomrung S (2022) GC × GC-TOFMS metabolomics analysis identifies elevated levels of plasma sugars and sugar alcohols in diabetic mellitus patients with kidney failure. J Biol Chem 298(10):102445. https://doi.org/10.1016/j.jbc.2022.102445

Dwivedi P, Schultz AJ, Jr HHH (2010) Metabolic profiling of human blood by high-resolution ion mobility mass spectrometry (IM-MS). Int J Mass Spectrom 298(1):78–90. https://doi.org/10.1016/j.ijms.2010.02.007

Escolà Casas M, Matamoros V (2022) Linking plant-root exudate changes to micropollutant exposure in aquatic plants (*Lemna minor* and *Salvinia natans).* A prospective metabolomic study. Chemosphere 287:132056. https://doi.org/10.1016/j.chemosphere.2021.132056

Evans AM, DeHaven CD, Barrett T, Mitchell M, Milgram E (2009) Integrated, nontargeted ultrahigh performance liquid chromatography/electrospray ionization tandem mass spectrometry platform for the identification and relative quantification of the small-molecule complement of biological systems. Anal Chem 81(16):6656–6667. https://doi.org/10.1021/ac901536h

Farag MA, Ryu C-M, Sumner LW, Pare PW (2006) GC-MS SPME profiling of rhizobacterial volatiles reveals prospective inducers of growth promotion and induced systemic resistance in plants. Phytochemistry 67(20):2262–2268. https://doi.org/10.1016/j.phytochem.2006.07.021

Fiehn O (2016) Metabolomics by gas chromatography-mass spectrometry: the combination of targeted and untargeted profiling. Curr Protoc Mol Biol 114:30.4.1–30.4.32. https://doi.org/10.1002/0471142727.mb3004s114

Forcisi S, Moritz F, Kanawati B, Tziotis D, Lehmann R, Schmitt-Kopplin P (2013) Liquid chromatography–mass spectrometry in metabolomics research: mass analyzers in ultra high pressure liquid chromatography coupling. J Chromatogr A 1292:51–65. https://doi.org/10.1016/j.chroma.2013.04.017

François I, de Villiers A, Tienpont B, David F, Sandra P (2008) Comprehensive two-dimensional liquid chromatography applying two parallel columns in the second dimension. J Chromatogr A 1178(1):33–42. https://doi.org/10.1016/j.chroma.2007.11.032

Fu Q, Ke Y, Jiang D, Jin Y (2020) Chemical separation and characterization of complex samples with herbal medicine. TrAC Trends Anal Chem 124:115775. https://doi.org/10.1016/j.trac.2019.115775

Furdíková K, Machyňáková A, Drtilová T, Špánik I (2020) Comparison of different categories of Slovak Tokaj wines in terms of profiles of volatile organic compounds. Molecules 25(3):669. https://doi.org/10.3390/molecules25030669

Gabelica V (2022) Ion mobility-mass spectrometry: an overview. In: Ashcroft AE, Sobott F (eds) Ion mobility-mass spectrometry: fundamentals and applications. Royal Soc Chemistry, Cambridge, pp 1–25

Gamboa-Becerra R, Desgarennes D, Molina-Torres J, Ramirez-Chavez E, Kiel-Martinez AL, Carrion G, Ortiz-Castro R (2022) Plant growth-promoting and non-promoting rhizobacteria from avocado trees differentially emit volatiles that influence growth of *Arabidopsis thaliana*. Protoplasma 259(4):835–854. https://doi.org/10.1007/s00709-021-01705-2

Gasilova N, Gassner A-L, Girault HH (2012) Analysis of major milk whey proteins by immunoaffinity capillary electrophoresis coupled with MALDI-MS. Electrophoresis 33(15):2390–2398. https://doi.org/10.1002/elps.201200079

Gazárková T, Plachká K, Svec F, Nováková L (2022) Current state of supercritical fluid chromatography-mass spectrometry. TrAC Trends Anal Chem 149:116544. https://doi.org/10.1016/j.trac.2022.116544

Ghafari N, Sleno L (2024) Challenges and recent advances in quantitative mass spectrometry-based metabolomics. Analytic Sci Adv 5(5–6):e2400007. https://doi.org/10.1002/ansa.202400007

Ghassemi KM, Demelenne A, Crommen J, Servais A-C, Fillet M (2022) Improvement of chemo- and stereoselectivity for phosphorothioate oligonucleotides in capillary electrophoresis by addition of cyclodextrins. J Chromatogr A 1676:463270. https://doi.org/10.1016/j.chroma.2022.463270

Gómez-Romero M, Segura-Carretero A, Fernández-Gutiérrez A (2010) Metabolite profiling and quantification of phenolic compounds in methanol extracts of tomato fruit. Phytochemistry 71(16):1848–1864. https://doi.org/10.1016/j.phytochem.2010.08.002

Gordillo R (2021) Supercritical fluid chromatography hyphenated to mass spectrometry for metabolomics applications. J Sep Sci 44(1):448–463. https://doi.org/10.1002/jssc.202000805

Gosset-Erard C, Lechner A, Wolff P, Aubriet F, Leize-Wagner E, Chaimbault P, Francois Y-N (2022) Optimization of nucleotides dephosphorylation for RNA structural characterization by tandem mass spectrometry hyphenated with separation methods. J Chromatogr B 1208:123396. https://doi.org/10.1016/j.jchromb.2022.123396

Gosset-Erard C, Aubriet F, Leize-Wagner E, François Y-N, Chaimbault P (2023) Hyphenation of Fourier transform ion cyclotron resonance mass spectrometry (FT-ICR MS) with separation methods: the art of compromises and the possible—a review. Talanta 257:124324. https://doi.org/10.1016/j.talanta.2023.124324

Gruebner M, Dunkel A, Steiner F, Hofmann T (2021) Systematic evaluation of liquid chromatography (LC) column combinations for application in two-dimensional LC Metabolomic studies. Anal Chem 93(37):12565–12573. https://doi.org/10.1021/acs.analchem.1c01857

Guerrero Esperanza M, Wrobel K, Gomez Ojeda A, Eugenia Garay Sevillo M, Corrales Escobosa AR, Barrientos EY, Wrobel K (2020) Liquid chromatography-mass spectrometry untargeted metabolomics reveals increased levels of tryptophan indole metabolites in urine of metabolic syndrome patients. Eur J Mass Spectrom 26(6):379–387. https://doi.org/10.1177/1469066720964632

Guillarme D, Ruta J, Rudaz S, Veuthey J-L (2010) New trends in fast and high-resolution liquid chromatography: a critical comparison of existing approaches. Anal Bioanal Chem 397(3):1069–1082. https://doi.org/10.1007/s00216-009-3305-8

Guo R, Zhong Q, Liu J, Bai P, Wang Z, Kou J, Chen P, Zhang J, Zhang B (2023) Polarity-extended liquid chromatography-triple quadrupole mass spectrometry for simultaneous hydrophilic and hydrophobic metabolite analysis. Anal Chim Acta 1277:341655. https://doi.org/10.1016/j.aca.2023.341655

Han S, Guiberson ER, Li Y, Sonnenburg JL (2024) High-throughput identification of gut microbiome-dependent metabolites. Nat Protoc 19(7):2180–2205. https://doi.org/10.1038/s41596-024-00980-6

Harada K, Fukusaki E (2009) Profiling of primary metabolite by means of capillary electrophoresis-mass spectrometry and its application for plant science. Plant Biotechnol 26(1):47–52. https://doi.org/10.5511/plantbiotechnology.26.47

Hazrati H, Kudsk P, Ding L, Uthe H, Fomsgaard IS (2022) Integrated LC-MS and GC-MS-based metabolomics reveal the effects of plant competition on the Rye metabolome. J Agric Food Chem 70(9):3056–3066. https://doi.org/10.1021/acs.jafc.1c06306

He L, Hu Q, Zhang J, Xing R, Zhao Y, Yu N, Chen Y (2023) An integrated untargeted metabolomic approach reveals the quality characteristics of black soybeans from different geographical origins in China. Food Res Int 169:112908. https://doi.org/10.1016/j.foodres.2023.112908

Hemmer S, Manier SK, Wagmann L, Meyer MR (2024) Comparison of reversed-phase, hydrophilic interaction, and porous graphitic carbon chromatography columns for an untargeted toxicometabolomics study in pooled human liver microsomes, rat urine, and rat plasma. Metabolomics 20(3):49. https://doi.org/10.1007/s11306-024-02115-0

Hemmler D, Heinzmann SS, Woehr K, Schmitt-Kopplin P, Witting M (2018) Tandem HILIC-RP liquid chromatography for increased polarity coverage in food analysis. Electrophoresis 39(13):1645–1653. https://doi.org/10.1002/elps.201800038

Hofstetter R, Fassauer GM, Link A (2018) Supercritical fluid extraction (SFE) of ketamine metabolites from dried urine and on-line quantification by supercritical fluid chromatography and single mass detection (on-line SFE–SFC–MS). J Chromatogr B 1076:77–83. https://doi.org/10.1016/j.jchromb.2018.01.024

Hooshmand K, Xu J, Simonsen AH, Wretlind A, de Zawadzki A, Sulek K, Hasselbalch SG, Legido-Quigley C (2021) Human cerebrospinal fluid sample preparation and annotation for integrated Lipidomics and metabolomics profiling studies. Mol Neurobiol 61(4):2021–2032. https://doi.org/10.1007/s12035-023-03666-4

Hu B, Ouyang G (2021) *In situ* solid phase microextraction sampling of analytes from living human objects for mass spectrometry analysis. TrAC Trends Anal Chem 143:116368. https://doi.org/10.1016/j.trac.2021.116368

Hu C, Li Z, Hu Z, Li Q, Fu Y, Chen Z (2022) Synthesis of multifunctional crown ether covalent organic nanospheres as stationary phase for capillary electrochromatography. J Chromatogr A 1677:463323. https://doi.org/10.1016/j.chroma.2022.463323

Hurtado-Fernández E, Pacchiarotta T, Longueira-Suárez E, Mayboroda OA, Fernández-Gutiérrez A, Carrasco-

Pancorbo A (2013) Evaluation of gas chromatography-atmospheric pressure chemical ionization-mass spectrometry as an alternative to gas chromatography-electron ionization-mass spectrometry: avocado fruit as example. J Chromatogr A 1313:228–244. https://doi.org/10.1016/j.chroma.2013.08.084

Ieritano C, Lee A, Crouse J, Bowman Z, Mashmoushi N, Crossley PM, Friebe BP, Campbell JL, Hopkins WS (2021) Determining collision cross sections from differential ion mobility spectrometry. Anal Chem 93(25):8937–8944. https://doi.org/10.1021/acs.analchem.1c01420

Iguiniz M, Corbel E, Roques N, Heinisch S (2019) Quantitative aspects in on-line comprehensive two-dimensional liquid chromatography for pharmaceutical applications. Talanta 195:272–280. https://doi.org/10.1016/j.talanta.2018.11.030

Jalili V, Barkhordari A, Ghiasvand A (2020) A comprehensive look at solid-phase microextraction technique: a review of reviews. Microchem J 152:104319. https://doi.org/10.1016/j.microc.2019.104319

Jankech T, Gerhardtova I, Stefanik O, Chalova P, Jampilek J, Majerova P, Kovac A, Piestansky J (2024) Current green capillary electrophoresis and liquid chromatography methods for analysis of pharmaceutical and biomedical samples (2019–2023)—a review. Anal Chim Acta 1323:342889. https://doi.org/10.1016/j.aca.2024.342889

Jemmali Z, Chartier A, Dufresne C, Elfakir C (2016) Optimization of the derivatization protocol of pentacyclic triterpenes prior to their gas chromatography–mass spectrometry analysis in plant extracts. Talanta 147:35–43. https://doi.org/10.1016/j.talanta.2015.09.026

Kanani HH, Klapa MI (2007) Data correction strategy for metabolomics analysis using gas chromatography–mass spectrometry. Metab Eng 9(1):39–51. https://doi.org/10.1016/j.ymben.2006.08.001

Kanu AB (2021) Recent developments in sample preparation techniques combined with high-performance liquid chromatography: a critical review. J Chromatogr A 1654:462444. https://doi.org/10.1016/j.chroma.2021.462444

Kaplitz AS, Mostafa ME, Calvez SA, Edwards JL, Grinias JP (2021) Two-dimensional separation techniques using supercritical fluid chromatography. J Sep Sci 44(1):426–437. https://doi.org/10.1002/jssc.202000823

Kartsova LA, Makeeva DV (2023) Biomedical applications of capillary electrophoresis. J Anal Chem 78(10):1362–1377. https://doi.org/10.1134/S1061934823100118

Kasicka V (2024) Recent developments in capillary and microchip electroseparations of peptides (2021-mid-2023). Electrophoresis 45(1–2):165–198. https://doi.org/10.1002/elps.202300152

Kataoka H, Lord HL, Pawliszyn J (2000) Applications of solid-phase microextraction in food analysis. J Chromatogr A 880(1):35–62. https://doi.org/10.1016/S0021-9673(00)00309-5

Khodadadi M, Pourfarzam M (2020) A review of strategies for untargeted urinary metabolomic analysis using gas chromatography–mass spectrometry. Metabolomics 16(6):66. https://doi.org/10.1007/s11306-020-01687-x

Kirk AT, Bohnhorst A, Raddatz C-R, Allers M, Zimmermann S (2019) Ultra-high-resolution ion mobility spectrometry—current instrumentation, limitations, and future developments. Anal Bioanal Chem 411(24):6229–6246. https://doi.org/10.1007/s00216-019-01807-0

Kiseleva O, Kurbatov I, Ilgisonis E, Poverennaya E (2022) Defining blood plasma and serum metabolome by GC-MS. Meta 12(1):15. https://doi.org/10.3390/metabo12010015

Koek MM, Jellema RH, van der Greef J, Tas AC, Hankemeier T (2011) Quantitative metabolomics based on gas chromatography mass spectrometry: status and perspectives. Metabolomics 7(3):307–328. https://doi.org/10.1007/s11306-010-0254-3

Kohoutek KM, Harrington P d B (2023) Electrospray ionization ion mobility mass spectrometry. Crit Rev Anal Chem 53(3):483–497. https://doi.org/10.1080/10408347.2021.1964938

Konya Y, Izumi Y, Hamase K, Bamba T (2022) Ultrafast simultaneous chiral analysis of native amino acid enantiomers using supercritical fluid chromatography/tandem mass spectrometry. J Chromatogr A 1677:463305. https://doi.org/10.1016/j.chroma.2022.463305

Kuehnbaum NL, Britz-McKibbin P (2013) New advances in separation science for metabolomics: resolving chemical diversity in a post-genomic era. Chem Rev 113(4):2437–2468. https://doi.org/10.1021/cr300484s

Kumar A, Malik AK, Singh B (2021) Recent advances in the analysis of plant alkaloids by capillary electrophoresis and micellar electrokinetic chromatography. Sep Sci Plus 4(11–12):401–412. https://doi.org/10.1002/sscp.202100040

Kurilung A, Limjiasahapong S, Kaewnarin K, Wisanpitayakorn P, Jariyasopit N, Wanichthanarak K, Sartyoungkul S, Wong SCC, Sathirapongsasuti N, Kitiyakara C, Sirivatanauksorn Y, Khoomrung S (2024) Measurement of very low-molecular weight metabolites by traveling wave ion mobility and its use in human urine samples. J Pharm Anal 14(5):100921. https://doi.org/10.1016/j.jpha.2023.12.011

Laboureur L, Ollero M, Touboul D (2015) Lipidomics by supercritical fluid chromatography. Int J Mol Sci 16(6):13868–13884. https://doi.org/10.3390/ijms160613868

Lang J, Ramos SE, Smohunova M, Bigler L, Schuman MC (2024) Screening of leaf extraction and storage conditions for eco-metabolomics studies. Plant Direct 8(4):e578. https://doi.org/10.1002/pld3.578

Lebanov L, Ghiasvand A, Paull B (2021) Data handling and data analysis in metabolomic studies of essential oils using GC-MS. J Chromatogr A 1640:461896. https://doi.org/10.1016/j.chroma.2021.461896

Lei Z, Huhman DV, Sumner LW (2011) Mass spectrometry strategies in metabolomics. J Biol Chem 286(29): 25435–25442. https://doi.org/10.1074/jbc.R111.238691

Lerner R, Baker D, Schwitter C, Neuhaus S, Hauptmann T, Post JM, Kramer S, Bindila L (2023) Four-dimensional trapped ion mobility spectrometry lipidomics for high throughput clinical profiling of human blood samples. Nat Commun 14(1):937. https://doi.org/10.1038/s41467-023-36520-1

Levy AJ, Oranzi NR, Ahmadireskety A, Kemperman RHJ, Wei MS, Yost RA (2019) Recent progress in metabolomics using ion mobility-mass spectrometry. TrAC Trends Anal Chem 116:274–281. https://doi.org/10.1016/j.trac.2019.05.001

Li C, Li M, Zhao Z, Khan A, Zhao T, Liu Y, Wang Z, Cheng G (2024a) The influences of extraction methods on the chemical constituents of *Lyonia ovalifolia* (wall.) Drude and intracellular protective effects based on metabolomics analysis. Food Chem 456: 140031. https://doi.org/10.1016/j.foodchem.2024.140031

Li Y, Miao S, Tan J, Zhang Q, Chen DDY (2024b) Capillary electrophoresis: a three-year literature review. Anal Chem 96(20):7799–7816. https://doi.org/10.1021/acs.analchem.4c00857

Liu D, Liu Y, Shen A, Li X, Yu L, Wang C, Liang X (2022) Analysis of alkaloids in *Gelsemium elegans* Benth using an online heart-cutting + comprehensive RPLC×RPLC system tandem mass spectrometry. Talanta 239:123069. https://doi.org/10.1016/j.talanta.2021.123069

Liu H, Wu S, Liu M, Sun Y, Li J (2024) Exploring flavour profile changes of strawberry fruits under different organic fertilizer strategies by combining HS-SPME-GC-MS and metabolomics. J Food Compos Anal 135: 106653. https://doi.org/10.1016/j.jfca.2024.106653

López-López Á, Ciborowski M, Niklinski J, Barbas C, López-Gonzálvez Á (2022) Optimization of capillary electrophoresis coupled to negative mode electrospray ionization-mass spectrometry using polyvinyl alcohol coated capillaries. Application to a study on non-small cell lung cancer. Anal Chim Acta 1226:340259. https://doi.org/10.1016/j.aca.2022.340259

Losacco GL, Marconetto E, Nicoli R, Kuuranne T, Boccard J, Rudaz S, Veuthey J-L, Guillarme D (2020) Supercritical fluid chromatography–mass spectrometry in routine anti-doping analyses: estimation of retention time variability under reproducible conditions. J Chromatogr A 1616:460780. https://doi.org/10.1016/j.chroma.2019.460780

Lou J, Xu X, Xu B, Wang H, Li X, Sun H, Zheng X, Zhou J, Zou Y, Wu H, Wang Y, Yang W (2024) Comprehensive metabolome characterization and comparison between two sources of Dragon's blood by integrating liquid chromatography/mass spectrometry and chemometrics. Anal Bioanal Chem 416(7): 1571–1587. https://doi.org/10.1007/s00216-024-05159-2

Luo M, Yin Y, Zhou Z, Zhang H, Chen X, Wang H, Zhu Z-J (2023) A mass spectrum-oriented computational method for ion mobility-resolved untargeted metabolomics. Nat Commun 14(1):1813. https://doi.org/10.1038/s41467-023-37539-0

Lv W, Shi X, Wang S, Xu G (2019) Multidimensional liquid chromatography-mass spectrometry for metabolomic and lipidomic analyses. TrAC Trends Anal Chem 120:115302. https://doi.org/10.1016/j.trac.2018.11.001

Ly R, Ly N, Sasaki K, Suzuki M, Kami K, Ohashi Y, Britz-McKibbin P (2022) Nontargeted Serum Lipid Profiling of Nonalcoholic Steatohepatitis by Multisegment Injection–Nonaqueous Capillary Electrophoresis–Mass Spectrometry: A Multiplexed Separation Platform for Resolving Ionic Lipids. J Proteome Res 21(3):768–777. https://doi.org/10.1021/acs.jproteome.1c00682

Ma B, Qi H, Li J, Xu H, Chi B, Zhu J, Yu L, An G, Zhang Q (2015) Triptolide disrupts fatty acids and peroxisome proliferator-activated receptor (PPAR) levels in male mice testes followed by testicular injury: a GC–MS based metabolomics study. Toxicology 336:84–95. https://doi.org/10.1016/j.tox.2015.07.008

Maier TV, Schmitt-Kopplin P (2016) Capillary electrophoresis in metabolomics. In: SchmittKopplin P (ed) Capillary electrophoresis: methods and protocols, 2nd edn. Humana Press Inc, Totowa, pp 437–470

Mamani-Huanca M, Villaseñor A, Gonzalez-Riano C, López-López Á, López-Gonzálvez Á, Barbas C (2023) Capillary electrophoresis mass spectrometry-based untargeted metabolomics to approach disease diagnosis. TrAC Trends Anal Chem 162:117049. https://doi.org/10.1016/j.trac.2023.117049

Marinaki M, Mouskeftara T, Arapitsas P, Zinoviadou KG, Theodoridis G (2023) Metabolic fingerprinting of Muscat of Alexandria grape musts during industrial alcoholic fermentation using HS-SPME and liquid injection with TMS Derivatization GC-MS methods. Molecules 28(12):4653. https://doi.org/10.3390/molecules28124653

Martakos IC, Tzavellas IF, Dasenaki ME, Thomaidis NS (2024) Food Lipidomics: development and application of a cutting-edge untargeted 4D HRMS workflow for the lipid profiling of food of animal origin. J Food Compos Anal 131:106232. https://doi.org/10.1016/j.jfca.2024.106232

Maruyama A, Sato Y, Nakayama J, Murai J, Ishikawa T, Soga T, Makinoshima H (2021) De novo deoxyribonucleotide biosynthesis regulates cell growth and tumor progression in small-cell lung carcinoma. Sci Rep 11(1):13474. https://doi.org/10.1038/s41598-021-92948-9

Masike K, Stander MA, de Villiers A (2021) Recent applications of ion mobility spectrometry in natural product research. J Pharm Biomed Anal 195:113846. https://doi.org/10.1016/j.jpba.2020.113846

Mhlongo M, Piater LA, Dubery IA (2022) Profiling of volatile organic compounds from four plant growth-promoting Rhizobacteria by SPME-GC-MS: a metabolomics study. Meta 12(8):763. https://doi.org/10.3390/metabo12080763

Misra BB, Olivier M (2020) High resolution GC-Orbitrap-MS metabolomics using both electron ionization and chemical ionization for analysis of human plasma. J Proteome Res 19(7):2717–2731. https://doi.org/10.1021/acs.jproteome.9b00774

Młynarczyk M, Belka M, Hewelt-Belka W (2024) Novel materials and approaches for solid-phase (micro) extraction in LC-MS-based metabolomics. TrAC Trends Anal Chem 180:117941. https://doi.org/10.1016/j.trac.2024.117941

Mondello L, Herrero M, Kumm T, Dugo P, Cortes H, Dugo G (2008) Quantification in comprehensive two-dimensional liquid chromatography. Anal Chem 80(14):5418–5424. https://doi.org/10.1021/ac800484y

Muñoz-Redondo JM, Ruiz-Moreno MJ, Puertas B, Cantos-Villar E, Moreno-Rojas JM (2020) Multivariate optimization of headspace solid-phase microextraction coupled to gas chromatography-mass spectrometry for the analysis of terpenoids in sparkling wines. Talanta 208:120483. https://doi.org/10.1016/j.talanta.2019.120483

Murphy RE, Schure MR, Foley JP (1998) Effect of sampling rate on resolution in comprehensive two-dimensional liquid chromatography. Anal Chem 70(8):1585–1594. https://doi.org/10.1021/ac971184b

Mushtaq MY, Choi YH, Verpoorte R, Wilson EG (2014) Extraction for metabolomics: access to the metabolome. Phytochem Anal 25(4):291–306. https://doi.org/10.1002/pca.2505

Nordström A, O'Maille G, Qin C, Siuzdak G (2006) Nonlinear data alignment for UPLC−MS and HPLC−MS based metabolomics: quantitative analysis of endogenous and exogenous metabolites in human serum. Anal Chem 78(10):3289–3295. https://doi.org/10.1021/ac060245f

Ovbude ST, Sharmeen S, Kyei I, Olupathage H, Jones J, Bell RJ, Powers R, Hage DS (2024) Applications of chromatographic methods in metabolomics: a review. J Chromatogr B 1239:124124. https://doi.org/10.1016/j.jchromb.2024.124124

Ovchinnikov DV, Ul'yanovskii NV, Kosyakov DS, Pokrovskiy OI (2022) Some aspects of additives effects on retention in supercritical fluid chromatography studied by linear free energy relationships method. J Chromatogr A 1665:462820. https://doi.org/10.1016/j.chroma.2022.462820

Owczarzy A, Kulig K, Piordas K, Piśla P, Sarkowicz P, Rogóż W, Maciążek-Jurczyk M (2024) Solid-phase microextraction—a future technique in pharmacology and coating trends. Anal Methods 16(20):3164–3178. https://doi.org/10.1039/D4AY00187G

Paglia G, Smith AJ, Astarita G (2022) Ion mobility mass spectrometry in the omics era: challenges and opportunities for metabolomics and lipidomics. Mass Spectrom Rev 41(5):722–765. https://doi.org/10.1002/mas.21686

Park KH, Kim MS, Baek SJ, Bae IH, Seo S-W, Kim J, Shin YK, Lee Y-M, Kim HS (2013) Simultaneous molecular formula determinations of natural compounds in a plant extract using 15 T Fourier transform ion cyclotron resonance mass spectrometry. Plant Methods 9(1):15. https://doi.org/10.1186/1746-4811-9-15

Pavagadhi S, Swarup S (2020) Metabolomics for evaluating flavor-associated metabolites in plant-based products. Meta 10(5):197. https://doi.org/10.3390/metabo10050197

Peng J, St Laurent CD, Befus AD, Zhou R, Li L (2014) Metabolomic profiling of bronchoalveolar lavage fluids by isotope labeling liquid chromatography mass spectrometry: a promising approach to studying experimental asthma. Metabolomics 10(6):1305–1317. https://doi.org/10.1007/s11306-014-0667-5

Pérez-Jiménez M, Sherman E, Pozo-Bayón MA, Pinu FR (2021) Application of untargeted volatile profiling and data driven approaches in wine flavoromics research. Food Res Int 145:110392. https://doi.org/10.1016/j.foodres.2021.110392

Popov RS, Ivanchina NV, Kicha AA, Malyarenko TV, Dmitrenok PS, Stonik VA (2015) LC-ESI MS/MS profiling of polar steroid metabolites of the far eastern starfish Patiria (=Asterina) pectinifera. Metabolomics 12(2):21. https://doi.org/10.1007/s11306-015-0939-8

Poria V, Jhilta P, Adhuna KP, Somvanshi VS, Rana A, Singh S (2024) Multistress-tolerant *Pseudomonas aeruginosa* SSVP3 uses multiple strategies to control *Meloidogyne incognita* in tomato. Plant Pathol. https://doi.org/10.1111/ppa.13974

Purohit K, Reddy N, Sunna A (2024) Exploring the potential of bioactive peptides: from natural sources to therapeutics. Int J Mol Sci 25(3):1391. https://doi.org/10.3390/ijms25031391

Quaranta A, Zöhrer B, Revol-Cavalier J, Benkestock K, Balas L, Oger C, Keyes GS, Wheelock ÅM, Durand T, Galano J-M, Ramsden CE, Hamberg M, Wheelock CE (2022) Development of a chiral supercritical fluid chromatography–tandem mass spectrometry and reversed-phase liquid chromatography–tandem mass spectrometry platform for the quantitative metabolic profiling of octadecanoid oxylipins. Anal Chem 94(42):14618–14626. https://doi.org/10.1021/acs.analchem.2c02601

Queiroz MEC, de Souza ID, de Oliveira IG, Grecco CF (2022) *In vivo* solid phase microextraction for bioanalysis. TrAC Trends Anal Chem 153:116656. https://doi.org/10.1016/j.trac.2022.116656

Queiroz EF, Guillarme D, Wolfender J-L (2024) Advanced high-resolution chromatographic strategies for efficient isolation of natural products from complex biological matrices: from metabolite profiling to pure chemical entities. Phytochem Rev. https://doi.org/10.1007/s11101-024-09928-w

Rodeo AJD, Mitcham EJ (2023) Chilling temperatures and controlled atmospheres alter key volatile compounds implicated in basil aroma and flavor. Front Plant Sci 14:1218734. https://doi.org/10.3389/fpls.2023.1218734

Rodríguez-Sánchez S, Hernández-Hernández O, Ruiz-Matute AI, Sanz ML (2011) A derivatization procedure for the simultaneous analysis of iminosugars and other low molecular weight carbohydrates by GC–MS in mulberry (*Morus sp.*). Food Chem 126(1):353–359. https://doi.org/10.1016/j.foodchem.2010.10.097

Ryparova Kvirencova J, Navratilova K, Hrbek V, Hajslova J (2023) Detection of botanical adulterants in saffron powder. Anal Bioanal Chem 415(23):5723–5734. https://doi.org/10.1007/s00216-023-04853-x

Saigusa D, Okamura Y, Motoike IN, Katoh Y, Kurosawa Y, Saijyo R, Koshiba S, Yasuda J, Motohashi H, Sugawara J, Tanabe O, Kinoshita K, Yamamoto M (2016) Establishment of protocols for global metabolomics by LC-MS for biomarker discovery. PLoS One 11(8):e0160555. https://doi.org/10.1371/journal.pone.0160555

Salzer L, Schmitt-Kopplin P, Witting M (2023) Capillary electrophoresis-mass spectrometry as a tool for Caenorhabditis elegans metabolomics research. Metabolomics 19(7):61. https://doi.org/10.1007/s11306-023-02025-7

Sana TR, Waddell K, Fischer SM (2008) A sample extraction and chromatographic strategy for increasing LC/MS detection coverage of the erythrocyte metabolome. J Chromatogr B 871(2):314–321. https://doi.org/10.1016/j.jchromb.2008.04.030

Sauer F, Sydow C, Trapp O (2020) A robust sheath-flow CE-MS interface for hyphenation with Orbitrap MS. Electrophoresis 41(15):1280–1286. https://doi.org/10.1002/elps.202000044

Scherling C, Roscher C, Giavalisco P, Schulze E-D, Weckwerth W (2010) Metabolomics unravel contrasting effects of biodiversity on the performance of individual plant species. PLoS One 5(9). https://doi.org/10.1371/journal.pone.0012569

Schmitt-Kopplin P (2008) Capillary electrophoresis: methods and protocols. Springer Science & Business Media

Seyfinejad B, Jouyban A (2022) Capillary electrophoresis-mass spectrometry in pharmaceutical and biomedical analyses. J Pharm Biomed Anal 221:115059. https://doi.org/10.1016/j.jpba.2022.115059

Shahid M, Singh UBB, Khan MS (2023) Metabolomics-based mechanistic insights into revealing the adverse effects of pesticides on plants: An interactive review. Metabolites 13(2):246. https://doi.org/10.3390/metabo13020246

Shulaev V, Isaac G (2018) Supercritical fluid chromatography coupled to mass spectrometry—a metabolomics perspective. J Chromatogr B 1092:499–505. https://doi.org/10.1016/j.jchromb.2018.06.021

Soga T (2023) Advances in capillary electrophoresis mass spectrometry for metabolomics. TrAC Trends Anal Chem 158:116883. https://doi.org/10.1016/j.trac.2022.116883

Srivastava A, Mishra A, Shrivastava A (2023) Optimizing solvents and Derivatizing agents for Metabolomic profiling of human plasma using GC–MS. Chromatographia 86(7):523–534. https://doi.org/10.1007/s10337-023-04266-z

Stefanik O, Majerova P, Kovac A, Mikus P, Piestansky J (2024) Capillary electrophoresis in the analysis of therapeutic peptides-a review. Electrophoresis 45(1–2):120–164. https://doi.org/10.1002/elps.202300141

Stephan S, Jakob C, Hippler J, Schmitz OJ (2016) A novel four-dimensional analytical approach for analysis of complex samples. Anal Bioanal Chem 408(14):3751–3759. https://doi.org/10.1007/s00216-016-9460-9

Strehmel N, Kopka J, Scheel D, Böttcher C (2014) Annotating unknown components from GC/EI-MS-based metabolite profiling experiments using GC/APCI(+)-QTOFMS. Metabolomics 10(2):324–336. https://doi.org/10.1007/s11306-013-0569-y

Šuklje K, Carlin S, Antalick G, Blackman JW, Deloire A, Vrhovsek U, Schmidtke LM (2019) Regional discrimination of Australian shiraz wine Volatome by two-dimensional gas chromatography coupled to time-of-flight mass spectrometry. J Agric Food Chem 67(36):10273–10284. https://doi.org/10.1021/acs.jafc.9b03563

Tagliaro F, Manetto G, Crivellente F, Smith FP (1998) A brief introduction to capillary electrophoresis. Forensic Sci Int 92(2):75–88. https://doi.org/10.1016/S0379-0738(98)00010-3

Tang D-Q, Zou L, Yin X-X, Ong CN (2016) HILIC-MS for metabolomics: An attractive and complementary approach to RPLC-MS. Mass Spectrom Rev 35(5):574–600. https://doi.org/10.1002/mas.21445

Tang W, Lu Y, Row KH, Baeck SH, Zhang Y, Sun G (2022) Novel bovine serum album and β-cyclodextrin-based mixed chiral stationary phase for the enantioseparation in capillary electrochromatography. Microchem J 181:107763. https://doi.org/10.1016/j.microc.2022.107763

Theodoridis GA, Gika HG, Want EJ, Wilson ID (2012) Liquid chromatography–mass spectrometry based global metabolite profiling: a review. Anal Chim Acta 711:7–16. https://doi.org/10.1016/j.aca.2011.09.042

Tian H, Bai J, An Z, Chen Y, Zhang R, He J, Bi X, Song Y, Abliz Z (2013) Plasma metabolome analysis by integrated ionization rapid-resolution liquid chromatography/tandem mass spectrometry. Rapid Commun Mass Spectrom 27(18):2071–2080. https://doi.org/10.1002/rcm.6666

Uchida T, Kisugi T, Ishii H, Yamada M, Kinoshita K, Leung GN-W (2024) Broad-spectrum and sensitive screening of more than 1000 compounds in equine urine using liquid chromatography/high-resolution mass spectrometry. Rapid Commun Mass Spectrom 38(17):e9856. https://doi.org/10.1002/rcm.9856

van de Velde B, Guillarme D, Kohler I (2020) Supercritical fluid chromatography—mass spectrometry in metabolomics: past, present, and future perspectives.

J Chromatogr B 1161:122444. https://doi.org/10.1016/j.jchromb.2020.122444

Venkatramani CJ, Girotti J, Wigman L, Chetwyn N (2014) Assessing stability-indicating methods for coelution by two-dimensional liquid chromatography with mass spectrometric detection. J Sep Sci 37(22):3214–3225. https://doi.org/10.1002/jssc.201400590

Vernocchi P, Vannini L, Gottardi D, Del Chierico F, Serrazanetti DI, Ndagijimana M, Guerzoni ME (2012) Integration of datasets from different analytical techniques to assess the impact of nutrition on human metabolome. Front Cell Infect Microbiol 2. https://doi.org/10.3389/fcimb.2012.00156

Verpoorte R, Kim HK, Choi YH (2022) Trivialities in metabolomics: artifacts in extraction and analysis. Front Mol Biosci 9. https://doi.org/10.3389/fmolb.2022.972190

Vuckovic D, Pawliszyn J (2011) Systematic evaluation of solid-phase microextraction coatings for untargeted Metabolomic profiling of biological fluids by liquid chromatography-mass spectrometry. Anal Chem 83(6):1944–1954. https://doi.org/10.1021/ac102614v

Wan L, Li Y, Wang H, Wang Y, Song L, Liang W (2024) Rapid detection of markers in green coffee beans with different primary processing treatments of Coffea arabica L. from Yunnan. Food Chem 455:139942. https://doi.org/10.1016/j.foodchem.2024.139942

Wang Z, Zhu H, Xiong W (2023) Advances in mass spectrometry-based multi-scale metabolomic methodologies and their applications in biological and clinical investigations. Science Bulletin 68(19):2268–2284. https://doi.org/10.1016/j.scib.2023.08.047

Wang Y, Huang Y, Song L, Wang H, Wan L, Pang J, Liang W (2024) The characteristic VOCs of different parts of *Artocarpus heterophyllus* fruit based on HS-SPME-GC-MS and PTR-TOF-MS. Food Chem 459:140431. https://doi.org/10.1016/j.foodchem.2024.140431

Weiller F, Diniz I, Pimentel D, Erban A, Reis P, Soares F, Rego C, Kopka J, Fortes AM (2024) Metabolomic analysis of grapes and leaves from symptomatic and asymptomatic *Vitis vinifera* grapevines with esca disease. Curr Plant Biol 40:100378. https://doi.org/10.1016/j.cpb.2024.100378

West C, Lemasson E, Bertin S, Hennig P, Lesellier E (2016) An improved classification of stationary phases for ultra-high performance supercritical fluid chromatography. J Chromatogr A 1440:212–228. https://doi.org/10.1016/j.chroma.2016.02.052

Winnike JH, Wei X, Knagge KJ, Colman SD, Gregory SG, Zhang X (2015) Comparison of GC-MS and GC×GC-MS in the analysis of human serum samples for biomarker discovery. J Proteome Res 14(4):1810–1817. https://doi.org/10.1021/pr5011923

Wolfender J-L, Marti G, Thomas A, Bertrand S (2015) Current approaches and challenges for the metabolite profiling of complex natural extracts. J Chromatogr A 1382:136–164. https://doi.org/10.1016/j.chroma.2014.10.091

Wolrab D, Chocholoušková M, Jirásko R, Peterka O, Holčapek M (2020) Validation of lipidomic analysis of human plasma and serum by supercritical fluid chromatography–mass spectrometry and hydrophilic interaction liquid chromatography–mass spectrometry. Anal Bioanal Chem 412(10):2375–2388. https://doi.org/10.1007/s00216-020-02473-3

Wong YF, Perlmutter P, Marriott PJ (2017) Untargeted metabolic profiling of eucalyptus spp. leaf oils using comprehensive two-dimensional gas chromatography with high resolution mass spectrometry: expanding the metabolic coverage. Metabolomics 13(5):46. https://doi.org/10.1007/s11306-017-1173-3

Xiao M, Qian K, Wang Y, Bao F (2020) GC-MS metabolomics reveals metabolic differences of the farmed mandarin fish Siniperca chuatsi in recirculating ponds aquaculture system and pond. Sci Rep 10(1):6090. https://doi.org/10.1038/s41598-020-63252-9

Xin L, Ren M, Lou Y, Yin H, Qin F, Xiong Z (2024) Integrated UHPLC-MS untargeted metabolomics and gut microbe metabolism pathway-targeted metabolomics to reveal the prevention mechanism of Gushudan on kidney-yang-deficiency-syndrome rats. J Pharm Biomed Anal 242:116062. https://doi.org/10.1016/j.jpba.2024.116062

Yang L, Nie H, Zhao F, Song S, Meng Y, Bai Y, Liu H (2020) A novel online two-dimensional supercritical fluid chromatography/reversed phase liquid chromatography–mass spectrometry method for lipid profiling. Anal Bioanal Chem 412(10):2225–2235. https://doi.org/10.1007/s00216-019-02242-x

Yu Z, Huang H, Reim A, Charles PD, Northage A, Jackson D, Parry I, Kessler BM (2017) Optimizing 2D gas chromatography mass spectrometry for robust tissue, serum and urine metabolite profiling. Talanta 165:685–691. https://doi.org/10.1016/j.talanta.2017.01.003

Zhang X, Kew K, Reisdorph R, Sartain M, Powell R, Armstrong M, Quinn K, Cruickshank-Quinn C, Walmsley S, Bokatzian S, Darland E, Rain M, Imatani K, Reisdorph N (2017) Performance of a high-pressure liquid chromatography-ion mobility-mass spectrometry system for metabolic profiling. Anal Chem 89(12):6384–6391. https://doi.org/10.1021/acs.analchem.6b04628

Zhong X, Zhang Z, Jiang S, Li L (2014) Recent advances in coupling capillary electrophoresis-based separation techniques to ESI and MALDI-MS. Electrophoresis 35(9):1214–1225. https://doi.org/10.1002/elps.201300451

Zhou J, Yin Y (2016) Strategies for large-scale targeted metabolomics quantification by liquid chromatography-mass spectrometry. Analyst 141(23):6362–6373. https://doi.org/10.1039/C6AN01753C

Zhou J, Zhong L (2022) Applications of liquid chromatography-mass spectrometry based metabolomics in predictive and personalized medicine. Front Mol Biosci 9. https://doi.org/10.3389/fmolb.2022.1049016

Zhou B, Xiao JF, Tuli L, Ressom HW (2012) LC-MS-based metabolomics. Mol Biosyst 8(2):470–481. https://doi.org/10.1039/C1MB05350G

Zhou Z, Shen X, Chen X, Tu J, Xiong X, Zhu Z-J (2019) LipidIMMS analyzer: integrating multi-dimensional information to support lipid identification in ion mobility—mass spectrometry based lipidomics. Bioinformatics 35(4):698–700. https://doi.org/10.1093/bioinformatics/bty661

Zuberovic A, Wetterhall M, Hanrieder J, Bergquist J (2009) CE MALDI-TOF/TOF MS for multiplexed quantification of proteins in human ventricular cerebrospinal fluid. Electrophoresis 30(10):1836–1843. https://doi.org/10.1002/elps.200800714

Metabolomics: Fundamentals, Methods, Analysis, Limits, and Recommendations

6

Ayush Praveen, Tripti Nair, Biplab Singha, Sumit Murmu, Murugesh Padmanarayana, Aditya Upadhyay, Yogita Soni, and Vijay Soni

Abstract

Metabolomics has emerged as a powerful discipline for characterizing the small molecules that define cellular physiology, environmental responses, and disease states. As technologies advance, researchers face an expanding landscape of analytical platforms, data-processing strategies, and integrative approaches that require clear guidance for effective application. This chapter was written to provide a comprehensive and accessible resource for students, clinicians, and researchers entering or advancing in the field. We outline the fundamentals of metabolomics, describe major analytical methodologies—including MS, NMR, chromatography, and imaging—and summarize key considerations for experimental design, data preprocessing, statistical analysis, and functional interpretation. We also address current challenges related to metabolite identification, reproducibility, and multi-omic integration, and highlight emerging innovations such as stable-isotope tracing, spatial metabolomics, and AI-driven analytics. Together, these elements offer a detailed roadmap for conducting robust, reproducible, and insightful metabolomic studies.

Keywords

Small molecules · Metabolome · Chemoinformatic · Network modeling · Microbiome · Mass analyzing · Metabolite imaging · Chromatography · Capillary electrophoresis · Heteroscedasticity · Cross-Omics

Ayush Praveen, Tripti Nair and Biplab Singha contributed equally with all other contributors.

A. Praveen
Elucidata Data Consulting Pvt Ltd, New Delhi, Delhi, India

T. Nair
Leonard Davis School of Gerontology, University of Southern California, Los Angeles, CA, USA

B. Singha
Department of Microbiology and Physiological Systems, University of Massachusetts Chan Medical School, Worcester, MA, USA

S. Murmu
Regional Centre of Biotechnology, Faridabad, India

M. Padmanarayana
Department of Biochemistry, Stanford University, Beckman Centre for Molecular and Genetic Medicine, Stanford, Palo Alto, CA, USA

A. Upadhyay
Hospital for Special Surgery, New York, NY, USA

Y. Soni
Department of Chemical and Biological Engineering, Tufts University, Medford, MA, USA

V. Soni (✉)
Division of Infectious Diseases, Weill Department of Medicine, Weill Cornell Medicine, New York, NY, USA
e-mail: vis2032@med.cornell.edu

M. Gama-Carvalho et al. (eds.), *Omics Approaches in Biomedicine and Biotechnology*, Advances in Experimental Medicine and Biology 1504, https://doi.org/10.1007/978-3-032-18966-0_6

6.1 Introduction

What impact do the foods we eat, the medications we take, and the environments we inhabit influence our health at the molecular level? Metabolomics seeks to answer these questions by examining the complete set of metabolites—small molecules (molecular mass less than 1500 Da) that play critical roles in our metabolism and overall well-being (Wishart et al. 2007). Metabolites, ranging from amino acids to lipids, act as essential substrates and products of metabolic processes, influencing cellular functions such as energy production, storage, signal transduction, and epigenetic regulation (Demain 1980; Sperber et al. 2015; Karlic et al. 2015; Yanes et al. 2010; Lodish 2008; Alberts et al. 2022). Understanding the metabolome—the vast array of metabolites within an organism—has profound implications for fields ranging from medicine to environmental science.

Metabolomics has emerged as a powerful tool for comprehensively analyzing these metabolites. From its early beginnings to today's high-throughput technologies, metabolomics has revolutionized our ability to identify, quantify, and characterize metabolites across diverse biological samples. The term "metabolome" was first introduced by Oliver et al. in 1998 to describe changes in relative metabolite concentrations resulting from gene deletions or overexpression (Oliver et al. 1998). The metabolome is vast and intricate, with about 114,000 chemicals identified in *Homo sapiens* and up to 1,000,000 in the plant kingdom (Wishart et al. 2018; Hounsome et al. 2008; Saito and Matsuda 2010). Compared to the human genome (around 20,000 genes) or proteome (over 620,000 protein molecules) (Ponomarenko et al. 2016; Salzberg 2018), the metabolome is larger and more challenging to measure, leading to the development of "metabolomics," a field coined by Fiehn in 2001 for the high-throughput identification, quantification, and characterization of metabolites (Fiehn 2002).

The complexity of metabolomics arises from the diversity of metabolites and the need for advanced bioinformatics to integrate this data with other omic information. In a single biological sample, thousands of different metabolites can exist, each with varying concentrations, chemical structures, and functions. Advanced bioinformatics tools are employed to identify and quantify these metabolites, map them to metabolic pathways, and integrate this information with genomic, transcriptomic, and proteomic data. This integration provides a comprehensive understanding of biological processes, including how genetic mutations (genomics) affect gene expression (transcriptomics), protein function (proteomics), and the overall metabolic state of the cell (metabolomics). Such insights are crucial for understanding disease mechanisms and identifying potential therapeutic targets.

The metabolome reflects the interplay between genome, epigenome, proteome, and environment, encompassing a complete set of metabolites influenced by these factors (Fig. 6.1). Genetic variations can affect enzyme activity and, consequently, metabolite production (Smith et al. 2022). Epigenomic modifications, such as DNA methylation and histone acetylation, impact gene expression and thus influence metabolic pathways (Wu et al. 2023). Posttranslational modifications to proteins also affect enzyme activity and expression, further influencing metabolic processes (Zhong et al. 2020). Environmental factors, including diet, stress, toxins, and microbial interactions, impact nutritional availability and metabolic activity (Amarnath et al. 2023). This complex interplay provides a comprehensive view of an organism's physiological state.

Metabolomics encompasses a wide array of samples, including tissues, cells, and biofluids, each requiring specific analytical techniques for comprehensive analysis. The main methodologies used to identify metabolites include untargeted (global) and targeted mass spectrometry (MS)-based metabolomics. Untargeted metabolomics aims to measure the broadest range of metabolites in an extracted sample without prior knowledge, while targeted metabolomics involves the analysis of specific metabolites based on prior information about the metabolome (Christ et al. 2018; Roberts et al. 2012). Additionally, imaging metabolomic

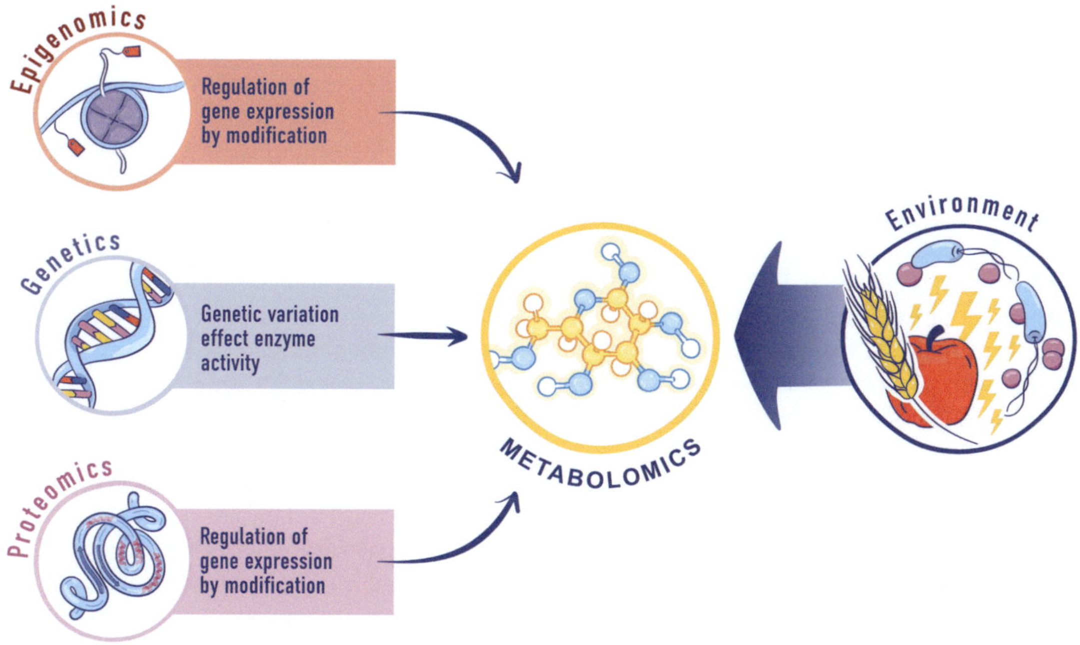

Fig. 6.1 The metabolome: a reflection of the genome, epigenome, proteome, and environment. The diagram illustrates the interconnected nature of the metabolome with various biological and environmental factors. Genetic variations in the genome can affect enzyme activity, influencing metabolic processes and metabolite composition

techniques such as matrix-assisted laser desorption ionization (MALDI) and nanostructure-imaging mass spectrometry (NIMS) play a critical role in characterizing metabolite localization within tissue samples (Tanaka et al. 1988; Wiseman et al. 2006). Nuclear magnetic resonance (NMR) spectroscopy provides structural information, while gas and liquid chromatography techniques (GC-MS, LC-MS) offer high sensitivity, resolution, reproducibility, and versatility in metabolite analysis (Dettmer et al. 2007; Zhang et al. 2012).

These analytical methods are complemented by a new surge of chemoinformatic and bioinformatic tools that support data acquisition, analysis, and integration (Gasteiger 2016). Comprehensive metabolite databases such as METLIN, HMDB, MassBank, and GMD aid in identifying metabolites based on accurate mass-to-charge ratios from MS data (Wishart et al. 2007; Montenegro-Burke et al. 2020; Horai et al. 2010). Network modeling and pathway mapping tools help to understand the roles of metabolites in biological perturbations and can integrate this information with upstream genes and proteins for mechanistic investigations (Swainston et al. 2013; Kanehisa et al. 2016). These advancements in analytical techniques and bioinformatics have revolutionized metabolomics, enabling the discovery of metabolic pathways, biomarkers, and personalized medicine approaches based on metabolic profiles.

Metabolomics has expanded its applications beyond traditional fields, significantly impacting areas such as biomarker discovery, pharmacology, toxicology, and environmental health. In biomarker discovery, metabolomics is pivotal in diagnostics, identifying biomarkers for various diseases, including cancer, cardiovascular diseases, and metabolic disorders. For instance, in cancer research, metabolic profiling has shown elevated total choline (tCho) levels due to an increase in phosphocholine and reduced glucose levels in breast cancer tissues compared to benign and healthy tissues (Spratlin et al. 2009; Sitter et al. 2002).

In pharmacology, metabolomics helps elucidate drug mechanisms of action and toxicity. It provides insights into how drugs are metabolized and how they affect metabolic pathways, facilitating personalized medicine approaches (Clayton et al. 2006). Moreover, metabolomics is pivotal in toxicology, assessing the impact of toxins, pollutants, and environmental exposures on metabolic pathways and identifying potential biomarkers of exposure (Nicholson et al. 2012). This capability is essential in environmental health, where metabolomics is used to study how environmental factors influence human health through metabolic perturbations.

Additionally, metabolomics is advancing in areas such as microbiome research, where it aids in understanding host-microbiome interactions and their impact on health and disease. It also supports agricultural sciences by enhancing crop quality and yield through metabolic profiling (Li et al. 2016). Overall, metabolomics continues to broaden its impact across disciplines, providing comprehensive insights into biological systems at the molecular level, from health and disease to environmental and agricultural contexts. Metabolomics faces significant challenges despite advancements in recent years. Biomarker validation and identification remain significant hurdles due to subtle differences in metabolite concentrations between control and aberrant conditions, as well as overcoming inter-individual metabolite variation; the metabolome is highly sensitive to genetic and environmental factors (Johnson and Gonzalez 2012; Johnson et al. 2012).

Understanding the biological roles of metabolites and their interactions with metabolic pathways remains a major hurdle. Current pathway maps, which cover approximately 2000 metabolites, are incomplete, necessitating a multi-omic approach to elucidate disease mechanisms and metabolic pathway interactions (Johnson et al. 2015a; Zamboni et al. 2015). Another major challenge lies in the choice of analytical platforms, namely, nuclear magnetic resonance (NMR) spectroscopy and mass spectrometry (MS), each with specific strengths and limitations. NMR spectroscopy offers advantages such as the ability to detect various nuclides and larger molecules but lacks the sensitivity of MS (Johnson and Gonzalez 2012).

The future of metabolomics holds great promise with ongoing advancements in technology and methodology. Integrating multi-omic approaches is crucial for elucidating complex disease mechanisms and identifying novel biomarkers (Fiehn 2002; Johnson et al. 2016). Bioinformatics is advancing with sophisticated algorithms and machine learning techniques that will significantly enhance metabolite identification, quantification, and pathway analysis (Kanehisa et al. 2016; Johnson et al. 2015a). Continuous expansion and refinement of metabolite databases, such as METLIN and HMDB, are essential for accurate and extensive metabolite identification (Wishart et al. 2018; Montenegro-Burke et al. 2020). Stable isotope-assisted analysis will allow precise tracking of metabolic changes under various conditions, offering deeper insights into metabolic pathways (Johnson et al. 2016; Bueschl et al. 2017). Innovations in single-cell metabolomics and spatial metabolomics will provide unprecedented insights into cellular heterogeneity and high-resolution mapping of metabolite distributions within tissues (Hu et al. 2023; Alexandrov 2020; Zenobi 2013). These advancements collectively promise to expand the scope of metabolomics, providing comprehensive insights into biological systems at a molecular level and paving the way for personalized medicine and precision health initiatives.

Overall, metabolomics represents a powerful approach to studying metabolism at a global scale, offering insights into biochemical pathways, disease mechanisms, and environmental influences. By unraveling the complexities of the metabolome, metabolomics contributes to personalized medicine, biomarker discovery, and our understanding of metabolic dysregulations in health and disease. As technological capabilities and analytical methods continue to advance, metabolomics is poised to make even greater contributions to biomedical research and clinical applications in the coming years.

6.2 Metabolomic Methodologies: An Overview

Metabolomics is the identification, characterization, and quantification of all small molecules, especially metabolites, and intermediates of metabolic pathways present in a sample. Metabolomics aids in the estimation of the biochemical activity of the organism at a systems level. Metabolomics is an ever-evolving and transitioning field. Currently, it complements other omic data, such as DNA sequencing, transcriptome, and proteome data. It is highly dynamic and rapidly affected by changes in the physical and chemical environments. Metabolomics is divided into different categories depending on the analytical techniques used and the range of metabolite analysis conducted. Each analysis method has advantages and disadvantages; however, they are sufficiently customizable to cater to specific metabolomic questions. There is plethora of factors, including the nature of the sample, the desired outcome, the sensitivity of the analysis technique, the depth of throughput, and the accessibility of data analysis, which help to choose and tailor the approach (Hartman and Lees 2023). To ensure optimal results, the following points must be considered when selecting an analytical technique:

Throughput High throughput approaches, such as nuclear magnetic resonance (NMR) and Fourier-transform ion cyclotron resonance mass spectrometry (FT-ICR-MS), are preferred for bulk analysis of samples quickly and accurately.

Sensitivity The sensitivity of the analytical technique is a critical point of consideration. Mass spectrometry-based techniques including liquid chromatography-mass spectrometry (LC-MS) and gas chromatography-mass spectrometry (GC-MS) are highly sensitive and accurate and provide high resolution in detecting metabolites even at very low concentrations.

Quantification The desired outcome of the research question determines the appropriate analytical technique. Targeted techniques such as tandem mass spectrometry (MS/MS) are typically used if quantitative analysis is required.

Sample Type The sample type being evaluated plays an instrumental role in determining the analytical technique to proceed with. For example, polar metabolites are typically analyzed using LC-MS (liquid chromatography-mass spectrometry), while nonpolar metabolites are better suited for GC-MS.

Other Considerations Apart from the aforementioned, other factors must be considered when selecting a metabolomic analysis technique. Cost-effectiveness, data reproducibility, statistical power, and technical feasibility are noteworthy factors that could influence the decision-making process.

Cost-effectiveness: Evaluate the total cost of the technique, including reagents, instruments, and personnel, against the expected outcomes.

Data reproducibility: The chosen technique must be reproducible and yield consistent, reliable results, ensuring high confidence in the data.

Statistical power: Choosing a technique that delivers the required statistical power and resolution to detect differences in metabolites across biological samples is essential.

Technical feasibility: It denotes the suitability and practicality of the technique in terms of available equipment, laboratory space, and staff expertise. A technique may not be considered suitable if it demands specialized equipment that is unavailable or it is a challenge to operate routinely. Thus, selecting the appropriate metabolomic technique is vital for ensuring accurate, precise, and reliable results, ultimately leading to valuable scientific insights. We will further describe the analytical methods most commonly used in metabolomics, including the primary approaches like the MS and NMR. A brief account of the various metabolomic approaches has been provided below.

6.2.1 Mass Spectrometry

Mass spectrometry is a prevalent technique in metabolomics that measures the mass-to-charge ratio of ions to identify and quantify a vast array of small molecules, including sugars, amino acids, and lipids. Due to its massive analytical potential, MS has become the most frequently sought-after tool for analyzing a myriad of biological systems. Mass spectrometers comprise of three major elements: an ion source, a mass analyzer, and a detector. The mass spectrometry workflow is as follows:

1. *Sample Introduction*: This can be performed via (1) direct injection, (2) gas-phase sampling, or (3) sample vaporization.
2. *Ionization:* In an ionization chamber, the sample is bombarded with electrons, causing the ejection of electrons from the sample molecules and resulting in the formation of charged parent ions ([M+] or [M-]).
3. *Acceleration and Separation*: Electric fields further accelerate the ionized sample ions, which are then separated based on their mass-to-charge ratio (*m/z*). Subsequently, the ions are resolved based on their indigenous mass-to-charge ratio (*m/z*).
4. *Detection*: After separation, the ions are detected by a detector system that measures the intensity or abundance of ions at different *m/z* values, thereby generating a mass spectrum.
5. *Data Analysis*: The mass spectrum is analyzed to determine the presence of different ions and their relative abundances providing information about the sample molecules' structure, molecular composition, and fragmentation patterns.
6. *Identification and Quantification*: The identified ions are matched to known compounds by comparing the mass spectrum with available databases or reference spectra. Additionally, the relative abundance of ions can be used to quantify the concentration of specific compounds.

In general mass spectrometry has three crucial steps: ionization, detection, and chromatography.

6.2.1.1 Ionization

The different ionization techniques include electron ionization (EI), matrix-assisted laser desorption/ionization (MALDI, etc.). Hard ionization (~70 eV) breaks the covalent bonds in the molecule and is commonly used for nonpolar, non-thermolabile molecules, making them less suitable for most biomolecules. Soft ionization techniques can accomplish the ionization of polar molecules within metabolic pathways without fragmentation with a gentler approach. The methods used are electrospray ionization (ESI), chemical ionization (CI), and MALDI are employed for metabolomics. The most commonly used is ESI; the sample liquid is pumped through the needle at 220–500 °C and 2–4 kV, resulting in protonated/deprotonated molecules. CI technique includes atmospheric pressure chemical ionization, combining the sample with ionized gas to form a plasma. Different gases alter the charge on the molecule accordingly $[M + H]^+$ or $[M - H]^-$. MS can also be performed on intact specimens (such as tissue samples) with the help of MALDI, where a laser is pulsed tangentially onto the sample surface, ablating molecules as singly charged ions. Molecular structure determination can be performed using hard ionization techniques in which the fragmentation pattern of the molecule of interest can be interpreted to deduce a likely structure. Soft ionization techniques are suited for biomolecule identification and quantification in sample solutions; MALDI signals can be used for biomolecule mapping in a tissue slice, furthermore, in microbial identification (Hartman and Lees 2023). Following are some recent advancements in MS ionization techniques:

6.2.1.1.1 Desorption Electrospray Ionization Mass Spectrometry (DESI-MS)

DESI-MS is a "soft" ionization technique suitable for the study of intact metabolites. DESI-MS ionizes metabolites via an electrospray plume directed to the sample surface creating a thin film on the surface where solid phase analytes are extracted. Primary droplets collide with the film, leading to the formation of secondary droplets that contain the analyte in liquid form. Ionization occurs through mechanisms similar to those in electrospray ionization (ESI). The composition ratio of organic to aqueous solvents and the type of organic solvent used in DESI-MS significantly influence the metabolites desorbed from the surface. Additionally, the ionic strength and pH of the desorption solvent play crucial roles in determining ionization efficiency, which in turn affects the intensity and stability of the metabolite signal. The inclusion of additional reagents in the desorption solvent can modify the ionization patterns of analytes or metabolites, a technique referred to as "reactive" DESI-MS (Clendinen et al. 2017).

6.2.1.1.2 (EASI)-MS

Easy ambient sonic-spray ionization (EASI)-MS is a variant of DESI that utilizes a supersonic cloud of charged droplets for analyte desorption and ionization. This technique generates both positive and negative ions, allowing for the analysis of a wide range of compounds. However, the technique has some limitations including lower sensitivity and the need for a high gas flux through the system (Clendinen et al. 2017).

6.2.1.1.3 DART

Direct analysis in real time (DART) is a direct sampling plasma ionization technique designed for high-throughput analysis of samples in various states including solids, liquids, and gases. In DART, the sample is exposed to a heated gas stream of metastable species typically nitrogen (N_2) or helium (He). These metastable species ionize thermally desorbed analytes, which are then introduced into a mass spectrometer for analysis (Clendinen et al. 2017).

6.2.1.1.4 PESI

Probe electrospray ionization (PESI) is a two-step technique involving a solid-sampling electrospray probe inserted into a sample, picking up material with ample water content in the picoliter range sufficient for electrospray ionization upon applying high voltage during analysis. The needle sampling depth has a profound effect on the reproducibility. PESI has high salt tolerance with minimal ion suppression (Clendinen et al. 2017).

6.2.1.1.5 LAESI

Laser ablation electrospray ionization (LAESI), a two-step laser-based desorption ablation technique, is coupled to ESI. It allows lateral mapping of metabolites on sample surfaces. In LAESI, the infrared radiation (IR)-ablated sample electrospray plume is combined with an electrospray droplet. The analytes are ionized via ESI mechanisms. IR lasers excite the OH vibrations from endogenous H_2O molecules in the sample, which in turn act as an ionization matrix to facilitate the desorption or ablation of neutral molecules from the surface (Clendinen et al. 2017).

6.2.1.1.6 REIMS

Rapid evaporative ionization mass spectrometry (REIMS) has contributed to the field of surgical metabolomics. During the process of electrosurgery, a high-frequency current is applied to surgical blades, and the resultant pyrolyzed tissue plume produces aerosols and charged species that are suctioned from the surgical site subjected to mass spectrometry (Clendinen et al. 2017).

6.2.1.1.7 PSMS

Paper spray mass spectrometry places a sample onto a paper positioned in front of the mass spectrometer. The samples are-dried, and after drying, analytes are eluted by solvent drop placed on the paper. The samples are simultaneously ionized at the paper tip via the application of high voltage and further subjected to MS analyses (Clendinen et al. 2017).

After ionization, the mass analyzer moves ions from the source to the detector while filtering them based on *m/z*. It is done by focusing the ion beam through an electrostatic or strong magnetic field. There are five major mass analyzers: (i) the time of fight (TOF) mass analyzer, (ii) the quadrupole mass analyzer, (iii) the Orbitrap, (iv) the ion trap mass analyzer, and (v) the ion cyclotron resonance (ICR). The quadrupole MS consists of four parallel electrodes in a vacuum chamber for ions to be shuttled between them based on the current or radiofrequency (RF) applied to the electrodes. These parameters can adjust to allow only the ions within a specific *m/z* range to pass through the quadrupole. Ions outside the range are lost in a vacuum. Thus, a quadrupole mass filter is useful to identify and/or quantify specific molecular species. Further advancements are triple quadrupole (QqQ) systems wherein the "q" denotes the middle quadrupole functioning as a collision cell, and ions are fragmented upon the introduction of a neutral gas that transfers translational energy to the ions via collision (collision-induced dissociation). Thus, establishing the tandem mass spectrometry (or MS/MS) process. MS/MS is useful in determining a compound's structure or distinguishing amongst molecules with a similar *m/z* value. The QqQ is often preferred while quantifying biomolecules within complex samples like blood, serum samples, cerebrospinal fluid (CSF), or urine. The other important feature of QqQ is selected reaction monitoring (SRM). This mode improves sensitivity; the quadrupoles filter out interfering ion signals, leading to noise-free quantification. Another add-on is multiple reaction monitoring (MRM) to quantify multiple ion species efficiently (Hartman and Lees 2023).

6.2.1.2 Detection

The detector makes an electronic signal upon registering an ion collision. Although Faraday cups (FCs) are currently still used, the detectors now are usually electron multipliers (EMs), photomultiplier conversion dynodes, or array detectors. The FC is a metal cup that intercepts the ion beam passing through the mass analyzer. Upon impact with an ionic beam, the cup induces a positive charge. That positive charge induces a current in the FC equivalent (+1.602 × 10–19 C, i.e., current/charge carried by a proton), thereby measuring the number of ions. In the prototype FC design, there is no scope for amplification—so these detectors have the limitation of low sensitivity. However, now EMs consisting of a series of aluminum oxide dynodes increase the potential of current generated. When an ion strikes the first dynode, it sets a cascade of electrons ejected that amplifies the ion current, resulting in a magnitude of the signal intensity and profoundly increasing the sensitivity. In the native state, the EM does not detect negative ions. Thus, a conversion dynode of positive voltage is placed in front of the EM releasing electrons down the electron multiplier chain. Another modification includes the photomultiplier conversion dynode detector. In this configuration, the ions from the analyzer strike the conversion dynode, releasing secondary electrons. These electrons eventually strike a phosphor screen, which releases photons that pass through EM (Hartman and Lees 2023).

6.2.1.2.1 High-Resolution Mass Spectrometry

The most used instruments for current metabolomics are the QqQ (500–1500 unit resolution), the Orbitrap (60,000–500,000 at m/z 200), the Q-TOF (20,000–40,000 FWHM), and the FTIR (5,000,000–10,000,000 resolving power). The ion trap has a three dimensional path compared to a quadrupole system's linear pattern. The ion trap has multiple configurations, of which the Orbitrap is the most common. The two configurations of ion trap apart from Orbitrap are (i) Paul's ion trap consisting of a ring electrode located between a pair of end-cap electrodes, and an oscillating RF field is intrinsically generated, ions orbit about the center of the trap until the orbit destabilizes. The voltage applied to the electrodes determines the point at which ions of different *m/z* form a stable orbit. (ii) Penning's ion trap that combines axial electromagnetic fields with quadrupole electric fields to trap/stabilize the ions for up to several days; this offers sensitivity to measure subatomic particulate species.

In the Kingdom's ion trap design, i.e., the Orbitrap, there is a high-field compact trap chamber with a central electrode sandwiched by electrode cups. Upon entering the trap, the ions are deflected toward the detector based on the electric field applied. The ions orbit around the central electrode core until they hit the amplifier and, subsequently, the detector.

The Q-TOF is a hybrid instrument; ions are separated based on the time an ion or packet of ions takes to travel through the vacuum "fight tube" to the detector. Within the vacuum tube, a high-voltage pulse accelerates the ions toward an electrostatic mirror, which deflects them to the detector. The travel time of an ion, i.e., time of flight, is determined by the intrinsic kinetic energy of the ions and is affected by the distance through the fight tube. Orbitraps and Q-TOFs offer to deliver high-resolution data and are suited for the recent metabolomic workflows.

Once the samples are analyzed, the output files comprise the mass spectrum detected by each scan over a run. After each scan, the mass spectrum is represented as a data file for that sample. The data file often provides itself as a table or in graphical form (the *m/z* ratios of the ions are plotted against their intensities). Each peak in a mass spectrum confirms the presence of a component of unique *m/z* in the sample, and the heights of the peaks indicate the relative abundance of the components in the sample. Verifying the peak on the MS data with the help of the quality check (QC) step in metabolomic studies is crucial. This is accomplished by spiking the sample matrix with the standard purified molecule, hence subjecting it to the identical analytical conditions as the sample. If the same fragment pattern appears in the standard as well as in the complex sample mixture, this gives data confidence regarding the identity of the molecule (Hartman and Lees 2023).

6.2.1.2.2 Metabolite Imaging/Mass Spectrometry Imaging

Mass spectrometry imaging (MSI) was noted for high-spatial-resolution imaging (between 0.5 and 5 μm) at even single cell level. Recent advancements are being made to resolve the issues regarding small sampling sizes and the low ionization yields of metabolites. Several on-tissue chemical derivatization approaches have been reported to augment MSI signals of the targeted compounds, especially in (MALDI)-MSI. Various combinations of chemical derivatization to selectively augment metabolite signals of categorical functional group for each consecutive tissue section have been adopted. The three most common on-tissue derivatization methods are (1) coniferyl aldehyde derivatization for primary amines, (ii) Girard's reagent T derivatization in case of carbonyl groups, and (iii) 2-picolylamine derivatization for carboxylic acids. These strategies enabled the detection of hundreds of novel unique metabolite features, compared to the results obtained without derivatization modification. Statistical analysis is indicative of quantitative variation between metabolites within tissue cross-sections, while MS images provided further resolution in the localization of these metabolites (Dueñas et al. 2019).

6.2.1.3 Chromatography

Chromatography is an undeniably versatile separation technique, enabling the efficient resolution of complex mixtures of compounds based on their innate chemical properties, such as polarity or size. However, not all metabolomic techniques require chromatography; there are various forms of chromatography, including LC (liquid chromatography), GC (gas chromatography), and IC (ion chromatography) (Hartman and Lees 2023).

6.2.1.3.1 Liquid Chromatography

Liquid chromatography separates metabolites based on physicochemical properties of polarity and hydrophobicity, using (i) a liquid mobile phase and (ii) a stationary phase. The mobile phase comprises two solvents carefully chosen based on the targeted molecules' solubility and compatibility with the assigned ionization technique. The sample is injected into a stream of the first mobile phase solvent (within which the sample is soluble). The mixture is then guided to the chromatography column, where it adsorbs to the stationary phase material. The second solvent is introduced at concentration gradients, slowly eluting the sample from the column and navigating it to the ion source. In the stationary phase, a solid matrix comprises tightly packed silica or alumina-based adsorbent. During the process, the solvents must be of high purity grade, not containing impurities that might appear on the mass spectrum. There are different LC modes, such as reverse-phase liquid chromatography (RPLC), HILIC (hydrophilic interaction chromatography), and ion-exchange chromatography (IEC). RPLC works on differential metabolite partitioning between polar mobile and nonpolar stationary phases. It is useful for separating moderately polar and nonpolar metabolites. Highly polar and ionic metabolites are compatible and not resolved in RPLC (e.g., some sugars, phosphates, and carboxylic acids). HILIC employs a categorically polar stationary phase and a less polar mobile phase. It is ideal for resolving highly polar metabolites. IEC resolves metabolites based on the charge with the help of the charged stationary phase while also adjusting the ionic strength and pH of the mobile phase. Both cationic and anionic resins are available to separate the respective charged ions. This technique is useful for separating proteins, amino acids, or even nucleotides (Hartman and Lees 2023).

LC-MS couples liquid chromatography and mass spectrometry. A separation procedure prior to the MS analysis is preferred in most metabolomic studies. As an efficient separation process, liquid chromatography allows the separation of chemical entities with wide-ranging polarity using the isocratic elution method (water-solvent ratio kept constant) or gradient elution method (water-solvent ratio kept variant). Thus, LC-MS effectively resolves metabolites based on their innate chemical properties before utilizing MS to identify and quantify the individual species. LC-MS offers notable advantages, for example, the ability to analyze polar and nonvolatile subsequently a wider range of metabolites, including lipids, peptides, and proteins. The associated reduction in sample complexity reduces matrix interference in the ionization process. Furthermore, as background signal/noise is reduced, reliable MS results are obtained. However, this technique is a complex, time-consuming endeavor (Munjal et al. 2022).

6.2.1.3.2 Gas Chromatography

Gas chromatography is a separation technique designed for thermally stable and volatile metabolites. Metabolites are vaporized and separated using a gaseous mobile phase. Ions are separated based on their volatility and affinity to the stationary phase. The sample is injected into an injection port which is heated. The sample is vaporized and moved forth by an inert gas through a capillary column. Metabolites in the sample interact with the stationary phase, and their resolution is based on differences in their chemical interactions in the stationary phase and their boiling points. The retention time of individual ions in the capillary is well standardized, with existing validated databases for the annotation of signals from the samples with great data confidence, thus making GC a robust method (Hartman and Lees 2023).

GC-MS segregates the metabolites based on their respective innate chemical properties, which are analyzed utilizing mass spectrometry for further resolution in identification and quantification. GC-MS is sensitive and analyzes a wide array of metabolites while providing reproducible results. GC-MS is instrumental in areas of volatile metabolites analysis. Metabolite samples are chemically derivatized (Derivatization using reagents like N,O-bis(trimethylsilyl), trifluoroacetamide (BSTFA) and trimethylsilyl (TMS)) to increase their volatility. Derivatization increases the coverage of experiments. GC-MS has limitations; it can only analyze volatile and semi-volatile compounds; thus, examining larger biomolecules such as lipids or proteins is not feasible via GC-MS (Munjal et al. 2022).

6.2.1.3.3 Capillary Electrophoresis (CE)

Capillary electrophoresis (CE) utilizes an electric field to separate metabolites based on charge and size. It uses a negatively charged hollow glass capillary. Voltage is applied to both ends of the capillary tube, and metabolite ions migrate and resolve based on their *m/z* ratio (due to consequent differences in electrophoretic mobility). CE offers short analysis time with minimal sample volume requisites. CE suits small polar charged metabolites like amino acids and other charged compounds (Hartman and Lees 2023).

CE-MS is an innovative and multifaceted technique. Metabolites, subjected to CE separation, are electrically resolved based on their characteristic charge and size, followed by the identification and quantification with the help of MS. CE-MS has several advantages, including its high sample resolution efficiency, the exceptional accuracy in analyzing charged metabolites.

CE-MS has its own limitations of low sensitivity compared to GC-MS/LC-MS, and it is restricted to the analysis of small molecules (Munjal et al. 2022).

6.2.2 Nuclear Magnetic Resonance Spectroscopy

NMR is a nondestructive approach; the samples can be rerun using multiple pulse programs. The advantages of NMR are as follows: it deals with limited sample preparation; it is quantitative; the sample run is cost-efficient; it is advisable for comprehensive untargeted analysis, thereby detecting a wide array of metabolites in a single experimental run; and it allows for the detection and characterization of novel compounds. It can also provide positional information in the case of isotopomers. Additionally, it is instrumental in analyzing molecules that cannot ionize well, require derivatization, or cannot be resolved based on mass. There are several different types of NMR; the major ones are (1) proton NMR (^1H-NMR), (2) carbon-13 NMR (^13C-NMR), and (3) two-dimensional NMR. The most common type of NMR is ^1H-NMR, providing in-depth information on hydrogen atoms present within a molecule. NMR experiments are reproducible, with negligible batch effects, hence increased precision and no carry-over between samples, therefore increased accuracy. These benefits, in combination with the short periods of experimental analyses and additional ease of automation, make NMR an ideal procedure for high-throughput metabolomic studies. A few noteworthy limitations of this technique are (1) low sensitivity and associated lowered number of identified biomarkers; (2) sample complexity, i.e., resonances from highly abundant or higher-molecular-weight molecules (e.g., lipoproteins or lipids) overlap with and obscure peaks from other molecules, thereby hindering NMR interpretations; (3) requirement of higher sample volume; (4) instrument costs are high; and (5) requirement of extensive lab space. NMR is employed in combination with techniques like MS and chromatography, making it more efficacious (Hartman and Lees 2023).

6.2.3 Stable Isotope-Resolved Metabolomics

Stable isotope-resolved metabolomics (SIRM) has a unique ability of isotope editing (i.e., distinguishing molecules containing the ^{13}C, ^{15}N or ^{31}P isotope labels). SIRM helps understand a compound's positional enrichment ion species, thereby providing further resolution into the associated metabolic flux. The stable isotopes are compatible within live cells and organisms, making it possible to measure metabolic processes in vivo. To delve deeper into the world of metabolomics, MS is frequently coupled with liquid chromatography (LC-MS), gas chromatography (GC-MS), or capillary electrophoresis (CE-MS), enabling the efficient separation and analysis of the metabolites in a given sample (Lane and Fan 2017).

6.2.4 Fourier Transform Infrared Spectroscopy (FTIR)

FTIR is a vibrational spectroscopy technique. IR (infrared radiation) generated from the radiation source (solid inert material heated to 1000–1800 °C) is passed through the sample. Some radiation gets absorbed, while some are reflected through to a monochromator, which then disperses the radiation to the detector. Either a thermal or photon detector analyzes the generated heat the IR produces and generates an electric interferogram plot. The IR spectrum can be instrumental for gleaning information about the sample composition (including the presence of functional groups and their exact quantification, all achieved through precision and accuracy). Generally, a solvent blank is run in the background, and the spectrum generated can be subtracted as a baseline from the sample run for a cleaner spectrum. Data analysis matches the frequency bands of the sample spectrum, Further, which is converted through a fast Fourier transform algorithm into high-resolution molecular fingerprinting data. FTIR can be utilized to analyze nonvolatile and nonpolar compounds with minimal sample preparation; however, it is constrained by low sensitivity (Hartman and Lees 2023).

6.3 Analysis of Metabolomic Data

High-throughput metabolomic profiling using methodologies such as NMR and MS has significantly advanced our understanding of how metabolite regulates phenotype in a biological system. While any single omic method offers a perspective on a biological system, none can provide a comprehensive view on its own. Although the same is true for metabolomics, it comes closest to offering a detailed view of regulation of phenotype. The road from generated metabolomic data to actual insights traverses through the series of data processing and analysis methodologies using bioinformatics. Employing incorrect data processing, statistical methods, and downstream analysis can often result in inconclusive or erroneous findings. In this section, we will focus on the bioinformatics processing of metabolomic data, explore the tools used for this purpose, and apply appropriate statistical approaches and downstream methods for confident findings. The typical bioinformatics workflow for the metabolomic data can be divided into the following steps.

1. Data preprocessing
2. Data post-processing
3. Statistical analysis
4. Functional analysis

In addition to these four steps, metabolomic data can be optionally integrated with other omic data to provide a comprehensive view of the system.

6.3.1 Data Preprocessing

Data preprocessing ensures that the raw signals from metabolomic profiling are translated into the quantified metabolite composition of different samples in the experiment. The usual steps in data pre-processing are chromatographic alignment, noise reduction, retention time correction, peak detection, isotope detection, etc. (Chen et al. 2022). Various software suites are available for performing these data preprocessing tasks in metabolomics (Table 6.1).

Typically, different suites have the following common steps to perform data preprocessing.

6.3.1.1 Data Import

Data import functionalities of these tools are used to select the raw mass spectrometry data to be pre-processed in the relevant format.[1] One should be mindful if the data is generated through profile acquisition or centroid acquisition and should select the appropriate mode in different suites. The profile mode data can be normalized to centroid mode by the preprocessing suite.

6.3.1.2 Peak Preprocessing

The broader step has multiple steps such as peak baseline setting and peak filtering based on different thresholds (mz threshold, frequency threshold, signal-to-noise ratio, etc.), peak grouping, and smoothing. If it is labeled data using tandem MS, isotope detection can also be performed.

6.3.1.3 Peak and Compound Detection

The metabolomic analysis can be targeted or untargeted. The targeted metabolomic analysis requires a defined chemical library that can be used to select a defined list of compounds from the profiled peaks. The peak detection step itself requires a defined chemical library in addition to the peak properties for selection. The untargeted analysis is a selection based on the peak properties. This requires selection of novel compounds and doesn't depend on a defined limited size library. The identified peaks can later be annotated as different compounds after selection. The unannotated compounds can be treated as novel compounds and studied further.

---

[1] Note: Different vendors and organizations have different data format standards for generating raw metabolomic data from their instruments. One can use tools like MSConvert (from the ProteoWizard toolkit) (Chambers et al. 2012) to convert data into formats supported by different data processing suites and applications. MSConvert supports both reading and writing open formats such as mzML 1.1, mzML 1.0, mzXML, MGF, MS2/CMS2/BMS2, mzIdentML and proprietary format (vendor formats require Windows with vendor libraries) such as RAW, T2D, .d and many others.

Table 6.1 Different preprocessing suites available for metabolomic data

XC-MS	Framework for processing and visualization of chromatographically separated and single-spectra mass spectral data available in the form of an R package (Smith et al. 2006)
Maven	MAVEN is an open-source software program for interactive processing of LC-MS- based metabolomic data (Clasquin et al. 2012)
MZMine	MZmine is an open-source, multiplatform, user-friendly, graphical software for mass spectrometry data processing (Katajamaa et al. 2006)
El-Maven	El-MAVEN is an open-source analysis platform that extends MAVEN and provides fast, powerful, and interactive analysis capabilities especially for datasets containing over 100 samples (Agrawal et al. 2019)
MetaboAnalyst	MetaboAnalyst is the most popular metabolomic data analysis platform available freely as a web tool and an R package. The platform supports all the different pre- processing steps required of a good metabolomic platform (Pang et al. 2024)

6.3.1.4 Data Export

The *m/z* ratios, RT, and other measured properties for different compounds/peaks in different groups are exported and are ready for data post-processing.

6.3.2 Data Post-processing

Data post-processing steps ensure that metabolomic data from different groups is free from systematic bias or technical variation (Sun and Xia 2024). The process first involves data normalization to remove any technical source of sample-to-sample variation. Post-data normalization, centering, scaling, and transformation of data is performed to adjust the variance of the different metabolites.

6.3.2.1 Data Normalization

Metabolomic data is high dimensional and has noise. Normalization ensures that metabolites across samples are comparable, and the effect of noise is minimized. In other words, normalization is done sample wise to adjust for across sample distribution making the profiles of one sample comparable to the other. There are many normalization methods employed for metabolomic data. We are discussing a few of the most common methods followed:

1. *Internal Standard Normalization (ISN):* ISN uses a reference metabolite or a set of metabolites for normalization. The reference is supposed to have similar concentration across all the samples (Kohler et al. 2016). The normalization involves dividing the concentration of each metabolite with the reference.
2. *Constant Sum Normalization (CSN):* CSN is total spectral area normalization or integral normalization that adjusts spectra to a constant sum by dividing each metabolite intensity or bin by the total peak areas (Warrack et al. 2009). This transforms each metabolite into a fraction of the total spectrum intensity.
3. *Quantile normalization (QN):* QN assumes that the broader metabolite distribution of each sample (as measured by quantiles) should roughly be equal and adjusting these distributions should normalize the distribution (Bolstad et al. 2003). The overall distribution being equal for samples would mean that a subset of metabolites will be differently enriched in samples.

6.3.2.2 Centering, Scaling and Transformation

As normalization corrects for sample-to-sample variations, centering, scaling, and transformation ensure that metabolites within one sample are comparable to each other. If the metabolites which are lowly expressed have a very different scale from the ones which are highly expressed, then it is difficult to compare them. Scaling and transformation ensure that metabolites across different levels of expression are comparable with each other.

6.3.2.2.1 Centering

Centering methods such as mean centering reduce the overall offset of values supporting data with a better fit for different applications. Mean centering subtracts the value of each metabolite measured for a sample with the mean for that metabolite value.

6.3.2.2.2 Scaling

Scaling is applied to adjust for differences in scale for different metabolites, adjust the heteroscedasticity of the data, and allow for different sizes of subsets of data (Bro and Smilde 2003). Scaling involves dividing the values with a scaling factor. Based on the calculation of scaling factors, the scaling methods can be (1) *Dispersion-based*: Uses deviation as a measure for calculating scaling factors, examples include unit variance scaling. (2) *Average-based:* Uses mean as a measure for calculating scaling factors, examples include level scaling.

6.3.2.2.3 Transformation

Transformation of the data concludes as the final step to make the metabolites comparable within a sample. The overall goals of transformations are (van den Berg et al. 2006):

1. Correct for heteroscedasticity.
2. Convert multiplicative relations into additive relations.
3. Make skewed distributions more symmetric.

There are many ways to transform the data, however, the most common ones follow one of the following types for calculations:

1. *Log transformation*: Log transformation turns the values into their logs (log(x, base)). The base can be natural (typical) or otherwise as well. Log transformation tries to bring the distribution of the values closer to the normal distribution.
2. *Power transformation*: Power transformation utilizes the power function (x^y) to control skewness and bring the data closer to normal distribution. Many different power transformations such as square root or inverse transformations.

6.3.3 Statistical Analysis

Statistical analysis is used to identify key features and assign statistical scores and significance to them. Typical statistical analyses of metabolomic data include univariate and multivariate approaches. They enable the evaluation of the input metabolomic dataset and identification of metabolites that change under different conditions.

6.3.3.1 Univariate Analysis

The univariate analysis focuses on one feature at a time and typically consists of nonparametric or parametric methods such as t-test, one-way ANOVA, and Wilcoxon-rank sum test to test for one feature at a time and identify differences between groups.

6.3.3.2 Multivariate Analysis

Multivariate analysis focuses on employing multiple features of relevance to study the different groups. This typically includes methods such as multiple-way variance analysis (ANOVA), multiple regression analysis, factor analysis, principal component analysis (PCA), partial least squares discriminant analysis (PLS), cluster analysis, and machine learning methods.

6.3.4 Functional Analysis

Functional analysis involves identifying functional differences between various biological states based on their biomolecular profiles. In metabolomics, this means determining the enrichment, either positive or negative, of different pathways based on changes in metabolite levels between states. It also helps to summarize these changes by highlighting a few significant pathways from many differentially enriched metabolites. The summarization can be done using different methods such as over-representation analysis (ORA), functional class scoring (FCS), or topology-based methods (Wieder et al. 2021). Typically, most methods involve scoring the overlap between the different pathways/metabolite sets and the differentially enriched metabolites. There are different tools which support the functional analysis of metabolomic data using well defined methods (Table 6.2). MetaboAnalyst (Pang et al. 2024), a popular tool for the overall analysis of metabolomic data, is routinely used for functional analysis. It supports two different functional analysis methods, Metabolic pathway analysis (MePA) and metabolite set enrichment analysis (MSEA). In addition, there are several different other tools and packages available which support functional analysis of metabolomic data.

Besides above abovementioned analysis steps, many times data from different omics can also be used and combined for better systematic level interpretation. Such multi-omic data integration involves using more than one omic data to get conclusive findings of a biological phenotype or effect (Jendoubi 2021).

There are many ways in which multi-omic analysis can be done (Table 6.3). There are ways to classify them in different ways:

1. Data types:
   (a) *Horizontal or Meta-analysis*: It typically involves using one data type. Multiple datasets of a given data type are analyzed to find common patterns across datasets.
   (b) *Vertical*: The vertical integration involves integration of more than one data type. Vertical analysis typically requires a stricter experimental design across different data types for measurement.
2. Data integration strategies:
   (a) *Early integration*: In early integration, all different omic data tables are concatenated at raw or processed levels to get a single dataset. The difference in scales is a constant challenge faced with early integration methodologies.
   (b) *Intermediate integration*: Intermediate integration methods transform the data into a form before individual datasets/ types can be combined with others. Many network-based methods employ intermediate integration for multi-omic analysis.
   (c) *Late integration*: For late integration, each data type is modeled separately contributing to their own separate scores. The results in the form of the scores are used to build a new high-level model, offering insights from different levels of omic data.

Table 6.2 Different tools available for functional analysis of metabolomic data

MetaboAnalyst	MetaboAnalyst is the most popular metabolomic data analysis platform available freely as a web tool and an R package. The platform also supports dedicated functional analysis over metabolomic data (Pang et al. 2024)
MetFlow	MetFlow is a web-based tool for comprehensive analysis of metabolomic data. MetFlow also supports functional analysis (Shen and Zhu 2019)
PaintOmics3	PaintOmics3 support integrated omic analysis and is available as a web-based tool (Hernandez-de-Diego et al. 2018)
3Omics	3Omics can be used in the single omic analysis mode or in multi-omic mode to do comprehensive analysis including integrative omic functional analysis of the omic data (Kuo et al. 2013)

Table 6.3 Different tools supporting multi-omic analysis

Shiny GATOM	Shiny GATOM is an R Shiny application that takes DE metabolites and DE genes/proteins data to generate a network of perturbed pathways and key points (Emelianova et al. 2022)
PaintOmics3	PaintOmics3 supports integrated omic analysis and is available as a web-based tool (Hernandez-de-Diego et al. 2018)
3Omics	3Omics can be used in the single omic analysis mode or in multi-omic mode to do comprehensive analysis including integrative omic functional analysis of the omic data (Kuo et al. 2013)
GAIT-GM	GAIT-GM is available as a python package for modeling metabolites as a function of gene expression (McIntyre et al. 2022)
iOmicsPASS	iOmicsPass is a network-based analysis tool that supports multi-omic predictive feature selection (Koh et al. 2019)

6.4 Constraints and Considerations in Metabolomic Research

One of the key challenges in metabolomics is the vast diversity of the chemical structures and the abundance of metabolites. This is further complicated by the need to accurately identify and quantify these metabolites with high precision. Most of the metabolites present in the living system have a short half-life, are volatile or unstable in nature, and are present at very low concentrations (Vuckovic 2018).

The lack of a single technology capable of analyzing the entire metabolome (Johnson and Gonzalez 2012) or a single metabolite (Beale et al. 2018) necessitates the use of diverse extraction and detection methods to efficiently and precisely characterize the metabolites. Researchers often employ a combination of complementary approaches, such as GC-MS, LC-MS, and NMR spectroscopy, to expand metabolite coverage (Johnson and Gonzalez 2012; Beale et al. 2018; Dudzik et al. 2018; Roessner and Bowne 2009; Clish 2015). Each of these methods comes with their limitations which can be reduced with careful consideration and other normalization processes.

6.4.1 Analytical and Reproducibility Challenges in Metabolomics

The use of metabolomics to evaluate clinical samples has become a key determinant in identifying specific regulators for conditions that can be used as clinical biomarkers. The analytical variability in clinical samples contributes significantly to conducting valid comparisons between different populations and conditions. Improper collection, handling, and preparation of data without employing a standard operating protocol leads to the generation of unreliable data. The differences are more prominent during untargeted metabolomics where a large amount of semi-quantitative data of unknown analytes are collected whose prior information is not available (Kohler et al. 2016).

A large number of metabolic studies are performed using chromatography and MS. The efficacy of these analytical techniques is affected over a period of time and requires frequent cleaning and calibrations. The division of samples in batches and randomization of samples increases the reproducibility and confidence in the results. Proper quality controls (QC) with standard operating procedures (SOP) are vital for minimizing the analytical error. The cleaning and maintenance of instruments before every batch with proper standards and control is essential for increasing the confidence of the output generated (Cajka and Fiehn 2016; Dunn et al. 2012). The QCs help in controlling and normalizing the intensity between batches and help in proper comparison of control and samples (Cajka and Fiehn 2016; Naz et al. 2014). The QC also help in checking the reproducibility of the instrument and normalizing accordingly.

The sample collection and storage are a very important part of metabolic profiling to generate reproducible data. A variety of metabolites are unstable and have very short half-lives which affect the reproducibility in the sample sets. Variation in sample collection even to a minute degree can largely affect the metabolic profiles. This heterogeneity in analytical methods across laboratories can lead to issues when comparing data and integrating results, as differences in precision, metabolite coverage, and data processing can hinder reproducibility. Efforts to address these challenges and improve metabolomics reliability have been ongoing. Standardized methods and SOPs have been proposed to harmonize experimental workflows and data analysis practices. Automation in sample collection and sample preparation can help to reduce the variability in data collected and provide high-quality data. The sample preparation is also crucial for generating reliable data, on the physiochemical properties of metabolite changes in different solvents and it becomes challenging to compare similar studies done by different practicing groups. Using a simple sample preparation method (enrichment, filtration, dilution, precipitation, etc.) can reduce the number of steps involved and decrease the effects of human perturbations and errors specially during untargeted metabolomics.

6.4.2 Metabolic Coverage and Sensitivity Limitation

Metabolomics provides a snapshot of all the cellular processes occurring in a cell at a particular time point. Metabolite capturing and identification is the key step in understanding the physiology of the cell. The range of metabolites in the cell can be from g/L to ng/L which makes it difficult to cover all the metabolites (Teoh et al. 2015). Generally, a separation step is required to separate all the metabolites during capture prior to their identification. High-sensitive mass spectrometry (MS) or nuclear magnetic resonance (NMR) is the choice of capturing metabolites for untargeted and targeted studies. But due to the wide variety and complexity of diverse metabolites generated inside a cell, it is impossible to determine all the metabolites using a single technique, therefore a combination of techniques is employed to capture and analyze all the available metabolites in the cell. Hyphenation or combination of various techniques, such as Spectroscope-LC-MS/MS or NMR-LC-MS/MS, along with other separation methods, are employed to efficiently separate and characterize all metabolites. Since the concentration of the metabolites differs vastly in cells, optimization of the sensitivity of the instrument is required to develop a comprehensive method to detect all metabolites. Every method has its superiority in certain aspects. For instance, in a study conducted by Fang et al. the authors suggested that there were around 40% of metabolites were degraded due to the sample preparation step in GC-MS which was not the case in LC-MS (Fang et al. 2015). Many of the techniques require derivatization of the samples for sharper signals during ion acquisition (Kloos et al. 2014; Fiehn et al. 2000), but it is difficult to achieve 100% derivatization of biological complex samples. Experts in different fields are required to develop novel methods to enhance the coverage of the metabolome (Pinu et al. 2019; Wishart 2016). Ionization methods and the sensitivity of the detectors greatly impact the data collection and identification. The choice of ionization technique is important which greatly depends upon some pre-existing data about the samples. There are a wide variety of ionization techniques like electron ionization (EI), chemical ionization (CI), electrospray ionization (ESI), and atmospheric pressure chemical ionization (APCI). The EI and CI are more suitable for GC-MS, ESI can be used with liquid chromatography (LC) and capillary electrophoresis (CE), APCI can be used with either GC, LC, or CE. The advantage of using MS-based analyzer is they can be employed either for high selectivity and sensitivity when using triple quadrupole (TQ) or quadrupole linear ion trap or for wide scanning of different metabolites using quadrupole time-of-flight (qTOF). Combination of the various ionization techniques and mass analyzer is employed to increase the coverage of metabolite as well as spectral information for the identification of the compound. The identification of metabolites cannot be solely done on the MS spectra, and hence tandem MS (MS/MS) is required for identification of metabolites from the sample. The use of additives such as ethylenediamine-tetra acetic acid (EDTA) can enhance or suppress the ionization potential of many metabolites. Nuclear magnetic resonance (NMR) is a nondestructive technique for sample identification. NMR-based analyzer can be used in tandem with LC for the identification of metabolites (Daykin et al. 2001; Akira et al. 2010). NMR is frequently utilized in clinical samples to detect metabolites which give good reproducible data. The main limitation is that NMR typically has lower sensitivity compared to MS. This is because the magnetic dipole in small populations of metabolites is very low, resulting in less pronounced differences in metabolite patterns. This can be improved by using high-power electromagnets and cryoprobes.

6.4.3 Complexity of Data Management Interpretation and Validation

Advancements in the sensitivity of the analytical techniques have increased the robustness of detecting thousands of metabolites from as little as 50 μl of urine sample or 10 mg of tissues (Johnson et al. 2015b; Yu et al. 2020; Warth et al. 2018). A small batch of 30 metabolic samples can generate over 10GB of data required for analysis. This huge amount of data cannot be processed and analyzed manually. There are four major aspects of metabolite data challenges are (1) data acquisition, (2) feature extraction, (3) quantitative and statistical analysis, and (4) metabolite annotation.

There are three basic modes for data acquisition in metabolic profiling: (1) MS full scan, (2) data-dependent acquisition (DDA), and (3) data-independent acquisition (DIA) (Benton et al. 2015; Roemmelt et al. 2014; Rost et al. 2014). MS full scan is applied for absolute quantification of MS data but cannot be used for annotating the unknown metabolite. DDA is used to acquire *m/z* (mass/charge) for a particular precursor ion in a narrow range of mass for pure MS/MS spectrum to generate high-quality data for annotation for an abundant precursor ion but is not suitable for low abundant ions. The DIA on the other hand is used to acquire *m/z* for a wide range of precursor ions but requires intensive deconvolution to assign the mass for the precursor ion. A closer inspection of all three methods revealed that DDA is less suitable for the discovery of new metabolites, but it is the most convenient method to obtain high-resolution MS/MS spectra for metabolite annotation. The hyphenation of DDA and DIA, i.e., data-dependent assisted data-independent acquisition (DaDIA) is much more suitable for the analysis of biological data samples (Guo et al. 2021).

Feature extraction or peak picking, i.e., separation of the real peak from the background noise is a challenge, especially during untargeted metabolomics (Sindelar and Patti 2020). Open-source automated software is available to for peak picking and annotation (Tautenhahn et al. 2008; Trevino et al. 2015). Enhancing the sensitivity of feature extraction also increases the risk of false positives (Sindelar and Patti 2020). To avoid the risk of false positives, the extracted ion chromatogram (EIC) of the metabolites must be verified individually, but the process is time-consuming, as the number of metabolites generated by a single experiment is huge. The advent of multiple artificial intelligence programs with proper training with existing datasets has led to the generation of multiple neural networks that can perform efficient feature extraction and validation of the peaks for large datasets.

For absolute quantification and comparison of the metabolomic data, normalization of the samples is important (Warrack et al. 2009; Bliss et al. 1999; Karu et al. 2018; Silva et al. 2013). Normalization can be used pre- or post-sample acquisition. If using pre-, the concentrations of the samples are adjusted to calculate the valid fold change, if done (post) the data are analyzed with some assumptions to normalize the values. Interpreting metabolic data and applying the correct statistical parameters present significant challenges because the data generated from ESI spectra are inherently nonlinear. Since the distribution pattern of the data is often diverse and complicated in metabolomic data transformation is done at various levels to change the data from non-normalized data to normalized data for easier statistical responses (van den Berg et al. 2006) changes applied to data during data transformation can affect the overall data during comparison and can lead to changes in principal component analysis.

Metabolite validation or metabolite annotation is the last step in metabolic study. The first step of metabolite annotation is comparing the retention time, MS-peak, and MS/MS spectra of the unknown with the known standards (Schymanski et al. 2014). Since the majority of the data in untargeted metabolomics is novel, proper annotation and validation require a lot of time by the researchers which is also called as dark matter in the field of metabolomics (Silva et al. 2013; Wang et al. 2016). The strategies used for the identification and validation of the metabolite are (1) known-to-unknown-based propagated annotation (Wang et al. 2016), (2) de novo annotations (Duhrkop et al. 2019), and

(3) in-silico fragmentation-based annotation (Huan et al. 2015). Annotation of the novel metabolites can be done by comparing the structurally similar metabolites which will show similar types of MS/MS spectra. Already available algorithms like Natural Products Social Molecular Networking (GNPS) and NIST Hybrid Similarity Search (HSS) are available that compare the MS/MS precursor ion and fragmentation ions and predict the structure of the metabolite (Moorthy et al. 2017; Falkner et al. 2008; Watrous et al. 2012).

6.4.4 Ethical Consideration in Metabolic Profiling

The metabolome of a biological system is the most observable state that can describe the phenotype of that biological system at that particular time point. The evolution of metabolomics stems from the fact that it is not only helpful in biomarker discovery but also can be used for the selection of therapeutic molecules for the intervention of particular diseases. Since metabolite identification is a noninvasive procedure, it has garnered the attention of many clinical and medical researchers. Many of the drugs that are used for activating or deactivating a signaling cascade inside a biological system are controlled by a diverse array of drugs or metabolite (Hagan et al. 2015).

The identification of drug and their targets through metabolomics requires extensive reproducible investigation with proper guidelines and SOPs (Chau et al. 2008). The quality check (QC) and quality assurance (QA) are defined parameters by the Food and Drug Association (FDA) for the development of any high-throughput analytical assay for drug or biomarker development. These definitions have been endorsed by the Cooperation on International Traceability in Analytical Chemistry (CITAC) and EuraChem (A Focus for Analytical Chemistry in Europe). Laboratories running the targeted and semi-targeted metabolomic study must follow the guidelines laid down by the FDA. There are no complete guidelines for untargeted metabolomics. Proper sample collection guidelines are also made for different sample types for targeted metabolomics. Since sample collection and storage greatly impact the metabolomic data, all the regulatory guidelines of patient confidentiality and proper labeling methods are to be followed. The sample maintenance and the data acquisition process should be followed with proper internal controls.

6.5 Advancements and Strategic Insights

The field of metabolomics has advanced significantly through the development of cutting-edge technologies and methodological improvements. With high-resolution mass spectrometry (HRMS) and nuclear magnetic resonance (NMR) spectroscopy, metabolite identification and quantification have become more precise and sensitive. In addition to providing comprehensive metabolite coverage, multiplatform approaches, such as LC-MS, GC-MS, and NMR, also improve detectability through chemical derivatization. Metabolomics of single cells, enabled by microfluidics and mass cytometry, reveals cellular heterogeneity and metabolic diversity. Dynamic and longitudinal metabolic profiling is possible with real-time and in vivo metabolomics, such as real-time mass spectrometry and magnetic resonance spectroscopy (MRS). Through advanced bioinformatics platforms and artificial intelligence, as well as metabolomics integrated with genomics, transcriptomics, and proteomics, biological systems can be understood holistically. Using guidelines from the Metabolomics Standards Initiative (MSI) to improve sample preparation, automation, and standardization ensures reproducibility and comparability. Personalized medicine and systems biology are being driven by these advancements, providing deeper insights into metabolic processes and disease mechanisms.

6.5.1 Cross-Omic Strategies: Integrating Metabolomics with Genomic and Proteomic Data

One of the major advancements of metabolomic study would be the cross-omic integration. The cross-omic integration approach combines data from different omic disciplines, such as genomics, proteomics, and metabolomics. Through this comprehensive approach, we can gain a deeper understanding of biological systems and discover new biomarkers, therapeutic targets, and insights into diseases. Identifying pathogenic genetic variations in patients suspected of genetic diseases requires DNA sequencing methods such as exome sequencing (ES) and whole genome sequencing (WGS). In contrast, WGS can produce millions of variants per patient, while ES can produce tens of thousands (Neu et al. 2019). Even with filtering strategies such as Combination Annotation Dependent Depletion (CADD), investigators must manually assess many variants, leading to clinically unsatisfactory classifications based on rare or novel variants. Integrating metabolomics, proteomics, and genomics requires interpreting results so that metabolite abnormalities and protein dysfunctions can be linked to potentially deficient reactions or enzymes and their corresponding genes. For example, consider a patient (various neurological diseases and gastric cancer) with elevated levels of methylmalonic acid, a valuable diagnostic marker for vitamin B12 deficiency (Lee et al. 2019). This abnormality suggests a deficiency in the enzyme methylmalonyl-CoA mutase, responsible for converting methylmalonyl-CoA to succinyl-CoA in the metabolism of certain amino acids and fatty acids. Proteomic analysis can further reveal reduced levels or dysfunction of the methylmalonyl-CoA mutase protein. Genetic analysis can then focus on the MUT gene, which encodes this enzyme. Identifying a pathogenic variant in the MUT gene confirms the diagnosis of methylmalonic acidemia (MMA, a rare genetic disorder of the liver) (Head et al. 2023), an inborn error of metabolism (IEM). This integrated approach, combining metabolomics, proteomics, and genomics, provides a comprehensive understanding of the disease mechanism and can guide more effective treatment strategies.

6.5.2 Proposals for Standardized Practices and Collaborative Efforts

Metabolomics, proteomics, and genomics should be integrated by creating unified data standards, creating integrated databases, and fostering interdisciplinary collaboration. It is essential to develop computational tools and standardize functional validation protocols. In order to speed discoveries and improve clinical outcomes, open data sharing, support for training, public-private partnerships, and implementation of clinical guidelines are crucial. This will enhance our understanding and treatment of metabolic and genetic disorders through regular monitoring and evaluation. There are a few essential practices that can help to understand data more comprehensive manner and solve the disease better way.

(a) *Establish Unified Data Standards*: Ensure consistency and reproducibility across studies by developing standard protocols for metabolomics, proteomics, and genomic data collection, processing, and analysis. The process includes defining common data formats, quality control measures, and reporting guidelines.
(b) *Create Integrated Databases:* Combine genomic, proteomic, and metabolomic data to create comprehensive databases. To link metabolite abnormalities, protein dysfunctions, and genetic variants, these databases must be easily accessible to researchers and clinicians.
(c) *Promote Interdisciplinary Collaboration:* Collaboration between researchers from different fields, such as biochemistry, molecular biology, bioinformatics, and clinical medicine, is encouraged. To gain a more

holistic understanding of disease mechanisms, multidisciplinary teams should be established to facilitate the integration of diverse expertise and perspectives.

(d) *Develop Computational Tools:* The development of advanced computational tools and algorithms will enable large metabolomic, proteomic, and genomic datasets to be integrated and analyzed. The tools should be user-friendly and should be effective at identifying potential pathogenic variants and their biological effects.

(e) *Standardize Functional Validation:* Develop standardized protocols for functional validation studies to assess genetic variant pathogenicity. In addition, this involves developing assays for measuring enzyme activity, protein expression, and metabolic flux, as well as guidelines for interpreting these results.

(f) *Encourage Data Sharing*: Promote open data sharing among researchers and institutions to accelerate discoveries and reduce redundancy. Establish data repositories and sharing platforms with appropriate privacy and ethical guidelines to protect patient information.

(g) *Support Training and Education:* To develop training programs and educational resources to equip researchers and clinicians with the skills necessary to integrate metabolomics, proteomics, and genomics. This includes various workshops, online courses, and collaborative research projects.

(h) *Foster Public-Private Partnerships:* For the advancement of integrated omic research, academic institutions, healthcare organizations, and the biotechnology industry should cooperate to leverage resources, expertise, and technology.

(i) *Implement Clinical Guidelines:* Develop clinical guidelines for using integrated omic data in diagnosing and treating metabolic and genetic disorders. These guidelines should be based on the latest research and consensus among experts in the field.

(j) *Monitor and Evaluate Progress:* Establish mechanisms to regularly monitor and evaluate the progress of standardized practices and collaborative efforts. This includes setting benchmarks, conducting periodic reviews, and making necessary adjustments to ensure continuous improvement and adaptation to new scientific advancements.

6.5.3 Recommendations for Infrastructure and Skill Development in Metabolomics

Although metabolomics is becoming more accessible, there are still two main challenges for researchers. The first thing to note about metabolomics is that it is a highly multidisciplinary science. To conduct a metabolomic study from beginning to end, you will need expertise in molecular biology, analytical chemistry, physical chemistry, quality assurance/control, cheminformatics, statistics and data science, bioinformatics, and biochemistry. Second, the metabolome is chemically very diverse (in terms of physicochemical properties, concentration range, and stability) and complex, since metabolites are highly correlated and interdependent. It is therefore difficult to standardize metabolomic analytical and data science workflows since there is often no single correct approach and every analysis is context-specific. As a result, effective metabolomic researchers are required to have a broad range of knowledge in all aspects of the workflow, as well as specific skills for some areas, as well as cross-disciplinary objective knowledge and subjective judgment based on experience. To overcome these challenges, we need to advance in skill development and infrastructure development.

(a) *Facilities and Infrastructure*: Invest in advanced laboratories equipped with high-quality metabolomic core facilities, including high-resolution mass spectrometers, nuclear magnetic resonance (NMR) spectrometers, Bruker Elute Ultra-High Performance Liquid Chromatograph, Bruker timsTOF, Pro Mass Spectrometer, Hamilton STARlet Automated Liquid Handling System, Thermo Orbitrap Fusion Tribrid Mass Spectrometer, Thermo Orbitrap Fusion Lumos Tribrid Mass Spectrometer, Thermo Vanquish Horizon Ultra-High-Performance Liquid Chromatographs, Sciex 7500 QTrap Ready Mass Spectrometer, Shimadzu Nexera Ultra-High Performance Liquid Chromatograph, and other essential analytical instruments. Ensure regular maintenance and updates to keep up with technological advancements.

(b) *Comprehensive Data Management Systems:* The development of robust data management and storage systems capable of handling large datasets generated by metabolomic studies is essential. Ensure data integrity and reproducibility by implementing standardized data formats and quality control protocols.

(c) *Interdisciplinary Training Programs*: Research in metabolomics requires interdisciplinary training programs that equip researchers with diverse skills. Hands-on experience with cutting-edge technologies and data analysis techniques should be provided by these programs in biochemistry, bioinformatics, and analytical chemistry. For example, a comprehensive training program might cover sample preparation, analysis, and interpretation using mass spectrometry, as well as practical applications in clinical and environmental research. Through this holistic approach, researchers can integrate metabolomic data with genomic and proteomic information, fostering innovation and collaboration across fields.

(d) *Bioinformatics and Computational Support*: To strengthen bioinformatics and computational support, researchers need access to high-performance computing resources and specialized software tools, such as XCMS for data processing, MetaboAnalyst for statistical analysis, and KEGG for pathway mapping. In addition, advanced statistical methods and machine learning techniques need to be learned. This resource can be used by a research group to preprocess mass spectrometry data, identify significant changes in metabolites, and map these changes onto metabolic pathways, thereby facilitating efficient data handling and actionable insights into the metabolic effects of a new drug.

6.6 Conclusions

Over the past two decades, metabolomics has advanced significantly. However, several aspects of this field are still in development and limit its application in a variety of areas. Some of these limitations include limited detection ranges, bulk molecular analysis, lack of chromatographic methods for improved resolution, and high costs associated with analytical devices like mass spectrometers and NMR. Despite these limitations, metabolomics remains a valuable tool in translational biology, pharmacological medicine, biomarker discovery, and diagnosis. It is possible to identify potential therapeutic targets based on the metabolic profiles of patients with different diseases. Additionally, it can monitor disease progression, treatment response, and side effects of therapies. Metabolomics is also enabling research in other fields, such as agriculture, environmental surveillance, and nutritional biology.

Metabolomics will emerge as a field of great promise in the coming years as advances in basic research and healthcare technologies surpass our current understanding of living organisms. It has the potential to improve precision medicine, personalized nutrition, disease diagnosis, and biomarker discovery. Further, the development of AI- and machine-learning-based tools for data integration and their applications in translational biology highlight its potential. Identifying biomarkers associated with heart disease, such as specific lipids and amino acids, can aid in early diagnosis and disease monitoring. There is evidence that elevated levels of trimethylamine-N-oxide (TMAO) may increase the risk of cardiovascular disease (Zheng and He 2022). New therapeutic strategies can be developed to reduce this risk by targeting metabolic pathways that produce TMAO. Metabolomics has proven to be valuable in diabetes research, providing insight into metabolic changes that occur in individuals with type 2 diabetes. Branched-chain amino acids (BCAAs) have been associated with insulin resistance (Yoon 2016), for example. Researchers can develop more effective treatments to manage blood glucose levels and improve insulin sensitivity by gaining a better understanding of these metabolic changes. Furthermore, metabolomics has played an important role in microbiome research. Scientists can gain a deeper understanding of the interaction between the microbiome and the host's metabolism by analyzing the metabolites produced by gut bacteria. In response to this knowledge, probiotics and dietary interventions aimed at modulating the microbiome have been developed to improve health outcomes, such as in the management of inflammatory bowel disease and obesity.

As metabolomics advances, it is expected that metabolite biomarkers will be used to diagnose diseases earlier and with greater accuracy. An individual's metabolic profile will be taken into consideration when tailoring treatment, which will significantly enhance personalized medicine. Using metabolic profiling, certain tumors can be identified by specific nutrients (glutamine) or metabolic pathways (Li et al. 2021). Tumor growth can be inhibited by therapies targeting glutamine metabolism. The use of metabolomics in neurological diseases enables the identification of biomarkers for the early detection of Alzheimer's disease. The presence of specific metabolites in the brain (Tau protein and amyloid-beta peptide) and cerebrospinal fluid (soluble amyloid precursor protein, neurogranin) can signal the onset of neurodegenerative processes before significant symptoms appear, enabling earlier intervention and better disease management (McKeen et al. 1994). Finally, machine learning and artificial intelligence are increasingly being applied to metabolomic data to uncover patterns and correlations previously unknown. It will lead to significant medical and scientific breakthroughs, including improved diagnostics, more effective treatments, and a deeper understanding of different diseases, ultimately improving patient outcomes.

References

Agrawal S et al (2019) El-MAVEN: a fast, robust, and user-friendly mass spectrometry data processing engine for metabolomics. Methods Mol Biol 1G78: 301–321

Akira K et al (2010) LC-NMR identification of a novel taurine-related metabolite observed in (1)H NMR-based metabonomics of genetically hypertensive rats. J Pharm Biomed Anal 51(5):1091–1096

Alberts B et al (2022) Molecular biology of the cell, 7th edn. W. W. Norton C Company, New York, NY

Alexandrov T (2020) Spatial metabolomics and imaging mass spectrometry in the age of artificial intelligence. Annu Rev Biomed Data Sci 3:61–87

Amarnath K et al (2023) Stress-induced metabolic exchanges between complementary bacterial types underly a dynamic mechanism of inter-species stress resistance. Nat Commun 14(1):3165

Beale DJ et al (2018) Review of recent developments in GC-MS approaches to metabolomics-based research. Metabolomics 14(11):152

Benton HP et al (2015) Autonomous metabolomics for rapid metabolite identification in global profiling. Anal Chem 87(2):884–891

Bliss DZ et al (1999) Comparison of subjective classification of stool consistency and stool water content. J Wound Ostomy Continence Nurs 26(3):137–141

Bolstad BM et al (2003) A comparison of normalization methods for high density oligonucleotide array data based on variance and bias. Bioinformatics 1G(2): 185–193

Bro R, Smilde AK (2003) Centering and scaling in component analysis. J Chemom 17(1):16–33

Bueschl C et al (2017) MetExtract II: a software suite for stable isotope-assisted untargeted metabolomics. Anal Chem 8G(17):9518–9526

Cajka T, Fiehn O (2016) Toward merging untargeted and targeted methods in mass spectrometry-based metabolomics and Lipidomics. Anal Chem 88(1): 524–545

Chambers MC et al (2012) A cross-platform toolkit for mass spectrometry and proteomics. Nat Biotechnol 30(10):918–920

Chau CH et al (2008) Validation of analytic methods for biomarkers used in drug development. Clin Cancer Res 14(19):5967–5976

Chen Y, Li EM, Xu LY (2022) Guide to metabolomics analysis: a bioinformatics Workffow. Meta 12(4)

Christ B et al (2018) Contribution of untargeted metabolomics for future assessment of biotech crops. Trends Plant Sci 23(12):1047–1056

Clasquin MF, Melamud E, Rabinowitz JD (2012) LC-MS data processing with MAVEN: a metabolomic analysis and visualization engine. Curr Protoc Bioinformatics. Chapter 14: p. Unit14 11

Clayton TA et al (2006) Pharmaco-metabonomic phenotyping and personalized drug treatment. Nature 440(7087):1073–1077

Clendinen CS, Monge ME, Fernández FM (2017) Ambient mass spectrometry in metabolomics. Analyst 142(17):3101–3117

Clish CB (2015) Metabolomics: an emerging but powerful tool for precision medicine. Cold Spring Harb Mol Case Stud 1(1):a000588

Daykin CA et al (2001) Application of directly coupled HPLC NMR to separation and characterization of lipoproteins from human serum. Anal Chem 73(6): 1084–1090

Demain AL (1980) Microbial production of primary metabolites. Naturwissenschaften 67(12):582–587

Dettmer et al. 2007 Mass spectrometry-based metabolomics.pdf

Dudzik D et al (2018) Quality assurance procedures for mass spectrometry untargeted metabolomics. A review. J Pharm Biomed Anal 147:149–173

Dueñas ME, Larson EA, Lee YJ (2019) Toward mass spectrometry imaging in the metabolomics scale: increasing metabolic coverage through multiple on-tissue chemical modifications. Front Plant Sci 10: 860

Duhrkop K et al (2019) SIRIUS 4: a rapid tool for turning tandem mass spectra into metabolite structure information. Nat Methods 16(4):299–302

Dunn WB et al (2012) The importance of experimental design and QC samples in large-scale and MS-driven untargeted metabolomic studies of humans. Bioanalysis 4(18):2249–2264

Emelianova M et al (2022) Shiny GATOM: omics-based identification of regulated metabolic modules in atom transition networks. Nucleic Acids Res 50(W1): W690–W696

Falkner JA et al (2008) A spectral clustering approach to MS/MS identification of post-translational modifications. J Proteome Res 7(11):4614–4622

Fang M et al (2015) Thermal degradation of small molecules: a global metabolomic investigation. Anal Chem 87(21):10935–10941

Fiehn O (2002) Metabolomics–the link between genotypes and phenotypes. Plant Mol Biol 48(1–2):155–171

Fiehn O et al (2000) Identification of uncommon plant metabolites based on calculation of elemental compositions using gas chromatography and quadrupole mass spectrometry. Anal Chem 72(15): 3573–3580

Gasteiger J (2016) Chemoinformatics: achievements and challenges, a personal view. Molecules 21(2):151

Guo J et al (2021) DaDIA: hybridizing data-dependent and data-independent acquisition modes for generating high-quality metabolomic data. Anal Chem G3(4): 2669–2677

Hagan SO et al (2015) A 'rule of 0.5' for the metabolite-likeness of approved pharmaceutical drugs. Metabolomics 11(2):323–339

Hartman TE, Lees HJ (2023) Introduction of metabolomics: an overview. In: Soni V, Hartman TE (eds) Metabolomics: recent advances and future applications. Springer International Publishing, Cham, pp 1–37

Head PE, Meier JL, Venditti CP (2023) New insights into the pathophysiology of methylmalonic acidemia. J Inherit Metab Dis 46(3):436–449

Hernandez-de-Diego R et al (2018) PaintOmics 3: a web resource for the pathway analysis and visualization of multi-omics data. Nucleic Acids Res 46(W1):W503–W509

Horai H et al (2010) MassBank: a public repository for sharing mass spectral data for life sciences. J Mass Spectrom 45(7):703–714

Hounsome N et al (2008) Plant metabolites and nutritional quality of vegetables. J Food Sci 73(4):R48–R65

Hu T et al (2023) Single-cell spatial metabolomics with cell-type specific protein profiling for tissue systems biology. Nat Commun 14(1):8260

Huan T et al (2015) MyCompoundID MS/MS search: metabolite identification using a library of predicted fragment-ion-spectra of 383,830 possible human metabolites. Anal Chem 87(20):10619–10626

Jendoubi T (2021) Approaches to integrating metabolomics and multi-omics data: a primer. Meta 11(3)

Johnson CH, Gonzalez FJ (2012) Challenges and opportunities of metabolomics. J Cell Physiol 227(8): 2975–2981

Johnson CH et al (2012) Xenobiotic metabolomics: major impact on the metabolome. Annu Rev Pharmacol Toxicol 52:37–56

Johnson CH et al (2015a) Bioinformatics: the next frontier of metabolomics. Anal Chem 87(1):147–156

Johnson CH et al (2015b) Metabolism links bacterial biofilms and colon carcinogenesis. Cell Metab 21(6): 891–897

Johnson CH, Ivanisevic J, Siuzdak G (2016) Metabolomics: beyond biomarkers and towards mechanisms. Nat Rev Mol Cell Biol 17(7):451–459

Kanehisa M et al (2016) KEGG as a reference resource for gene and protein annotation. Nucleic Acids Res 44 (D1):D457–D462

Karlic H et al (2015) Inhibition of the mevalonate pathway affects epigenetic regulation in cancer cells. Cancer Genet 208(5):241–252

Karu N et al (2018) A review on human fecal metabolomics: methods, applications and the human fecal metabolome database. Anal Chim Acta 1030:1–24

Katajamaa M, Miettinen J, Oresic M (2006) MZmine: toolbox for processing and visualization of mass spectrometry based molecular profile data. Bioinformatics 22(5):634–636

Kloos DP et al (2014) Comprehensive gas chromatography-electron ionisation mass spectrometric analysis of fatty acids and sterols using sequential one-pot silylation: quantification and isotopologue analysis. Rapid Commun Mass Spectrom 28(13): 1507–1514

Koh HWL et al (2019) iOmicsPASS: network-based integration of multiomics data for predictive subnetwork discovery. NPJ Syst Biol Appl 5:22

Kohler I et al (2016) Analytical pitfalls and challenges in clinical metabolomics. Bioanalysis 8(14):1509–1532

Kuo TC, Tian TF, Tseng YJ (2013) 3Omics: a web-based systems biology tool for analysis, integration and visualization of human transcriptomic, proteomic and metabolomic data. BMC Syst Biol 7:64

Lane AN, Fan TW (2017) NMR-based stable isotope resolved metabolomics in systems biochemistry. Arch Biochem Biophys 628:123–131

Lee SM et al (2019) Methylmalonic acid and homocysteine as indicators of vitamin B12 deficiency in patients with gastric cancer after gastrectomy. Nutrients 11(2)

Li H, He J, Jia W (2016) The inffuence of gut microbiota on drug metabolism and toxicity. Expert Opin Drug Metab Toxicol 12(1):31–40

Li T, Copeland C, Le A (2021) Glutamine metabolism in cancer. Adv Exp Med Biol 1311:17–38

Lodish HF (2008) Molecular Cell Biology. W. H. Freeman

McIntyre LM et al (2022) GAIT-GM integrative cross-omics analyses reveal cholinergic defects in a C. elegans model of Parkinson's disease. Sci Rep 12(1):3268

McKeen ES, Feniuk W, Humphrey PP (1994) Mediation by SRIF1 receptors of the contractile action of somato statin in rat isolated distal colon; studies using some novel SRIF analogues. Br J Pharmacol 113(2): 628–634

Montenegro-Burke JR, Guijas C, Siuzdak G (2020) METLIN: a tandem mass spectral library of standards. Methods Mol Biol 2104:149–163

Moorthy AS et al (2017) Combining fragment-ion and neutral-loss matching during mass spectral library searching: a new general purpose algorithm applicable to illicit drug identification. Anal Chem 8G(24): 13261–13268

Munjal Y, Tonk R, Sharma R (2022) Analytical techniques used in metabolomics: a review. Systematic Reviews in Pharmacy 13:550–556

Naz S et al (2014) Method validation strategies involved in non-targeted metabolomics. J Chromatogr A 1353:99–105

Neu MB, Bowling KM, Cooper GM (2019) Clinical utility of genomic sequencing. Curr Opin Pediatr 31(6): 732–738

Nicholson JK et al (2012) Metabolic phenotyping in clinical and surgical environments. Nature 4G1(7424): 384–392

Oliver SG et al (1998) Systematic functional analysis of the yeast genome. Trends Biotechnol 16(9):373–378

Pang Z et al (2024) MetaboAnalyst c.0: towards a unified platform for metabolomics data processing, analysis and interpretation. Nucleic Acids Res 52(W1):W398–W406

Pinu FR et al (2019. G(4)) Systems biology and multi-omics integration: viewpoints from the metabolomics research community. Meta 9

Ponomarenko EA et al (2016) The size of the human proteome: the width and depth. Int J Anal Chem 2016:7436849

Roberts LD et al (2012) Targeted metabolomics. Curr Protoc Mol Biol. Chapter 30: p. Unit 30 2 1–24

Roemmelt AT et al (2014) Liquid chromatography, in combination with a quadrupole time-of-flight instrument (LC QTOF), with sequential window acquisition of all theoretical fragment-ion spectra (SWATH) acquisition: systematic studies on its use for screenings in clinical and forensic toxicology and comparison with information-dependent acquisition (IDA). Anal Chem 86(23):11742–11749

Roessner U, Bowne J (2009) What is metabolomics all about? BioTechniques 46(5):363–365

Rost HL et al (2014) OpenSWATH enables automated, targeted analysis of data-independent acquisition MS data. Nat Biotechnol 32(3):219–223

Saito K, Matsuda F (2010) Metabolomics for functional genomics, systems biology, and biotechnology. Annu Rev Plant Biol 61:463–489

Salzberg SL (2018) Open questions: how many genes do we have? BMC Biol 16(1):94

Schymanski EL et al (2014) Identifying small molecules via high resolution mass spectrometry: communicating confidence. Environ Sci Technol 48(4):2097–2098

Shen X, Zhu ZJ (2019) MetFlow: an interactive and integrated workffow for metabolomics data cleaning and differential metabolite discovery. Bioinformatics 35(16):2870–2872

Silva LP et al (2013) Measurement of DNA concentration as a normalization strategy for metabolomic data from adherent cell lines. Anal Chem 85(20):9536–9542

Sindelar M, Patti GJ (2020) Chemical discovery in the era of metabolomics. J Am Chem Soc 142(20):9097–9105

Sitter B et al (2002) High-resolution magic angle spinning MRS of breast cancer tissue. NMR Biomed 15(5): 327–337

Smith CA et al (2006) XCMS: processing mass spectrometry data for metabolite profiling using nonlinear peak alignment, matching, and identification. Anal Chem 78(3):779–787

Smith CJ et al (2022) Integrative analysis of metabolite GWAS illuminates the molecular basis of pleiotropy and genetic correlation. elife 11:11

Sperber H et al (2015) The metabolome regulates the epigenetic landscape during naive-to-primed human

embryonic stem cell transition. Nat Cell Biol 17(12): 1523–1535
Spratlin JL, Serkova NJ, Eckhardt SG (2009) Clinical applications of metabolomics in oncology: a review. Clin Cancer Res 15(2):431–440
Sun J, Xia Y (2024) Pretreating and normalizing metabolomics data for statistical analysis. Genes Dis 11(3):100979
Swainston N, Mendes P, Kell DB (2013) An analysis of a 'community-driven' reconstruction of the human metabolic network. Metabolomics G(4):757–764
Tanaka K et al (1988) Protein and polymer analyses up to m/z 100 000 by laser ionization time-of-flight mass spectrometry. Rapid Commun Mass Spectrom 2(8): 151–153
Tautenhahn R, Bottcher C, Neumann S (2008) Highly sensitive feature detection for high resolution LC/MS. BMC Bioinformatics 9G:504
Teoh ST et al (2015) A metabolomics-based strategy for identification of gene targets for phenotype improvement and its application to 1-butanol tolerance in Saccharomyces cerevisiae. Biotechnol Biofuels 8:144
Trevino V et al (2015) GridMass: a fast two-dimensional feature detection method for LC/MS. J Mass Spectrom 50(1):165–174
van den Berg RA et al (2006) Centering, scaling, and transformations: improving the biological information content of metabolomics data. BMC Genomics 7:142
Vuckovic D (2018) Improving metabolome coverage and data quality: advancing metabolomics and lipidomics for biomarker discovery. Chem Commun (Camb) 54(50):6728–6749
Wang M et al (2016) Sharing and community curation of mass spectrometry data with global natural products social molecular networking. Nat Biotechnol 34(8): 828–837
Warrack BM et al (2009) Normalization strategies for metabonomic analysis of urine samples. J Chromatogr B Analyt Technol Biomed Life Sci 877(5–6):547–552
Warth B et al (2018) Metabolomics reveals that dietary Xenoestrogens Alter cellular metabolism induced by Palbociclib/Letrozole combination cancer therapy. Cell Chem Biol 25(3):291–300 e3
Watrous J et al (2012) Mass spectral molecular networking of living microbial colonies. Proc Natl Acad Sci USA 10G(26):E1743–E1752
Wieder C et al (2021) Pathway analysis in metabolomics: recommendations for the use of over-representation analysis. PLoS Comput Biol 17(9):e1009105
Wiseman JM et al (2006) Tissue imaging at atmospheric pressure using desorption electrospray ionization (DESI) mass spectrometry. Angew Chem Int Ed 45(43):7188–7192
Wishart DS (2016) Emerging applications of metabolomics in drug discovery and precision medicine. Nat Rev Drug Discov 15(7):473–484
Wishart DS et al (2007) HMDB: the human metabolome database. Nucleic Acids Res 35(Database issue): D521–D526
Wishart DS et al (2018) HMDB 4.0: the human metabolome database for 2018. Nucleic Acids Res 46 (D1):D608–D617
Wu YL et al (2023) Epigenetic regulation in metabolic diseases: mechanisms and advances in clinical study. Signal Transduct Target Ther 8(1):98
Yanes O et al (2010) Metabolic oxidation regulates embryonic stem cell differentiation. Nat Chem Biol 6(6): 411–417
Yoon MS (2016) The emerging role of branched-chain amino acids in insulin resistance and metabolism. Nutrients 8(7)
Yu H et al (2020) Parallel metabolomics and lipidomics enables the comprehensive study of mouse brain regional metabolite and lipid patterns. Anal Chim Acta 1136:168-1143–168-1177
Zamboni N, Saghatelian A, Patti GJ (2015) Defining the metabolome: size, ffux, and regulation. Mol Cell 58(4): 699–706
Zenobi R (2013) Single-cell metabolomics: analytical and biological perspectives. Science 342(6163):1243259
Zhang A et al (2012) Modern analytical techniques in metabolomics analysis. Analyst 137(2):293–300
Zheng Y, He JQ (2022) Pathogenic mechanisms of trimethylamine N-oxide-induced atherosclerosis and cardiomyopathy. Curr Vasc Pharmacol 20(1):29–36
Zhong Q et al (2020) Protein posttranslational modifications in health and diseases: functions, regulatory mechanisms, and therapeutic implications. MedComm 4(3):e261

Interactomics: Dissecting Protein Networks in Living Systems

7

Carlos M. Farinha, Francisco Rodrigues Pinto, and João A. I. Miranda

Abstract

Interactomics aims at identifying and characterizing in a comprehensive way the complete set of protein-protein interactions that occur within a cell, tissue, or organism. In this chapter, we cover the most relevant experimental and computational approaches that are used in the field to assess the interactome of a specific protein or of all the proteins in a defined compartment. We address their applicability as well as their advantages and disadvantages. We then describe the different databases available to researchers interested in dissecting protein-protein interactions and finally briefly summarize the several attempts that have been made to identify the human interactome, i.e., the complete set of protein interactions that occur within a human cell.

Keywords

Protein-protein interactions · Interactome · Interactomics · Affinity purification · Yeast two-hybrid · Proximity labeling · Protein interaction databases

C. M. Farinha (✉) · F. R. Pinto · J. A. I. Miranda
BioISI—Biosystems & Integrative Sciences Institute & Department of Chemistry and Biochemistry, Faculty of Sciences, Universidade de Lisboa, Lisbon, Portugal
e-mail: cmfarinha@fc.ul.pt

7.1 Introduction

Proteins are responsible for performing most of the key processes that occur in living organisms. In each cell, there is an average of 10^9 protein molecules that function as catalysts (enzymes), as scaffolding to the cell and its internal compartments (structural proteins), as receptors responsible for communication between the outside and inside of the cell, as gene expression regulators, as transporters across membranes, and as motors in cell (and organismal) movement and performing many other more specific functions.

To perform their specific functions in a precise and coordinated manner, proteins interact with other proteins (and biomolecules) in finely tuned interaction networks (Braun and Gingras 2012). The full set of protein-protein interactions that occur in a system is called the interactome, a term that was coined in 1999 to describe interactions reported as relevant in Drosophila development (Sanchez et al. 1999). Interactomics thus focuses on the study of the interactome, being at the intersection of biology and bioinformatics—by studying interactions among proteins and the consequences of those interactions and often comparing such networks of interactions (the interactomes) between and within species or physiological conditions (e.g., healthy or diseased) in order to characterize how these protein traits are preserved or altered. These latter aspects are often referred to as "comparative

M. Gama-Carvalho et al. (eds.), *Omics Approaches in Biomedicine and Biotechnology*, Advances in Experimental Medicine and Biology 1504, https://doi.org/10.1007/978-3-032-18966-0_7

interactomics." Interactomics is an example of "top-down" systems biology, as the sets of genome-wide and proteomics data are collected and used to infer correlations between different molecules. Over several other "omics" methodologies, interactomics by studying interactions brings a dynamic view to the processes occurring within cells.

In this chapter, we cover the more relevant experimental and computational methods to study (identify) interactomes and provide an overview of the most used databases in the field and how these developments shaped the identification of the human interactome.

7.2 Experimental Methods to Assess Interactomes

Experimentally assessing an interactome is a complex task, as each of the techniques available does not have the ability to capture all of the interactions occurring within a cell. There are three major sets of methods (Fig. 7.1) that have been used to define an interactome: affinity purification coupled to mass spectrometry, yeast two-hybrid (Y2H) (and related) methods, and proximity labeling coupled to mass spectrometry. When selecting the most adequate method to apply in a specific situation, several aspects need to be considered (Sharifi Tabar et al. 2022):

- Localization of the interaction(s) to be assessed
- Molecular weight of the protein of interest (bait)
- Need to include epitope tags or fusion proteins.
- Composition of the cell culture medium.
- Inclusion of appropriate controls (to minimize false positives or negatives).
- Quality of the beads (in methods that include pull-down steps).
- Quantification method (when appropriate).

Below we describe these three sets of methods, taking into consideration these critical aspects.

7.2.1 Affinity Purification Coupled to Mass Spectrometry (AP-MS)

Probably the most used experimental procedure to isolate and then identify the set of proteins interacting with a given protein of interest is affinity purification coupled to mass spectrometry (Phizicky and Fields 1995). Although the term "affinity purification" refers in general to the use of a well-known biological interaction and thus to a set of related approaches, the most common type of affinity explored is the antibody-antigen interaction, as the basis for an immunoprecipitation (IP) approach. IP has been used for a long time as a pivotal technique in biochemistry and molecular biology to detect and study protein-protein interactions (Markham et al. 2007). It explores the affinity and specificity of the interaction between an antigen and an antibody and under specific conditions allows the isolation (also termed pull down) of not only the antigen but also of its interacting partners from a homogenate (a cell or tissue lysate). When used to isolate the protein together with the other proteins with which it is complexed, the procedure is normally called co-immunoprecipitation (co-IP).

A co-IP experiment includes several key steps (Fig. 7.1a), as detailed below:

(i) *Preparation of the lysate*—this corresponds to the lysis of the cells (or tissue) from which the target will be isolated. The lysis protocol will release the intracellular component into the buffer from which the target can be isolated. Thus, the composition of the lysis buffer is critical for the success of the procedure and depends on the protein to be isolated, its intracellular location and abundance, and on the nature of the interactions to be detected. One of the most used buffers is the radioimmunoprecipitation assay (RIPA) buffer, which contains several detergents to lyse the membrane and solubilize membrane-embedded proteins, but most of the time, co-IP buffers contain only milder detergents, such as NP-40/IGEPAL

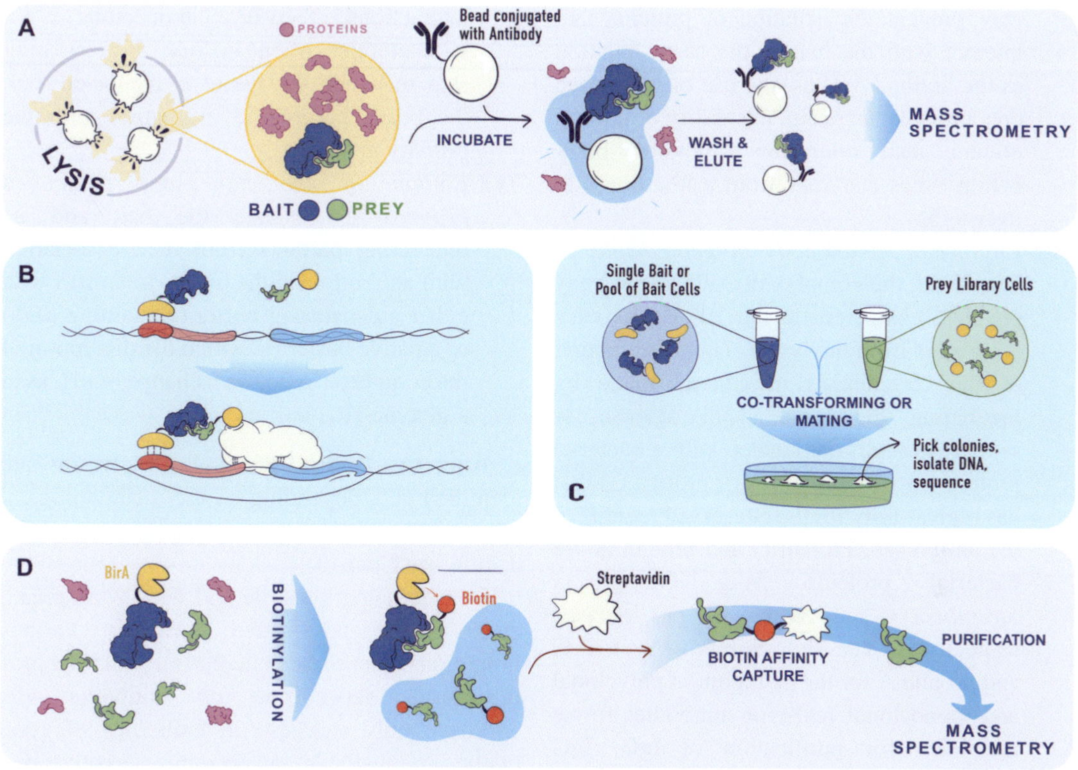

Fig. 7.1 Most common experimental approaches to identify the interactome. (**a**) Co-immunoprecipitation. After cell lysis, an antibody is used to selectively capture the bait (protein of interested) from the protein lysate. The experimental conditions allow the protein-protein interactions to be preserved. Subsequently, protein A/G-conjugated beads are added to capture the antibodies together with the protein assemblies. This is followed by washing and elution to release the bait and its interactors that are then identified/quantified by MS. (**b**) Yeast two-hybrid principle. The method relies in the activation of a downstream reporter gene by a transcription factor binding to an upstream activating sequence (UAS). This transcription factor is split into two separate fragments—the DNA binding domain (DBD) and the activation domain (AD). The DBD is fused to the bait and the AD is fused to the prey. When bait and prey interact, the transcription factor is reactivated by the close proximity of DBD and AD, and the reporter gene is activated. (**c**) For large-scale Y2H interactomics approaches, a prey library is transformed with bait constructs, and selection is made based on growth reporter genes. Prey plasmids are then recovered from each colony and identity of the interactors determined by sequencing analysis. (**d**) Proximity labeling (with BioD). In this approach, the bait protein is fused with a biotin ligase that will promiscuously tag lysine residues with biotin in proteins that are within a 10 nm radius of the bait. After cell lysis, biotinylated proteins are pulled down with streptavidin-conjugated beads and then identified/quantified by MS

or Triton X-100, as these allow the preservation of protein complexes. The lysis buffer should also contain protease inhibitors to prevent protein degradation during the procedure. After lysis, there is usually a clearing step to remove cellular debris and obtain a cleared lysate.

(ii) *Incubation with the primary antibody*—the cleared lysate is incubated with the antibody raised against the protein of interest (the bait). The choice (and quality) of the antibody is probably the most critical step the procedure, as it should have high specificity and affinity for the target protein. As an alternative, a mixture of more than one antibody against the bait can be used—this is particularly for cases in which the interaction of the antigen with a single antibody may prevent the isolation of proteins that interact with the bait at the same location as the antibody. Thus, the use of more than one antibody increases the ability of pulling down as many interactors as possible. Incubation times can vary from a few hours to overnight.

(iii) *Capture of the antibody-antigen complex*—in general, the complex doesn't possess any physical characteristic that allows its easy separation from the lysate. Thus, the capture of the complex is in general made by incubating the lysate with agarose or sepharose beads conjugated with a bacterial protein (such as protein A or protein G) that has high affinity for the conserved regions of the antibodies. Protein G and protein A are bacterial proteins from Group G *Streptococci* and *Staphylococcus aureus*, respectively. The high affinity of protein G and protein A for the Fc region of polyclonal and monoclonal IgG-type antibodies forms the basis for purification of IgG, IgG fragments containing the Fc region, and IgG subclasses. When coupled to sepharose, protein G and protein A create extremely useful, easy-to-use chromatography media for routine purification of antibodies. For co-IP procedures followed by mass spectrometry identification of complexes, it is common to precede the incubation by cross-linking the protein A/G beads to the antibody. This will prevent the antibody elution along with the protein complexes of interest, affecting their detection and identification. Incubation times typically vary from 1 to 4/5 h.

(iv) *Washing*—after incubation period, the beads are washed several times with the aim of removing nonspecifically bound proteins. The washing is performed with lysis buffer, usually with adjusted salt concentration or detergent type, to modulate the stringency of the procedure—and thus minimize the nonspecific binding but retaining the targeted complex (bearing in mind that some of the interactions being picked are weak and/or transient). To minimize the nonspecific interactions, co-IP procedures for interactomics often include a preclearing step in which the lysate is incubated with the beads before the incubation with the primary antibody.

(v) *Elution of the bait and associated proteins*—To isolate the bait and its interacting partners from their association with antibody and the beads, treatment with either a denaturing buffer (containing SDS) or a native buffer (in which the disruption of the complexes relies on a change of pH, as in a glycine-HCl buffer.

After the elution, the isolated proteins are identified and, depending on the methodological approach, quantified by mass spectrometry (for details, see Chap. 3).

Despite being versatile and used widespread, co-IP approaches have however various limitations (Markham et al. 2007), which include the large dependence on antibodies—the approach only succeeds if antibodies of good quality are available and recognize a native epitope; the dynamic range as transient and weak interactions are difficult to capture and may be lost during the procedure—in these cases, a cross-link agent can be added to live cells before the co-IP procedure and will thus covalently link the proteins within the complex (in vivo cross-linking); the protein abundance, as the procedure requires a relatively high amount of the target protein and is more difficult for proteins that exist in reduced levels; the technical variability that affects antibody binding, washing, and elution conditions and thus may hamper the reproducibility and make the use of proper controls (no antibody, no beads, no lysate) critical to obtain reliable information.

7.2.2 Yeast Two-Hybrid Assay and Related Methods

Another method—or set of methods—commonly used to determine the interactome of a specific protein is the yeast two-hybrid (Y2H) assay, first described by Fields and Song in 1989 (Fields and

Song 1989). Over the years, the technique—which is performed in vivo—has undergone several adaptations and modifications, and it is at the basis of several related methodologies that can not only assess binary protein-protein interactions but also unravel new ones.

The technique makes use of the knowledge that transcription factors are generally composed of two domains—an activation domain (AD) and a DNA-binding domain (DBD). The DBD binds to a specific sequence in the DNA molecule and triggers the AD to recruit the machinery needed to start transcription, enabling the expression of a downstream reporter gene (Fig. 7.1b). The original Y2H method is based on the yeast transcription factor GAL4 with a "bait" protein fused to the DBD and a "prey" protein fused to the AD. If the bait and prey proteins interact, the two domains are brought into proximity, reform a functional GAL4 factor, and induce the expression of a reporter gene—usually a gene needed for cells to proliferate and survive. Thus, when grown on a selective medium, cells will only proliferate if the two proteins of interest interact. So, the process involves the simultaneous transfection of the yeast cell with both the constructs Bait-DBD and Prey-AD plus a plasmid encoding a reporter gene (or cDNA) under the control of a promoter which contains an upstream activating sequence (UAS) recognized by the GAL4 transcription factor. The Y2H has been widely used due to its ability to screen libraries of potential interacting proteins in a high-throughput manner.

A standard Y2H assay usually involves the following steps:

(i) *Design of the fusion proteins*—with the bait protein usually fused to a DBD and the prey protein fused to the AD.
(ii) *Transformation of yeast cells*—with insertion of both fusion constructs into a yeast strain that has a reporter gene under the control of a promoter recognized by the DBD.
(iii) *Assessment of protein interaction in yeast*—if bait and prey proteins interact, the DBD and AD are brought together, reconstituting the functional transcription factor.
(iv) *Activation of the reporter gene*—the reconstituted transcription factor activates the reporter gene, producing a detectable signal (e.g., color change, growth on selective media).
(v) *Readout and analysis*—with detection of the interaction by checking reporter activity (e.g., colony color on X-gal plates or growth on selective medium).

Considering that both genome-scale clone repositories such as the human open reading frame (ORF) clone collection or complex cDNA libraries covering most expressed genes are available, these libraries are used fused to the cDNA encoding the GAL4 AD and co-transfected into cells with a plasmid encoding the bait protein fused to the cDNA for GAL4 DBD (Fig. 7.1c). In this approach, generally termed forward Y2H, cells will only grow when one prey protein interacts with the bait protein—allowing the discovery of new interactions and the construction of proteome-scale PPI networks (Rolland et al. 2014).

Despite its extensive use and relative ease of application, the Y2H system has some limitations that include:

- The occurrence of false positives, mainly due to the overexpression of the two proteins leading to interactions that do not occur in vivo.
- The occurrence of false negatives, due to the possible perturbation of the folding of bait or prey caused by their fusion to DBD and AD, or when the interaction is dependent on post-translational modifications that do not occur in yeast cells.
- The difficulty to detect interactions involving membrane proteins or those d is particularly adequate.

Thus, the method has been adapted or modified to circumvent these limitations, generating some variants, including:

- Split ubiquitin system (SUS), designed for membrane proteins. Also called Membrane Yeast Two-Hybrid (MYTH), the system is based on the splitting of the ubiquitin molecule into two halves (Johnsson and Varshavsky 1994). The C-terminal part is fused to the bait and anchored to the membrane (with a transmembrane domain). The N-terminal half is fused to the prey protein. One of the halves (most commonly the C-terminal part) is fused with a transcription factor. When bait and prey interact, the two ubiquitin halves come together and reconstitute a functional ubiquitin that is recognized by ubiquitin-specific proteases leading to the release of the transcription factor. This will translocate to the nucleus and induce the expression of the reporter gene. The most used reporter genes are ADE2, HIS3, and lacZ. ADE2 and HIS3 allow qualitative assessment of the interactions through growth on selective medium, whereas lacZ enables a semiquantitative readout through colony color (blue/white).

- Bimolecular fluorescence complementation (BiFC) is based on the splitting of two halves of a fluorescent protein (the most used are green or yellow fluorescent protein, GFP/YFP), each of which is fused with the proteins of interest (bait and prey). When these interact, the two halves come together and reconstitute a functional fluorescent protein. This approach was first used in 2000 (Ghosh et al. 2000) and has now been extensively perfected and optimized, using over 15 different fluorescent proteins. Although relying on overexpressed proteins (similarly to Y2H or SUS), BiFC allows the detection of interactions in their native cellular context (not favoring nuclear or membrane interactions), thus providing more reliable spatial and temporal information on the interaction.
- Mammalian protein-protein interaction trap (MAPPIT) is a two-hybrid assay based on type I cytokine signal transduction. The assay is based on a signaling-deficient cytokine I receptor. The extracellular domain is based on the homodimeric erythropoietin receptor (EpoR) which is fused to a transmembrane domain and a cytoplasmic region of the leptin receptor and leptin receptor. This chimeric version is signaling-deficient as three relevant tyrosine residues in the cytoplasmic region are mutated into phenylalanine. In the presence of an extracellular ligand, Janus kinases (JAKs) are activated, but the lack of residues to be phosphorylated precludes the recruitment of signal transducer and activator of transcription (STAT) proteins. In the two-hybrid approach, the bait protein is C-terminally fused to the deficient receptor, and the prey protein(s) is (are) coupled to a domain of the glycoprotein 130 receptor (gp130), which is another cytokine receptor and contains tyrosine motifs that can be JAK-phosphorylated. If bait and prey interact, interacting JAKs can phosphorylate gp130 and recruit STAT3, which migrate to the nucleus and induce the expression of a reporter gene (Eyckerman et al. 2001). In this assay, the interaction occurs in intact mammalian cells, and the compartmentalization that separates interaction and readout (cytoplasm vs nucleus) reduces background. MAPPIT has been shown to be complementary to other two-hybrid-based methods and to be scalable to large-scale interactomics approaches (Braun et al. 2009).

7.2.3 Proximity Labeling Coupled to Mass Spectrometry

A third set of methods regularly used to assess interactomics is the so-called proximity labeling ones. These methods identify proteins that are near the protein under study within living cells and can include not only direct interactors but proteins that are within a very small distance, even if not interacting directly—having with this the ability to capture dynamic and transient interactions that many times will be missed by other approaches.

Proximity labeling uses engineered enzymes which are fused to the protein of interest. These enzymes will thus promote the attachment of a tag (biotin or other small molecules) to proteins that are in proximity (typically less than 10–20 nm) to the bait (Fig. 7.1d). The labeling reaction occurs in living cells, that are subsequently lysed, being the labeled proteins pulled down, identified (and quantified) by mass spectrometry. The most common techniques are BioID (Roux et al. 2012), Apex2 (Lam et al. 2015), and TurboID (Branon et al. 2018).

A standard proximity labeling approach includes the following steps:

(i) *Design of the fusion protein*—by fusing the protein of interest (bait) to the enzyme that catalyzes labeling (a biotin ligase or an engineered peroxidase).
(ii) *Expression in cells*—the fusion protein is transfected into living cells to allow it to localize naturally. It is particularly important to ensure that fusion with the labeling enzyme does not alter the physiological location and role of the bait protein.
(iii) *Initiation of the labeling reaction*—by adding the labeling substrate to activate the enzyme, thus allowing nearby proteins get covalently tagged (usually with biotin) within a defined radius.
(iv) *Cell lysis and protein capture*—with cell lysis and use of streptavidin beads to capture the biotinylated (labeled) proteins.
(v) *Mass spectrometry analysis*—with elution, digestion of the captured proteins, and analysis of the peptides by liquid chromatography-mass spectrometry (LC-MS/MS) to identify and quantify the proximal interactors.

BioID (for proximity-dependent biotin identification) was the first technique widely adopted for proximity labeling and is based on a mutated version of BirA, a biotin ligase from *Escherichia coli* that biotinylates lysine residues on proteins in the presence of biotin. Whereas the wild-type counterpart BirA requires a specific recognition sequence to promote biotinylation of the target, the mutated version BirA* can biotinylate any protein that is within a 10 nm radius. Since its development, the advantages of the method were obvious: BioID can be performed in live cells, preserving the native environment; the duration of the labeling can be controlled, and since then it has been used to a wide diversity of cellular structures, including the nuclear pore complexes, chromatin, and different organelles. However, the labeling time is quite long (usually requiring 18–24 h for efficient labeling), and the method has been shown to result in a significant number of false positives.

The main limitations of BioID were overcome by the more recent and efficient version coined TurboID (Branon et al. 2018). TurboID was developed from BirA* using yeast display-based directed evolution, which catalyzes proximity labeling with much greater efficiency than BioID and enables 10 min labeling in cells with nontoxic and easily deliverable biotin. This faster kinetics enables the study of rapid and transient interactions, missed by slower methods, increasing the sensitivity of detecting low-abundance proteins, maintain (as with BioID) the integrity of the cellular context. The rapid activity of TurboID comes with its downfalls, as it can lead to excessive labeling if not carefully controlled.

APEX2 (for enhanced ascorbate peroxidase) differs from BioID and TurboID in its use of an engineered peroxidase enzyme rather than a biotin ligase. APEX2 catalyzes the oxidation of biotin-phenol by hydrogen peroxide, resulting in the formation of biotin-phenoxyl radicals. As these radicals are highly reactive, they have the ability to attach covalently to tyrosine residues in proteins which are a few nanometers from the peroxidase. Due to its smaller labeling radius (1–2 nm), APEX2 provides a more precise mapping of the microenvironment surrounding the baits. It is also able to label proteins rapidly (from seconds to minutes) and is compatible with electron microscopy, as APEX-2 can generate electron-dense products, thus allowing combination of proteomic-based data with ultrastructural information. However, the use of hydrogen peroxide is likely to cause oxidative damage affecting cell physiology. The high spatial resolution (frequently listed as an advantage) can also be a limitation in cases where broader proximity is sought.

Proximity labeling techniques have advanced the ability to assess the spatial organization of protein complexes and have been applied to mapping of organelle proteomes (e.g., endoplasmic reticulum, mitochondria, nucleus), to the investigation of signal transduction pathways (which many times relying on transient/weak interactions) or the exploration of subcellular microenvironments (e.g., the synaptic clef in neurons or the intermembrane space in mitochondria (Rhee et al. 2013; Hung et al. 2014; Loh et al. 2016)). However, some challenges remain including:

- Background noise, caused by nonspecific labeling and making it difficult to distinguish true from false positives. The experimental conditions need to be finely tuned, and the use of appropriate controls is imperative.
- Expansion to other organisms, as most of the studies have been performed in mammalian cells. Application to plants, yeast, or bacteria will be of great interest.
- Integration with other techniques, including quantitative MS, live cell imaging, or structural biology, which is still limited.
- Development of new proximity labeling enzymes, with better specificity, faster kinetics, or alternative labeling reactions that could circumvent the current limitations.

7.3 Computational Methods to Predict Interactomes

Experimental techniques for detecting protein-protein interactions (PPIs) are costly and time-consuming. Despite the accumulation of experimental evidence regarding physical interactions among millions of protein pairs, comprehensive coverage of most interactomes remains incomplete, even in extensively studied model species (Hart et al. 2006). Researchers have developed computational methods to predict protein physical interactions, which generally entail reduced costs and time compared to experimental approaches.

Protein interactions occur through the formation of non-covalent bonds between atoms in distinct proteins. For bonds to form, the geometric configuration of the two surfaces in the structure of each interacting protein must permit the involved atoms to approach each other closely enough. The primary categories of inter-protein non-covalent bonds include (Soleymani et al. 2022; Stollar and Smith 2020):

1. Electrostatic interactions, which result from attractions between positively and negatively charged amino acid residues.
2. Hydrogen bonds, when a hydrogen atom is shared between a donor group, such as -NH or -OH, and an acceptor group, such as -C=O.
3. Van der Waals, weak, short-range interactions that result from transient or permanent dipole moments in atoms.
4. Hydrophobic interactions, when nonpolar residues aggregate to reduce their exposure to water, thereby promoting interactions in aqueous environments with a favorable entropic effect.

Determining whether interactions occur between two proteins requires knowledge of the amino acid residues present in each protein and their spatial arrangement when the proteins are in proximity. Some computational methods for predicting protein physical interactions use either protein structure or protein sequence as input data. Some use indirect evidence, including protein functional annotations or the context within established protein interaction networks. Regardless of the input data, many of these methods use various machine learning models to identify patterns that predict protein physical interactions (Casadio et al. 2022). Examples of methods for each type of input data are provided in Table 7.1. More extensive lists of methods can be found in recent reviews (Casadio et al. 2022; Kewalramani et al. 2023; Grassmann et al. 2024) or in the Bio. Tools website (https://bio.tools/).

7.3.1 Supervised Machine Learning Models

Machine learning models are functions that integrate multiple input predictor variables, which characterize a protein pair, to compute an output variable that facilitates predictions regarding the interaction of the protein pair (Casadio et al.

Table 7.1 Examples of computational methods to predict protein physical interactions

Method	Input data	Principle	Reference
MEGADOCK	Protein structures	Molecular docking	Ohue et al. (2014)
SGPPI	Protein structures	Train graph convolutional neural betwork with graph representations of protein structures predicted with AlphaFold	Huang et al. (2023)
AlphaFold2	Protein sequences	Fold and dock prediction for the two proteins simultaneously	Bryant et al. (2022)
DPPI	Protein sequences	Train neural networks with one-hot encodings of the two protein sequences	Hashemifar et al. (2018)
DeepPPI	Protein sequences	Train neural networks with physicochemical properties estimated from protein sequences	Du et al. (2017)
BIPS	Protein sequences	Interolog predictions	Garcia-Garcia et al. (2012)
ADVICE	Protein sequences	Detect coevolution in protein pairs	Tan et al. (2004)
SFCN	PPI network	Neighbor similarity	Li et al. (2018)
RNM	PPI network	L3 principle	Kovacs et al. (2019)
KGsim2vec	GO annotations	Knowledge graph semantic similarity	Sousa et al. (2024)

2022; Kewalramani et al. 2023; Hu et al. 2021). The functions are designed to be flexible, featuring internal parameters that can be adjusted to optimize predictions when applied to a training dataset with known labels (interacting or noninteracting). The adjustment of these parameters with training data corresponds to the learning stage. This process is classified as supervised learning, as it uses a training dataset in which the true labels to be predicted are already known.

The training dataset for predicting protein physical interactions comprises a set of protein pairs identified as known physical interactors (positive examples) and a set of protein pairs that do not interact (negative examples). While positive examples are readily available in public databases of interactomic data, there is a lack of systematic documentation for protein pairs that do not interact. To address this limitation, researchers typically create random pairs of proteins that have not been previously recognized as interacting pairs.

The entirety of the collected training data is not used for model training. The dataset is typically divided into three components: (1) training set, (2) testing set, and (3) validation set. The training set may be used multiple times with various machine learning models or different versions of the same model. The variations are compared in relation to the testing set. Cross-validations are commonly conducted to enhance the robustness of selecting the optimal model. The data in the training and testing sets is divided multiple times, either randomly or using a k-fold approach, facilitating the evaluation of the model with a variety of testing subsets. The optimal model is assessed using the validation set, which was not used in the training of any model. The predictive performance of binary classification models, applicable to protein physical interaction prediction, is assessed using metrics such as precision, recall, F1-measure, and Matthew's correlation coefficient (MCC). All measures are derived from the counts of true positives (TP—instances where the model correctly predicts protein interactions), true negatives (TN—instances where the model correctly predicts noninteractions), false positives (FP—instances where the model incorrectly predicts interactions), and false negatives (FN—instances where the model incorrectly predicts non-interactions). Precision quantifies the ratio of true positives (TP) among all positive predictions, calculated as TP/(TP + FP). In contrast, recall assesses the ratio of true positives within all actual positive instances, expressed as TP/(TP + FN). The F1-measure represents the geometric mean of precision and recall, serving as an effective performance metric when the training data is balanced, indicating a similar quantity of positive and negative examples. Imbalanced datasets often need the use of the Matthews correlation coefficient (MCC) as a performance metric, as it evaluates the true positives, true negatives, false positives, and false negatives against the expected values of a random classifier. Should the model demonstrate strong performance on the validation set, it is applicable for predicting novel protein pairs. If a pre-trained model is used to make new predictions, it is essential to previously check which data has been used during model training. The performance measures of the pre-trained model can only be trusted if the input data that is to be used (new protein pairs) has properties that are well represented in the training data (species, structural types, amino acid composition, etc.).

7.3.2 Methods Using Protein Structures

The forecast of whether two proteins can develop a specific physical relationship relies on the ability to measure how many non-covalent contacts, and of what sort, are produced when they come into contact. Knowing the tridimensional structure of two proteins allows anticipating the occurrence of inter-protein interactions using two major approaches (Grassmann et al. 2024): (1) molecular docking and (2) molecular dynamic (MD) simulations. Molecular docking algorithms iteratively evaluate various contact orientations and determine the success of the connection by evaluating the non-covalent bonds that may form. However, one shortcoming of docking techniques is that proteins are represented as rigid structures. MD simulations overcome these constraints by allowing atoms to move in response to the forces acting upon them. These forces are created by electrostatic and Van der Waals interactions with other atoms within or between molecules, including solvents, and constrained by covalent interactions within the same molecule (due to bond stretching, angle bending, and torsional rotation). Making an MD simulation of two proteins and solvent molecules allows tracking the spatial trajectories of all atoms over time and seeing if a physical interaction between the two proteins is possible. Because protein structures are flexible, when two proteins come close together, they can induce conformational changes in each other, changing the shape of the contact surfaces between them. These changes may allow the establishment of more stabilizing non-covalent contacts that would otherwise go undetected during rigid structure docking. In addition to predicting the occurrence of an interaction, both approaches can be used to predict the interacting surfaces, involved amino acid residues, and binding affinity.

Both molecular docking and MD simulations need a significant amount of computer time and resources. However, the greater detail produced with MD simulations comes at a higher computational cost. Nevertheless, the availability of experimentally characterized protein structures is the greatest barrier to the widespread use of these approaches for interaction prediction. The Protein Data Bank (PDB) contains approximately 230,000 structures, but the UniProt database contains entries for over 250 million proteins. For proteins with high sequence homology to a known structure, homology modeling can be used to infer the unknown structure. The recent success of the AlphaFold algorithm (Jumper et al. 2021), which predicts protein structure based solely on sequence, promises to address this issue, and the scientific community now has access to a database of 200 million protein structure predictions. Indeed, AlphaFold-based approaches for predicting protein physical interactions have recently been developed (Bryant et al. 2022).

7.3.3 Methods Using Protein Sequences

The sequences of two proteins provide information that can be used to predict whether they interact—e.g., two proteins may share sequence motifs that have been detected in previously identified pairs of interacting proteins. However, most machine learning algorithms do not accept sequences directly as data input. The simplest way is to employ one-hot encoding (Casadio et al. 2022). Briefly, each amino acid is represented as a binary vector of length equal to the number of potential amino acids. The vector for a specific amino acid contains a 1 at the corresponding place and 0 s elsewhere. To encode a full protein sequence, each amino acid's one-hot encoding is stacked to create a matrix. Alternatively, some approaches convert the amino acid sequences in each protein into vectors of quantitative variables by measuring characteristics such as hydrophobicity, electrical charge, and polarity for various protein subsequences (Kewalramani et al. 2023). These matrices can then be used to train supervised machine learning algorithms.

Protein sequences can also be used to predict protein interactions using an evolutionary perspective. If a homologous protein A interacts with a homologous protein B in a closely related species, then proteins A and B are likely to interact. A protein-protein interaction that is conserved across related species is known as an interolog (Garcia-Garcia et al. 2012). Furthermore, if an interaction is conserved across species, the interacting proteins are likely to exhibit coevolutionary characteristics (Tan et al. 2004). This indicates that when a sequence variation is found by comparison of homologous protein sequences from different species, compensating variations in the interacting protein sequences should also be searched for. These compensatory changes are required to maintain the physical connection between the two proteins. As a result, protein-protein interactions can be predicted by looking for evidence of coevolution in protein sequence pairs from different species.

7.3.4 Methods Using PPI Networks

As interactomics data expanded, researchers began to investigate the possibility of predicting novel protein interactions using patterns from previously known interaction networks. The first approaches assumed that if two proteins shared many interactors, they were more likely to interact physically (Hu et al. 2021). These approaches have been moderately successful. They take advantage of the fact that biological networks have a modular structure, which means that proteins are more likely to interact with proteins in the same module than with proteins in different modules. Furthermore, a substantial subset of protein interactions is associated with stable protein complexes, in which each protein interacts directly with the majority of the other members of the complex.

More recent approaches improved prediction performance by using a slightly different assumption. If protein A and protein B have many common neighbors, this could indicate that they have comparable interaction interfaces or domains. Proteins must interact through complementary interaction domains, which are not necessarily the same. As a result, if protein A interacts with protein C and protein D shares numerous interactors with C, A is likely to interact with D. The argument is as follows: C has a complementary interacting domain with A; C and D should have similar interacting domains (since they share many interactors); and consequently, D should have complementary interacting domains with A. A practical method for identifying candidate protein interacting pairs (such as proteins A and D in this example), denominated as the L3 principle (Kovacs et al. 2019), is to select protein pairs that are connected by a large number of network paths of length 3 (A—C—(common interactor of C and D)—D, where — represents a known protein physical interaction). The performance of network-based approaches predicting novel protein physical interactions has been systematically compared, and methods integrating the L3 principle are among the top performers across the evaluated interactomes (Wang et al. 2023). Naturally, the absolute performance levels are higher for interactomes that have been more extensively researched. In these circumstances, the network properties of the unknown entire interactomes are more accurately estimated.

7.3.5 Methods Using Protein Annotations

In order for a protein physical interaction to occur, the two proteins must be expressed at the same time and in the same cell compartment, in addition to having complementary interacting domains. Furthermore, empirical investigations into known protein physical interaction networks have revealed that two proteins are more likely to interact if they are involved in the same cellular process or pathway (Sharan et al. 2007). These extra annotations regarding proteins or protein pairs, such as co-expression (usually at the protein or mRNA level) and Gene Ontology terms (which include cellular component and biological process annotations), can help to predict protein physical interactions (Wu et al. 2006; Zhong and Rajapakse 2020). The current state-of-the-art strategy for including these various annotations is to construct a knowledge graph (Sousa et al. 2024). Knowledge graphs are networks in which nodes represent various entities. In the context of protein interaction prediction, nodes can be proteins, cellular components, biological processes, or other protein annotations. Interactions between nodes can potentially take numerous forms. Interactions between proteins can be established if the proteins are known to interact, are highly co-expressed, or have other features in common. Nodes indicating annotation terms are linked to proteins that have been annotated with that term. If the annotation terms are part of an ontology, the ontology graph, which represents the relationships between concepts, is also included. The knowledge graphs can then be used to create graph neural network (GNN) models that forecast missing interactions (Zhang et al. 2021). Most GNNs accomplish this by computing an edge embedding, which entails determining a low-dimensional vector representation for each edge while preserving the similarity between edges in the graph. In a process known as aggregation, the information associated with neighboring edges or nodes influences an edge's embedding vector. Given the embedding vectors, a classifier model can be trained to distinguish between positive edges (when proteins physically interact) and negative edges.

7.4 Protein-Protein Interaction Databases

Most experimentally identified protein physical interactions (PPI) are documented in scientific journals. This acquired knowledge is then entered into databases by human curation of scientific publications or direct submissions by researchers. PPI databases are beneficial for researchers interested in studying a specific protein, as well as those investigating the organization of molecular networks and how they affect cellular function.

Primary PPI databases focus on curating experimentally proven interactions. These databases prioritize high-quality evidence to substantiate reported interactions. Examples include IntAct (Del Toro et al. 2022), MINT (Licata et al. 2012), and BioGRID (Oughtred et al. 2021). Secondary PPI databases, also known as meta-databases, combine and consolidate interaction data from many primary sources. Secondary databases are intended to provide a more comprehensive and integrated view of PPI networks. Examples include STRING (Szklarczyk et al. 2023) and APID (Alonso-Lopez et al. 2019; Alonso-Lopez et al. 2016). They provide larger coverage than primary databases since primary databases typically have small overlaps between them.

PPI databases are curated to varying degrees of detail and stringency. At one end of the scale, some databases use "light curation," which focuses on extracting essential information like interacting partners and detection techniques. In contrast, "deep curation," as demonstrated by the IMEx consortium (Porras et al. 2020) members such as IntAct and MINT, includes additional experimental information such as assay settings, interaction types, and evidence for direct binding. This thorough curation provides high data quality and compliance with community standards such as the PSI-MI format and ontology (Deutsch et al. 2023), but it may limit the number of interactions gathered.

Interactions from small-scale studies are frequently manually curated from the literature, whereas those from high-throughput screens are more likely to be provided directly by researchers. Interactions collected from the literature may suffer from selection biases, as proteins having biological or biotechnological significance are more frequently examined. Sets of interactions detected in high-throughput screens have a lower selection bias, but they may have technical bias because all interactions are detected using the same experimental method, which can fail more frequently for some classes of proteins (such as membrane proteins in yeast-two-hybrid methods) (Rolland et al. 2014).

Some protein physical interaction databases include computationally predicted interactions to broaden the range of accessible PPI data. STRING is a notable example, combining predictions and experimental data to provide a comprehensive perspective of PPI networks. While these predictions can provide useful insights, they also add a greater degree of uncertainty than experimentally derived interactions. Therefore, it is critical to examine the prediction algorithms and accompanying confidence scores when analyzing data from these databases.

Protein physical interaction databases can also include additional information layers to increase the value and usefulness of PPI data. These databases frequently combine data from multiple sources, including cellular compartments (Veres et al. 2015), tissue specificity (Kotlyar et al. 2016), and signaling roles (Turei et al. 2021; Lo Surdo et al. 2023). By combining many data types, these tools allow exploring PPIs in a broader biological context, understanding their functional consequences, and proposing ideas regarding their roles in cellular processes and disease.

7.5 The Human Interactome(s)

The experimental and computational approaches described above have been extensively applied to several biological models and allowed the identification of global and protein-specific interactomes in many species, from bacteria to man. The term "human interactome" encompasses the complete set of interactions among proteins within human cells. Identification (and subsequent characterization) of PPIs provides extended knowledge of processes such as signal transduction, gene regulation, structural organization, and metabolism and elucidates mechanisms relevant to human health (and disease). Many studies have addressed the human interactome over the last two decades, leading to the development of extensive, and thorough, PPI networks.

7.5.1 Previous Efforts

One of the earliest efforts to map the human interactome was the Human Protein Reference Database, first published in 2003 by Peri et al. (2003). This database was created by manual curation of PPIs described in the literature, and by 2009 it contained 30,000 interactions.

In 2005, the first large-scale Y2H screen of the human proteome was published by Rual et al. (2005), identifying close to 3000 high-confidence binary interactions among over 7000 human proteins. Despite being a ground-breaking work, this study also evidenced the need for adequate controls when performing large-scale PPI mapping to reduce the occurrence of false positives.

Following these initial attempts, several large-scale projects expanded and refined the human interactome. In 2013, the HI-II-14 (Human Interactome Version 2) project used an extensive Y2H screen and identified 14,000 binary interactions involving 4000 proteins—and clearly highlighted the importance of performing adequate validation using different experimental methods (Rolland et al. 2014). In 2015, the BioPlex was another relevant milestone in the field. BioPlex used affinity purification coupled to mass spectrometry and identified more than 23,000 proteins among over 10,000 proteins (Huttlin et al. 2015). Two years later, BioPlex 2.0 expanded this network to over 56,000 proteins among 11,000 proteins (Huttlin et al. 2017).

The contributions from different experimental approaches to the development of a global and complete human interactome made clear that no single technique would be able to capture all the relevant interactions. As described above, each technique offers unique advantages and limitations—and thus a more comprehensive and accurate map of "all" human protein-protein interactions can only be derived from a combination approach. The Human Reference

Interactome Mapping Project was launched in 2017 with the aim of creating a reference map of the human interactome, based on the several studies published to that date. The project, led by Anne-Claude Gingras, integrated over 52,000 binary interactions (Luck et al. 2020)—on top of the expansion of the interactome, this study also improved the reliability of the data through the cross-validation of interactions identified with different methods. Its current version reports over 64,000 interactions (http://www.interactome-atlas.org).

As described above, computational methods have also played a relevant role in predicting PPIs, using machine learning, network analysis, or comparative genomics—and databases such as STRING (Szklarczyk et al. 2019) and IntAct (Orchard et al. 2014) include information from both experimentally reported and predicted interactions.

Identification of the human interactome has given insight into the organization and function of many cellular processes, facilitating discovery of novel drug targets and identifying changes associated with several pathologies (Sahni et al. 2015), including cancer and neurodegenerative and rare disorders such as cystic fibrosis (Pankow et al. 2015; Canato et al. 2018). The application to the study of cancer is particularly notable, with the integration of PPI data with information from studies in genomics and transcriptomics leading to the identification of cancer-specific interaction networks. As a central player in several tumorigenic processes, the interactome of the tumor suppressor p53 has been a focus of several studies, which highlighted its involvement in a wide diversity of cellular processes, ranging from DNA repair to apoptosis and cell cycle regulations (Kussie et al. 1996).

7.5.2 Future Directions

Although many advances have been made toward the development of a comprehensive map of the human interactome, there are still some challenges that persist. The first is the completeness of the interactome—estimates suggest that only a fraction has been uncovered, probably due to the inherent limitations of the techniques available. The missing interactions have even been coined the "dark interactome" (Sharifi Tabar et al. 2022). Another challenge is the context-dependence of the interactions, with many specific to certain cell types, environmental conditions, or development stages. Single-cell proteomics is a technique that may bring significant advances in overcoming this stage.

Integration of interactomics data with other omics data will also enable a deeper understanding of how cellular networks function (in health and disease). Finally, development of improved computational tools will help to refine our knowledge and understanding of the human interactome.

Acknowledgments Work supported by center Grant UID/04046/2025 from FCT to BioISI—Portugal (https://doi.org/10.54499/UID/04046/2025). Work at CM Farinha's lab is supported by grants from Cystic Fibrosis Foundation (FARINH24G0), from Fondazione Fibrosi Cistica Ricerca (grant FFC#2/2023), and from Emily's Entourage.

References

Alonso-Lopez D, Gutierrez MA, Lopes KP, Prieto C, Santamaria R, De Las Rivas J (2016) APID interactomes: providing proteome-based interactomes with controlled quality for multiple species and derived networks. Nucleic Acids Res 44(W1):W529–W535. https://doi.org/10.1093/nar/gkw363

Alonso-Lopez D, Campos-Laborie FJ, Gutierrez MA, Lambourne L, Calderwood MA, Vidal M, De Las Rivas J (2019) APID database: redefining protein-protein interaction experimental evidences and binary interactomes. Database (Oxford) 2019. https://doi.org/10.1093/database/baz005

Branon TC, Bosch JA, Sanchez AD, Udeshi ND, Svinkina T, Carr SA, Feldman JL, Perrimon N, Ting AY (2018) Efficient proximity labeling in living cells and organisms with TurboID. Nat Biotechnol 36(9): 880–887. https://doi.org/10.1038/nbt.4201

Braun P, Gingras AC (2012) History of protein-protein interactions: from egg-white to complex networks. Proteomics 12(10):1478–1498. https://doi.org/10.1002/pmic.201100563

Braun P, Tasan M, Dreze M, Barrios-Rodiles M, Lemmens I, Yu H, Sahalie JM, Murray RR, Roncari L, de Smet AS, Venkatesan K, Rual JF, Vandenhaute J, Cusick ME, Pawson T, Hill DE,

Tavernier J, Wrana JL, Roth FP, Vidal M (2009) An experimentally derived confidence score for binary protein-protein interactions. Nat Methods 6(1):91–97. https://doi.org/10.1038/nmeth.1281

Bryant P, Pozzati G, Elofsson A (2022) Improved prediction of protein-protein interactions using AlphaFold2. Nat Commun 13(1):1265. https://doi.org/10.1038/s41467-022-28865-w

Canato S, Santos JD, Carvalho AS, Aloria K, Amaral MD, Matthiesen R, Falcao AO, Farinha CM (2018) Proteomic interaction profiling reveals KIFC1 as a factor involved in early targeting of F508del-CFTR to degradation. Cell Mol Life Sci 75(24):4495–4509. https://doi.org/10.1007/s00018-018-2896-7

Casadio R, Martelli PL, Savojardo C (2022) Machine learning solutions for predicting protein-protein interactions. Wires Comput Mol Sci 12(6):e1618. https://doi.org/10.1002/wcms.1618

Del Toro N, Shrivastava A, Ragueneau E, Meldal B, Combe C, Barrera E, Perfetto L, How K, Ratan P, Shirodkar G, Lu O, Meszaros B, Watkins X, Pundir S, Licata L, Iannuccelli M, Pellegrini M, Martin MJ, Panni S, Duesbury M, Vallet SD, Rappsilber J, Ricard-Blum S, Cesareni G, Salwinski L, Orchard S, Porras P, Panneerselvam K, Hermjakob H (2022) The IntAct database: efficient access to fine-grained molecular interaction data. Nucleic Acids Res 50(D1):D648–D653. https://doi.org/10.1093/nar/gkab1006

Deutsch EW, Vizcaino JA, Jones AR, Binz PA, Lam H, Klein J, Bittremieux W, Perez-Riverol Y, Tabb DL, Walzer M, Ricard-Blum S, Hermjakob H, Neumann S, Mak TD, Kawano S, Mendoza L, Van Den Bossche T, Gabriels R, Bandeira N, Carver J, Pullman B, Sun Z, Hoffmann N, Shofstahl J, Zhu Y, Licata L, Quaglia F, Tosatto SCE, Orchard SE (2023) Proteomics standards initiative at twenty years: current activities and future work. J Proteome Res 22(2):287–301. https://doi.org/10.1021/acs.jproteome.2c00637

Du X, Sun S, Hu C, Yao Y, Yan Y, Zhang Y (2017) DeepPPI: boosting prediction of protein-protein interactions with deep neural networks. J Chem Inf Model 57(6):1499–1510. https://doi.org/10.1021/acs.jcim.7b00028

Eyckerman S, Verhee A, der Heyden JV, Lemmens I, Ostade XV, Vandekerckhove J, Tavernier J (2001) Design and application of a cytokine-receptor-based interaction trap. Nat Cell Biol 3(12):1114–1119. https://doi.org/10.1038/ncb1201-1114

Fields S, Song O (1989) A novel genetic system to detect protein-protein interactions. Nature 340(6230): 245–246. https://doi.org/10.1038/340245a0

Garcia-Garcia J, Schleker S, Klein-Seetharaman J, Oliva B (2012) BIPS: BIANA Interolog Prediction Server. A tool for protein-protein interaction inference. Nucleic Acids Res 40 (Web Server issue):W147–W151. https://doi.org/10.1093/nar/gks553

Ghosh I, Hamilton AD, Regan L (2000) Antiparallel leucine zipper-directed protein reassembly: application to the Green fluorescent protein. J Am Chem Soc 122(23):5658–5659. https://doi.org/10.1021/ja994421w

Grassmann G, Miotto M, Desantis F, Di Rienzo L, Tartaglia GG, Pastore A, Ruocco G, Monti M, Milanetti E (2024) Computational approaches to predict protein-protein interactions in crowded cellular environments. Chem Rev 124(7):3932–3977. https://doi.org/10.1021/acs.chemrev.3c00550

Hart GT, Ramani AK, Marcotte EM (2006) How complete are current yeast and human protein-interaction networks? Genome Biol 7(11):120. https://doi.org/10.1186/gb-2006-7-11-120

Hashemifar S, Neyshabur B, Khan AA, Xu J (2018) Predicting protein-protein interactions through sequence-based deep learning. Bioinformatics 34(17): i802–i810. https://doi.org/10.1093/bioinformatics/bty573

Hu L, Wang X, Huang YA, Hu P, You ZH (2021) A survey on computational models for predicting protein-protein interactions. Brief Bioinform 22(5). https://doi.org/10.1093/bib/bbab036

Huang Y, Wuchty S, Zhou Y, Zhang Z (2023) SGPPI: structure-aware prediction of protein-protein interactions in rigorous conditions with graph convolutional network. Brief Bioinform 24(2). https://doi.org/10.1093/bib/bbad020

Hung V, Zou P, Rhee HW, Udeshi ND, Cracan V, Svinkina T, Carr SA, Mootha VK, Ting AY (2014) Proteomic mapping of the human mitochondrial intermembrane space in live cells via ratiometric APEX tagging. Mol Cell 55(2):332–341. https://doi.org/10.1016/j.molcel.2014.06.003

Huttlin EL, Ting L, Bruckner RJ, Gebreab F, Gygi MP, Szpyt J, Tam S, Zarraga G, Colby G, Baltier K, Dong R, Guarani V, Vaites LP, Ordureau A, Rad R, Erickson BK, Wuhr M, Chick J, Zhai B, Kolippakkam D, Mintseris J, Obar RA, Harris T, Artavanis-Tsakonas S, Sowa ME, De Camilli P, Paulo JA, Harper JW, Gygi SP (2015) The BioPlex network: a systematic exploration of the human interactome. Cell 162(2):425–440. https://doi.org/10.1016/j.cell.2015.06.043

Huttlin EL, Bruckner RJ, Paulo JA, Cannon JR, Ting L, Baltier K, Colby G, Gebreab F, Gygi MP, Parzen H, Szpyt J, Tam S, Zarraga G, Pontano-Vaites L, Swarup S, White AE, Schweppe DK, Rad R, Erickson BK, Obar RA, Guruharsha KG, Li K, Artavanis-Tsakonas S, Gygi SP, Harper JW (2017) Architecture of the human interactome defines protein communities and disease networks. Nature 545(7655):505–509. https://doi.org/10.1038/nature22366

Johnsson N, Varshavsky A (1994) Split ubiquitin as a sensor of protein interactions in vivo. Proc Natl Acad Sci USA 91(22):10340–10344. https://doi.org/10.1073/pnas.91.22.10340

Jumper J, Evans R, Pritzel A, Green T, Figurnov M, Ronneberger O, Tunyasuvunakool K, Bates R, Zidek A, Potapenko A, Bridgland A, Meyer C, Kohl SAA, Ballard AJ, Cowie A, Romera-Paredes B, Nikolov S, Jain R, Adler J, Back T, Petersen S,

Reiman D, Clancy E, Zielinski M, Steinegger M, Pacholska M, Berghammer T, Bodenstein S, Silver D, Vinyals O, Senior AW, Kavukcuoglu K, Kohli P, Hassabis D (2021) Highly accurate protein structure prediction with AlphaFold. Nature 596(7873):583–589. https://doi.org/10.1038/s41586-021-03819-2

Kewalramani N, Emili A, Crovella M (2023) State-of-the-art computational methods to predict protein-protein interactions with high accuracy and coverage. Proteomics 23(21–22):e2200292. https://doi.org/10.1002/pmic.202200292

Kotlyar M, Pastrello C, Sheahan N, Jurisica I (2016) Integrated interactions database: tissue-specific view of the human and model organism interactomes. Nucleic Acids Res 44(D1):D536–D541. https://doi.org/10.1093/nar/gkv1115

Kovacs IA, Luck K, Spirohn K, Wang Y, Pollis C, Schlabach S, Bian W, Kim DK, Kishore N, Hao T, Calderwood MA, Vidal M, Barabasi AL (2019) Network-based prediction of protein interactions. Nat Commun 10(1):1240. https://doi.org/10.1038/s41467-019-09177-y

Kussie PH, Gorina S, Marechal V, Elenbaas B, Moreau J, Levine AJ, Pavletich NP (1996) Structure of the MDM2 oncoprotein bound to the p53 tumor suppressor transactivation domain. Science 274(5289): 948–953. https://doi.org/10.1126/science.274.5289.948

Lam SS, Martell JD, Kamer KJ, Deerinck TJ, Ellisman MH, Mootha VK, Ting AY (2015) Directed evolution of APEX2 for electron microscopy and proximity labeling. Nat Methods 12(1):51–54. https://doi.org/10.1038/nmeth.3179

Li S, Huang J, Zhang Z, Liu J, Huang T, Chen H (2018) Similarity-based future common neighbors model for link prediction in complex networks. Sci Rep 8(1): 17014. https://doi.org/10.1038/s41598-018-35423-2

Licata L, Briganti L, Peluso D, Perfetto L, Iannuccelli M, Galeota E, Sacco F, Palma A, Nardozza AP, Santonico E, Castagnoli L, Cesareni G (2012) MINT, the molecular interaction database: 2012 update. Nucleic Acids Res 40(Database issue):D857–D861. https://doi.org/10.1093/nar/gkr930

Lo Surdo P, Iannuccelli M, Contino S, Castagnoli L, Licata L, Cesareni G, Perfetto L (2023) SIGNOR 3.0, the SIGnaling network open resource 3.0: 2022 update. Nucleic Acids Res 51(D1):D631–D637. https://doi.org/10.1093/nar/gkac883

Loh KH, Stawski PS, Draycott AS, Udeshi ND, Lehrman EK, Wilton DK, Svinkina T, Deerinck TJ, Ellisman MH, Stevens B, Carr SA, Ting AY (2016) Proteomic analysis of unbounded cellular compartments: synaptic clefts. Cell 166(5):1295–1307. e1221. https://doi.org/10.1016/j.cell.2016.07.041

Luck K, Kim DK, Lambourne L, Spirohn K, Begg BE, Bian W, Brignall R, Cafarelli T, Campos-Laborie FJ, Charloteaux B, Choi D, Cote AG, Daley M, Deimling S, Desbuleux A, Dricot A, Gebbia M, Hardy MF, Kishore N, Knapp JJ, Kovacs IA, Lemmens I, Mee MW, Mellor JC, Pollis C, Pons C, Richardson AD, Schlabach S, Teeking B, Yadav A, Babor M, Balcha D, Basha O, Bowman-Colin C, Chin SF, Choi SG, Colabella C, Coppin G, D'Amata C, De Ridder D, De Rouck S, Duran-Frigola M, Ennajdaoui H, Goebels F, Goehring L, Gopal A, Haddad G, Hatchi E, Helmy M, Jacob Y, Kassa Y, Landini S, Li R, van Lieshout N, MacWilliams A, Markey D, Paulson JN, Rangarajan S, Rasla J, Rayhan A, Rolland T, San-Miguel A, Shen Y, Sheykhkarimli D, Sheynkman GM, Simonovsky E, Tasan M, Tejeda A, Tropepe V, Twizere JC, Wang Y, Weatheritt RJ, Weile J, Xia Y, Yang X, Yeger-Lotem E, Zhong Q, Aloy P, Bader GD, De Las Rivas J, Gaudet S, Hao T, Rak J, Tavernier J, Hill DE, Vidal M, Roth FP, Calderwood MA (2020) A reference map of the human binary protein interactome. Nature 580(7803):402–408. https://doi.org/10.1038/s41586-020-2188-x

Markham K, Bai Y, Schmitt-Ulms G (2007) Co-immunoprecipitations revisited: an update on experimental concepts and their implementation for sensitive interactome investigations of endogenous proteins. Anal Bioanal Chem 389(2):461–473. https://doi.org/10.1007/s00216-007-1385-x

Ohue M, Matsuzaki Y, Uchikoga N, Ishida T, Akiyama Y (2014) MEGADOCK: an all-to-all protein-protein interaction prediction system using tertiary structure data. Protein Pept Lett 21(8):766–778. https://doi.org/10.2174/09298665113209990050

Orchard S, Ammari M, Aranda B, Breuza L, Briganti L, Broackes-Carter F, Campbell NH, Chavali G, Chen C, del-Toro N, Duesbury M, Dumousseau M, Galeota E, Hinz U, Iannuccelli M, Jagannathan S, Jimenez R, Khadake J, Lagreid A, Licata L, Lovering RC, Meldal B, Melidoni AN, Milagros M, Peluso D, Perfetto L, Porras P, Raghunath A, Ricard-Blum S, Roechert B, Stutz A, Tognolli M, van Roey K, Cesareni G, Hermjakob H (2014) The MIntAct project--IntAct as a common curation platform for 11 molecular interaction databases. Nucleic Acids Res 42(Database issue):D358–D363. https://doi.org/10.1093/nar/gkt1115

Oughtred R, Rust J, Chang C, Breitkreutz BJ, Stark C, Willems A, Boucher L, Leung G, Kolas N, Zhang F, Dolma S, Coulombe-Huntington J, Chatr-Aryamontri-A, Dolinski K, Tyers M (2021) The BioGRID database: a comprehensive biomedical resource of curated protein, genetic, and chemical interactions. Protein Sci 30(1):187–200. https://doi.org/10.1002/pro.3978

Pankow S, Bamberger C, Calzolari D, Martinez-Bartolome S, Lavallee-Adam M, Balch WE, Yates JR 3rd (2015) F508 CFTR interactome remodelling promotes rescue of cystic fibrosis. Nature 528(7583): 510–516. https://doi.org/10.1038/nature15729

Peri S, Navarro JD, Amanchy R, Kristiansen TZ, Jonnalagadda CK, Surendranath V, Niranjan V, Muthusamy B, Gandhi TK, Gronborg M, Ibarrola N,

Deshpande N, Shanker K, Shivashankar HN, Rashmi BP, Ramya MA, Zhao Z, Chandrika KN, Padma N, Harsha HC, Yatish AJ, Kavitha MP, Menezes M, Choudhury DR, Suresh S, Ghosh N, Saravana R, Chandran S, Krishna S, Joy M, Anand SK, Madavan V, Joseph A, Wong GW, Schiemann WP, Constantinescu SN, Huang L, Khosravi-Far R, Steen H, Tewari M, Ghaffari S, Blobe GC, Dang CV, Garcia JG, Pevsner J, Jensen ON, Roepstorff P, Deshpande KS, Chinnaiyan AM, Hamosh A, Chakravarti A, Pandey A (2003) Development of human protein reference database as an initial platform for approaching systems biology in humans. Genome Res 13(10):2363–2371. https://doi.org/10.1101/gr.1680803

Phizicky EM, Fields S (1995) Protein-protein interactions: methods for detection and analysis. Microbiol Rev 59(1):94–123. https://doi.org/10.1128/mr.59.1.94-123.1995

Porras P, Barrera E, Bridge A, Del-Toro N, Cesareni G, Duesbury M, Hermjakob H, Iannuccelli M, Jurisica I, Kotlyar M, Licata L, Lovering RC, Lynn DJ, Meldal B, Nanduri B, Paneerselvam K, Panni S, Pastrello C, Pellegrini M, Perfetto L, Rahimzadeh N, Ratan P, Ricard-Blum S, Salwinski L, Shirodkar G, Shrivastava A, Orchard S (2020) Towards a unified open access dataset of molecular interactions. Nat Commun 11(1):6144. https://doi.org/10.1038/s41467-020-19942-z

Rhee HW, Zou P, Udeshi ND, Martell JD, Mootha VK, Carr SA, Ting AY (2013) Proteomic mapping of mitochondria in living cells via spatially restricted enzymatic tagging. Science 339(6125):1328–1331. https://doi.org/10.1126/science.1230593

Rolland T, Tasan M, Charloteaux B, Pevzner SJ, Zhong Q, Sahni N, Yi S, Lemmens I, Fontanillo C, Mosca R, Kamburov A, Ghiassian SD, Yang X, Ghamsari L, Balcha D, Begg BE, Braun P, Brehme M, Broly MP, Carvunis AR, Convery-Zupan D, Corominas R, Coulombe-Huntington J, Dann E, Dreze M, Dricot A, Fan C, Franzosa E, Gebreab F, Gutierrez BJ, Hardy MF, Jin M, Kang S, Kiros R, Lin GN, Luck K, MacWilliams A, Menche J, Murray RR, Palagi A, Poulin MM, Rambout X, Rasla J, Reichert P, Romero V, Ruyssinck E, Sahalie JM, Scholz A, Shah AA, Sharma A, Shen Y, Spirohn K, Tam S, Tejeda AO, Trigg SA, Twizere JC, Vega K, Walsh J, Cusick ME, Xia Y, Barabasi AL, Iakoucheva LM, Aloy P, De Las RJ, Tavernier J, Calderwood MA, Hill DE, Hao T, Roth FP, Vidal M (2014) A proteome-scale map of the human interactome network. Cell 159(5):1212–1226. https://doi.org/10.1016/j.cell.2014.10.050

Roux KJ, Kim DI, Raida M, Burke B (2012) A promiscuous biotin ligase fusion protein identifies proximal and interacting proteins in mammalian cells. J Cell Biol 196(6):801–810. https://doi.org/10.1083/jcb.201112098

Rual JF, Venkatesan K, Hao T, Hirozane-Kishikawa T, Dricot A, Li N, Berriz GF, Gibbons FD, Dreze M, Ayivi-Guedehoussou N, Klitgord N, Simon C, Boxem M, Milstein S, Rosenberg J, Goldberg DS, Zhang LV, Wong SL, Franklin G, Li S, Albala JS, Lim J, Fraughton C, Llamosas E, Cevik S, Bex C, Lamesch P, Sikorski RS, Vandenhaute J, Zoghbi HY, Smolyar A, Bosak S, Sequerra R, Doucette-Stamm L, Cusick ME, Hill DE, Roth FP, Vidal M (2005) Towards a proteome-scale map of the human protein-protein interaction network. Nature 437(7062):1173–1178. https://doi.org/10.1038/nature04209

Sahni N, Yi S, Taipale M, Fuxman Bass JI, Coulombe-Huntington J, Yang F, Peng J, Weile J, Karras GI, Wang Y, Kovacs IA, Kamburov A, Krykbaeva I, Lam MH, Tucker G, Khurana V, Sharma A, Liu YY, Yachie N, Zhong Q, Shen Y, Palagi A, San-Miguel A, Fan C, Balcha D, Dricot A, Jordan DM, Walsh JM, Shah AA, Yang X, Stoyanova AK, Leighton A, Calderwood MA, Jacob Y, Cusick ME, Salehi-Ashtiani K, Whitesell LJ, Sunyaev S, Berger B, Barabasi AL, Charloteaux B, Hill DE, Hao T, Roth FP, Xia Y, Walhout AJM, Lindquist S, Vidal M (2015) Widespread macromolecular interaction perturbations in human genetic disorders. Cell 161(3):647–660. https://doi.org/10.1016/j.cell.2015.04.013

Sanchez C, Lachaize C, Janody F, Bellon B, Roder L, Euzenat J, Rechenmann F, Jacq B (1999) Grasping at molecular interactions and genetic networks in Drosophila melanogaster using FlyNets, an internet database. Nucleic Acids Res 27(1):89–94. https://doi.org/10.1093/nar/27.1.89

Sharan R, Ulitsky I, Shamir R (2007) Network-based prediction of protein function. Mol Syst Biol 3:88. https://doi.org/10.1038/msb4100129

Sharifi Tabar M, Parsania C, Chen H, Su XD, Bailey CG, Rasko JEJ (2022) Illuminating the dark protein-protein interactome. Cell Rep Methods 2(8):100275. https://doi.org/10.1016/j.crmeth.2022.100275

Soleymani F, Paquet E, Viktor H, Michalowski W, Spinello D (2022) Protein-protein interaction prediction with deep learning: a comprehensive review. Comput Struct Biotechnol J 20:5316–5341. https://doi.org/10.1016/j.csbj.2022.08.070

Sousa RT, Silva S, Pesquita C (2024) Explaining protein-protein interactions with knowledge graph-based semantic similarity. Comput Biol Med 170:108076. https://doi.org/10.1016/j.compbiomed.2024.108076

Stollar EJ, Smith DP (2020) Uncovering protein structure. Essays Biochem 64(4):649–680. https://doi.org/10.1042/EBC20190042

Szklarczyk D, Gable AL, Lyon D, Junge A, Wyder S, Huerta-Cepas J, Simonovic M, Doncheva NT, Morris JH, Bork P, Jensen LJ, Mering CV (2019) STRING v11: protein-protein association networks with increased coverage, supporting functional discovery in genome-wide experimental datasets. Nucleic Acids Res 47(D1):D607–D613. https://doi.org/10.1093/nar/gky1131

Szklarczyk D, Kirsch R, Koutrouli M, Nastou K, Mehryary F, Hachilif R, Gable AL, Fang T, Doncheva

NT, Pyysalo S, Bork P, Jensen LJ, von Mering C (2023) The STRING database in 2023: protein-protein association networks and functional enrichment analyses for any sequenced genome of interest. Nucleic Acids Res 51(D1):D638–D646. https://doi.org/10.1093/nar/gkac1000

Tan SH, Zhang Z, Ng SK (2004) ADVICE: automated detection and validation of interaction by co-evolution. Nucleic Acids Res 32(Web Server issue):W69–W72. https://doi.org/10.1093/nar/gkh471

Turei D, Valdeolivas A, Gul L, Palacio-Escat N, Klein M, Ivanova O, Olbei M, Gabor A, Theis F, Modos D, Korcsmaros T, Saez-Rodriguez J (2021) Integrated intra- and intercellular signaling knowledge for multicellular omics analysis. Mol Syst Biol 17(3):e9923. https://doi.org/10.15252/msb.20209923

Veres DV, Gyurko DM, Thaler B, Szalay KZ, Fazekas D, Korcsmaros T, Csermely P (2015) ComPPI: a cellular compartment-specific database for protein-protein interaction network analysis. Nucleic Acids Res 43 (Database issue):D485–D493. https://doi.org/10.1093/nar/gku1007

Wang XW, Madeddu L, Spirohn K, Martini L, Fazzone A, Becchetti L, Wytock TP, Kovacs IA, Balogh OM, Benczik B, Petervari M, Agg B, Ferdinandy P, Vulliard L, Menche J, Colonnese S, Petti M, Scarano G, Cuomo F, Hao T, Laval F, Willems L, Twizere JC, Vidal M, Calderwood MA, Petrillo E, Barabasi AL, Silverman EK, Loscalzo J, Velardi P, Liu YY (2023) Assessment of community efforts to advance network-based prediction of protein-protein interactions. Nat Commun 14(1):1582. https://doi.org/10.1038/s41467-023-37079-7

Wu X, Zhu L, Guo J, Zhang DY, Lin K (2006) Prediction of yeast protein-protein interaction network: insights from the Gene Ontology and annotations. Nucleic Acids Res 34(7):2137–2150. https://doi.org/10.1093/nar/gkl219

Zhang XM, Liang L, Liu L, Tang MJ (2021) Graph neural networks and their current applications in bioinformatics. Front Genet 12:690049. https://doi.org/10.3389/fgene.2021.690049

Zhong X, Rajapakse JC (2020) Graph embeddings on gene ontology annotations for protein-protein interaction prediction. BMC Bioinform 21(Suppl 16):560. https://doi.org/10.1186/s12859-020-03816-8

Functional Genomics: Principles and Methods

8

Ana Neves-Costa and Luís Ferreira Moita

Abstract

Functional genomics studies address how genes, gene products, and noncoding sequences dictate phenotypes. Unlike traditional gene-by-gene approaches, functional genomics screens comprehensively interrogate genomes. Continuous advances in nucleic acid manipulation and gene-targeting tools, combined with the availability of complete genomes, the variety of phenotypes that can be screened, and the expanding capacity of omics data analysis, contribute to the lasting success of functional genomics in biological discovery. This chapter focuses on designing, implementing, and analyzing functional genetic screens in model organisms and human cells. It also reviews recent findings with profound impacts on biomedical research. Additionally, current limitations that still compromise the use of functional genomics screens across all model organisms and cells are examined, and future applications are anticipated. Given the extensive work presented in this chapter, there is little doubt that functional genomics screens will continue to be a powerful driving force in understanding how genomes regulate health and disease.

Keywords

Functional genomics screen · Forward screen · Reverse screen · Arrayed library · Pooled library · siRNA · shRNA · CRISPR/Cas9

A. Neves-Costa · L. F. Moita (✉)
Innate Immunity and Inflammation Laboratory, Instituto Gulbenkian de Ciência, Oeiras, Portugal

Faculdade de Medicina da Universidade de Lisboa, Lisbon, Portugal
e-mail: lmoita@medicina.ulisboa.pt

8.1 Introduction

Understanding how genomes determine biological processes is fundamental to explaining cell, organ, and organismal behavior and, therefore, fighting diseases. Functional genomics studies attempt to explain how genes, gene products, or even noncoding sequences in an organism, system, organ, cell, or pathway interact to produce a particular phenotype or trait. Ultimately, functional genomics contributes to understanding the function of every gene in a defined time and location (Bunnik and Le Roch 2013).

Functional genomics methods are based on perturbing, modifying, or abolishing gene function to study the resulting phenotypes. What distinguishes these approaches from traditional gene-by-gene molecular biology techniques is that they interrogate, or screen the genomes, in a more or less comprehensive way, ranging from targeted approaches—for example, by focusing on all known genes with kinase activity—to whole genome coverages. The variety of

M. Gama-Carvalho et al. (eds.), *Omics Approaches in Biomedicine and Biotechnology*, Advances in Experimental Medicine and Biology 1504, https://doi.org/10.1007/978-3-032-18966-0_8

phenotypes that can be screened, combined with the continuous advances in nucleic acid manipulation, has contributed to the success of functional genomics throughout the years. In addition, the more recent availability of complete genome sequences, including of the human genome (for two discussions on this topic separated by two decades, see Watson (2001), Amaral et al. (2023)), and the impressive development of omics technologies, unquestionably established functional genomic screens as a powerful moving force behind major discoveries that relate gene function to the observed phenotypes in many cell types and organisms. The latest advances in clustered regularly interspaced short palindromic repeats (CRISPR)/Cas tools are expected to prompt many more breakthroughs that will expand our understanding of the complex biology governing cells, organs, and organisms (Pickar-Oliver and Gersbach 2019; Przybyla and Gilbert 2022).

In particular, biomedical research profits greatly from functional genomics studies. Novel knowledge of the function of genes and gene families often contributes to advances in understanding human health and pathophysiology. Some examples include microbiome composition, drug sensitivity and resistance, cancer susceptibility and progression, and immune responses, in particular to specific infectious agents. Functional genomics studies have also focused on the impacts of human genomic variation, which is a key aspect to consider when developing novel therapies, especially those treatments involving gene therapy and regenerative medicine (Luca et al. 2018; Przybyla and Gilbert 2022; Wang et al. 2022; Shi et al. 2023; Vredevoogd and Peeper 2023). This chapter will outline the principles and methods employed in functional genomics screens, including the experimental design, the platforms and technical approaches, and the choice of readouts. Significant examples, mostly contributing to a better understanding of human health, will be discussed to highlight how functional genomic screens can be fundamental to the understanding of the dynamic nature of gene expression, taking into consideration the intricate network of contributing factors such as cell type, developmental stage, metabolism, cell activation status, response to stimuli, infection, and other diseases.

8.2 Forward Versus Reverse Genetic Screens

Classically, functional genomics screens aim to identify the genes and gene products responsible for a phenotype of interest. Such approaches, aiming to determine the genetic basis for a specific trait, are considered forward screens and rely on phenotype-to-genotype methodologies (Fig. 8.1). In forward screens, gene expression of a series of genes, or even genome-wide gene

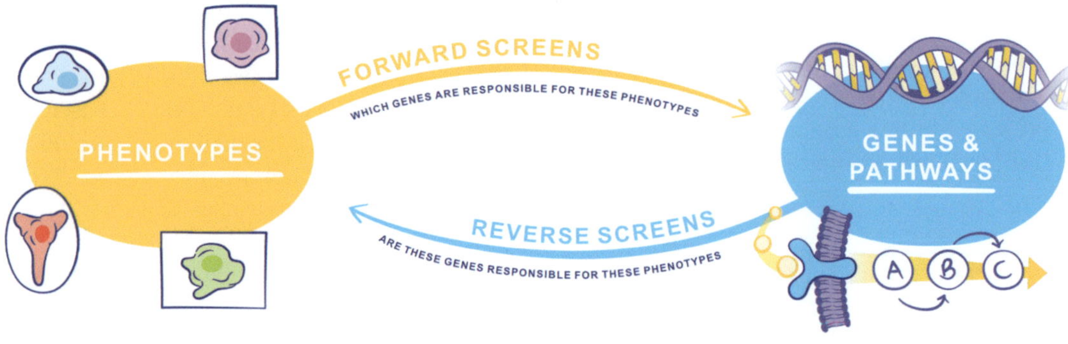

Fig. 8.1 Forward and reverse genetic screens. Whereas forward genetic screens aim to identify genes responsible for particular phenotypes of interest, reverse screens start with the manipulation of genes and pathways of interest aim to identify the phenotypes that are controlled by particular genetic sequences

expression, is altered in a more or less random way, and one or more phenotypes are assessed in that diverse population of cells or individuals. By tracking down which genetic changes induced the observed phenotype, novel players that are involved in the biological pathways or processes being studied can be identified. Importantly, forward genetic screens have led to considerable progress in complex and experimentally challenging biological questions, such as the identification of human core fitness genes and how they vary across cell types (Hart et al. 2015).

In contrast, reverse genetic screens interrogate a known set of genes of interest in gene-to-phenotype approaches (Fig. 8.1). Typically, these genes are part of molecular pathways being studied, and the aim is to characterize the phenotypes that emerge from the modulation of that particular pathway. Reverse genetics approaches or smaller reverse genetic screens are often used to validate and/or expand results from forward genetic screens, once a gene or pathway of interest is identified. The ever-evolving tools based on CRISPR/Cas9 system have transformed reverse genetics by allowing precise editing of theoretically any gene or noncoding genome element in a wide variety of cell types and model organisms. As discussed below, these technical advances contributed immensely to clarifying how genotypes shape phenotypes. Moreover, the ability to successfully edit selected genes has allowed the understanding of otherwise puzzling observations, such as those arising from redundancy in biological networks and from molecular mechanisms operating to compensate for gene loss-of-function (discussed, e.g., in Salanga and Salanga (2021)).

8.3 Targeting Libraries

One central aspect of functional genomics studies is the choice of the gene-targeting screening library. The chosen option may be a whole-genome library or, instead, a library that allows for specific knockout or knockdown of a collection of genes of interest. Impressive technical developments in nucleic acid manipulation, together with the current understanding of how genomes work and are regulated, have considerably expanded the toolkit of genomic libraries used in screens, from the early days of chromosome-specific DNA hybridizations to the latest optimizations of genome-wide human CRISPR/Cas9 tools (Krumlauf et al. 1982; Gonçalves et al. 2021). In the last 50 years, screening and target identification became significantly more precise and reliable, mostly due to substantial progress in the library reagents available for a wide variety of organisms. Originally, functional genomics screens relied on random mutagenesis approaches with the use of chemical mutagens, of viruses, and of transposons to induce gene inactivation. However, only screens that use defined targeting libraries will be discussed here.

The main types of libraries that are currently available for sequence-specific targeting of mammalian genomes make use of RNA molecules to do the targeting: siRNAs (small interfering RNAs), including shRNAs (short hairpin RNAs), and CRISPR/Cas9 sgRNAs (single-guide RNAs). The delivery methods for the plasmids carrying these libraries and the associated molecular tools may vary, depending on the type of DNA constructs where the libraries can be found and also on the cell or organism where the targeting is expected to occur. Whereas some plasmids can be delivered directly to the cells in the screen, others may require viral packaging. In addition to deciding what type of library to use, which is closely linked to the intent of the targeting—either genomic inactivation leading to knockout or gene expression modulation, for example—one critical aspect to consider is the decision between arrayed or pooled screening formats. Various technical aspects of working with targeting libraries are discussed in the next sections.

8.3.1 Arrayed Libraries Versus Pooled Libraries

The screen is said to be arrayed when only one gene or genomic sequence is targeted per well (or tube, plate, flask, other) in a multi-well plate format experiment. If, on the contrary, gene perturbations occur in bulk in the same well (or tube, plate, flask, other), then it is a pooled screen (see, e.g., Canver et al. (2018) and So et al. (2019) for the technical differences between the two types of screens using CRISPR/Cas9 tools). Since arrayed screens target each gene separately, gene identification following phenotypic characterization is simple and direct, for example, in different wells of a 96-well plate, which constitutes the major advantage of using arrayed libraries (Fig. 8.2). However, when a large number of genes need to be analyzed, these screens are very labor- and reagent-consuming, which can drastically increase the costs associated and the time to conduct the screen. Frequently, arrayed screens are performed in automated or semiautomated ways using plate-handling devices.

On the contrary, pooled screens are a more obvious choice when a large collection of genes needs to be analyzed, as when testing all genes of an entire genome. This format has been developed more recently and is more versatile because, by pooling several target genes into the same assay well, pooled libraries allow for less expensive and more expeditious experimental approaches. Pooled libraries are usually supplied as a mixed collection of plasmids in a single tube. Independent of the source of the library, all constructs are built with the same backbone, and only the gene targeting region differs. Either a gift from another laboratory, a commercially available library, or one from a non-profit repository such as Addgene (https://www.addgene.org), good practices dictate that libraries should be sequenced to verify that all targeting sequences, corresponding to the genes believed to be included, can be detected. Due to the advantages of setting up a pooled screen, with lower initial costs and faster results, these are frequently the first options for unbiased discovery. The identification of gene hits from the screen when using pooled libraries relies on DNA or RNA sequencing of the final populations of cells, and therefore, sequencing capacity and costs should also be considered when designing a screen (Fig. 8.3). At the end, candidate genes that were selected from the vast list of targets often become part of a second screen to validate the results and possibly further expand to other members of the same family. These smaller confirmatory screens are then arrayed and can be designed to contain more than one target per candidate gene to increase the specificity of the results.

Although not true in all settings, the accuracy of the phenotypic responses can increase in arrayed screens compared to pooled screens. It has been reported that higher hit rates can be expected, with fewer false-positive and false-negative hits (Kim et al. 2018). Modulation of the expression of a particular gene may result in a very mild phenotype, which can go unnoticed when only a few cells in a well have had that gene targeted, as could happen in a pooled screen. In contrast, if most of the cells in a well—and after selective pressure, it should be the vast majority of the cells—have the same gene knocked out, then the phenotype may become clearer (Kim et al. 2018). Another limitation of pooled screens is that cells with different responses due to different genes being targeted can influence one another in the same assay well. These effects may be especially impactful in certain cell types, such as primary cells. Cell death, senescence, or the production of inflammatory mediators, such as cytokines and chemokines, are examples of possible outcomes resulting from gene targeting in functional genomics screens. If the screen is pooled, such responses from cells where a particular gene was targeted may affect neighboring cells and induce a response that is not caused by the targeting of a different gene in that cell but a consequence of the external stimuli (discussed for primary neurons in So et al. (2019)). As explained, such confounding screen results are mostly a matter of concern in pooled screens but can also occur in arrayed screens in cases with high levels of cell death or inflammation due to the experimental treatment or the selection method.

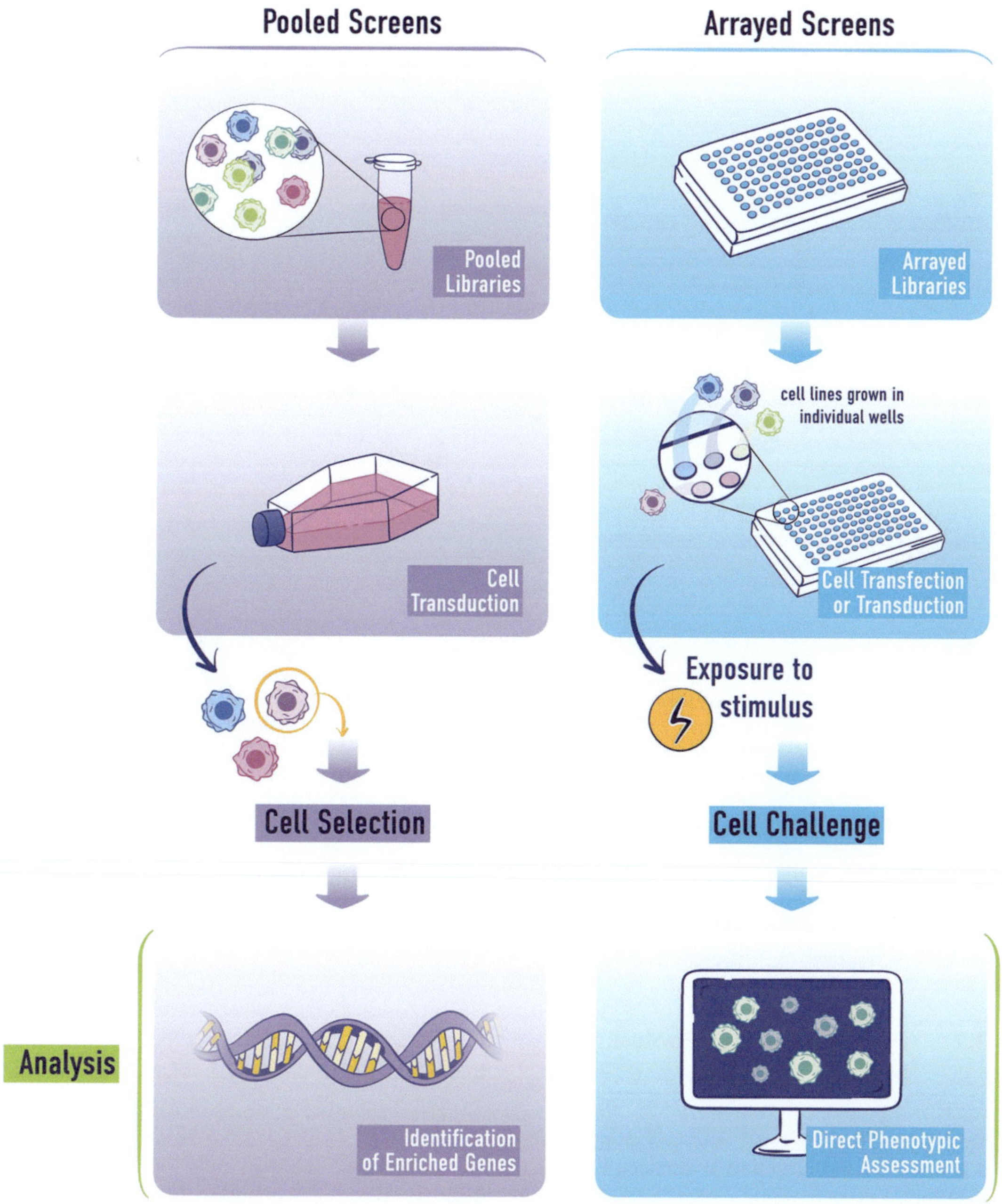

Fig. 8.2 Pooled versus arrayed screens. In pooled screens, as shown on the left, the genetic perturbations are introduced in a population of cells as a "pool," meaning that the library used is composed by a mixture of different gene-targeting sequences. Cells are then selected based on a phenotype of interest, and the population is finally assessed to determine which genes have been favored by the treatment. The analysis can both determine which genes are present in the final population, and the proportion in which each gene is present in relation to the others. In this way, the screen will reveal which genes have been selected or depleted by the treatment, depending on the design of the screen. In arrayed screens, as shown on the right, different genetic perturbations are introduced separately, targeting, for example, one gene per well in 96-well plates. Following cell growth or exposure to a chosen stimulus, the resulting phenotypes can be assessed directly in each well, using technical approaches such as high-throughput microscopy. Finally, the genes that were perturbed in the wells where phenotypes of interest were observed are directly identified

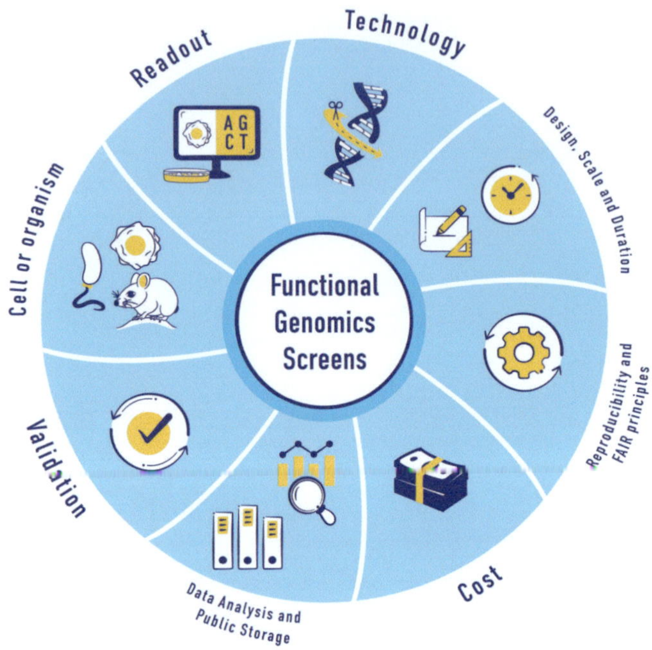

Fig. 8.3 Key considerations in the design and execution of functional genomics screens. Before performing a functional genomics screen, the model organism or cell type of interest needs to be carefully considered, firstly taking into consideration the biological question, but also for the ease of growth, maintenance, and manipulation. The choice of a suitable readout is one of the main challenges of a screen, as many aspects must contribute to the final choice, such as feasibility, availability, specificity, and the amount and type of information provided. The platform chosen for gene editing or RNA interference is central to the screen and must be chosen according to the biological model and the readout. The design of each specific screen should incorporate aspects dedicated to the scale of the screen, both in the number of samples and replicates, but also the steps involved, which will dictate the duration of the screen, to be estimated before the beginning of the experimental procedures. One important aspect that should be assessed before the start of the screen is its reproducibility and therefore robustness of the observations, which should comply with FAIR (findability, accessibility, interoperability, and reuse of digital assets) principles. The costs must be calculated taking into consideration all steps of the screen, including the analysis and validation stages. Data should be deposited in publicly available databases as soon as possible

8.3.2 Particularities of Pooled Screens

Pooled libraries can be of different sizes, from just a few selected targeting sequences to a vast collection of gene-targeting sequences, such as whole-genome libraries. One aspect that is very important for all functional genomic screens, but is particularly central when performing pooled screens, is the ratio of targeting plasmids per cultured cells. Obviously, this ratio should be optimized for different cell types and is method-dependent, as some screens may use transfection reagents to deliver the libraries to cells. In contrast, others may require viral packaging of the libraries and transduction of cells. However, one fundamental rule in every pooled screen is that the number of plasmids used per each cultured cell should be as low as possible so that, theoretically, each cell receives only one plasmid. As different plasmids encode different targeting sequences, the ideal scenario in a pooled screen would be to have only one gene being targeted in each cell. Consequently, the number of cells used should be adjusted to the size of the library to increase the odds of every plasmid being represented in the cell population assayed (Hart et al. 2015). In addition, the design of pooled screens should also take into consideration the need for biological replicates, which, in this case, are different cells carrying the same sequence-targeting plasmid. Whereas in arrayed screens there can be as many biological replicates as desired, with only the need of adding the same plasmid to more than one well in a multi-well culture plate, in pooled screens the biological replicates become possible when a large enough number of cells are infected. Again, to ensure an adequate number of theoretical biological replicates, the number of cells targeted should be adjusted accordingly, always maintaining a low ratio of plasmids to cells. In addition to increasing the possibility of hit identification, including biological replicates in the screen also minimizes false-positive and false-negative results.

Another important step when performing pooled screens is the selection method or readout for hit identification when cell death or any treatment compromises viability. Examples of these types of selection are drug sensitivity, susceptibility to infection, and cell proliferation (Lin et al. 2024). In arrayed screens, if cells die or do not proliferate due to gene knockdown or knockout, the gene targeted in that "empty" well can still be identified, as each well receives only one gene-targeting plasmid. On the contrary, if cells die in a pooled screen, it is impossible to identify which plasmid they had received. To overcome such limitations, the design of a pooled screen should start with the decision of performing a positive or a negative screen. In positive screens, the phenotype of interest can be found in the cells that survive the treatment (the readout is drug resistance, for example). The selection pressure should be strong enough that only a small proportion of the cells can be selected as hits. In contrast, negative screens lead to the death of cells with the phenotype of interest (e.g., when assaying drug sensitivity), and, for that reason, two "identical" screens need to be done in parallel so that in one of the replicate screens, no selection or treatment is applied. Negative screens are therefore more complex and expensive.

Both for positive and negative pooled screens, hit identification requires DNA or RNA sequencing after selection. In pooled positive screens, the collection of gene hits in the final selected population of cells is sequenced after plasmid amplification (or amplification of plasmid-carried specific sequences from the genome in the case of stable transfections). The sequences retrieved correspond to the candidate genes revealed by the pooled screen. When the pooled screen is a negative screen, both the selected and treated cells, and also the replicate set of untreated cells, must be sequenced. In this case, the gene hits that are revealed by the screen are the ones that have been depleted when comparing treated with untreated cell populations. Details of sequencing and sequencing analysis required for library deconvolution and hit identification will not be mentioned here but have been extensively discussed in the literature, including the choice between custom barcode microarrays or next-generation sequencing (see, e.g., Hu and Luo (2012); Bock et al. (2022)).

8.3.3 siRNA Libraries

RNA interference (RNAi) was first characterized in *Caenorhabditis elegans* but is also found in many biological systems, including mammals (Fire et al. 1998; Elbashir et al. 2001). This evolutionarily conserved phenomenon has been harnessed in laboratorial research for suppression of gene expression through posttranscriptional gene-silencing. In RNAi, short double-stranded RNA species (dsRNA) mediate the breakdown of mRNA molecules in a sequence-specific manner through the action of the multiprotein RNA-induced silencing complex (RISC), which directs mRNA molecules to degradation (Pratt and MacRae 2009). The targeting dsRNAs, known as small interfering RNAs (siRNAs, of 19 to 24 base pairs in length), are complementary to the mRNA molecules to be degraded and can have several origins: (1) derived from a long dsRNA by ribonuclease cleavage in cells; (2) be chemically synthesized; or (3) be expressed from a vector with an RNA polymerase promoter (Elbashir et al. 2001). Long dsRNAs, once believed to be of viral origin exclusively and their presence in eukaryotic cells attributed solely to infection, are now known to also be produced by vertebrate cells from endogenous sources (Chen and Hur 2022). In contrast with the introduction of long dsRNAs in mammalian cells, siRNAs are not associated with a strong antiviral response by mammalian cells and therefore are useful tools for gene silencing (Maillard et al. 2019). Only dsRNAs used in functional genomic screens will be discussed here.

Synthetic and plasmid-encoded siRNA libraries have been widely used in functional genomics screens, commonly known as siRNA libraries. When using synthesized siRNAs, gene silencing can be very transient, usually peaking around 48 to 72 hours after transfection and disappearing after 5 to 7 days in culture for actively dividing cells. It is, therefore, important to have a screen readout that is compatible with this timeframe. For more stable suppression of gene expression, targeting libraries composed of sequence-specific siRNA fragments cloned into expression vectors are the obvious choice. For particular nonessential genes, knockdown can be maintained even when mRNA levels are assessed 2 months after transfection (Brummelkamp et al. 2002). One particular type of vectors that have proved to be very successful in eliciting RNAi in mammalian cells is the one encoding hairpin-type short RNAs, as discussed in the next section (for the earliest versions of these expression vectors, see Paddison et al. (2002), Brummelkamp et al. (2002), Yu et al. (2002)). As explained, vector-based libraries should be expanded and sequenced before use to ensure that the targeting sequences correspond to the full set of genes to be studied.

Several vectors and strategies for encoding siRNAs have been designed for the stable expression of siRNAs in mammalian cells. Vector-based siRNA libraries have the advantage that the transfection efficiency can be assessed through a marker (fluorescent markers for microscopy assessment are commonly used, for example) in the backbone of the vector chosen for cloning of targeting sequences. Selection markers can also be included, as, for example, when the vector is engineered to contain antibiotic-resistant genes so that untransfected cells can be negatively selected and removed from the experiment with the use of antibiotics in the culture medium. As with any experimental approach relying on transfection for DNA delivery into cells, reagents should be carefully selected and transfection conditions optimized according to cell type, plating setup, and vector size, to name just a few of possible interfering factors. In difficult-to-transfect cells, other delivery methods can be used to overcome the need for transfection and instead rely on viral transduction, as discussed below.

8.4 shRNA Libraries and Viral Delivery

When first experimenting with designs that would be suitable for siRNA expression from a single RNA molecule, short hairpin RNA (shRNA)-based cloning strategies were the preferred choice (Paddison et al. 2002; Brummelkamp et al. 2002; Yu et al. 2002). In shRNA vectors, both siRNA strands are encoded in a segment of DNA with 50 to 60 bp in length, which is transcribed into a single RNA that folds like a hairpin due to sequence complementarity within the RNA molecule. In each shRNA, the sequence for each strand of the hairpin is usually connected by just a few nucleotides. Similarly to endogenous siRNAs, the transcribed and folded shRNA is incorporated into the RISC complex and directs sequence-specific mRNA degradation (reviewed in Chen et al. (2005)). Other strategies for encoding siRNAs have been developed and include, for example, the use of two promoters on opposing sites of a short DNA molecule to drive the expression of complementary RNA strands. Dual-promoter vectors have also been developed to be used in tandem, where the two siRNA strands are transcribed from two separate DNA sequences. Most of the vectors used in siRNA libraries have in common the use of Pol III promoters, which are active in a wide variety of cells and can be designed to be inducible. Details of technical aspects related to the production of siRNA libraries, including the choice of promoters for efficient siRNA expression and the selection of the targeting sequences, have been extensively discussed in the literature (an overview can be found in Chen et al. (2005)).

Cell transfection for delivery of siRNA vectors can be very inefficient in some mammalian cell types, including suspension cells and primary cells. To overcome these limitations, libraries have been cloned into viral vectors such as lentivirus, retrovirus, and adenovirus, as viral transduction can considerably increase nucleic acid delivery efficiency into mammalian cells (Abbas-Terki et al. 2002). Over the past 20 years, the use of viral packaging for library delivery has revolutionized RNAi-based genomic screens, including the possibility of genome-wide scale-up in virtually all cell types (work pioneered, e.g., in Moffat et al. (2006)). Early loss-of-function screens using large libraries to target a wide selection of human genes employed shRNA-expressing retroviral vectors and, shortly after, used lentiviral vectors (Abbas-Terki et al. 2002; Stewart et al. 2003; Berns et al. 2004; Paddison et al. 2004). This viral genetic material can integrate into the genome, thus allowing durable transgene expression. The advantage of using lentiviruses is that, unlike other retroviruses, lentiviruses are capable of infecting non-dividing cells in addition to dividing cells, which is particularly relevant for screens using primary terminally differentiated cells. Adenoviruses can also be used for packaging RNAi libraries because they infect both dividing and non-dividing cells too and have high transduction efficiencies and usually high levels of gene expression. However, adenoviral vectors cannot ensure long-lasting transgene expression.

8.5 Lentiviral-Based Libraries

In functional genomics screens, lentiviral vectors for shRNA library expression have been popular since their initial developments due to the high efficiencies of transduction and the relative ease of cloning, as the size of the sequences to be cloned is usually not a limiting factor in these vectors (Abbas-Terki et al. 2002; Stewart et al. 2003; Moffat et al. 2006). However, library production is not as straightforward as in other expression systems and can be more time-consuming, which is an important consideration when designing a screen. Lentiviral vector systems are usually composed of three independent plasmids, namely, the envelope, the packaging, and the transfer plasmids. The envelope plasmid usually originates from a virus with wide infectivity in mammalian cells. Commonly, envelope plasmids contain the glycoprotein G of the vesicular stomatitis virus (VSV-G), known to mediate cell attachment and membrane fusion in a variety of cell types. For safety reasons, the packaging plasmid has been engineered so that self-packaging is prevented due to missing packaging proteins. The transfer or cloning plasmid contains, in addition to the mammalian expression cassette and targeting sequences, elements to ensure genomic integration. Biosafety measures should be observed during viral plasmid handling, as there is a very remote possibility of generating a replication-competent virus in case of recombination between the packaging and the transfer plasmids. Many iterations of lentiviral-producing plasmids have been developed over the years to increase handling safety while maintaining high viral titers. Many of these vectors and libraries are available to the scientific community either commercially or free of charge from repositories, and the most suitable options must be selected according to each screen.

For viral production, the three plasmids need to be co-transfected into packaging HEK293T cells. After a few days in culture, cells release transgene-containing lentiviral particles into the culture media (Moffat et al. 2006). These particles are then collected and can be used to transduce cells of interest. For convenience, these viral particles can be stored at −80 °C without considerably compromising viability. Nevertheless, long-term storage and many freezing and thawing cycles should be avoided. This experimental set-up is amenable to high-throughput arrayed screens because viral production can be made in an arrayed format in microplates, and then the viral particles can be used to transduce microplate-cultured cells in the same arrayed positions (Moffat et al. 2006). Also important, small volumes of lentiviral particles are usually required due to the high efficiencies of transduction (as long as good viral titers are harvested from the supernatants of HEK293T cells), therefore allowing screens to be conducted in plate formats that are compatible with small volumes of culture medium. The need for small volumes also allows many screens to be performed from one same batch of viral particles, including, for example, screens done in parallel using different treatments and/or different readouts or screens to be used as replicates.

Although viral transduction with lentiviral vectors is usually highly efficient in most cell types, optimization of the conditions for viral delivery is still required. It is central to carefully monitor cell density and to adjust transduction confluency according to each cell type. As mentioned, high viral titers and good storage conditions are also essential. Finally, the transduction protocol should be carefully designed. For example, one particularly successful approach is the transduction of cells while centrifuging, or spinfection. Another important reagent to include is a polycation—usually polybrene—that improves viral delivery into cells by neutralizing the charge repulsion created when the viral particles approach cell membranes. The balance between cell seeding density, polybrene concentration, and spinfection time and conditions will contribute decisively to the success of viral delivery and library expression. The authors of this chapter have developed and optimized protocols for functional genomic screens in several cell types, mostly using lentiviral-mediated shRNA libraries in murine and human innate immune cells. The screens performed led to the identification of novel cellular players in antigen presentation, cytokine production, exosome secretion, and responses to systemic inflammation, among other observations with physiological impact (Ostrowski et al. 2010; Rato et al. 2010; Colvin et al. 2010; Moura-Alves et al. 2011; Moita et al. 2012; Neves-Costa and Moita 2013; Figueiredo et al. 2013).

8.6 Gene-Editing Libraries

As discussed so far, large-scale high-throughput screens have allowed the interrogation of the function of a large number of genes in parallel. The technical approaches introduced above rely on the modulation of gene expression through mRNA degradation through siRNA, but functional genomic screens can also take advantage of genome editing to link genotypes to phenotypes. The first editing tools available harnessed the power of cellular DNA-damage repair mechanisms. By using DNA endonucleases to introduce DNA double-strand breaks (DSB) in the genomes, and then providing an exogenous DNA donor sequence to be used for the repair of the DSB by homologous recombination (HR), genes could be edited although in a very inefficient way. For genomic specificity, nucleases were engineered to become sequence-specific. The design of zinc-finger DNA-binding domains fused to endonucleases, known as zinc-finger nucleases (ZFNs), allowed the targeting of these ZFNs to specific genomic domains and the consequent introduction of DSBs in selected DNA sequences. Other tools involved in early genome editing protocols were the transcription activator-like effector nucleases (TALENs), which also consisted of a DNA-binding domain fused to a DNA endonuclease. TALENs proved to be easier to customize and had higher cleavage efficiencies than ZFNs. Although genome editing by induction of DSBs followed by DNA damage repair (DDR) was a breakthrough in functional genomics, it was always associated with low efficiencies of targeted editing, high risks of off-target effects, and considerable increase in DNA instability, which compromised their use in large-scale functional genomics (reviewed in Urnov et al. (2010), Joung and Sander (2013), Yeh et al. (2019)).

The development of CRISPR/Cas9 tools for gene editing, which have been in use for less than 10 years, has already revolutionized functional genomics. The toolset currently available is vast, and Cas9 is part of an expanding family of Cas endonucleases that can be used in CRISPR. Only Cas9 will be discussed here to exemplify the wide range of possibilities that using CRISPR/Cas endonucleases brings to functional genomics screens. Of note, some Cas proteins target RNA instead of DNA (Przybyla and Gilbert 2022). To guide Cas9 to selected genomic sequences, 20-nucleotide-long single-guide RNAs (sgRNAs) must be provided, which recognize the target DNA by sequence pairing. Once at the target sequence, Cas9 further recognizes a DNA domain known as the protospacer adjacent motif (PAM) and cleaves the DNA upstream of the PAM, thus introducing a DSB. Cas9 variants with altered PAM-recognizing capacities have been developed to expand the targeting capacity of the enzyme (Kleinstiver et al. 2015). DDR pathways are then activated, and if non-homologous end joining (NHEJ) operates to repair the damage, this often introduces small insertion–deletion mutations (indels). Frameshifts due to indels usually originate transcripts with premature stop codons, which are degraded and cause specific gene knockout (Sternberg et al. 2014).

Loss-of-function libraries of CRISPR/Cas9 have been designed to make use of the gene-inactivating ability of Cas9, associated with the targeting specificity of sgRNAs. These novel functional genomics platforms co-express the Cas9 protein with a library of sgRNAs targeting the genes to be studied. In addition to gene genome-editing by gene knockout, other effector activities have been developed using CRISPR/Cas9 tools that can be used with sgRNA libraries, such as gain-of-function libraries and epigenome-targeting libraries. Some examples will be discussed next.

8.7 CRISPR Libraries

Cas9 is often used in its catalytically inactive, or dead, form (dCas9), to solely function as a DNA-targeting protein without the endonuclease cleaving activity, whereas the fused domains then perform the effector functions. Theoretically, dCas9 can be engineered to perform all sorts of activities depending on the fusion domain. The many existing dCas9 fusions have become invaluable tools for functional genomic studies, considerably expanding the range of biological questions that can be addressed. Some examples include variants designed for gene activation (CRISPRa), for inhibition through epigenetic regulation (CRISPRi), for base editing, and for regulation of DNA methylation. By using dCas9 fused to effector domains that regulate transcriptional activation and chromatin organization, gene modulation occurs at the transcriptional level without the need to permanently modify genomic sequences (reviewed in Ford et al. (2019)).

Cas9 and sgRNA delivery into cells follows the same experimental procedures as explained above for RNAi-mediated libraries (more examples of delivery platforms can be found in Ford et al. 2019). Although sgRNA-containing vectors are small and are easily taken up by cells, dCas9 fusion sequences can be long, and therefore their expression plasmids become more difficult to transfect. Co-transfection of a plasmid containing Cas9 (or dCas9 fusions) and another vector containing the sgRNA of interest is an obvious choice, but viral-based platforms, such as those discussed above for shRNAs, can also be used for difficult-to-transfect cells or when the plasmids are too big. In case of co-transfection or co-transduction, Cas9 levels should be uniformized between cells so that targeting efficiency is not compromised by the lack of Cas9 proteins. To overcome delivery challenges of Cas9-containing plasmids, and especially if cell lines are used in the screen, Cas9 can be constitutively expressed from a genomic location, increasing assay reproducibility as all cells will, in theory, express comparable amounts of the enzyme. If Cas9-expressing cells are used in the screen, then only the sgRNA library must be delivered to cells.

Similarly, when working with mouse primary cells, ex vivo screens can be performed using either constitutive or inducible CRISPR/Cas9 knock-in (KI) mice (Platt et al. 2014). In these animals, all cells or induced cells express Cas9, and therefore only the sgRNAs need to be delivered. Cas9 KI animals are also suitable for in vivo

CRISPR/Cas9 screens. Library delivery to in vivo models is of course organism-dependent, but in the case of the mouse model, viral delivery through adeno-associated viruses (AAVs) is a common choice (for protocols on mouse in vivo screens using CRISPR/Cas9, see Braun et al. (2022)).

One advantage of CRISPR/Cas9 screens over siRNA screens is the possibility of functionally interrogating the entire genome and not only transcribed regions. Many genome-wide association studies (GWAS) have identified genomic single-nucleotide polymorphisms (SNPs) that correlate with risks for several diseases. It is not uncommon for SNPs to be located in noncoding regions of the human genome, and when the closest genes are studied, no disease-modulating effects can be identified. By using CRISPR/Cas9 screens to directly interrogate those SNP-containing regions, it may be possible to gain further insights on their functional relevance (discussed in Przybyla and Gilbert (2022)). This is just one example of the possible applications of CRISPR/Cas9 screens to the study of noncoding genomes, which can be extended to the study of gene regulatory elements and spatial regulatory elements.

8.8 Choice of Readouts

Possibly the simplest and most used readouts of functional genomics screens are cell viability and cell proliferation assays following the introduction of a perturbation, commonly described as fitness assays. In screens with positive selection—also known as enrichment screens—cells that proliferate and can be detected at the end of the screen are the ones with the phenotype of interest. A widespread application of enrichment screens is for identifying genes conferring resistance to cytotoxic compounds, including antibiotic-resistant genes. These genes, which make cells refractory to particular drugs, can influence therapies, and their identification is therefore critical for clinical applications (Kurata et al. 2018). It is now common in drug development to perform preclinical CRISPR/Cas9 resistance screens. The perturbation introduced, which can be a drug treatment as just mentioned, incubation with an external stimulus, or an infection, for example, should be adjusted in dose and duration so that it allows the clear distinction between the nonviable cells and the cells that survive at the end of the screen. On the contrary, in screens with negative selection—also known as dropout screens—cells bearing the phenotype of interest are depleted from the final population. These screens are particularly useful for the study of oncogenes and the discovery of putative targets for drug development (Wirth et al. 2022). When proliferating cancer cells are screened with loss-of-function libraries, it is possible to detect novel oncogenes in the pool of cells that ceased to proliferate. For the identification of the hits, a wide variety of technical approaches can be chosen, from simple observation of cells by microscopy in arrayed screens to complex next-generation sequencing (NGS) of sgRNA in the case of pooled CRISPR/Cas9 screens (Bock et al. 2022) (Table 8.1). These more complex readouts will be discussed at the end of this section.

Another common readout in functional genomics screens is based on molecular reporter systems, so that the activity of the reporter correlates with the biological phenotype of interest. When choosing the reporter assay detection method, either by fluorescence-activated cell sorting (FACS), fluorescence microscopy coupled to high-content image analysis, luminescence, absorbance, or others, the signal-to-noise ratio should be sufficient for the readout of the screen to be unambiguous, so that hits can not only be detected but, if possible, also ranked. Fluorescent readouts in screens can also be due to dyes that are incorporated into cells to mark specific cell domains or organelles, to quantify not only the number of such structures but also their morphology, composition, and location. Dyes can also be used to record functional cellular processes, such as apoptotic induction or cell division, as in T cell proliferation assays (Moita et al. 2012).

Readouts do not need to be confined to the cells of interest but can instead be detected in the culture media. For example, the detection of protein secretion by immune cells, either cytokines or chemokines, can be performed with

Table 8.1 Examples of readouts used in functional genomics screens. Details and references can be found throughout the text

Readouts	Advantages	Limitations
Cell viability	Commonly used in pooled screens; simple to scale-up	Should include reliable nonselected controls
Cell proliferation	Suitable for cancer research	Confounding competition effects
Drug sensitivity	Commonly used in pooled screens; simple to scale-up	Choice of doses and incubation times can limit discovery; confounding compensatory effects
Synthetic lethality	For cells and unicellular organisms	Need for well-characterized mutations; confounding compensatory effects
Cell migration	Provides information about location and kinetics	Requirement for kinetic evaluation
Differentiation	Of particular interest in immune and developmental research	Relies on well-characterized end-points
Phagocytosis	Of interest in infection and immune studies; requires labeled bacteria	Applied only to a limited number of cells and microorganisms
Cell and organelle morphology	Used in cells and unicellular organisms	Subtle changes are difficult to assess
Protein localization and trafficking	Specificity; provides information about location and kinetics	Need for reliable labeling of proteins of interest and of cellular structures
Quantitative microscopy-based techniques	Used mostly in arrayed screens; adaptable to complex cellular characterization	Automated equipment required when scaling up; need for reliable labeling of factors and structures of interest; the limits of detection should be established
Biochemical assay (colorimetric, fluorescence, luminescence)	Easy to quantify in plate readers; simple to scale-up	Need for high specificity of probes; the limits of detection should be established
FACS-based	Suitable for pooled-screens; simple to scale-up	Requires optimization of the photoactive signal
Immunoassays	Easy to quantify in plate readers; possibility of multiplexing; mostly used in small and middle scale arrayed screens	The antibodies chosen should have high specificity; the limits of detection should be established
Single cell transcriptional profile	Allows for single cell resolution	Strong computational pipeline required
In vivo assays	Possibility of use in behavioral research; commonly used in tumor growth and suppressing assays	Limitations associated with in vivo work, such as number of individuals, type of procedure, choice of operators, reproducibility, costs, ethics concerns
Omics: Transcriptomics, epigenomics, proteomics, metabolomics, lipidomics	Highly informative; mostly used in small scale arrayed screens	Very dependent on the availability of the technological platforms; strong computational pipelines required; costs

relative ease by enzyme-linked immunosorbent assays (ELISAs) if the screen is arrayed (Figueiredo et al. 2013). As explained before, paracrine effects can influence cellular responses in pooled screens. For example, pro-inflammatory cytokines released by a subpopulation of cells with a particular gene targeted can activate neighboring cells and produce pro-inflammatory mediators even if the knockout of the neighboring population is not directly related to cytokine regulation. For this reason, screens with readouts that require the analysis of secreted factors should be restricted to arrayed formats.

8.9 Outputs of Pooled Screens

As previously discussed, using pooled screens allows large targeting libraries to be tested in the same population of cells to the point that genome-wide targeting can ultimately be achieved in one single petri dish. The obvious advantage is that pooled screens save reagents and time, but on the other hand, this screen format limits the types of readouts that can be used, as mentioned for protein secretion. In the same petri dish or assay well in a microplate, each cell ideally receives a unique targeting sequence, such as a unique sgRNA in the case of CRISPR/Cas9 screens. Therefore, sequencing capacity needs to be available for the detection of the sgRNAs present in the final population and their relative abundance when compared with the other sgRNAs of the library. Massively parallel sequencing with the possibility to quantitate reads has converted NGS into the preferred choice for hit identification and stratification in pooled CRISPR/Cas9 screens (reviewed in Morris et al. (2024)). This can be a complex readout and a dissuasive aspect when considering pooled CRISPR/Cas9 screens, but the possibilities for biological discoveries are immense. The primary aim is to determine, from the set of sequencing reads representing sgRNA abundances in the selected population of cells at the end of the screen, which genes were enriched and which were depleted because of the perturbation introduced. Independent of further bioinformatics analysis, a preliminary list of hits can be obtained from counting which sgRNAs are abundant, and a good indication that a gene is a hit from the screen is if multiple sgRNAs targeting the same gene rank high in the final list of sequences detected. Bioinformatic analysis of the screening results can then be performed using a series of packages developed for different needs, which have been discussed and compared in the literature (Ford et al. 2019).

The availability of sequencing capacity also expands the functional analysis that can be performed from the data collected in pooled screens. If single-cell sequencing (sc-RNA seq) is employed, the results may show that more than one targeting event occurred in each cell, which may be a very relevant biological finding in case more than one gene (or gene knockout) contributed to the phenotype observed. So far, functional genomics screens rely on the modification of gene expression at single-gene level, but it may be possible to target multiple genes if pooled screens are designed with that intent, which would open the door to the understanding of how complex phenotypes are driven by multigene interactions. In addition, the corresponding transcriptomic and epigenomic profiles of independent cells can also be determined when using sc-RNA-seq as a readout. Screens that combine pooled CRISPR/Cas9 targeting with single-cell sequencing are known as sc-CRISPR-seq (Morris et al. 2024).

Finally, one aspect where single-cell sequencing combined with pooled screens can be of particular interest is toward a better understanding of chromatin domains and genomic accessibility. At its various orders of organization, chromatin is central for the dynamic regulation of coding and noncoding genomic regions, affecting all DNA-based nuclear activities. However, mostly due to the challenges of finding adequate readouts to study how chromatin components are themselves regulated, the genetic basis of chromatin organization, which may also be responsible for critical chromatin dynamics, is very poorly understood. By combining pooled CRISPR/Cas9 screens with chromatin assays for transposase-accessible chromatin (ATAC), for transposase-accessible chromatin with visualization (ATAC-see), and for chromosome conformation capture (3C), to name just a few recently developed examples, sc-CRISPR-seq now allows high-throughput, single-cell identification of genetic perturbations with impact on chromatin organization and regulation (Rubin et al. 2019; Liscovitch-Brauer et al. 2021; Liu et al. 2023; Ishii et al. 2024).

8.10 Advantages and Limitations of Common Readouts

Readouts used in functional genomics screens are not thoroughly discussed here due to space limitations and because this is a rapidly evolving field where new outputs are constantly developed according to the biological questions to be addressed and potentiated by considerable advances in the use of novel technologies.

Readouts used in screens addressing the immune system are discussed in the next section, and a list of commonly used readouts in functional genomics screens can be found in Table 8.1. In addition, multimodal screens, where a series of independent readouts are recorded in the same cells, often at a single-cell level, are already in use. In these multiplexed readouts, information concerning distinct cellular aspects, such as abundance of selected proteins, genomic recombination, and transcriptomes, is combined to better understand the biological responses to experimental perturbations (Mimitou et al. 2019). The correct choice of a readout can determine the success of the screen, as discussed below in the summary of screen design, where the most important features of planning, executing, and analyzing a functional genomics screen are highlighted.

8.11 Screens in Biomedical Research: Key Findings in Immunology

The human immune system is comprised of a series of different cell types tailored for immune surveillance and recognition, effector activities, and regulatory processes, which are central in the development and control of infection, inflammation, and cancer. Immune cell function is dynamically driven both by intracellular regulators and by external stimuli. In an unprecedented way, CRISPR/Cas9 functional genomic screens have been and continue to be used for the systematic modulation of immune responses because they are compatible with the complexity of the immune system and also because they are suitable for use in primary cells and in vivo models of disease. These screens identified novel genes with roles in immune responses and immune cell fates but also previously unknown links between immunity-driven processes and biological adaptations controlling host-microbe interactions, cancer progression, and metabolism (Shi et al. 2023).

In arrayed immune screens using CRISPR/Cas9-based perturbations, the functional readouts commonly rely on flow cytometry, imaging, proteomics, and metabolomics. In addition, cell separation of selected immune subsets with the use of membrane cell markers or intracellular cell markers allows for increased specificity of downstream cell or molecular analysis and has the advantage of separating cell-autonomous effects directly related to gene perturbations from the effects due to perturbations in neighboring cells. However, high-throughput may not be adequate when performing arrayed screens, and therefore pooled immune screens were from the beginning the preferred choice (Parnas et al. 2015). The initial CRISPR/Cas9 screens using mouse primary immune cells, even if focused on well-studied innate immune pathways, still led to the identification of novel functional regulators of cytokine production and antimicrobial responses (Parnas et al. 2015). A second pooled screen, with selected hits from the original unbiased screen but with a library consisting of a more thorough gene coverage, is usually also employed in immune functional screens to validate the results and narrow the gene selection (Parnas et al. 2015). The exceptions to the use of pooled libraries are validation studies, where smaller, targeted arrayed screens are still the best option (Shi et al. 2023).

T cell activation, proliferation, and adhesion have also been the focus of a series of CRISPR/Cas9 screens, contributing to a better understanding of the adaptive immune system and further exploring the role of these cells in cancer immunotherapy (reviewed in Shi et al. (2023)). In a mouse in vivo screen, Cas9-expressing primary T cells (CD8+ cytotoxic T cells, CTLs) were transduced with a genome-wide CRISPR library and, by adoptive cell transfer, delivered into tumor-bearing mice (Zhou et al. 2023). After 1 week, tumor-infiltrating T cells were collected and analyzed by single-cell sgRNA sequencing, which was also combined with single-cell transcriptomic analysis. The results showed that tumor-specific CTLs are a diverse population and highlighted the molecular players contributing to the heterogeneity of T cell responses within tumors. The newly identified factors with roles in T cell differentiation into proliferative cell subsets are candidate targets for the improvement of cancer immunotherapies through functional reprogramming of intratumor CTL proliferating capacity (Zhou et al. 2023).

In addition to requiring conventional immune signaling—T cell receptor engagement, co-stimulation, and cytokine environment—T cell activation also relies on cellular metabolic programs (reviewed in Chapman et al. (2020)). A CRISPR/Cas9 screening performed when there was little understanding of how nutrients control T cell fate decisions strongly contributed to establishing a functional link between metabolism and the adaptive immune system (Long et al. 2021). Like in the study above, primary mouse Cas9-expressing T cells were transduced with a genome-wide lentiviral-based CRISPR library. As is often the case for genome-wide studies, the validation of the candidate hits, both positive and negative regulators of T cells, relied on a secondary, more targeted screen focused on the original hits. This screen showed that nutrient transporters, sensors, and transducers regulate immune signaling in T cells and can now be targeted in therapies for infectious diseases and other immune-related diseases, such as cancer and autoimmunity (Long et al. 2021).

8.12 Summary of Screen Design

Functional genomics screens are more than tools for gene function discovery; they enable researchers to address complex questions and to integrate observations from multiple biological pathways. As discussed, well-planned and executed screens can lead to breakthroughs at the cellular and organismal level, potentially revealing new players and unexpected regulatory networks with implications for physiological responses and disease. What successful functional genomic screens all have in common is that they thoughtfully address a series of aspects (Fig. 8.3), namely, (1) choice of biological model; (2) readout, which can rely on the acquisition of more than one biological feature; (3) type of targeting library to induce perturbations in gene expression; (4) experimental design of the biological challenges leading to cell selection and the kinetics associated; (5) reproducibility and the inclusion of FAIR principles (findable, accessible, interoperable, and reusable) in data acquisition, treatment, and storage; (6) costs associated with all steps of the screen; (7) complexity and extent of data analysis and the possibility of data sharing; and, finally, (8) validation of the hits. Given the availability of completely sequenced genomes (of humans and many model organisms) and due to considerable advances in library preparations and the wide range of platforms available (shRNA, CRISPR/Cas9 for example), high-throughput screens targeting entire genomes are now possible. In the recent years, technical limitations, such as robust robotic plate-handling, reliable automation in image acquisition and processing, and expeditious DNA and RNA sequencing, have also been overcome to a great extent. As a result of the impressive advances in libraries and technologies, the current main challenges to functional genomics screens are perhaps linked to library delivery and targeting specificity, as discussed next in the final section of this chapter.

8.13 Current Limitations and Perspectives

Plasmid transfection can lead to toxicity in several cell types. Therefore, this simple nucleic-acid delivery method cannot be used in many screens, which is why other delivery methods, such as viral delivery, have been used instead, as discussed. Another reason is the need for in vivo delivery. However, and especially when studying the immune system, viral factors themselves can induce immune responses. Viral components can also become integrated into the target genome with unintended consequences. Of no less concern, even if engineered to be safe to human health, viral vectors still maintain a very remote capacity of generating replication-competent particles (viral delivery for in vivo CRISPR/Cas9 was reviewed in Xu et al. (2019)). In the case of CRISPR/Cas9 screens, both with plasmid transfection and with viral-mediated delivery, it can be difficult to determine the best time to assess the chosen phenotype, as the production of Cas9 enzyme and the production and targeting of gRNAs may not be synchronous. To bypass these limitations, ribonucleoproteins (RNPs), composed of recombinant Cas9 enzymes and chemically synthesized gRNA, can be an option. When pre-formed RNPs are introduced in cells, editing at the target site is faster as both the enzyme and the gRNA are already in their effector forms. Another advantage of RNPs is that genomic integration is no longer a concern. A series of cell- and organ-specific protocols have been developed recently for difficult-to-transfect systems, including in vivo (for brain delivery, for example, see Chen et al. (2024)). In the near future, efficient systemic CRISPR/Cas9 delivery may instead rely on nanoparticles (Nie et al. 2022).

Off-target effects (OTEs) of modulating gene expression are also a matter of concern in functional genomics screens. In the case of CRISPR/Cas9, OTEs are a consequence of unexpected editing of genomic sequences different from the desired target sites due to sequence similarity. The propensity for OTEs to occur increases in cases where the Cas9 enzyme and the gRNA are present in cells for long periods of time, which is more likely with plasmid delivery than with the use of RNPs for example. Efforts to increase precision led to improvements in CRISPR/Cas9 specificity, but a series of challenges remain (reviewed in Guo et al. (2023)). Computational prediction tools have become very reliable but mostly rely on the analysis of sequence similarities, whereas other genomic features possibly leading to OTEs, such as epigenetic modifications and spatial organization, are still largely neglected. DISCOVER-seq is an experimental approach for the unbiased identification of off-target CRISPR/Cas-induced DNA breaks (Wienert et al. 2019). This technique can be used to assess the extent to which OTEs occur in the system chosen for the CRISPR/Cas9 screen using a series of representative gRNAs. To minimize OTEs, more than one biological model, type of library, or delivering method, for example, can be compared by DISCOVER-seq before performing the screen.

Due to the wide range of potential applications discussed throughout this chapter, functional genomic screens, in particular CRISPR/Cas9-based, will certainly result in groundbreaking discoveries with the promise of novel therapies and the prospect of a better quality of life. One important biological question where functional genomics screens are expected to bring considerable advances to the current knowledge is in host-microbe interactions. A series of cutting-edge works in the early days of CRISPR/Cas9 screens elucidated host pathways involved in bacterial toxicity and susceptibility to toxins in various cell types (see, e.g., Koike-Yusa et al. (2014), Zhou et al. (2014), Gilbert et al. (2014)). Shortly after, CRISPR/Cas9 screens were also performed in bacteria (Peters et al. 2016). However, it was only when a system for the transfer of modular CRISPR/Cas9 components that are functional throughout a wide variety of bacterial species was developed that it became possible to thoroughly interrogate gene function in microbes (Peters et al. 2019). This technological development, called Mobile-CRISPRi, led to the expansion of functional genomic screens in non-model bacteria (reviewed in Rousset and Bikard (2020)). To better elucidate bacterial gene function, bacterial functional screens have also been performed in living animal models, as gene requirement can be different when assessed in vitro or in vivo (some bacterial genes may not be essential for in vitro growth but are required in the animal gut for example). One of the pioneering in vivo bacterial genomic screens was performed in a murine pneumonia model and identified bacterial features contributing to pathogenesis (Liu et al. 2021). Importantly, variations were detected when screens were performed in different animals despite being genetically identical, highlighting the significant influence of the host and potentially the broader microbial community on disease progression. Microbial functional genomics screens are expected to become essential tools for investigating the diversity of microbiomes, particularly in the context of disease.

References

Abbas-Terki T, Blanco-Bose W, Déglon N, Pralong W, Aebischer P (2002) Lentiviral-mediated RNA interference. Hum Gene Ther 13:2197–2201. https://doi.org/10.1089/10430340232098788

Amaral P, Carbonell-Sala S, De La Vega FM, Faial T, Frankish A, Gingeras T, Guigo R, Harrow JL, Hatzigeorgiou AG, Johnson R, Murphy TD, Pertea M, Pruitt KD, Pujar S, Takahashi H, Ulitsky I, Varabyou A, Wells CA, Yandell M, Carninci P, Salzberg SL (2023) The status of the human gene catalogue. Nature 622:41–47. https://doi.org/10.1038/s41586-023-06490-x

Berns K, Hijmans EM, Mullenders J, Brummelkamp TR, Velds A, Heimerikx M, Kerkhoven RM, Madiredjo M, Nijkamp W, Weigelt B, Agami R, Ge W, Cavet G, Linsley PS, Beijersbergen RL, Bernards R (2004) A large-scale RNAi screen in human cells identifies new components of the p53 pathway. Nature 428:431–437. https://doi.org/10.1038/nature02371

Bock C, Datlinger P, Chardon F, Coelho MA, Dong MB, Lawson KA, Lu T, Maroc L, Norman TM, Song B, Stanley G, Chen S, Garnett M, Li W, Moffat J, Qi LS, Shapiro RS, Shendure J, Weissman JS, Zhuang X (2022) High-content CRISPR screening. Nat Rev Methods Primers 2:8. https://doi.org/10.1038/s43586-021-00093-4

Braun CJ, Adames AC, Saur D, Rad R (2022) Tutorial: design and execution of CRISPR in vivo screens. Nat Protoc 17:1903–1925. https://doi.org/10.1038/s41596-022-00700-y

Brummelkamp TR, Bernards R, Agami R (2002) A system for stable expression of short interfering RNAs in mammalian cells. Science (1979) 296:550–553. https://doi.org/10.1126/science.1068999

Bunnik EM, Le Roch KG (2013) An introduction to functional genomics and systems biology. Adv Wound Care (New Rochelle) 2:490–498. https://doi.org/10.1089/wound.2012.0379

Canver MC, Haeussler M, Bauer DE, Orkin SH, Sanjana NE, Shalem O, Yuan G-C, Zhang F, Concordet J-P, Pinello L (2018) Integrated design, execution, and analysis of arrayed and pooled CRISPR genome-editing experiments. Nat Protoc 13:946–986. https://doi.org/10.1038/nprot.2018.005

Chapman NM, Boothby MR, Chi H (2020) Metabolic coordination of T cell quiescence and activation. Nat Rev Immunol 20:55–70. https://doi.org/10.1038/s41577-019-0203-y

Chen YG, Hur S (2022) Cellular origins of dsRNA, their recognition and consequences. Nat Rev Mol Cell Biol 23:286–301. https://doi.org/10.1038/s41580-021-00430-1

Chen M, Du Q, Zhang H-Y, Wahlestedt C, Liang Z (2005) Vector-based siRNA delivery strategies for high-throughput screening of novel target genes. J RNAi Gene Silencing 1:5–11

Chen K, Stahl EC, Kang MH, Xu B, Allen R, Trinidad M, Doudna JA (2024) Engineering self-deliverable ribonucleoproteins for genome editing in the brain. Nat Commun 15:1727. https://doi.org/10.1038/s41467-024-45998-2

Colvin RA, Means TK, Diefenbach TJ, Moita LF, Friday RP, Sever S, Campanella GSV, Abrazinski T, Manice LA, Moita C, Andrews NW, Wu D, Hacohen N, Luster AD (2010) Synaptotagmin-mediated vesicle fusion regulates cell migration. Nat Immunol 11:495–502. https://doi.org/10.1038/ni.1878

Elbashir SM, Harborth J, Lendeckel W, Yalcin A, Weber K, Tuschl T (2001) Duplexes of 21-nucleotide RNAs mediate RNA interference in cultured mammalian cells. Nature 411:494–498. https://doi.org/10.1038/35078107

Figueiredo N, Chora A, Raquel H, Pejanovic N, Pereira P, Hartleben B, Neves-Costa A, Moita C, Pedroso D, Pinto A, Marques S, Faridi H, Costa P, Gozzelino R, Zhao JL, Soares MP, Gama-Carvalho M, Martinez J, Zhang Q, Döring G, Grompe M, Simas JP, Huber TB, Baltimore D, Gupta V, Green DR, Ferreira JA, Moita LF (2013) Anthracyclines induce DNA damage response-mediated protection against severe sepsis. Immunity 39:874–884. https://doi.org/10.1016/j.immuni.2013.08.039

Fire A, Xu S, Montgomery MK, Kostas SA, Driver SE, Mello CC (1998) Potent and specific genetic interference by double-stranded RNA in Caenorhabditis elegans. Nature 391:806–811. https://doi.org/10.1038/35888

Ford K, McDonald D, Mali P (2019) Functional genomics via CRISPR–Cas. J Mol Biol 431:48–65. https://doi.org/10.1016/j.jmb.2018.06.034

Gilbert LA, Horlbeck MA, Adamson B, Villalta JE, Chen Y, Whitehead EH, Guimaraes C, Panning B, Ploegh HL, Bassik MC, Qi LS, Kampmann M, Weissman JS (2014) Genome-scale CRISPR-mediated control of gene repression and activation. Cell 159:647–661. https://doi.org/10.1016/j.cell.2014.09.029

Gonçalves E, Thomas M, Behan FM, Picco G, Pacini C, Allen F, Vinceti A, Sharma M, Jackson DA, Price S, Beaver CM, Dovey O, Parry-Smith D, Iorio F, Parts L, Yusa K, Garnett MJ (2021) Minimal genome-wide human CRISPR-Cas9 library. Genome Biol 22:40. https://doi.org/10.1186/s13059-021-02268-4

Guo C, Ma X, Gao F, Guo Y (2023) Off-target effects in CRISPR/Cas9 gene editing. Front Bioeng Biotechnol 11. https://doi.org/10.3389/fbioe.2023.1143157

Hart T, Chandrashekhar M, Aregger M, Steinhart Z, Brown KR, MacLeod G, Mis M, Zimmermann M, Fradet-Turcotte A, Sun S, Mero P, Dirks P, Sidhu S, Roth FP, Rissland OS, Durocher D, Angers S, Moffat J (2015) High-resolution CRISPR screens reveal fitness genes and genotype-specific cancer liabilities. Cell 163:1515–1526. https://doi.org/10.1016/j.cell.2015.11.015

Hu G, Luo J (2012) A primer on using pooled shRNA libraries for functional genomic screens. Acta Biochim Biophys Sin Shanghai 44:103–112. https://doi.org/10.1093/abbs/gmr116

Ishii S, Kakizuka T, Park S-J, Tagawa A, Sanbo C, Tanabe H, Ohkawa Y, Nakanishi M, Nakai K, Miyanari Y (2024) Genome-wide ATAC-see screening identifies TFDP1 as a modulator of global chromatin accessibility. Nat Genet 56:473–482. https://doi.org/10.1038/s41588-024-01658-1

Joung JK, Sander JD (2013) TALENs: a widely applicable technology for targeted genome editing. Nat Rev Mol Cell Biol 14:49–55. https://doi.org/10.1038/nrm3486

Kim HS, Lee K, Kim S-J, Cho S, Shin HJ, Kim C, Kim J-S (2018) Arrayed CRISPR screen with image-based assay reliably uncovers host genes required for coxsackievirus infection. Genome Res 28:859–868. https://doi.org/10.1101/gr.230250.117

Kleinstiver BP, Prew MS, Tsai SQ, Topkar VV, Nguyen NT, Zheng Z, Gonzales APW, Li Z, Peterson RT, Yeh J-RJ, Aryee MJ, Joung JK (2015) Engineered CRISPR-Cas9 nucleases with altered PAM specificities. Nature 523:481–485. https://doi.org/10.1038/nature14592

Koike-Yusa H, Li Y, Tan E-P, Velasco-Herrera MDC, Yusa K (2014) Genome-wide recessive genetic screening in mammalian cells with a lentiviral CRISPR-guide RNA library. Nat Biotechnol 32:267–273. https://doi.org/10.1038/nbt.2800

Krumlauf R, Jeanpierre M, Young BD (1982) Construction and characterization of genomic libraries from specific human chromosomes. Proc Natl Acad Sci 79:2971–2975. https://doi.org/10.1073/pnas.79.9.2971

Kurata M, Yamamoto K, Moriarity BS, Kitagawa M, Largaespada DA (2018) CRISPR/Cas9 library screening for drug target discovery. J Hum Genet 63:179–186. https://doi.org/10.1038/s10038-017-0376-9

Lin K, Chang Y-C, Billmann M, Ward HN, Le K, Hassan AZ, Bhojoo U, Chan K, Costanzo M, Moffat J, Boone C, Bielinsky A-K, Myers CL (2024) A scalable platform for efficient CRISPR-Cas9 chemical-genetic screens of DNA damage-inducing compounds. Sci Rep 14:2508. https://doi.org/10.1038/s41598-024-51735-y

Liscovitch-Brauer N, Montalbano A, Deng J, Méndez-Mancilla A, Wessels H-H, Moss NG, Kung C-Y, Sookdeo A, Guo X, Geller E, Jaini S, Smibert P, Sanjana NE (2021) Profiling the genetic determinants of chromatin accessibility with scalable single-cell CRISPR screens. Nat Biotechnol 39:1270–1277. https://doi.org/10.1038/s41587-021-00902-x

Liu X, Kimmey JM, Matarazzo L, de Bakker V, Van Maele L, Sirard J-C, Nizet V, Veening J-W (2021) Exploration of bacterial bottlenecks and Streptococcus pneumoniae pathogenesis by CRISPRi-Seq. Cell Host Microbe 29:107–120.e6. https://doi.org/10.1016/j.chom.2020.10.001

Liu Z, Chen Y, Xia Q, Liu M, Xu H, Chi Y, Deng Y, Xing D (2023) Linking genome structures to functions by simultaneous single-cell Hi-C and RNA-seq. Science (1979) 380:1070–1076. https://doi.org/10.1126/science.adg3797

Long L, Wei J, Lim SA, Raynor JL, Shi H, Connelly JP, Wang H, Guy C, Xie B, Chapman NM, Fu G, Wang Y, Huang H, Su W, Saravia J, Risch I, Wang Y-D, Li Y, Niu M, Dhungana Y, KC A, Zhou P, Vogel P, Yu J, Pruett-Miller SM, Peng J, Chi H (2021) CRISPR screens unveil signal hubs for nutrient licensing of T cell immunity. Nature 600:308–313. https://doi.org/10.1038/s41586-021-04109-7

Luca F, Kupfer SS, Knights D, Khoruts A, Blekhman R (2018) Functional genomics of host–microbiome interactions in humans. Trends Genet 34:30–40. https://doi.org/10.1016/j.tig.2017.10.001

Maillard P V, van der Veen AG, Poirier EZ, Reis e Sousa C (2019) Slicing and dicing viruses: antiviral <scp>RNA</scp> interference in mammals. EMBO J 38. https://doi.org/10.15252/embj.2018100941

Mimitou EP, Cheng A, Montalbano A, Hao S, Stoeckius M, Legut M, Roush T, Herrera A, Papalexi E, Ouyang Z, Satija R, Sanjana NE, Koralov SB, Smibert P (2019) Multiplexed detection of proteins, transcriptomes, clonotypes and CRISPR perturbations in single cells. Nat Methods 16:409–412. https://doi.org/10.1038/s41592-019-0392-0

Moffat J, Grueneberg DA, Yang X, Kim SY, Kloepfer AM, Hinkle G, Piqani B, Eisenhaure TM, Luo B, Grenier JK, Carpenter AE, Foo SY, Stewart SA, Stockwell BR, Hacohen N, Hahn WC, Lander ES, Sabatini DM, Root DE (2006) A Lentiviral RNAi library for human and mouse genes applied to an arrayed viral high-content screen. Cell 124:1283–1298. https://doi.org/10.1016/j.cell.2006.01.040

Moita CF, Chora Â, Hacohen N, Moita LF (2012) <scp>RNA</scp> i screen for kinases and phosphatases that play a role in antigen presentation by dendritic cells. Eur J Immunol 42:1843–1849. https://doi.org/10.1002/eji.201142167

Morris JA, Sun JS, Sanjana NE (2024) Next-generation forward genetic screens: uniting high-throughput perturbations with single-cell analysis. Trends Genet 40:118–133. https://doi.org/10.1016/j.tig.2023.10.012

Moura-Alves P, Neves-Costa A, Raquel H, Pacheco TR, D'Almeida B, Rodrigues R, Cadima-Couto I, Chora Â, Oliveira M, Gama-Carvalho M, Hacohen N, Moita LF (2011) An shRNA-based screen of splicing regulators identifies SFRS3 as a negative regulator of IL-1β secretion. PLoS One 6:e19829. https://doi.org/10.1371/journal.pone.0019829

Neves-Costa A, Moita LF (2013) TET1 is a negative transcriptional regulator of IL-1β in the THP-1 cell line. Mol Immunol 54:264–270. https://doi.org/10.1016/j.molimm.2012.12.014

Nie D, Guo T, Yue M, Li W, Zong X, Zhu Y, Huang J, Lin M (2022) Research progress on nanoparticles-based CRISPR/Cas9 system for targeted therapy of tumors. Biomolecules 12:1239. https://doi.org/10.3390/biom12091239

Ostrowski M, Carmo NB, Krumeich S, Fanget I, Raposo G, Savina A, Moita CF, Schauer K, Hume AN, Freitas RP, Goud B, Benaroch P, Hacohen N, Fukuda M, Desnos C, Seabra MC, Darchen F, Amigorena S, Moita LF, Thery C (2010) Rab27a and Rab27b control different steps of the exosome secretion pathway. Nat Cell Biol 12:19–30. https://doi.org/10.1038/ncb2000

Paddison PJ, Caudy AA, Bernstein E, Hannon GJ, Conklin DS (2002) Short hairpin RNAs (shRNAs) induce sequence-specific silencing in mammalian cells. Genes Dev 16:948–958. https://doi.org/10.1101/gad.981002

Paddison PJ, Silva JM, Conklin DS, Schlabach M, Li M, Aruleba S, Balija V, O'Shaughnessy A, Gnoj L, Scobie K, Chang K, Westbrook T, Cleary M, Sachidanandam R, Richard McCombie W, Elledge SJ, Hannon GJ (2004) A resource for large-scale RNA-interference-based screens in mammals. Nature 428:427–431. https://doi.org/10.1038/nature02370

Parnas O, Jovanovic M, Eisenhaure TM, Herbst RH, Dixit A, Ye CJ, Przybylski D, Platt RJ, Tirosh I, Sanjana NE, Shalem O, Satija R, Raychowdhury R, Mertins P, Carr SA, Zhang F, Hacohen N, Regev A (2015) A genome-wide CRISPR screen in primary immune cells to dissect regulatory networks. Cell 162:675–686. https://doi.org/10.1016/j.cell.2015.06.059

Peters JM, Colavin A, Shi H, Czarny TL, Larson MH, Wong S, Hawkins JS, Lu CHS, Koo B-M, Marta E, Shiver AL, Whitehead EH, Weissman JS, Brown ED, Qi LS, Huang KC, Gross CA (2016) A comprehensive, CRISPR-based functional analysis of essential genes in bacteria. Cell 165:1493–1506. https://doi.org/10.1016/j.cell.2016.05.003

Peters JM, Koo B-M, Patino R, Heussler GE, Hearne CC, Qu J, Inclan YF, Hawkins JS, Lu CHS, Silvis MR, Harden MM, Osadnik H, Peters JE, Engel JN, Dutton RJ, Grossman AD, Gross CA, Rosenberg OS (2019) Enabling genetic analysis of diverse bacteria with Mobile-CRISPRi. Nat Microbiol 4:244–250. https://doi.org/10.1038/s41564-018-0327-z

Pickar-Oliver A, Gersbach CA (2019) The next generation of CRISPR–Cas technologies and applications. Nat Rev Mol Cell Biol 20:490–507. https://doi.org/10.1038/s41580-019-0131-5

Platt RJ, Chen S, Zhou Y, Yim MJ, Swiech L, Kempton HR, Dahlman JE, Parnas O, Eisenhaure TM, Jovanovic M, Graham DB, Jhunjhunwala S, Heidenreich M, Xavier RJ, Langer R, Anderson DG, Hacohen N, Regev A, Feng G, Sharp PA, Zhang F (2014) CRISPR-Cas9 Knockin mice for genome editing and cancer modeling. Cell 159:440–455. https://doi.org/10.1016/j.cell.2014.09.014

Pratt AJ, MacRae IJ (2009) The RNA-induced silencing complex: a versatile gene-silencing machine. J Biol Chem 284:17897–17901. https://doi.org/10.1074/jbc.R900012200

Przybyla L, Gilbert LA (2022) A new era in functional genomics screens. Nat Rev Genet 23:89–103

Rato S, Maia S, Brito PM, Resende L, Pereira CF, Moita C, Freitas RP, Moniz-Pereira J, Hacohen N, Moita LF, Goncalves J (2010) Novel HIV-1 knockdown targets identified by an enriched kinases/phosphatases shRNA library using a Long-term iterative screen in Jurkat T-cells. PLoS One 5:e9276. https://doi.org/10.1371/journal.pone.0009276

Rousset F, Bikard D (2020) CRISPR screens in the era of microbiomes. Curr Opin Microbiol 57:70–77. https://doi.org/10.1016/j.mib.2020.07.009

Rubin AJ, Parker KR, Satpathy AT, Qi Y, Wu B, Ong AJ, Mumbach MR, Ji AL, Kim DS, Cho SW, Zarnegar BJ, Greenleaf WJ, Chang HY, Khavari PA (2019) Coupled single-cell CRISPR screening and epigenomic profiling reveals causal gene regulatory networks. Cell 176:361–376.e17. https://doi.org/10.1016/j.cell.2018.11.022

Salanga CM, Salanga MC (2021) Genotype to phenotype: CRISPR gene editing reveals genetic compensation as a mechanism for phenotypic disjunction of morphants and mutants. Int J Mol Sci 22:3472. https://doi.org/10.3390/ijms22073472

Shi H, Doench JG, Chi H (2023) CRISPR screens for functional interrogation of immunity. Nat Rev Immunol 23:363–380. https://doi.org/10.1038/s41577-022-00802-4

So RWL, Chung SW, Lau HHC, Watts JJ, Gaudette E, Al-Azzawi ZAM, Bishay J, Lin LT-W, Joung J, Wang X, Schmitt-Ulms G (2019) Application of CRISPR genetic screens to investigate neurological diseases. Mol Neurodegener 14:41. https://doi.org/10.1186/s13024-019-0343-3

Sternberg SH, Redding S, Jinek M, Greene EC, Doudna JA (2014) DNA interrogation by the CRISPR RNA-guided endonuclease Cas9. Nature 507:62–67. https://doi.org/10.1038/nature13011

Stewart SA, Dykxhoorn DM, Palliser D, Mizuno H, Yu EY, An DS, Sabatini DM, Chen ISY, Hahn WC, Sharp PA, Weinberg RA, Novina CD (2003) Lentivirus-delivered stable gene silencing by RNAi in primary cells. RNA 9:493–501. https://doi.org/10.1261/rna.2192803

Urnov FD, Rebar EJ, Holmes MC, Zhang HS, Gregory PD (2010) Genome editing with engineered zinc finger nucleases. Nat Rev Genet 11:636–646. https://doi.org/10.1038/nrg2842

Vredevoogd DW, Peeper DS (2023) Heterogeneity in functional genetic screens: friend or foe? Front Immunol 14. https://doi.org/10.3389/fimmu.2023.1162706

Wang T, Antonacci-Fulton L, Howe K, Lawson HA, Lucas JK, Phillippy AM, Popejoy AB, Asri M, Carson C, Chaisson MJP, Chang X, Cook-Deegan R, Felsenfeld AL, Fulton RS, Garrison EP, Garrison NA, Graves-Lindsay TA, Ji H, Kenny EE, Koenig BA, Li D, Marschall T, McMichael JF, Novak AM,

Purushotham D, Schneider VA, Schultz BI, Smith MW, Sofia HJ, Weissman T, Flicek P, Li H, Miga KH, Paten B, Jarvis ED, Hall IM, Eichler EE, Haussler D (2022) The Human Pangenome Project: a global resource to map genomic diversity. Nature 604:437–446. https://doi.org/10.1038/s41586-022-04601-8

Watson JD (2001) The human genome revealed. Genome Res 11:1803–1804. https://doi.org/10.1101/gr.211601

Wienert B, Wyman SK, Richardson CD, Yeh CD, Akcakaya P, Porritt MJ, Morlock M, Vu JT, Kazane KR, Watry HL, Judge LM, Conklin BR, Maresca M, Corn JE (2019) Unbiased detection of CRISPR off-targets in vivo using DISCOVER-Seq. Science (1979) 364:286–289. https://doi.org/10.1126/science.aav9023

Wirth A-K, Wange L, Vosberg S, Henrich K-O, Rausch C, Özdemir E, Zeller CM, Richter D, Feuchtinger T, Kaller M, Hermeking H, Greif PA, Senft D, Jurinovic V, Bahrami E, Jayavelu AK, Westermann F, Mann M, Enard W, Herold T, Jeremias I (2022) In vivo PDX CRISPR/Cas9 screens reveal mutual therapeutic targets to overcome heterogeneous acquired chemo-resistance. Leukemia 36:2863–2874. https://doi.org/10.1038/s41375-022-01726-7

Xu CL, Ruan MZC, Mahajan VB, Tsang SH (2019) Viral delivery systems for CRISPR. Viruses 11:28. https://doi.org/10.3390/v11010028

Yeh CD, Richardson CD, Corn JE (2019) Advances in genome editing through control of DNA repair pathways. Nat Cell Biol 21:1468–1478. https://doi.org/10.1038/s41556-019-0425-z

Yu J-Y, DeRuiter SL, Turner DL (2002) RNA interference by expression of short-interfering RNAs and hairpin RNAs in mammalian cells. Proc Natl Acad Sci 99: 6047–6052. https://doi.org/10.1073/pnas.092143499

Zhou Y, Zhu S, Cai C, Yuan P, Li C, Huang Y, Wei W (2014) High-throughput screening of a CRISPR/Cas9 library for functional genomics in human cells. Nature 509:487–491. https://doi.org/10.1038/nature13166

Zhou P, Shi H, Huang H, Sun X, Yuan S, Chapman NM, Connelly JP, Lim SA, Saravia J, KC A, Pruett-Miller SM, Chi H (2023) Single-cell CRISPR screens in vivo map T cell fate regulomes in cancer. Nature 624:154–163. https://doi.org/10.1038/s41586-023-06733-x

Part II

From Large Scale Data to Biological Insights

FAIR Omics Data Management: Overview, Challenges, and Best Practices

9

Francisco Pina-Martins, Gil Poiares-Oliveira, Jorge Oliveira, and Catia Pesquita

Abstract

This chapter provides a comprehensive overview of FAIR principles applied to omics data management, addressing key challenges in handling large-scale, heterogeneous datasets while promoting reproducibility, collaboration, and open science practices. It covers essential strategies including Data Management Plans (DMPs), file organization, metadata standards, ontologies, repository selection, interoperability protocols, and security measures for privacy and ethics compliance. Best practices are outlined for achieving Findable, Accessible, Interoperable, and Reusable data, with recommendations for continuous improvement to support cross-omics integration and long-term data stewardship.

Keywords

Findable · Accessible · Interoperable · Reusable · Data management

9.1 Introduction

Omics data in biotechnology and biomedicine is growing at an unprecedented rate, with increasing volume, diversity, and complexity, creating both immense opportunities and significant challenges (Namsaraev et al. 2023). Efficient management of omics data is now critical for enabling meaningful analyses, facilitating data sharing and reproducibility, and accelerating scientific discovery.

Omics data management encompasses a range of activities, from data organization, storage, curation, and sharing. These processes are essential for harnessing the full potential of omics data, ensuring that researchers can interpret findings accurately, draw meaningful conclusions, and compare results across studies (Giannattasio et al. 2023). These systems also foster collaboration across disciplines, facilitating cross-omics integration and empowering interdisciplinary teams to explore complex biological questions.

A carefully designed strategy for omics data management is fundamental to support

F. Pina-Martins (✉)
Computational Biology and Population Genomics Group, cE3c—Center for Ecology, Evolution and Environmental Changes & CHANGE—Global Change and Sustainability Institute, Faculdade de Ciências, Universidade de Lisboa, Lisbon, Portugal

Departamento de Engenharia Química e Biológica, Escola Superior de Tecnologia Do Barreiro, Instituto Politécnico de Setúbal, Rua Américo da Silva Marinho, Lavradio, Portugal
e-mail: frmartins@ciencias.ulisboa.pt

G. Poiares-Oliveira · J. Oliveira
INESC-ID, Instituto Superior Técnico, Universidade de Lisboa, LisboaLisbon, Portugal

BioData.pt | ELIXIR Portugal, Oeiras, Portugal

C. Pesquita
LASIGE, Faculdade de Ciências da Universidade de Lisboa, Lisboa, Portugal

M. Gama-Carvalho et al. (eds.), *Omics Approaches in Biomedicine and Biotechnology*, Advances in Experimental Medicine and Biology 1504, https://doi.org/10.1007/978-3-032-18966-0_9

reproducibility and promote collaborative research. As the scientific community embraces open science practices, data management systems ensure that omics datasets are aligned with the FAIR data principles (Wilkinson et al. 2016). The principles were established by the authors with the objective of acting as a guideline that enables researchers to enhance the dissemination and repurposing of their datasets by themselves or others. To fulfil this goal, it was proposed that data should be *Findable, Accessible, Interoperable, and Reusable* (Box 9.1). A set of guidelines accompanies each of these principles to help implement each one in a generic context. Quickly after their publication, these concepts were widely adopted by communities of multiple scientific disciplines to a point that can be considered almost ubiquitous (Brinson et al. 2024; Nicholson et al. 2023). Despite its widespread presence, the adoption of these principles is still an ongoing process (Garcia et al. 2019) as FAIRness itself evolves with time.

This chapter overviews the basic principles and technologies that are essential to designing an effective omics data management strategy. It explores the challenges associated with storing and processing massive, heterogeneous datasets, discusses tools and standards for data storage, annotation, and integration, and examines best practices for ensuring data security and compliance.

9.2 Data Management Plans

In the lifecycle of omics data, the importance of proper data structure and organization cannot be overstated. The implementation of best practices, guided by a robust Data Management Plan (DMP), facilitates effective data handling and promotes transparency, reproducibility, and collaboration within the scientific community (Vitale and Sandy 2019).

From the moment data is generated, careful management is essential to ensure data quality, accessibility, and interoperability. This foundation is crucial for achieving FAIR data status.

A Data Management Plan (DMP) is a formal document that outlines how data should be handled during and after a research project. It serves as a roadmap for data collection, organization, storage, and sharing. A DMP should cover topics such as ownership, management responsibilities, sharing policies, and retention plans. Additionally, data storage requirements should be stated in the document. DMPs should approach the following stages of a research project:

- *Collection*: Description of data types, formats, and methodologies.
- *Storage and Backup*: Strategies for storing and backing up data.
- *Documentation and Metadata*: Procedures for documenting data and creating comprehensive metadata to facilitate understanding and (re)use.
- *Sharing and Access*: Plans for making data publicly available, including where and when.
- *Preservation*: Long-term preservation strategies to ensure data remains accessible and usable over time.

The level of detail a DMP should have must be appropriate to the institution that develops it. Too many details may lead technically challenged users to fail to comply, while too little detail may lead to undefined and inconsistent practices. Finally, a DMP document should not be static but rather a "living document" that is updated as circumstances change. Therefore, it is crucial that it is versioned from the first draft onwards. The following subsections can be used independently or as guidelines for drafting a DMP.

Furthermore, Data Management Plans are not only useful for data organization and results reproducibility, but many funding agencies also require them for grant submission, such as the H2020 programme, or the NSF, and usually provide official templates. Several tools exist, such as ARGOS/OpenDMP (OpenAIRE n.d.), Data Stewardship Wizard (Pergl et al. 2019), and DMPonline (Donnelly et al. 2010), among others, that guide users step-by-step through developing their DMP. Those tools present several advantages, namely, the fact that they're able to output DMPs in a computer-actionable format, and have extensive template collections based on funders' official guidelines.

9.3 Data Collection and Organization

Proper data structure and organization are critical components in the lifecycle of omics data, serving as the foundation for ensuring data quality, accessibility, and interoperability. Before data can be made publicly available and achieve the status of FAIR, it must be managed from the moment it is generated by scientific instruments. This pre-publication phase is essential for subsequent data sharing, analysis, and reuse (Michener 2015).

Adhering to systematic file-naming conventions, establishing a logical directory structure, and employing version control and comprehensive documentation practices are key strategies in maintaining data integrity. Moreover, integrating quality control measures and generating detailed metadata further enhance the usability and reliability of omics data (Michener 2015). Proactive and accurate organization of omics data can be considered a critical research process step, by streamlining data sharing, analysis, and long-term preservation. Embracing these practices, even when they are not a requirement, is an asset that should not be underestimated.

This section provides a guide on best practices for handling omics data immediately after it is produced, through systematic and consistent data organization. This step is not just a technical requirement but rather a pivotal step to facilitate data management and enhance the reproducibility value of scientific data. These practices aim to enable researchers to ensure their data is FAIR in a local context and well-prepared for eventual public sharing. Proper data management at this stage not only benefits individual research projects but also contributes to more efficient data sharing and access within local research groups, promoting transparency, reproducibility, and collaboration.

9.3.1 File-Naming Conventions

Consistent file naming is fundamental for effective data organization. Clear and descriptive file names enhance data discoverability and reduce the error risk. Effectively managed file names should have:

- *Descriptive Names*: Use meaningful names that reflect the file's content and origin (and possibly version).
- *Standardized Format*: Define which elements compose a filename, in which order, and the separator character. For example, include "project name", "sample ID", "date (DDMMYYYY)", and "version" (vXX—where XX is an incremental number between 01 and 99) using *PascalCase* (each word starts with a capital letter, without separators) for each element, and "_" as an element separator. The filename should end with ".format".
- *Avoid Special Characters*: Use underscores or hyphens instead of spaces, and avoid special characters that may not be supported by all systems.

An example filename using the above recommendations could look like the one in Fig. 9.1.

9.3.2 Directory Structure and Hierarchy

A logical and consistent directory structure facilitates data management by organizing files systematically. An effective directory hierarchy includes:

- *Top-Level Folders*: Create main folders based on projects or experiments.
- *Subfolders*: Organize subfolders by data type, or project stage (in this case, folders can be numbered). For example, use "001-raw_data", "002-filtered_data", "003-processed data", and "004-results".
- *Sub-subfolders*: You can go deeper into the hierarchy and further organize each data type

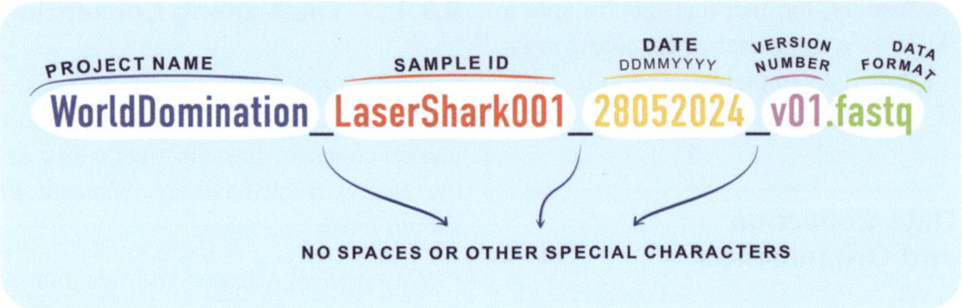

Fig. 9.1 An example file-naming scheme following the guidelines recommended in this chapter. Different colours denote different elements

or project stage. This should be decided on a case-by-case basis, but it should be consistent (and documented) within each project.

9.3.3 Documentation and Metadata

Documentation is of capital importance for effective data storage and reuse. It is recommended that, at least, there is a README file (a simple text file that describes the content and/or intent of files in a directory) to accompany each dataset. This should be a simple text file, but it can be made expressive using Markdown, or ReStructured text. This file should contain a description of the data type, what it refers to, how it was obtained, when it was first stored/ obtained, what project it is a part of, and which files are included (and what each contains). It should also indicate where sample metadata can be found.

Metadata provides context to make data understandable and reusable. Good metadata uses standard fields defined by standard schemas. A deeper dive into metadata best practices is elaborated later.

9.3.4 Data Quality Control and Validation

Ensuring the quality of omics data is crucial for reliable interpretation of analysis and results. Quality control measures should be integrated into the data organization workflow. Establish QC protocols for each handled data type, including the tools used, their versions, and any necessary parameters. Ideally, this process should be automated and run whenever new data is added to the local repository.

9.3.5 Version Control and Tracking Changes

Version control, typically used for code, can also be effectively applied to data. It allows for maintaining a comprehensive history of data changes and ensuring reproducibility. Implementing a version control system (VCS) such as Git can provide features such as change tracking (recording changes to data, including who made the changes and why) and version tags (marking significant milestones), facilitating easier navigation through data history. However, VCSs have a steep learning curve and can be a high barrier to entry. Consider who will be managing and using the data before mandating data VCS. Sometimes, simply adding a version number to the filename will suffice.

9.4 Data Annotation and Curation

Effective processes for data annotation and curation are very important to ensure research data adheres to the FAIR principles. Such practices involve the use of detailed metadata to describe datasets and their respective curation to maximize usability. The adoption of standardized curation techniques allows researchers to maximize data value and reuse potential and support systematic interoperability.

9.4.1 Metadata

Metadata, meaning "data about data", is a collection of highly structured information that describes a dataset (ELIXIR 2021). Although classifying a particular set of information as "data" or "metadata" is context-dependent, metadata should always provide the necessary information to reproduce (upstream) and interpret (downstream) the research data it describes. Metadata is key to making data FAIR, since it is what enables the application of these principles:

- Findable: Through the use of unique and persistent identifiers, allowing it to be indexed and discovered view keywords.
- Accessible: Metadata should include availability and licensing information which dictate how accessible the data is. Persistent identifiers also ensure data remains accessible over time.
- Interoperable: By using a standard compliant ontology, metadata ensures the data stays compatible with other systems and datasets.
- Reusable: Metadata provides context and provenance information, ensuring data is understandable over space and time, maximizing its reuse potential.

Interoperability between different usage contexts of metadata is facilitated by metadata standards. A metadata standard is a set of guidelines and specifications that define the structure, format, and content of metadata, with the intent of ensuring common understanding of the data's meaning by its creator and others (Mosha and Ngulube 2023). Like all technical standards, metadata standards enable information originating from one context to be used in another in an automated manner (Erik Duval 2021).

Different metadata standards may be suitable for different contexts. More general models, like Dublin Core Metadata Terms (DCMT) (DCMI Metadata Terms n.d.), offer a more reduced set of terms that are able to describe a broad range of resources (Weibel 1997). Domain-specific models, like the EDAM ontology for bioscientific data analysis and data management (Black et al. 2022), provide a set of more specific terms for a particular subject or field of knowledge.

FAIRsharing (Sansone et al. 2019), an online resource on data management and FAIR practices, lists on its database, as of July 2024, more than 1700 different metadata standards. Domain-specific repositories often support a subset of those standards that relate to the information contained in them. As such, an important criterion for selecting a repository is precisely which metadata standards it supports, and which of them are useful to ensure the FAIRness of the information being shared. Table 9.1 shows some of the most commonly used standards for omics data.

9.4.2 Curation

Different repositories have different levels of data curation, which is the process that ensures that data are fit for purpose and available for discovery and reuse, a foundational practice in realizing the FAIR goals (Johnston et al. 2024).

In this angle, domain-specific repositories may also present a significant advantage, as they are able to employ metadata standardization to boost resource findability and may apply extra efforts to ensure the data adheres to field-specific standards (Lin et al. 2024).

Domain-specific repositories are also able to leverage the standard ontologies and controlled vocabularies employed by that domain. Ontologies convey the biological and biomedical meaning of experiments and data in a computer-actionable format and allow for the integration of data and knowledge from various sources (Noy et al. 2008). They also contribute positively to the findability of datasets as they provide a standardized set of keywords to be used in searches.

As such, it can prove very useful to familiarize oneself with the domain ontologies that can be

Table 9.1 Some of the most used metadata standards in omics

Name	Website	Data type	Organization/ Consortium	Short description
BioSample	BioSample	Multi-omics	NCBI (National Center for Biotechnology Information)	Metadata standard for biological samples across omics disciplines
MIABE (Minimum Information About a Bioactive Entity)	MIABE	Molecular interaction	PSI (Proteomics Standards Initiative)	For bioactive entities, aiding in sharing bioactivity data
MIAME (Minimum Information About a Microarray Experiment)	MIAME	Transcriptomics (Microarray)	FGED (Functional Genomics Data Society)	Specifies requirements for annotating and sharing microarray data
MIAPE (Minimum Information About a Proteomics Experiment)	MIAPE	Proteomics	PSI (Proteomics Standards Initiative)	A set of guidelines for proteomics experiments, supporting data reproducibility
MINIML (Minimum Information about a Microarray and RNA-seq Experiment)	MINiML	Transcriptomics	NCBI (National Center for Biotechnology Information)	XML format for metadata exchange in GEO for gene expression experiments
MINSEQE (Minimum Information About a High-throughput Nucleotide Sequencing Experiment)	MINSEQ	Genomics, Transcriptomics	FGED (Functional Genomics Data Society)	Ensures detailed reporting for next-generation sequencing data
MIxS (Minimum Information about any (x) Sequence)	MixS	Genomics, Metagenomics	GSC (Genomics Standards Consortium)	A suite of checklists and environmental packages that standardize the reporting of genomic data metadata

used to annotate an experiment or dataset to facilitate its curation and increase its findability and computer-actionability. To that end, BioPortal (Whetzel et al. 2011) offers a comprehensive resource in finding biomedical ontologies, allowing users to quickly find the ontologies adequate for their needs, browse their respective terms, and retrieve entries' unique identifiers. Some of the most used ontologies in omics can be found in Table 9.2.

9.5 Data Accessibility, Interoperability, and Reuse

Compliance with the FAIR principles, discussed earlier, requires much more effort than just tossing all the datasets in a public storage drive and adding the URL to a research paper. The data must be made available in a repository that exposes a set of metadata that allows potential users to easily find it and reuse it for their own experiments.

When choosing a data repository, it is important to reflect on how a particular solution would facilitate compliance with the FAIR principles. Some points of reflection:

1. Would the repository assure the *findability* of my data? How easy would it be for a college to discover this dataset using a select number of relevant keywords? How about a machine? Would automated algorithms easily find the content I published?
2. Is my data *accessible*? Would it be available free of charge? Are access permissions as broad as they can be without infringing any laws or regulations? Would someone in another jurisdiction be able to access the data? What if they don't have an institutional account? If access must be restricted, is there a defined process for requesting permission?

Table 9.2 Useful ontologies in omics data management

Ontology	Domain	Description
Gene Ontology (GO)	Cellular and Molecular Biology	Describes gene functions in terms of molecular function, biological processes, and cellular components.
Sequence Ontology (SO)	Cellular and Molecular Biology	Provides a structured vocabulary for describing sequence features and annotations.
Protein Ontology (PRO)	Cellular and Molecular Biology	Defines relationships among proteins, protein isoforms, and complexes.
Chemical Entities of Biological Interest (ChEBI)	Cellular and Molecular Biology	Standardizes chemical compounds relevant to metabolomics and biochemical studies.
Cell Ontology (CL)	Cellular and Molecular Biology	Defines types of cells and their characteristics across species, aiding in single-cell omics.
Ontology for Biomedical Investigations (OBI)	Experimental Data and Methods	Standardizes terms for experimental procedures, protocols, and study designs.
Experimental Factor Ontology (EFO)	Experimental Data and Methods	Provides a comprehensive vocabulary for experimental variables and conditions.
MetaboLights Ontology (MLO)	Experimental Data and Methods	Specializes in metabolomics datasets, describing sample types, data processing, and analytical methods.
BioAssay Ontology (BAO)	Experimental Data and Methods	Focuses on high-throughput screening and omics assay descriptions.
Human Phenotype Ontology (HPO)	Phenotype Annotation	Captures phenotypic abnormalities associated with human diseases, useful for genotype-phenotype mapping.
Plant Trait Ontology (TO)	Phenotype Annotation	Defines terms for plant phenotypes and traits in the context of genomics and breeding studies.
Ontology for Microbial Phenotypes (OMP)	Phenotype Annotation	Describes phenotypes of microorganisms, aiding in microbiome and metagenomics studies.
Disease Ontology (DO)	Disease and Clinical	Aids in annotating datasets related to diseases and associated biomarkers.
National Cancer Institute Thesaurus (NCIT)	Disease and Clinical	Covers cancer-related terms, pathways, and molecular events.
Mondo Disease Ontology (MONDO)	Disease and Clinical	Integrates multiple disease ontologies into a unified framework for disease annotation.

3. Is the repository Designed with *interoperability* in mind? Does it support the established metadata standards employed in my field? What other fields of knowledge would benefit from access to this dataset? What information about my assay would a machine gather just by interpreting the metadata?
4. Does the information the repository exposes facilitate the *reuse* of the data? Would someone be able to easily use my data in their studies? Does the repository facilitate the sharing of data under open licenses?

9.5.1 Interoperability

The "I" in FAIR stands for *Interoperable*. The concept, defined as early as 1991, means "The ability of two or more systems or components to exchange information and to use the information that has been exchanged" (Geraci et al. 1991). Although revisited and refined over the years (J. Park and Ram 2004), its apparently simple meaning remains crucial yet challenging to implement in omics data management. Here, interoperability extends to ensuring that data formats, structures, and terminologies are compatible and standardized for integration, sharing, and reuse across research institutions, clinical settings, and computational tools. This encompasses three types of interoperability:

- *Syntactic*: Standardized and compatible *data formats* and structures.
- *Semantic*: Data *meanings* preserved and understood across different systems.
- *Technical*: Effective data *exchange* between communicating systems.

To address these interoperability challenges in omics data management, various standards, ontologies, and protocols have been developed.

9.5.2 Syntactic Interoperability

Standard data formats address syntactic *interoperability* by ensuring consistent schema for the same type of information. For example, sequence data and the associated error rates can be stored in many ways, but the FASTQ format (Cock et al. 2010) is a widely recognized standard, ensuring compatibility across most systems (Box 9.2). Other widely used standard formats include the Variant Calling Format (VCF) (Danecek et al. 2011) for variant calls, and SAM/BAM (Li et al. 2009) for genome assemblies (Box 9.3).

9.5.3 Semantic Interoperability

Controlled vocabulary protocols, such as the Gene Ontology (GO) (The Gene Ontology Consortium et al. 2023), the Human Phenotype Ontology (HPO) (Gargano et al. 2024), and the Sequence Ontology (SO) (Moore et al. 2010), ensure *semantic interoperability* by defining data meanings so all systems interpret data consistently. These ontologies are hierarchical structures based on domain-specific *concepts* and their *relationships* (Almeida and Barbosa 2009). *Concepts* can refer to any real-world phenomenon, whereas *relations* are the pre-set ways in which they interact. They use controlled, standardized vocabulary to define possibilities and are collaboratively developed (but see Poveda-Villalón et al. 2020) and versioned to track their changes over time. Many are collected in the Open Biological and Biomedical Ontology Foundry (Jackson et al. 2021).

9.5.4 Technical Interoperability

Data exchange protocols enable *technical interoperability*, allowing different systems to connect and communicate effectively. These protocols often come in the form of application programming interfaces (APIs), which are sets of rules and protocols enabling communication between different systems and programs (Stylos 2009). For example, APIs allow bioinformatics pipelines to retrieve data from public web services like the National Center for Biotechnology Information (NCBI)'s Entrez (Sayers et al. 2020) and return results to laboratory management or electronic health record systems.

9.5.5 Importance of Interoperability in Genomics Data Management

Interoperability in genomics data management is not just a technical requirement but a crucial enabler of scientific progress and research. Sharing, reusing, and integrating genomic data across different systems, institutions, and topics is foundational to advancements in biology, including biotechnology and medicine (Amorim et al. 2022; Byrd et al. 2020). Despite its undeniable technical costs (implementing interoperability is a challenge, and maintaining it can be an even harder one (Canuel et al. 2015), interoperability addresses key challenges in collaborative, interdisciplinary research, offering several benefits:

- *Collaboration*: Without a good implementation of interoperability, scientific collaboration becomes much more difficult, or even impossible.
- *Data Reusability*: It is interoperability that allows the same data to be used in different contexts, to support different research.
- *Efficiency*: Interoperability streamlines data workflows, reducing redundancies.
- *Time gains*: Interoperability ensures there is no need to "reinvent the wheel" whenever a specific scientific problem is being dealt with.

The 1000 Genomes Project (Auton et al. 2015) is arguably one of the largest and emblematic cases of collaboration in biology and exemplifies the importance of interoperability in genomics. Its extensive data and collaborator base required stringent interoperability practices, making it a model for best practices in the field. Clear guidelines on data formats and analysis methods were key to its success, resulting in the timely completion of its scientific goals (Sudmant et al. 2015) and the public release of all data and tools (Clarke et al. 2012). The project's FAQ highlights the significance of interoperability, and its open data policy has enabled numerous other research projects (see Tanjo et al. (2021) for details). Additionally, many widely used bioinformatics tools, such as bwa (Li and Durbin 2009) and samtools (Li 2011), were developed for this project, benefiting the genomics community by providing efficient, standardized tools. This was paramount to a community that, before these tools were developed, was very fragmented in terms of the software used.

Implementing interoperability involves adopting best practices, such as selecting appropriate standards and tools, and keeping them up-to-date with evolving protocols. Institutions should invest in training, infrastructure, and staff to support these practices effectively (Canuel et al. 2015).

9.5.6 Interoperability: The Most Important Part of FAIR or a Stifle to Innovation?

For data to be truly FAIR, a comprehensive "data ecosystem" must exist where data flows seamlessly across different platforms, systems, and institutions, enabling comprehensive analysis and efficient collaboration. This makes interoperability arguably the most crucial part of FAIR data principles. Without interoperability, data cannot be *Findable, Accessible, or Reusable*. Strong interoperability practices are essential to harness the full potential of genomic data (Berrios et al. 2018). Achieving this requires a concerted effort to conceptualize, adopt, and implement interoperability standards, develop interoperable tools, and foster collaborative networks. While progress has been made, much work still remains ahead.

Despite the clear advantages of interoperability—the ability of diverse systems to work together seamlessly—which are often celebrated as a catalyst for efficiency, collaboration, and user convenience, there is a lesser-discussed downside: its potential to stifle innovation (Hodapp and Hanelt 2022). While interoperability promotes standardization, it can inadvertently constrain the creative freedom of developers, forcing them to design within the confines of existing frameworks rather than exploring groundbreaking paradigms. The very standards meant to facilitate compatibility often become chains that tie new ideas to the old, hindering progress.

One of the critical ways interoperability stifles innovation is through its reliance on uniform standards. When communities coalesce around specific protocols or formats, there is little room for deviation. While such standards undeniably simplify collaboration, they often lock stakeholders into incremental improvements rather than transformative breakthroughs. Scientists may find themselves pressured to prioritize compatibility with legacy systems over pioneering new, potentially disruptive technologies that don't fit established moulds. As a result, the drive for conformity can stymie the exploration of novel approaches that could redefine entire fields (Rust 2023).

In essence, while interoperability is undeniably valuable in fostering connectivity and user convenience, its dark side cannot be overlooked. The demand for compatibility can lead to technological stagnation, limiting experimentation and disincentivizing bold innovation (M. Park et al. 2023). To counteract this, scientists must strike a balance: championing interoperability where it enhances collective progress while encouraging the freedom to disrupt and diverge where it matters most. Only by navigating this tension thoughtfully can we ensure that interoperability serves as a bridge to the future rather than a tether to the past.

To illustrate this point, we can provide the example of standard file formats—FASTQ files can technically be replaced with "unmapped BAM" (uBAM) files. The newer format has various advantages over FASTQ, such as fields for metadata, (considerably) smaller size, and the fact that paired-end reads can be kept in the same file. The standardization around FASTQ as a "raw read" format, however, keeps uBAM from being more widely adopted.

9.6 Data Storage

Data repositories are a central part of research data management and stewardship. As science is built on the sharing of knowledge, it is good scientific practice, and often a requirement by funders and publishers, to share research data publicly (ELIXIR 2021). However, not all repositories are built alike, and deciding where to deposit research data is a nontrivial task. Ahead, we highlight the main deciding factors in choosing a repository.

9.6.1 On-Premises Versus the Cloud

Although data may, at first, appear to be an abstract concept, it very much has a material presence, as it relies on physical media to be stored, be it on spinning disks, memory cells, tape, or other storage technologies. Deciding where to physically store the data generated by an experiment is an unavoidable, yet easily overlooked, decision for researchers.

As it relates to storage location, data repositories can broadly be classified into two categories: (i) on-premises, meaning the servers hosting the data are located on-site at the organization responsible for its generation, and (ii) in the cloud, meaning the data is stored in data centers operated by third-party providers.

Storing data on-premises allows for a higher degree of control over the data and its hosting conditions. In cases where the data is extremely sensitive, such as with datasets coming from human patients, that high degree of control is a requirement to ensure the privacy and security of the records. This approach is less reliant on third parties, which can be beneficial, as public repositories, particularly ones that store data free of charge, do sometimes cease to exist, and the information they store can be permanently lost.

There are, however, disadvantages to this approach, particularly to smaller institutions which may not have the infrastructure, the staff, or the funding required to operate and maintain computational resources to host the data publicly or otherwise for an indefinite amount of time. Regardless of the conditions of the responsible organization, computational biology resources do frequently shut down, most likely because the project's funding has been depleted or because the responsible researchers have switched teams or shifted priorities (Veretnik et al. 2008).

Cloud computing is generally considered to be a loosely defined term but can generally be understood as a computing technique which provides IT services that rely on a vast network of low-cost computing units (Qian et al. 2009). Large cloud providers maintain networks of those units spread across the globe and provide users access to those resources, usually by metering the resources (such as compute time or storage capacity) used by a customer and then charging according to their usage.

The major benefit of these services is that they have a very low cost of entry. In fact, publishing relatively small datasets can usually be done without any cost. For more sizable datasets, which is frequently the case with omics data, researchers can expect recurring fees for as long as they need the data stored in those facilities. Storing sensitive data adds another degree of complexity, as there's a need to ensure compliance with applicable laws and regulations. And, although some cloud services now offer Trusted Research Environments as a service (Estevez et al. 2022), there's always some degree of trust involved in resorting to a third party to be the safeguard of highly sensitive data.

9.6.2 Data Security, Privacy, Legal, and Ethical Concerns

Omics data management must consider the ethical and legal context. When working with sensitive human data, failing to secure your data or disrespecting privacy and ethical laws can have severe consequences, not only reputational but also legal, including severe administrative fines or sanctions.

The first step would be to be aware of the local law regarding sensitive data. Many jurisdictions have enacted or adopted data protection laws inspired by the GDPR (https://eur-lex.europa.eu/eli/reg/2016/679/oj) to govern the collection, use, and sharing of personal data, some examples are the Health Insurance Portability and Accountability Act (https://www.hhs.gov/hipaa/for-professionals/privacy/laws-regulations/index.html) in the USA or the Data Protection Act (https://www.gov.uk/data-protection) in the UK.

While many of these laws share GDPR's core principles—such as data minimization, purpose limitation, and user rights—they often differ in scope, enforcement mechanisms, and penalties. Unlike GDPR, some laws may not apply extraterritorially or may focus on specific types of data, like health or financial information.

These regulations collectively highlight a global trend towards stricter data privacy and protection standards, with GDPR serving as a model for many.

The GDPR, for instance, states that "National authorities can or must assess fines for specific data protection violations [...]". Additionally, other measures can be applied including a ban on data processing. The next sections cover the most common measures that should be taken into consideration. However, it is important to talk to the Data Protection Officer (DPO) of your institution and take into consideration your institution's guidelines and policies, together with the national law.

9.6.3 Data Security

Encryption is a foundational element, ensuring that data is protected both in transit and at rest, using robust algorithms to safeguard against unauthorized access. An option for genomics, for instance, is Crypt4GH (https://samtools.github.io/hts-specs/crypt4gh.pdf), developed by the Global Alliance for Genomics and Health (https://www.ga4gh.org/), which has a set of characteristics that allow handling large, complex genomic datasets, ensuring efficient encryption without compromising performance.

Regular certification or audits of data management systems are essential to ensure compliance with international standards such as ISO/IEC 27001 (https://www.iso.org/standard/27001), reinforcing trust and identifying vulnerabilities proactively.

Standard Operating Procedures (SOPs) should be meticulously developed and include clear protocols for routine data handling, as well as detailed response plans in case of data breaches, such as immediate notification, containment strategies, and corrective measures.

Formation and training of personnel are equally crucial; all team members must understand best practices for secure data handling, emerging threats, and compliance requirements.

Together, these measures create a multilayered defence system, safeguarding omics data while fostering ethical and secure research environments.

9.6.4 Data Privacy and Ethics

Protecting the privacy of the data subjects is fundamental, fostering trust, and ensuring compliance with legal frameworks.

Apart from the already mentioned laws on the protection of natural persons with regard to the processing of personal data and on the free movement of such data, that need to be respected, privacy is often a delicate task, requiring a balance between having rich and useful data and the risk of subject identification.

The next step is usually the decision between anonymization and pseudonymization, two essential techniques for protecting individual privacy in research. Anonymization involves permanently removing or altering identifiers in a dataset so that individuals cannot be directly or indirectly identified. This process is irreversible, meaning it is impossible to re-link the data to the original participants. It is particularly useful when the goal is to share data publicly or across institutions, as it ensures compliance with stringent privacy regulations like GDPR. Pseudonymization, on the other hand, replaces identifiers with artificial identifiers, such as codes or aliases, while retaining the possibility of re-linking the data to individuals using a *separate* key. This approach is valuable for research scenarios where re-linking data might be necessary, such as longitudinal studies or validation of findings. However, pseudonymized data requires careful management of the re-identification key and robust access controls to ensure it remains secure.

While anonymization offers stronger privacy protection, pseudonymization keeps a linkage that is essential in scenarios where discoveries or findings may have direct implications for a participant's health. For example, if a study uncovers a genetic marker associated with a treatable condition or a significant risk factor, pseudonymization enables researchers to re-identify the individual and inform them or their healthcare provider, ensuring timely intervention. Both methods reduce privacy risks, but the choice between them depends on the specific research needs. Data usage consents must explicitly state the type of anonymization or pseudonymization that will be applied to the data, and the data subject should be aware of the risks inherent to the chosen approach.

Rich metadata improves most aspects of FAIRness, with all the advantages it comes with, but it must come with an evaluation of the risk of re-identification. The risk of re-identification is a critical concern in research, particularly when working with anonymized or pseudonymized datasets. Even when direct identifiers (like names or IDs) are removed, combining seemingly innocuous data points (e.g., age, location, or unique behaviours) can inadvertently reveal a participant's identity, especially when cross-referenced with external datasets (e.g., genome). This risk should be calculated, and if necessary critical metadata fields should be removed.

9.6.5 Informed Consents

Transparency is one of the main takeaway messages when talking about privacy respect. Consent forms should allow patients to specify their preferences, ensuring that their decisions are respected throughout the research or clinical process.

Before participating in a study or undergoing genetic testing, patients should be clearly informed about the possibility of discovering secondary findings (unrelated to the primary purpose of the test or study), the nature of such findings, particularly if they involve actionable health risks (e.g., predisposition to hereditary cancers), and their right to opt in or out of being informed about these findings.

This aspect, in particular, also enters the ethics field. There is a dilemma between offering patients the opportunity to receive actionable findings, particularly if interventions can prevent serious harm while measuring the potential psychological distress or unwanted knowledge about their health.

9.6.6 Ethics

Ethics in omics data management requires careful consideration of several factors, especially when dealing with sensitive datasets. Ethical committee approval is fundamental, serving as the first safeguard to ensure that research involving omics data adheres to established ethical principles, respects participant rights, and complies with legal frameworks and institutional policies. A key challenge lies in the possible representation of ethnic minorities, vulnerable populations, and underrepresented groups in omics studies. Their inclusion might help avoid biases that could perpetuate health disparities but carries inherent risks, particularly regarding re-identification and potential legal or ethical violations. These risks stem from the unique and often rare genetic markers or population-specific traits present in such groups, which can increase the likelihood of individuals being re-identified even in anonymized datasets. Additionally, Data Access Committees (DACs) play a pivotal role in overseeing omics research. Their responsibilities include evaluating whether the intended use of data respects participant consent, protects privacy, and aligns with the purpose for which the data was originally collected. DACs also play a pivotal role in preventing misuse or exploitation of data by ensuring that access is granted only to qualified researchers and institutions with a legitimate and scientifically sound purpose. Together, these mechanisms uphold the integrity of omics research while balancing scientific advancement with the protection of individual and group rights.

9.6.7 Openness

Laws like GDPR and others promote data sharing by emphasizing secure and ethical practices. They recognize the potential of shared data to drive societal benefits, particularly in areas like healthcare, education, and research.

The balance between ethics, privacy, and open data is dynamic and context-dependent. It requires a deep study of the specific case we are dealing with, and it should also rely on robust frameworks, technological innovation, and an ongoing commitment to transparency, fairness, and respect for participant rights. By carefully navigating this balance, researchers can uphold ethical principles while advancing open science and maximizing the societal benefits of their work. In a simple way, follow the open science principle of "as open as possible, as closed as necessary".

9.7 Recommendations for Designing an Effective Omics Data Management Strategy

Establishing an effective omics data management strategy is crucial for maximizing the value of data generated in research, ensuring reproducibility, facilitating collaboration, and supporting regulatory compliance. The following recommendations (summarized in Fig. 9.2) aim to guide researchers, labs, and institutions in creating efficient, robust, and sustainable omics data management practices.

9.7.1 Defining Clear Data Management Goals

- *Identify Key Objectives*: Determine the primary goals of data management in the project or institution, such as data sharing, reproducibility, regulatory compliance, or future data reuse.
- *Assess Regulatory and Ethical Requirements*: For sensitive data, particularly in human studies, ensure compliance with regulations like

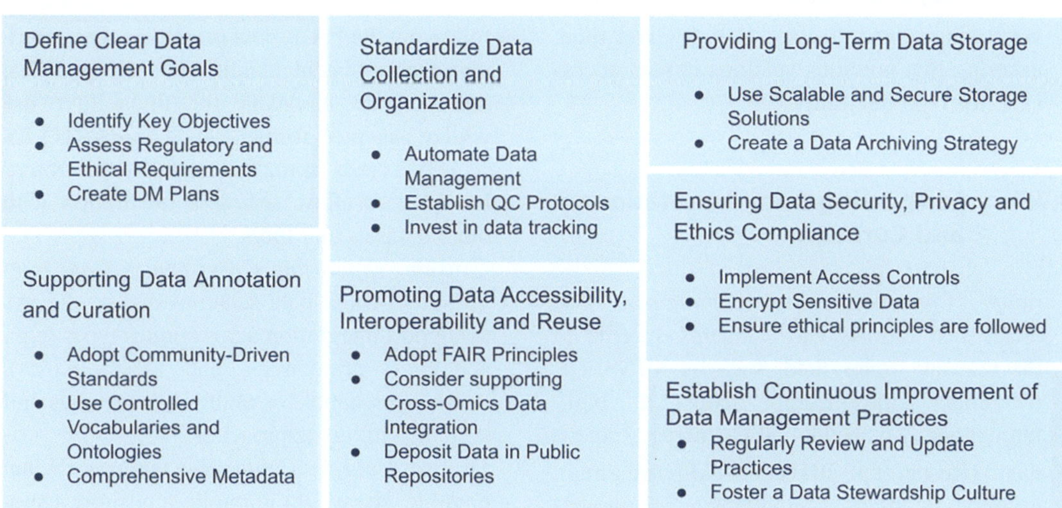

Fig. 9.2 Recommendations for designing an effective omics data management strategy

the General Data Protection Regulation (GDPR) and any institutional data protection policies.
- *Create Data Management Plans (DMPs)*: Develop a data management plan at the outset of each project to define data handling, storage, sharing, and retention strategies. Many funding agencies require a DMP as part of grant proposals.

9.7.2 Standardizing Data Collection and Organization

- *Automate Data Management where possible.* Use Laboratory Information Management Systems (LIMS) to streamline data capture, sample tracking, and metadata collection, which is especially useful for high-throughput omics labs. Consider specialized data management platforms such as Galaxy, BaseSpace, or the Cancer Genomics Cloud, which offer workflows tailored to omics data handling and analysis.
- *Establish Quality Control Protocols*: Design Quality Control steps to perform consistency checks, identify and correct errors, and ensure data completeness and consistency.
- *Invest in Data Tracking*: Implement a versioning system to track changes over time, ensuring that previous versions remain accessible for reproducibility.

9.7.3 Supporting Data Annotation and Curation

- *Adopt Community-Driven Standards*: Use established formats and standards specific to the relevant omics field, such as MINSEQE (Minimum Information about a high-throughput Nucleotide SEQuencing Experiment) (Rustici et al. 2021) or MIAPE (minimum information about a proteomics experiment) (https://www.nature.com/articles/nbt1329).
- *Use Controlled Vocabularies and Ontologies*: Employ standardized terminologies, such as the Gene Ontology (GO) or Disease Ontology (DO), to ensure consistent annotation across datasets and improve data interoperability.
- *Annotate with Comprehensive Metadata*: Include detailed metadata covering sample information, experimental conditions, data processing steps, and biological context. Metadata should be complete, unambiguous, and follow standardized formats.

9.7.4 Providing Long-Term Data Storage

- *Use Scalable and Secure Storage Solutions*: Invest in scalable storage solutions, such as cloud storage or high-performance computing clusters, to handle large volumes of omics data.
- *Create a Data Archiving Strategy*: For long-term storage, develop an archiving plan that includes data retention periods, secure storage for sensitive data, and backup solutions.

9.7.5 Promoting Data Accessibility, Interoperability, and Reuse

- *Adopt FAIR Principles*: Make data *Findable, Accessible, Interoperable*, and *Reusable* by following the FAIR data principles. For example, use persistent identifiers, such as DOIs, and provide metadata for findability; use widely accepted formats, such as FASTQ for sequence reads or mzML for mass spectrometry data, to allow seamless integration with other datasets and tools.
- *Consider supporting Cross-Omics Data Integration*: Design data schemas and workflows to support integration across omics types (e.g., genomics, proteomics, and metabolomics), which is essential for multi-omics studies and systems biology approaches.
- *Deposit Data in Public Repositories*: When possible, share data in public repositories specific to your omics domain, such as the Gene Expression Omnibus (GEO) for genomics or PRIDE for proteomics, to support transparency and reproducibility.

9.7.6 Ensuring Data Security, Privacy, and Ethics Compliance

- *Implement Access Controls*: Use role-based access control (RBAC) to restrict access to sensitive data, ensuring only authorized users have access.
- *Encrypt Sensitive Data*: Apply encryption both at rest and in transit for datasets containing sensitive information, such as patient identifiers or clinical data.
- *Ensure ethical principles are followed*: ethical omics data management requires strict privacy protections, compliance with regulatory standards, and transparency in data handling.

9.7.7 Establish Continuous Improvement of Data Management Practices

- *Regularly Review and Update Practices*: Data management needs evolve, so periodically assess and refine your practices to incorporate new standards, technologies, and regulatory requirements.
- *Foster a Data Stewardship Culture*: Promote a culture of responsibility around data management and sharing, emphasizing the value of high-quality data for the scientific community. Regularly train researchers and data managers on best practices, data standards, software tools, and ethical guidelines related to omics data.

Box 9.1 The FAIR Principles

Findable: Data should have unique identifiers, complete metadata, and be registered to search resources.

Accessible: Data should be retrievable using standard protocols, with clear access policies.

Interoperable: Data should be described using standard vocabulary, format, and ontologies so it can be integrated with other datasets.

Reusable: Clear licensing, provenance, and restrictions should accompany the data to maximize its reuse by the scientific community.

Box 9.2 The FASTQ Format

The FASTQ format is a de facto standard in bioinformatics, meaning it became standardized through popular use rather than official mandate. As a result, it has various issues, such as multiple incompatible variants (*fastq-sanger*, *fastq-solexa*, *fastq-illumina*). For a comprehensive history of the format and how its issues were addressed, see Cock et al. (2010).

Box 9.3 The VCF Format

Unlike the FASTQ format, the VCF and SAM/BAM formats were well-defined from the start. They are versioned for improvements and updates by the Global Alliance for Genomics & Health (GA4GH), an open body. That's not to say there are no issues with these standards. For example, the 1000 Genomes VCF "fork" was later merged back into the standard specification (Garrison et al. 2022). The latest specifications of VCF and SAM/BAM are continually updated.

Acknowledgements This work was partially funded by FCT through the LASIGE Research Unit (ref. UID/00408/2025, DOI: https://doi.org/10.54499/UID/00408/2025), through CE3C ref. UID/00329/2025 (DOI: https://doi.org/10.54499/UID/00329/2025) and through INESC-ID ref. UID/50021/2025 (DOI: https://doi.org/10.54499/UID/50021/2025) and UID/PRR/50021/2025 (DOI: https://doi.org/10.54499/UID/PRR/50021/2025). It was also partially supported by the CancerScan project by the EU's HORIZON Europe research and innovation programme under grant agreement No 101186829.

This work was further supported by BioData.pt | ELIXIR Portugal, the national research infrastructure for life and health data.

References

Almeida MB, Barbosa RR (2009) Ontologies in knowledge management support: a case study. J Am Soc Inf Sci Technol 60(10):2032–2047. https://doi.org/10.1002/asi.21120

Amorim M, Silva S, Machado H, Teles EL, Baptista MJ, Maia T, Nwebonyi N, de Freitas C (2022) Benefits and risks of sharing genomic data for research: comparing the views of rare disease patients, informal caraers and healthcare professionals. Int J Environ Res Public Health 19(14):8788. https://doi.org/10.3390/ijerph19148788

Auton A, Abecasis GR, Altshuler DM, Durbin RM, Abecasis GR, Bentley DR, Chakravarti A, Clark AG, Donnelly P, Eichler EE, Flicek P, Gabriel SB, Gibbs RA, Green ED, Hurles ME, Knoppers BM, Korbel JO, Lander ES, Lee C et al (2015) A global reference for human genetic variation. Nature 526(7571):68–74. https://doi.org/10.1038/nature15393

Berrios DC, Beheshti A, Costes SV (2018) FAIRness and usability for open-access omics data systems. AMIA Ann Symp Proc 2018:232–241

Black M, Lamothe L, Hager Eldakroury, Kierkegaard M, Ankita Priya, Machinda A, Khanduja US, Drashti Patoliya, Rashika Rathi, Tawah Peggy Che Nico, Umutesi G, Blankenburg C, Op A, Chieke P, Omodolapo Babatunde, Laurie S, Neumann S, Schwämmle V, Kuzmin I et al (2022) EDAM: The bioscientific data analysis ontology (update 2021). F1000 Research Limited. https://doi.org/10.7490/F1000RESEARCH.1118900.1

Brinson LC, Bartolo LM, Blaiszik B, Elbert D, Foster I, Strachan A, Voorhees PW (2024) Community action on FAIR data will fuel a revolution in materials research. MRS Bull 49(1):12–16. https://doi.org/10.1557/s43577-023-00498-4

Byrd JB, Greene AC, Prasad DV, Jiang X, Greene CS (2020) Responsible, practical genomic data sharing that accelerates research. Nat Rev Genet 21(10):615–629. https://doi.org/10.1038/s41576-020-0257-5

Canuel V, Rance B, Avillach P, Degoulet P, Burgun A (2015) Translational research platforms integrating clinical and omics data: a review of publicly available solutions. Brief Bioinform 16(2):280–290. https://doi.org/10.1093/bib/bbu006

Clarke L, Zheng-Bradley X, Smith R, Kulesha E, Xiao C, Toneva I, Vaughan B, Preuss D, Leinonen R, Shumway M, Sherry S, Flicek P (2012) The 1000 Genomes Project: data management and community access. Nat Methods 9(5):459–462. https://doi.org/10.1038/nmeth.1974

Cock PJA, Fields CJ, Goto N, Heuer ML, Rice PM (2010) The Sanger FASTQ file format for sequences with quality scores, and the Solexa/Illumina FASTQ variants. Nucleic Acids Res 38(6):1767–1771. https://doi.org/10.1093/nar/gkp1137

Danecek P, Auton A, Abecasis G, Albers CA, Banks E, DePristo MA, Handsaker RE, Lunter G, Marth GT, Sherry ST, McVean G, Durbin R, Group, 1000 Genomes Project Analysis (2011) The variant call format and VCFtools. Bioinformatics 27(15):2156–2158. https://doi.org/10.1093/bioinformatics/btr330

DCMI Metadata Terms (n.d.) DCMI. Retrieved July 9, 2024, from https://www.dublincore.org/specifications/dublin-core/dcmi-terms/

Donnelly M, Jones S, Pattenden-Fail JW (2010) DMP online: a demonstration of the digital curation centre's web-based tool for creating, maintaining and exporting data management plans. In: Lalmas M, Jose J, Rauber A, Sebastiani F, Frommholz I (eds) Research and advanced technology for digital libraries. Springer, pp 530–533. https://doi.org/10.1007/978-3-642-15464-5_74

Duval E (2021) Metadata standards: what, who & why. JUCS 7(7):591–601. https://doi.org/10.3217/JUCS-007-07-0591

ELIXIR (2021) Research data management kit. A deliverable from the EU-funded ELIXIR-CONVERGE project (grant agreement 871075). https://rdmkit.elixir-europe.org

Estevez M, Li R, Paul B, Daneshvar K, Mullen AC, Romerio F, Addepalli B (2022) Identification and mapping of post-transcriptional modifications on the HIV-1 antisense transcript *Ast* in human cells. RNA 28(5):697–710. https://doi.org/10.1261/rna.079043.121

Garcia L, Bolleman J, Gehant S, Redaschi N, Martin M (2019) FAIR adoption, assessment and challenges at UniProt. Sci Data 6(1):175. https://doi.org/10.1038/s41597-019-0180-9

Gargano MA, Matentzoglu N, Coleman B, Addo-Lartey EB, Anagnostopoulos AV, Anderton J, Avillach P, Bagley AM, Bakštein E, Balhoff JP, Baynam G, Bello SM, Berk M, Bertram H, Bishop S, Blau H, Bodenstein DF, Botas P, Boztug K et al (2024) The Human Phenotype Ontology in 2024: phenotypes around the world. Nucleic Acids Res 52(D1):D1333–D1346. https://doi.org/10.1093/nar/gkad1005

Garrison E, Kronenberg ZN, Dawson ET, Pedersen BS, Prins P (2022) A spectrum of free software tools for processing the VCF variant call format: Vcflib, bio-VCF, cyvcf2, hts-nim and slivar. PLoS Comput Biol 18(5):e1009123. https://doi.org/10.1371/journal.pcbi.1009123

Geraci A, Katki F, McMonegal L, Meyer B, Lane J, Wilson P, Radatz J, Yee M, Porteous H, Springsteel F (1991) IEEE standard computer dictionary: compilation of IEEE standard computer glossaries. IEEE Press

Giannattasio S, Heil KF, Hermjakob H, Hooft RWW, Klapa MI, Mastrorocco F, Neumann S, O'Donovan C, Payne T, Perez-Riverol Y, Prigent S, Thévenot EA (2023) The next challenge: data management, submission and FAIRness in multi-omics experiments. F1000Res 12(1379):1379. https://doi.org/10.7490/f1000research.1119646.1

Hodapp D, Hanelt A (2022) Interoperability in the era of digital innovation: an information systems research agenda. J Inf Technol 37(4):407–427. https://doi.org/10.1177/02683962211064304

Jackson R, Matentzoglu N, Overton JA, Vita R, Balhoff JP, Buttigieg PL, Carbon S, Courtot M, Diehl AD, Dooley DM, Duncan WD, Harris NL, Haendel MA, Lewis SE, Natale DA, Osumi-Sutherland D, Ruttenberg A, Schriml LM, Smith B et al (2021) OBO Foundry in 2021: operationalizing open data principles to evaluate ontologies. Database 2021: baab069. https://doi.org/10.1093/database/baab069

Johnston LR, Curty R, Braxton SM, Carlson J, Hadley H, Lafferty-Hess S, Luong H, Petters JL, Kozlowski WA (2024) Understanding the value of curation: a survey of US data repository curation practices and perceptions. PLoS One 19(6):e0301171. https://doi.org/10.1371/journal.pone.0301171

Li H (2011) Improving SNP discovery by base alignment quality. Bioinf (Oxf) 27(8):1157–1158. https://doi.org/10.1093/bioinformatics/btr076

Li H, Durbin R (2009) Fast and accurate short read alignment with Burrows–Wheeler transform. Bioinformatics 25(14):1754–1760. https://doi.org/10.1093/bioinformatics/btp324

Li H, Handsaker B, Wysoker A, Fennell T, Ruan J, Homer N, Marth G, Abecasis G, Durbin R (2009) The sequence alignment/map format and SAMtools. Bioinformatics (Oxford, England) 25(16):2078–2079

Lin D, McAuliffe M, Pruitt KD, Gururaj A, Melchior C, Schmitt C, Wright SN (2024) Biomedical data repository concepts and management principles. Sci Data 11(1):622. https://doi.org/10.1038/s41597-024-03449-z

Michener WK (2015) Ten simple rules for creating a good data management plan. PLoS Comput Biol 11(10): e1004525. https://doi.org/10.1371/journal.pcbi.1004525

Moore B, Fan G, Eilbeck K (2010) SOBA: sequence ontology bioinformatics analysis. Nucleic Acids Res 38(suppl_2):W161–W164. https://doi.org/10.1093/nar/gkq426

Mosha NF, Ngulube P (2023) Metadata standard for continuous preservation, discovery, and reuse of research data in repositories by higher education institutions: a systematic review. Information 14(8):8. https://doi.org/10.3390/info14080427

Namsaraev ZB, Korzhenkov AA, Fedosov DY, Patrushev MV (2023) Digitalization of genetic information: prospects and challenges. Nanobiotechnol Rep 18(3): 320–328. https://doi.org/10.1134/S2635167623700192

Nicholson C, Kansa S, Gupta N, Fernandez R (2023) Will it ever be FAIR?: making archaeological data findable, accessible, interoperable, and reusable. Adv Archaeol Pract 11(1):63–75. https://doi.org/10.1017/aap.2022.40

Noy NF, Tudorache T, de Coronado S, Musen MA (2008) Developing biomedical ontologies collaboratively. AMIA Ann Symp Proc 2008:520–524

OpenAIRE (n.d.) ARGOS [Computer software]. https://argos.openaire.eu

Park J, Ram S (2004) Information systems interoperability: what lies beneath? ACM Trans Inf Syst 22(4): 595–632. https://doi.org/10.1145/1028099.1028103

Park M, Leahey E, Funk RJ (2023) Papers and patents are becoming less disruptive over time. Nature 613(7942):138–144. https://doi.org/10.1038/s41586-022-05543-x

Pergl R, Hooft R, Suchánek M, Knaisl V, Slifka J (2019) "Data stewardship wizard": a tool bringing together researchers, data stewards, and data experts around data management planning. Data Sci J 18(1). https://doi.org/10.5334/dsj-2019-059

Poveda-Villalón M, Espinoza-Arias P, Garijo D, Corcho O (2020) Coming to terms with FAIR ontologies. In: Keet CM, Dumontier M (eds) Knowledge engineering and knowledge management. Springer International Publishing, pp 255–270. https://doi.org/10.1007/978-3-030-61244-3_18

Qian L, Luo Z, Du Y, Guo L (2009) Cloud computing: an overview. In: Jaatun MG, Zhao G, Rong C (eds) Cloud computing. Springer, pp 626–631. https://doi.org/10.1007/978-3-642-10665-1_63

Rust R (2023) Rebooting disruptive science: exploring the challenges and potential solutions. Eur J Clin Investig 53(8):e13988. https://doi.org/10.1111/eci.13988

Rustici G, Williams E, Barzine M, Brazma A, Bumgarner R, Chierici M, Furlanello C, Greger L, Jurman G, Miller M, Ouellette BFF, Quackenbush J, Reich M, Stoeckert CJ, Taylor RC, Trutane SC, Weller J, Wilhelm B, Winegarden N (2021) Transcriptomics data availability and reusability in the transition from microarray to next-generation sequencing (p. 2020.12.31.425022). bioRxiv. https://doi.org/10.1101/2020.12.31.425022

Sansone S-A, McQuilton P, Rocca-Serra P, Gonzalez-Beltran A, Izzo M, Lister AL, Thurston M (2019) FAIRsharing as a community approach to standards, repositories and policies. Nat Biotechnol 37(4): 358–367. https://doi.org/10.1038/s41587-019-0080-8

Sayers EW, Beck J, Bolton EE, Bourexis D, Brister JR, Canese K, Comeau DC, Funk K, Kim S, Klimke W, Marchler-Bauer A, Landrum M, Lathrop S, Lu Z, Madden TL, O'Leary N, Phan L, Rangwala SH, Schneider VA et al (2020) Database resources of the National Center for biotechnology information. Nucleic Acids Res 49(D1):D10–D17. https://doi.org/10.1093/nar/gkaa892

Stylos J (2009) Making apis more usable with improved api designs, documentation and tools [Doctoral dissertation, Carnegie Mellon University]. http://www.cs.cmu.edu/~NatProg/papers/CMU-CS-09-130-Stylos-Dissertation.pdf

Sudmant PH, Rausch T, Gardner EJ, Handsaker RE, Abyzov A, Huddleston J, Zhang Y, Ye K, Jun G,

Hsi-Yang Fritz M, Konkel MK, Malhotra A, Stütz AM, Shi X, Paolo Casale F, Chen J, Hormozdiari F, Dayama G, Chen K et al (2015) An integrated map of structural variation in 2,504 human genomes. Nature 526(7571):75–81. https://doi.org/10.1038/nature15394

Tanjo T, Kawai Y, Tokunaga K, Ogasawara O, Nagasaki M (2021) Practical guide for managing large-scale human genome data in research. J Hum Genet 66(1):39–52. https://doi.org/10.1038/s10038-020-00862-1

The Gene Ontology Consortium, Aleksander SA, Balhoff J, Carbon S, Cherry JM, Drabkin HJ, Ebert D, Feuermann M, Gaudet P, Harris NL, Hill DP, Lee R, Mi H, Moxon S, Mungall CJ, Muruganugan A, Mushayahama T, Sternberg PW, Thomas PD et al (2023) The gene ontology knowledgebase in 2023. Genetics 224(1):iyad031. https://doi.org/10.1093/genetics/iyad031

Veretnik S, Fink JL, Bourne PE (2008) Computational biology resources lack persistence and usability. PLoS Comput Biol 4(7):e1000136. https://doi.org/10.1371/journal.pcbi.1000136

Vitale CH, Sandy HM (2019) Data management plans a review. DESIDOC J Libr Inf Technol *39*(06):06. https://doi.org/10.14429/djlit.39.06.15086

Weibel S (1997) The Dublin core: a simple content description model for electronic resources. Bull Am Soc Inf Sci Technol 24(1):9–11. https://doi.org/10.1002/bult.70

Whetzel PL, Noy NF, Shah NH, Alexander PR, Nyulas C, Tudorache T, Musen MA (2011) BioPortal: enhanced functionality via new Web services from the National Center for Biomedical Ontology to access and use ontologies in software applications. Nucleic Acids Res 39(suppl_2):W541–W545. https://doi.org/10.1093/nar/gkr469

Wilkinson MD, Dumontier M, Aalbersberg IJ, Appleton G, Axton M, Baak A, Blomberg N, Boiten J-W, da Silva Santos LB, Bourne PE, Bouwman J, Brookes AJ, Clark T, Crosas M, Dillo I, Dumon O, Edmunds S, Evelo CT, Finkers R et al (2016) The FAIR Guiding Principles for scientific data management and stewardship. Sci Data 3(1):1. https://doi.org/10.1038/sdata.2016.18

Challenges and Opportunities for Statistics in Omics Data Analysis

10

Eunice Carrasquinha, Lisete Sousa, and Carina Silva

Abstract

Omics data, comprising a diverse array of high-throughput molecular datasets, present substantial statistical challenges due to their intrinsic heterogeneity and variability. Effectively distinguishing biologically meaningful variations from random noise requires the application and development of robust statistical approaches. Interdisciplinary collaboration plays a pivotal role in refining these methodologies and enhancing the understanding of intricate biological systems. This chapter reviews the importance of statistical methods in omics data analysis, highlighting the need for ongoing advancements to address key challenges, including experimental design, preprocessing, dimensionality reduction, statistical modeling of complex datasets, and the interpretation of results. The pursuit of improved reliability in biological insights creates opportunities for the development and refinement of advanced statistical methodologies.

Keywords

Statistics · Omics · High-dimensional data · Experimental design · Machine learning

10.1 Experimental Design

Designing experiments for omics data analysis presents several challenges due to the complexity and high dimensionality of the data, as well as the inherent variability in biological systems. These challenges must be carefully considered during the experimental design phase to ensure valid and interpretable results.

10.1.1 Major Challenges and Considerations in Experimental Design

In experimental design, careful planning is essential to ensure that the experiment accurately answers the research question, minimizes bias, and produces valid and reliable results. Some primary considerations and major challenges commonly encountered in experimental design are sample size determination, biological

E. Carrasquinha · L. Sousa (✉)
Faculdade de Ciências, Universidade de Lisboa, Lisbon, Portugal

CEAUL—Centro de Estatística e Aplicações, Universidade de Lisboa, Lisbon, Portugal
e-mail: eitrigueirao@ciencias.ulisboa.pt; lmsousa@ciencias.ulisboa.pt

C. Silva
CEAUL—Centro de Estatística e Aplicações, Universidade de Lisboa, Lisbon, Portugal

ESSL—Escola Superior de Saúde, Instituto Politécnico de Lisboa, Lisbon, Portugal
e-mail: carina.silva@essl.ipl.pt

M. Gama-Carvalho et al. (eds.), *Omics Approaches in Biomedicine and Biotechnology*, Advances in Experimental Medicine and Biology 1504, https://doi.org/10.1007/978-3-032-18966-0_10

variability and replication, technical variability, complex experimental designs, cost and resource constraints, integration of multiple omics data types, and ethical and regulatory considerations.

Omics experiments often involve large numbers of variables (e.g., genes, proteins, metabolites) and limited sample sizes.

The primary objective of experimental design in omics research is to meticulously plan and organize studies to enable accurate, dependable, and reproducible analyses of biological data, ultimately producing significant and unbiased findings (Baggerly and Coombes 2009). Determining an appropriate sample size to achieve sufficient statistical power for detecting meaningful effects can be challenging, especially considering the potential for multiple hypothesis testing. Too small a sample size relative to the dimensionality of the data (the $p \gg n$ dilemma) fosters inflated effect sizes and poor reproducibility. To mitigate this, designs should be grounded in empirical evidence, ideally via pilot studies or reanalyzes of similar datasets, and incorporate power simulations tailored to high-dimensional settings. In many cases, a minimum of six biological replicates per condition is recommended, though the optimal number is context-dependent (Schurch et al. 2016).

Biological variability, often underestimated in study planning, can overshadow the very effects one aims to detect. This variability may arise from genetic background, epigenetic drift, environmental exposure, or even stochastic fluctuations within systems biology. Hence, the inclusion of genuine biological replicates (distinct individuals or experimental units subjected to the same condition) is essential. These replicates allow estimates of within-group variability and ensure that conclusions extend beyond a singular biological instance. Importantly, they should not be conflated with technical replicates, which serve distinct purposes: while technical replicates assess procedural consistency, they do not capture biological heterogeneity and should never substitute true biological replication. Equally important is the control of technical variability. Omics workflows are inherently multistep, each phase susceptible to introducing artifactual variation, from sample handling and storage to library preparation, instrument calibration, or batch processing. These can culminate in batch effects, systematic, non-biological differences that confound interpretation. Effective designs implement randomization at key points (e.g., sample processing order, lane, or plate assignment) and, where feasible, adopt blocking strategies to distribute known sources of technical variation evenly across experimental groups. This is particularly crucial when sample throughput necessitates processing in multiple batches or runs. The inclusion of control materials (e.g., pooled reference samples, spike-in standards) may further support quality assurance and enable post hoc correction where needed, though these should complement, not replace, sound design choices.

Contemporary omics studies often explore multifactorial hypotheses or integrate data across molecular layers. In the former, factorial designs are particularly powerful, allowing the disentanglement of main effects and interactions—provided there is sufficient replication across factor combinations. However, such comprehensive designs are resource-intensive, and trade-offs are inevitable. Often, increasing replication within a single omics layer yields greater interpretative clarity than diluting resources across too many assays. Accordingly, design decisions must reflect both the scientific questions and the practical constraints of the study. Resource limitations, in fact, shape much of the experimental landscape. Sequencing costs, sample availability, and analytical throughput impose natural boundaries. Designing within these requires strategic prioritization: is it preferable to profile more samples at moderate depth, or fewer samples with maximum coverage? Should one expand the number of conditions tested, or increase replication for a subset?

Ethical and regulatory considerations are very important to omics study design. In human research, informed consent must address data sharing, future analyses, and recontact (Knoppers 2014), while data privacy and compliance with regulations such as the General Data Protection Regulation (GDPR) must be planned from the outset (Shabani and Borry 2018). Sample size has both statistical and ethical implications. When working with genetically modified organisms or pathogens, prior regulatory approval and biosafety measures are required (OECD 2016; European Commission 2009). Integrating ethical foresight into the design safeguards participants, researchers, and scientific integrity.

Some considerations can be made to overcome some issues related with experimental designs: (i) use common data formats, metadata standards (e.g., Minimum Information about a high-throughput Nucleotide SeQuencing Experiment (MINSEQE) a proposal for standards in functional genomic data reporting (1.0) (Brazma et al. 2001; Rustici et al. 2021), Human Proteome Organization Proteomics Standards Initiative (HUPO PSI) for proteomics (Deutsch et al. 2023)), and reference ontologies (e.g., gene ontology (Ashburner et al. 2000)) to facilitate compatibility across platforms; (ii) develop data harmonization pipelines that handle pre-processing, normalization, and alignment of datasets. Specialized tools, like Multi-Omics Factor Analysis (MOFA) (Argelaguet et al. 2018) or OmicsIntegrator (Kedaigle and Fraenkel 2018), can assist in integrating multi-omics data; (iii) advanced statistical frameworks, such as Bayesian modeling and machine learning algorithms, to manage discrepancies in data structure, scale, and noise across platforms; (iv) conduct pilot studies to identify potential technical issues or compatibility constraints between platforms, allowing for adjustments in experimental design; (v) use internal or external standards across platforms (e.g., control groups or synthetic controls) for benchmarking, ensuring consistency and comparability across datasets; and (vi) integrating ethical dimensions into the design phase supports legal compliance, participant protection, and the overall integrity of the research.

To guide design, biostatisticians, bioinformaticians, molecular biologists, and ethicists should work together, and since this field is constantly evolving, it is very important to keep teams updated on new tools, methods, and ethical regulations to improve data integration and compliance.

10.1.2 Sample Design and Sample Size

Careful planning of sample design and sample size is a fundamental component of any experimental study and particularly critical in omics research. Due to the complexity, cost, and high dimensionality of omics data, poor choices at this stage can compromise the entire analysis. For this reason, we highlight sample design as a distinct and essential element in the planning process.

Sample design refers to the strategy used to select experimental units (e.g., patients, tissues, cell cultures) that will provide the data for analysis. In omics studies, the method of sampling, whether random, stratified, or otherwise, has a significant impact on the validity, reproducibility, and generalizability of the findings. However, these concepts are often borrowed from statistical theory and may not always appear intuitive to researchers with a life sciences background. We therefore provide a practical example: suppose one is conducting an RNA-seq experiment to investigate how gene expression changes in cultured cells following exposure to a drug. A random sampling approach could involve randomly assigning available cell culture replicates to either control or treatment groups, ensuring that each sample has an equal chance of being selected for each condition. This minimizes selection bias and helps distribute unknown confounding factors evenly across groups. Alternatively, a stratified sampling strategy might be more appropriate if the cell cultures vary by a known characteristic, such as passage number or donor origin (in the case of primary cells). In this case, one would divide the cell cultures into strata (e.g., early vs. late passage) and then sample equally or proportionally from each stratum in both treatment and control groups. This ensures that the comparison between conditions is not confounded by systematic differences in cell characteristics. Such an approach is particularly useful when studying human subjects, where stratification by age, sex, disease stage, or ethnicity may be necessary to ensure representativeness and reduce variance. In population-based studies using patient samples, cluster sampling or multistage sampling may also be employed to address logistical constraints or to ensure coverage across geographical or clinical strata. In environmental omics, spatial or temporal sampling strategies may be required to capture the relevant biological variability.

Determining the appropriate sample size is equally essential. It directly influences the precision, statistical power, and robustness of downstream analyses. Given the substantial cost and effort associated with omics data collection, researchers must find a balance between available resources and the desired level of statistical certainty. Larger sample sizes generally improve the reliability of the results and allow for more nuanced analyses, such as subgroup comparisons or interaction effects. In omics, where the number of measured variables often vastly exceeds the number of samples, sample size planning should ideally incorporate pilot data or variance estimates from previous studies and be informed by power calculations that account for multiple testing and high dimensionality. Although there are no universal rules for sample size, several guidelines and recommendations have been proposed by leading organizations in the field. For example, the ENCODE consortium recommends using at least six biological replicates for gene expression studies in order to detect moderate fold changes with acceptable false discovery rates (ENCODE 2017). When pilot data are available, they can be used to estimate biological variability and inform more accurate power calculations, particularly for high-throughput settings. In Sect. 4.3 of this chapter, we provide a more detailed discussion on power analysis approaches adapted to omics data, including strategies for estimating effect sizes and addressing the multiple testing burden. We refer the reader to that section for methodological details and practical recommendations.

As Baggerly and Coombes (2009) argue, the reliability of high-throughput data analysis depends not only on analytical techniques but also, critically on how the data were obtained. A carefully constructed sample design and a justified sample size are the foundation for meaningful, reproducible, and unbiased discovery in omics research.

10.2 Preprocessing

Preprocessing omics data poses numerous challenges due to the intrinsic complexity and variability of biological datasets. Several critical aspects must be carefully addressed prior to analyzing this type of data. Below, some of the most common issues are outlined.

10.2.1 Missing Values

As with many other types of data, omics data often contain missing values due to experimental limitations or technical issues. These missing values can be handled using various techniques. However, it is important to consider the underlying reasons for the missing data. Three mechanisms can be identified: missing completely at random (MCAR), missing at random (MAR), and missing not at random (MNAR) (Little and Rubin 1987). When the data are MCAR, the missingness is independent of both the observed and unobserved data. In contrast, for MAR, the missingness is systematically related to the observed data but not to the unobserved data. Finally, in the case of MNAR, the missingness is systematically related to the unobserved data, meaning that the missingness is connected to events or factors not measured by the researcher.

Handling missing values in a dataset is a crucial preprocessing step. Several imputation methods are available in the literature to address the issue of missing values. Simple strategies include mean, median, or mode imputation, as well as regression imputation. A more advanced approach involves using the Expectation-Maximization (EM) algorithm to generate a new dataset, where all missing values are imputed with values estimated using maximum likelihood (Kang 2013). However, these methods can sometimes lead to a reduction in the amount of available data for subsequent analyses. One solution to mitigate this issue is to replace missing values with multiple predictions from an imputation model. Techniques such as k-nearest neighbors (k-NN) presented by Batista and Monard (2003) and multiple imputation by chained equations (MICE), proposed by van Buuren (2018), are examples of methods that provide multiple imputations.

Another approach to handling missing data is through deletion methods. In this case, observations or features with missing values are removed from the dataset. Deletion methods include listwise deletion and pairwise deletion (Kim and Curry 1977).

Both R (R Core Team 2024) and Python (Python Software Foundation 2024) provide numerous tools and libraries for identifying, managing, and imputing missing values. In R, commonly used packages include `mice` and `DMwR`, while in Python, libraries such as `pandas`, `NumPy`, and `scikit-learn` offer extensive functionality for handling missing data.

10.2.2 Normalization and Scaling

Normalization and scaling are essential preprocessing steps in data analysis, particularly in omics data, in order to remove technical variation while preserving biological signals. The choice of normalization or scaling technique depends on the type of omics data and the goal of the analysis. It is important to note that the selection of techniques for both normalization and scaling influences the final results.

Normalization methods for transcriptomics data, such as reads per million (RPM), fragments per kilobase per million (FPKM), quantile normalization, trimmed mean of M-values (TMM), and DESeq normalization, are crucial for reducing technical variability and ensuring accurate comparative analysis (Abbas-Aghababazadeh et al. 2018). RPM and FPKM are techniques used to normalize RNA-seq data by adjusting for sequencing depth and gene length, enabling precise comparisons of gene expression levels between samples. Quantile normalization ensures the distribution of gene expression levels is uniform across all samples, minimizing technical variability and improving comparability. The TMM method normalizes RNA-seq data by accounting for compositional differences between samples, thus improving the accuracy of differential expression analysis. DESeq normalization adjusts RNA-seq data for variations in sequencing depth and library size, enhancing the precision of differential expression analysis.

For proteomic data, normalization involves standardizing protein abundance measurements to mitigate technical variability and ensure accurate comparisons and reliable biological interpretations. The most common normalization methods for proteomics are total ion current (TIC) normalization, median normalization, and spike-in normalization (Välikangas et al. 2018). TIC normalization scales protein abundance measurements according to the overall intensity of ions detected in a sample, ensuring consistent comparisons across different experimental conditions. Median normalization scales protein abundance measurements by the median intensity observed across samples, improving consistency in comparative analyses of protein expression levels. Spike-in normalization in proteomics involves introducing standardized amounts of reference proteins (spike-ins) into samples before analysis, ensuring precise quantification and facilitating consistent comparisons of protein abundance across different experimental settings.

In metabolomic data, normalization refers to the process of standardizing metabolomic measurements to mitigate technical biases, thereby facilitating valid comparisons between samples and improving the consistency of metabolomic analyses. Two common normalization techniques for metabolomic data are normalization to internal standards and probabilistic quotient normalization (PQN) (Gotsmy et al. 2022). Normalization to internal standards in metabolomics involves using predefined reference compounds that are consistently added to samples, while PQN standardizes metabolomic measurements using the distribution of internal standards, which helps account for differences in sample size and composition. Both methods enhance the reliability of comparative analyses.

Scaling omics data involves adjusting measurements to a uniform scale or distribution, promoting equitable comparisons and enhancing the interpretability of diverse biological datasets (Walach et al. 2018). Below are the scaling techniques for different types of omics data. In genomics, Z-score normalization is used to standardize single nucleotide polymorphism (SNP) data, ensuring consistent comparison of variations, while log transformation is applied to sequencing read counts to address data skewness and improve analytical precision. The log2 transformation is used to normalize RNA-seq read counts, aiding in variance stabilization and data interpretability. Quantile normalization is employed in transcriptomics to standardize microarray intensities across samples, minimizing systematic biases and ensuring reliable comparative analysis. In proteomics, total ion current normalization is used to adjust mass spectrometry proteomic data for variations in sample loading and instrument sensitivity. Z-score normalization is then applied to standardize the data distribution, enabling reliable comparisons and precise quantitative analysis. Finally, in metabolomic data, log transformation is used to enhance data distribution and stabilize variance.

A variety of software tools and libraries are available for normalization and scaling of omics data. In R (R Core Team 2024), widely used packages include `limma`, `DESeq2`, `edgeR`, and `preprocessCore`, which offer extensive functionality for handling high-throughput data. Python (Python Software Foundation 2024) provides similar capabilities through libraries such as `scikit-learn`, `statsmodels`, and `pandas`, facilitating robust normalization and scaling for diverse omics datasets.

10.2.3 Batch Effect Correction

Successfully integrating and analyzing high-throughput data, particularly gene expression and proteomic profiling data, holds the potential to uncover groundbreaking clinical insights and therapeutic possibilities. However, technical variability, often referred to as batch effects, stemming from differences in experimental timing, personnel, or reagent batches, presents significant challenges. Batch effects in omics data result from inconsistencies introduced at various stages of the experimental workflow and can severely compromise the validity of study outcomes (Goh et al. 2017). These variations, when correlated with batch-specific factors, can distort analyses, leading to a high prevalence of false positives or false negatives and, ultimately, to misleading conclusions. In the era of high-dimensional and multi-omics data, the issue of batch effects has become increasingly prominent.

Effectively addressing batch effects in omics data necessitates a combination of robust correction algorithms, standardized experimental protocols, and advanced statistical and computational techniques to ensure data consistency and reliability. A variety of correction algorithms have been developed to mitigate batch effects in omics datasets, including ComBat (Zhang et al. 2020), surrogate variable analysis (SVA) (Leek et al. 2012), and empirical Bayes methods (Johnson et al. 2007). These methods are designed to standardize datasets and minimize systematic biases arising from technical variability.

Advanced statistical and computational techniques are also widely employed to manage batch effects. Techniques such as principal component analysis (PCA) are used to reduce dimensionality and identify sources of variation (Yu et al. 2023). Mixed-effects models account for batch-related variability by modeling it as a random effect, while machine learning approaches, including neural networks, are increasingly applied to accurately detect and correct batch-related biases in large datasets.

Several software packages are available for addressing batch effects, offering tools for both statistical and machine learning-based approaches. Software R (R Core Team 2024) provides several widely used packages for batch effect correction. These include: `ComBat` for harmonizing data between batches, `SVA` for removing batch-related variability, and `limma` for adjusting batch effects in high-throughput genomic analyses. Python (Python Software Foundation 2024) also offers powerful libraries for addressing batch effects, including `scikit-learn` for machine learning approaches, `limma`, and `combat-py`.

10.2.4 Exploratory Data Analysis

Exploratory analysis of omics data involves an initial examination aimed at uncovering patterns, trends, and relationships within large-scale biological datasets, serving as a foundation for hypothesis generation and data-driven insights.

Before conducting data analysis, it is essential to assess various aspects related to the nature of the data. Exploratory analysis should be performed only after addressing all the considerations outlined in the preceding sections. To ensure a comprehensive understanding and meaningful interpretation of the dataset, exploratory analysis should begin with an examination of the dataset's dimensions and the nature of its features. Descriptive statistics play a pivotal role in summarizing and analyzing the distribution, variability, and central tendencies of biological measurements, providing crucial insights for data exploration and interpretation.

Data visualization is a fundamental component of exploratory analysis, as it provides clear graphical insights into complex biological datasets, revealing patterns and relationships for further investigation. Numerous graphical representations are available, but here we highlight those most commonly employed in omics studies. Distribution plots, such as histograms and density plots, are invaluable tools in exploratory analysis, providing a visual depiction of the distribution and characteristics of omics data. To compare the distribution of features across different groups, boxplots are a suitable option. This graphical representation is also effective in identifying outlying observations. Scatter plots are particularly useful for visualizing correlations and patterns between features, facilitating the identification of relationships and potential outliers within biological datasets. Heatmaps are often favored by biologists as effective visual tools for presenting high-dimensional data derived from high-throughput omics experiments. Figure 10.1 illustrates a heatmap with an accompanying dendrogram, a graphical representation produced through hierarchical clustering analysis. The interpretation of a heatmap with a dendrogram entails examining the clustered patterns within the heatmap, which reveal hierarchical relationships and similarities among features or observations in the omics dataset. The color scheme reflects the relative measurement values, visually highlighting differences and patterns across features and observations within the dataset.

Another widely used multivariate technique is principal component analysis (PCA) (Jolliffe 2002), which serves as an effective exploratory tool for omics data. PCA simplifies complex datasets by reducing their dimensions, thereby uncovering significant patterns, trends, and anomalies. More details regarding this multivariate technique can be found in Sect. 10.3.

For high-dimensional datasets, the Uniform Manifold Approximation and Projection (UMAP), proposed by McInnes et al. (2018), is another powerful visualization technique. UMAP is instrumental in revealing complex biological relationships by transforming data into a lower-dimensional space, making patterns and clusters more apparent and easier to interpret.

In the R (R Core Team 2024) software environment, packages such as `ggplot2`, `EDAseq`, `pheatmap`, and `umap` are widely employed for the exploratory analysis of omics data, offering robust capabilities for visualization, clustering, and dimensionality reduction. Both R and Python provide powerful tools for exploratory analysis of omics data. R includes packages like `ggplot2`, `EDAseq`, `pheatmap`, and `umap`, while Python (Python Software Foundation 2024) offers packages such as `pandas`, `seaborn`, and `umap-learn`, which deliver comprehensive functionality for data visualization, clustering, and dimensionality reduction.

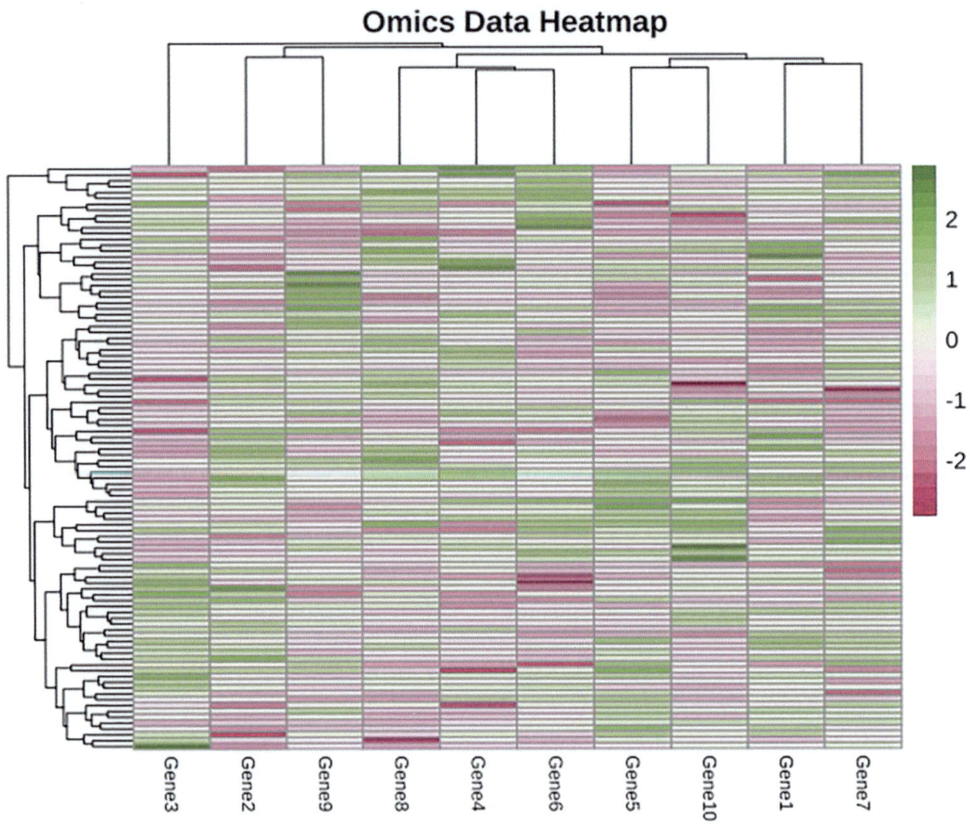

Fig. 10.1 Graphical representation of a heatmap and dendrogram of simulated normalized data considering 100 samples and 10 features (genes)

10.2.5 Outliers

During the exploratory analysis of omics data, particularly in the interpretation of graphical representations, abnormal observations can often be identified. In statistics, these abnormal observations are referred to as *outliers*, which are data points that significantly deviate from the majority of observations in a dataset. In life sciences, outliers are often removed from datasets; however, this approach should be applied with caution. Not all extreme values are errors; some may represent true biological variability. Therefore, it is essential to assess each outlier in context, considering both domain knowledge and statistical criteria. When appropriate, robust statistical methods can be used to mitigate their influence without excluding potentially meaningful observations. In the context of omics data, an outlier might represent, for instance, an unusually high or low expression level of a gene compared to the rest of the dataset. Identifying such outliers is crucial as they can substantially influence overall results and their potential biological significance. The presence of outliers may arise due to various reasons, including measurement errors, data entry mistakes, or inherent variability in the data. Identifying outliers in omics data is vital for uncovering potential errors, detecting rare biological phenomena, and ensuring the reliability of subsequent analyses.

Depending on the type of analysis and the nature of the omics data, various approaches for detecting outliers are described in the literature. Graphical representations are often the primary tool for identifying outliers, as they provide a clear visualization of unusual biological variations or extreme measurements, facilitating their detection and interpretation. Among these, boxplots are the most commonly used visualization tool for identifying abnormal observations.

In statistical modeling, outliers can also be identified based on the residuals of a given model. This involves analyzing the discrepancies between observed data points and their predicted values, thereby identifying data points that significantly deviate from the expected pattern. For example, in the context of survival analysis, Carrasquinha et al. (2018) employed martingale residuals through a rank product test to identify long- or short-term survivors of ovarian cancer, potentially uncovering new prognostic factors. Similarly, Carrasquinha et al. (2018) proposed an ensemble approach to classify triple-negative breast cancer data.

After identifying anomalous observations, researchers can employ various strategies to handle them. Common techniques include data transformations such as logarithmic transformations, square root transformations, or normalization. Robust statistical methods, including the median absolute deviation (MAD) and robust principal component analysis (RPCA), are effective in minimizing the impact of outliers on overall analyses. More recently, machine learning algorithms have gained prominence in outlier detection. Examples include Isolation Forest (Liu et al. 2008), local outlier factor (LOF) (Breunig et al. 2000), and autoencoders (Chen et al. 2017).

Most statistical software platforms offer tools for outlier detection. In R (R Core Team 2024), the popular visualization package `ggplot2` provides several graphical options for identifying outlying observations. For more advanced analyses, the `ROSIE` package, introduced by Jensch et al. (2022), implements an ensemble classification approach for outlier detection and gene selection in cancer omics data. Recently, a software program named `EnsMOD`, available in github, was proposed by Manes et al. (2023) for outlier detection in omics data. `EnsMOD` includes a stand-alone Rmarkdown implementation and an R Shiny application with a user-friendly graphical interface. Similarly, Python (Python Software Foundation 2024) offers packages such as `scikit-learn`, `pyod`, and `statsmodels`, which provide a wide range of tools and algorithms for efficient outlier detection.

10.2.6 Data Splitting

In the context of omics data analysis, data splitting is a critical step that involves partitioning the dataset into training, validation, and test sets to ensure robust model development and evaluation. The importance of data splitting has increased in recent years, as it is essential for evaluating the applicability of statistical models, validating research findings, and minimizing biases, thereby promoting trustworthy and reproducible research outcomes (Badillo et al. 2020). It is important to note that data splitting strategies are typically applicable in large-scale omics studies, such as those involving hundreds of samples from consortia like TCGA. In small-scale experiments with very limited replicates, such as three to five samples per condition, data splitting is generally not feasible. Nevertheless, understanding these strategies remains important as omics datasets grow in size and researchers increasingly integrate public or multi-study data sources.

The following is a comprehensive guide outlining how to effectively perform data splitting. Before partitioning the dataset, it is necessary to shuffle the data points. Randomization, in this context, refers to reordering the dataset entries in a randomized sequence, ensuring that the dataset receives an unbiased distribution of data points.

After completing the preprocessing steps, including randomization, the omics dataset is divided into three subsets: the training set, the validation set, and the test set. The training set, which constitutes the largest portion of the dataset, is used to teach the statistical model, allowing it to identify patterns and relationships that enable accurate predictions or classifications. The validation set, a smaller portion of the dataset, is used to refine and optimize the model's performance, ensuring its robustness and applicability before the final evaluation. Finally, the test set is reserved for assessing the model's ability to generalize to new, unseen data after the training and validation phases.

Following the initial data splitting process, cross-validation (Arlot and Celisse 2010) is employed to rigorously evaluate the model's performance across various subsets, ensuring the consistency and accuracy of predictive analyses.

Thus, the data splitting procedure plays a vital role in promoting reproducibility in omics research, as it establishes standardized subsets for training, validation, and testing, ensuring consistent and reliable results across different analyses.

In R (R Core Team 2024), data splitting can be performed using packages such as `caret`, `rsample`, or `tidymodels`. In Python (Python Software Foundation 2024), data splitting can be implemented using libraries like `scikit-learn`, `TensorFlow`, or `Keras`. These packages provide functions for partitioning datasets and performing cross-validation, which are essential for effective model development and evaluation.

10.3 Dimensionality Reduction

Rapid advances in technologies have led to the generation of high-dimensional omics data, such as genomics, transcriptomics, and metabolomics, in many biomedical studies (Cavill et al. 2016). At the same time, these advances present many analytical challenges.

Modeling omics data constitutes a complex statistical problem when inferring models, as the number of covariates greatly outnumbers the observations. Bühlmann and Van De Geer (2011) present several approaches to tackle this $N << p$ challenge. Over the past few years, machine learning and deep learning algorithms have demonstrated their ability to address this issue by simplifying data representation while retaining the most important information. In this context, machine learning and deep learning techniques provide powerful methods for dimensionality reduction in omics data analysis, enabling researchers to efficiently extract meaningful information from large-scale molecular datasets. By reducing dimensionality while preserving essential features, these techniques facilitate visualization, interpretation, and analysis of omics data, leading to an improved understanding of biological processes and enhanced capabilities for biomedical research and clinical applications.

Next, some methodologies used in machine learning and deep learning are presented.

10.3.1 Machine Learning Algorithms

Machine learning algorithms can be divided into two groups: supervised and unsupervised learning. Supervised learning relies on labeled data, while unsupervised learning seeks to uncover patterns or structures within unlabeled data. The choice between these approaches depends on the nature of the problem, the availability of labeled data, and the desired outcomes.

The most common algorithms for supervised and unsupervised learning, used to reduce the dimensionality of the data, are presented next.

10.3.1.1 Supervised Learning

Regularization techniques are commonly used for dimensionality reduction tasks. Elastic net, proposed by Zou and Hastie (2005), is a regularization method that combines the lasso and ridge regression penalties. The elastic net penalty provides a high-performance, flexible regularization approach that can be used with any linear model, whether regression, classification, or survival data, and has been widely applied to omics data in many different contexts. Another example is the sparse group Lasso, proposed by Simon et al. (2013), which is an extension of Lasso regularization that adds an additional group penalty to encourage sparsity at the group level.

While these methods lead to more interpretable models, they do not fully leverage the relationships between features, especially when these relationships can be represented through graphs (Veríssimo et al. 2016). Cell regulatory networks are highly interconnected, meaning that any gene expressed in disease-relevant cells can affect the regulation or function of core disease-related genes. This fosters the need for new research directions that target not only differentially expressed genes but also those that, although non-differentially expressed, play a role in the molecular mechanisms of complex biological phenomena. Proposed alternatives, such as HubCox and OrphanCox, proposed by Veríssimo et al. (2018), incorporate genomic network data into the elastic net penalization term, improving the interpretability of the models.

When dealing with a classification problem, linear discriminant analysis (LDA), presented by Fisher (1936), could be a solution. LDA aims to find the linear combinations of features that best separate different classes in the data. It is a very useful classification tool, as it maximizes the between-class scatter while minimizing the within-class scatter. The data is projected into a lower-dimensional subspace spanned by the discriminative directions. For normally distributed data with well-separated classes, the performance of the model is very high. Another option is using canonical correlation analysis (CCA) (Hotelling 1936), which is a multivariate statistical technique that finds linear combinations of two sets of features that are maximally correlated. It considers features and class labels as two sets of variables. The data is projected into a lower-dimensional subspace where the correlations between features and class labels are maximized.

In multivariate statistics, another option to reduce the dimensionality of the data is to use supervised principal component analysis (supervised PCA), proposed by Bair and Hastie (2003), which is an extension of the well-known multivariate statistical exploratory technique PCA by incorporating class information into the dimensionality reduction process. The goal is to maximize the variance of the projected data while also maximizing the separation between different classes. It is an optimization problem, where the objective is to find a transformation matrix that optimally separates the classes while minimizing reconstruction error.

The last approach presented is regarding supervised kernel methods. Kernel methods can be applied to linearize nonlinear decision boundaries in high-dimensional space. Kernel discriminant analysis (KDA) (Mika et al. 1999) and kernel PCA (KPCA) (Schölkopf et al. 1998) are examples of supervised kernel methods. The class labels are incorporated into the kernel space to learn a discriminative low-dimensional representation. These methods are particularly useful when the data is not linearly separable.

Several R (R Core Team 2024) libraries are available for implementing the supervised methods presented above. For elastic net regularization, examples include caret and glmnet. For network-based regularization, glmSparseNet is a good option. For LDA, the penalized library can be used. Regarding CCA, the PMA library is available. Supervised PCA can be performed using the superpc library. For kernel methods, KDA does not have a specific R library dedicated solely to this method, but it can be implemented using existing libraries that support kernel methods and discriminant analysis, such as kknn. For KPCA, the best option is the kernlab library. In Python (Python Software Foundation 2024), the well-known machine learning library scikit-learn can be used for most of the methods described above. One exception is supervised PCA, for which there is no dedicated library in Python. KDA can be implemented using the kernel trick in combination with LDA, in the milk library.

10.3.1.2 Unsupervised Learning

Unsupervised learning is another type of machine learning algorithm where a model searches for patterns in a dataset without labels and with minimal human supervision. In this context, the training data consists of an unlabeled dataset, and the goal is to learn the inherent structure from the input data. There are several unsupervised learning techniques in the literature. Below, the most commonly used techniques in the context of omics data are presented.

Cluster analysis, introduced by Breiman et al. (1984), is probably the most well-known unsupervised learning technique used across all scientific fields, particularly in the context of omics data. Cluster analysis is an exploratory multivariate technique that aims to group objects or features together according to their similarity, thus reducing the row space. There are various methods for grouping objects/features, and many clustering algorithms are available, such as k-means (McQueen 1967), hierarchical clustering (Rencher 2002), or biclustering (Madeira and Oliveira 2004).

Among all the cluster analysis algorithms, the most commonly used in omics data is the k-means algorithm, a non-hierarchical clustering method. In the first step, an initial partition of n objects into k groups is selected, and then each object's displacement from its own group to each of the others is evaluated, recording the distance of each object to the centroid of the new group. The objects are moved in a way that each object is placed in the group with the closest centroid. The centroids of the newly formed groups are recalculated, and the process is repeated until no further moves are possible.

Another multivariate statistical technique used for data reduction in omics data is principal component analysis (PCA), with details of this approach available in Jolliffe (2002). The idea of PCA is to replace the p original features with a smaller number, q, of derived features, the principal components, which are linear combinations of the original features. This facilitates visualization and interpretation of the data. Often, it is possible to retain most of the variability in the original features, with q much smaller than p. Despite its apparent simplicity, PCA has several important features and many uses and extensions. However, PCA does not account for the nonlinear structure of the data, which is common in omics contexts. One possible way to overcome this problem is t-distributed stochastic neighbor embedding (t-SNE), proposed by Cieslak et al. (2020). The t-SNE approach is complex mathematically, but the idea is simple. Like PCA, t-SNE maps high-dimensional data to a lower-dimensional space while preserving the neighborhood of each point. By minimizing the Kullback–Leibler divergence (KL divergence) between the two distributions with respect to the locations of the points on the map, it tries to preserve the local structure of the data. t-SNE can be used for different purposes, but it is primarily used for cluster visualization and identifying potential outliers. The main drawbacks are its computational expense, sensitivity to hyperparameters, and lack of generalizability.

These techniques can be applied individually or in combination, depending on the specific characteristics of the omics dataset and the objectives of the analysis. It is important to assess the performance and interpretability of each technique in the context of the biological questions being addressed.

In R (R Core Team 2024), several libraries offer functionality for cluster analysis, depending on the problem and the characteristics of the data. The most popular ones are `cluster`, `MClust`, and `fclust`. Both k-means and PCA are available in the base R package `stats`. t-SNE can be performed using the `Rtsne` package. Regarding Python (Python Software Foundation 2024), there are also several packages available for performing cluster analysis. The most commonly used packages are `scikit-learn` and `KMeans`, a simple library specifically for k-means clustering. Both PCA and t-SNE are available in the `scikit-learn` package.

10.3.2 Deep Learning

Deep learning techniques have emerged as powerful tools for dimensionality reduction in omics data analysis (Min et al. 2017). By leveraging neural network architectures with multiple layers of abstraction, deep learning models can automatically learn hierarchical representations of complex omics datasets, capturing intricate patterns and relationships between molecular features.

The most popular approach is the use of autoencoders, which are neural networks trained to reconstruct input data from a compressed latent space (Hinton and Salakhutdinov 2006). Autoencoders, in the context of omics data, can effectively reduce high-dimensional feature spaces while preserving relevant biological information. This approach also facilitates visualization, interpretation, and downstream analysis tasks such as clustering, classification, and anomaly detection (Tan et al. 2015).

Additionally, deep learning techniques such as convolutional neural networks (CNNs) (Zeng et al. 2016) and recurrent neural networks (RNNs) (Nguyen et al. 2019) can be adapted to analyze omics data in its raw format, such as DNA sequences or gene expression time series. These models can automatically extract hierarchical features from raw data, enabling comprehensive analysis and interpretation of complex biological phenomena. Overall, deep learning techniques offer promising avenues for data reduction in omics data analysis.

In R (R Core Team 2024), packages like `Keras`, `TensorFlow`, and `deepnet` provide powerful capabilities for utilizing deep learning methods to achieve dimensionality reduction in omics data. Additionally, Python (Python Software Foundation 2024) offers versatile tools for applying deep learning techniques, such as `TensorFlow`, `Keras`, and `PyTorch`.

10.4 Statistics and Data Complexity

Omics fields generate large, multidimensional datasets that require sophisticated statistical techniques to analyze. However, many researchers face challenges in understanding and implementing appropriate statistical approaches, leading to common mistakes that can compromise the integrity of their findings (Rahnenführer et al. 2023).

The multifaceted nature of high-dimensional biomedical data research emphasizes the importance of the collaboration between data and computational scientists, including statisticians, computational biologists, bioinformaticians, and computer scientists, along with biomedical scientists such as clinicians and biologists, which is deemed essential for the effective generation, management, processing, analysis, and interpretation of high-dimensional biomedical data. The versatility of data, which can be sourced from multiple origins and represented in diverse formats, is a continuous challenge for every intervenient but also contributes to the advances in statistical methodology and machine learning to enhance data mining, statistical inference, and prediction within the context of high-dimensional data (HDD) settings (Tarazona et al. 2021). Different disciplines may have varied vocabularies and basic concepts, necessitating researchers to understand and communicate across disciplines. This requires patience and practice but becomes easier with experience. Establishing strong working relationships across disciplines is vital for effective scientific development.

This section discusses three critical aspects: the challenges associated with omics data, common mistakes in omics statistical analysis, and the essential statistical principles that researchers should adhere to when working with omics data.

10.4.1 Challenges in Omics Data Analysis

Omics data is characterized by its high dimensionality, referring to the large number of variables (e.g., genes, proteins, metabolites) measured relative to the relatively small number of samples. For instance, a typical transcriptomics experiment may measure the expression levels of 20,000 genes across 12 samples. This disproportionate ratio poses several challenges, notably the *curse of dimensionality*. This term describes the difficulties inherent in analyzing datasets with many variables but relatively few observations, which include the following: (i) when there are more features (e.g., genes or proteins) than samples, statistical models can easily overfit, capturing noise instead of meaningful patterns; (ii) omics studies often involve thousands of simultaneous statistical tests (e.g., identifying differentially expressed genes), leading to an increased likelihood of false positives. Correcting for multiple comparisons is critical but adds complexity to the analysis. Biological systems involve intricate networks of interactions, with multiple layers of regulation, and this complexity is further compounded by technical variability (e.g., differences in sample preparation, batch effects) and biological variability (e.g., genetic diversity, environmental factors, disease states), leading to data that are inherently heterogeneous. Understanding how to disentangle true biological signals from noise is a significant challenge. Missing data and sparsity are common issues in omics datasets. Some data points may be missing due to technical failures or limitations in detection sensitivity. In some omics technologies, such as single-cell RNA sequencing, sparsity (i.e., many zeros in the data) can occur due to the low capture efficiency of transcripts. Missing or sparse data can distort statistical analyses if not properly accounted for. To gain a holistic view of biological processes, researchers increasingly integrate multiple layers of omics data (e.g., combining genomics, transcriptomics, and proteomics). However, integrating data from different platforms introduces additional challenges. Each type of omics data has its own characteristics, measurement scales, and noise profiles. Statistically modeling the relationships between these layers while accounting for their distinct features is a daunting task (Rahnenführer et al. 2023).

10.4.2 Common Mistakes in Omics Data Analysis

Given the complexity of omics data, it is not uncommon for researchers to make mistakes during statistical analysis. These mistakes can undermine the validity of the study's findings, leading to false conclusions or missed opportunities for discovery. Neglecting proper data normalization is a common and critical mistake in omics data analysis. Normalization adjusts for technical biases, such as differences in sequencing depth, signal intensity, or batch effects, and ensures comparability across samples, thereby preserving the integrity of downstream analyses. Failure to apply appropriate normalization techniques can result in misleading findings (see Sect. 10.2.2). Many researchers rely heavily on *p*-values as a measure of significance, and in high-dimensional data, *p*-values can be misleading due to multiple testing issues. Without proper adjustments, such as controlling for the false discovery rate (FDR) (see Sect. 10.4.3), researchers may report false positives—results that appear significant but are actually due to random chance. Over-reliance on *p*-values also risks ignoring biologically relevant findings that may not meet stringent thresholds but still hold biological significance. Inadequate handling of batch effects (see Sect. 10.2.2), which arise from differences in experimental conditions (e.g., different days of sample processing or different reagent batches), can introduce substantial variability into omics data. Ignoring batch effects can lead to biased results. Failing to properly account for factors such as age, sex, or genetic background can lead to confounded results. Omics studies often pool together data from diverse populations, and ignoring these biological covariates can result in associations driven by population stratification or other confounding factors. Predictive modeling, such as machine learning, is commonly applied to omics data for tasks like disease classification or biomarker discovery. However, when there are far more features than samples, predictive models can easily overfit the training data which occurs when a model becomes too complex and captures noise rather than generalizable patterns. As a result, the model performs well on training data but poorly on unseen test data, where proper model validation using techniques like cross-validation is essential to avoid this pitfall.

10.4.3 Statistical Analysis Principles in Omics Data

To address the challenges outlined and avoid common pitfalls, researchers must adhere to fundamental statistical principles when analyzing omics data. These principles provide a foundation for conducting robust and reproducible research.

The high-throughput nature of omics data introduces a critical challenge: the proliferation of multiple hypothesis tests. Without appropriate adjustments, the risk of false positives, or Type I errors, increases exponentially. A well-known illustration of this issue is the "dead salmon" experiment, in which researchers falsely identified patterns of brain activity in a deceased fish, an outcome stemming from the failure to account for the vast number of comparisons made (Røislien 2023). In transcriptomics, for instance, researchers often test tens of thousands of genes for differential expression. Without correcting for multiple testing, such analyses can result in an unacceptably high false positive rate, leading to spurious findings. These false discoveries can misdirect research efforts, resulting in the pursuit of biological pathways that do not exist, wasting valuable resources, and undermining the credibility of scientific conclusions.

To mitigate this issue, two principal correction methods are commonly employed: the family-wise error rate (FWER) and the false discovery rate (FDR). FWER control ensures that the probability of making at least one false positive remains below a specified threshold, while FDR control focuses on limiting the proportion of false positives among all rejected null hypotheses. The FDR approach, exemplified by the Benjamini-Hochberg procedure (Benjamini and Hochberg 1995), is particularly suited for omics data due to the large number of simultaneous tests involved. By focusing on the proportion of false positives rather than eliminating them entirely, the FDR method strikes a balance between discovery and statistical rigor, making it ideal for exploratory, high-dimensional datasets characteristic of omics research.

Several tools are available for implementing these corrections. Software R (R Core Team 2024) provides the p.adjust function, while specialized R packages such as qvalue and multtest offer additional capabilities. Integrated workflows for differential expression analysis, such as those provided by the `limma` package (designed for microarray data) and `edgeR` or `DESeq2` (optimized for next-generation sequencing data), also incorporate multiple testing corrections. For a detailed exploration of multiple testing procedures and their specific applications in omics data analysis, refer to the foundational work of Storey (2002).

In addition to p-value adjustments, some methods provide corrected confidence intervals to account for multiple testing, offering a more comprehensive interpretation of effect sizes under conditions of multiplicity. For hierarchical models, adjustments can be applied at various levels to address dependencies between tests, thereby improving the control of false positives. Conditional hypothesis testing approaches, which incorporate covariates or prior knowledge, are also gaining popularity. These methods enable more tailored corrections in complex experimental designs, such as those frequently encountered in omics studies.

Power analysis represents another critical yet often overlooked aspect of omics research. It ensures that a study is adequately powered to detect meaningful effects, allowing researchers to determine the probability of identifying true effects given a specific sample size and significance level. In the context of omics data, power analysis is crucial for ensuring that studies are equipped to detect significant biological insights within complex, high-dimensional datasets. This increases the reliability of research outcomes and their potential impact on advancing biological understanding. Strategies such as focusing on a subset of hypotheses or combining related hypotheses can enhance statistical power without unnecessarily increasing the complexity of testing procedures.

Normalization is another essential step in omics data analysis, ensuring that technical biases do not overshadow biological signals. Neglecting proper normalization or transformation can lead to biased parameter estimates and flawed inferences (see Sect. 10.2.2).

Finally, external variables, known as confounders, pose a significant risk to the validity of omics studies. Confounders, such as batch effects (see Sect. 10.2.3), demographic variables, or environmental factors, can influence both the genetic exposure and the outcome of interest, leading to spurious associations or incorrect interpretations. Statistical models must adequately account for these known confounders to ensure that observed relationships between genetic variants and outcomes, such as diseases or phenotypic traits, reflect true biological associations rather than artifacts of confounding influences.

10.4.4 Summary

The interpretation of omics data demands meticulous attention to both statistical principles and the inherent complexity of the datasets. Researchers face significant challenges, including high dimensionality, multiple testing, and confounding factors, while also needing to avoid common pitfalls such as inadequate normalization or over-reliance on *p*-values. By adhering to best practices in statistical analysis, such as applying corrections for multiple testing, ensuring proper data normalization, and appropriately addressing confounding variables, researchers can produce findings that are robust, reproducible, and biologically meaningful. The following section delves deeper into these challenges and addresses several methods designed to address them effectively.

10.5 Interpretation of Results

As the field of omics expands and becomes increasingly complex, the need for accurate, reliable, and biologically meaningful interpretations has never been greater. The interpretation of results in the context of omics data has multiple aspects to consider, as the entire process from preprocessing to knowledge translation involves several stages, each dependent on the previous and on various factors. We will focus on three significant challenges in the interpretation of omics data: balancing statistical and biological significance, managing the complexity of multiple methods and tools that yield different results, and dealing with confounding factors that can obscure the true biological signal.

10.5.1 Balancing Statistical and Biological Significance

Interpreting statistical results from omics data analyses requires careful consideration of the underlying biological context, experimental design, and statistical methods employed. Data analysis can be conducted through numerical results and graphical results. Many graph visualization methods exist and have their own algorithm or style for visualization (Silva-Fortes et al. 2012; Silva-Fortes et al. 2014). The interpretation of a graph, although it may be an authentic work of art, cannot have the same subjectivity that a painting can bring. The challenge lies in creating visualizations that are clear, meaningful, and well-integrated, which can convey biological insights without being drowned out by the complexity of the data (Mougin et al. 2018). Some common mistakes in the interpretation of omics statistical results include interpreting statistical findings without considering the biological context, which can lead to misinterpretation. Small but biologically relevant changes may not always be statistically significant in large-scale omics datasets.

10.5.2 Complexity of Multiple Methods and Tools

As seen in previous sections, statistical significance is often determined by *p*-values or other measures, such as false discovery rate (FDR), to control for the probability of finding false positives due to multiple testing. While this approach can effectively highlight associations that are unlikely to have occurred by chance, these findings may not necessarily represent biologically important insights. For example, a gene might exhibit a strong statistical association with a disease but may not play a direct or functional role in the underlying biological process. The challenge for researchers is to avoid over-reliance on statistical measures and ensure that findings are biologically validated. This involves integrating existing biological knowledge, such as pathways, gene ontology, or network-based approaches, to evaluate whether the statistically significant findings are aligned with known biology. Biological experiments, such as functional assays or follow-up validation studies, are crucial in confirming whether the identified associations have real-world relevance. Another common issue is the risk of using overly conservative statistical corrections, such as Bonferroni adjustments, which may exclude potentially biologically relevant signals. On the other hand, liberal thresholds may result in false positives. Researchers must strike a delicate balance, considering both the statistical rigor and the biological implications of their findings. By incorporating prior biological knowledge into the interpretation process, researchers can refine their results and ensure that their findings contribute meaningfully to the understanding of the biological systems in question. It is important to provide a comprehensive overview of all analyzed results, including both significant and nonsignificant findings. Selectively reporting statistically significant findings while ignoring nonsignificant results can lead to biased interpretations.

10.5.3 Dealing with Confounding Factors

Omics research is inherently interdisciplinary, involving a broad array of computational and statistical methods. With the rapid advancement of technologies and tools, researchers now have access to a wide range of algorithms for analyzing omics data. While this diversity enhances the flexibility and innovation in the field, it also presents a significant challenge, as different tools often yield different results. Consequently, the same dataset analyzed by different methods may produce overlapping yet distinct sets of DEGs. In some cases, applying different analysis pipelines to the same dataset, such as using DESeq2 versus limma for RNA-seq differential expression, can lead to noticeably different sets of results. This methodological variability adds complexity to the interpretation and reproducibility of omics findings. This variability in results presents a considerable challenge for researchers. How should they interpret results when different methods point to different conclusions? One approach is to conduct a comparative analysis across several methods to identify robust findings that are consistent regardless of the approach used. However, this can be computationally expensive and requires expertise in multiple bioinformatics tools. Benchmarking studies, where the performance of different tools is systematically compared, can also help researchers identify the most appropriate method for their specific dataset (Chen et al. 2023). Moreover, the development of consensus or ensemble methods, which combine results from multiple algorithms, is emerging as a promising approach to reconcile differences and improve the reliability of omics analyses. In addition, it is important to remember that different tools are often optimized for specific types of data or experimental designs. As such, the choice of method should be guided by the characteristics of the data, including sample size, technical variability, and the biological question of interest. When possible, researchers should conduct sensitivity analyses to test the robustness of their findings to potential confounding effects. These efforts, combined with rigorous statistical techniques, can help to ensure that omics results reflect true biological signals rather than confounding artifacts.

10.6 Final Remarks

Distinguishing between statistical and biological significance remains a central challenge in omics data analysis, compounded by the potential for different methods and tools to produce inconsistent results. Confounding factors further complicate analyses, often obscuring genuine biological relationships within the data. Addressing these issues requires thoughtful study design, robust statistical methodologies, and rigorous biological validation to enhance the interpretability of omics data and foster meaningful insights into complex biological systems.

Failure to provide comprehensive documentation of statistical methods, data preprocessing steps, and interpretation criteria undermines reproducibility and transparency. Clear and transparent reporting of analytical workflows and interpretation frameworks is paramount to ensuring the reliability and integrity of omics data analyses.

Despite these challenges, the field of omics analysis holds tremendous promise. The obstacles faced today offer opportunities to drive significant advancements and contribute to transformative discoveries in the exploration of biological systems.

References

Abbas-Aghababazadeh F, Li Q, Fridley BL (2018) Comparison of normalization approaches for gene expression studies completed with high-throughput sequencing. PLoS One 13(10). https://doi.org/10.1371/journal.pone.0206312

Argelaguet R, Velten B, Arnol D, Dietrich S, Zenz T, Marioni JC, Buettner F, Huber W, Stegle O (2018) Multi-Omics Factor Analysis—a framework for unsupervised integration of multi-omics data sets. Mol Syst Biol 14:e8124. https://doi.org/10.15252/msb.20178124

Arlot S, Celisse A (2010) A survey of cross-validation procedures for model selection. Stat Surveys 4:40–79. https://doi.org/10.1214/09-SS054

Ashburner M, Ball C, Blake J et al (2000) Gene ontology: tool for the unification of biology. Nat Genet 25:25–29. https://doi.org/10.1038/75556

Badillo S, Banfai B, Birzele F, Davydov II, Hutchinson L, Kam-Thong T, Siebourg-Polster J, Steiert B, Zhang JD (2020) An introduction to machine learning. Clin Pharmacol Ther 107(4):871–885. https://doi.org/10.1002/cpt.1796

Baggerly KA, Coombes KR (2009) Deriving chemosensitivity from cell lines: forensic bioinformatics and reproducible research in high-throughput biology. Ann Appl Stat 3(4):1309–1334. https://doi.org/10.1214/09-AOAS291

Bair E, Hastie T (2003) Semi-supervised principal component analysis. J R Stat Soc 65(3):565–586. https://doi.org/10.1198/016214505000000628

Batista GEAPA, Monard MC (2003) An analysis of four missing data treatment methods for supervised learning. Appl Artif Intell 17:519–533. https://doi.org/10.1080/713827181

Benjamini Y, Hochberg Y (1995) Controlling the false discovery rate: a practical and powerful approach to multiple testing. J R Stat Soc Ser B 57(1):289–300. https://doi.org/10.1111/j.2517-6161.1995.tb02031.x

Brazma A, Hingamp P, Quackenbush J, Sherlock G, Spellman P, Stoeckert C, Aach J, Ansorge W, Ball CA, Causton HC, Gaasterland T, Glenisson P, Holstege FC, Kim IF, Markowitz V, Matese JC, Parkinson H, Robinson A, Sarkans U, Schulze-Kremer S, Stewart J, Taylor R, Vilo J, Vingron M (2001) Minimum information about a microarray experiment (MIAME)-toward standards for microarray data. Nat Genet 29(4):365–371. https://doi.org/10.1038/ng1201-365

Breiman L, Friedman J, Stone CJ, Olshen RA (1984) Classification and regression trees. CRC Press

Breunig MM, Kriegel HP, Ng RT, Sander J (2000) LOF: identifying density-based local outliers. In: Proceedings of the 2000 ACM SIGMOD International Conference on Management of Data, pp 93–104. https://doi.org/10.1145/342009.335388

Bühlmann P, Van De Geer S (2011) Statistics for high-dimensional data: methods, theory and applications. Springer Berlin Heidelberg. https://doi.org/10.1007/978-3-642-20192-9

Carrasquinha E, Veríssimo A, Lopes MB, Vinga S (2018) Identification of influential observations in high-dimensional cancer survival data through the rank product test. BioData Min 11(1). https://doi.org/10.1186/s13040-018-0162-z

Cavill R, Jennen D, Kleinjans J, Briedé JJ (2016) Transcriptomic and metabolomic data integration. Brief Bioinform 17(5):891–901. https://doi.org/10.1093/bib/bbv090

Chen J, Sathe S, Aggarwal C, Turaga D (2017) Outlier detection with autoencoder ensembles. In: Proceedings of the 2017 SIAM International Conference on Data Mining (SDM), pp 90–98. https://doi.org/10.1137/1.9781611974973.11

Chen C, Wang J, Pan D, Wang X, Xu Y, Yan J, Wang L, Yang X, Yang M, Liu GP (2023) Applications of multi-omics analysis in human diseases. MedComm 4(4):e315. https://doi.org/10.1002/mco2.315

Cieslak MC, Castelfranco AM, Roncalli V, Lenz PH, Hartline DK (2020) t-Distributed Stochastic Neighbor Embedding (t-SNE): a tool for eco-physiological transcriptomic analysis. Mar Genomics 51. https://doi.org/10.1016/j.margen.2019.100723

Deutsch EW, Vizcaíno JA, Jones AR, Binz PA, Lam H, Klein J, Bittremieux W, Perez-Riverol Y, Tabb DL, Walzer M, Ricard-Blum S, Hermjakob H, Neumann S, Mak TD, Kawano S, Mendoza L, Van Den Bossche T, Gabriels R, Bandeira N, Carver J, Pullman B, Sun Z, Hoffmann N, Shofstahl J, Zhu Y, Licata L, Quaglia F, Tosatto SCE, Orchard SE (2023) Proteomics standards initiative at twenty years: current activities and future work. J Proteome Res 22(2):287–301. https://doi.org/10.1021/acs.jproteome.2c00637

ENCODE Project Consortium (2017) ENCODE guidelines for experiments generating RNA-seq data. https://www.encodeproject.org/documents/

European Commission (2009) Directive 2009/41/EC on the contained use of genetically modified micro-organisms

Fisher RA (1936) The use of multiple measurements in taxonomic problems. Ann Eugenics 7(2):179–188

Goh WWB, Wang W, Wong L (2017) Why batch effects matter in omics data, and how to avoid them. Trends Biotechnol 35(6):498–507. https://doi.org/10.1016/j.tibtech.2017.02.012

Gotsmy M, Brunmair J, Büschl C, Gerner C, Zanghellini J (2022) Probabilistic quotient's work and pharmacokinetics' contribution: countering size effect in metabolic time series measurements. BMC Bioinform 23:379. https://doi.org/10.1186/s12859-022-04918-1

Hinton GE, Salakhutdinov RR (2006) Reducing the dimensionality of data with neural networks. Science 313(5786):504–507. https://doi.org/10.1126/science.1127647

Hotelling H (1936) Relations between two sets of variates. Biometrika 28(3/4):321–377

Jensch A, Lopes MB, Vinga S, Radde N (2022) ROSIE: RObust sparse ensemble for outlIEr detection and gene selection in cancer omics data. Stat Methods Med Res 31(5):947–958. https://doi.org/10.1177/09622802211072456

Johnson WE, Li C, Rabinovic A (2007) Adjusting batch effects in microarray expression data using empirical Bayes methods. Biostatistics 8(1):118–127. https://doi.org/10.1093/biostatistics/kxj037

Jolliffe IT (2002) Principal component analysis, Springer Series in Statistics, 2nd edn. Springer-Verlag

Kang H (2013) The prevention and handling of the missing data. Korean J Anesthesiol 64(5):402–406. https://doi.org/10.4097/kjae.2013.64.5.402

Kedaigle AJ, Fraenkel E (2018) Discovering altered regulation and signaling through network-based integration of transcriptomic, epigenomic, and proteomic tumor data. Methods Mol Biol 1711:13–26. https://doi.org/10.1007/978-1-4939-7493-1_2

Kim JO, Curry J (1977) The treatment of missing data in multivariate analysis. Sociol Methods Res 6:215–241

Knoppers BM (2014) Framework for responsible sharing of genomic and health-related data. HUGO J 8(1):3. https://doi.org/10.1186/1877-6566-8-3

Leek JT, Johnson WE, Parker HS, Jaffe AE, Storey JD (2012) The sva package for removing batch effects and other unwanted variation in high-throughput experiments. Bioinformatic 28:882–883. https://doi.org/10.1093/bioinformatics/bts034

Little RJA, Rubin DB (1987) Statistical analysis with missing data. John Wiley & Sons, New York

Liu FT, Ting KM, Zhou Z (2008) Isolation forest. In: 2008 Eighth IEEE International Conference on Data Mining. IEEE, pp 413–422. https://doi.org/10.1109/ICDM.2008.17

Madeira SC, Oliveira AL (2004) Biclustering algorithms for biological data analysis: a survey. IEEE ACM Trans Comput Biol Bioinform 1(1):24–45. https://doi.org/10.1109/TCBB.2004.2

Manes NP, Song J, Nita-Lazar A (2023) EnsMOD: a software program for omics sample outlier detection. J Comput Biol 30(6):726–735. https://doi.org/10.1089/cmb.2022.0243

McInnes L, Healy J, Melville J (2018) UMAP: Uniform Manifold Approximation and Projection for dimension reduction. arXiv. https://arxiv.org/abs/1802.03426

McQueen J (1967) Some methods for classification and analysis of multivariate observations. Comput Chem 4: 257–272

Mika S, Rätsch G, Weston J, Schölkopf B, Müller KR (1999) Fisher discriminant analysis with kernels. In: Neural Networks for Signal Processing IX: Proceedings of the 1999 IEEE Signal Processing Society Workshop, pp 41–48

Min S, Lee B, Yoon S (2017) Deep learning in bioinformatics. Brief Bioinform 18(5):851–869. https://doi.org/10.1093/bib/bbw068

Mougin F, Auber D, Bourqui R, Diallo G, Dutour I, Jouhet V, Thiessard F, Thiébaut R, Thébault P (2018) Visualizing omics and clinical data: which challenges for dealing with their variety? Methods 132:3–18. https://doi.org/10.1016/j.ymeth.2017.08.012

Nguyen T, Tran D, Kavuluru R, Nguyen T (2019) A comprehensive analysis of deep learning approaches for biomarker discovery in clinical omics data. IEEE Access 7:143231–143247. https://doi.org/10.1109/ACCESS.2019.2944428

OECD (2016) Considerations for the design and implementation of biosafcty and bioscourity oversight systems. https://www.oecd.org/sti/emerging-tech/biosafety-biosecurity.htm

Python Software Foundation (2024) Python Language Reference, version 3. https://www.python.org

R Core Team (2024) R: a language and environment for statistical computing. R Foundation for Statistical Computing, Vienna. https://www.R-project.org/

Rahnenführer J, De Bin R, Benner A et al (2023) Statistical analysis of high-dimensional biomedical data: a gentle introduction to analytical goals, common approaches and challenges. BMC Med 21:182. https://doi.org/10.1186/s12916-023-02858-y

Rencher AC (2002) Methods of multivariate analysis, 2nd edn. Wiley, New York

Røislien J (2023) Statistics lessons from a salmon. Medicine in Numbers - Tidsskr Nor Legeforen, 143

Rustici G, Williams E, Barzine M, Brazma A, Bumgarner R, Chierici M, Furlanello C, Greger L, Jurman G, Miller M, Ouellette BFF, Quackenbush J, Reich M, Stoeckert CJ Jr, Taylor RC, Trutane SC, Weller J, Wilhelm B, Winegarden N (2021) Transcriptomics data availability and reusability in the transition from microarray to next-generation sequencing. bioRxiv. https://doi.org/10.1101/2020.12.31.425022

Schölkopf B, Smola A, Müller KR (1998) Nonlinear component analysis as a kernel eigenvalue problem. Neural Comput 10(5):1299–1319

Schurch NJ, Schofield P, Gierliński M, Cole C, Sherstnev A, Singh V, Wrobel N, Gharbi K, Simpson GG, Owen-Hughes T, Blaxter M, Barton GJ (2016) How many biological replicates are needed in an RNA-seq experiment and which differential expression tool should you use? RNA 22(6):839–851. https://doi.org/10.1261/rna.053959.115

Shabani M, Borry P (2018) Rules for processing genetic data for research purposes in view of the new EU General Data Protection Regulation. Eur J Hum Genet 26:149–156. https://doi.org/10.1038/s41431-017-0045-7

Silva-Fortes C, Amaral Turkman MA, Sousa L (2012) Arrow plot: a new graphical tool for selecting up and down regulated genes and genes differentially expressed on sample subgroups. BMC Bioinform 13: 147. https://doi.org/10.1186/1471-2105-13-147

Silva-Fortes C, Freitas A, Roque S, Sousa L (2014) Arrow plot and CA maps on microarray prepossessing methods. In: 14th International Conference on Computational Science and Its Applications, Guimaraes, Portugal, pp 229–232. https://doi.org/10.1109/ICCSA.2014.52

Simon N, Friedman J, Hastie T, Tibshirani R (2013) A sparse-group lasso. J Comput Graph Stat 22(2): 231–245. https://doi.org/10.1080/10618600.2012.681250

Storey JD (2002) A direct approach to false discovery rates. J R Stat Soc Ser B 64(3):479–498. https://doi.org/10.1111/1467-9868.00346

Tan J, Ung M, Cheng C, Greene CS (2015) Unsupervised feature construction and knowledge extraction from genome-wide assays of breast cancer with denoising autoencoders. Pac Symp Biocomput 20:132–143. https://doi.org/10.1142/9789814644730_0014

Tarazona, A., Arzalluz-Luque, A. and Conesa, A. (2021). Undosclosed, unmet and neglected challenges in multi-omics studies. Nat Comput Sci, 1: 395–402..1038/s43588–021-00086-z

Välikangas T, Suomi T, Elo LL (2018) A systematic evaluation of normalization methods in quantitative label-free proteomics. Brief Bioinform 19(1):1–11. https://doi.org/10.1093/bib/bbw100

van Buuren S (2018) Flexible imputation of missing data, 2nd edn. CRC Press, Boca Raton

Veríssimo A, Oliveira A, Sagot M-F, Vinga S (2016) DegreeCox—a network-based regularization method for survival analysis. BMC Bioinform 17(16):449. https://doi.org/10.1186/s12859-016-1310-4

Veríssimo A, Carrasquinha E, Lopes MB, Oliveira AL, Sagot M-F, Vinga S (2018) Sparse network-based regularization for the analysis of patientomics high-dimensional survival data. bioRxiv:403402. https://doi.org/10.1101/403402

Walach J, Filzmoser P, Hron K (2018) Chapter seven - Data normalization and scaling: consequences for the analysis in omics sciences. In: Jaumot J, Bedia C, Tauler R (eds) Comprehensive analytical chemistry, vol 82. Elsevier, pp 165–196. https://doi.org/10.1016/bs.coac.2018.06.004

Yu Y, Zhang N, Mai Y et al (2023) Correcting batch effects in large-scale multiomics studies using a reference-material-based ratio method. Genome Biol 24:201. https://doi.org/10.1186/s13059-023-03047-z

Zeng H, Edwards MD, Liu G, Gifford DK (2016) Convolutional neural network architectures for predicting DNA–protein binding. Bioinformatics 32(12):i121–i127. https://doi.org/10.1093/bioinformatics/btw255

Zhang Y, Parmigiani G, Johnson WE (2020) ComBat-seq: batch effect adjustment for RNA-seq count data. NAR Genom Bioinform 2(3). https://doi.org/10.1093/nargab/lqaa078

Zou H, Hastie T (2005) Regularization and variable selection via the elastic net. J R Stat Soc Ser B 67(2): 301–320. https://doi.org/10.1111/j.1467-9868.2005.00503.x

Key Concepts in Transcriptomics Data Analysis in the Era of Next-Gen Sequencing

11

Saumya Kumar

Abstract

With the rise of next-generation sequencing technologies, gaining new insights into the molecular profiles of cells and tissues has become easier than ever. These technologies enable the investigation of the complete transcriptome, encompassing both known and unknown RNA molecules at single-base pair resolution and at much lower costs. Since comprehensive proteome measurements for biological samples remain challenging, transcriptome sequencing offers valuable insights and new avenues for exploring biological samples. These RNA-sequencing (RNA-seq) experiments generate high-throughput data, making them useful for examining unique molecular profiles alongside biological pathways and networks in various tissues and cellular subsets. Moreover, with the increasing ease of conducting single-cell transcriptomics studies, there has been a surge in large datasets, which greatly enhance our understanding of biological systems in human health and disease. In this chapter, fundamental and advanced analysis methods and concepts for both bulk and single-cell transcriptomics are discussed.

Keywords

Transcriptomics · RNA-seq · Bulk RNA-seq · Single-cell RNA-seq · Data analysis

11.1 Next-Generation Sequencing Technologies for Transcriptomics Studies

RNA-seq assays require the generation of sequencing libraries that capture transcripts in the biological sample, including ligation of sequencing adaptors and conversion into cDNA (although direct RNA-seq methods have recently been developed). Various library generation protocols exist to target different types of transcripts. Most commonly, these protocols focus on polyadenylated messenger RNA (mRNA), but protocols examining total RNA in ribosomal RNA-depleted samples or small non-coding RNAs are also frequently used (Kukurba and Montgomery 2015; RNA sequencing methods collection n.d.). The starting material for RNA-seq data is typically total RNA isolated from tissue samples or enriched cell populations. For transcriptomics of very few cells or low-abundance samples, specialized library

S. Kumar (✉)
Centre for Individualised Infection Medicine (CiiM), a joint venture between the Helmholtz Centre for Infection Research (HZI) and Hannover Medical School (MHH), Hannover, Germany

TWINCORE, a joint venture between the Helmholtz-Centre for Infection Research (HZI) and the Hannover Medical School (MHH), Hannover, Germany
e-mail: saumya.dileepkumar@helmholtz-hzi.de

M. Gama-Carvalho et al. (eds.), *Omics Approaches in Biomedicine and Biotechnology*, Advances in Experimental Medicine and Biology 1504, https://doi.org/10.1007/978-3-032-18966-0_11

preparation steps, such as additional PCR amplification, are necessary. The transcriptome data produced by these approaches offer insights at a bulk RNA level. These methods are useful in mapping the common or different modules of gene expression, as well as in defining the cellular processes unique to specific samples. Recent advancements enable efficient isolation of RNA from single cells, facilitating the study of minor cell populations and cellular states that may reveal underlying heterogeneity. Moreover, single-cell approaches provide greater precision in investigating developmental features and identification of transient cellular states within cell populations compared to bulk RNA-seq.

11.2 Bulk Sequencing Versus Single-Cell Transcriptomics

Transcriptome studies from both bulk and single-cell techniques have their advantages and disadvantages. Among the various considerations required for a sequencing experiment (Nguyen et al. 2018), three major issues, namely choice of sequencing methods, the sample size, and sequencing read depth, are important and also linked with the high experimental costs.

The choice of sequencing methods depends mainly on the biological question of interest. In bulk RNA-seq, RNA quantified from a population of homogenous cells or tissue provides an average read-out of the sample. Typically, the number of cells used for RNA extraction in bulk RNA-seq ranges from 50,000 to 100,000, employing well-established sequencing techniques. These produce gene expression data that cover most of the coding and noncoding transcriptomes of the samples. Some modifications of bulk RNA-seq are necessary when access to sufficient cell numbers is limited, requiring low-input sequencing protocols adapted from single-cell experiments, such as SMART-Seq2, an index-sorting method designed to capture full-length cDNA molecules (Picelli et al. 2014). A consequence of this protocol is that the resulting transcriptome may experience dropout events due to low RNA content. Since single-cell experiments measure gene expression at an individual cell level, they also inherently face the challenge of lower RNA quantities, resulting in sparse data and biased PCR amplification. However, PCR bias can be mitigated through techniques like Unique Molecular Identifier (UMI) barcoding (Kivioja et al. 2011) or External RNA Controls Consortium (ERCC) spikes (External RNA Controls Consortium | NIST n.d.). A primary challenge with index-sorting single-cell methods is their high cost, which limits the number of cells studied. In contrast, droplet microfluidics technology enables sequencing thousands of cells per experiment at comparable costs (Svensson et al. 2018, 2020; Zhang et al. 2019). While the datasets may contain fewer genes compared to deep sequencing with index-sorting, microfluidics is valuable for detecting rare cell populations and previously unrecognized cellular states. Recent advancements in microfluidics technology, along with its increased flexibility for users, are driving its growing adoption for generating single-cell data.

11.3 Sampling and Sequencing Depth

For any transcriptomics experiment, sequencing many biological replicates and obtaining sufficient read depth for each sample also incurs very high costs. However, these two parameters are crucial for assessing the power of the transcriptomics study. Various studies have described different strategies for appropriate power analysis in bulk datasets, ranging from the use of biological and technical replicates to power analysis methods employing different statistical distribution models for analysis and paired or unpaired sample studies (Ching et al. 2014). Although there is no "one-size-fits-all" solution, certain trends have emerged that could be useful for understanding and implementing better data generation. Ching et al. evaluated the performance of simulated datasets modeled over different real-time biological datasets and compared the performance of various differential expression analysis packages to estimate the expected power in the data. They demonstrated that high-power estimates can be achieved with sample sizes of ≤5 per condition, which show low dispersion, using a sequencing depth of only up to 20 million reads per sample. Comparing the performance of various packages, Deseq2 and EdgeR showed a power estimate of up to 0.8 with these options. For samples with higher dispersion, better power estimates were observed with multifactor studies, such as paired samples, but these would also require larger sample sizes. Using these options, most of the protein-coding RNA is captured, making it suitable for estimating log2 fold change; however, expression changes of lowly expressed long intergenic noncoding RNA (lincRNA) are estimated incorrectly. Therefore, a higher read depth may be necessary to better capture these lowly expressed RNA molecules.

Power analysis for single-cell sequencing related to cost factors involves considering the number of samples, cells, read depth per cell, and cell type. Similar to bulk RNA-seq analysis, methods for power analysis in single-cell datasets have used negative binomial linear regression (Ching et al. 2014). Recommendations for plate-based SMART-seq2 single-cell transcriptomics state that a read depth of 500,000 to 1,000,000 reads per cell is needed for better gene coverage. However, the number of cells is usually limited in this method, as sequencing more cells increases costs. For droplet-based methods, the power improves by increasing parameters in the following order: number of cells, sample size, and read depth. Schmid et al. suggest that increasing the number of cells per sample with shallow read depth (<10,000 reads per cell) provides greater power for transcriptome capture and analysis than increasing the number of individuals sequenced. For rare cell types, this effect is more noticeable because the frequency of the cell type is strongly linked to an individual, and increasing sample size only slightly enhances the power (Schmid et al. 2020). Large-scale single-cell sequencing projects, such as the Human Cell Atlas (HCA), have employed a "Skydive" approach, which involves an iterative and strategic design based on prior biological knowledge (Paper 2017). As single-cell sequencing datasets and methods continue to emerge rapidly, better methodologies and guidelines for power analysis are likely to be developed over time (See et al. 2018). Analysis of both bulk and single transcriptomics datasets generally shares initial steps, diverging downstream based on the biological questions and methods used to explore those questions (Fig. 11.1).

11.4 Bulk Transcriptome Data Analysis

11.4.1 Quality Control and Filtering

Analysis of transcriptome datasets involves several steps to distinguish biologically meaningful information from noise, which often arises from technical variations. Once the data is generated by sequencing facilities, the dataset is available in fastq format. This dataset contains the reads sequenced for the samples and is usually large, up to a few gigabytes per sample. Depending on the protocol, these datasets can be paired-end or single-end reads. Quality control steps mainly involve examining the occurrence of sequencing artifacts within the datasets. One of the most widely used tools, FASTQC (Andrews 2010), reports on multiple parameters including the quality of sequenced reads, nucleotide bias if any, in the reads, the distribution of reads across the dataset, estimated duplication levels, and any over-represented sequences. QC analysis offers insights into biases, if present, which should be considered for downstream analysis. Although most datasets arrive with pre-processed reads from sequencing facilities, after removing adaptors and other trailing sequences, tools like cutadapt (Martin 2011) and other adaptor- and primer-trimming utilities can be employed to eliminate artifact sequences that are not part of the biological data.

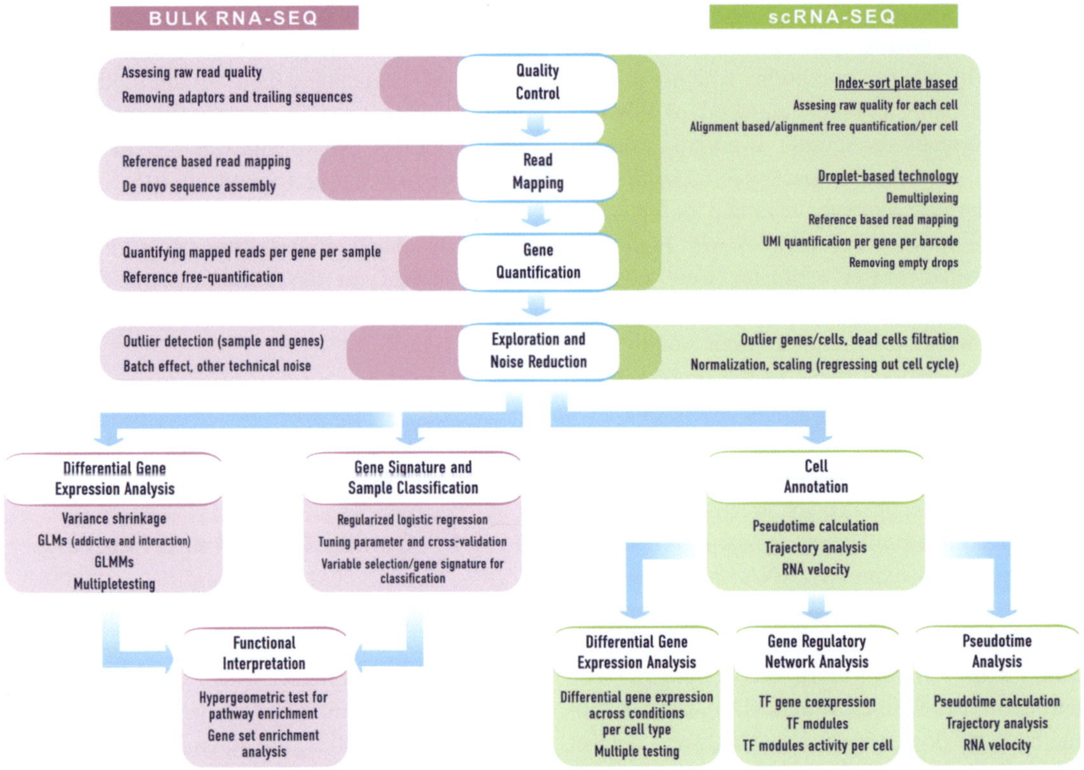

Fig. 11.1 Schematic highlighting common and different steps in bulk and single-cell RNA-seq datasets analysis

11.4.2 Read Mapping

In most cases, when sequenced genomes of model organisms are available, the fastq reads are aligned to these references. When the sequenced genome is not accessible, de novo sequence assembly is necessary to produce contigs. In reference-assisted alignment, the reference sequences used may differ depending on the study type. Most studies examining gene expression profiles require alignment solely against a transcriptome reference, while studies investigating alternative splicing, different isoform usage, and noncoding RNA require alignment against a genome reference. Reference genome and transcriptome files for most model organisms are easily accessible from online data resources, such as NCBI and ENSEMBL. These sites maintain curated and updated sequence information along with various data types and formats needed for downstream alignment and other genomic or transcriptomics analyses.

Various alignment tools have evolved, enabling confident and unique alignment of reads against reference sequences (Canzar and Salzberg 2017; Hatem et al. 2013; Li and Homer 2010). Key advances in RNA-seq alignment methods have been made to improve speed, increase read mapping accuracy, and reduce reliance on high memory usage (Musich et al. 2021). Although most alignment tools aim to address at least two of these issues, a trade-off among the three variables always exists. Spliced Transcripts Alignment to a Reference (STAR), a widely used RNA-seq aligner, outperforms most other tools with high accuracy and ultra-fast speed, but it is memory-intensive, requiring ideally up to 32 GB RAM, to run alignments for human and mouse genomes. STAR is a splice-aware aligner that employs a two-step method to accurately align reads. Briefly, STAR first splits each read to identify the Maximal Mappable Prefixes (MMPs), which map accurately to the reference. It then attempts to extend these MMPs, allowing mismatches and short indels that may be part of the read. If, however, these extensions do not produce good alignments, due to poor-quality sequences or synthetic sequences (e.g., adaptor sequences), they are soft-clipped. Next, the MMPs are clustered and stitched together to create a complete read and are scored for the best alignment. For most annotated gene models, this approach allows for precise detection of read mapping to splice junctions and enables the identification of fusion transcripts (Dobin et al. 2013). STAR was developed as part of the ENCODE project and functions well for large-scale sequencing projects involving from a few million to billions of reads. However, smaller sequencing projects or laboratories lacking high-performance computing systems may encounter difficulties in using these computationally intensive tools.

Alternative methods, such as "pseudoalignment", "lightweight alignment", and "quasi-mapping", have recently been developed, offering remarkable speed and direct quantification of reads to transcripts and genes. These algorithms are based on the key insight that transcript quantification does not require base-to-base alignment of reads. Pseudoalignment, introduced in Kallisto (Bray et al. 2016), focuses on determining the membership or origin of a read in a transcript without determining the location or alignment of the read to the transcript. Lightweight alignment (Patro et al. 2017) further scores matches between reads and transcripts, considering the position and orientation of the mapped reads. Quasi-mapping, inspired by both lightweight alignment and pseudoalignment, reports the potential loci of a read's origin (Patro et al. 2017; Srivastava et al. 2016). Salmon provides the flexibility to use either quasi-mapping or previously aligned reads (bam file) for quantification, employing its "rich model" which accounts for sample-specific parameters and biases typical of RNA-seq experiments (Patro et al. 2017). These methods have enabled researchers to bypass the computationally intensive and time-consuming processes of sequence alignment and transcript quantification with high accuracy. The computational resources required for these tools are generally modest, making them accessible to most laboratories, as they use minimal memory and are incredibly fast.

11.4.3 Quantification of Gene and Transcript Expression

Post-alignment processes involve read count or transcript quantification. Various tools using different approaches to quantify transcripts for each sample have been evaluated and compared (Zhang et al. 2017). As mentioned earlier, Kallisto, Salmon, and RapMap directly produce transcript quantification data without requiring alignment, whereas most alignment-based analyses depend on quantification from the alignments using tools like RSEM, Cufflinks (Li and Dewey 2011; Trapnell et al. 2010). Salmon can also quantify transcripts from an aligned transcriptome BAM file. The most efficient quantification tools (e.g., RSEM, Salmon, Kallisto, Cufflinks, RapMap) employ iterations of the Expectation-Maximization (EM) algorithm and Maximum Likelihood (ML) estimates to assign reads to transcripts. Common issues affecting accurate quantification include complex gene structures, especially for genes with multiple highly similar transcripts. Additionally, transcripts with low expression levels often face challenges in accurate quantification. However, increasing sequencing read depth generally improves the accuracy of transcript quantification, while the use of either read-mapping tools (e.g., Salmon, Kallisto) or alignment-based tools (e.g., STAR) during upstream analysis has little or no effect on the overall quantification accuracy. Zhang et al. suggest that Salmon and RSEM are the most accurate and time-efficient methods for transcript or isoform quantification, performing well across a variety of datasets (Zhang et al. 2017).

Transcriptome quantification using the above methods or tools for each sample involves dealing with sample-specific biases, such as varied sequencing read depth across different samples that should be normalized. Normalization for gene length is also necessary, as longer genes tend to have more reads mapped to them compared to shorter genes. Many different normalization methods have been described for various purposes. Count per million (CPM) adjusts for sequencing depth and is useful for gene count comparisons between replicates. Reads/Fragments per Kilobase Million (RPKM and FPKM) both adjust for sequencing depth and gene length, with RPKM used for single-end read datasets, while FPKM is used for paired-end read datasets, and can be applied for gene count comparisons within a sample. A more sensitive correction method called Transcripts Per Million (TPM) also adjusts for sequencing depth and gene length, making it useful for gene count comparisons within a sample or between samples. Since most RNA-seq experiments involve comparing samples, TPM is generally preferred over RPKM and FPKM for normalization. However, none of these methods are recommended for direct differential expression testing. As described, employing these methods allows for the exploration and estimation of differences that may arise due to technical artifacts.

A recent study has highlighted the benefit of additionally introducing average expressed transcript length offsets or normalizing for expressed transcript length, alongside sequencing depth, as a better estimate for downstream differential gene expression testing. This is achieved by using transcript-level abundance data and summarizing it at the gene level, while also calculating the average expressed transcript length across samples. Normalizing in this way helps to reduce inflation in false discovery rate calculations used in downstream differential gene expression testing. The inflated false discovery rate without this normalization may be attributed to genes displaying different transcript usage across conditions, although this only affects a small percentage of genes in most realistic datasets. Additionally, abundance estimation at the gene level after normalization can also lessen the impact of technical biases on expression data (Soneson et al. 2016). Tximport, a tool introduced in the same study, enables this normalization across various quantification methods, making the output easily accessible for most downstream differential gene expression testing approaches (Soneson et al. 2016).

11.4.4 Exploratory Data Analysis and Noise Reduction

Data exploration of quantified data can be an iterative process. It requires bidirectional evaluation of different variables, which may result from technical artifacts or biologically relevant signals. Although the normalization methods described above are reasonable for exploring certain technical artifacts, such as sequencing depth and transcription length correction, other issues like batch effects and additional technical noise remain that cannot be entirely eliminated. Instead, it is possible to estimate their impact on the data and separate or correct these biases to prioritize the biologically relevant signal. One such effect is prominently observed in bulk RNA-seq datasets generated using the Smart-seq2 protocol, where the datasets often contain more zeros than those produced by other library preparation methods. As mentioned earlier, this protocol is used for sequencing rare cell populations with limited cell numbers. Carefully evaluating these datasets involves filtering out very low or very high signals. An initial, relaxed filtering criterion focuses on removing genes with no counts in any samples. Another approach involves filtering genes based on a threshold expression level (normalized to sequencing depth) across biological replicates from the same condition, particularly in pairwise analyses. The filtering step of the RNA-seq dataset is important to prevent irrelevant zero-count data from occupying computer memory and to reduce spurious calculations, which can often lead to false positives.

Another major unintended variable introduced during data generation is the batch effect. During the data generation process, care should be taken to minimize the creation of batch effects. However, with large-scale sequencing projects, batch effects are unavoidable. For this task, certain tools can estimate the effect caused by different batches and adjust the data to counteract them (Johnson et al. 2007; Zhang et al. 2018). Some differential expression analysis tools include batch information as a covariate, a method generally preferred and recommended over modifying the data. Nevertheless, even after modeling the different batches, the data still experiences underlying latent variation, which hampers the proper investigation of the biologically relevant signal. The cause of this unwanted noise, although difficult to identify, can be modeled, enabling further "cleaning" of the count dataset and prioritization of biologically relevant information. Surrogate Variable Analysis (SVA), particularly SVA-seq, allows modeling of unwanted noise by borrowing information from genes that do not align with the phenotype or batch effect. The tool attempts to identify the portion of the dataset affected solely by artifacts that cannot be explained by the phenotype of interest or the batch effect covariate in the modeling steps. This unwanted noise signal is subtracted from the baseline effect, and the resulting matrix is decomposed using either principal component analysis or singular value decomposition to estimate the "surrogates" for the noise (Leek 2014). These surrogate variables can then be included as covariates, alongside batch effects and biological variables of interest, in linear models for expression data.

11.4.5 Differential Expression Analysis

Most widely used differential expression analysis methods employ Generalized Linear Models (GLMs) to model the expression data. These models essentially assess the expression data resulting from the combination of different variables, which can be biological, such as different treatments and conditions, or technical, such as batch effects and surrogate variables, as described above. Based on data exploration, previous knowledge, and the biological question of interest, the relationships among the variables and the genes influenced by those relationships can be examined using linear models. Most small-sample-size experiments evaluating differential gene expression due to different treatments involve simple models with only one variable. However, linear models also allow evaluation of complex relationships and multifactor variables of interest, for example, pairwise samples taken before and after treatment, or analyzing genes responsible for different independent variables. Limma, a tool initially designed for microarray experiments, has extended and adapted the use of linear models to analyze complete experiments with complex relationships among the variables (Ritchie et al. 2015).

Effective detection of differential gene expression between conditions requires consistent signals among biological replicates of the same condition. However, if the inter-sample variability within each condition is high, the resulting differential expression may lack sufficient statistical significance. Inter-sample variability can arise from many factors, including biological differences in genetically identical mice or variations in human genetic backgrounds. Differential gene expression methods address this variability by modeling the variance either of the entire dataset or for each gene individually. Deseq2, a widely used tool for differential expression analysis, models RNA-seq data using a negative binomial distribution and has been successfully used for small sample size experiments, as it efficiently controls inter-sample variability (Love et al. 2014). It uses the empirical Bayes method to shrink dispersion and log-fold-change estimates while accounting for the "heteroscedasticity" present in such datasets. The tool also implements Cook's distance to flag outlier genes with unusual expression values, which may not be considered for downstream differential expression testing if the sample size is very small ($n = 3$–5). If the sample size is larger, Deseq2 attempts to impute outlier values, permitting the gene to be included in the differential expression analysis. Deseq2 provides either the "Wald significance test" or the "likelihood ratio test" for differential expression testing, with multiple testing correction via the Benjamini and Hochberg procedure.

Limma employs a gene-wise model and fits a mean-variance trend across samples for each gene, along with a "moderated t-statistic" for differential expression testing and the Benjamini and Hochberg procedure for multiple testing. It converts count data to a log scale and empirically estimates the mean–variance relationship. Limma also uses the empirical Bayes method to borrow information between genes and estimate a "compromise" between gene-wise and global variance estimates. This use of empirical Bayes was an innovation of the Limma package, especially for small-sample-size experiments (Ritchie et al. 2015). Limma provides two methods for estimating the mean–variance trend: Limma trend, which is suitable for datasets that are not highly variable and applies the relationship at the gene level, and Voom, short for "variance modeling at the observational level", which applies this relationship to each observation using precision weights. Voom demonstrates greater sensitivity in managing data with high variance.

Complex experimental designs, especially those involving multiple samples from the same individuals (e.g., longitudinal studies or different stimulations), also introduce variance due to inter-individual differences. In generalized linear mixed models (GLMMs), controlling for repeated measurements is usually achieved by adding a random effect term (e.g., individual ID) alongside the fixed effect (e.g., time point, stimulation). Recent RNA-seq analysis methods have enabled the inclusion of the random effect in linear mixed models by examining the dataset's correlation structure. Limma calculates duplicate correlation to estimate the data variance caused by inter-individual differences. However, the limma duplicate correlation model assumes a single value for all genes, which can lead to false-positive results. A recent advancement to address this issue was introduced in Dream, where random effects are estimated separately for each gene, along with precision weights and other limma–voom features. This inter-individual variance calculation also offers the opportunity to identify the genes most affected by genetic differences (Hoffman and Roussos 2021).

The concerns listed above in RNA-seq data and the methods used by various tools to address them are some of the most well-known major issues. Alongside considering these issues, assessing their impact and whether these effects are present or absent requires careful data exploration and visualization by plotting different aspects of the dataset. Differential analysis tools provide functions to evaluate these effects within the datasets, along with other sample plotting options in R. These can include simple box or violin plots to compare library sizes before and after sample correction and correlation plots in R. Both Deseq2 and limma–voom offer functions to estimate gene expression dispersion across samples, trends in log two-fold change before and after shrinkage, and the influence of data transformation before and after normalization. Principal Component Analysis (PCA) and Multi-dimensional Scaling (MDS) can be used to identify and examine the largest sources of variation, as well as to detect outliers among the samples. Therefore, combining different plots with an awareness of various artifact types is a crucial step in data cleaning and preparing data for differential expression analysis. Additionally, heatmaps can be used to visualize differentially expressed genes, highlighting the most significant genes for each comparison or patterns of expression across different samples.

The results from differential expression testing usually provide statistical relevance, such as p value, adjusted p value, log two-fold change, and average expression across all samples. Additionally, limma also provides log odds (B-statistic) between the conditions, which can be used to prioritize and select the most differentially expressed genes. While prioritizing top genes using user-defined thresholds is a helpful approach, it is not sufficient, as even after this filtering, a large number of genes—up to a few hundred—still appear as differentially expressed. The next section discusses methods for interpreting large lists of differentially expressed genes.

11.4.6 Gene Signatures for Sample Classification

Traditional analysis of differential expression data aims to identify genes that are likely responsible for or affected by different phenotypes. The use of generalized linear models works quite well in these cases. However, there are other situations involving classifying an unknown biological sample into different phenotype categories based on its transcriptome profile, for example, classifying transcriptome data obtained from cancer patients into different cancer subtypes (Algamal and Lee 2015; Wu et al. 2018). The typical approach for this analysis requires previously collected transcriptome samples with known phenotypes. Statistical methods like logistic regression can then be used to train a model (classifier) on these samples and subsequently predict the phenotype of test datasets. Generally, logistic regression works well with low-dimensional datasets, where the number of samples significantly exceeds the number of variables. However, with transcriptome datasets, where the number of variables (genes) far surpasses the number of samples, additional strategies such as shrinkage methods called "regularization" are employed for improved model interpretation. Three common methods—ridge, lasso, and elastic net regularization—are available, and thus the entire analysis is referred to as "regularized logistic regression".

The use of regularization methods enables the selection of variables (genes) that are employed to accurately classify a sample as belonging to a phenotypic category. Ridge utilizes all variables for classification, whereas Lasso selects only a subset by setting nonselected variables to zero. This "variable selection" process used by Lasso regularization facilitates easier model interpretation. These chosen variables can serve as a "transcriptional signature" sufficient to accurately classify a test sample into its corresponding phenotype. Elasticnet offers a hybrid between Ridge and Lasso, resulting in the selection of a subset of variables along with their coefficients based on their relevance in classification. Each of these regularizations requires a tuning parameter "λ" that determines the degree of shrinkage of the coefficients toward zero or near-zero values. Selecting the optimal "λ" value is a critical step for classification and can be achieved through cross-validation. Therefore, combining regularized logistic regression with cross-validation to identify the best tuning parameter can be employed for classifying new samples. Additionally, this approach can be utilized to derive a transcriptional signature for the related phenotypes from the training datasets.

11.4.7 Functional Interpretation

A functional interpretation of these gene lists could provide a more biologically relevant way to prioritize genes for further analysis by biologists. One such method involves investigating the enrichment of gene lists against a background dataset of known functional classifications of genes. This classification may refer to different pathways or ontologies based on gene function, although a variety of annotations are now widely available. Gprofiler is an online tool offering a broad range of data resources annotated from ontologies, such as Gene Ontology, and biological pathways including KEGG and Reactome. Additionally, it includes resources for annotated regulatory DNA motifs, protein databases, and Human Phenotype Ontology. Like most gene enrichment tools, Gprofiler uses hypergeometric tests to measure the enrichment and significance of the annotated gene set relative to the query gene list. The extensive data resources provided by GProfiler enable accessible and straightforward association of gene lists with various biological categories (Raudvere et al. 2019; Reimand et al. 2007).

An unbiased method for associating genes with a phenotype has been described by the Gene Set Enrichment Analysis (GSEA) framework. This framework considers the entire gene expression profile of the dataset and does not require a user-defined gene list based on a threshold. Although this method was initially designed for microarray data analysis, it can also be applied to RNA-seq datasets by arranging the genes in a pre-ranked list. The pre-ranked list can reflect the log two-fold change alongside the adjusted p-value, with the most significantly upregulated genes, having the highest log two-fold change, placed at the top and the most significantly downregulated genes at the bottom. Other statistics may also be considered depending on the data analysis method used. GSEA determines whether the members of a gene set (annotated list of known biological significance) are enriched at either the top or bottom of this list. It employs a running-sum statistic, akin to a weighted Kolmogorov–Smirnov-like statistic, to calculate the enrichment score. This score is obtained by walking down the gene list and matching genes with the gene set, with each match increasing the score and each mismatch decreasing it. The enrichment score indicates the maximum deviation from zero during this "random walk." Additionally, the genes that match from the two lists to produce the significant enrichment score can provide further interpretation as the core of the gene set. These leading-edge genes may offer additional insights as biologically relevant signals from a comprehensive list of differentially expressed genes (Subramanian et al. 2005).

11.5 Single-Cell Transcriptomics Analysis

11.5.1 Pipeline Overview

Single-cell transcriptomics analysis requires specific considerations, in addition to the issues mentioned above. Although there are considerably more technical variabilities in these datasets, the biological insights gained from them have transformed the understanding of the continuum in biological processes and discoveries in rare cell populations. Consequently, the analysis steps necessitate careful evaluation of various variabilities at each stage.

Datasets generated using the index-sort plate-based Smart-seq2 protocol produce individual fastq files for each cell. These files are processed similar to bulk RNA-seq datasets, where each sample corresponds to a single cell. As previously described, these fastq files first undergo quality control with multiQC, which provides a summarized assessment of all sequenced single cells (Ewels et al. 2016). The QC considerations at this stage are similar to those for bulk RNA-seq datasets. These sequence files are then aligned to the reference genome using STAR and quantified with Salmon to generate gene-level counts, or directly using Salmon's quasi-mapping combined with the gene count estimation method. The latter generally offers a faster and accurate approach, producing results comparable to STAR alignment while being less computationally intensive. These datasets may require a customized reference genome if ERCC spike-in sequences have been added during data generation. As previously mentioned, these spike-in sequences are used to estimate PCR bias during sequencing. Therefore, manual addition of ERCC sequences to the reference genome, along with a customized GTF file, is essentially required. However, newer protocols for index-based single-cell sequencing have replaced the need for ERCC spike-in sequences with UMI counting.

11.5.2 Demultiplexing of 10× Datasets

The most widely available single-cell sequencing datasets result from commercially available droplet-based technology, such as those generated by 10X Chromium. This technology enables the capture of RNA from a large number of cells (in thousands) per sample. The sequenced data are available either as base call files (BCLs) or most commonly as fastq files. Unlike index-based single-cell sequencing data, the data generated using 10X technology does not consist of separate fastq files for each cell. Freely available pipelines from 10X Genomics, such as "Cell Ranger", facilitate various initial analysis steps of single-cell datasets. Two commands, "mkfastq" and "count", are primarily used to generate a data matrix of gene counts per cell. The "mkfastq" command allows demultiplexing of BCL files into fastq files, while "count" creates a data matrix and feature files from the fastq files generated by "mkfastq".

The Cell Ranger "count" command involves two main steps for generating matrix files. These include read trimming to remove nontemplate sequences, followed by aligning reads to the reference genome. Cell Ranger uses STAR for sequence alignment. The authors of STAR also offer a pipeline called "STARsolo," specifically designed for mapping, demultiplexing, and gene quantification in single-cell transcriptomics datasets from droplet-based single-cell techniques (Kaminow et al. 2021). However, Cell Ranger does not incorporate STARsolo into its pipeline. The alignment of reads by both Cell Ranger and STARsolo is performed using a standard splice-aware method. Unlike bulk RNA-seq, all RNA molecules from droplet technology are tagged with UMI sequences and a barcode unique to each cell. Therefore, the quantification process used by Cell Ranger first involves correcting sequencing errors in UMI sequences. It then groups reads with shared barcode, UMI, and gene annotation. If multiple read groups share the same barcode and UMI but have different gene annotations (due to multiple aligned reads), the gene annotation with the most supporting reads is retained for UMI counting. If the maximum support from read groups is similar, all such read groups are discarded as a confident gene assignment is not possible. The resulting dataset is considered as a UMI count, comprising each observed combination of barcode, UMI, and gene.

One of the crucial steps after quantifying reads is calling cell barcodes. Ideally, each barcode represents a single cell; however, this is not always the case. Drop-based technology also produces empty droplets without encapsulating a cell. These empty droplets do not contain cells but still contain "ambient RNA". Ambient RNA consists of cell-free transcripts found in the cell suspension solution. It results from cells that experience stress during dissociation and microfluidics or undergo apoptosis. Therefore, removing these empty droplets before downstream analysis is essential. Usually, empty droplets with ambient RNA have very low UMI counts compared to droplets containing cells. Most methods, including early versions of Cell Ranger, implement simple filtration during cell barcode calling based on a threshold number of UMI counts. However, this filtration method may cause the removal of rare cells with low RNA content that are not necessarily from empty droplets. EmptyDrops addresses this issue by modeling the ambient RNA pool and comparing each barcode against this model to test for deviations. It employs a Dirichlet multinomial distribution along with the Good-Turing algorithm for posterior estimates of count proportions in modeling the ambient RNA pool. Additionally, it uses a knee point method to call cell barcodes. The resulting barcode calls include cells with high UMI counts—mainly droplets containing cells with high RNA content—as well as barcodes with low RNA content that are not ambient RNA (Lun et al. 2019). Cell Ranger version 3.0 and above initially uses a cutoff based on the total UMI count per barcode. It first identifies the primary mode of high RNA content cells, followed by the implementation of EmptyDrops for cell barcode calling.

The gene-quantified information for each cell from either of two methods (index-sort or droplet technology) is essentially a data matrix where each column represents a single cell, and each row represents a quantified gene. Although creating this matrix from index-sort data involves manual curation, Cell Ranger mainly generates files ready to be processed and used by single-cell analysis tools such as Seurat, Scran, and Scanpy.

11.5.3 Quality Control of Single-Cell Data

Post-quantification, quality control is necessary to remove low-quality cells before downstream analysis. These cells are often a result of cell rupture during the lysis stage of the protocol or cell doublets caused by cells sticking together. Although it is not feasible to image each cell prior to sequencing to assess their quality, certain characteristics previously evaluated can be useful for filtering them out from the data matrix (Ilicic et al. 2016). Such characteristics include high mitochondrial DNA expression. Other criteria involve examining the number of genes expressed per cell, sequencing depth, as well as uniquely mapped and unmapped reads. There are no strict cut-offs for these filtering steps; instead, these variables should be assessed based on their distribution plots across cells for each study. An exceptionally high gene count in a small subset of cells may indicate doublets, whereas very low gene numbers might primarily reflect low-quality cells. Index-sorted samples may include ERCC counts, which offer insight into PCR bias across cells and assist in selecting high-quality cells. Since droplet-based samples use UMI counts, this bias is less prominent in these samples. Therefore, careful selection of datasets for downstream analysis—by removing low-quality cells and maximizing the number of usable cells—is essential.

11.5.4 Preprocessing Single-Cell Data

Essential and standard steps in the analysis of single-cell datasets include normalization, identifying the most variable features (genes) across all cells, and scaling data before examining biological heterogeneity. However, technical factors can confound biological variation. Therefore, preprocessing steps should aim to remove or reduce the influence of technical factors, leaving the observed variability primarily biological. Two main approaches are used for preprocessing: first, calculating a scale factor either through log normalization or using spike-ins, such as ERCC, to distinguish technical noise from biological variability; second, employing a probabilistic framework using a negative binomial distribution that relies on accurate per-gene error model estimates (Lopez et al. 2018; Eraslan et al. 2019). Hafemeister et al. conducted a detailed comparison of these approaches on various single-cell datasets, suggesting that applying a single scaling factor does not normalize highly and lowly expressed genes equally, while using a per-gene error model can lead to overfitting. Instead, they recommend a regularized negative binomial regression method, where the sequencing depth of each cell is used as a covariate in a generalized linear model fitted for each gene. Additional constraints are then calculated, integrating information from genes with similar expression patterns, contrasting with independent gene-level estimates. A second iteration of NB regression is performed with these constraints, and the residuals are interpreted as normalized expression levels. This data-driven approach has demonstrated robust normalization for both highly and lowly expressed genes while avoiding overfitting (Hafemeister and Satija 2019) and is implemented in sctransform within the Seurat single-cell toolkit.

A second essential step in preprocessing single-cell datasets involves calculating highly variable genes across all cells. Seurat uses the "variance stabilization transformation" (vst) method to select the top 2000 variable genes. This selection of highly variable genes reflects the heterogeneity within the single-cell dataset and is therefore useful for dimensionality reduction steps. The vst method standardizes gene expression using the observed mean and expected variance, followed by the calculation of gene variance on unnormalized log counts. This approach ensures that highly variable features represent genuine biological heterogeneity (Stuart et al. 2019). When applying the negative binomial regression method for normalization, the Pearson residuals effectively correspond to vst values (Hafemeister and Satija 2019). Finally, scaling the dataset through a z-score transformation for each gene across all cells is a standard procedure before conducting downstream dimensionality reduction. This step also allows regressing out unwanted effects, such as the percentage of mitochondrial DNA expression, from all cells against each gene. The resulting residuals are then scaled and centered for each gene. For datasets generated by Smart-seq2, it is recommended to regress out sequencing depth as well, provided the data is not UMI count-based.

11.5.5 Exploratory Data Analysis

The high dimensionality of single-cell transcriptomics datasets allows for the examination of biological heterogeneity that is lost in bulk RNA datasets. Data exploration for this purpose requires a low-dimensional representation of these datasets. Principal Component Analysis (PCA) is a linear dimensionality reduction method used for this purpose. PCA preserves the pairwise distances among all data points (cells) and can capture the underlying biological variability. It is performed on the 2000 highly variable features identified in previous steps, which should effectively reflect the biological characteristics of the dataset. The principal components calculated must be evaluated for their significant representation of different biological dimensions of the data. Here, biological dimension refers to a list of genes known for their roles in various biological functions, for example, cell cycle, cell type-specific genes, etc. Evaluating the PCs is essential, as irrelevant PCs may obscure biological variance of interest. Several methods have been suggested to select relevant PCs, including enrichment analysis of genes from different PCs to associate them with biological functions, previously known markers for different cell subsets, evaluation based on elbow plots, and resampling tests such as the JackStraw procedure (Chung and Storey 2015). Usually, a combination of all three methods provides a good indication of the relevant PCs to consider. Depending on the biological question, if the PCs indicate a high proportion of cell cycle genes, scoring and regressing out these effects might be necessary before proceeding to subsequent steps.

11.5.6 Mapping Cell Diversity

One of the most exciting applications of single-cell transcriptomics is its ability to identify rare cell populations and cellular states. Clustering cells enables exploration of the high dimensionality of scrna-seq datasets, effectively segregating different cell subsets based on gene expression similarities among cells. For high-dimensional data such as single-cell transcriptomics datasets, the Shared Nearest Neighbor (SNN) graph approach is widely used to find similarities between nodes (cells) (Houle et al. 2010). A nearest-neighbor graph comprises nodes (cells) connected by edges, with weights indicating similarity between the two nodes. To construct the SNN, a graph of k-nearest neighbors (k- NN), i.e., a set of the k-closest neighbours to each node, is first created based on Euclidean distance in the PCA space. The number of relevant Principal Components (PCs) identified earlier determines the PCA space. The next step involves refining the edge weights between two nodes based on the overlap of shared nearest neighbors identified in the previous step, using the Jaccard similarity score. Multiple clustering methods for both scrna-seq and mass cytometry datasets have demonstrated the successful use of SNN graphs (Xu and Su 2015; Levine et al. 2015). The shared nearest-neighbor graph is further used to define modules (clusters of cells) iteratively with the Louvain algorithm (Blondel et al. 2008). The Louvain algorithm is widely employed for community detection in large and complex datasets. Cell clusters can be viewed as communities of cells sharing similar expression patterns. The Louvain algorithm identifies these clusters by initially moving nodes locally between communities and evaluating the quality function. Once it determines the separation between communities, each community becomes a node. This process repeats until there is no significant improvement in the quality function (Blondel et al. 2008). Another approach, the Leiden algorithm, has also been proposed for community detection. It is said to overcome some community disconnection issues that can occur with the Louvain algorithm (Traag et al. 2019).

Both methods require a resolution input, which determines the granularity of clustering. A larger resolution value results in many small clusters, whereas a smaller value produces fewer larger clusters of cells. Choosing the appropriate resolution for clustering depends on empirical knowledge of the cell subsets involved and the markers that define them. However, a resolution range of 0.4–1.2 is generally recommended for up to 3000 cells. The optimal resolution also depends on the specific research question. If the dataset includes very different cell types (e.g., T cells, B cells, myeloid cells) and the goal is to broadly identify these subsets, a lower resolution may be sufficient. Conversely, if the aim is to distinguish subpopulations within T and B cells, iterative clustering using various resolutions combined with differential expression tests to identify unique markers for each cluster should be performed. This approach requires some empirical knowledge of known markers for the subpopulations and is best used alongside other methods to determine the most suitable clustering. With the advancements in community-driven single-cell research, reference mapping algorithms are also available, facilitating the assignment of cellular identities. These algorithms can map unlabeled cells onto well-annotated reference datasets, which are widely accessible. Tools such as Azimuth and celltypist enable the mapping of new, unlabeled scRNA-seq datasets to existing reference datasets for both human and mouse (Hao et al. 2021; Domínguez Conde et al. 2022). This facilitates rapid cell identification and annotation, reducing the need for clustering and somewhat overcoming the requirement for prior marker knowledge for subpopulations.

11.5.7 Data Visualization

As resolving cells in different clusters is an essential step for scrna-seq data analysis, visualizing these datasets to explore their underlying data structure is equally important. Previously performed PCA captures the larger distances between cells; however, it can lead to overcrowding of cells that are closer to each other. Nonlinear dimensionality reduction methods have been particularly useful in capturing both the overall global and local structures of such datasets. t-distributed stochastic neighbor embedding (t-SNE) is an algorithm widely used for nonlinear dimensionality reduction and in machine learning applications across various fields (Van Der Maaten and Hinton 2008). The method uses previously generated k-NN graphs and positions the cells in two dimensions such that cells that are close neighbors remain close, while cells far apart based on the kNN graph stay far apart. This process can effectively capture local variability between clusters of cells, but the global structure, influenced by the larger variability of distant clusters, may appear arbitrary within the two- dimensional space. This loss of global structure poses significant issues in single-cell transcriptomics datasets, where cell subsets inherently exhibit hierarchical structures. For example, t-SNE can correctly cluster cells from different subpopulations of T cells; however, the placement of these clusters in a global context—relative to B cells or other myeloid cells—tends to be arbitrary. Such disparities between the local and global positions of cell clusters may lead to misinterpretations of the data. Additionally, performing t-SNE on larger datasets is computationally intensive, resulting in different visualizations on each run—an inherent characteristic of t-SNE. Recent work suggests certain steps to generate more informative t-SNE visualizations (Kobak and Berens 2019). Another nonlinear dimensionality reduction method, Uniform Manifold Approximation and Projection (UMAP) (McInnes et al. 2018), belongs to the same category as t-SNE; it also uses k-NN graphs and has been widely recommended over t-SNE for various biological datasets, including those from flow cytometry, Cytof, and single-cell transcriptomics. A comparison between UMAP and t-SNE indicated that UMAP performs better, offering improved representation and interpretation of datasets, and mitigates some issues highlighted above in t-SNE visualizations (Becht et al. 2018). UMAP can preserve local structures between cells similar to t-SNE; additionally, it better preserves the relative global structure in higher dimensions. Consequently, UMAP projections tend to align more closely with the clustering of cells resulting from previously mentioned clustering methods, enabling more meaningful data representation and visualization. UMAP is also less computationally demanding than t- SNE, with shorter run times and more reproducible visualizations.

UMAP assumes an underlying manifold structure in the dataset and uses this to identify the data structure despite noise. As more data is sampled, this structure becomes more apparent because noise tends to decrease. Therefore, caution should be exercised when interpreting data from small datasets (McInnes et al. 2018). It is important to note that, unlike PCA where each principal component allows interpretation of the data, the two dimensions of t-SNE or UMAP are arbitrary and lack direct interpretability. Both t-SNE and UMAP focus on finding local structure precisely, rather than global structure. If the goal of the analysis is to examine the global structure, other methods such as MDS or PHATE should be used; however, these methods are more computationally intensive. Consequently, interpretation of data based on visualization, along with cell clustering, should be carefully evaluated in relation to the specific questions of interest and the limitations of each method.

11.5.8 Differential Expression Analysis

As previously mentioned, differential gene expression testing helps define the identity of each cell cluster by identifying both known and unknown markers. This process can be used to find genes that are up- or downregulated compared to another cluster. Additionally, genes that are uniquely upregulated in each cluster can also be identified. Because sparsity is an inherent feature of single-cell datasets, certain parameters should be considered during differential expression testing. These include the minimum percentage of cells expressing a gene in each comparison, the log-fold change between the clusters, and the minimum number of cells required for testing. The choice of test type may also be important for identifying differentially expressed genes in each cluster. If the goal is to find unique markers for each cluster, the logistic regression method can be employed. Integrating approaches such as clustering, dimensionality reduction, and differential gene testing provides new insights from single-cell datasets, particularly in elucidating the heterogeneity and identity of rare cell populations.

11.5.9 Inferring Dynamic Processes from Static Data

Biological systems comprise various dynamical processes, such as those seen in developmental biology, the cell cycle processes, cellular maturation, and disease progression. Before the advent of single-cell technologies, time course studies combined with bulk RNA sequencing were often employed to explore these processes. However, bulk RNA-seq measures the average expression across all cells in a sample. It has been noted that most cells do not undergo transitions in a synchronized manner; therefore, capturing a snapshot with bulk RNA-seq of a mixture of different cell subsets or cellular states does not accurately reflect these dynamic processes. Single-cell transcriptomics, although still providing a snapshot of RNA activity at a specific time point, offers the advantage of single-cell resolution, enabling the deconvolution of various cell subsets and states. Due to this powerful capability, single-cell transcriptomics datasets have become widely popular for investigating not only differentiation but also many other dynamic biological processes. This is achieved through the application of the concept of "pseudotime" ordering of single cells. Pseudotime is a quantitative measure of biological progression through a dynamic process (Trapnell 2015; Trapnell et al. 2014). Pseudotemporal ordering, based on changes in gene expression, allows the reconstruction of a developmental trajectory, projecting cells onto this continuum. Pseudotime should not be mistaken for actual time during a dynamical process. However, collecting time-series single-cell data and constructing pseudotime can enable an estimation of progression that aligns with real time. This does not imply that, at any given real time, all cells are in a particular state, especially if the cellular state is transient. Various studies have used time series measurements and described asynchronous cellular states within each time point (Proserpio et al. 2016; Lönnberg et al. 2017). Instead, identifying and analyzing different cellular states to construct the dynamical process trajectory using pseudotime provide a more effective approach.

Pseudotime analysis involves arranging cells based on how gene expression changes continually over time. Single-cell data contains both biological signals and noise. To correctly order cells in pseudotime, methods must be able to extract the true biological signals from the noise. Various tools employ dimensionality reduction techniques to identify these hidden biological signals and organize cells in fewer dimensions.

Principal Component Analysis (PCA), Independent Component Analysis (ICA), and other linear dimensionality reduction methods are commonly used for this purpose. However, nonlinear dimensionality reduction techniques are often better at capturing the low-dimensional structures within high-dimensional data. It should be noted that, unlike t-SNE and UMAP—which aim to find discrete clusters of cells—pseudotime analysis seeks to model a continuous process. These clustering algorithms focus on local structures within datasets, often neglecting the global structure, and thus may not be ideal for representing continuous cell organization. Lonnberg and Svensson et al. described using the Bayesian Gaussian Process Latent Variable (BGPLVM) method, a nonlinear dimensionality reduction technique, to derive low-dimensional representations of cells and infer pseudotime with *per-cell* priors, making it semi-supervised (Lönnberg et al. 2017; Titsias and Lawrence 2010). However, in many cases, the latent variables from BGPLVM are sufficient for pseudotime inference even without prior information (Miragaia et al. 2019). Different latent variables in BGPLVM may represent various, sometimes overlapping, transcriptional programs linked to cellular states, and typically two or three latent variables are enough to visualize cell ordering. Another nonlinear technique, Diffusion Maps, measures "diffusion distance" between cells—defined as the average length of paths connecting two cells in an N-dimensional space, where N corresponds to the number of genes. Furthermore, the use of transition probabilities between cell states across multiple paths makes this method robust to noise. Diffusion distance is particularly relevant for analyzing biological datasets demonstrating dynamic processes, as it mimics the stochastic diffusion-like dynamics observed in such systems (Coifman and Lafon 2006; Haghverdi et al. 2015). Similar to BGPLVM, three to four dimensions identified through Diffusion Maps can effectively help visualize cell ordering.

In addition to low-dimensional continuous visualization of cell ordering, it is essential to define the trajectory, which enables the identification of different trajectories leading to various lineages in the dynamic process (Fig. 11.2).

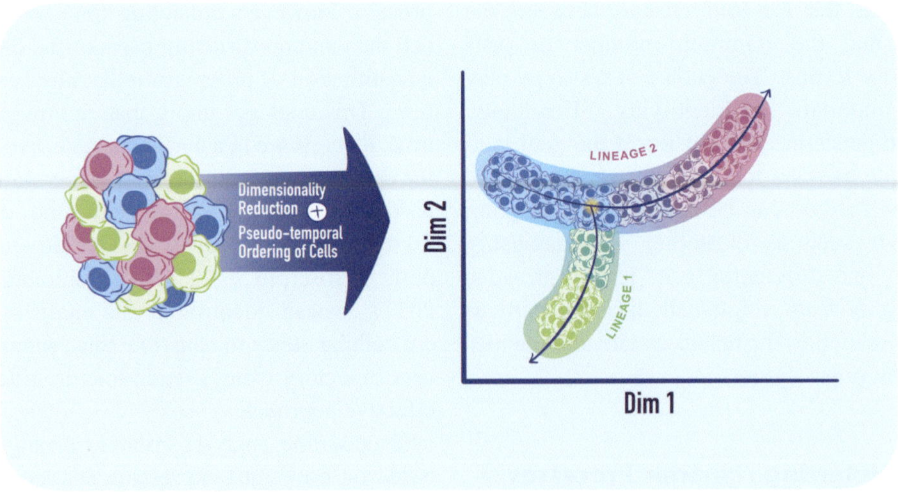

Fig. 11.2 Schematic illustrating pseudotime analysis in single-cell data, where multiple trajectories correspond to distinct cellular lineages. Single cells from the major cell type, such as T cells, consist of distinct cellular subsets, indicated by multicolored cells; that exist as transitional states rather than discrete cell subset categories. By arranging single cells based on their continous changing gene expression, pseudotime algorithms place them along a continuum representing dynamic cellular progression, as shown in the X-Y plot. This allows the reconstruction of different trajectories of these transitional states along different lineages. These methods are valuable for understanding diverse biological processes, such as cellular differentiation and proliferation, and can identify key genes involved in these processes

Processes with linear trajectories are usually straightforward; however, it becomes challenging in most differentiation studies where multiple end stages or cyclic trajectories may be possible. There are various sets of trajectory inference methods currently available, many of which require prior information. The two main categories of these methods are distinguished based on whether they use fixed topology and the type of topology they can detect. An extensive study by Saelens et al. has examined up to 45 different trajectory inference methods across a wide range of both real and synthetic datasets and evaluated these methods according to different benchmarking metrics. They conclude that although no single method works for every dataset, it is crucial to investigate the data dimensions and trajectory topology. In cases where there is no prior knowledge of the expected trajectory topology, multiple trajectory inference methods should be employed and data dimensions assessed before identifying a definitive trajectory (Saelens et al. 2019). Slingshot, recognized as one of the most feasible and accurate methods, is a tree-based approach capable of predicting various types of tree-like trajectories. It does not depend on any specific dimensionality reduction method, as it initially uses a cluster-based minimum spanning tree (MST) to capture the overall lineage structure. It further employs simultaneous principal curves to smooth the inference of pseudotime variables for cells throughout the lineage. Although Slingshot does not require prior information about the starting or terminal cells, it allows user input for the initial or terminal clusters. This user input does not influence the global structure inference, indicating that Slingshot is robust and capable of inferring different types of trajectories (Street et al. 2018).

11.5.10 Dynamic Modelling of Gene Expression Processes

Cellular dynamic processes retraced using pseudotime inference offer new insights into these processes; however, they all require some user input and are not always consistent when evaluated using different methods. New methods and concepts are developing daily to better understand and model these processes, including different types of single-cell dataset generation. RNA velocity is one such concept, adapted from bulk-RNA-seq for single-cell RNA-seq, and has proven to be of great significance in studying these processes. RNA velocity uses real transcription kinetics, analyzing the abundance of spliced and unspliced RNA molecules for each gene in each cell to estimate the future state of the cell. Single-cell transcriptomics datasets generated from either Smart-seq2 or droplet technology have both shown a percentage of RNA that is unspliced. Therefore, by combining information from spliced and unspliced reads and modeling the steady-state for each gene based on transcription, splicing, and degradation rates, it is possible to estimate the upregulation or downregulation of genes. The resulting data on gene regulation allows prediction of the future state of individual cells over a timescale of a few hours (La Manno et al. 2018). Although the authors suggest that RNA velocity can be visualized using Principal Components or other dimensionality reduction methods, recent studies have proposed that visualization methods specifically embedding RNA velocity offer a better solution (Atta et al. 2021). Further developments in RNA velocity modeling have led to dynamical modeling, replacing the previously used steady-state model. This dynamical modelling addresses transcriptional events such as transcription, splicing, and degradation within a probabilistic framework, rather than linear regression. It calculates the full dynamics of splicing kinetics for each gene, unlike the steady-state model, which assumes a constant splicing rate for all genes. This advancement better captures transient cellular states, which might be missed when assuming constant splicing dynamics across genes in most datasets (Bergen et al. 2020). However, in some cases, RNA velocity may not fully reflect the known biology and requires further consideration, such as variable kinetic rates of genes across different cell states, which can also provide insights into various post-transcriptional processes important for studying gene dynamics. Additionally, the use of multimodal datasets, including protein expression dynamics or chromatin accessibility, might offer improved solutions when transcriptional dynamics alone are insufficient. A combined approach using RNA velocity along with Markov chain transitions can also be effective in modeling cellular dynamics (La Manno et al. 2018; Bergen et al. 2021). All these methods are already being utilized by the scientific community to investigate previously understood dynamic processes with new emerging insights. With the exponential increase in single-cell datasets and collaborative efforts, these technologies are likely to generate significant insights, especially in humans, where experimental exploration poses certain challenges.

11.5.11 Functional Annotation and Regulatory Networks

Single-cell transcriptomics analysis typically identifies various known or novel cell clusters based on lists of differentially expressed genes across those clusters. As mentioned in earlier sections, interpreting dynamic processes is one of the most valuable applications of this type of dataset, aside from exploring heterogeneity. However, the lists of differentially expressed genes also include previously unknown genes that may play a role in the function of specific cell clusters. A gene set enrichment analysis of significant genes from each cluster can reveal the involved biological pathways, similar to methods used for bulk RNA-seq (Raudvere et al. 2019; Reimand et al. 2007).

An alternative and more detailed analysis examines the regulatory networks active within each cell cluster. Attempts have been made to adapt co- expression network tools originally designed for bulk RNA-seq to single-cell transcriptomics data (Cha and Lee 2020; Bacher and Kendziorski 2016; Xue et al. 2013). However, these approaches require careful handling to manage the sparsity of such datasets and tend to be computationally demanding in terms of memory. Few methods have been developed specifically for exploring expression networks from single-cell transcriptomics data, and these networks alone do not directly reveal regulation relationships between genes and transcription factors (Iacono et al. 2019; Guo et al. 2015). A recent method, Single- Cell rEgulatory Network Inference and Clustering (SCENIC), utilizes multiple techniques to infer regulatory networks, called "regulons," from single-cell datasets (Aibar et al. 2017; Van de Sande et al. 2020). SCENIC operates through a workflow comprising three main steps. It initially employs GENIE 3 or GRNBoost/GRNBoost2 algorithms to infer relationships between transcription factors and their targets (Huynh-Thu et al. 2010; Natekin and Knoll 2013; Moerman et al. 2019). GENEI3 uses random forest regression to construct decision trees based on gene expression, treating expression levels as response variables and transcription factor expression as predictors. The random forest approach effectively captures complex, nonlinear relationships. GRNBoost performs a similar function but utilizes Gradient Boosting Machines (GBMs) as its algorithm. This part of the SCENIC workflow is computationally intensive, especially when using GRNBoost. Nonetheless, since both methods employ tree-based algorithms, they do not suffer from sparsity and multicollinearity problems, which are common in single-cell data. The inferred interactions are weighted values for each transcription factor–target gene pair, which are then used to build modules. The SCENIC workflow incorporates the cisTarget database, providing motif information associated with each transcription factor (Herrmann et al. 2012; Imrichová et al. 2015). This enables motif enrichment analysis at the transcriptional start sites of target genes, adding an additional regulatory layer often overlooked by co-expression analysis alone. However, the regulons derived from motif enrichment may exclude many genes from the original co-expression modules if they do not meet certain thresholds. Despite this limitation, this approach has successfully identified various regulatory networks across multiple single-cell studies. Furthermore, integrating multi-omics datasets such as single-cell ATAC-seq or Chip-seq could enhance the power of this method and assist in retaining genes that might be missed due to stringent enrichment thresholds.

Resulting regulons are essentially lists of genes where each regulatory transcription factor, along with its target genes, is identified. Since these regulons are derived from the analysis of all cells, their specificity can vary across individual cells. SCENIC employs methods like AUCell, which assess the activity of these regulons in each cell using Area Under Curve (AUC) analysis (Aibar et al. 2017). AUCell performs enrichment analysis to determine whether a regulon is overrepresented within the expressed genes of each cell. It achieves this by ranking all expressed genes based on their decreasing expression values and calculating the AUC for the regulons against the background ranked gene list of the cell. This approach is independent of gene expression units and does not require normalization, which is both an advantage and a limitation because single cells from different experiments cannot be directly compared. AUCell provides a score for each regulon, reflecting its relative biological activity within each cell. Gene regulatory networks should be similar in related cells and different in distinct cell types. The AUCell scores for individual cells can be used for clustering, where cells are grouped based on the expression patterns of regulons as units, rather than individual genes. These scores are also valuable for assessing the specificity of regulons across different cell clusters, highlighting transcription factors that may drive key cellular processes. Regulon Specificity Score (RSS) is a useful metric that identifies highly specific regulons for each cell cluster (Suo et al. 2018). This score can be calculated using any user-defined clustering, providing crucial insights into gene regulatory networks that are specific to and important for the cellular phenotype. The SCENIC workflow offers a flexible and modular approach to investigating gene regulatory networks for various purposes, such as cell clustering, identifying key transcription factors driving specific cell populations, and determining the genes they regulate. Furthermore, integrating additional omics data, such as single-cell ATAC-seq, can provide further insights into gene regulation, especially for noncoding genomic regions. Therefore, multi-omics approaches should also be considered when studying gene regulatory networks.

References

Aibar S, González-Blas CB, Moerman T, Huynh-Thu VA, Imrichova H, Hulselmans G, Rambow F, Marine J-C, Geurts P, Aerts J, van den Oord J, Atak ZK, Wouters J, Aerts S (2017) SCENIC: single-cell regulatory network inference and clustering. Nat Methods 14:1083–1086

Algamal ZY, Lee MH (2015) Regularized logistic regression with adjusted adaptive elastic net for gene selection in high dimensional cancer classification. Comput Biol Med 67:136–145

Andrews S (2010) FastQC: a quality control tool for high throughput sequence data., babraham bioinformatics. http://www.bioinformatics.babraham.ac.uk/projects/

Atta L, Sahoo A, Fan J (2021) VeloViz: RNA velocity-informed embeddings for visualizing cellular trajectories. Bioinformatics. https://doi.org/10.1093/BIOINFORMATICS/BTAB653

Bacher R, Kendziorski C (2016) Design and computational analysis of single-cell RNA-sequencing experiments. Genome Biol 17:1–14

Becht E, McInnes L, Healy J, Dutertre C-A, Kwok IWH, Ng LG, Ginhoux F, Newell EW (2018) Dimensionality reduction for visualizing single-cell data using UMAP. Nat Biotechnol 37(1):38–44

Bergen V, Lange M, Peidli S, Wolf FA, Theis FJ (2020) Generalizing RNA velocity to transient cell states through dynamical modeling. Nat Biotechnol 38(12):1408–1414

Bergen V, Soldatov RA, Kharchenko PV, Theis FJ (2021) RNA velocity—current challenges and future perspectives. Mol Syst Biol 17:e10282

Blondel VD, Guillaume J-L, Lambiotte R, Lefebvre E (2008) Fast unfolding of communities in large networks. J Stat Mech: Theory Exp 2008:P10008

Bray NL, Pimentel H, Melsted P, Pachter L (2016) Near-optimal probabilistic RNA-seq quantification. Nat Biotechnol 34:525–527

Canzar S, Salzberg SL (2017) Short read mapping: an algorithmic tour. Proc IEEE Inst Electr Electron Eng 105:436

Cha J, Lee I (2020) Single-cell network biology for resolving cellular heterogeneity in human diseases. Exp Mol Med 52(11):1798–1808

Ching T, Huang S, Garmire LX (2014) Power analysis and sample size estimation for RNA-Seq differential expression. RNA 20:1684–1696

Chung NC, Storey JD (2015) Statistical significance of variables driving systematic variation in high-dimensional data. Bioinformatics 31:545

Coifman RR, Lafon S (2006) Diffusion maps. Appl Comput Harmon Anal 21:5–30

Dobin A, Davis CA, Schlesinger F, Drenkow J, Zaleski C, Jha S, Batut P, Chaisson M, Gingeras TR (2013) STAR: ultrafast universal RNA-seq aligner. Bioinformatics 29:15–21

Domínguez Conde C, Xu C, Jarvis LB, Rainbow DB, Wells SB, Gomes T, Howlett SK, Suchanek O, Polanski K, King HW, Mamanova L, Huang N, Szabo PA, Richardson L, Bolt L, Fasouli ES, Mahbubani KT, Prete M, Tuck L, Richoz N, Tuong ZK, Campos L, Mousa HS, Needham EJ, Pritchard S, Li T, Elmentaite R, Park J, Rahmani E, Chen D, Menon DK, Bayraktar OA, James LK, Meyer KB, Yosef N, Clatworthy MR, Sims PA, Farber DL, Saeb-Parsy K, Jones JL, Teichmann SA (2022) Cross-tissue immune cell analysis reveals tissue-specific features in humans. Science 376:eabl5197

Eraslan G, Simon LM, Mircea M, Mueller NS, Theis FJ (2019) Single-cell RNA-seq denoising using a deep count autoencoder. Nat Commun 10:1–14

Ewels P, Magnusson M, Lundin S, Käller M (2016) MultiQC: summarize analysis results for multiple tools and samples in a single report. Bioinformatics 32:3047–3048

External RNA Controls Consortium | NIST. https://www.nist.gov/programs-projects/external-rna-controls-consortium

Guo M, Wang H, Potter SS, Whitsett JA, Xu Y (2015) SINCERA: a pipeline for single-cell RNA-Seq profiling analysis. PLoS Comput Biol 11:e1004575

Hafemeister C, Satija R (2019) Normalization and variance stabilization of single-cell RNA-seq data using regularized negative binomial regression. Genome Biol 20:1–15

Haghverdi L, Buettner F, Theis FJ (2015) Diffusion maps for high-dimensional single-cell analysis of differentiation data. Bioinformatics 31:2989–2998

Hao Y, Hao S, Andersen-Nissen E, Mauck WM, Zheng S, Butler A, Lee MJ, Wilk AJ, Darby C, Zager M, Hoffman P, Stoeckius M, Papalexi E, Mimitou EP, Jain J, Srivastava A, Stuart T, Fleming LM, Yeung B, Rogers AJ, McElrath JM, Blish CA, Gottardo R, Smibert P, Satija R (2021) Integrated analysis of multimodal single-cell data. Cell 184:3573–3587.e29

Hatem A, Bozdağ D, Toland AE, Çatalyürek ÜV (2013) Benchmarking short sequence mapping tools. BMC Bioinform 14:1–25

Herrmann C, Van de Sande B, Potier D, Aerts S (2012) i-cisTarget: an integrative genomics method for the prediction of regulatory features and cis-regulatory modules. Nucleic Acids Res 40:e114

Hoffman GE, Roussos P (2021) Dream: powerful differential expression analysis for repeated measures designs. Bioinformatics 37:192–201

Houle ME, Kriegel H-P, Kröger P, Schubert E, Zimek A (2010) Can shared-neighbor distances defeat the curse of dimensionality? In: International conference on scientific and statistical database management. Springer Berlin Heidelberg, Berlin, Heidelberg

Huynh-Thu VA, Irrthum A, Wehenkel L, Geurts P (2010) Inferring regulatory networks from expression data using tree-based methods. PLoS One 5:e12776

Iacono G, Massoni-Badosa R, Heyn H (2019) Single-cell transcriptomics unveils gene regulatory network plasticity. Genome Biol 20:1–20

Ilicic T, Kim JK, Kolodziejczyk AA, Bagger FO, McCarthy DJ, Marioni JC, Teichmann SA (2016) Classification of low quality cells from single-cell RNA-seq data. Genome Biol 17

Imrichová H, Hulselmans G, Kalender Atak Z, Potier D, Aerts S (2015) i-cisTarget 2015 update: generalized cis-regulatory enrichment analysis in human, mouse and fly. Nucleic Acids Res 43:W57

Johnson WE, Li C, Rabinovic A (2007) Adjusting batch effects in microarray expression data using empirical bayes methods. Biostatistics 8:118–127

Kaminow B, Yunusov D, Dobin A (2021) STARsolo: accurate, fast and versatile mapping/quantification of single-cell and single-nucleus RNA-seq data. bioRxiv. https://doi.org/10.1101/2021.05.05.442755

Kivioja T, Vähärautio A, Karlsson K, Bonke M, Enge M, Linnarsson S, Taipale J (2011) Counting absolute numbers of molecules using unique molecular identifiers. Nat Methods 9:72–74

Kobak D, Berens P (2019) The art of using t-SNE for single-cell transcriptomics. Nat Commun 10:1–14

Kukurba KR, Montgomery SB (2015) RNA sequencing and analysis. Cold Spring Harb Protoc 2015:951–969

La Manno G, Soldatov R, Zeisel A, Braun E, Hochgerner H, Petukhov V, Lidschreiber K, Kastriti ME, Lönnerberg P, Furlan A, Fan J, Borm LE, Liu Z, van Bruggen D, Guo J, He X, Barker R, Sundström E, Castelo-Branco G, Cramer P, Adameyko I, Linnarsson S, Kharchenko PV (2018) RNA velocity of single cells. Nature 560:494–498

Leek JT (2014) Svaseq: removing batch effects and other unwanted noise from sequencing data. Nucleic Acids Res 42:e161

Levine JH, Simonds EF, Bendall SC, Davis KL, Amir EA, Tadmor MD, Litvin O, Fienberg HG, Jager A, Zunder ER, Finck R (2015) Data-driven phenotypic dissection of AML reveals progenitor-like cells that correlate with prognosis. Cell 162:184–197

Li B, Dewey CN (2011) RSEM: accurate transcript quantification from RNA-Seq data with or without a reference genome. BMC Bioinform 12:1–16

Li H, Homer N (2010) A survey of sequence alignment algorithms for next-generation sequencing. Brief Bioinform 11:473

Lönnberg T, Svensson V, James KR, Fernandez-Ruiz D, Sebina I, Montandon R, Soon MSF, Fogg LG, Nair AS, Liligeto UN, Stubbington MJT, Ly L-H, Bagger FO, Zwiessele M, Lawrence ND, Souza-Fonseca-Guimaraes F, Bunn PT, Engwerda CR, Heath WR, Billker O, Stegle O, Haque A, Teichmann SA (2017) Single-cell RNA-seq and computational analysis using temporal mixture modeling resolves $T_H 1/T_{FH}$ fate bifurcation in malaria. Sci Immunol 2:eaal2192

Lopez R, Regier J, Cole MB, Jordan MI, Yosef N (2018) Deep generative modeling for single-cell transcriptomics. Nat Methods 15:1053–1058

Love MI, Huber W, Anders S (2014) Moderated estimation of fold change and dispersion for RNA-seq data with DESeq2. Genome Biol 15:550

Lun ATL, Riesenfeld S, Andrews T, Dao TP, Gomes T, Marioni JC (2019) EmptyDrops: distinguishing cells from empty droplets in droplet-based single-cell RNA sequencing data. Genome Biol 20:1–9

Martin M (2011) Cutadapt removes adapter sequences from high-throughput sequencing reads. EMBnetjournal 17:10

McInnes L, Healy J, Melville J (2018) UMAP: uniform manifold approximation and projection for dimension reduction

Miragaia RJ, Gomes T, Chomka A, Jardine L, Riedel A, Hegazy AN, Whibley N, Tucci A, Chen X, Lindeman I, Emerton G, Krausgruber T, Shields J, Haniffa M, Powrie F, Teichmann SA (2019) Single-cell transcriptomics of regulatory T cells reveals trajectories of tissue adaptation. Immunity 50:493–504.e7

Moerman T, Aibar Santos S, Bravo González-Blas C, Simm J, Moreau Y, Aerts J, Aerts S (2019) GRNBoost2 and Arboreto: efficient and scalable inference of gene regulatory networks. Bioinf (Oxf) 35:2159–2161

Musich R, Cadle-Davidson L, Osier MV (2021) Comparison of short-read sequence aligners indicates strengths and weaknesses for biologists to consider. Front Plant Sci 12:692

Natekin A, Knoll A (2013) Gradient boosting machines, a tutorial. Front Neurorobot 7:21

Nguyen QH, Pervolarakis N, Nee K, Kessenbrock K (2018) Experimental considerations for single-cell RNA sequencing approaches. Front Cell Dev Biol 6:108

Paper W (2017) The human cell atlas. Elife 6:e27041

Patro R, Duggal G, Love MI, Irizarry RA, Kingsford C (2017) Salmon: fast and bias-aware quantification of transcript expression using dual-phase inference. Nat Methods 14:417

Picelli S, Faridani OR, Björklund ÅK, Winberg G, Sagasser S, Sandberg R (2014) Full-length RNA-seq from single cells using Smart-seq2. Nat Protoc 9:171–181

Proserpio V, Piccolo A, Haim-Vilmovsky L, Kar G, Lönnberg T, Svensson V, Pramanik J, Natarajan KN, Zhai W, Zhang X, Donati G, Kayikci M, Kotar J, McKenzie ANJ, Montandon R, Billker O, Woodhouse S, Cicuta P, Nicodemi M, Teichmann SA (2016) Single-cell analysis of CD4+ T-cell differentiation reveals three major cell states and progressive acceleration of proliferation. Genome Biol 17:103

Raudvere U, Kolberg L, Kuzmin I, Arak T, Adler P, Peterson H, Vilo J (2019) g:Profiler: a web server for functional enrichment analysis and conversions of gene lists (2019 update). Nucleic Acids Res 47:W191–W198

Reimand J, Kull M, Peterson H, Hansen J, Vilo J (2007) g:Profiler—a web-based toolset for functional profiling of gene lists from large-scale experiments. Nucleic Acids Res 35:W193

Ritchie ME, Phipson B, Wu D, Hu Y, Law CW, Shi W, Smyth GK (2015) Limma powers differential expression analyses for RNA-sequencing and microarray studies. Nucleic Acids Res 43:e47

RNA sequencing methods collection. https://www.illumina.com/content/dam/illumina-marketing/documents/products/research_reviews/rnasequencing-methods-review-web.pdf

Saelens W, Cannoodt R, Todorov H, Saeys Y (2019) A comparison of single-cell trajectory inference methods. Nat Biotechnol 37(5):547–554

Schmid KT, Cruceanu C, Böttcher A, Lickert H, Binder EB, Theis FJ, Heinig M (2020) Design and power analysis for multi-sample single cell genomics experiments. bioRxiv

See P, Lum J, Chen J, Ginhoux F (2018) A single-cell sequencing guide for immunologists. Front Immunol 9:2425

Soneson C, Love MI, Robinson MD (2016) Differential analyses for RNA-seq: transcript-level estimates improve gene-level inferences. F1000Res 4:1521

Srivastava A, Sarkar H, Gupta N, Patro R (2016) RapMap: a rapid, sensitive and accurate tool for mapping RNA-seq reads to transcriptomes. Bioinformatics 32:i192–i200

Street K, Risso D, Fletcher RB, Das D, Ngai J, Yosef N, Purdom E, Dudoit S (2018) Slingshot: cell lineage and pseudotime inference for single-cell transcriptomics. BMC Genomics 19:1–16

Stuart T, Butler A, Hoffman P, Hafemeister C, Papalexi E, Mauck WM, Hao Y, Stoeckius M, Smibert P, Satija R (2019) Comprehensive integration of single-cell data. Cell 177:1888–1902.e21

Subramanian A, Tamayo P, Mootha VK, Mukherjee S, Ebert BL, Gillette MA, Paulovich A, Pomeroy SL, Golub TR, Lander ES, Mesirov JP (2005) Gene set enrichment analysis: a knowledge-based approach for interpreting genome-wide expression profiles. Proc Natl Acad Sci 102:15545–15550

Suo S, Zhu Q, Saadatpour A, Fei L, Guo G, Yuan G-C (2018) Revealing the critical regulators of cell identity in the mouse cell atlas. Cell Rep 25:1436

Svensson V, Vento-Tormo R, Teichmann SA (2018) Exponential scaling of single-cell RNA-seq in the past decade. Nat Protoc 13:599–604

Svensson V, da Veiga Beltrame E, Pachter L (2020) A curated database reveals trends in single-cell transcriptomics. Database 2020

Titsias MK, Lawrence ND (2010) Bayesian Gaussian process latent variable model. Proceedings of the thir-

teenth international conference on artificial intelligence and statistics. JMLR Workshop and Conference Proceedings
Traag VA, Waltman L, van Eck NJ (2019) From Louvain to Leiden: guaranteeing well-connected communities. Sci Rep 9:1–12
Trapnell C (2015) Defining cell types and states with single-cell genomics. Genome Res 25:1491–1498
Trapnell C, Williams BA, Pertea G, Mortazavi A, Kwan G, van Baren MJ, Salzberg SL, Wold BJ, Pachter L (2010) Transcript assembly and quantification by RNA-Seq reveals unannotated transcripts and isoform switching during cell differentiation. Nat Biotechnol 28:511–515
Trapnell C, Cacchiarelli D, Grimsby J, Pokharel P, Li S, Morse M, Lennon NJ, Livak KJ, Mikkelsen TS, Rinn JL (2014) The dynamics and regulators of cell fate decisions are revealed by pseudotemporal ordering of single cells. Nat Biotechnol 32(4):381–386
Van de Sande B, Flerin C, Davie K, De Waegeneer M, Hulselmans G, Aibar S, Seurinck R, Saelens W, Cannoodt R, Rouchon Q, Verbeiren T, De Maeyer D, Reumers J, Saeys Y, Aerts S (2020) A scalable SCENIC workflow for single-cell gene regulatory network analysis. Nat Protoc 15:2247–2276
Van Der Maaten L, Hinton G (2008) Visualizing data using t-SNE. J Mach Learn Res 9:2579–2605
Wu S, Jiang H, Shen H, Yang Z (2018) Gene selection in cancer classification using sparse logistic regression with L1/2 regularization. Appl Sci (Switz) 8:1–12
Xu C, Su Z (2015) Identification of cell types from single-cell transcriptomes using a novel clustering method. Bioinformatics 31:1974–1980
Xue Z, Huang K, Cai C, Cai L, Jiang CY, Feng Y, Liu Z, Zeng Q, Cheng L, Sun YE, Liu JY (2013) Genetic programs in human and mouse early embryos revealed by single-cell RNA sequencing. Nature 500:593–597
Zhang C, Zhang B, Lin L-L, Zhao S (2017) Evaluation and comparison of computational tools for RNA-seq isoform quantification. BMC Genom 18:1–11
Zhang Y, Jenkins DF, Manimaran S, Johnson WE (2018) Alternative empirical bayes models for adjusting for batch effects in genomic studies. BMC Bioinform 19:1–15
Zhang X, Li T, Liu F, Chen Y, Yao J, Li Z, Huang Y, Wang J (2019) Comparative analysis of droplet-based ultra-high-throughput single-cell RNA-Seq systems. Mol Cell 73:130–142.e5

Data Analysis in Extreme Resolution Mass Spectrometry Untargeted Metabolomics

12

Marta Sousa Silva, Francisco Traquete, João Luz, António E. N. Ferreira, and Carlos Cordeiro

Abstract

Mass spectrometry (MS)-based metabolomics is a powerful tool for understanding the complexity of biochemical processes and to identify biomarkers across diverse biological systems. The vast amount of data generated by extreme resolution mass spectrometers poses significant data processing challenges, requiring robust computational approaches and workflows for meaningful data interpretation. This chapter provides a comprehensive overview of current methodologies in MS-based metabolomics data analysis, with a focus on data preprocessing and pretreatment, m/z extraction and annotation, univariate and multivariate statistical approaches, as well as data visualization. We discuss key considerations for ensuring data quality and the growing role of bioinformatics in pathway analysis and metabolite identification. We highlight the transforming role of extreme resolution and mass accuracy enabled by FT-ICR mass spectrometers, and finally, we explore emerging trends, including artificial intelligence-driven insights and real-time data processing, to guide future developments in this rapidly evolving field.

Keywords

Untargeted metabolomics · FT-ICR-MS · Data processing and treatment · Compound identification

12.1 Introduction

Metabolomics is the comprehensive analysis and quantification of all the metabolites in a biological system (Fiehn 2002; Nicholson et al. 2007). Metabolites are endogenous and exogenous small molecules whose ensemble constitutes the system's metabolome. They are the product of all biochemical reactions and are, consequently, informative of the biochemical activity of the system. In an untargeted metabolomics workflow, the goal is to comprehensively profile and analyze the broad spectrum of metabolites present in a biological sample, without prior knowledge of their identity or abundance. This approach aims to capture as much of the metabolome as possible, enabling the discovery of novel metabolites, metabolic pathways, and biomarker candidates associated with specific biological states, diseases, or environmental conditions (Bartel et al. 2013; Roberts et al. 2012). For untargeted metabolomics, mass spectrometry (MS) is the method of choice, being far superior to nuclear

M. Sousa Silva · F. Traquete · J. Luz · A. E. N. Ferreira · C. Cordeiro (✉)
FT-ICR and Structural Mass Spectrometry Laboratory, Biosystems and Integrative Sciences Institute (BioISI), Faculdade de Ciências, Universidade de Lisboa, Lisbon, Portugal
e-mail: cacordeiro@fc.ul.pt

M. Gama-Carvalho et al. (eds.), *Omics Approaches in Biomedicine and Biotechnology*, Advances in Experimental Medicine and Biology 1504, https://doi.org/10.1007/978-3-032-18966-0_12

magnetic resonance (NMR). MS has much higher sensitivity and dynamic range, allowing the detection of metabolites at much lower concentrations (nanomolar to picomolar range), enabling the identification of low-abundance compounds in trace amounts of material. It also provides greater metabolite coverage because of its higher dynamic range and can easily be coupled with separative techniques, like liquid chromatography (LC)–MS or gas chromatography (GC)–MS. Finally, the ability to perform gas-phase fragmentation allows the identification of unknown metabolites and the elucidation of their structures (reviewed by Ferreira et al. 2021). In untargeted metabolomics, the highest mass accuracy and resolution are essential, and these parameters directly impact the reliability, precision, and depth of the metabolomics analysis. Presently, Fourier Transform Ion Cyclotron Resonance Mass Spectrometry (FT-ICR-MS) is the most powerful technique to perform such analysis, with over 10,000,000 resolution and a mass accuracy in ppb (Brown et al. 2005). This extreme mass accuracy and resolution allow the identification of metabolites based on their *m/z* value which enables unambiguous molecular formula assignments in low-mass metabolites (Brown et al. 2005). Moreover, it allows direct infusion through electrospray ionization, without chromatographic separation, thus allowing the simultaneous detection of tens of thousands of compounds in a single run. Furthermore, MS^n experiments can be performed in FT-ICR-MS using a vast plethora of fragmentation methods (Ghaste et al. 2016). To fully exploit these advantages, innovative computational and statistical tools are required for data analysis and to extract meaningful biological information.

12.2 Data Preprocessing

The first step in data analysis of an untargeted metabolomics experiment is data preprocessing. This step is critical for converting raw mass spectra data into a suitable format for downstream statistical analysis and biological interpretation.

Data preprocessing deals directly with the spectral raw data with the aim of generating a 2D data matrix, with m/z represented in one dimension and samples in the other (Karaman 2017; Katajamaa and Oresic 2007). For MS data, preprocessing includes spectra deconvolution with noise filtering and baseline correction procedures, peak detection and alignment, m/z values' filtering, and gap filling (Castillo et al. 2011). There are currently many available platforms and software, both commercial and freely available for metabolomics data preprocessing, such as XCMS (Tautenhahn et al. 2012) and MZMine (Schmid et al. 2023).

Noise reduction and baseline correction consist in the removal of the background noise or low-intensity signals. Noise can arise from the instrument or from solvents and sample contaminants (Trygg et al. 2009). A predefined intensity threshold is normally applied to remove signals below that value, bearing in mind that setting the threshold too high can exclude low-abundance metabolites and setting it too low can retain noisy and meaningless signals.

Peak alignment is a critical step in metabolomics data preprocessing, especially in untargeted metabolomics. It usually refers to the alignment of *m/z* values with a defined tolerance (either by peak alignment or spectral alignment), ensuring that the same metabolite peaks from different samples are matched, and can also include the alignment of retention time when LC-MS is performed (Karaman 2017). Of course, large datasets with hundreds of samples can make the alignment more difficult, due to extensive variability, and the occurrence of different charge states as well as adducts of the same metabolite ($[M + H]^+$, $[M + Na]^+$, $[M + K]^+$) make alignment a challenging task.

At this stage, filtering steps can be applied to remove peaks with very low variance between samples (since these are noninformative), or with low reproducibility between samples based on quality-control samples if possible, or peaks that appear below a chosen threshold of number of samples in the dataset, among others (Schiffman et al. 2019). Data filtering aims to discard uninformative variables, reduce noise in the dataset while reducing the computational load of applying different statistical methodologies further down the processing. Thus, it is important to adapt data filtering procedures to the data to analyze (Schiffman et al. 2019).

Gap filling or missing value imputation is a necessity. Since missing values arise in metabolomics datasets after peak or spectral alignments when a compound detected in a sample is not detected in another. The imputation consists of replacing those missing values by a certain value to facilitate the different kinds of data analysis performed downstream in the workflow (which do not adequately account for the presence of missing values), while maintaining the overall structure of the data. Missing values can be values missed completely at random (MCAR), missed at random (MAR), and missed not at random (MNAR), (Wei et al. 2018; Di Guida et al. 2016). For metabolomics data, MCAR values can result from instrumental factors, such as stochastic fluctuations during data acquisition, while MAR values can be, for example, due to incorrect signal detection by the chosen algorithm (Wei et al. 2018). MNAR values are, however, caused by the signal of some compounds being under the detection limit (Di Guida et al. 2016), either present at a low concentration or absent in samples, justified by the metabolic context. There are strategies to reduce the number of values that have to be imputed by filtering the number of variables used in further analyses. These can be done by imposing a maximum percentage of missing values, filtering those that exceed this limit (Wei et al. 2018). This threshold can be set on the overall dataset, or it can be set on a group of technical replicates. There are several different strategies to impute the missing values, some favoring the imputation of MNAR and others the imputation of MCAR/MAR (Wei et al. 2018). One of the strategies used by methods that favor the imputation of MNAR, considering all missing values as such, consists in replacing all missing values by a constant small value, that is usually zero, ½, or 1/5th of the minimum intensity value on the dataset. Another strategy, known as Quantile Regression Imputation of Left-Censored data (QRILC), randomly imputes values from a small-value distribution, estimated by quantile regression (Wei et al. 2018; Di Guida et al. 2016; Lazar 2015). Concerning the methods that favor the imputation of MAR/MCAR considering all missing values as such, these are usually replaced by the mean or median of the corresponding variable or by using more complex methods, like k-Nearest Neighbors (kNN) (Troyanskaya et al. 2001) or Random Forest (RF) imputation (Stekhoven and Bühlmann 2012), among others (Wei et al. 2018; Di Guida et al. 2016). As always, the choice of method has a considerable effect on the data matrix and on the results of downstream analysis. The choice of missing value imputation depends on the data analysis method that will be used subsequently, meaning that there is no single one-size-fits-all method (Wei et al. 2018; Di Guida et al. 2016).

12.3 Data Pretreatment

After the preprocessing steps, a clean dataset is obtained that is ready to be treated. Data pretreatment has the objective of highlighting relevant biological information within the dataset while reducing the effect of undesired variation (van den Berg et al. 2006) due to measurement or technical errors, slight changes in temperature, batch or operator variation, etc. Data pretreatment normally includes normalization (which can often be considered a preprocessing step), transformation, centering, and scaling. Each category contemplates multiple options of data treatment and may also be applied in combination with other categories, exponentially increasing the number of available options (Xi et al. 2014). A thoughtful deliberation is required over which combination of pretreatment methods should be made, considering the data characteristics and the downstream performed analysis.

12.3.1 Normalization

The first step of pretreatment in untargeted metabolomics is usually normalization. This step has the objective of removing unwanted variation between samples to eliminate the systematic bias between them (Katajamaa and Oresic 2007). This is achieved by multiplying or dividing the intensity values of the samples by a certain normalization factor, which is specific to each sample, allowing a quantitative comparison analysis. Several methods can be used to normalize metabolomics data: normalization by a reference compound, normalization using optimal selection of multiple internal standards, normalization by the total peak area sum of a sample, quantile normalization, and probabilistic quotient normalization.

The normalization by a reference compound, which can be an internal or external standard present in every sample, considers the normalization factor as the peak area or intensity of that reference compound in each sample, or that value multiplied by a constant. Since this standard has a known concentration across all samples, by equalizing its intensity values, the global comparison of all samples becomes more reliable. A possible disadvantage may come from the normalization depending only on one metabolite intensity, which, as mentioned previously, may have issues in reproducibility (Misra 2020). The Normalization using Optimal selection of Multiple Internal Standards (NOMIS) addresses this latter issue by using multiple internal standards for normalization (Sysi-Aho et al. 2007). This method requires the use of multiple internal standards, and the intensity variation of a metabolite can be modeled based on the variation of the different standards (Sysi-Aho et al. 2007). Both methods require a careful selection of internal standards that are chemically similar to the metabolites of interest.

In the normalization by the total peak area, each sample's normalization factor is the sum of intensities of all its signals. This method assumes that the total metabolite concentration in each sample is identical. Hence, metabolites with a higher concentration will contribute more to the normalization factor. So, considerable concentration changes in these metabolites will affect the normalization factor, and its efficacy will be reduced, a possible disadvantage of this method (Karaman 2017).

Quantile normalization aims to make all samples to have the same peak intensity distribution. It is different from the other methods as it does not use a conventional normalization factor. Instead, it creates a "reference spectrum" from the data and uses it to replace all the values in the dataset. This is done by first ranking all intensities in each sample. Then, by calculating the mean or median of all intensities with the same rank, the "reference spectrum" is obtained. Finally, all values are replaced by the intensity value in the "reference spectrum" with the same rank. After normalization, a dataset where each sample has the same set of intensity values but distributed between the features differently is obtained (Bolstad et al. 2003). This method achieves the goal of having all samples with the same distribution, although it is problematic in the tails of the distribution where, if a metabolite is the most intense in all samples, for example, it will have the same value for all of them, and in datasets with considerable amounts of missing values or missing values that were imputed with constant values, since it creates samples with many identical values in the lowest ranks.

The Probabilistic Quotient Normalization (PQN) is a method that assumes that the difference in intensity in most peaks is due to different dilutions. This method, therefore, starts by performing a normalization by the total peak area, scaling all samples to the same magnitude, and then proceeds to calculate another normalization factor based on a "reference spectrum". This "reference spectrum" can either be calculated as the mean or median intensity of each signal, or it can be a separate sample from the study or from a database. All samples are divided by this "reference spectrum", and the median value of the quotients of all its intensity values is taken as the normalization factor (dilution factor) for each sample (Dieterle et al. 2006). Since it assumes that most metabolites do not vary significantly between samples, which may not always hold true, this can be a limitation of this method.

Again, there is no single method for data normalization in untargeted metabolomics, as the choice depends on the dataset, experimental design, and biological questions being addressed. Different methods should be evaluated to choose a good normalization method that minimizes technical variation while preserving biological differences.

12.3.2 Transformations

Transformations are a set of nonlinear treatments whose main objective is to reduce heteroscedasticity and to make the data less skewed. This affects the reliability of the statistical analysis since most methods assume that the dataset is homoscedastic and that it has a symmetrical distribution (Karaman 2017). Transformations can have a "pseudo-scaling" effect since they reduce the intensity of the higher values more than that of the lower values, thus reducing their differences (Karaman 2017). However, they do not replace the proper scaling methods presented below. The most common transformation methods are logarithmic or power transformations.

Logarithmic transformation is a straightforward transformation that usually applies either natural base or base 2 logarithms to each point of the dataset. This transformation, besides the previously mentioned effects, also turns multiplicative relations into additive ones, which might be biologically relevant. It should be noted that this transformation cannot be directly applied on a dataset with null values. Another limitation is the dominance of small values in downstream analyses if the dataset has many low-abundance compounds (van den Berg et al. 2006).

Generalized Logarithmic Transformation (*glog*) is a variation of the logarithmic transformation that introduces a transformation parameter λ, with the objective of stabilizing variable variance. This transformation is made by applying Eq. 12.1 to the dataset (Rocke and Durbin 2001; Rocke and Durbin-Johnson 2003). The standard logarithmic transformation stabilizes variance for most of its variables, except for those with low intensity, since as they get closer to zero their variance increases dramatically. The *glog* transformation aims to correct this drawback with the λ factor. Ideally, $\lambda = b/a$, where b and a are normal distributions (with mean 0) of the error/variance, dependent (b) and independent (a) of the intensity in the dataset in a model where the variance of the variable x is given by Eq. 12.2 (Rocke and Durbin 2001; Rocke and Durbin-Johnson 2003):

$$\tilde{x}_{ij} = \log_2\left(\frac{x_{ij} + \sqrt{\left({x_{ij}}^2 + \lambda^2\right)}}{2}\right) \quad (12.1)$$

$$\text{Variance}(x_i) = b^2 \times {x_i}^2 + a^2 + \alpha \quad (12.2)$$

where $\tilde{x}$ is the transformed intensity x of the metabolite i of sample j, λ is a transformation parameter, and α is background noise. By incorporating the λ parameter, *glog* reduces the need for arbitrary constant addition to zero values.

The *power* transformation, as the name entails, is another common transformation that transforms the values to their n-power. A common choice is the square root ($\sqrt{x_{ij}}$). It is a simple transformation that, despite not transforming multiplicative relations into additive relations, it tends to reduce heteroscedasticity, improve symmetry, and can be used in datasets with null values (Karaman 2017).

The general recommendation is to use log or *glog* transformation for simpler datasets, choosing power transformation for datasets with complex distributions and wide dynamic ranges.

12.3.3 Mean Centering and Scaling

The final steps of untargeted metabolomics pretreatment are mean centering and scaling. These procedures, often coupled together, aim to reduce the difference between high- and low-intensity biologically important metabolites to prevent skewness toward higher magnitude signals in downstream analysis.

Mean centering makes every metabolite have a mean of zero by subtracting the average intensity of each metabolite in every sample (Eq. 12.3):

$$\tilde{x}_{ij} = x_{ij} - \bar{x}_i \quad (12.3)$$

In Eq. 12.3, $\tilde{x}$ is the scaled intensity x of the metabolite i of sample j, $\bar{x}$ is the average intensity of metabolite i across all samples, respectively.

Higher magnitude metabolites will, however, also have greater absolute differences in intensity (or concentration) than lower magnitude metabolites (especially in datasets where transformations were not applied to reduce heteroscedasticity) (Karaman 2017).

Scaling complements mean centering since it specifically aims to transform these absolute differences into relative differences, such as fold changes. These methods also include mean centering, subtracting the mean of the corresponding variable to each value and performing the scaling by then dividing it by a scaling factor, which varies between the different methods. These scaling factors are mostly based on a dispersion measure, such as standard deviation, with size measures (mean, median) being a possible alternative. Although these very useful methods help emphasize the importance of low-concentration biologically important metabolites, they can also lead to an amplification of the error in these metabolites' intensity, assuming that the errors are relatively large in small values (van den Berg et al. 2006). Common scaling methods are auto scaling, Pareto scaling, range scaling, vast scaling, level scaling, and weighted scaling. For each method's equation, $\tilde{x}$ is the scaled intensity x of the metabolite i of sample j, s and $\overline{x}$ are the standard deviation and average intensity of metabolite i across all samples, respectively.

Auto scaling, also known as unit variance or standard scaling, uses the standard deviation of each metabolite as a scaling factor (Eq. 12.4). This directly achieves the proposed aim of centering and scaling data by giving equal weight to all features for further statistical analysis. However, this also best represents the aforementioned increase of error that scaling might lead to in uninformative features affected by noise that can then be interpreted as important. To apply this method, a considerable analytical effort has to be made during peak filtering to remove noisy and uninformative features of the dataset to minimize its shortcomings (Karaman 2017):

$$\tilde{x}_{ij} = \frac{x_{ij} - \overline{x}_i}{s_i} \tag{12.4}$$

Pareto scaling is a slight modification to auto scaling that uses the square root of the standard deviation as the scaling factor (Eq. 12.5). This combines elements of both mean centering and auto scaling and still increases the importance of low-concentration metabolites while limiting the inflation of measurement errors and preserving the data structure. This way, the method limits the influence of the drawbacks of mean centering and auto scaling, but it is still not entirely free from them. While sensitive to large fold changes, it does not present any major drawback due to the compromise taken (van den Berg et al. 2006; Karaman 2017):

$$\tilde{x}_{ij} = \frac{x_{ij} - \overline{x}_i}{\sqrt{s_i}} \tag{12.5}$$

Range scaling, or min–max scaling, uses the range between the maximum and the minimum value a variable has across all samples as a scaling factor (Eq. 12.6), thus aiming to remove instrumental response factors (Smilde et al. 2005). The logic behind the scaling factor is that it represents the biological range of each metabolite (van den Berg et al. 2006). A disadvantage of the method is that its scaling factor depends only on the maximum and minimum values of a metabolite, making it very sensitive to outliers on any extreme (the minimum tends to be an imputed value given to missing values):

$$\tilde{x}_{ij} = \frac{x_{ij} - \overline{x}_i}{x_{i\,(\max)} - x_{i\,(\min)}} \tag{12.6}$$

Vast scaling, or VAriable STability scaling, is another extension of auto scaling that multiplies its result by the ratio between the mean and the standard deviation of metabolite i (prior to auto scaling) (Eq. 12.7). The purpose of this extra factor is to diminish the inflation of measurement errors—the major drawback of auto scaling—and focus on more "stable" features. This way, the low-abundance metabolites (low average) with a high relative error (high standard deviation) will have a low $\overline{x}_i/s_i$ ratio and will have less importance (Keun et al. 2003). However, a problem that arises is that biologically relevant features that have considerable fold changes between samples will also be considered unstable (low $\overline{x}_i/s_i$ ratio) and may be overlooked. Thus, this method focuses on features that change subtly between samples—stable features—which is the opposite of most other strategies:

$$\tilde{x}_{ij} = \frac{x_{ij} - \overline{x}_i}{s_i} \times \frac{\overline{x}_i}{s_i} \tag{12.7}$$

Level scaling uses the mean of a metabolite (or the median as an alternative)—a size measure—as a scaling factor (Eq. 12.8). It changes each value to the intensity percentage change from the mean of the metabolite, which helps see fold changes between samples (van den Berg et al. 2006):

$$\tilde{x}_{ij} = \frac{x_{ij} - \overline{x}_i}{\overline{x}_i} \tag{12.8}$$

Weighted scaling uses a weighting function as a scaling factor (Biswas et al. 2023). This scaling factor has the objective of being robust to the outliers by approaching their weights to zero by how far they are from the median values. This scaling can be made by applying the weighting function (Eq. 12.9) to the values and then performing auto scaling on the transformed values (Eq. 12.11 and 12.12). Here, $z_{0.05}^2$ is the upper 5% critical value of a standard normal distribution and mad is the median absolute deviation (Eq. 12.10) (Biswas et al. 2023):

$$w_{ij} = \min\left(1, \frac{z_{0.05}^2}{\left(x_{ij} - \mathrm{median}(x_i)\right)/\mathrm{mad}(x_i))^2}\right) \tag{12.9}$$

$$\mathrm{mad}(x_i) = \frac{1}{0.6754} \underset{j}{\mathrm{median}} \left| x_{ij} - \underset{j}{\mathrm{median}}\left(x_{ij}\right) \right| \tag{12.10}$$

$$x_{ij}{}^{(w)} = w_{ij} \times x_{ij} \tag{12.11}$$

$$\tilde{x}_{ij} = \frac{x_{ij}{}^{(w)} - \overline{x_i{}^{(w)}}}{s_i{}^{(w)}} \tag{12.12}$$

Each of these scaling methods provides a different outlook to scaling the data and brings its own sets of advantages and disadvantages, showing how the choice of pretreatment methods depends on the objective of the analysis and should be regarded as a challenging and key task in data analysis. The choice depends on the biological question and the analysis method: auto scaling can be used for equal contribution across all compounds, and Pareto scaling can be used if more intense signals are biologically significant. Nevertheless, the data before and after scaling should be always visualized to confirm that the transformation is appropriate.

12.3.4 Metabolite Occurrence-Based Methodologies

Besides this more common workflow in metabolomics, other approaches focused on metabolite occurrence have been suggested. For these approaches, the sole presence or absence of a metabolite in a sample is considered informative (Traquete et al. 2021). The rationale behind this idea is the lower reproducibility of signal intensity in comparison with occurrence. The lack of reproducibility in intensity values between laboratories has been reported, even when analyzing the same samples and following identical protocols, despite a large fraction of the same metabolites being identified (Lin et al. 2020). Thus, substituting intensity values with occurrence data may lead to a more robust analysis (Traquete et al. 2021). The information that can be deduced from occurrence depends on the fraction of missing values. These methods are better suited to extreme resolution methodologies such as FT-ICR-MS with sub-1 ppm mass accuracy.

These treatments are markedly different from the more commonly used workflow, replacing the data pretreatment step (and missing value imputation processing step). There are currently two approaches in this category: binary simplification and sample mass-difference networks (sMDiNs).

Binary simplification (BinSim) is a straightforward form of pretreatment tailored for qualitative analysis, providing a robust yet simplified representation of metabolomics data. This treatment transforms the data matrix by replacing non-missing intensities with 1 and all missing values with 0 (Traquete et al. 2021). The resulting matrix can then be analyzed by all the different statistical analysis methods as it is done with a usual data matrix. Thus, it is a complementary approach to the usual workflow since it will highlight a different set of metabolites focusing especially on those that are exclusive of one or a specific set of biological classes (Traquete et al. 2021). It is also suited for datasets with substantial intensity variability, being insensitive to variations in sample preparation, instrument sensitivity, or experimental conditions.

The application of sample Mass-Difference Networks (sMDiNs) is far more complex, as it uses the mass differences between values of *m/z* to infer potential biochemical transformations in a sample (Traquete et al. 2022). In Mass-Difference Networks (MDiNs) (Breitling et al. 2006), the masses of metabolic signals are nodes in a graph, connected by edges representing mass differences from a list of defined mass-differences related to neutral losses or enzymatic transformations (Tziotis et al. 2011; Moritz et al. 2017). They are mostly used for formula annotation or data visualization but have also been used to identify chemical transformation with different prevalence between classes (Moritz et al. 2017). sMDiNs use this framework by building an MDiN for each sample analyzed. These networks can then be analyzed by graph-based methodologies such as profiling the degree or closeness centrality of every node in each network or by the prevalence of each chemical transformation, making up a data matrix that can be used for statistical analysis (Traquete et al. 2022). Constructing sMDiNs requires extreme mass accuracy to reliably detect small differences (sub-1 ppm) and has the advantage of allowing metabolic pathway discovery, as it helps to infer biochemical relationships without requiring metabolite identification (Traquete et al. 2022).

12.4 Univariate and Multivariate Statistical Analyses

Data preprocessing and pretreatment cleaned and highlighted the induced biological variation in the data to enhance the results of statistical analysis of the metabolomics data. The statistical analysis step of an untargeted metabolomics workflow has a plethora of different approaches and types of analyses available. As already discussed, metabolomics is characterized by high-dimensional data, with the number of measured variables being much higher than the number of samples. This poses specific issues to the statistical methods due to multicollinearity (Palermo et al. 2009), which cannot be corrected by the previously described pretreatments other than the removal of highly collinear signals. The latter approach also causes a problem, since relevant predictors can be highly correlated due to their similar variations between biological classes. Statistical analysis can be broadly classified into univariate or multivariate analysis which can be used individually or in conjunction. Briefly, univariate analysis involves analyzing each individual metabolite separately, one at a time, to assess its behavior, while multivariate analysis considers each sample a multivariate vector and thus takes into account the interactions and relationships between compounds (Saccenti et al. 2013).

12.4.1 Univariate Analysis

Univariate analysis is applied to investigate whether there is a significant difference or relationship between a single metabolite and the classes or groups being studied. These methods have the disadvantage of considering a multivariate dataset as multiple individual variables, which will therefore fail to consider the interactions between different metabolites expected from a dynamic system. They counterbalance this disadvantage by being relatively simple to use and easily interpretable (Saccenti et al. 2013). The choice of the statistical test is tied to the assumptions on the characteristics of the data regarding their distribution and variance, as well as the design of the experiment (Vinaixa et al. 2012).

When the univariate tests are applied iteratively to all compounds in a dataset, the likelihood (and number) of false-positives increases and they become almost inevitable when thousands of tests are being performed—the problem of multiple testing. Therefore, multiple test correction procedures need to be applied to control the number of false-positives while trying to prevent missing true-positives. Common correction methods are the Bonferroni correction and the False Discovery Rate (FDR). The Bonferroni correction is a conservative approach to minimize false-positives (increasing false-negatives) that changes the threshold to reject the null hypothesis by dividing the predefined threshold by the number of tests performed, $\text{threshold}_{\text{new}} = \text{threshold}_{\text{old}}$ / number of tests (Benjamini and Hochberg 1995; Vinaixa et al. 2012). Minimizing the FDR is a less conservative approach that aims to minimize the ratio of false-positives to true-positives instead of just reducing the absolute number of false-positives. It transforms the set thresholds (that indicate that $x\%$ of all tests will result in false-positives) into a q-value that indicates that $x\%$ of all significant tests will be false-positives (Benjamini and Hochberg 1995; Vinaixa et al. 2012). The Benjamini–Hochberg procedure is the most common among the FDR-based methods. This procedure orders every p-value obtained from the statistical test for every feature from the lowest to highest and considers features significant (rejecting null hypothesis) if the condition indicated in Eq. 12.13 is verified (Benjamini and Hochberg 1995):

$$p-\text{value} \le \frac{\text{rank of feature by lowest } p-\text{value}}{\text{number of features}} \times q-\text{value} \tag{12.13}$$

12.4.2 Multivariate Analysis

As previously mentioned, multivariate analysis considers metabolomics data by focusing on samples as an ensemble of multivariate vectors. As such, these methods use mathematical concepts, like distance and covariance, to establish models capable of finding global structural patterns to characterize data or of making predictions based on such data patterns. Unlike univariate analysis, metabolites are not treated independently, which means that possible interactions are implicitly considered (Saccenti et al. 2013). However, due to the thousands of metabolic variables in metabolomics datasets compared to the number of samples, aggravated by the metabolites that can be biologically correlated and connected and biological transformations, multicollinearity of the data is unavoidable. This is an issue for multivariate methodologies that rely on building models. Multivariate analysis methods are roughly subdivided into unsupervised or supervised learning methods. Unsupervised methods analyze data without information regarding the biological classes of the samples and thus detect intrinsic patterns within the data (Cuperlovic-Culf 2018). Supervised analysis methods use the biological classes' information of *training* data to build predictive models suitable to classify unknowns.

12.4.2.1 Unsupervised Statistical Analysis

Unsupervised statistical analysis is a type of multivariate analysis that aims to detect intrinsic patterns within the data to group or separate them purely on a data-driven way (Cuperlovic-Culf 2018). They can be used as data integrity check or, more commonly, for data visualization. The most widely used methodologies are Principal Component Analysis (PCA) and Clustering, with the Hierarchical Clustering Analysis (HCA) and K-means clustering being the most popular.

Principal Component Analysis (PCA) is a dimension reduction algorithm that focuses on reducing the dataset dimension from its number of variables (n) to a small number of orthogonal Principal Components (PCs), (Hotelling 1933; Jolliffe 2011; Jolliffe and Cadima 2016; Hansen 2007). These PCs are directions in the n-dimensional space designed to represent a new coordinate system computed by maximizing the variance of the projections of the data point on these new coordinates. They are linear combinations of the original variables. Therefore, the first PC is defined as the direction in the *n*-dimensional space, which corresponds to the largest variance in the data; the second PC is defined as the direction of the largest variance of the data orthogonal to the first PC; the third PC is orthogonal to the first two, and henceforth (Jolliffe and Cadima 2016; Hansen 2007). The new coordinates of our samples in these PCs are called PC *scores,* and the coordinates of variables in these PCs are called *loadings*. There are theoretically as many PCs as there are variables; however, the first PCs model most of the variation in the data. The dimension reduction algorithm comes from selecting only those PCs that explain the most variance which is usually much lower than *n*. The number of components to select is thus a crucial topic. Strategies include setting a threshold for the minimum accumulated explained variance by the different components (e.g., 80–90%) or observing when the variance explained by a new component compared to the variance explained by the prior component does not meaningfully improve the PCA model (Jolliffe 2011). In untargeted metabolomics, PCA as a descriptive technique is mainly used for data visualization. In this case, only two or three components that are selected represent the sample in a 2D or 3D PC projection plot, called a *scores plot* (such as the one in Fig. 12.1a). Since PC calculation depends on the magnitudes of the variables, prior centering and scaling are mandatory.

Clustering methods seek for proximity relationships among samples or variables. These relationships can be used to establish a predefined number of disjoint clusters or a hierarchy of progressively larger clusters in the whole dataset. There are currently several clustering methods, such as hierarchical clustering (agglomerative HCA, for example) and partitioning clustering (like the K-means clustering) (Andreopoulos et al. 2009).

In agglomerative HCA, all samples start in their own separate cluster and are progressively clustered together based on a similarity measure. The closest clusters are joined together first, and this process is performed iteratively until one cluster with all samples is obtained. The similarity is based on a distance metric, commonly the Euclidian distance between the samples, although others can be used depending on the nature of the

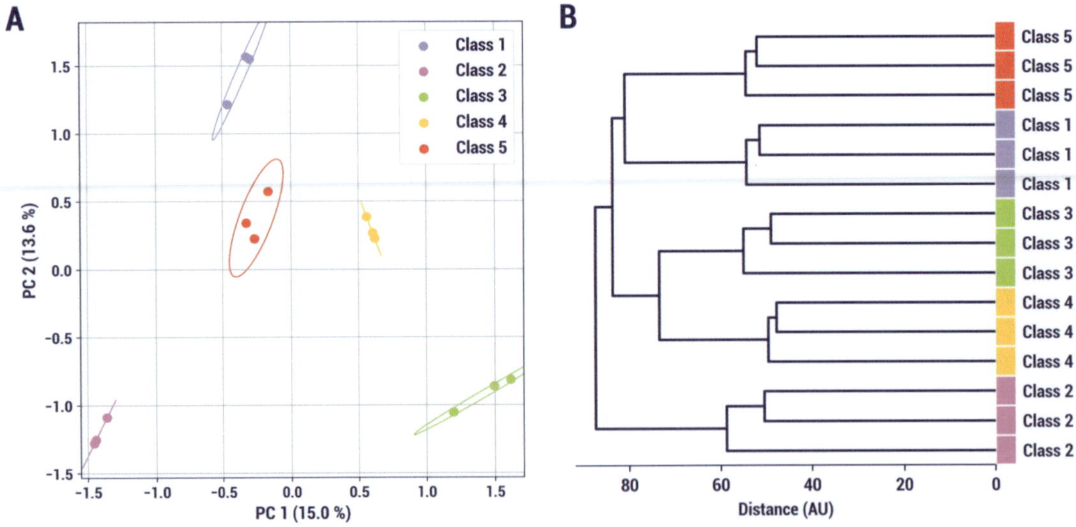

Fig. 12.1 Analysis results from PCA and hierarchical clustering. (**a**) Sample projection on the first two Principal Components with the explained variance of each Principal Component (PC) indicated in the respective axis. (**b**) Dendrogram representation of Hierarchical Clustering. Both representations were made from an example dataset with three samples of five different classes. (Data from Luz et al. 2021)

data and the research question. When data is normalized or standardized (mean-centered and scaled to unit variance), Euclidean distance effectively captures the magnitude of differences between variables, pairing effectively with the commonly used linkage criteria (like average, complete, or single linkage) to build dendrograms (Elmore and Richman 2001). Linkage criteria are needed for calculation of the distances between multi-sample clusters or multi-sample clusters and single samples. Several such criteria are commonly used. In average linkage or UPGMA (Unweighted Pair Group Method with Arithmetic Mean), the distance between two clusters is the average distance between all pairs of points (one from each cluster) (Sokal and Rohlf 1962). Ward's linkage, or Ward's minimum variance, merges clusters based on the minimum increase in total within-cluster variance (minimizes variance after merging) (Ward 1963) and is commonly used in untargeted metabolomics due to its focus on minimizing variance. The single linkage, also known as Nearest Neighbor, considers that the distance between two clusters is defined as the shortest distance between any two points in the clusters. The complete linkage, or Farthest Neighbor, considers that the distance between two clusters is defined as the largest distance between any two points in the clusters (Andreopoulos et al. 2009; Hastie et al. 2009; Hansen 2007). Like PCA, this unsupervised statistical method is widely used for data visualization and interpretation since it can be neatly represented as a dendrogram after applying a chosen distance metric and linkage method combination (Fig. 12.1b).

K-means clustering is a type of partitioning-based clustering method where k clusters are made with a centroid being representative of each cluster (k is defined by the user), (MacQueen 1967). Samples are attributed to the cluster defined by the closest centroid in the n-dimensional space (n being the number of variables in the studied dataset), and the centroids are updated to minimize the distance to the samples that are in their cluster which can be negatively affected by outliers. These two steps are repeated until the samples in each cluster do not change between iterations or their shift is below a predefined threshold. Since the starting centroids are chosen randomly, this method does not always generate the same results, and the algorithm may stop at a local minimum instead of a global minimum (Andreopoulos et al. 2009; Hastie et al. 2009). Thus, repeating the algorithm may help find the best estimation of the global minimum. When analyzing an untargeted metabolomics dataset, it is important to test different linkage criteria and evaluate the results to help ensure robust clustering outcomes.

Independent of the chosen clustering method, it is important to assess its quality. However, defining what is a good clustering is difficult, making its evaluation also difficult (Kleinberg 2002; Palacio-Niño and Berzal 2019). Two types of analyses can be performed: internal (unsupervised) validation, looking at the intrinsic patterns of the data without any external validation (like clustering tendency and number), with an example being the Silhouette Coefficient, or external (supervised) validation, using outside information defining a *ground truth* (such as class membership) to compare and evaluate the results (Palacio-Niño and Berzal 2019), with two examples being the Rand Index and the Correct First Clustering Percentage. The Silhouette coefficient is a metric that aims to evaluate cluster validity by comparing the distances of samples, inter- and intra-clusters (Rousseeuw and Rousseeuw 1987). It measures how well data points are clustered by comparing how similar a data point is to its own cluster (cohesion) versus other clusters (separation). For each sample in a dataset, the Silhouette coefficient (s) for a datapoint i is given by Eq. 12.14:

$$s(i) = \frac{b(i) - a(i)}{\text{maximum } \{a(i), b(i)\}} \tag{12.14}$$

where $a(i)$ and $b(i)$ are the mean distance of sample i to the samples in its cluster (cohesion) and to the nearest cluster (separation), respectively. Thus, it is a value between -1 and 1, where a value closer to 1 means that points are well-clustered and far from neighboring clusters, and a negative value means that points are likely misclassified into the wrong cluster (Rousseeuw and Rousseeuw 1987).

The Rand index evaluates how similar the clustering made is to the natural group separation of the samples on the dataset (defined a priori of the analysis) (Rand 1971). This index compares *k* clusters (each is a group) to the original class memberships. Thus, each pair of samples is tested to see if their group relation by the original class memberships and by the clusters made from clustering is in agreement. They are in agreement if the two samples that belong (or do not belong) to the same original class memberships were or were not clustered in the same cluster. For example, considering the samples shown in Fig. 12.1b, the class memberships and clustering performed are in agreement if two samples of the same class were clustered in the same cluster and are also in agreement if a sample from a class and another from another were clustered in different clusters. The Rand index is then the ratio of every pair of samples that is in agreement by the total amount of pairs of samples (Hubert and Arabie 1985; Santos and Embrechts 2009). The Adjusted Rand Index is a slight variation of the Rand Index that accounts for the chance, that is, the expected ratio of pairs in agreement if the clustering had happened at random (Hubert and Arabie 1985). By doing this, zero becomes an expected value if the clustering is random, becoming progressively better the closer it is to 1 (Hubert and Arabie 1985; Santos and Embrechts 2009). Note that slightly negative values can also occur—worse than the average random. This metric is one of the most popular ones for evaluating clustering; however it requires that the clusters are predefined, that is, for HCA, we must define a threshold or a cut-off line to define the clusters in the agglomerative procedure.

The Correct First Clustering Percentage is a metric tailored to Hierarchical Clustering Analysis where the number of clusters is not predefined (Traquete et al. 2021). As indicated by the name, it is the ratio of samples which are correctly clustered in the first time they cluster in the agglomerative HCA. Correctly clustered is defined as a sample clustering with only the sample of their own class. This allows an evaluation without definition of number of clusters surpassing that issue with Rand Index. However, it comes with their own issue of only being a shallow analysis at the level of the first few clusterings without considering the full structure of the clustering.

12.4.2.2 Supervised Statistical Analysis

Supervised methodologies are methods that build models based on metabolomics data with a priori knowledge of biological class memberships (discrete) or of the response variables (continuous). In the former case, the models built will be classifiers, while in the latter case, the models will be regressions. In both cases, these models aim to make predictions about new data that could undergo the same experimental and data analysis procedures but lacking information about class membership or response variable values. In general, the quality of these predictions must be assessed either by applying the model to new testing data by extending the metabolomics study results or, more commonly, by splitting the already obtained data into a *training* set, used to develop the model and a *testing* set, used to assess the performance of model predictions.

Regressions attempt to predict the best fit of an "unknown" (to the model) sample to the continuous response variable. Thus, prediction quality can be assessed by continuous metrics such as the root-mean-squared error. On the other hand, classifiers will predict the discrete class memberships of the unknown test samples. Their performance has then to be assessed with different metrics such as accuracy, F-beta score, precision, or recall that focus on correct (or incorrect) predictions. In untargeted metabolomics, classification problems such as distinguishing between different biological classes are more common than regressions; thus, here, more emphasis will be put on classifiers. Common supervised methodologies for classification are Partial Least Squares (or Projection to Latent Structures)—Discriminant Analysis (PLS-DA) and Random Forests (RF).

Partial Least Squares Discriminant Analysis (PLS-DA)

Partial Least Squares—Discriminant Analysis (PLS-DA) is a popular classification method in untargeted metabolomics since it can analyze efficiently high-dimensional data with multicollinearity and does not make assumptions on the distribution of the underlying data. Sample class membership is numerically encoded, and the numerical regression predictions are converted to categorical class membership predictions by the use of decision rules (Mateos-Aparicio 2011). Like PCA, PLS is a dimension reduction algorithm. However, instead of maximizing the variance of the projections of the samples on the principal components, PLS maximizes the covariance between the dataset (the samples)—X—and a vector or a matrix representing class memberships—Y [116,122] (Lee et al. 2018; Lee et al. 2019). The PLS algorithm used is slightly different depending on whether there are two classes (PLS1-DA) or more than two (PLS2-DA) to predict (Lee et al. 2018; Brereton and Lloyd 2014). For two classes, class membership is encoded in a vector of 0 s and 1 s, where each number represents a categorical class. For more than two categorical classes, a dummy matrix is constructed with one-hot encoding, where each column represents a class with a sample having a 1 or a 0 in that column whether they belong to the class or not, respectively. PLS components are projections called Latent Variables (LVs) and fullfil the same role as the principal components of PCA, being a new coordinate system where original data can be projected, thus defining *scores* and *loadings* as in PCA (while adding the computation of weights) (Lee et al. 2018; Lee et al. 2019; Mehmood et al. 2012; Wold et al. 1993). The number of LVs in the model should be restricted to those which contribute to its predictive significance using validation methods (Wold et al. 1993). Although all the general measures of performance assessment for classifiers can be applied to PLS-DA, a popular measure of performance that is specific to this method is 1—Predictive Residual Sum of Squares (PRESS)/SS), where SS is the residual sum of squares of Y corrected for the mean—Q^2—a value that the closer to one the better. Furthermore, the number of LVs used in the model can be chosen by looking for a maximum or convergence of Q^2.

To either select important compounds for class discrimination or for building a model with only more meaningful variables, we can apply different strategies to PLS-DA models (as reviewed by Mehmood et al. 2012). Some of these methods (filter methods) aim to identify important features based on a certain measure. These measures can be a set of parameters obtained by the PLS model, specifically the absolute values of X-weights or the regression coefficients (measure of association between X and Y) (Mehmood et al. 2012). An ordered list of their absolute values provides an indication of the most important features to build the PLS-DA model. Variable Importance/Influence in Projection (VIP) score (Wold et al. 1993) is another method that estimates the importance of each compound to explain the variance in Y using Eq. 12.15, which takes into account the variance explained of each LV in the model in comparison to the total variance in Y and the contribution of each compound to each LV (represented as the X-weights) to select those that are most important to the model (Mehmood et al. 2012; Galindo-Prieto et al. 2014). Since the average of all squared VIP scores is 1, this number is usually used as a minimum threshold to consider a feature as important or relevant. However, a careful consideration of a suitable threshold based on the results obtained is recommended.

$$\mathrm{VIP}_i = \sqrt{n \times \frac{\sum_{a=1}^{n^\circ \mathrm{LV}} \left(q_a^2 \times t'_a \times t_a\right) \times \left(w_{ai}/\|w_a\|\right)^2}{\sum_{a=1}^{A} q_a^2 \times t'_a \times t_a}} \tag{12.15}$$

In Eq. 12.15 (adapted from Mehmood et al. 2012), q_a and t_a are Y-loadings and X-scores of the ath LV, w_{ai} is the x-weight of the ath LV for feature I, and n is the total number of features.

Many different variants of PLS-DA have been developed since the start of PLS-DA use in statistical analysis coming from the different variants of PLS, such as the orthogonal PLS-DA (O-PLS-DA) or the sparse PLS-DA (sPLS-DA), the natural extensions from O-PLS (Trygg et al. 2002) and sPLS (Lê Cao et al. 2008).

Random Forest (RF)

Random Forest is a classifier that consists of an ensemble of independent decision trees, each built with a subset of the training data. Each tree makes a decision about which class a tested sample belongs to. The global result for the Random Forest classifier is then based on the majority vote of all de-correlated individual trees (Breiman 2001; Probst and Boulesteix 2017; Hastie et al. 2009). This allows for a better performance, resulting in less variance (more robust) compared to using single-decision trees (Probst and Boulesteix 2017), and shows less susceptibility to overfitting (Breiman 2001). A decision tree is a model that makes binary decisions, each internal node on the tree representing one of these decisions so that it can split the samples in their respective classes in the terminal nodes or leaves (Rokach and Maimon 2005), (Fig. 12.2). With each decision made, the data subset is split into two different subsets. The rules (decisions) are chosen to minimize an impurity measure such as the Gini Index/Impurity (Hastie et al. 2009; Louppe et al. 2013). The Gini Index is a measure of the probability of a random sample being incorrectly labeled if randomly labeled by the node distribution of the training samples' classes where that sample is. Each binary decision is then made so that misclassification of samples used to build the tree is decreased as much as possible until all subsets of samples only have samples of one class or until a maximum tree depth predefined by the user is reached. Thus, tree size is an important parameter to control the complexity of the tree and minimize overfitting (Hastie et al. 2009). A test sample, after navigating the tree, will be classified as belonging to the majority group of the terminal node that it reaches. For random forest, each tree is independently built based on a random bootstrap sampling of the training data, with the forest being a set of de-correlated trees (Probst and Boulesteix 2017; Hastie et al. 2009). The number of trees to build in a random forest is thus an important parameter. With more trees added to the forest, the generalization error of the prediction converges, which means that adding more trees will not lead to overfitting the model (Breiman 2001; Probst and

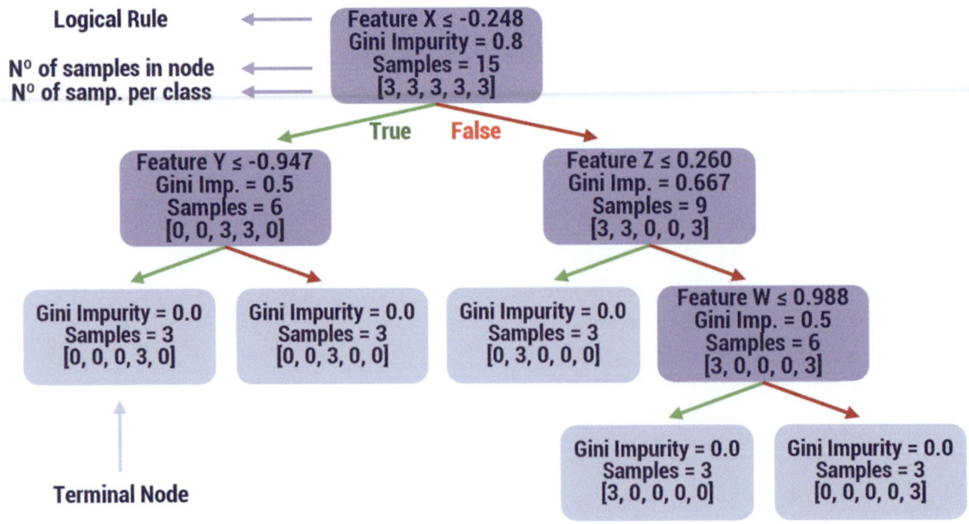

Fig. 12.2 Example of a small decision tree in a Random Forest. All samples are part of the training set used to build the example decision tree

Boulesteix 2017). This means that an appropriate number of trees to use is one that is sufficient to approximate the converged generalized error and, therefore, the convergence of the model accuracy without being too computationally expensive. Other important parameters include tree size or defining the complexity of individual trees to avoid them being overfitted (see the available parameters of the algorithm used, e.g., RandomForestClassifier with the Python package scikit-learn (Pedregosa et al. 2011) or the randomForest package in R language (Liaw and Wiener 2002)).

Due to being robust to overfitting and making discrete decisions without many assumptions about the data, Random Forest can be applied to great effect with the collinear nature of metabolomics data. When the impurity measure used is the Gini Index, this is called the Gini Importance or Mean Decrease Gini (Louppe et al. 2013). Other approaches include the Mean Decrease Accuracy, which assesses importance by assessing its impact on the performance of the model (Louppe et al. 2013).

Model Validation Methods

An essential and ubiquitous step for all supervised methodologies is the validation of the models built. This is crucial to evaluate how robust the model is and to avoid overfitting. Model validation is usually external validation by holding out test samples, internal validation bootstrapping, or permutation testing (Lee et al. 2018). Internal validation tends to be used when the number of samples is low and the loss of a few samples to test the model would have a significant impact on the model built, while external validation is better used when there is a greater number of samples (Lee et al. 2018).

Internal validation techniques include *k*-fold cross-validation (CV), stratified *k*-fold CV, and leave-one-out cross-validation (LOOCV). This consists of iteratively splitting the training data that will be used for the final model in a further subset with training sets and test sets in different ways (depending on the method), respectively, and train and test the performance of the model until all samples were in the test set once. The mean of the performance metrics of the models built for each test set is an estimate of the performance of the prediction model built with the full dataset. *K*-fold CV randomly splits the samples of a dataset into *k* different sample sets, with 5, 7, or 10 being the most common, where one of them is iteratively the test set and all others are the training set. Stratified k-fold CV follows the exact same approach but guarantees that each biological class has its samples evenly split between the different sample sets. The higher the number of folds, the higher the computational intensity and the lower the number of folds, the more pessimistic our estimation is (Kohavi 1995). Leave-one-out CV is this logic pushed to the extreme by having each sample set comprised by a single sample—in each iteration, a model is built with all samples, except one.

External validation works by holding out a set of samples that is only used to test the model, while the rest of the samples are used to build the model. The two main parameters are the number of the samples excluded, around 30% of the total samples being the norm, and the procedure to select the samples to leave out, such as random sampling or Kennard–Stone (Kennard and Stone 1969) sampling used to have a uniform distribution between training and test sets (Lee et al. 2018).

Bootstrapping (Efron 1979) is a wide family of methods based on the idea of sampling *n* instances from the data with replacement, that is, the same sample can be sampled multiple times (Kohavi 1995). Thus, samples are chosen in this way for the training set, and the remaining samples which were not picked are selected for the test set. The training and test sets have the same function as in CV.

Permutation tests determine the probability of observing an equal or better performance of the predictive model compared to pure-chance predictions, that is, the significance of the predictive accuracy of the model. This is achieved by randomly permuting the class labels of each sample in the training set, building the model (with the same parameters) with the permuted labels' dataset, and comparing the performance of its prediction model. The rationale is that if there is some intrinsic structure within the groups, the prediction model should have a greater accuracy compared to when the data is jumbled. Thus, many label permutations are repeated, and if our original prediction model has a better performance $x\%$ of the times, with x normally being equal to 95% or 99%, the prediction model's performance is significantly better than the models built with randomly labeled datasets (Westerhuis et al. 2008).

12.5 Chemical Diversity Visualization

For biological interpretation, it is useful for the complex chemical diversity of different samples to be represented in simplified and visually clear ways. Some common representations of a system's chemical diversity are Van Krevelen, Kendrick Mass Defect (KMD) plots, and Mass-Difference Networks (MDiNs).

Van Krevelen plots represent the carbon, hydrogen, and oxygen (CHO) chemical space by using the identified formulas to plot the ratio of the number of hydrogen to carbon atoms (H/C) over the ratio of the number of oxygen to carbon atoms (O/C) (Fig. 12.3a). The areas the different formulas occupy can be used to estimate to which groups or categories they belong, such as amino acids, carbohydrates, lipids, lignins, amino acid-like, or carbohydrate-like compounds, among others (Kim et al. 2003; Roullier-Gall et al. 2014). For example, this representation has been used in several works that analyzed the chemical diversity of wines (Gougeon et al. 2009; Roullier-Gall et al. 2014). This rough estimation is far from perfect, leading to many inaccuracies since it only uses the information of O/C and H/C ratios (Rivas-Ubach et al. 2018). Moreover, there is a lack of concretely defined boundaries for each category and compound class and considerable overlap among these. These factors associated with the difficulty in categorizing compounds solely based on their H/C and O/C ratios make the classification often inefficient and not reliable (Rivas-Ubach et al. 2018).

KMD plots represent the Kendrick Mass Defect over the Kendrick nominal masses (Fig. 12.3b). The Kendrick mass of a compound is given by standardizing the group CH_2 to exactly 14 (Kendrick mass) by Eq. 12.16. The Kendrick Mass Defect (KMD) is the difference between the nominal Kendrick mass and the exact Kendrick mass (Hughey et al. 2001). Thus, compounds with the same heteroatoms and same amount of double bonds plus rings (changes by H_2 or exactly by 0.0134 KMD) will be represented in a horizontal line, with their position on the x-axis being based on the number of CH_2 groups. Therefore, it allows the visualization of the different classes and types of compounds in the samples due to their characteristic KMD in a 2D space (Hughey et al. 2001). Color coding of different formulas based on a characteristic further helps the visualization of the chemical diversity. For example, in another work, when analyzing the chemical diversity in grapevines, formulas were colored based on the chemical elements they possess, CHO, CHON, CHOS, CHOP, CHONS, CHONP, and CHONSP (Adrian et al. 2017).

$$m_{\text{Kendrick}} = m_{\text{IUPAC}} \times 14 \div 14.01565 \text{ (IUPAC mass of the group } CH_2) \quad (12.16)$$

MDiNs is a more recent approach to represent chemical diversity. They were developed specifically for metabolomics data. As stated earlier, they were developed by Breitling et al. (2006). Their rational is to represent a structure similar to a metabolic network where the nodes are the detected metabolites which are connected by edges representing specific mass differences between these metabolites, which correspond to common biochemical transformations (Moritz et al. 2017; Tziotis et al. 2011). Furthermore, these networks consider possible spontaneous nonenzymatic reactions and are not skewed by the known metabolic pathways (Breitling et al. 2006; Moritz et al. 2017). Thus, it represents the diversity in the data although it becomes more

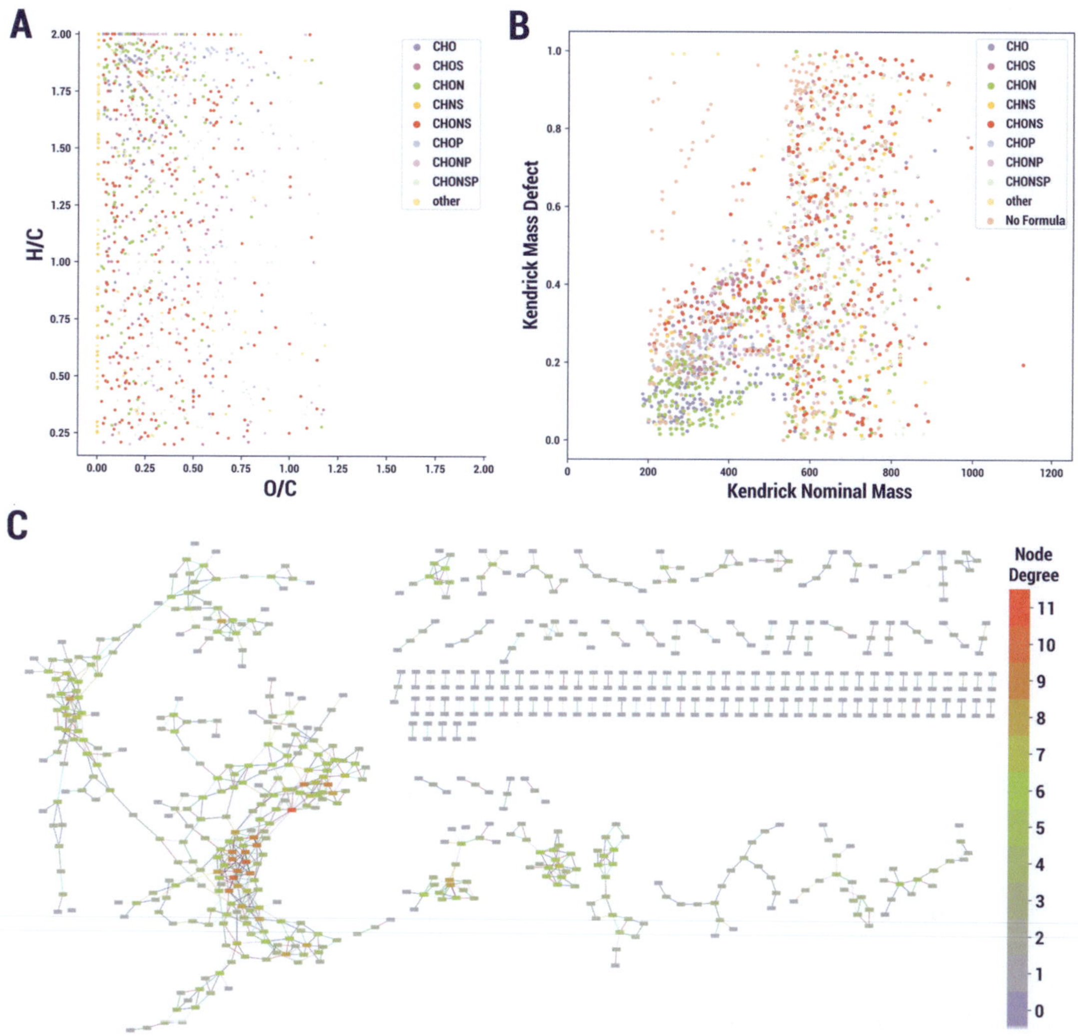

Fig. 12.3 Example of a Van Krevelen diagram (**a**), a Kendrick Mass Defect plot (**b**) and a Mass-Difference Network (**c**) of metabolomics data. Dots are coloured according to their elemental formula composition in (**a** and **b**). Network nodes are coloured by their degree (indicated by a color gradient from purple to red) and edge colours represent each MDB. Network representations were made with the Cytoscape 3.8.1 (Shannon et al. 2003). (Data from Luz et al. 2021)

difficult to interpret, as represented in Fig. 12.3c. The exact mass differences correspond to the changes in the elemental formula of metabolites due to common biochemical reactions such as: methylations (total change of CH_2 after substitution of –H hydrogen atom by a $-CH_3$ group), hydrogenations/dehydrogenations (H_2), and phosphorylations ($PO_3(H)$), among many other possibilities (such as amino acids for bigger groups) that have specific masses, for example, H_2 = 2.01565 Da and CH_2 = 14.01565 Da (Breitling et al. 2006; Tziotis et al. 2011). MDBs can be chosen a priori by selecting a set of common biochemical reactions or by the most common mass differences in the dataset, where a specific chemical transformation can be assigned (Moritz et al. 2017). Other than data visualization (as well as the pretreatment mentioned earlier),

MDiNs can also be used with two further objectives commonly. They can be used for formula annotation, where one single reliable annotated formula/compound in a network component can be propagated to all other nodes by subtracting or adding the corresponding elemental composition based on the mass difference of the edges while keeping in mind certain characteristics of metabolites such as acceptable ratios between the different elements (Tziotis et al. 2011). On one hand, this makes formula assignment biased to formulas similar to the starting set of formulas, instead of considering the complete metabolite chemical space; on the other hand, these formulas belong to known metabolites that have a high probability of being present in the sample, and, therefore, it could be argued that formulas similar to these, especially when the difference can be explained by common biochemical reactions, also have a higher probability of being present in the same sample. Since it depends on very specific mass differences, it is better suited to high or extreme resolution methodologies that have a higher mass accuracy, leading to nonrandom and informative networks.

12.6 Compound Identification

This is an essential step in untargeted metabolomics data analysis, as it translates *m/z* data into meaningful biological entities such as metabolites. Despite being essential for revealing important metabolic pathways, identifying biomarkers, and gaining insights into the underlying biochemistry of the samples, it is also perhaps the major bottleneck in metabolomics data analysis. Challenges in compound identification in an untargeted metabolomics workflow arise from the complexity of metabolomes, with the detection of thousands of metabolites with diverse chemical structures, functional groups, and physicochemical properties, many with identical nominal masses (isobaric) or identical formulas (isomers), making them difficult to distinguish. Also, lower mass accuracy and resolution data can make difficult identification, making extreme resolution and mass accuracy instruments, like FT-ICR-MS, essential to unequivocally assign a molecular formula. Another problem comes from metabolite databases, which do not cover all naturally occurring or modified compounds, and this is particularly crucial for rare or unknown metabolites from secondary metabolism. The main strategies for compound identification will be presented.

The simplest way to perform compound identification is through mass-based annotation, by comparing the measured *m/z* values to theoretical masses of known metabolites in databases. It requires preferably the use of extreme resolution MS data to achieve accurate mass measurements (<1 ppm deviation). Different databases can be used for searching, like the broad chemical database Chemspider (Pence and Williams 2010), or the more specific databases such as the Human Metabolome Database (HMDB, Wishart et al. 2022), or the natural product LOTUS database (Rutz et al. 2022). In the search, possible ion adducts must be considered, for example, H^+, Na^+, or K^+ in positive ionization mode analysis, or H^- and Cl^- in negative ionization mode analysis. Afterward, if the theoretical and experimental mass values are within a chosen maximum mass deviation (normally expressed as parts per million—ppm), the signal is assigned to that compound.

In extreme mass accuracy MS data (1 ppm range), presently only achieved through FT-ICR mass spectrometers, through the monoisotopic compound mass (actually, the *m/z*) and the analysis of the isotopic distribution, it is possible to assign an unequivocal formula. At an extreme resolving power and high dynamic range, FT-ICR mass spectrometers have also the ability to detect the fine isotopic structure of more intense signals, making it possible to infer the elemental composition. Formula assignment starts with the definition of several parameters, like the elemental composition based on biological relevance (usually C, H, O, N, P, S), limiting the number of elements to avoid generating implausible formulas. Additionally, elements' ratio to carbon ranges, the presence of multiple heteroatoms, with respect to Senior and Lewis chemical rules, and the detection of the isotopic pattern for the metabolite are also parameters to consider (Kind and Fiehn 2007). With the development of powerful instrumentation like FT-ICR-MS, the definition of these rules must evolve, since they are no longer the primary decision-making criteria but act as secondary checks. In FT-ICR-MS metabolomics data, sub-ppm mass deviation drastically reduces the number of possible formulas, and the isotopic pattern detection improves confidence in elemental composition, making most of these legacy rules useless.

Challenges include the identification of isomeric compounds, with the same exact mass but different structures, and ambiguity in formula assignment, as multiple plausible formulas may fit a single *m/z* value, requiring additional information for a reliable annotation. For these cases, MS/MS (or MS^n) analysis can be performed, by selecting a specific peak in the sample and fragmenting it to re-analyze by MS, thus obtaining an MSMS pattern that can be used to provide a structural identification of the compound. This can be achieved by comparing the experimental MS/MS spectrum to a spectral library of previously acquired MSMS data (Xiao et al. 2012) and calculating a similarity score. Despite the continued growth of these libraries, there is still a lack of reference data and coverage (Xiao et al. 2012; Schrimpe-Rutledge et al. 2016). To further capitalize on MSMS data, molecular networking is a recent computational approach that allows visualization of the MSMS data in a network representation where a similarity score between MSMS spectra is calculated, making the MSMS spectra the network nodes, linked if they have a high similarity (Watrous et al. 2012). Molecular networking has gained popularity, especially with the publicly available GNPS (Global Natural Products Social Molecular Networking) (Wang et al. 2016) that allows both the representation of these networks with also access to large amounts of reference spectra that can be used for annotation of compounds, allowing grouping of spectra.

12.7 Conclusion

Metabolomics is continuing its fast evolution to grow closer and closer to the ideal objective of unambiguous identification and quantification of all metabolites in a biological system. This is still a far-off proposition; however, the past success of metabolomics is bringing more and more researchers into this field, accelerating this growth. It holds a unique potential to represent a biological system's phenotype at a certain time point, but it comes at a price of high complexity in data acquisition and analysis. Thus, this led to a concerted effort for the development of newer and better methodologies and techniques to use in every single step of the metabolomics workflow. A few examples include the resurgence of deep learning methodologies for both preprocessing steps such as peak picking as well as for supervised classifier model building, the automatization of data acquisition steps to minimize human variation, batch correction algorithms, the continued development of new in silico fragmentation and spectral matching tools to increase annotation coverage and accuracy of MS/MS data, and the increasing size of reference spectra databases. Although LC–MS has been the most popular approach for metabolomics mass spectrometry, these tools have been developed for LC–MS, GC–MS and extreme-resolution direct infusion MS, each with their own characteristics that can be capitalized on.

After data analysis, biological interpretation begins, and this is the most critical and complex step in metabolomics, transforming data into meaningful insights about biochemical pathways, physiological states, and disease mechanisms. Its importance lies in its potential to unravel novel biomarkers, elucidate metabolic networks, and provide a holistic understanding of biological systems. Additionally, the integration of metabolomics data with other omics layers and clinical information requires sophisticated computational tools and domain expertise. Advancements in pathway analysis algorithms, machine learning models, and database curation are progressively addressing these limitations. By overcoming these challenges, biological interpretation will continue to drive innovation in systems biology, precision medicine, and translational research, unlocking the full potential of untargeted metabolomics.

Acknowledgments This work has been funded by the Fundação para a Ciência e a Tecnologia (Portugal) through the project EXPL/BTA-BTA/1389/2021 and the PhD grants 2021.06370.BD to F. Traquete and 2023.05150. BDANA to J. Luz. We also acknowledge the support from the research center BioISI, grant UIDB/04046/2020, DOI 10.54499/UIDB/04046/2020. This work was conducted in the framework of the Research Infrastructure project European Network of Fourier-Transform Ion-Cyclotron-Resonance Mass Spectrometry Centers (EU FT-ICR-MS) funded by the European Union's Horizon 2020 research and innovation programme, Grant Agreement No 731077.

References

Adrian M, Lucio M, Roullier-Gall C, Héloir MC, Trouvelot S, Daire X, Kanawati B, Lemaître-Guillier C, Poinssot B, Gougeon R et al (2017) Metabolic fingerprint of PS3-induced resistance of grapevine leaves against plasmopara viticola revealed differences in elicitor-triggered defenses. Front Plant Sci 8:1–14. https://doi.org/10.3389/fpls.2017.00101

Andreopoulos B, An A, Wang X, Schroeder M (2009) A roadmap of clustering algorithms: finding a match for a biomedical application. Brief Bioinform 10:297–314. https://doi.org/10.1093/bib/bbn058

Bartel J, Krumsiek J, Theis FJ (2013) Statistical methods for the analysis of high-throughput metabolomics data. Comput Struct Biotechnol J 4:e201301009–e201301009. https://doi.org/10.5936/csbj.201301009

Benjamini Y, Hochberg Y (1995) Controlling the false discovery rate - a practical and powerful approach to multiple testing. J R Stat Soc Ser B 57:289–300. https://doi.org/10.2307/2346101

Biswas B, Kumar N, Hoque MA, Alam MA (2023) Weighted scaling approach for metabolomics data analysis. Jpn J Stat Data Sci. https://doi.org/10.1007/s42081-023-00205-2

Bolstad BM, Irizarry RA, Åstrand M, Speed TP (2003) A comparison of normalization methods for high density oligonucleotide array data based on variance and bias. Bioinformatics 19:185–193. https://doi.org/10.1093/bioinformatics/19.2.185

Breiman L (2001) Random forests. Mach Learn 45:5–32. https://doi.org/10.1023/A:1010933404324

Breitling R, Ritchie S, Goodenowe D, Stewart ML, Barrett MP (2006) Ab initio prediction of metabolic networks using Fourier transform mass spectrometry data. Metabolomics 2:155–164. https://doi.org/10.1007/s11306-006-0029-z

Brereton RG, Lloyd GR (2014) Partial least squares discriminant analysis: taking the magic away. J Chemom 28:213–225. https://doi.org/10.1002/cem.2609

Brown SC, Kruppa G, Dasseux J-L (2005) Metabolomics applications of FT-ICR mass spectrometry. Mass Spectrom Rev 24:223–231. https://doi.org/10.1002/mas.20011

Castillo S, Gopalacharyulu P, Yetukuri L, Orešič M (2011) Algorithms and tools for the preprocessing of LC-MS metabolomics data. Chemom Intell Lab Syst 108:23–32. https://doi.org/10.1016/j.chemolab.2011.03.010

Cuperlovic-Culf M (2018) Machine learning methods for analysis of metabolic data and metabolic pathway modeling. Meta 8. https://doi.org/10.3390/metabo8010004

Di Guida R, Engel J, Allwood JW, Weber RJM, Jones MR, Sommer U, Viant MR, Dunn WB (2016) Non-targeted UHPLC-MS metabolomic data processing methods: a comparative investigation of normalisation, missing value imputation, transformation and scaling. Metabolomics 12:93. https://doi.org/10.1007/s11306-016-1030-9

Dieterle F, Ross A, Schlotterbeck G, Senn H (2006) Probabilistic quotient normalization as robust method to account for dilution of complex biological mixtures. Application in 1H NMR metabonomics. Anal Chem 78:4281–4290. https://doi.org/10.1021/ac051632c

Efron B (1979) Bootstrap methods: another look at the jackknife. Ann Stat 7:1–26

Elmore KL, Richman MB (2001) Euclidean distance as a similarity metric for principal component analysis. Mon Weather Rev 129(3):540–549. https://doi.org/10.1175/1520-0493(2001)129<0540:EDAASM>2.0.CO;2

Ferreira AEN, Sousa Silva M, Cordeiro C (2021) Metabolic network inference from time series. In: Wolkenhauer O (ed) Systems medicine: integrative, qualitative and computational approaches. Elsevier, Oxford, pp 127–133

Fiehn O (2002) Metabolomics—the link between genotypes and phenotypes. Plant Mol Biol 48:155–171. https://doi.org/10.1023/A:1013713905833

Galindo-Prieto B, Eriksson L, Trygg J (2014) Variable influence on projection (VIP) for orthogonal projections to latent structures (OPLS). J Chemom 28:623–632. https://doi.org/10.1002/cem.2627

Ghaste M, Mistrik R, Shulaev V (2016) Applications of Fourier transform ion cyclotron resonance (FT-ICR) and orbitrap based high resolution mass spectrometry in metabolomics and lipidomics. Int J Mol Sci 17. https://doi.org/10.3390/ijms17060816

Gougeon RD, Lucio M, Frommberger M, Peyron D, Chassagne D, Alexandre H, Feuillat F, Voilley A, Cayot P, Gebefügi I et al (2009) The chemodiversity of wines can reveal a metabologeography expression of cooperage oak wood. Proc Natl Acad Sci USA 106: 9174–9179. https://doi.org/10.1073/pnas.0901100106

Hansen MAE (2007) Data analysis. In: Metabolome analysis. Wiley, pp 146–187

Hastie T, Tibshirani R, Friedman J (2009) The elements of statistical learning: data mining, inference, and prediction, 2nd edn. Springer. ISBN 9780387848570

Hotelling H (1933) Analysis of a complex of statistical variables into principal components. J Educ Psychol 24:417–441. https://doi.org/10.1037/h0071325

Hubert L, Arabie P (1985) Comparing partitions. J Classif 2:193–218. https://doi.org/10.1007/BF01908075

Hughey CA, Hendrickson CL, Rodgers RP, Marshall AG, Qian K (2001) Kendrick mass defect spectrum: a compact visual analysis for ultrahigh-resolution broadband mass spectra. Anal Chem 73:4676–4681. https://doi.org/10.1021/ac010560w

Jolliffe IT (2011) Principal component analysis. In: BT - international encyclopedia of statistical science, pp 1094–1096

Jolliffe IT, Cadima J (2016) Principal component analysis: a review and recent developments. Philos Trans A Math Phys Eng Sci 374:20150202. https://doi.org/10.1098/rsta.2015.0202

Karaman I (2017) Preprocessing and pretreatment of metabolomics data for statistical analysis. Adv Exp Med Biol 965:145–161. https://doi.org/10.1007/978-3-319-47656-8_6

Katajamaa M, Oresic M (2007) Data processing for mass spectrometry-based metabolomics. J Chromatogr A 1158:318–328. https://doi.org/10.1016/j.chroma.2007.04.021

Kennard RW, Stone LA (1969) Computer aided design of experiments. Technometrics 11:137–148. https://doi.org/10.2307/1266770

Keun H, Ebbels T, Antti H, Bollard M, Beckonert O, Holmes E, Lindon J, Nicholson J (2003) Improved analysis of multivariate data by variable stability scaling: application to NMR-based metabolic profiling. Anal Chim Acta 490:265–276. https://doi.org/10.1016/S0003-2670(03)00094-1

Kim S, Kramer RW, Hatcher PG (2003) Graphical method for analysis of ultrahigh-resolution broadband mass spectra of natural organic matter, the Van Krevelen diagram. Anal Chem 75:5336–5344. https://doi.org/10.1021/ac034415p

Kind T, Fiehn O (2007) Seven golden rules for heuristic filtering of molecular formulas obtained by accurate mass spectrometry. BMC Bioinformatics 8:1–20. https://doi.org/10.1186/1471-2105-8-105

Kleinberg J (2002) An impossibility theorem for clustering. In: Proceedings of the 15th international conference on neural information processing systems. MIT Press, Cambridge, MA, pp 463–470

Kohavi R (1995) A study of cross-validation and bootstrap for accuracy estimation and model selection. In: Proceedings of the 14th international joint conference on artificial intelligence, vol 2. Morgan Kaufmann Publishers Inc., San Francisco, CA, pp 1137–1143

Lazar C (2015) ImputeLCMD: a collection of methods for left-censored missing data imputation version 2.0. Available online: https://rdrr.io/cran/imputeLCMD/

Lê Cao K-A, Rossouw D, Robert-Granié C, Besse P (2008) A sparse PLS for variable selection when integrating omics data. Stat Appl Genet Mol Biol 7:35. https://doi.org/10.2202/1544-6115.1390

Lee LC, Liong C-Y, Jemain AA (2018) Partial least squares-discriminant analysis (PLS-DA) for classification of high-dimensional (HD) data: a review of contemporary practice strategies and knowledge gaps. Analyst 143:3526–3539. https://doi.org/10.1039/c8an00599k

Lee LC, Liong C-Y, Jemain AA (2019) Statistical comparison of decision rules in PLS2-DA prediction model for classification of blue gel pen inks according to pen brand and pen model. Chemom Intell Lab Syst 184:94–101. https://doi.org/10.1016/j.chemolab.2018.11.014

Liaw A, Wiener M (2002) Classification and regression by RandomForest. R News 2:18–22

Lin Y, Caldwell GW, Li Y, Lang W, Masucci J (2020) Inter-laboratory reproducibility of an untargeted metabolomics GC–MS assay for analysis of human plasma. Sci Rep 10:1–11. https://doi.org/10.1038/s41598-020-67939-x

Louppe G, Wehenkel L, Sutera A, Geurts P (2013) Understanding variable Importances in forests of randomized trees. Adv Neural Inf Process Syst 26:431–439

Luz J, Pendão AS, Sousa Silva M, Cordeiro C (2021) FT-ICR-MS based untargeted metabolomics for the discrimination of yeast mutants. Figshare. Dataset. https://doi.org/10.6084/m9.figshare.15173559.v1

MacQueen JB (1967) Some methods for classification and analysis of multivariate observations. In: Le Cam LM, Neyman J (eds) Proceedings of the fifth Berkeley symposium on mathematical statistics and probability, vol 1. University of California Press, pp 281–297

Mateos-Aparicio G (2011) Partial least squares (PLS) methods: origins, evolution, and application to social sciences. Commun Stat Theory Methods 40:2305–2317. https://doi.org/10.1080/03610921003778225

Mehmood T, Liland KH, Snipen L, Sæbø S (2012) A review of variable selection methods in partial least squares regression. Chemom Intell Lab Syst 118:62–69. https://doi.org/10.1016/j.chemolab.2012.07.010

Misra BB (2020) Data normalization strategies in metabolomics: current challenges, approaches, and tools. Eur J Mass Spec 26(3):165–174. https://doi.org/10.1177/1469066720918446

Moritz F, Kaling M, Schnitzler JP, Schmitt-Kopplin P (2017) Characterization of poplar metabotypes via mass difference enrichment analysis. Plant Cell Environ 40:1057–1073. https://doi.org/10.1111/pce.12878

Nicholson JK, Holmes E, Lindon JC (2007) Chapter 1 - Metabonomics and metabolomics techniques and their applications in mammalian systems. In: Lindon JC, Nicholson JK, Holmes E (eds) The handbook of metabonomics and metabolomics. Elsevier Science B. V., Amsterdam, pp 1–33. ISBN 978-0-444-52841-4.

Palacio-Niño J-O, Berzal F (2019) Evaluation metrics for unsupervised learning algorithms. arXiv. https://doi.org/10.48550/arXiv.1905.05667

Palermo G, Piraino P, Zucht H-D (2009) Performance of PLS regression coefficients in selecting variables for each response of a multivariate PLS for omics-type data. Adv Appl Bioinform Chem 2:57–70. https://doi.org/10.2147/aabc.s3619

Pedregosa F, Varoquaux G, Gramfort A, Michel V, Thirion B, Grisel O, Blondel M, Prettenhofer P, Weiss R, Dubourg V et al (2011) Scikit-learn: machine learning in python. J Mach Learn Res 12:2825–2830

Pence HE, Williams A (2010) ChemSpider: an online chemical information resource. J Chem Educ 87: 1123–1124. https://doi.org/10.1021/ed100697w

Probst P, Boulesteix A-L (2017) To tune or not to tune the number of trees in random forest. J Mach Learn Res 18:6673–6690

Rand WM (1971) Objective criteria for the evaluation of clustering methods. J Am Stat Assoc 66:846–850. https://doi.org/10.2307/2284239

Rivas-Ubach A, Liu Y, Bianchi TS, Tolić N, Jansson C, Paša-Tolić L (2018) Moving beyond the van Krevelen diagram: a new stoichiometric approach for compound classification in organisms. Anal Chem 90:6152–6160. https://doi.org/10.1021/acs.analchem.8b00529

Roberts LD, Souza AL, Gerszten RE, Clish CB (2012) Targeted metabolomics. Curr Protoc Mol Biol 30: 30.2.1-30.2.24. https://doi.org/10.1002/0471142727.mb3002s98

Rocke DM, Durbin B (2001) A model for measurement error for gene expression arrays. J Comput Biol 8:557–569. https://doi.org/10.1089/106652701753307485

Rocke D, Durbin-Johnson B (2003) Approximate variance-stabilizing transformations for gene-expression microarray data. Bioinformatics 19:966–972. https://doi.org/10.1093/bioinformatics/btg107

Rokach L, Maimon O (2005) Decision trees. In: The data mining and knowledge discovery handbook, vol 6, pp 165–192

Roullier-Gall C, Witting M, Gougeon RD, Schmitt-Kopplin P (2014) High precision mass measurements for wine metabolomics. Front Chem 2:1–9. https://doi.org/10.3389/fchem.2014.00102

Rousseeuw P, Rousseeuw PJ (1987) Silhouettes: a graphical aid to the interpretation and validation of cluster analysis. J Comput Appl Math 20:53–65. https://doi.org/10.1016/0377-0427(87)90125-7

Rutz A, Sorokina M, Galgonek J, Mietchen D, Willighagen E, Gaudry A, Graham JG, Stephan R, Page R, Vondrášek J et al (2022) The LOTUS initiative for open knowledge management in natural products research. elife 11. https://doi.org/10.7554/eLife.70780

Santos JM, Embrechts M (2009) On the use of the adjusted rand index as a metric for evaluating supervised classification. In: Alippi C, Polycarpou M, Panayiotou C, Ellinas G (eds) Proceedings of the Artificial Neural Networks—ICANN 2009. Springer, Berlin, Heidelberg, pp 175–184

Saccenti E, Hoefsloot H, Smilde A, Westerhuis J, Hendriks M (2013) Reflections on univariate and multivariate analysis of metabolomics data. Metabolomics 10. https://doi.org/10.1007/s11306-013-0598-6

Schiffman C, Petrick L, Perttula K, Yano Y, Carlsson H, Whitehead T, Metayer C, Hayes J, Rappaport S, Dudoit S (2019) Filtering procedures for untargeted LC-MS metabolomics data. BMC Bioinformatics 20: 334. https://doi.org/10.1186/s12859-019-2871-9

Schmid R, Heuckeroth S, Korf A, Smirnov A, Myers O, Dyrlund TS, Bushuiev R, Murray KJ, Hoffmann N, Lu M et al (2023) Integrative analysis of multimodal mass spectrometry data in MZmine 3. Nat Biotechnol 41: 447–449. https://doi.org/10.1038/s41587-023-01690-2

Schrimpe-Rutledge AC, Codreanu SG, Sherrod SD, McLean JA (2016) Untargeted metabolomics strategies-challenges and emerging directions. J Am Soc Mass Spectrom 27:1897–1905. https://doi.org/10.1007/s13361-016-1469-y

Shannon P, Markiel A, Ozier O et al (2003) Cytoscape: a software environment for integrated models of biomolecular interaction networks. Genome Res 13(11): 2498–2504. https://doi.org/10.1101/gr.1239303

Smilde AK, Van Der Werf MJ, Bijlsma S, Van Der Werff-Van Der Vat BJC, Jellema RH (2005) Fusion of mass spectrometry-based metabolomics data. Anal Chem 77:6729–6736. https://doi.org/10.1021/ac051080y

Sokal R, Rohlf F (1962) The comparison of dendrograms by objective methods. Taxon 11:33–40. https://doi.org/10.2307/1217208

Stekhoven DJ, Bühlmann P (2012) Missforest-non-parametric missing value imputation for mixed-type data. Bioinformatics 28:112–118. https://doi.org/10.1093/bioinformatics/btr597

Sysi-Aho M, Katajamaa M, Yetukuri L, Orešič M (2007) Normalization method for metabolomics data using optimal selection of multiple internal standards. BMC Bioinformatics 8:93. https://doi.org/10.1186/1471-2105-8-93

Tautenhahn R, Patti GJ, Rinehart D, Siuzdak G (2012) XCMS online: a web-based platform to process untargeted metabolomic data. Anal Chem 84:5035–5039. https://doi.org/10.1021/ac300698c

Traquete F, Luz J, Cordeiro C, Sousa Silva M, Ferreira AEN (2021) Binary simplification as an effective tool in metabolomics data analysis. Meta 11. https://doi.org/10.3390/metabo11110788

Traquete F, Luz J, Cordeiro C, Sousa Silva M, Ferreira AEN (2022) Graph properties of mass-difference networks for profiling and discrimination in untargeted metabolomics. Front Mol Biosci 9. https://doi.org/10.3389/fmolb.2022.917911

Troyanskaya O, Cantor M, Sherlock G, Brown P, Hastie T, Tibshirani R, Botstein D, Altman RB (2001) Missing value estimation methods for DNA microarrays. Bioinformatics 17:520–525. https://doi.org/10.1093/bioinformatics/17.6.520

Trygg J, Wold S (2002) Orthogonal projections to latent structures (O-PLS). J Chemom 16(3):119–128. https://doi.org/10.1002/cem.695

Trygg J, Gabrielsson J, Lundstedt T (2009) Data preprocessing: background estimation, denoising, and preprocessing. In: Brown SD, Tauler R, Walczak B (eds) Comprehensive chemometrics. Elsevier, pp 1–8

Tziotis D, Hertkorn N, Schmitt-Kopplin P (2011) Kendrick-analogous network visualisation of ion cyclotron resonance Fourier transform mass spectra: improved options for the assignment of elemental compositions and the classification of organic molecular complexity. Eur J Mass Spectrom 17:415–421

van den Berg RA, Hoefsloot HCJ, Westerhuis JA, Smilde AK, van der Werf MJ (2006) Centering, scaling, and transformations: improving the biological information content of metabolomics data. BMC Genomics 7:142. https://doi.org/10.1186/1471-2164-7-142

Vinaixa M, Samino S, Saez I, Duran J, Guinovart JJ, Yanes O (2012) A guideline to univariate statistical analysis for LC/MS-based untargeted metabolomics-derived data. Meta 2:775–795. https://doi.org/10.3390/metabo2040775

Wang M, Carver JJ, Phelan VV, Sanchez LM, Garg N, Peng Y, Nguyen DD, Watrous J, Kapono CA, Luzzatto-Knaan T et al (2016) Sharing and community curation of mass spectrometry data with global natural products social molecular networking. Nat Biotechnol 34:828–837. https://doi.org/10.1038/nbt.3597

Ward JH (1963) Hierarchical grouping to optimize an objective function. J Am Stat Assoc 58:236–244. https://doi.org/10.1080/01621459.1963.10500845

Watrous J, Roach P, Alexandrov T, Heath BS, Yang JY, Kersten RD, van der Voort M, Pogliano K, Gross H, Raaijmakers JM et al (2012) Mass spectral molecular networking of living microbial colonies. Proc Natl Acad Sci 109:E1743–E1752. https://doi.org/10.1073/pnas.1203689109

Wei R, Wang J, Su M, Jia E, Chen S, Chen T, Ni Y (2018) Missing value imputation approach for mass spectrometry-based metabolomics data. Sci Rep 8:1–10. https://doi.org/10.1038/s41598-017-19120-0

Westerhuis JA, Hoefsloot HCJ, Smit S, Vis DJ, Smilde AK, van Velzen EJJ, van Duijnhoven JPM, van Dorsten FA (2008) Assessment of PLSDA cross validation. Metabolomics 4:81–89. https://doi.org/10.1007/s11306-007-0099-6

Wishart DS, Guo AC, Oler E, Wang F, Anjum A, Peters H, Dizon R, Sayeeda Z, Tian S, Lee BL et al (2022) HMDB 5.0: the human metabolome database for 2022. Nucleic Acids Res 50:D622–D631. https://doi.org/10.1093/nar/gkab1062

Wold S, Johansson A, Cochi M (1993) PLS-partial least squares projections to latent structures. ESCOM Science Publishers, Leiden, pp 523–550

Xi B, Gu H, Baniasadi H, Raftery D (2014) Statistical analysis and modeling of mass spectrometry-based metabolomics data. Methods Mol Biol 1198:333–353. https://doi.org/10.1007/978-1-4939-1258-2_22

Xiao JF, Zhou B, Ressom HW (2012) Metabolite identification and quantitation in LC-MS/MS-based metabolomics. Trends Analyt Chem 32:1–14. https://doi.org/10.1016/j.trac.2011.08.009

Primer on Modelling Approaches for Omics Data

13

Guillem A. Santamaria, João Miranda, Margarida Carrolo, Sara Silva, and Francisco Rodrigues Pinto

Abstract

Omics data allow us to study complex biological systems. Mathematical models can help us understand these systems and extract more knowledge from the data. In this chapter, we introduce three types of mathematical models that can be used to potentiate the interpretation of omics datasets: statistical, machine learning and mechanistic models. We delve into the characteristics of these modelling approaches, the essential stages involved in their development and their potential applications to omics datasets.

G. A. Santamaria
BioISI – Biosystems & Integrative Sciences Institute, Faculty of Sciences, University of Lisboa, Lisbon, Portugal

Luxembourg Centre for Systems Biomedicine (LCSB), University of Luxembourg, Esch-sur-Alzette, Luxembourg

J. Miranda · F. R. Pinto (✉)
BioISI – Biosystems & Integrative Sciences Institute, Faculty of Sciences, University of Lisboa, Lisbon, Portugal
e-mail: frpinto@ciencias.ulisboa.pt

M. Carrolo
Centro Hospitalar Universitário Lisboa Norte, Hospital de Santa Maria, Lisbon, Portugal

S. Silva
LASIGE, Faculdade de Ciências da Universidade de Lisboa, Lisbon, Portugal

Keywords

Statistical model · Machine learning model · Mechanistic model · Omics data

13.1 Introduction

A model is a simplified representation of a real-world system (Torres and Santos 2015). Models can be used to understand and predict the behaviour of the real-world systems they represent. Models can be real-world systems themselves. For example, a yeast cell can be used as a model of all eukaryotic cells. But models can also be abstract representations of real-world systems. Within abstract models, we can identify conceptual models and mathematical models. The former can be diagrams representing the components and variables of the real-world system and how they interact with each other. Diagrams in biochemistry textbooks that represent metabolic or signalling pathways can be considered conceptual models. Conceptual models can evolve to mathematical models that use some type of mathematical formalism (algebraic equations, differential equations, difference equations, logical expressions, etc.) that allows the prediction of the values of some variables of the system. These mathematical expressions use parameters and other variables of the system to compute the predicted values.

M. Gama-Carvalho et al. (eds.), *Omics Approaches in Biomedicine and Biotechnology*, Advances in Experimental Medicine and Biology 1504, https://doi.org/10.1007/978-3-032-18966-0_13

Omics technologies allow the measurement of a large number of variables in biological systems under different experimental conditions. Therefore, mathematical models that aim to represent such biological systems can be induced and validated with omics data. Models based on omics data improve our understanding of the system and allow us to make predictions about its behaviour. In other words, mathematical models can be powerful tools to extract knowledge from omics data.

We will discuss three main types of mathematical models that can be applied to omics data: (1) statistical models, (2) machine learning models and (3) mechanistic models. First, we will present the common structure of omics data and the typically measured variables. Then, we will explain how the three modelling types can be applied to quantify the relationship between those variables and how they can be used to better understand the underlying biological systems and predict their behaviours.

13.2 Common Characteristics of Omics Studies

Omics assays identify and quantify a large number of molecular entities in biological samples. The type of entity varies with the omics technology. Transcriptomics measures mRNA; proteomics measures proteins; and metabolomics measures metabolites. The number of identified entities is ideally large enough to cover most entities of that type that exist in the biological system under study.

Omics studies are composed of a set of omics assays, each of which is applied to a different biological sample. A given omics study, like any other research study, can be observational or experimental. In observational studies, biological samples are selected because they have some properties of interest, but samples are not experimentally manipulated by researchers. Common examples are clinical observational studies where samples from a group of individuals with a particular disease are compared with samples collected from healthy individuals. In experimental studies, the biological samples are all similar or equivalent prior to the experiment. Then, subgroups of samples are exposed to different experimental treatments. One possible example would be an experiment with cell line cultures, where half of the cultures would be grown in simple growth media while the other half would be grown in the same media supplemented with a drug.

Both observational and experimental studies need to have some sort of replication (Blainey et al. 2014). In observational studies, replicates correspond to different biological samples sharing the same property of interest. As samples are not manipulated in observational studies, these replicate samples can differ in other aspects. In experimental studies, replicates correspond to biological samples that were exposed to the same experimental treatment. Replication is needed to determine if the differences that are observed between samples are correlated with their different properties (disease vs. healthy, drug vs. no drug) or, alternatively, are more easily explained by random fluctuations that are observable even between replicate samples (also known as noise or interindividual variability (Altman and Krzywinski 2015)). If the differences due to changes in sample properties (or experimental treatments) are larger than differences between replicates, then we can infer that the property (or treatment) has a significant impact on the measured variables.

13.3 Omics Data and Design Tables

Given this description, all omics studies give rise to large data tables, where each row corresponds to a unique molecular entity, and each column corresponds to a biological sample. Columns are organised in replicate groups (Fig. 13.1a). Each entry of the table contains the abundance of a molecular entity (row) in one sample (column).

For the complete analysis of omics data, we also need an additional table, sometimes called a design table, with the number of rows

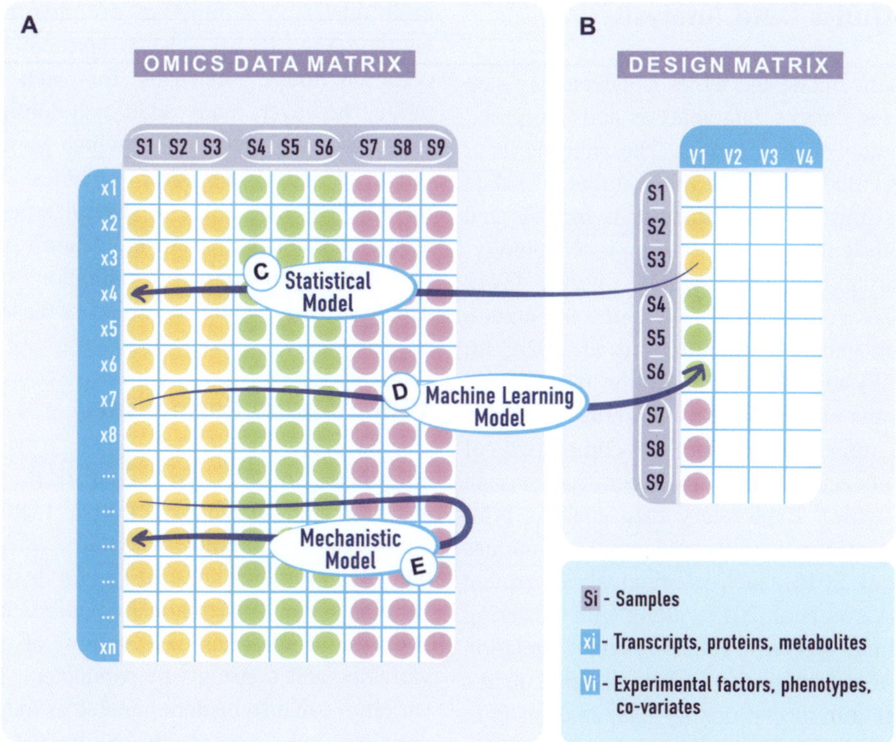

Fig. 13.1 Modelling for omics data analysis. Omics data usually generates a data table (**a**) where each column corresponds to a sample and each row corresponds to an omics-specific entity (transcript, protein or metabolite, for example). The samples are usually organised in replicate groups (here identified by different colours). In each replicate group, all samples received the same experimental treatment or share the same phenotype (in observational studies). Each sample can also be characterised by additional attributes, called covariates. The information about the levels (values) of each experimental factor (or phenotype) and covariate values for each sample is organised in a design table (**b**). Model approaches to analyse omics data consist in finding a function that encodes the quantitative relation between values in both data and design tables. Models can be of different types according to the variables used and the type of function and how it is adjusted to the data. Typical modelling approaches include Statistical models (**c**) that explain the values in each row of the data table through the values of the column variables in the design table, Machine Learning models (**d**) that predict the value of one column in the design table using the values of multiple (or all) rows in the data table and Mechanistic models (**e**) that predict the values in one (or several) row(s) of the data table using the values of other rows

corresponding to the number of biological samples (matching the columns of the data table), but with rows corresponding to the sample properties (disease status in the observational study example and drug exposure in the experimental study example) (Fig. 13.1b).

In many studies, there is only one main sample property of interest. But more complex studies compare samples that differ in the values of multiple properties simultaneously. In experimental design language, each property of interest is an experimental factor (Krzywinski and Altman 2014). The various values that a property can take across samples are referred to as the levels of the experimental factor. The design table may include additional columns for other properties that characterise each sample. These properties, called covariates, although not being variables of interest for the hypothesis being studied, should be included if we suspect that their values may influence the outcome measurements.

13.4 Omics Data Analysis

Having data and design tables is a necessary step to allow the correct data analysis and interpretation of omics studies results. The analysis itself can be divided into three main stages: (1) data preprocessing, (2) exploratory data analysis and (3) modelling. Data preprocessing is technology-dependent and includes quality filtering (Wen et al. 2022; Yamada et al. 2021), the imputation of missing values (Chilimoniuk et al. 2024; Jin et al. 2021) and data normalisation (Chua et al. 2023; Evans et al. 2017), which aims to reduce technical noise and enable the comparison of values between samples (between columns in our data table). Exploratory data analysis typically uses dimensionality reduction techniques (Meng et al. 2016), such as principal component analysis (Lever et al. 2017), along with clustering (Altman and Krzywinski 2017) and correlation analyses to compare the various samples in the study. The aim of exploratory data analysis is to determine whether the replicates are highly similar to each other and whether there are large dissimilarities between samples from different replicate groups.

Finally, modelling can be employed: (i) to check if the data supports the hypotheses under study, (ii) to potentiate data interpretation or (iii) to make new predictions of the system's behaviour. Different types of models can be used, each with different aims and methodological characteristics. Statistical models are commonly applied to explain the abundances of molecular entities across samples (in the data table) as a function of the experimental factors and covariates (in the design table) (Fig. 13.1c). Machine learning models, on the other hand, are typically used to predict factor levels (in the design table) as a function of the abundances of multiple molecular entities (Fig. 13.1d). Finally, mechanistic models use previous knowledge about how the molecular entities interact and the physical or chemical laws that rule those interactions to predict how molecular entity abundances change (with time or relatively to each other) as a function of other molecular entities (Fig. 13.1e). These represent the most common usage scenarios for each type of model; however, many additional configurations exist that involve changing which variables are predicted based on other variables. Different model types can differ in several aspects, such as mathematical formalism, feature selection, parameter estimation or validation strategies. The following sections will discuss these aspects.

13.5 Model Formalisms

All three model types (statistical, machine learning and mechanistic) use mathematical equations to predict variable values. The predicted variables are also known as dependent variables. The mathematical equations that compute the predictions use values of predictor variables and constants or parameters. Predictor variables can also be denominated as independent variables or features. In biological systems characterised through omics data, both predicted and predictor variables can be: (i) the abundances of the measured molecular entities or (ii) factors, phenotypes or covariates that distinguish the different biological samples (Fig. 13.1). The types of mathematical equations vary according to the modelling approach.

13.5.1 Statistical Models

The focus of statistical models is not the mathematical equation used to make predictions. Instead, statistical models focus on the errors or residuals, that is, on the difference between the observed values of the dependent variable and the values predicted by the model. Inferences about the model's quality and parameter values are based on the residual values' magnitude and variability. We can apply the statistical approach using many different mathematical formalisms, but usually general and simple functions are used, such as linear models (Eq. 13.1):

$$Y = \beta_0 + \beta_1 X_1 + \beta_2 X_2 + \ldots + \beta_n X_n + \varepsilon \quad (13.1)$$

These functions only capture linear relationships between predictors and dependent variables, but they are easy to understand and flexible enough to adapt to the use of many predictor variables with either continuous or categorical scales.

Linear models can be applied when the dependent variable is continuous and the residuals are independent of each other and follow a normal distribution. Predictor variables can either be continuous or categorical. An extension of linear models, called Generalised Linear Models (GLMs) (Dobson and Barnett 2018), allows for dependent variables with different non-normal distributions such as Poisson (for count data), binomial (for binary outcomes) or negative-binomial distributions (for overdispersed count data, with applications in the analysis of mRNAseq data).

Both GLMs and traditional linear models can accommodate some non-linear relationships with the introduction of interaction terms between predictors (bold term in Eq. 13.2):

$$Y = \beta_0 + \beta_1 X_1 + \ldots + \beta_n X_n + \boldsymbol{\beta_{12} X_1 X_2} + \ldots + \varepsilon \quad (13.2)$$

The interaction terms introduce new coefficients that encode the influence of the combination of the values of two (or more) predictor variables. Although interaction terms allow the discovery of more complex relationships between the predictor and the dependent variables, they demand more complete factorial designs (where all the combinations of values of the predictor variables are tested and replicated) and larger sample sizes. Besides these extra needs, model interpretability can also become more difficult. Other methods for capturing non-linear relationships in statistical models have been developed, including generalised additive models (GAM) (Hastie and Tibshirani 2017), which utilise smooth, non-linear functions (such as smoothing splines and local regression smoothers) estimated from the data itself while maintaining the overall structure of a linear model.

Although it is expected that real systems have variables that are related through complex non-linear functions, within the range of values explored in a given experiment, the linear approximation can be sufficiently accurate to determine which predictor variables have a significant influence on the dependent variable. Therefore, linear formalisms are still the most commonly used in statistical models.

13.5.2 Machine Learning Models

Machine learning models (Bzdok et al. 2017; Greener et al. 2022) can be roughly classified according to their learning strategy as (i) supervised learning (Bzdok et al. 2018), (ii) unsupervised learning (James et al. 2023) or (iii) reinforcement learning (Sutton and Barto 2018). For the analysis of omics data, the most commonly applied machine learning models use one of the first two learning strategies. Supervision means that the models can be induced on datasets that have been labelled with the correct values to predict. Unsupervised learning includes clustering and dimensionality reduction algorithms that are mainly used in the exploratory analysis of omics data. They are not so comparable with mechanistic or statistical models, as they do not explicitly make predictions of the values of a dependent variable. In this chapter, when we refer to machine learning models, we are referring to supervised machine learning. Most supervised machine learning models are either classification or regression models. In classification models, the dependent variable is categorical, while in regression models, the dependent variable has a continuous scale. However, most commonly used mathematical formalisms can be adapted for either classification or regression. As with statistical models, machine learning models can use very different mathematical formalisms. Independently of the formalism, the main aim of machine learning models is to minimise the prediction error while avoiding overfitting (Lever et al.

2016a). Compared with mechanistic or statistical models, machine learning is traditionally less concerned with model interpretability—the ability to understand which predictor variables, under which specific conditions, have a greater influence on the dependent variable. Despite increasing efforts to provide explanations for the model predictions, one can apply machine learning using a black-box approach, which encapsulates the mathematical formalism of the model within a closed system. We provide input data to our black box, and it returns predictions as outputs. We may not be interested in the details of how the black box works. This approach means that we can use any mathematical formalism for our model, as long as it is able to produce accurate predictions for our input data. In fact, linear models, like the ones used in statistical models, can be used with a machine learning approach. More complex formalisms can also be used to capture more complex non-linear relationships, such as support vector machines (Vanneschi and Silva 2023a; Yang 2004), decision trees (Kingsford and Salzberg 2008; Vanneschi and Silva 2023a), random forests (Touw et al. 2013; Vanneschi and Silva 2023a), gradient boosting (Boldini et al. 2023; Vanneschi and Silva 2023b), genetic programming (Poli et al. 2008; Vanneschi and Silva 2023b) or deep learning/artificial neural networks (Sapoval et al. 2022; Vanneschi and Silva 2023a). The choice of which type of model to use should take into account the nature and the quantity of data available for inducing such a model. For example, deep learning models are nowadays the most powerful for learning very complex relationships among variables, but this advantage is only possible when very large datasets (thousands of individual samples) are available.

13.5.3 Mechanistic Models

Mechanistic models use expressions that are derived from chemical or physical laws that govern the interactions between biomolecules. For example, to interpret transcriptomic data, one can use a dynamical model of gene expression where each transcript is synthesised and degraded according to biochemical rate laws (that may depend on the abundance of transcription factors, parameters and independent variables that may represent environmental factors or signals). Synthesis and degradation rate laws are combined in differential equations that can be numerically solved to predict the evolution of transcript abundances after some experimental perturbation. Differential equations are the most common mathematical formalism for mechanistic models of biological systems (Novère 2015), but they require previous knowledge of the rate laws for each process in the system and the associated parameters. Systems that imply the determination of hundreds or thousands of parameters may become practically unfeasible due to the huge amount of data that would be necessary to adjust all parameter values.

Some alternative mathematical formalisms for mechanistic models are available that have lower requirements for parameter values and previous knowledge of the system. One such formalism is that of qualitative or logical models (Naldi et al. 2018; Niarakis and Helikar 2020), where each variable (such as a gene, protein or metabolite) is modelled as a Boolean variable (meaning that it can only have two states: 0 (inactive) and 1 (active)) or multivalued discrete variables (that may have more than two ordinal values of activity levels). The next values of each variable are computed with logical formulas that use the current values of other variables as inputs. Each variable has its own logical formula that encodes the regulatory interactions that influence it. This formalism does not need to define quantitative parameters. It needs the knowledge of which molecules (genes, proteins or metabolites) activate or inhibit which other molecules, that is, it needs to have a defined regulatory graph of the system. Additionally, it needs the logical rules that define how multiple incoming regulatory interactions combine to predict the future state of each variable. Due to these properties, this formalism is more appropriate for studying signalling and gene expression regulatory networks.

Genome-scale metabolic models (GMMs) serve as a third mechanistic modelling framework that can enhance our understanding of omics datasets (Hyduke et al. 2012; Orth et al. 2010). These models are based on in silico reconstructions of the full metabolic network of an organism or cell type. Functional annotations of metabolic and transport genes expressed in the target cell type are used to build the network of metabolic reactions that may occur in that cell. Instead of attempting to perform dynamical simulations of the metabolite concentrations, GMMs compute the flux values for each reaction that allow the metabolic network to reach a steady state and fulfil or maximise important functions such as biomass production (necessary if the cells are actively dividing) or maintenance of energetic (ATP) and redox (NAD(P)H) balances. GMMs do not need to consider rate laws and their respective parameters. Therefore, they are less data-demanding than differential equation models. Transcriptomic or proteomic data can be used to further constrain GMMs' metabolic flux predictions. In this way, GMMs can predict metabolic consequences of perturbations studied with some omics approaches.

13.6 Feature Selection

All mathematical models have independent or predictor variables. The choice of which variables are included as predictor variables is called feature selection. Feature selection (Theng and Bhoyar 2024) has been introduced within the field of machine learning, especially when machine learning models are applied to high-dimensional datasets, meaning that each individual/sample is characterised by a high number of variables (also known as features) (Altman and Krzywinski 2018). This is the case of omics data, where each sample has values for thousands of biomolecules. It is usually the case that only a small fraction of the available variables contains useful information for predictions. The majority of variables are unrelated to the dependent variable and only introduce noise to model predictions. Therefore, it is useful to filter out uninformative variables. In the context of omics data, feature selection corresponds to the selection of relevant rows in the omics data table or in the design table.

In statistical and machine learning models, feature selection can be done using different approaches. The most basic approach is to study all possible predictor variables separately, using correlation, mutual information or univariate models to evaluate whether each variable can potentially help predict the dependent variable. Only variables that are above a defined threshold (of correlation with the dependent variable, for example) are then used in the final model.

A second approach is based on iterative model building. For example, we start with a model with all predictor variables, and at each iteration, we select a predictor to discard that maximises the model performance. Other iterative heuristics are possible, but they do not search over all possible predictor combinations, which means that finding the best set of predictor variables is not guaranteed.

A third approach embeds the feature selection process with parameter optimisation. In statistical models, this is called regularisation (Lever et al. 2016b). Instead of minimising only the deviation between model prediction and observations, a penalty term is added to the function that is being minimised. This penalty term can be the sum of parameter absolute values (Lasso regression), the sum of the squares of parameter values (Ridge regression) or a mixture of the Lasso and Ridge penalty terms (Elastic Net regression). Inclusion of such a penalty term during parameter optimisation (or model induction, as explained in the next section) will favour the selection of models with fewer predictor variables. Uninformative variables will be discarded through zero-value or very low-value coefficients.

Some machine learning algorithms, such as tree-based ones, have implicit ways to perform feature selection. Within the induction process of tree-based models, individual predictors are selected to define splits in regression or classification trees. Variables are chosen for splits because the resulting branches have more homogeneous data labels. This increase in homogeneity is measured as a decrease in the impurity index that is usually the Gini index or an entropy-based measure. The order in which variables are chosen for splits can be interpreted as their importance in the tree. Random forests produce ensembles of trees using bootstrapped datasets and random subsets of predictor variables. The importance of each predictor variable in the final random forest can be computed as the average of variable importance in each of the individual trees. Therefore, we can induce a random forest model with all data and evaluate the variable importance statistics. Then, we can impose a threshold to select only the most important variables to use for inducing the final random forest model. Also, evolution-based methods like genetic programming evolve free-form models that may use only a subset of features.

In mechanistic models, the choice of variables to include in the model is based on our previous knowledge about the system. It is not sufficient to know that a variable may influence the system, but we must also know in which way and through which mechanism it influences the system. On the other hand, some variables known to be part of the modelled system may be excluded from the model because they are approximately constant under the conditions being studied. Generally, one should refrain from incorporating variables in the model that are not pertinent to the current research question.

13.7 Model Learning and Validation

All mathematical models have parameters. The mathematical formalism defines the type of formulas used, but the parameter values are the ones that adjust our model predictions to our data. Finding the best parameters for the model using available data is done during model induction. This process of inducing a model is the job of the learning algorithm. It is also called learning a model or, most commonly, training a model, a term that was initially used by the neural networks community and later widely (over) used to designate the induction of any machine learning model.

Once we have a model, we can validate it by testing whether its predictions are accurate on data that was not used for training it. The process of learning a model on a dataset and testing it on another dataset is called cross-validation, and it is essential to determine whether the model is overfitted to the training data (Lever et al. 2016a). Overfitting happens when the optimal parameters obtained during training allow for very accurate predictions with the training data but lead to inaccurate predictions when new input data is used. Overfitting is due to two main causes: (i) the model is overparameterised or (ii) the training data is not representative of the real system that is being modelled.

Overparameterisation means that our model has more parameters than it should. Usually, the number of parameters grows with the number of independent variables (predictors) in our models. So, overparameterisation happens when we are including independent variables that (i) are not bringing in useful information for our model or (ii) could add some useful information but we do not have sufficient data to correctly estimate the extra parameter values. Regarding overparameterisation, all datasets contain noise and measurement errors. Having an excess of parameters allows the model to capture not only the true relationships between variables but also the noise in the training dataset. When overparameterised models are used with new data, which has different noise values, predictions will fail.

Regarding the lack of representativeness of the training data, this happens when the training set does not contain samples that represent all the possible variations in the values of independent variables that could be observed in the real system that we want to model. If data comes from an observational study, this defect is probably due to a small sample size that is only capturing the most frequent values for all the independent variables. If the model is used with samples from under-represented subpopulations, it remains unaware of the true relationships between independent and dependent variables in that context, leading to predictions that are likely to be incorrect. If training data comes from experimental studies, the values of some or all of the independent values are determined by the researchers. But due to technical or resource limitations, the range of values explored for the different independent variables may also be limited. Thus, the model should only be used to predict samples with properties within the training ranges.

For some models, there may exist distinct parameter values that produce exactly the same model predictions. This property, called non-identifiability (Heinrich et al. 2025; Paulino and de Pereira 1994; Ran and Hu 2017), can be a direct consequence of the structure of the model equations. For example, if two parameters appear in the same equation multiplied by the same independent variables, there are infinite combinations of their values that lead to the same model predictions. Alternatively, if two parameters are multiplied by different independent variables, but those independent variables are highly correlated, a similar non-identifiability occurs. This collinearity between predictors can be avoided with a careful feature selection. Finally, non-identifiability can be dependent on the training data. With a limited training dataset, there could be two sets of parameter values that lead to identical predictions. Adding more training data, it is possible that a unique parameter set is able to make model predictions concordant with data.

All modelling approaches covered in this chapter require training and validation steps, but each has its own specificities.

13.7.1 Statistical Models

As previously said, residuals are the focus of statistical models. Most statistical models assume that the residuals follow a known probability distribution. Different parameter sets will give rise to different model predictions and, hence, different residual values. If we know the residual probability distribution, we can calculate the likelihood that a given residual value is observed. Maximum Likelihood Estimation (MLE) (Irizarry and Love 2016) is one of the main approaches for parameter estimation in statistical models and consists in searching the parameter values that maximise the likelihood of all the residuals obtained with those parameter values. The likelihood is given by the probability density (according to the probability distribution function) at the residual value. As each dataset contains multiple data points, the likelihood function to maximise is the product of the likelihood of all individual residual values. Depending on the assumed residual probability distribution, the calculus of the parameter values that maximise the likelihood function may have an analytical solution. If there is no analytical solution, a numerical optimisation algorithm must be used to search for the optimal parameter values. In the case of linear models, where the residual distribution is assumed to be normal, MLE is equivalent to the least-squares method. This method finds the parameters that minimise the sum of the squares of the residuals. For linear models, it has an analytical solution, which greatly facilitates the computational estimation of the parameters.

An alternative way to estimate model parameters is using a Bayesian approach (Puga et al. 2015). In this approach, a probability distribution for each parameter is estimated instead of point values. We must start with a prior distribution for the parameters. This can be a non-informative prior (a uniform distribution, for example), a distribution obtained from previous knowledge, or from the analysis of previous datasets. Then, the conditional probability distribution for parameters given the new data is computed using Bayes theorem. For most models, there is no explicit formula to compute this conditional probability, and numerical approaches such as Markov Chain Monte Carlo (MCMC) (Speagle 2019) or Approximate Bayesian Computation (ABC) (Sunnåker et al. 2013) methods must be used.

In many situations, statistical models are not developed for making new predictions but to better understand the current dataset. Therefore, in traditional statistical analysis, all data is used in model training, and no validation dataset is kept aside. The main goal is to identify which predictor variables have a significant effect on the dependent variable. In parallel with parameter estimation, statistical tests on model coefficients allow the detection of the influent predictor variables. But these statistical tests are only valid if the modelling assumptions about the data hold and if, globally, the model is significantly well adjusted to the data. This is done by performing diagnostic analyses, such as checking the probability distribution of the residuals, checking whether the residual values are independent of each other (that is, checking that there are no non-random patterns in the residuals), and performing goodness-of-fit tests. These procedures can be considered an internal validation of the statistical models, but they cannot guarantee that there is no overfitting. For that, one needs to compare model predictions with data that was not used in model training.

13.7.2 Machine Learning Models

The main goal of machine learning model training is to find the parameter and hyperparameter values that maximise predictive ability. Hyperparameters are set before adjusting parameter values to maximise model performance. This means that hyperparameters are not learnt from the data. Nevertheless, they influence model performance, as they may specify model structure (in random forests, for example, the number of trees and the maximum depth of each tree are important hyperparameters) or the way the parameter optimisation is done (defining how parameter optimisation algorithms work, for example).

To maximise model performance, it is important to define how this performance is measured (Lever et al. 2016c; Rainio et al. 2024). If the dependent variable is continuous, Mean Absolute Error (MAE) or Root-Mean-Squared Error (RMSE) can be the performance metrics to minimise. If the dependent variable is categorical, the most basic performance metric is accuracy, the fraction of samples whose class was correctly predicted by the model. When predicting categorical variables, it is common to have only two classes. This is the case when machine learning models are used to predict whether an individual has one disease from his transcriptomic profile, for example. In this type of classification problem, we can have two types of errors: false positives (the individual is healthy, but the model predicts he has the disease) and false negatives (the individual has the disease but the model fails to detect it). Machine learning models may specialise in avoiding just one type of error. Particularly, if the training data has a class imbalance (having more negative than positive cases, for example) and the learning maximises accuracy, then it is expectable that the model will specialise in classifying the majority class correctly. In some practical applications, having one error type may lead to worse consequences than the other type. Therefore, it may be useful to evaluate performance separately for each error type. Precision, which is the fraction of samples classified as positive that are true positives, is a measure that is penalised by false positives. On the other hand, recall, which is the fraction of all the real positive cases that are classified as positive, is a measure that is penalised by false negatives. Ideal machine learning models would have both high precision and high recall. Therefore, one commonly used performance metric to maximise during model learning is the F1-score, which is the harmonic mean of precision and recall. Other performance metrics are available, and the choice of which one to use depends on our goals (for what purpose we are developing the machine learning model) and characteristics of our training data, such as the class imbalance problem.

Due to the need to validate model predictions with data that was not used during model learning, available data should be split into training and validation datasets (Greener et al. 2022; Vanneschi and Silva 2023a). In the context of omics data, a data split corresponds to a split in both the omics data table and the design table, giving rise to two tables with complementary subsets of the samples of the original tables. A typical split is 80% of data to the training set and 20% to the validation set. Larger training sets allow for better parameter estimation, but the validation set should be large enough to be a representative sample of the available data. That representativity can be enforced through stratified sampling, especially in the presence of class imbalance.

Problems of dataset size can be partially solved by data augmentation techniques. Some of these techniques were specifically developed to solve the class imbalance problem, such as ADASYN (Adaptive Synthetic sampling) (He et al. 2008) or SMOTE (Synthetic Minority Over-Sampling Technique) (Chawla et al. 2002), despite some limitations (Blagus and Lusa 2013). Others use specialised artificial neural networks (such as GANs (Generative Adversarial Networks) (Cusworth et al. 2024) or VAEs (Variational Autoencoders) (Karlberg et al. 2024)) to generate artificial data points that are similar to input real data. Data augmentation should only be applied to the training data; otherwise, validation performance could be due to overfitting the artificial data characteristics.

If some feature selection procedure is applied previously to model learning (and it is not an integral part of it), it should be based only on the training set to avoid a data leakage problem. Data leakage occurs whenever some step during model development or data preprocessing is done including data from the validation set, and in that way, model performance may be inflated. Subsequently, all preprocessing steps, such as predictive variable scaling or outlier detection, should also be done using only the training data.

Cross-validation is crucial to monitor overfitting and the ability of models to make predictions with new data. But what if the specific data split we are using has issues? Maybe the validation set is not really representative of the entire dataset. This is why performing a simple cross-validation, with only one split into training and validation data, is seldom enough. A better method of cross-validation is called k-fold cross-validation, where the dataset is divided into k subsets of equal size. Each k subset is used as a validation set once for a model trained with the remaining k-1 data subsets. The average performance of the models in the k validation subsets is the global model performance measure, and its comparison with the average performance in the training sets will inform us about overfitting. Since we evaluate all subsets once as validation sets, the specific data splits we use do not influence the overall validation performance. Such a cross-validation is also a more robust way to compare different models and fine-tune hyperparameters. Hence, a typical data-splitting strategy is to perform k-fold cross-validation within the training set and perform a final evaluation with the validation set that was never used in model development.

13.7.3 Mechanistic Models

As the mathematical equations of mechanistic models reflect physical or chemical processes, it is usual that model parameters have a defined physicochemical meaning. This implies that other authors may have previously determined the value of these parameters. Even if the other author estimated the parameter value in a different cell type or species, its value may be a useful first guess value that can be optimised later. It is therefore normal to start by doing a literature search for parameter values. Some databases can also be useful for parameter value searches. The BioModels database (Malik-Sheriff et al. 2019) collects previously published mechanistic models, including the parameter values used. Even if we are developing a model for a system that has never been modelled before, there may be a part of a published model that is similar to ours, or some parameters may have the same physicochemical meaning. Enzyme catalysis is a common feature in biological system models. It is then useful to search databases such as BRENDA (Chang et al. 2020), SABIO-RK (Wittig et al. 2017) or STRENDA (Swainston et al. 2018) that store enzyme kinetic parameter values. A very useful and more general-purpose database to find parameter values is BioNumbers (Milo et al. 2010). The fact that mechanistic model parameters have a defined physicochemical meaning also opens the possibility to perform experiments targeted to estimate a particular parameter value. These experiments do not necessarily involve the full system under study, which can facilitate the experimental design.

But getting model parameter values through the literature, databases or targeted experiments is usually not enough to fully parameterise a mechanistic model. Parameter optimisation algorithms are then employed that adjust the parameter values in a way that minimises the differences between model predictions and experimental observations. In this case, the experimental observations must be obtained with the complete system being modelled. Omics technologies are a powerful experimental approach to get data for complex biological systems. As many mechanistic models of biological systems are dynamical in nature, the most common training data are of the time series type. Most biological systems in a constant environment reach a dynamical equilibrium or steady state. In such a steady state, all components (transcripts, proteins, metabolites, etc.) reach a constant concentration because their rates of synthesis and degradation are balanced. Experimentally measuring a time series of a biological system in steady state is useless. To produce useful time-series datasets, biological systems must be perturbed either through changing environmental conditions, changing gene expression (overexpression or knockdown of some genes), changing gene sequences (mutations or knockouts) or applying pharmacological inhibitors. An alternative to time-series datasets is to apply a perturbation and collect the omics data after the system reaches a new steady state. Data from multiple perturbations must be collected to gather enough information for parameter estimation.

Given the possibly large number of parameters to optimise and the complexity of mechanism-based mathematical expressions, these optimisations may be technically hard, with many local minima in the parameter space. GA (Genetic Algorithms) (Vanneschi and Silva 2023a), SA (Simulated Annealing) (Gonzalez et al. 2006) and PSO (Particle Swarm Optimisation) (Kennedy and Eberhart 1995) are global optimisation algorithms that succeed in such conditions, while classical gradient-based optimisation algorithms (such as Levenberg–Marquardt or Gauss–Newton) are more easily trapped in the local minima that are closest to the initial guess of parameter values. Hybrid algorithms that start with a global optimisation and refine the final solutions with local gradient-based algorithms are the most common choices for parameter optimisation in mechanistic models.

When doing a parameter search for mechanistic models, one must acknowledge that there is no guarantee that the optimal or the correct set of parameters is going to be found by the optimisation algorithms. Additionally, one should consider that distinct sets of parameter values can be equally good at reproducing the available training data (meaning that the model parameters are non-identifiable). There is always going to be some uncertainty about the model parameter values. Sensitivity analysis (Zi 2011)—studying how the model predictions change after applying small perturbations in parameter values—is a common approach to check whether there are non-identifiability problems and to estimate the impact of having badly estimated parameters.

For mechanistic models, model validation implies comparing model predictions with new data that was not used to optimise parameter values. If making new experiments is not a viable option, it means that some of the available datasets must be kept aside for validation and not used during model development.

13.8 Conclusion

Modelling is a fundamental part of omics data analysis, as it is the appropriate tool to extract knowledge from data. The previous sections presented three modelling approaches, their properties and how their models are developed, which are summarised in Table 13.1. However, how can a researcher conducting omics experiments or analysing omics data decide which type of model to develop?

If we are analysing data from an experiment using omics technologies, it is almost certain that we will use statistical modelling. In transcriptomics, most packages used for differential expression analysis in mRNAseq data use linear models, fitting one model for each transcript analysed (Law et al. 2014; Love et al. 2014; Robinson et al. 2009). Basically, if we

Table 13.1 Properties of different modelling approaches used in omics data analysis

Model type	Statistical	Machine learning	Mechanistic
Mathematical formalism	Linear models, generalised linear models, generalised additive models, among others	Linear models, decision trees / random forests, support vector machines, artificial neural networks, among others	Algebraic equations, differential equations, logical equations, among others These equations are based on the physical and chemical laws that govern biomolecular interactions
Typical predicted variables	Molecular entities	Sample factors	Molecular entities
Typical predictor variables	Sample factors and covariates	Molecular entities	Molecular entities
Model learning	Least squares, maximum likelihood estimation	Cross-validation within training data partition allows optimisation of parameters and hyperparameters to maximise model performance	Parameters from literature or adjusted to experimental data (error minimisations)
Model validation	Goodness-of-fit tests, analysis of residuals and parameter value significance tests	Predictive ability in validation dataset that was not used for model training	New predictions to be tested experimentally
Modelling aims	Explain the quantitative relation between variables, considering experimental variability	Make predictions with new data	Generate understanding about system behaviour, check if current hypothesis about physical and chemical processes affecting the system's molecular components allow accurate simulations

need to identify the transcripts, proteins or metabolites that significantly changed between the experimental groups, we will employ statistical modelling. Statistical models are particularly advantageous for experiments in which we are studying the effect of multiple experimental factors (rows in the omics design table) that will be used as predictor variables in the statistical model.

Proposals for new machine learning applications in biology are rapidly increasing. The power of machine learning models to learn complex patterns in data and to make accurate predictions justifies such interest. However, the generally small sample size limits its application in the analysis of individual omics experiments. Omics datasets may be large, but mainly due to their high dimensionality (number of variables per sample) and not due to the number of samples studied (Altman and Krzywinski 2018). Some large-scale studies, such as the TCGA project for cancer research, where thousands of samples were analysed using the same protocols, have sufficient data to induce and validate machine learning models. However, this does not preclude the usefulness of machine learning models in omics data analysis. Models can be developed to learn patterns across multiple datasets and then used to help in the analysis of a new dataset. For example, machine learning models have been developed to fill in missing data using data from different omics technologies (Cai et al. 2024).

From the molecular biology perspective, one of the aims of omics experiments is to gain knowledge about the molecular mechanisms responsible for key cellular functions and their regulation. In that sense, the use of mechanistic models to test our hypothesis about molecular mechanisms is an exciting approach. Omics data can be used to either train or validate mechanistic models. Since most of these models are dynamical, omics data should either be time-series data or used to evaluate the system's steady-state response after multiple perturbations.

One limitation of the use of mechanistic models with omics data is that it is not likely that models explicitly consider all molecular species monitored by the omics technology. This limitation is related to the difficulty in adjusting parameter values. More molecular species included in the model imply more parameters. More parameters imply more data for model development. Nevertheless, some modelling frameworks, such as genome-scale metabolic modelling and logical modelling, partially circumvent the need for parameters and allow for the development of models that include a larger number of molecular species.

Our aims should ultimately guide our choice of modelling approach. If our goal is to make predictions for new samples, machine learning provides a straightforward solution. If we are keen to discover which variables influence other variables (either experimental factors or molecular species), statistical models can estimate and test those effects. Finally, if we know one or multiple hypothetical molecular mechanisms that can explain a cellular function or phenotype, mechanistic models can help us identify which mechanism has a higher concordance with observed data and hence is probably the real mechanism.

References

Altman N, Krzywinski M (2015) Sources of variation. Nat Methods 12:5–6. https://doi.org/10.1038/nmeth.3224

Altman N, Krzywinski M (2017) Clustering. Nat Methods 14:545–546. https://doi.org/10.1038/nmeth.4299

Altman N, Krzywinski M (2018) The curse(s) of dimensionality. Nat Methods 15:399–400. https://doi.org/10.1038/s41592-018-0019-x

Blagus R, Lusa L (2013) SMOTE for high-dimensional class-imbalanced data. BMC Bioinform 14:106. https://doi.org/10.1186/1471-2105-14-106

Blainey P, Krzywinski M, Altman N (2014) Replication. Nat Methods 11:879–880. https://doi.org/10.1038/nmeth.3091

Boldini D, Grisoni F, Kuhn D, Friedrich L, Sieber SA (2023) Practical guidelines for the use of gradient boosting for molecular property prediction. J Cheminformatics 15:73. https://doi.org/10.1186/s13321-023-00743-7

Bzdok D, Krzywinski M, Altman N (2017) Machine learning: a primer. Nat Methods 14:1119–1120. https://doi.org/10.1038/nmeth.4526

Bzdok D, Krzywinski M, Altman N (2018) Machine learning: supervised methods. Nat Methods 15:5–6. https://doi.org/10.1038/nmeth.4551

Cai Z, Apolinário S, Baião AR, Pacini C, Sousa MD, Vinga S, Reddel RR, Robinson PJ, Garnett MJ, Zhong Q, Gonçalves E (2024) Synthetic augmentation of cancer cell line multi-omic datasets using unsupervised deep learning. Nat Commun 15:10390. https://doi.org/10.1038/s41467-024-54771-4

Chang A, Jeske L, Ulbrich S, Hofmann J, Koblitz J, Schomburg I, Neumann-Schaal M, Jahn D, Schomburg D (2020) BRENDA, the ELIXIR core data resource in 2021: new developments and updates. Nucleic Acids Res 49:D498–D508. https://doi.org/10.1093/nar/gkaa1025

Chawla NV, Bowyer KW, Hall LO, Kegelmeyer WP (2002) SMOTE: synthetic minority over-sampling technique. J Artif Intell Res 16:321–357. https://doi.org/10.1613/jair.953

Chilimoniuk J, Grzesiak K, Kała J, Nowakowski D, Krętowski A, Kolenda R, Ciborowski M, Burdukiewicz M (2024) Imputomics: web server and R package for missing values imputation in metabolomics data. Bioinformatics 40:btae098. https://doi.org/10.1093/bioinformatics/btae098

Chua AE, Pfeifer LD, Sekera ER, Hummon AB, Desaire H (2023) Workflow for evaluating normalization tools for omics data using supervised and unsupervised machine learning. J Am Soc Mass Spectrom 34:2775–2784. https://doi.org/10.1021/jasms.3c00295

Cusworth S, Gkoutos GV, Acharjee A (2024) A novel generative adversarial networks modelling for the class imbalance problem in high dimensional omics data. BMC Méd Inform Decis Mak 24:90. https://doi.org/10.1186/s12911-024-02487-2

Dobson AJ, Barnett AG (2018) An introduction to generalized linear models. Chapman and Hall/CRC. https://doi.org/10.1201/9781315182780

Evans C, Hardin J, Stoebel DM (2017) Selecting between-sample RNA-Seq normalization methods from the perspective of their assumptions. Brief Bioinform 19:776–792. https://doi.org/10.1093/bib/bbx008

Gonzalez OR, Küper C, Jung K, Naval PC, Mendoza E (2006) Parameter estimation using simulated annealing for S-system models of biochemical networks. Bioinformatics 23:480–486. https://doi.org/10.1093/bioinformatics/btl522

Greener JG, Kandathil SM, Moffat L, Jones DT (2022) A guide to machine learning for biologists. Nat Rev Mol Cell Biol 23:40–55. https://doi.org/10.1038/s41580-021-00407-0

Hastie TJ, Tibshirani RJ (2017) Generalized additive models. (Routledge). https://doi.org/10.1201/9780203753781

He H, Bai Y, Garcia EA, Li S (2008) ADASYN: adaptive synthetic sampling approach for imbalanced learning. In: 2008 IEEE International Joint Conference on Neural Networks (IEEE World Congress on Computational Intelligence), pp 1322–1328. https://doi.org/10.1109/ijcnn.2008.4633969

Heinrich M, Rosenblatt M, Wieland F-G, Stigter H, Timmer J (2025) On structural and practical identifiability: current status and update of results. Curr Opin Syst Biol 41:100546. https://doi.org/10.1016/j.coisb.2025.100546

Hyduke DR, Lewis NE, Palsson BØ (2012) Analysis of omics data with genome-scale models of metabolism. Mol BioSyst 9:167–174. https://doi.org/10.1039/c2mb25453k

Irizarry RA, Love MI (2016) Data analysis for the life sciences with R. CRC Press

James G, Witten D, Hastie T, Tibshirani R, Taylor J (2023) An introduction to statistical learning, with applications in python, Springer texts in statistics. Springer, Cham. https://doi.org/10.1007/978-3-031-38747-0

Jin L, Bi Y, Hu C, Qu J, Shen S, Wang X, Tian Y (2021) A comparative study of evaluating missing value imputation methods in label-free proteomics. Sci Rep 11: 1760. https://doi.org/10.1038/s41598-021-81279-4

Karlberg B, Kirchgaessner R, Lee J, Peterkort M, Beckman L, Goecks J, Ellrott K (2024) SyntheVAEiser: augmenting traditional machine learning methods with VAE-based gene expression sample generation for improved cancer subtype predictions. Genome Biol 25:309. https://doi.org/10.1186/s13059-024-03431-3

Kennedy J, Eberhart R (1995) Particle swarm optimization. In: Proceedings of ICNN'95 – International Conference on Neural Networks, vol 4, pp 1942–1948. https://doi.org/10.1109/icnn.1995.488968

Kingsford C, Salzberg SL (2008) What are decision trees? Nat Biotechnol 26:1011 1013. https://doi.org/10.1038/nbt0908-1011

Krzywinski M, Altman N (2014) Designing comparative experiments. Nat Methods 11:597–598. https://doi.org/10.1038/nmeth.2974

Law CW, Chen Y, Shi W, Smyth GK (2014) Voom: precision weights unlock linear model analysis tools for RNA-seq read counts. Genome Biol 15:R29. https://doi.org/10.1186/gb-2014-15-2-r29

Lever J, Krzywinski M, Altman N (2016a) Model selection and overfitting. Nat Methods 13:703–704. https://doi.org/10.1038/nmeth.3968

Lever J, Krzywinski M, Altman N (2016b) Regularization. Nat Methods 13:803–804. https://doi.org/10.1038/nmeth.4014

Lever J, Krzywinski M, Altman N (2016c) Classification evaluation. Nat Methods 13:603–604. https://doi.org/10.1038/nmeth.3945

Lever J, Krzywinski M, Altman N (2017) Principal component analysis. Nat Methods 14:641–642. https://doi.org/10.1038/nmeth.4346

Love MI, Huber W, Anders S (2014) Moderated estimation of fold change and dispersion for RNA-seq data with DESeq2. Genome Biol 15:550. https://doi.org/10.1186/s13059-014-0550-8

Malik-Sheriff RS, Glont M, Nguyen TVN, Tiwari K, Roberts MG, Xavier A, Vu MT, Men J, Maire M, Kananathan S, Fairbanks EL, Meyer JP, Arankalle C, Varusai TM, Knight-Schrijver V, Li L, Dueñas-Roca C, Dass G, Keating SM, Park YM, Buso N, Rodriguez N, Hucka M, Hermjakob H (2019) BioModels—15 years of sharing computational models in life science. Nucleic Acids Res 48:D407–D415. https://doi.org/10.1093/nar/gkz1055

Meng C, Zeleznik OA, Thallinger GG, Kuster B, Gholami AM, Culhane AC (2016) Dimension reduction techniques for the integrative analysis of multi-omics data. Brief Bioinform 17:628–641. https://doi.org/10.1093/bib/bbv108

Milo R, Jorgensen P, Moran U, Weber G, Springer M (2010) BioNumbers—the database of key numbers in molecular and cell biology. Nucleic Acids Res 38: D750–D753. https://doi.org/10.1093/nar/gkp889

Naldi A, Hernandez C, Abou-Jaoudé W, Monteiro PT, Chaouiya C, Thieffry D (2018) Logical modeling and analysis of cellular regulatory networks with GINsim 3.0. Front Physiol 9:86–16. https://doi.org/10.3389/fphys.2018.00646

Niarakis A, Helikar T (2020) A practical guide to mechanistic systems modeling in biology using a logic-based approach. Brief Bioinform 22:bbaa236. https://doi.org/10.1093/bib/bbaa236

Novère NL (2015) Quantitative and logic modelling of molecular and gene networks. Nat Rev Genet:1–13. https://doi.org/10.1038/nrg3885

Orth JD, Thiele I, Palsson BØ (2010) What is flux balance analysis? Nat Biotechnol 28:245–248. https://doi.org/10.1038/nbt.1614

Paulino CDM, de Pereira CAB (1994) On identifiability of parametric statistical models. J Ital Stat Soc 3:125 151. https://doi.org/10.1007/bf02589044

Poli R, Langdon WB, McPhee NF (2008) A field guide to genetic programming. Lulu Enterprises

Puga JL, Krzywinski M, Altman N (2015) Bayesian statistics. Nat Methods 12:377–378. https://doi.org/10.1038/nmeth.3368

Rainio O, Teuho J, Klén R (2024) Evaluation metrics and statistical tests for machine learning. Sci Rep 14:6086. https://doi.org/10.1038/s41598-024-56706-x

Ran Z-Y, Hu B-G (2017) Parameter Identifiability in statistical machine learning: a review. Neural Comput 29: 1151–1203. https://doi.org/10.1162/neco_a_00947

Robinson MD, McCarthy DJ, Smyth GK (2009) edgeR: a Bioconductor package for differential expression analysis of digital gene expression data. Bioinf (Oxf) 26: 139–140. https://doi.org/10.1093/bioinformatics/btp616

Sapoval N, Aghazadeh A, Nute MG, Antunes DA, Balaji A, Baraniuk R, Barberan CJ, Dannenfelser R, Dun C, Edrisi M, Elworth RAL, Kille B, Kyrillidis A,

Nakhleh L, Wolfe CR, Yan Z, Yao V, Treangen TJ (2022) Current progress and open challenges for applying deep learning across the biosciences. Nat Commun 13:1728. https://doi.org/10.1038/s41467-022-29268-7

Speagle JS (2019) A conceptual introduction to Markov chain Monte Carlo methods. arXiv. https://doi.org/10.48550/arxiv.1909.12313

Sunnåker M, Busetto AG, Numminen E, Corander J, Foll M, Dessimoz C (2013) Approximate Bayesian computation. PLoS Comput Biol 9:e1002803. https://doi.org/10.1371/journal.pcbi.1002803

Sutton RS, Barto AG (2018) Reinforcement learning: an introduction. A Bradford Book, Cambridge, MA

Swainston N, Baici A, Bakker BM, Cornish-Bowden A, Fitzpatrick PF, Halling P, Leyh TS, O'Donovan C, Raushel FM, Reschel U, Rohwer JM, Schnell S, Schomburg D, Tipton KF, Tsai M, Westerhoff HV, Wittig U, Wohlgemuth R, Kettner C (2018) STRENDA DB: enabling the validation and sharing of enzyme kinetics data. FEBS J 285:2193–2204. https://doi.org/10.1111/febs.14427

Theng D, Bhoyar KK (2024) Feature selection techniques for machine learning: a survey of more than two decades of research. Knowl Inf Syst 66:1575–1637. https://doi.org/10.1007/s10115-023-02010-5

Torres NV, Santos G (2015) The (mathematical) modeling process in biosciences. Front Genet 6:354. https://doi.org/10.3389/fgene.2015.00354

Touw WG, Bayjanov JR, Overmars L, Backus L, Boekhorst J, Wels M, van Hijum SAFT (2013) Data mining in the life sciences with random Forest: a walk in the park or lost in the jungle? Brief Bioinform 14:315–326. https://doi.org/10.1093/bib/bbs034

Vanneschi L, Silva S (2023a) Lectures on intelligent systems. Nat Comput Ser 283–288. https://doi.org/10.1007/978-3-031-17922-8_11

Vanneschi L, Silva S (2023b) Lectures on intelligent systems. Nat Comput Ser 161–204. https://doi.org/10.1007/978-3-031-17922-8_7

Wen B, Jaehnig EJ, Zhang B (2022) OmicsEV: a tool for comprehensive quality evaluation of omics data tables. Bioinformatics 38:5463–5465. https://doi.org/10.1093/bioinformatics/btac698

Wittig U, Rey M, Weidemann A, Kania R, Müller W (2017) SABIO-RK: an updated resource for manually curated biochemical reaction kinetics. Nucleic Acids Res 46:gkx1065. https://doi.org/10.1093/nar/gkx1065

Yamada R, Okada D, Wang J, Basak T, Koyama S (2021) Interpretation of omics data analyses. J Hum Genet 66:93–102. https://doi.org/10.1038/s10038-020-0763-5

Yang ZR (2004) Biological applications of support vector machines. Brief Bioinform 5:328–338. https://doi.org/10.1093/bib/5.4.328

Zi Z (2011) Sensitivity analysis approaches applied to systems biology models. IET Syst Biol 5:336–346. https://doi.org/10.1049/iet-syb.2011.0015

14 Functional Interpretation of Omics Data: From Pathways to Networks

Marina L. García-Vaquero, João A. I. Miranda, Margarida Carrolo, and Francisco Rodrigues Pinto

Abstract

Analysis of omics experiments usually identifies lists of molecular entities (genes, transcripts, other RNA molecules, proteins, or metabolites) that need to be functionally interpreted, not as individual molecules but as ensembles that are functionally related. Multiple computational methods have been developed to study the functions or pathways that are more relevant in those lists of biomolecules. In this chapter, we describe the principles behind these methods and discuss their use in the functional interpretation of omics data. First, we describe the two main approaches to identify the most relevant functional annotations in a list of biomolecules: Over-Representation Analysis (ORA) and Functional Class Scoring (FCS). Afterward, we introduce basic concepts of network biology and discuss how network analysis methods, such as Centrality Analysis, Network Clustering, and Network diffusion, can be used to potentiate the functional interpretation of omics data.

M. L. García-Vaquero · J. A. I. Miranda · F. R. Pinto (✉)
BioISI – Biosystems & Integrative Sciences Institute, Faculty of Sciences, University of Lisboa, Lisbon, Portugal
e-mail: frpinto@ciencias.ulisboa.pt

M. Carrolo
Centro Hospitalar Universitário Lisboa Norte, Hospital de Santa Maria, Lisbon, Portugal

Keywords

Over-Representation Analysis · Functional Class Scoring · Biological networks · Network centrality · Network clustering · Network diffusion

14.1 Introduction

The most common omics techniques detect and quantify molecular entities in biological samples. Experiments may use these techniques to compare multiple samples that were manipulated with different treatments or represent different phenotypes.

Data analysis of omics experiments leads to lists of biological entities that are differentially expressed across experimental conditions tested or were found to have a similar profile associated with a sample type. How can we further explore and interpret these lists?

For this task, we benefit from the knowledge accumulated during years of molecular biology research. A significant fraction of known genes, proteins, and metabolites have known functions (especially for *Homo sapiens (*del Toro et al. 2022*)* and other model species). Bioinformatics development also helped in (1) expanding this knowledge (many functional annotations are automatically inferred (Huang et al. 2009; Zhao et al. 2020)) and in (2) organizing it in accessible databases and ontologies (KEGG (Janjić and

M. Gama-Carvalho et al. (eds.), *Omics Approaches in Biomedicine and Biotechnology*, Advances in Experimental Medicine and Biology 1504, https://doi.org/10.1007/978-3-032-18966-0_14

Pržulj 2012), REACTOME (McGillivray et al. 2018), Wikipathways (Agrawal et al. 2023), Gene Ontology (Aleksander et al. 2023; Ashburner et al. 2000), and MSigDB (Milacic et al. 2023; Schwikowski et al. 2000)). These databases identify thousands of pathways and cellular processes, each composed of a set of entities (genes, proteins, or metabolites).

As recently reviewed by Maleki and colleagues, we can identify in the literature multiple functional enrichment strategies (Lv et al. 2021). One natural question is whether our experimental list contains a significant number of entities in common with some of these known pathways. Alternatively, we need to determine if our experimental perturbations have a specific impact on some of the established pathways. There are two fundamental approaches to answering these questions. One is denominated Over-Representation Analysis (ORA), while the other is Functional Class Scoring (FCS) (Fig. 14.1). However, both approaches consider pathways (or annotation terms) as sets where every gene (or protein or metabolite) has an identical contribution. Biological networks represent how different biomolecules interact with each other and provide a third approach to functionally explore omics results (Oughtred et al. 2018; Zitnik et al. 2024). Biological networks can help us distinguish between elements of the same pathway (some genes may be more peripheral or more central). We can also identify relevant genes (or proteins or metabolites) in our list that, although not annotated with a pathway or function, have direct interactions with known pathways. Alternatively, we can try to identify the most affected subnetworks before assigning genes to pathways. The remaining sections of this chapter describe these multiple ways to functionally explore the results of omics experiments. Although these functional analyses can be applied to several types of molecular entities (genes, proteins, or metabolites), for simplicity, in the remainder of this chapter, we will refer to the analysis of lists of differentially expressed genes as a representative example. The discussion about the methods remains valid if we exchange genes by either proteins or metabolites.

14.2 Sources of Functional Annotations

Cataloguing the molecular players and their functional capabilities is critical to investigating the different types of cell programs. With this objective, the "functionome" defines the compendium of biological functions a gene product (a protein or an RNA molecule) can exert, ranging from molecular activities through protein complexes to coordinated cell activities like phagocytosis. Therefore, even though it does not qualify precisely as an omics category, many researchers consider it as such.

The Gene Ontology (GO) consortium (geneontology.org/) provides the framework and the set of concepts for describing the functions of gene products from all organisms (Aleksander et al. 2023; Ashburner et al. 2000). Gene ontology is divided into three independent domains to distinguish the different aspects of a gene product function: Molecular Function (MF), Cellular Component (CC), and Biological Process (BP). MF terms describe the actions or activities of a gene product from a molecular mechanism perspective, and CC terms describe their relative localization in the cell. Biological Process (BP) is the most extended category and represents the ultimate biological programs accomplished by multiple molecular activities. Gene functional associations can be addressed using varied evidence protocols. Initially, GO terms were manually annotated based on published experimental evidence and so were biased toward processes with anticipated research interest. Later, thanks to the accumulation of multivariate biological data, prediction algorithms have notably improved annotations and are now pivotal for electronically inferring GO terms. These assist in providing a broader coverage of the functionome, albeit also introducing a large fraction of false-positive annotations (Aleksander et al. 2023).

Besides Gene Ontology, there are different pathway-centric functional annotation resources such as Reactome (McGillivray et al. 2018), KEGG (Janjić and Pržulj 2012), or WikiPathways

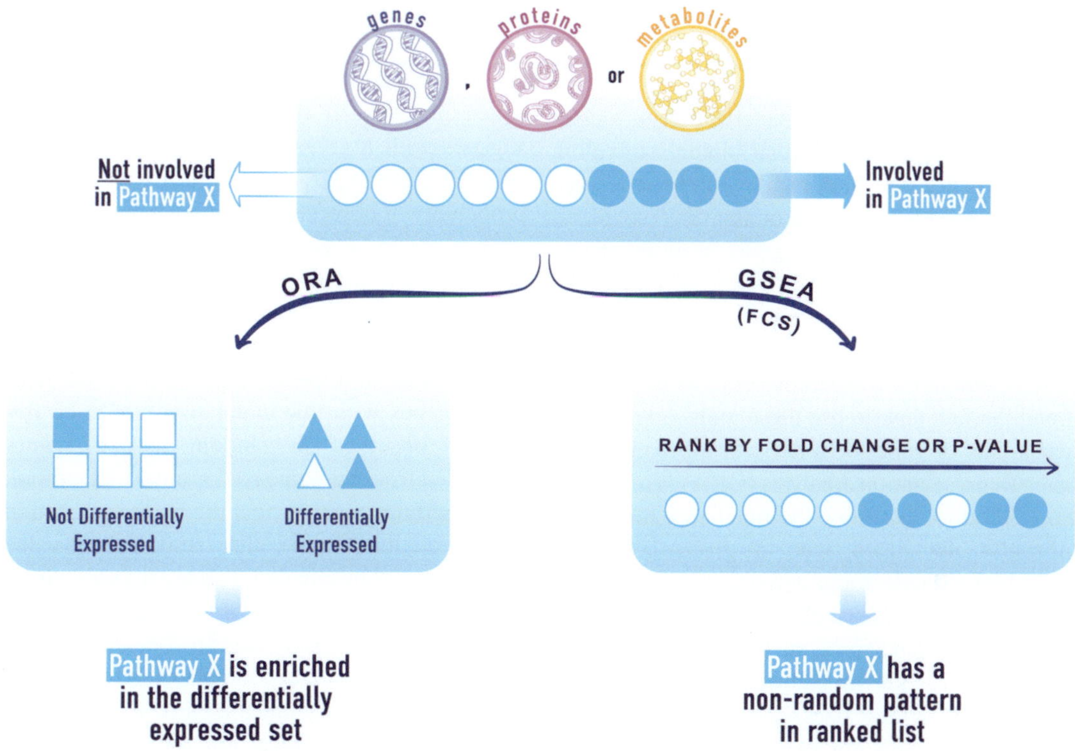

Fig. 14.1 Functional analysis of lists of molecular entities. Over-Representation Analysis (ORA): The analysis of omic data usually produces lists of molecular entities (genes, proteins, and metabolites) that were considered differentially expressed. These lists can be used as input sets in ORA. In this analysis, the frequency of functional annotations within the input set is compared with the frequency of functional annotations in a background universe. This background should correspond to all the molecular entities of the same type that were detected in the biological samples under study. In the figure, molecular entities are represented as circles, and orange color represents a functional annotation to Pathway X. After differential expression analysis, molecular entities are either differentially expressed (triangles) or not differentially expressed (squares). To evaluate if the input set is enriched in entities with a particular annotation, an hypergeometric test is used to estimate the probability of observing the same or a higher frequency of entities with that annotation in random sets (with the same size of the input set) taken from the background universe. Functional Class Scoring (FCS): As an alternative to ORA, the complete list of molecular entities detected in the omics data can be used for functional analysis. The most common FCS method is Gene Set Enrichment Analysis (GSEA). In this method, the detected entities are ranked according to relevant quantitative statistics (log fold-change and p-value of differential expression are common choices). The ranking positions of the entities with a given annotation are then compared with a random distribution across the ranking. Functional annotations associated with entities that are more concentrated in the top or the bottom of the ranking are considered to be significantly associated with the experimental perturbation (or phenotype) under study

(Agrawal et al. 2023) (benchmarking analysis of these databases for functional analysis is discussed by Mubeen and colleagues (Mootha et al. 2003)). These databases are more stable and contain fewer false positives than GO. However, gene ontology contains more annotations and captures the complexity of protein functional relationships better. An additional benefit of GO annotation is that it employs a common vocabulary across different species and thus facilitates cross-species knowledge integration.

14.3 Over-Representation Analysis (ORA)

Beyond the functional characterization of single gene products, most molecular biologists exploit GO annotations to biologically interpret results derived from high-throughput assays. To explore the annotations of lists of genes resulting from omics data analysis (lists of differentially expressed genes, for example), the most used approach is Over-Representation Analysis (ORA), due to the availability of tools to perform it and the fact that, to use these tools, one simply needs lists of gene identifiers. In a broad sense, Over-Representation Analysis (ORA) is a statistical method to determine the functional traits significantly associated with a given gene list (Fig. 14.1). These tools determine if the list of interest has overlaps with all the pathway gene sets (from a given source, such as GO). Due to the large number of known pathways, with associated gene sets of varying sizes, it is relatively easy to find overlaps with our experimental gene lists. The problem is to define how big the overlap must be for it to be considered significant.

Most tools implementing ORA employ the hypergeometric test (mathematically equivalent to Fisher's exact test). This test is applied in scenarios analogous to the following situation:

1. We have a large box containing 900 white balls and 100 black balls.
2. Without seeing the color of the balls, we randomly select 50 balls.
3. Out of the 50 balls, 25 are black.
4. What is the probability, by chance alone, that out of 50 balls we get 25 or more blacks?
5. If this probability (that is, the p-value of the test) is very small, it is unlikely that the 50 balls were randomly selected, and the set is significantly enriched in black balls.

In the interpretation of omics-derived gene lists, the 100 black balls represent the pathway gene set, and the 50 extracted balls represent the gene list of interest (that resulted from the analysis of omics experiments). If we obtain a low p-value, it means that the size of the overlap is bigger than expected by chance alone. Therefore, we can conclude that the experimental gene list is significantly enriched in genes associated with the tested pathway. If the list of interest contains x genes with a given annotation (GO term, for example), the hypergeometric test p-value is the probability of observing x or more genes with that annotation in a gene list that was randomly selected from a background list (all genes that were detected as being expressed in the transcriptomic experiment, for example). If the p-value is high, it means that the observed number of genes with that annotation would be frequently observed in random lists of genes (extracted from the same background). Therefore, we should not consider that the list is enriched in that annotation. On the other hand, if the p-value is very low, it means that it is highly unlikely to observe so many genes with that annotation in random lists. Therefore, we can conclude that the list of interest is enriched in genes with that annotation. The hypergeometric test can also be interpreted as a statistical test for the difference between the proportions of genes with that annotation in the gene list and in the remaining background list. Since a single experimental gene list can potentially generate thousands of tests—one for each pathway or biological process in the database—multiple correction testing is necessary (Newman 2006). This means that results should be filtered with an adjusted p-value.

ORA can be used without a profound understanding of bioinformatics or biostatistics (Glaab et al. 2012). However, GO annotations are not exempt from biases and limitations that affect ORA outcomes (discussed in (García-Campos et al. 2015)). The two most popular pitfalls when functionally characterizing a gene set are (i) the correct assignment of the background universe and (ii) the management of redundant outputs.

To compute the hypergeometric test, we need to know how many genes are in the background list and how many of them are annotated with each function. This set is known as the universe (or background) of the analysis and should contain all the genes that could eventually be part of our experimentally selected list. It should be noted that the GO and other knowledge bases collect the functions associated with all the genes in each genome, without considering tissue- or cell-specific contexts. The inexperienced users frequently dismiss that the default enrichment analysis (in most tools) uses the entire genome as the background gene list. However, the statistical enrichment must always be calculated against an experiment-specific gene universe. The genome contains all the genes that a cell of a given species can potentially express. But our experiments were probably done with a particular cell type exposed to a defined environment. Both the cell type and the environment will constrain the expression of certain genes. To avoid biases in the pathway enrichment results, we should restrict our universe according to our experimental settings. If we have performed transcriptomic experiments, it is usual to consider the universe as the set of genes that were detected as expressed (TPM>1 in RNA-seq data, for example) in at least one of the experimental conditions. Using background lists that include genes that have no chance of being selected to be part of the gene list of interest can lead to falsely enriched terms, either through inflated statistical significance or biases in the frequency of annotated terms within the background.

GO terms are organized in directed acyclic graphs. This organization represents the hierarchical relationship between the functions in such a way that the less specific terms connect the increasingly detailed functional annotations. Therefore, each gene is annotated with the most specific term and also with all the functional predecessors. Due to the hierarchical structure of GO data (further discussed in (García-Campos et al. 2015)), most ORA outputs return a very large list of enriched GO terms, among which there is a considerable fraction of uninformative or redundant terms. The lists of genes annotated with more general terms include all the genes that are also annotated with similar but more specific terms. If one GO term is enriched in the list of interest, probably the broader GO terms that contain it are also enriched. To simplify the ORA outputs, the user can discard uninformative or redundant GO terms by applying different strategies.

The hierarchical structure of GO enables the evaluation of the information content (Mazandu and Mulder 2013) of a functional term according to its position in the graph. For instance, top-positioned terms describe general functions with low information content, while their descendants gradually incorporate more specific information. Likewise, the terms descending in the same branch are restricted to the same function in increasing levels of detail (further discussed in (Yu 2020)). An objective way of accessing the information content of a GO term is through the frequency of the term in the background list. Terms that are very common in the background have a low information content, while terms that are very rare in the background have a high information content. Therefore, a simple way to simplify the list of enriched terms is to eliminate the terms that have an information content lower than a chosen threshold.

It is also advisable to eliminate enriched GO terms that, although significantly enriched, are associated with a small number of genes in our gene list of interest. The enrichment of these terms could be an artifact due to a few genes that were false positives in the omics data analysis or an extremely low number of associated genes in the background.

Semantic similarity algorithms (Matys et al. 2006) exploit the graph structure of the GO to identify whether two GO terms refer to very similar functions or not. Two GO terms are considered semantically similar if they both possess similar information content and share a first common ancestor GO term that also has comparable information content. If two GO terms were enriched in the gene list of interest and are semantically similar, they constitute a pair of redundant terms. To remove redundancy from the final list of enriched terms, we can eliminate one of the terms in the redundant pair, usually the one that is less significantly enriched. Alternatively, semantic similarity could also be used by clustering algorithms to generate clusters of similar GO terms. From each cluster, a representative GO term can be selected, generating a more compact list of enriched terms.

Another approach is to investigate if enriched terms are associated with the same genes in the list of interest. If two GO terms are redundant, it is expected that they share many genes in common. But two functionally unrelated GO terms may also share genes. In this case, one could argue that both terms can be enriched due to the presence of these common genes in the list of interest. Therefore, it may make sense to consider both terms as a single enrichment. We can evaluate the overlap of annotated genes between two GO terms using Jaccard's similarity index (Vorontsov et al. 2013). The Jaccard index is defined as the size of the intersection (common genes) divided by the size of the union of the sample sets (total genes annotated with at least one of the two GO terms). When the two GO term set sizes are very unbalanced, it is preferable to apply Simpson's similarity index since it divides the gene intersection by the size of the smallest set. Once the similarity index is calculated for the complete list, we can reduce the ORA result by merging the GO terms with high similarity.

Even after these simplifications of the list of enriched terms, the remaining terms could still be too many to fit in a legible plot or table. Many users opt to summarize the analysis by outlining the top enriched functional terms, either considering the total number of genes annotated with each term or the statistical p-value. However, the top GO terms do not necessarily describe the complete functional landscape of a given gene set. Even if we only provide a selected list of the top enriched terms, a complete list should always be provided in supplementary materials.

14.4 Functional Class Scoring (FCS)

Functional Class Scoring (FCS) is an alternative approach for the functional interpretation of omics results that avoids the definition of a gene list of interest (usually the differentially expressed genes). Instead, it estimates a single statistic for each set of genes annotated with the same pathway or function. This statistic will use the omics results of all genes in the pathway and will tell us if that pathway is strongly affected by the experimental perturbation under study (Garcia-Alonso et al. 2019; Kanehisa et al. 2024).

The most common approach in FCS is to rank all the genes (or other entity types) detected in our experiment (using expression fold-change or differential expression p-value, for example) (Fig. 14.1). Then, it is possible to characterize each pathway gene set by the ranks occupied by its elements. The pathways that are more affected by the experimental perturbation (or that are more involved in the molecular response/adaptation to the experimental perturbation) should have a nonrandom set of ranks (ranks concentrated in the top fold changes, either positive or negative), while the ranks of pathways not affected by the perturbation should be randomly scattered. This nonrandom pattern can be tested statistically with a Wilcoxon Rank Sum Test (Barry et al. 2005) or with the enrichment score running-sum statistic, developed for the Gene Set Enrichment Analysis (GSEA—https://www.gsea-msigdb.org/gsea/index.jsp tool) (Schwikowski et al. 2000). As with ORA results, FCS results need to be corrected for multiple testing. Depending on the source of gene annotations, FCS results may also suffer from the redundancy problem, and solutions similar to the ones described for ORA may need to be applied.

In the ORA analysis, a bad choice of background list may lead to erroneous results. In FCS analysis, there is no need for a background list, making FCS results more robust to user choices. Similarly, the definition of gene lists of interest (used in ORA) may produce different results depending on the statistical threshold used in the omics data analysis. For example, changing from FDR <0.05 to FDR <0.1 may lead to a large increase in the list size and consequent changes in the list of enriched functional terms, according to the ORA analysis. FCS analysis, by avoiding the use of these gene lists of interest, is not so susceptible to changes in statistical thresholds of the differential expression analysis.

Not using a hard threshold for differential expression statistical significance may also allow the detection of more subtle collective effects. A pathway may have only a few differentially expressed genes, but a larger set of genes in the pathway may have consistent expression changes, although not big enough to reach statistical significance.

FCS uses more information from our results, but that is only useful if the fold-change ranking distribution is informative and not dominated by noise. In many experiments, only a small fraction of genes change their expression after the experimental perturbation. This means that most fold-change values randomly fluctuate around 0 (in log scale), and their relative ranking is meaningless. In these datasets, some FCS results may be statistically significant but biologically irrelevant.

The size of the pathway gene sets may influence both ORA and FCS results. Large pathways in ORA may exhibit significant enrichment, despite only a modest fraction of the pathway being present in the list of interest. In FCS, large pathways may have a significant deviation of the ranks from a random distribution only due to a small number of pathway members. Small pathways can also be problematic. In this case, only one or two genes present in our list are sufficient to generate a significant enrichment in ORA. Similarly, in FCS, only one or two genes with top ranks are enough to make the pathway significant. As our data already originates from a previous analysis of omics data that can be naturally noisy, we must be aware that a small fraction of our list may be contaminated with false positives. Therefore, we cannot rule out the possibility that a pathway enrichment (or biased ranking) based on one or two genes could be an artifact.

14.5 Biological Networks

Both ORA and FCS approaches implicitly consider that cellular functions result from the concerted actions of multiple molecular entities. However, these approaches do not consider how these molecular entities interact with each other to implement the associated cellular function. They also ignore genes or proteins that still have an unknown function but interact closely with known pathway members. To incorporate this type of knowledge, new approaches and data sources must be explored.

Systems biology investigates how the relationship between the elements of a complex system gives rise to the biological phenomena observed in nature (Westerhoff and Palsson 2004). Thus, systems biology shifts the interest from studying single molecular elements to the characterization of their interactions. Graphs became cardinal to represent complex interactions and model the outcomes of these systems. A graph is a mathematical structure used to represent collections of discrete objects, represented as nodes, and the interactions between them (Koschützki and Schreiber 2008; Maleki et al. 2020). Although they are not strictly synonyms, for simplicity, the terms "graph" and "network" will be used here with the same meaning. The representation of molecular interactions in networks not only facilitates the integration of complex data but also the application of graph theory concepts to interpret biological phenomena. Graph theory was first introduced in 1736 by mathematician Leonhard Euler and aims to study the topological or structural properties of graphs. However, it was not until the early 2000s that the research community realized the potential benefits of using graph theory in biomolecular research (Barabási and Oltvai 2004).

Several types of biological networks can be used to potentiate the interpretation of omics data (Koutrouli et al. 2020; Mazandu et al. 2017). They vary in the types of molecular entities that are considered as nodes and in the meaning of the interactions between nodes. The molecular entities can be genes, transcripts, other RNA molecules (miRNAs, lncRNAs, circRNAs), proteins, and metabolites. Occasionally, the same node can simultaneously represent the protein, the transcript, and the gene that encodes them. The interactions may represent physical events or other functional relationships. Among the physical events, we can identify specific binding (most commonly between proteins, but also between proteins and DNA or between proteins and RNA molecules), post-translational modifications (phosphorylation, dephosphorylation, ubiquitination, and other types of modifications), or chemical reactions. Nonphysical interactions include coexpression correlations, belonging to the same pathway, genetic interactions (when mutations in one gene modify the phenotype induced by mutations in another gene), or even text-mining-based interactions.

To build these biomolecular networks, we can query interaction databases. There is more information available for protein physical interactions (in large databases such as Intact (Szklarczyk et al. 2023), String (Surdo et al. 2022), or Biogrid (Nowicki and Snijders 2001)), but there are also sources for TF–target interactions (TRANSFAC (Masuda and Porter 2017) and DOROTHEA (Feuermann et al. 2025)) or signaling interactions (OMNIPATH (Türei et al. 2016; Türei et al. 2021), Signalink (Csabai et al. 2021), and Signor (Subramanian et al. 2005)). The list of molecular interaction data sources is continually evolving, with new databases being created, existing resources updated frequently, and curation standards improving as scientific knowledge advances.

14.6 Basic Network Concepts

A network is a representation of relations between discrete objects within complex systems. In cellular networks, the nodes represent macromolecules such as DNA, RNA, proteins, or metabolites, and the edges depict their functional or physical relations. There are various types of networks according to the type of biological information we intend to represent. The edges can be directed or undirected depending on whether they represent directional interactions, such as signaling events (a protein kinase phosphorylating its target protein, a transcription factor regulating the gene expression of a target) or undirected relations (such as physical protein–protein or RNA–protein interactions). In parallel, the edges can be weighted or unweighted, depending on whether the interactions incorporate quantitative information. Edge weight can be a discrete variable representing categorical information as “repression” or “activation” (e.g., −1, 1) or a continuous variable to describe interaction strength or stoichiometry relations, among others. Coexpression or metabolic networks frequently incorporate quantitative data relative to expression correlation or stoichiometry data. Heterogeneous macromolecular interactions can also be portrayed in multilayered networks (Liu et al. 2020). In these networks, each layer is restricted to the interactions between the elements of a single class, while the interlayer edges represent relationships across the distinct types of macromolecules. However, it must be noted that the use of multilayered and/or weighted networks increases the complexity of network analysis.

In some cases, one edge might connect to the same node itself. These self-loops are frequently found in regulatory networks to represent autoregulatory feedback relations. In the case of PPI networks, these can depict homodimer complexes. However, most analyses using PPI networks discard these self-loops to retrieve the so-called simple graph. Simple graphs also eliminate multiple edges between the same pair of nodes.

The number of edges connected to a node is termed the node’s degree. In directed networks, we can further distinguish out-degree (number of edges leading away) and in-degree (number of edges incident to the node).

Nodes are directly connected by edges but can also be connected by paths. A path is the list of edges (interactions) necessary to connect one node to another. The shortest path length, as the name implies, is the minimum number of edges one must traverse to move from one node to another.

A network is said to be connected if for all pairs of nodes there is a path that connects the two nodes. When a network is unconnected, each subset of nodes that forms a connected subnetwork is a network component. In these cases, the largest connected subnetwork is commonly referred to as the main component. A network can also be completely connected, meaning that the main component is the full network. Complete connectivity is rare on large networks. When a subset of nodes forms a subnetwork where every node has an edge connecting it to every other node, that maximally connected subnetwork is denominated as a clique. Cliques in protein physical interaction networks denote the existence of tightly connected protein complexes or molecular machines.

Network topology describes the arrangement of nodes and their interactions in the network, and it provides valuable insights to interpret biological properties (Barabási and Oltvai 2004). Topology can be employed to describe overall network properties and infer mechanistic properties. On a local scale, network topology can be investigated to identify the decisive nodes for maintaining network functionality. The most basic parameters to describe a network topology are the number of nodes and the number of edges in the network. The network edge density is calculated by dividing the number of edges by the theoretical maximum number of edges possible for the given number of nodes (that is, if the full network were a clique). The network diameter, which is the maximum shortest path length in the network, can also be used to compare the topology of different networks. But single numbers are not efficient at describing the highly diverse structures that a network may have. A more complete characterization of network topology is possible through the study of the distribution of some parameters across all nodes in the network. The degree distribution is an archetype feature to describe the network topology. The degree distribution of a network quantifies the frequency of nodes in the network with degree k for all possible k values. Another determining network property is the clustering coefficient. The local clustering coefficient measures the ratio between the observed and theoretical maximum number of edges linking the direct neighbors of a given node. It gives an estimation of edge density around a node. In turn, the average clustering coefficient is a network-wide measure of the clustering coefficient, and it shows the tendency of a graph to have densely connected subnetworks. Lastly, a topological property that is particularly relevant in biological networks is modularity (Alcalá-Corona et al. 2021). A network is considered to be modular if it is possible to partition nodes into non-overlapping subsets called modules that are densely connected within each module but rarely connected across modules. Known biological networks tend to be modular, which is thought to reflect the organization of cellular functions, as genes or proteins belonging to the same module tend to have related functions. Modularity is evaluated by a coefficient that compares the number of edges within modules to the number of edges across different modules (Nanni et al. 2020). This coefficient is normalized by the randomly expected number of each type of edge, considering the number of modules and the size of each module. In this way, a modularity coefficient of 0 means that it is similar to the modularity of a random network. A negative modularity coefficient means that the network is less modular than a random network. A modularity coefficient of 1 means that each module is an isolated component, and no edges connect nodes in different modules.

The study of the topology of biological networks has attracted a lot of interest in the research community (Barabási and Oltvai 2004; Jalili et al. 2016), but it is not of direct utility in the functional interpretation of omics data. But several types of analyses that use the discussed network topological concepts can be adapted to potentiate the functional interpretation of omics data. In this chapter, we will discuss three of these approaches (Fig. 14.2): (1) centrality analysis, (2) network clustering, and (3) network diffusion.

14.7 Network Centrality

One common type of network analysis is to rank nodes according to their importance in the network structure (Fig. 14.2a). The importance of a node can be interpreted in different ways. It may refer to nodes that are necessary to maintain local pathways or network connectivity, or it may be related to the speed of information spread across the network. This information spread could correspond to regulatory events in signaling networks or biomass flow in metabolic networks.

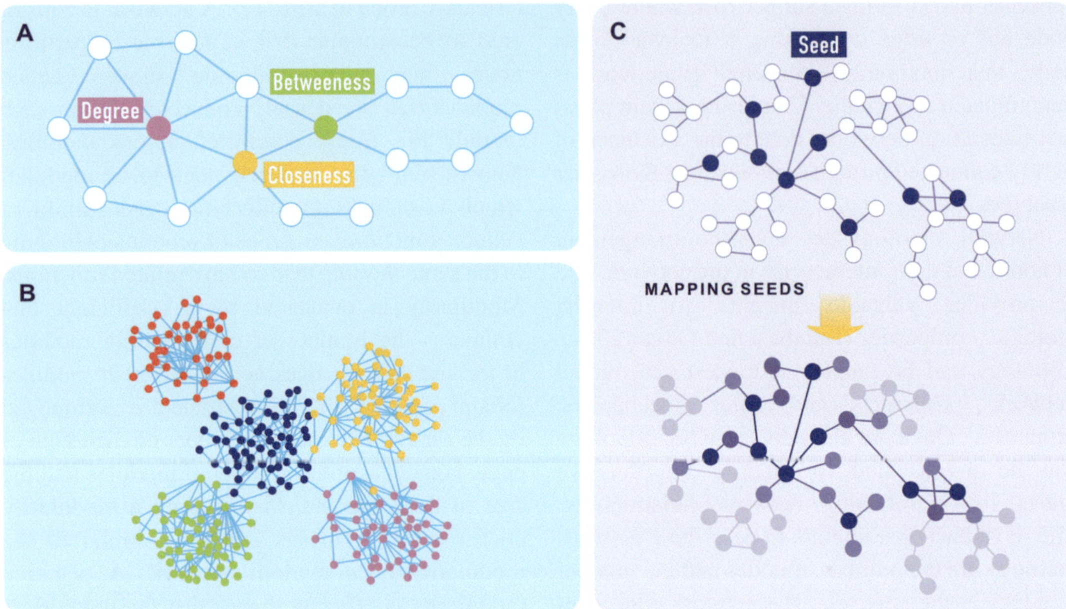

Fig. 14.2 Biological network analysis. (**a**) Centrality analysis: identifies nodes in the network that are potentially more influent in the network behavior. There are several centrality measures, the degree centrality (that counts the number of interactions that a node has) being the most common, betweenness centrality (the fraction of shortest paths (between all node pairs) that include the node), and closeness centrality (the inverse of the average distance (measured as the shortest path length) between the node and all the other nodes in the network). (**b**) Clustering analysis: divides the network into groups of nodes that have many interactions within the group and few interactions with nodes in other groups. (**c**) Network diffusion analysis: starting from a subset of seed nodes (that could be a set of differentially expressed genes, for example), these methods simulate a diffusion process throughout the entire network, enabling the discovery of other nodes that are significantly connected to the seed nodes. This significant connectivity is evidence that these nodes can influence or be influenced by the seed nodes

Currently, we have at our disposal a vast choice of network centrality measures, based on varying assumptions of node importance. Here, we will focus on the simplest measures of centrality. For a more detailed discussion on the topic, refer to (Ispano et al. 2023; Kondor and Lafferty 2002).

The simplest concept around centrality is that nodes with the largest degree (number of interactions) are central because they connect many elements. Therefore, the degree centrality of a node can be used to estimate its ability to spread a signal in its neighborhood. Nodes with the highest degree are commonly referred to as connectivity hubs. On the other hand, it can be argued that a molecular signal would preferentially propagate through the shortest paths. Therefore, it is expected that the most central nodes will be involved in many shortest paths between other node pairs. Betweenness centrality counts the number of shortest paths a node is involved in, estimating in this way the amount of information that runs through it. Nodes with the highest betweenness centrality also tend to have a high degree, but there are some nodes that can exhibit high betweenness while having a modest degree. These nodes are often regarded as connectivity bottlenecks. A slightly different perspective is used in the closeness centrality measure. Closeness considers the average shortest path length from one node to all the other nodes in the network. Nodes with a lower average shortest path length are considered to have higher closeness and to be more central in the network. A perturbation to a node with high closeness will propagate across the network more quickly.

Centrality analysis can be useful in the functional interpretation of omics experiments. It can be used to improve the interpretation of ORA results. For every pathway or functional annotation that is enriched in the gene list of interest (list of differentially expressed genes), we can build an interaction network using known interactions between pathway members. Then, we can assess the centrality of the genes in the list of interest for each enriched pathway. This can help to rank enriched pathways, selecting the ones that have genes with a higher centrality. A pathway may have many genes in the list of interest, but if their centrality is low, the overall impact on the pathway function may be minor. On the other hand, the differential expression of a few genes with high centrality may cause a higher impact on pathway function.

14.8 Network Clustering

Networks are useful tools to represent complex systems. One common strategy to address complex systems is to try to partition the full system into smaller subsystems. That partition is useful if the network is modular, meaning that there are many fewer interactions between subsystems than interactions within subsystems. This modularity facilitates the study of each module in isolation. Network clustering algorithms aim to determine the optimal partition of a network into subnetworks using different heuristics (Fig. 14.2b). The most direct approach is to try to maximize the modularity coefficient. At each iteration of this greedy algorithm, a node is chosen to change its module membership. The chosen membership change is the one that leads to a greater increase in modularity. Although simple, this heuristic does not guarantee the best result, and in some networks, it may lead to an excessive number of modules, each with a small number of components. It is advisable to test multiple clustering algorithms for each different network. Louvain (Blondel et al. 2008), Girvan-Newman (Gillis and Pavlidis 2012), or Stochastic Block Models (Noble 2009) are alternative algorithms commonly used for network clustering. To choose the best algorithm, the resulting partition must be evaluated at least in terms of modularity coefficient, number of modules, and size of modules. If external node data is available, one should also check if nodes within the same module also share other properties.

Biological processes are exerted through the interaction of molecular components. From this, it follows that the macromolecules interacting directly in the network are likely to be involved in similar processes. This assumption is commonly referred to as "guilt-by-association" and constitutes the foundation of numerous network-based methods designed for protein functional prediction (Gaudet and Dessimoz 2016; Paton et al. 2025). Following the same rationale, we can consider network clusters as naturally defined functional modules. By "naturally", we mean through the known molecular interaction patterns. This characteristic can be an advantage relative to the conventionally defined pathways or biological processes. Human decisions determine what genes are in a pathway, where it starts, and where it ends. Although these decisions were made with the best available knowledge, they end up being influenced by the historical process of how the pathway was discovered and characterized. Therefore, in the context of omics data analysis, ORA and FCS analyses could be performed with network clusters instead of pathway or functional annotation gene sets. In the case of ORA, this corresponds to identifying network clusters that are enriched in differentially expressed genes, for example. Using network clusters instead of GO annotations would solve the problem of redundancy in enriched terms and would alleviate the multiple testing problem (as the number of network clusters should be lower than the number of terms to test). But for the final interpretation, we would have to determine which GO terms (or other annotation terms) are associated with the enriched clusters. The main difficulty with applying this strategy is ensuring that it is based on valid network clustering.

14.9 Network Diffusion

Biological networks encode knowledge about how complex biological systems work. When, for example, a set of genes is found to be differentially expressed in a transcriptomic experiment, networks can help to identify other genes that interact closely with that set, unraveling pathways or subnetworks that can explain how the system responds (or adapts) to the experimental perturbation under study. Network diffusion (or network propagation) methods (Figure 14.2C) are the appropriate approach for this task (Cowen et al. 2017; Mubeen et al. 2019). These methods receive as input a set of seed nodes. Then, they apply an algorithm to simulate the diffusion (or propagation) of a signal originating in the seeds and spreading across the network. The output of that algorithm is a numerical score for every node in the network. The higher the node score, the stronger the connection between the seeds and that node.

Some of these methods apply diffusion kernels that model the spread of information across a network in a manner directly analogous to the physical diffusion of a substance (like heat or chemicals) across a medium (Köhler et al. 2008). In this analogy, nodes represent locations and edges represent routes along which the "substance" can travel. The value at each node in the steady state represents how much of the initial signal would have arrived there after an infinite time or how "influenced" it is by the seed nodes. Like physical diffusion, the kernels capture both direct and indirect paths through which information flows, resulting in measures that reflect both local and global network structures. Network diffusion kernels use a diffusion coefficient parameter that controls the information (or "substance") diffusion rate. Higher diffusion coefficients allow for measuring the signal spread to parts of the network that are more distant from the seed nodes, but they can also make the results more homogeneous across the network.

One of the most successful methods of network diffusion is Random Walk with Restarts (RWR, also known as the Personalized Page Rank algorithm) (Khatri et al. 2012; Mason and Verwoerd 2007). This method simulates a random walk across the network. In a simple random walk, the walker starts in one of the seed nodes and randomly selects the next node to visit among the direct neighbors of the present node. If the random walk has infinite steps and we analyze how frequently a given node is visited by the walker, we will get a probability of visit that is proportional to the node's degree. Put differently, a simple random walk does not assess the connectivity to the seed nodes. To solve that, RWR introduces a restart probability p. At each step of the random walk, the walker can return directly to a seed node with probability p. With probability 1-p, the walker will visit a direct neighbor of the current node. The probability of visiting a node in RWR is an efficient way to measure its connectivity with the seed nodes. The parameter p can be fine-tuned to achieve good discrimination among the nodes that are densely connected to the seeds. High values of p will give more weight to short-range connectivity (established through short paths).

Both kernel diffusion methods and RWR tend to produce similar results and are computationally easy to apply to biomolecular networks. Lists of genes (or other molecular entities) coming from omics data analysis can be used as seeds in the appropriate biological network (one that has the same type of entities as nodes). Numerical weights can also be associated with each of the seed nodes, allowing the inclusion of quantitative measurements from the omics data (abundances, log fold changes, etc.). To functionally interpret the results from network diffusion or RWR, we can compute an average score for nodes that are associated with known pathways or functional annotation terms. In this way, we can rank the different pathways and identify the ones that are more affected by the experimental perturbation that was studied in the omics experiment. This approach allows us to obtain meaningful functional interpretations, even if the genes identified in the omics data analysis have poor functional annotations, as long as they are present in the biomolecular network used. One should be aware that nodes with high centralities (high degree, for example) may have a higher chance of getting high network diffusion scores, no matter which seed sets are being analyzed. To correct this bias, the network diffusion process must be repeated a large number of times (100x–1000x, depending on the computational capacity) using either randomized seed sets or randomized networks (maintaining the same degree for each node). The results obtained in the randomized analysis will allow us to estimate a null distribution to perform permutation-based statistical tests or to compute z-scores that are not biased by the node centralities in each pathway. Several variations of this approach have been presented in the literature (Alexeyenko et al. 2012; Girvan and Newman 2002).

14.10 Conclusion

In this chapter, we presented several methods to functionally interpret the results of omics data analysis. We started from simpler methods, which use less information from the omics data (just lists of molecular entities) and just use annotation databases, to more complex methods that can also use quantitative data characterizing the molecular entities and use biological network data besides annotation databases. There are plenty of user-friendly tools to run the simplest method (ORA), although we should be careful about how these tools choose background lists, correct for multiple testing, or address enriched term redundancy. More complex methods require more computational skills, as most of them are available as packages/libraries in the R or Python programming languages. All these methods, independent of their level of complexity, should be used with caution because, in essence, they are exploratory analysis tools that should be used to generate hypotheses. If we discover, using one of the methods, that a particular pathway was significantly affected in our omics experiment, we should design new experiments to validate that finding. It is also difficult to evaluate in a single omics experiment if the changes in that pathway were a side effect of the experimental perturbation or were a fundamental step in the organism's response to the perturbation. Despite these limitations, applying one or more of these methods is essential for fully analyzing and interpreting omics data.

References

Agrawal A, Balcı H, Hanspers K, Coort SL, Martens M, Slenter DN, Ehrhart F, Digles D, Waagmeester A, Wassink I, Abbassi-Daloii T, Lopes EN, Iyer A, Acosta JM, Willighagen LG, Nishida K, Riutta A, Basaric H, Evelo CT, Willighagen EL, Kutmon M, Pico AR (2023) WikiPathways 2024: next generation pathway database. Nucleic Acids Res 52:D679–D689. https://doi.org/10.1093/nar/gkad960

Alcalá-Corona SA, Sandoval-Motta S, Espinal-Enríquez J, Hernández-Lemus E (2021) Modularity in biological networks. Front Genet 12:701331. https://doi.org/10.3389/fgene.2021.701331

Aleksander SA, Balhoff J, Carbon S, Cherry JM, Drabkin HJ, Ebert D, Feuermann M, Gaudet P, Harris NL, Hill DP, Lee R, Mi H, Moxon S, Mungall CJ, Muruganugan A, Mushayahama T, Sternberg PW, Thomas PD, Auken KV, Ramsey J, Siegele DA, Chisholm RL, Fey P, Aspromonte MC, Nugnes MV, Quaglia F, Tosatto S, Giglio M, Nadendla S, Antonazzo G, Attrill H, dos Santos G, Marygold S, Strelets V, Tabone CJ, Thurmond J, Zhou P, Ahmed SH, Asanitthong P, Buitrago DL, Erdol MN, Gage MC, Kadhum MA, Li KYC, Long M, Michalak A, Pesala A, Pritazahra A, Saverimuttu SCC, Su R, Thurlow KE, Lovering RC, Logie C, Oliferenko S, Blake J, Christie K, Corbani L, Dolan ME, Drabkin HJ, Hill DP, Ni L, Sitnikov D, Smith C, Cuzick A, Seager J, Cooper L, Elser J, Jaiswal P, Gupta P, Jaiswal P, Naithani S, Lera-Ramirez M, Rutherford K, Wood V, Pons JLD, Dwinell MR, Hayman GT, Kaldunski ML, Kwitek AE, Laulederkind SJF, Tutaj MA, Vedi M, Wang S-J, D'Eustachio P, Aimo L, Axelsen K, Bridge A, Hyka-Nouspikel N, Morgat A, Aleksander SA, Cherry JM, Engel SR, Karra K, Miyasato SR, Nash RS, Skrzypek MS, Weng S, Wong ED, Bakker E, Berardini TZ, Reiser L, Auchincloss A, Axelsen K, Argoud-Puy G, Blatter M-C, Boutet E, Breuza L, Bridge A, Casals-Casas C, Coudert E, Estreicher A, Famiglietti ML, Feuermann M, Gos A, Gruaz-Gumowski N, Hulo C, Hyka-Nouspikel N, Jungo F, Mercier PL, Lieberherr D, Masson P, Morgat A, Pedruzzi I, Pourcel L, Poux S, Rivoire C, Sundaram S, Bateman A, Bowler-Barnett E, Bye-A-Jee H, Denny P, Ignatchenko A, Ishtiaq R, Lock A, Lussi Y, Magrane M, Martin MJ, Orchard S, Raposo P, Speretta E, Tyagi N, Warner K, Zaru R, Diehl AD, Lee R, Chan J, Diamantakis S, Raciti D, Zarowiecki M, Fisher M, James-Zorn C, Ponferrada V, Zorn A, Ramachandran S, Ruzicka L, Westerfield M, Aleksander SA, Balhoff J, Carbon S, Cherry JM, Drabkin HJ, Ebert D, Feuermann M, Gaudet P, Harris NL, Hill DP, Lee R, Mi H, Moxon S, Mungall CJ, Muruganugan A, Mushayahama T, Sternberg PW, Thomas PD, Auken KV, Ramsey J, Siegele DA, Chisholm RL, Fey P, Aspromonte MC, Nugnes MV, Quaglia F, Tosatto S, Giglio M, Nadendla S, Antonazzo G, Attrill H, dos

Santos G, Marygold S, Strelets V, Tabone CJ, Thurmond J, Zhou P, Ahmed SH, Asanitthong P, Buitrago DL, Erdol MN, Gage MC, Kadhum MA, Li KYC, Long M, Michalak A, Pesala A, Pritazahra A, Saverimuttu SCC, Su R, Thurlow KE, Lovering RC, Logie C, Oliferenko S, Blake J, Christie K, Corbani L, Dolan ME, Drabkin HJ, Hill DP, Ni L, Sitnikov D, Smith C, Cuzick A, Seager J, Cooper L, Elser J, Jaiswal P, Gupta P, Jaiswal P, Naithani S, Lera-Ramirez M, Rutherford K, Wood V, Pons JLD, Dwinell MR, Hayman GT, Kaldunski ML, Kwitek AE, Laulederkind SJF, Tutaj MA, Vedi M, Wang S-J, D'Eustachio P, Aimo L, Axelsen K, Bridge A, Hyka-Nouspikel N, Morgat A, Aleksander SA, Cherry JM, Engel SR, Karra K, Miyasato SR, Nash RS, Skrzypek MS, Weng S, Wong ED, Bakker E, Berardini TZ, Reiser L, Auchincloss A, Axelsen K, Argoud-Puy G, Blatter M-C, Boutet E, Breuza L, Bridge A, Casals-Casas C, Coudert E, Estreicher A, Famiglietti ML, Feuermann M, Gos A, Gruaz-Gumowski N, Hulo C, Hyka-Nouspikel N, Jungo F, Mercier PL, Lieberherr D, Masson P, Morgat A, Pedruzzi I, Pourcel L, Poux S, Rivoire C, Sundaram S, Bateman A, Bowler-Barnett E, Bye-A-Jee H, Denny P, Ignatchenko A, Ishtiaq R, Lock A, Lussi Y, Magrane M, Martin MJ, Orchard S, Raposo P, Speretta E, Tyagi N, Warner K, Zaru R, Diehl AD, Lee R, Chan J, Diamantakis S, Raciti D, Zarowiecki M, Fisher M, James-Zorn C, Ponferrada V, Zorn A, Ramachandran S, Ruzicka L, Westerfield M (2023) The gene ontology knowledgebase in 2023. Genetics 224:iyad031. https://doi.org/10.1093/genetics/iyad031

Alexeyenko A, Lee W, Pernemalm M, Guegan J, Dessen P, Lazar V, Lehtiö J, Pawitan Y (2012) Network enrichment analysis: extension of gene-set enrichment analysis to gene networks. BMC Bioinformatics 13:226. https://doi.org/10.1186/1471-2105-13-226

Ashburner M, Ball CA, Blake JA, Botstein D, Butler H, Cherry JM, Davis AP, Dolinski K, Dwight SS, Eppig JT, Harris MA, Hill DP, Issel-Tarver L, Kasarskis A, Lewis S, Matese JC, Richardson JE, Ringwald M, Rubin GM, Sherlock G (2000) Gene ontology: tool for the unification of biology. Nat Genet 25:25–29. https://doi.org/10.1038/75556

Barabási A-L, Oltvai ZN (2004) Network biology: understanding the cell's functional organization. Nat Rev Genet 5:101–113. https://doi.org/10.1038/nrg1272

Barry WT, Nobel AB, Wright FA (2005) Significance analysis of functional categories in gene expression studies: a structured permutation approach. Bioinformatics 21:1943–1949. https://doi.org/10.1093/bioinformatics/bti260

Blondel VD, Guillaume J-L, Lambiotte R, Lefebvre E (2008) Fast unfolding of communities in large networks. arXiv. https://doi.org/10.48550/arxiv.0803.0476

Cowen L, Ideker T, Raphael BJ, Sharan R (2017) Network propagation: a universal amplifier of genetic associations. Nat Rev Genet 18:551–562. https://doi.org/10.1038/nrg.2017.38

Csabai L, Fazekas D, Kadlecsik T, Szalay-Bekő M, Bohár B, Madgwick M, Módos D, Ölbei M, Gul L, Sudhakar P, Kubisch J, Oyeyemi OJ, Liska O, Ari E, Hotzi B, Billes VA, Molnár E, Földvári-Nagy L, Csályi K, Demeter A, Pápai N, Koltai M, Varga M, Lenti K, Farkas IJ, Türei D, Csermely P, Vellai T, Korcsmáros T (2021) SignaLink3: a multi-layered resource to uncover tissue-specific signaling networks. Nucleic Acids Res 50:D701–D709. https://doi.org/10.1093/nar/gkab909

del Toro N, Shrivastava A, Ragueneau E, Meldal B, Combe C, Barrera E, Perfetto L, How K, Ratan P, Shirodkar G, Lu O, Mészáros B, Watkins X, Pundir S, Licata L, Iannuccelli M, Pellegrini M, Martin MJ, Panni S, Duesbury M, Vallet SD, Rappsilber J, Ricard-Blum S, Cesareni G, Salwinski L, Orchard S, Porras P, Panneerselvam K, Hermjakob H (2022) The IntAct database: efficient access to fine-grained molecular interaction data. Nucleic Acids Res 50:D648–D653. https://doi.org/10.1093/nar/gkab1006

Feuermann M, Mi H, Gaudet P, Muruganujan A, Lewis SE, Ebert D, Mushayahama T, Consortium GO, Aleksander SA, Balhoff J, Carbon S, Cherry JM, Drabkin HJ, Harris NL, Hill DP, Lee R, Logie C, Moxon S, Mungall CJ, Sternberg PW, Auken KV, CACAO/EcoliWiki, Ramsey J, Siegele DA, Chisholm RL, Fey P, Ontology E and C, Giglio M, Nadendla S, FlyBase, Antonazzo G, Attrill H, Brown NH, Garapati PV, Marygold S, UCL FGA, Ahmed SH, Asanitthong P, Buitrago DL, Erdol MN, Gage MC, Huang S, Kadhum MA, Li KYC, Long M, Michalak A, Pesala A, Pritazahra A, Saverimuttu SCC, Su R, Xu Q, Lovering RC, Informatics MG, Blake J, Christie K, Corbani L, Dolan ME, Ni L, Sitnikov D, Smith C, PomBase, Lera-Ramirez M, Rutherford K, Wood V, Reactome, D'Eustachio P, Database RG, Demos WM, Pons JLD, Dwinell MR, Hayman GT, Kaldunski ML, Kwitek AE, Laulederkind SJF, Smith JR, Tutaj MA, Vedi M, Wang S-J, Database SG, Engel SR, Karra K, Miyasato SR, Nash RS, Skrzypek MS, Weng S, Wong ED, Consortium S, Achsel T, Andres-Alonso M, Bagni C, Bayés À, Biederer T, Brose N, Chua JJE, Coba MP, Cornelisse LN, de Juan-Sanz J, Goldschmidt HL, Gundelfinger ED, Huganir RL, Imig C, Jahn R, Jung H, Kaeser PS, Kim E, Koopmans F, Kreutz MR, Lipstein N, MacGillavry HD, McPherson PS, O'Connor V, Pielot R, Ryan TA, Sala C, Sheng M, Smalla K-H, Smit AB, Toonen RF, van Weering JRT, Verhage M, Verpelli C, (TAIR) TAIR, Bakker E, Berardini TZ, Reiser L, Group) UC (Swiss-P, Auchincloss A, Axelsen K, Argoud-Puy G, Blatter M-C, Boutet E, Breuza L, Bridge A, Casals-Casas C, Coudert E, Estreicher A, Famiglietti ML, Gos A, Gruaz-Gumowski N, Hulo C, Hyka-Nouspikel N, Jungo F, Mercier PL, Lieberherr D, Masson P, Morgat A, Pedruzzi I, Pourcel L, Poux S, Rivoire C, Sundaram S, (EMBL-EBI) UC, Bowler-Barnett E, Bye-A-Jee H, Denny P, Ignatchenko A, Ishtiaq R, Lock A, Lussi Y, Magrane M, Martin MJ, Orchard S, Raposo P, Speretta E, Tyagi N, Warner K, Zaru R, WormBase, Chan J, Diamantakis S, Raciti D, Xenbase,

Fisher M, James-Zorn C, Ponferrada V, Zorn A, (ZFIN) ZIN, Ramachandran S, Ruzicka L, Westerfield M, Thomas PD (2025) A compendium of human gene functions derived from evolutionary modelling. Nature 640:146–154. https://doi.org/10.1038/s41586-025-08592-0

Garcia-Alonso L, Holland CH, Ibrahim MM, Turei D, Saez-Rodriguez J (2019) Benchmark and integration of resources for the estimation of human transcription factor activities. Genome Res 29:1363–1375. https://doi.org/10.1101/gr.240663.118

García-Campos MA, Espinal-Enríquez J, Hernández-Lemus E (2015) Pathway analysis: state of the art. Front Physiol 6:383. https://doi.org/10.3389/fphys.2015.00383

Gaudet P, Dessimoz C (2016) The gene ontology handbook. Methods Mol Biol 1446:189–205. https://doi.org/10.1007/978-1-4939-3743-1_14

Gillis J, Pavlidis P (2012) “Guilt by association” is the exception rather than the rule in gene networks. PLoS Comput Biol 8:e1002444. https://doi.org/10.1371/journal.pcbi.1002444

Girvan M, Newman M (2002) Community structure in social and biological networks. Proc Natl Acad Sci 99:7821

Glaab E, Baudot A, Krasnogor N, Schneider R, Valencia A (2012) EnrichNet: network-based gene set enrichment analysis. Bioinformatics 28:i451–i457. https://doi.org/10.1093/bioinformatics/bts389

Huang DW, Sherman BT, Zheng X, Yang J, Imamichi T, Stephens R, Lempicki RA (2009) Extracting biological meaning from large gene lists with DAVID. Curr Protoc Bioinformatics 27:13.11.1–13.11.13. https://doi.org/10.1002/0471250953.bi1311s27

Ispano E, Bianca F, Lavezzo E, Toppo S (2023) An overview of protein function prediction methods: A deep learning perspective. Curr Bioinforma 18:621–630. https://doi.org/10.2174/1574893618666230505103556

Jalili M, Salehzadeh-Yazdi A, Gupta S, Wolkenhauer O, Yaghmaie M, Resendis-Antonio O, Alimoghaddam K (2016) Evolution of centrality measurements for the detection of essential proteins in biological networks. Front Physiol 7:892–894. https://doi.org/10.3389/fphys.2016.00375

Janjić V, Pržulj N (2012) Biological function through network topology: a survey of the human diseasome. Brief Funct Genomics 11:522–532. https://doi.org/10.1093/bfgp/els037

Kanehisa M, Furumichi M, Sato Y, Matsuura Y, Ishiguro-Watanabe M (2024) KEGG: biological systems database as a model of the real world. Nucleic Acids Res 53:D672–D677. https://doi.org/10.1093/nar/gkae909

Khatri P, Sirota M, Butte AJ (2012) Ten years of pathway analysis: current approaches and outstanding challenges. PLoS Comput Biol 8:e1002375. https://doi.org/10.1371/journal.pcbi.1002375

Köhler S, Bauer S, Horn D, Robinson PN (2008) Walking the Interactome for prioritization of candidate disease genes. Am J Hum Genet 82:949–958. https://doi.org/10.1016/j.ajhg.2008.02.013

Kondor RI, Lafferty JD (2002) Diffusion kernels on graphs and other discrete input spaces. In: Proceedings of the nineteenth international conference on machine learning. Morgan Kaufmann Publishers Inc., San Francisco, pp 315–322

Koschützki D, Schreiber F (2008) Centrality analysis methods for biological networks and their application to gene regulatory networks. Gene Regul Syst Biol 2: GRSB.S702. https://doi.org/10.4137/grsb.s702

Koutrouli M, Karatzas E, Paez-Espino D, Pavlopoulos GA (2020) A guide to conquer the biological network era using graph theory. Front Bioeng Biotechnol 8:34. https://doi.org/10.3389/fbioe.2020.00034

Liu C, Ma Y, Zhao J, Nussinov R, Zhang Y-C, Cheng F, Zhang Z-K (2020) Computational network biology: data, model, and applications. Phys Rep:1–66. https://doi.org/10.1016/j.physrep.2019.12.004

Lv Y, Huang S, Zhang T, Gao B (2021) Application of multilayer network models in bioinformatics. Front Genet 12:664860. https://doi.org/10.3389/fgene.2021.664860

Maleki F, Ovens K, Hogan DJ, Kusalik AJ (2020) Gene set analysis: challenges, opportunities, and future research. Front Genet 11:654. https://doi.org/10.3389/fgene.2020.00654

Mason O, Verwoerd M (2007) Graph theory and networks in biology. IET Syst Biol 1:89. https://doi.org/10.1049/iet-syb:20060038

Masuda N, Porter MA (2017) Random walks and diffusion on networks. Phys Rep 716–717:1–58. https://doi.org/10.1016/j.physrep.2017.07.007

Matys V, Kel-Margoulis OV, Fricke E, Liebich I, Land S, Barre-Dirrie A, Reuter I, Chekmenev D, Krull M, Hornischer K, Voss N, Stegmaier P, Lewicki-Potapov-B, Saxel H, Kel AE, Wingender E (2006) TRANSFAC® and its module TRANSCompel®: transcriptional gene regulation in eukaryotes. Nucleic Acids Res 34:D108–D110. https://doi.org/10.1093/nar/gkj143

Mazandu GK, Mulder NJ (2013) Information content-based gene ontology semantic similarity approaches: toward a unified framework theory. Biomed Res Int 2013:292063. https://doi.org/10.1155/2013/292063

Mazandu GK, Chimusa ER, Mulder NJ (2017) Gene ontology semantic similarity tools: survey on features and challenges for biological knowledge discovery. Brief Bioinform 18:886–901. https://doi.org/10.1093/bib/bbw067

McGillivray P, Clarke D, Meyerson W, Zhang J, Lee D, Gu M, Kumar S, Zhou H, Gerstein M (2018) Network analysis as a grand unifier in biomedical data science. Ann Rev Biomed Data Sci 1:153–180. https://doi.org/10.1146/annurev-biodatasci-080917-013444

Milacic M, Beavers D, Conley P, Gong C, Gillespie M, Griss J, Haw R, Jassal B, Matthews L, May B, Petryszak R, Ragueneau E, Rothfels K, Sevilla C, Shamovsky V, Stephan R, Tiwari K, Varusai T,

Weiser J, Wright A, Wu G, Stein L, Hermjakob H, D'Eustachio P (2023) The Reactome pathway knowledgebase 2024. Nucleic Acids Res 52:D672–D678. https://doi.org/10.1093/nar/gkad1025

Mootha VK, Lindgren CM, Eriksson K-F, Subramanian A, Sihag S, Lehar J, Puigserver P, Carlsson E, Ridderstråle M, Laurila E, Houstis N, Daly MJ, Patterson N, Mesirov JP, Golub TR, Tamayo P, Spiegelman B, Lander ES, Hirschhorn JN, Altshuler D, Groop LC (2003) PGC-1α-responsive genes involved in oxidative phosphorylation are coordinately downregulated in human diabetes. Nat Genet 34:267–273. https://doi.org/10.1038/ng1180

Mubeen S, Hoyt CT, Gemünd A, Hofmann-Apitius M, Fröhlich H, Domingo-Fernández D (2019) The impact of pathway database choice on statistical enrichment analysis and predictive modeling. Front Genet 10: 1203. https://doi.org/10.3389/fgene.2019.01203

Nanni ND, Bersanelli M, Milanesi L, Mosca E (2020) Network diffusion promotes the integrative analysis of multiple omics. Front Genet 11:106. https://doi.org/10.3389/fgene.2020.00106

Newman M (2006) Modularity and community structure in networks. Proc Natl Acad Sci USA 103:8577–8582

Noble WS (2009) How does multiple testing correction work? Nat Biotechnol 27:1135–1137. https://doi.org/10.1038/nbt1209-1135

Nowicki K, Snijders TAB (2001) Estimation and prediction for stochastic Blockstructures. J Am Stat Assoc 96:1077–1087. https://doi.org/10.1198/016214501753208735

Oughtred R, Stark C, Breitkreutz B-J, Rust J, Boucher L, Chang C, Kolas N, O'Donnell L, Leung G, McAdam R, Zhang F, Dolma S, Willems A, Coulombe-Huntington J, Chatr-aryamontri A, Dolinski K, Tyers M (2018) The BioGRID interaction database: 2019 update. Nucleic Acids Res 47:D529–D541. https://doi.org/10.1093/nar/gky1079

Paton V, Türei D, Ivanova O, Müller-Dott S, Rodriguez-Mier P, Venafra V, Perfetto L, Garrido-Rodriguez M, Saez-Rodriguez J (2025) NetworkCommons: bridging data, knowledge, and methods to build and evaluate context-specific biological networks. Bioinformatics 41:btaf048. https://doi.org/10.1093/bioinformatics/btaf048

Schwikowski B, Uetz P, Fields S (2000) A network of protein–protein interactions in yeast. Nat Biotechnol 18:1257–1261. https://doi.org/10.1038/82360

Subramanian A, Tamayo P, Mootha VK, Mukherjee S, Ebert BL, Gillette MA, Paulovich A, Pomeroy SL, Golub TR, Lander ES, Mesirov JP (2005) Gene set enrichment analysis: a knowledge-based approach for interpreting genome-wide expression profiles. Proc Natl Acad Sci 102:15545–15550. https://doi.org/10.1073/pnas.0506580102

Surdo PL, Iannuccelli M, Contino S, Castagnoli L, Licata L, Cesareni G, Perfetto L (2022) SIGNOR 3.0, the SIGnaling network open resource 3.0: 2022 update. Nucleic Acids Res 51:D631–D637. https://doi.org/10.1093/nar/gkac883

Szklarczyk D, Kirsch R, Koutrouli M, Nastou K, Mehryary F, Hachilif R, Gable AL, Fang T, Doncheva NT, Pyysalo S, Bork P, Jensen LJ, von Mering C (2023) The STRING database in 2023: protein–protein association networks and functional enrichment analyses for any sequenced genome of interest. Nucleic Acids Res 51:D638–D646. https://doi.org/10.1093/nar/gkac1000

Türei D, Korcsmaros T, Saez-Rodriguez J (2016) OmniPath: guidelines and gateway for literature-curated signaling pathway resources. Nat Methods 13:966–967. https://doi.org/10.1038/nmeth.4077

Türei D, Valdeolivas A, Gul L, Palacio-Escat N, Klein M, Ivanova O, Ölbei M, Gábor A, Theis F, Módos D, Korcsmáros T, Saez-Rodriguez J (2021) Integrated intra- and intercellular signaling knowledge for multicellular omics analysis. Mol Syst Biol 17: MSB20209923. https://doi.org/10.15252/msb.20209923

Vorontsov IE, Kulakovskiy IV, Makeev VJ (2013) Jaccard index based similarity measure to compare transcription factor binding site models. Algorithms Mol Biol 8: 23. https://doi.org/10.1186/1748-7188-8-23

Westerhoff HV, Palsson BO (2004) The evolution of molecular biology into systems biology. Nat Biotechnol 22:1249–1252. https://doi.org/10.1038/nbt1020

Yu G (2020) Stem cell transcriptional networks, methods and protocols. Methods Mol Biol 2117:207–215. https://doi.org/10.1007/978-1-0716-0301-7_11

Zhao Y, Wang J, Chen J, Zhang X, Guo M, Yu G (2020) A literature review of gene function prediction by modeling gene ontology. Front Genet 11:400. https://doi.org/10.3389/fgene.2020.00400

Zitnik M, Li MM, Wells A, Glass K, Gysi DM, Krishnan A, Murali TM, Radivojac P, Roy S, Baudot A, Bozdag S, Chen DZ, Cowen L, Devkota K, Gitter A, Gosline SJC, Gu P, Guzzi PH, Huang H, Jiang M, Kesimoglu ZN, Koyuturk M, Ma J, Pico AR, Pržulj N, Przytycka TM, Raphael BJ, Ritz A, Sharan R, Shen Y, Singh M, Slonim DK, Tong H, Yang XH, Yoon B-J, Yu H, Milenković T (2024) Current and future directions in network biology. Bioinform Adv 4:vbae099. https://doi.org/10.1093/bioadv/vbae099

Multi-omics Data Integration

15

Saumya Kumar, Javier Botey-Bataller, Nhan Nguyen, Wenchao Li, Zhaoli Liu, Cheng-Jian Xu, and Yang Li

Abstract

Human diseases are multi-factorial, affecting multiple aspects of a homeostatic system. Recent advances in high-throughput technology have allowed the generation of various omics datasets from large cohorts at affordable costs and hence made it possible to study the complex dynamical systems perturbed in human diseases. Studying the complex perturbed systems offers a mechanistic understanding to identify druggable targets and offers new avenues for individualised medical intervention. Mechanisms driving complex human diseases cannot be explored merely by single omics-focused studies. In addition, the heterogeneity among the human populations adds additional complexity and limits the possibility for inferring the regulatory mechanisms underlying these diseases. Examining the disease or phenotype of interest through the lens of multiple omics layers may allow the dissection of the perturbed biological processes associated with the disease. Studying a complex disease through multiple omics layers providing vast information is quite a challenging task and, therefore, requires statistical frameworks to achieve integrative multi-omics analysis. In this chapter, we first

S. Kumar · J. Botey-Bataller · N. Nguyen · W. Li · Z. Liu
Centre for Individualised Infection Medicine (CiiM), a joint venture between the Helmholtz Centre for Infection Research (HZI) and Hannover Medical School (MHH), Hannover, Germany

TWINCORE, a joint venture between the Helmholtz-Centre for Infection Research (HZI) and the Hannover Medical School (MHH), Hannover, Germany

C.-J. Xu
Centre for Individualised Infection Medicine (CiiM), a joint venture between the Helmholtz Centre for Infection Research (HZI) and Hannover Medical School (MHH), Hannover, Germany

TWINCORE, a joint venture between the Helmholtz-Centre for Infection Research (HZI) and the Hannover Medical School (MHH), Hannover, Germany

Department of Internal Medicine and Radboud Center for Infectious Diseases, Radboud University Medical Center, Nijmegen, the Netherlands

Y. Li (✉)
Centre for Individualised Infection Medicine (CiiM), a joint venture between the Helmholtz Centre for Infection Research (HZI) and Hannover Medical School (MHH), Hannover, Germany

TWINCORE, a joint venture between the Helmholtz-Centre for Infection Research (HZI) and the Hannover Medical School (MHH), Hannover, Germany

Department of Internal Medicine and Radboud Center for Infectious Diseases, Radboud University Medical Center, Nijmegen, the Netherlands

Cluster of Excellence Resolving Infection Susceptibility (RESIST; EXC 2155), Hannover Medical School, Hannover, Germany

Lower Saxony center for artificial intelligence and causal methods in medicine (CAIMed), Hannover, Germany
e-mail: yang.li@helmholtz-hzi.de

M. Gama-Carvalho et al. (eds.), *Omics Approaches in Biomedicine and Biotechnology*, Advances in Experimental Medicine and Biology 1504, https://doi.org/10.1007/978-3-032-18966-0_15

summarise key characteristics of each of the omics layers and the various considerations important for the implementation of statistical methods. We then shed light on the most common statistical methods used for multi-omics integration studies and highlight various published examples showing the use of these methods for addressing key biological questions. For this, we show integration examples focused on at least two prime omics layers. We next focus on methods and examples showing multi-omics integration to study dynamical systems in large cohort studies. Finally, we discuss some of the multi-omics approaches and examples from single-cell multi-omics datasets.

Keywords

Multi-omics integration · Single-cell omics · Quantitative trait loci (QTL) · Dynamic biological processes · Latent factor analysis

15.1 Omics Dataset Characteristics

A wide range of high-throughput technologies are available to cover molecular and cellular phenotypes: genome, epigenome, transcriptome, proteome, metabolome, cellular phenotype (cell count) and cytokine production capacity. To get a holistic picture of biological changes in the system, the collection of multi-omics datasets is crucial. Table 15.1 summarises some of the high-throughput measurements widely used for large-scale multi-omics studies. Further, in this section, we describe the characteristics of each omics layer and the implications they have in the integration of the datasets across modalities.

15.1.1 Genomics: Genetic Variation

The human genome harbours a high level of variation. The genome of any individual differs, on average, at four to five million sites compared to the established human genome reference. These sites of variation include small variations including single-nucleotide polymorphisms (SNPs) or insertion/deletions (InDels), as well as large-scale mutations such as insertions, deletions and amplifications, which can lead to variation among individuals, for example, individual differences in immune function (Netea et al. 2016). Genetic information can be generated through PCR-based methods (low-throughput) or DNA microarrays and DNA sequencing (high-throughput). DNA microarrays, also known as genotyping chips, contain probes of allele-specific DNA sequences that hybridise with the complementary sequences in DNA samples and detect specific genetic variants. Depending on the chip design, this method can measure different ranges of known genetic variations (Verlouw et al. 2021). Although the number of variants that can be detected using genotyping chips has increased over the years, even the high-density chip (Illumina OMNI5) only covers a small fraction of the human genome (see Table 15.1). Applying imputation algorithms {Next-generation genotype imputation service and methods. Nature Genetics 48, 1284–1287 (2016) https://pubmed.ncbi.nlm.nih.gov/27571263/} based on reference panels, these numbers usually reach millions of SNPs, amounting to the estimated five million common (with a frequency of at least 5%) SNPs in the population. DNA sequencing, on the other hand, detects genetic variants of the whole genome (whole-genome sequencing) or within the exome regions (whole-exome sequencing), including known or novel SNPs/InDels and structural variations (Abel and Duncavage 2013). Genetic variation can also be used to assess the determinants of inter-individual variation at all different omics layers (gene expression, proteome, metabolome and other molecular readouts) which in turn maybe involved in regulating various cellular and molecular functions in a disease (Li et al. 2016a, b; Aguirre-Gamboa et al. 2016, 2020; Zhang et al. 2021; Sun et al. 2023; Karjalainen et al. 2024; Botey-Bataller et al. 2024).

15.1.2 Epigenomics

15.1.2.1 DNA Methylation

DNA methylation is one of the most extensively studied epigenetic modifications (Gupta et al. 2023b). It involves the transfer of a methyl

Table 15.1 High-throughput measurements widely used for large-scale multi-omics studies

Modality	Commonly used tissues	Technology	No of features	Measurement
Genome	Blood/cells	Microarray chips/ DNA-seq	600–700 k	Whole genome
Methylome	Blood, tissues, cells	Microarray chips/BS-seq	850–960 k/ 10–20mil	Whole genome
Open chromatin	Blood, tissues, cells of interest	Bulk or single cell ATAC-seq	30–150 k	Whole genome
Transcriptome	Blood, tissues, cells of interest	Bulk or single cell RNA-seq	30–100 k	Whole transcriptome
Proteome	Plasma, serum	PEA, SomaScan	3–10 k	Targeted circulating molecules
Metabolome	Plasma, serum	LC-MS	2–5 k	Targeted and un-targeted circulating molecules

group to the fifth carbon of cytosine residue (m^5C) catalysed by a family of methyltransferases, including DNMT3A, DNMT3B and DNMT1 (Leonhardt et al. 1992; Xie et al. 1999). DNA methylation modification can be reversed by demethylases, for example, ten–eleven translocation (TET) family (Wu and Zhang 2017). Consequently, DNA methylation levels reflect a dynamic balance between methylation gains and losses. DNA methylation possesses profound biological importance for several reasons: First, m^5C accounts for 1% of the DNA dinucleotides in human somatic cells and affects 70–80% CpG in human genome (Ehrlich et al. 1982; Bird 2002). Second, DNA methylation plays important roles in a variety of biological processes and mechanisms, including gene expression regulation, embryonic development, X chromosome inactivation, genomic imprinting, chromosome stability, genome organisation and silencing of endogenous retrovirus (Riggs 1975; Holliday and Pugh 1975; Sager and Kitchin 1975; Okano et al. 1999). For instance, DNA methylation is essential for repressing the expression of harmful genetic elements (Moore et al. 2013), and the methylation of CpG islands and CpG shores is highly correlated with reduced gene expression (Irizarry et al. 2009; Moore et al. 2013). One well-studied phenomenon is hypermethylation at the promoter regions of tumour suppressor genes, which leads to their silencing. Recent research has expanded to explore the role of DNA methylation in complex diseases (Xu et al. 2018; Qi et al. 2020). Moreover, DNA methylation is a dynamic process that can be influenced by both genetic and environmental factors like smoking (Joubert et al. 2016), air pollution (Gruzieva et al. 2019) and aging (Merid et al. 2020; Oltmanns et al. 2023).

15.1.2.2 Open Chromatin Regions

Assays for Transposase-Accessible Chromatin sequencing (ATAC-seq) is a widely used protocol for genome-wide identification of open chromatin (Li et al. 2019). DNase-seq and ATAC-seq are based on the use of cleavage enzymes (DNase-I and Tn5, respectively), which recognise and cleave DNA in open chromatin regions. Sequencing and the alignment of reads from these fragments allow the detection of open chromatin by identifying genomic intervals to which reads are mapped (Yan et al. 2020). The reads aligned to the reference genome with a higher frequency of reads correspond to 'open peaks' overlapping regulatory elements within the genome. These regulatory elements (RE) can be promoters, enhancers, silencers and insulators that regulate the gene expression profiles of the cells. ATAC-seq datasets have also been useful in identifying novel REs. One of the main analyses of this dataset is differentially accessible regions (DARs), which can provide information on regulatory differences in case/control studies. Additionally, the sequence of regulatory elements can be examined to identify transcription factor (TF)-specific motifs, providing evidence of differential TF regulation in case/control studies. To link RE to genes regulated by RE, usually nearest overlapping genes are considered, giving evidence of *cis*-regulation. When the gene expression from same samples is available, a correlation of peak intensity (reads overlapping open reads) with the expression of genes located nearby provides a statistically relevant prioritisation of gene regulation. ATAC-seq requires fewer cells (only needs a standard input of 50,000 cells) and is less laborious (Li et al. 2019), thereby providing whole-genome open chromatin landscape with minimal biomaterial, and now is also available at a single-cell level. Thus, we can also have scATAC-seq that measures chromatin accessibility in thousands of individual cells and can generate genome-wide profiles from 10 k to 100 k cells per experiment (Chen et al. 2019), capturing the cellular heterogeneity of gene regulation (Teo et al. 2024).

15.1.3 Transcriptomics

RNA-seq is one of the most widely used omics modality useful for providing evidence of differential gene expression with measurements encompassing whole transcriptome (Mortazavi et al. 2008). Since RNA-seq datasets cover whole transcriptomes, a magnitude of biological questions can be examined with it. These datasets have been widely used to explore novel transcripts, differential gene expression in case/control studies, different usage of transcripts of same genes and many more. Bulk RNA-seq, generated from either whole blood, tissues or purified cellular populations, provides gene expression profiles of the samples but does not uncover the heterogeneity underlying the samples. Since the first discovery of single-cell RNA-seq (scRNA-seq) study in 2009 (Tang et al. 2009), it has attracted a growing interest and significantly promoted exploration in all fields of biological sciences. Currently, a myriad of scRNA-seq technologies and protocols are widely used, such as SMART-seq (Goetz and Trimarchi 2012) and 10× Genomic (Zheng et al. 2017). Thereby, tens of thousands of cells could be sensitively captured at once, instead of only hundreds of samples, through conventional RNA-seq methods. This breakthrough profoundly enables researchers to explore the complexity and diversity of the biological system. While the in-depth characteristics and analysis of transcriptomic datasets are discussed in Chap. 11, in this chapter, we highlight how transcriptomic datasets can be integrated with other modalities with some examples.

15.1.4 Proteomics

Proteome refers to the measurement of abundance of proteins in the system/tissue or cells of interest. Even though proteins are generated from mRNA transcripts, other mechanisms, such as post-translational modifications and protein degradation, can affect the concordance between the transcriptome and proteome. Proteome can be measured via untargeted approaches, such as mass spectrometry (MS)-based technologies, or targeted approaches, such as antibody-, bead- or aptamer-based assays. Since proteins can interact with each other and have direct biological functions, this process can also be identified, quantified and verified through protein pathway array (PPA) or tissue microarray (TMA) (Cui et al. 2022). The range, coverage and sensitivity of protein measurements vary widely depending on the technologies. Large studies evaluating immune system often evaluate known cytokine panels or ELISA assays from either whole blood/PBMCs or plasma/serum samples collected. Studies examining immune response heterogeneity use the stimulation of samples as a surrogate of immune response, followed by the measurement of cytokines or proteins using ELISA from the supernatant of the stimulated samples. Recently, OLINK and SomaLogic have established large targeted panels evaluating protein expression of up to 10,000 proteins in circulation. Especially, the protein quantification, which are relative or absolute values, could affect the data normalisation, interpretation and translational application. Various studies have evaluated plasma OLINK panels to study proteome differences evaluating immune and inflammatory responses in infection (Filbin et al. 2021; Byeon et al. 2022), vaccination (Koeken et al. 2020; Kumar et al. 2024) and different tissues (Wimmers et al. 2023).

15.1.5 Metabolomics

Metabolome refers to the whole set of metabolites (small molecules), which are the substrates, intermediates and products of biochemical processes in organisms or biological samples. These metabolites are often the terminal products or the result of the interaction between proteins and environmental factors. Integrating metabolome allows the exploration of the interactions of multi-layered features in the biological system and the molecular regulatory mechanism of these features. Metabolites can be quantified using mass spectrometry (MS)-based approaches, such as liquid chromatography–MS (LC–MS) or gas chromatography–MS (GC–MS), and nuclear magnetic resonance (NMR) spectroscopy (Nguyen et al. 2022; Fu et al. 2023). MS-based approaches are more high-throughput, quantify metabolites in a non-targeted way and are commonly used in metabolomics data generation; however, there are limitations in annotating metabolites, as the technology detects the mass-to-charge ratio of the molecule, which may be overlapping for multiple molecules. Compared to MS-based technologies, NMR spectroscopy is based on the unique magnetic field of compounds; it can be used to identify biochemical molecules. NMR is less sensitive than MS, which means the number of detectable metabolites is generally lower; however, it requires minimal sample preparation, offers a high level of experimental reproducibility, and is non-destructive in nature (Chu et al. 2021b; Nguyen et al. 2022). The vast landscape of all metabolites makes the complete characterisation of the metabolites challenging. Depending on the research interests, metabolomics studies can also focus on a segment/class of metabolite such as lipidomics, glycomics, microbiome metabolomics, etc. A detailed analysis of metabolomics datasets with a focus on systems immunology can be found here (Fu et al. 2023).

15.2 Omics Integration Approaches and Interpretations

In this section, we discuss the statistical methods and considerations essential for the integration of different omics datasets, followed by examples of different omics integration studies. For each omic layer, first, a quality control (QC) evaluating the extreme outlier samples, measured variables, etc., should be performed. This is followed by normalisation, most suitable for each of the omic layer. Different frameworks/pipelines for the genomics, methylation and transcriptomics datasets are widely available and also discussed in the transcriptomics data analysis chapter (Wilhelm-Benartzi et al. 2013; Zhou et al. 2018; Zhao et al. 2018; Sahoo and Sundararajan 2024). However, proteomics and metabolomics datasets are usually controlled and normalised when received from the companies (e.g. NPX values from OLINK). Nevertheless, evaluating the distribution of each measured variable across samples and detection of outliers forms a crucial step before any downstream analysis.

15.2.1 Considerations for Omics Integration Analysis: Data Variables and Confounders

To dissect the biologically relevant signal in each of the omics layers, various confounders need to be evaluated and/or corrected. A confounder is an extraneous variable whose presence affects the variables being studied (e.g. gene/protein expression) such that the results do not reflect the actual relationship between the variables under study (Pourhoseingholi et al. 2012). In study designs, it is important to carefully think and record the well-known and measurable confounders, such as age, sex, body mass index, smoking, etc. Different tactics can be used such as: (1) randomisation, which randomly assigns study subjects to exposure categories to generate groups that are fairly comparable with respect to known and unknown confounding variables; (2) restriction, which eliminates variation in the confounder (focus on a selected age / sex /… group); and (3) matching, which is commonly used in case–control studies (Nørgaard et al. 2017). In addition to these, other omics-specific confounders consist of technical variability (different technology used for data generation, different library preparation) and biological technical variability (e.g. cell viability, protein stability).

However, the study design, data collection methods and technology used are predetermined and cannot be always controlled by researchers. In this case, if confounding factors are recognised and recorded, statistical methods can be used to adjust the effect of confounders. There are mostly two options for dealing with known confounders in the analysis stage: stratification and multivariate methods. Stratification of samples consists of splitting the data into groups of samples within which the confounders do not vary. Stratified analysis works best if only one or two confounders have to be controlled (Pourhoseingholi et al. 2012). Multivariate analysis/models can handle large numbers of covariates (and also confounders) simultaneously. Classical regression models can include many covariables at one time and clarify how much the confounders distort the relationship between the variable of interest (e.g. gene/protein expression) and biological outcome (e.g. diseased state) (Pourhoseingholi et al. 2012). Additionally, Analysis of Covariance (ANCOVA), which is a combination of Analysis of Variance (ANOVA) and linear regression, can increase the statistical power by comparing means across groups, after removing the variance for which quantitative covariates (confounders) account (Pourhoseingholi et al. 2012). Additional confounder identification and adjustment steps applicable for different modalities (e.g. datasets generated from next-generation sequencing technologies) should be considered, dependent on the data modality.

15.2.2 Multi-omics Integration with Genomics: Commonly Used Methods

The integration of genetics and phenotypes (GWAS) or quantitative traits (QTL) can be used to infer causality between traits (Fig. 15.1). Multivariate linear models are the most common and widely used approaches allowing the control of various cofounders and establishing the links between different modalities. Additionally, using Mendelian Randomisation (MR), a statistical method where genetic variants are considered instrumental variables and resembles a randomised controlled trial, can be used to establish a causal link. MR analyses are based on a series of assumptions that must be carefully checked to ensure a robust analysis (Burgess et al. 2023). R packages that have been implemented to perform MR are TwoSample MR (Hemani et al. 2018) and MendelianRandomisation (Patel et al. 2023). An example of this can be found in (Botey-Bataller et al. 2024) where MR was carried out to examine whether cytokine production upon stimulation had a causal effect on the susceptibility to Lyme borreliosis. For this, both the results of cytokine QTL mapping and GWAS of Lyme susceptibility were used. Selecting the variants that significantly altered cytokine responses to the bacteria causing Lyme (*Borrelia burgdorferi*), MR revealed that there was a causal link between higher cytokine responses and protection against *Borrelia* infection.

The vast majority of genetic association signals occur at the genome loci with a high density of correlated SNPs and in non-coding regions of the genome. Because of this, annotating and interpreting genetic associations require statistical techniques such as genetic co-localisation and fine mapping. Genetic co-localisation employs Bayesian statistics to establish whether a genetic variant (or group) is jointly associated with two genetic traits. This may aid, for instance, in linking a disease-associated variant to gene expression. The most widely used method to estimate co-localisation is

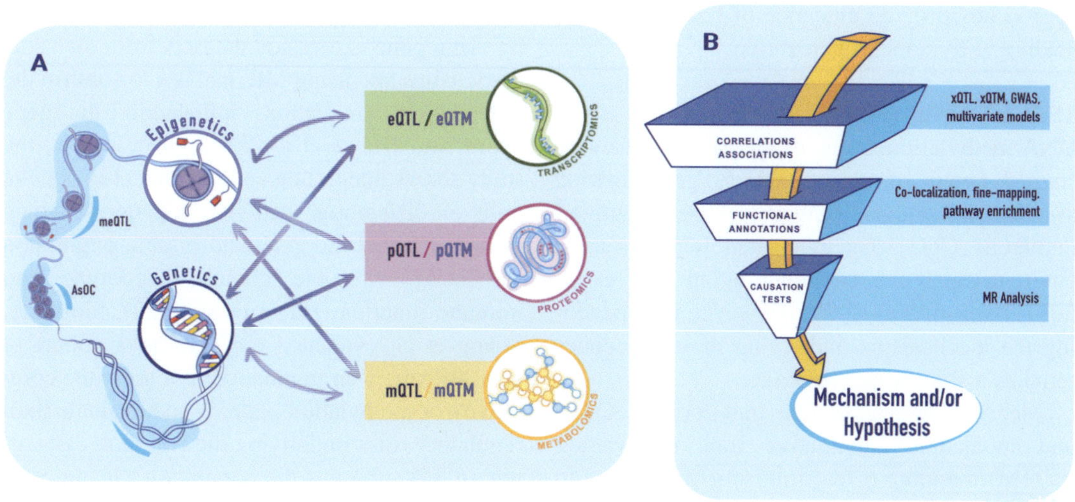

Fig. 15.1 Pairwise omics integration and methods. (**a**) Integration of different omics layers to study the impact of genetics and methylation at each omics layer. (**b**) Statistical methods useful for integration and hypothesis generation

the R package *coloc* (Wang et al. 2020), whose results include the posterior probability of four tested hypotheses: that there is no association with either trait in the region, that there is only an association with one of the traits, that both traits are associated with the region but differ in the causal variant and that both traits are associated with the region and share causal variants (proving that there is genetic co-localisation). In a study examining cytokine response to Lyme disease, co-localisation highlighted genetic loci that regulate cytokine production, immune-mediated diseases and pinpoint genetic variants associated with cytokine production and immune-mediated diseases (Botey-Bataller et al. 2024). This can help establish the effect of a disease-associated variant on immune function. This way, a link between one of the loci, the *TLR1-6-10* locus, and allergic disease could be established.

Genetic fine-mapping is based on feature selection methods and aims to determine which variant in a locus is 'causal' for a given genetic association. Commonly used methods include *Susie* (Wallace 2021) or *FINEMAP* (Benner et al. 2016), which rely on the correlation among SNPs (linkage disequilibrium) to infer this causal variant. Common resources reporting GWAS in population cohorts, such as FinnGen (Kurki et al. 2023), already report the causal variants identified using fine-mapping, allowing researchers to study the mechanisms of GWAS-associated variants.

The results of genome-wide association studies often require the use of publicly available resources to interpret the identified associations and move from variant to function. A breadth of major initiatives is available for this. To identify the effect of a variant in gene expression in a given tissue, eQTL studies such as GTEx (Lonsdale et al. 2013) or eQTLgen (Võsa et al. 2021) provide a resource. Other studies, such as the UK Biobank Pharma Proteomics Project (Sun et al. 2023), detail the effect of genetic variants on plasma protein concentrations (pQTL). Moreover, resources enable to link the variant to a disease or phenotype, such as the UK Biobank (Sudlow et al. 2015) or FinnGen (Kurki et al. 2023). Lastly, major studies also allow, depending on the area of research, to link genetic variants to epigenetic changes (methylation QTL (Oliva et al. 2023)), metabolites (metabolite QTL (Karjalainen et al. 2024)) or the microbiome (microbiome QTL (Zhernakova et al. 2016)).

In the sections below, we describe with examples integration of different omics layers with genetics and how these integration approaches provide evidence of biological links leading to hypothesis generation.

15.2.2.1 Methylation Integration

DNA methylation is one of the most extensively studied layers of epigenome. Epigenome-wide association studies (EWAS) have identified numerous CpG sites associated with diseases, environmental exposure and aging (Xu et al. 2017; Battram et al. 2022). However, understanding the mechanisms underlying these associates remains a significant challenge.

Methylation quantitative trait loci (MeQTL) and molecular quantitative trait methylation (xQTM) mapping offer promising approaches to connect methylation CpG sites and regions with molecular phenotypes (e.g. gene expression, protein levels, cytokine concentrations, metabolite levels, microbiome and cell count proportion), essentially providing a promising approach to examine the impact of DNA methylation in downstream regulation. Methyl modifications can mediate the genetic effects (Liu et al. 2013) and can also be affected by the environmental factors (Joubert et al. 2016). Various studies examining meQTL from human blood have shown associations to behavioural, diseased traits as well as changing patterns with age, smoking, ancestry and BMI, suggesting the influence of genetic traits in differential methylation patterns (Smith et al. 2014; Huan et al. 2019). Recently, Min et al. examined genetic and methylation associations within a cohort evaluating >32,000 participants showing polygenicity in DNA methylation data. Further, they evaluated the shared impact of genetic variants on both DNA methylation and complex diseases using bidirectional MR and found that DNA methylation had a causal relationship with the trait in only a minority of cases (Min et al. 2021).

Another study examining meQTLs across two ancestries—European and South Asian populations—has shown both *cis* and *trans* effects on DNA methylation associated with the genetic traits. Further, the study annotated the CpG sites overlapping the functional chromatin states from the available resources, for example, Roadmap Epigenomics Project for primary blood cells. Additionally, the study evaluated eQTMs by using gene expression datasets from the same participants and using MR analysis to confirm the methylation associations influencing the gene expression (Hawe et al. 2022). In this way, the study shows integration and the impact of genetic traits on differential methylation patterns, which in turn affected the gene and protein expression associated with the clinical phenotype and immune function (Hawe et al. 2022). Similarly, Shang et al. evaluated meQTLs in a cohort of African American population to dissect the complexity of methylation patterns and evaluate their regulatory roles underlying the common diseases within this subset of the population (Shang et al. 2023). In another study evaluating the brain health using methylome and proteome (pQTMs) and brain phenotypes, Gadd et al. reported new DNA methylation signatures from blood associated with proteins implicated in neurological processes, for example, Parkinson's disease. Although this study did not specifically evaluate brain DNA methylation patterns, the authors were able to show an association of cross-tissue DNA methylation patterns by evaluating the pQTMs with brain health metrics, for example, with the GWAS QTL region for cognitive ability and intelligence score as well as integrating with Hi-C datasets of brain and ChiP-seq database (ENCODE project) (Gadd et al. 2022).

15.2.2.2 Transcriptomics Integration

Genome-wide association studies (GWAS) are a widely used tool to identify genetic risk variants which are related to a specific disease. Gene expression patterns among individuals show heterogeneity (Fairfax and Knight 2014), and this variation may largely come from the genetic components (Barreiro et al. 2012; Battle and Montgomery 2014; Fairfax and Knight 2014). This genetic variation is associated with diseases and traits; therefore, deciphering the comprehensive genetic effects on genes which could provide substantial information towards therapy and drug development. Previous studies showed that disease-associated genetic variants primarily influence gene expression rather than affecting the protein function directly (Umans et al.

2021). Expression quantitative trait loci (eQTL) mapping, a framework to link genetics and transcriptomics by employing reliable statistical models. Is largely used to examine the genetic effects. Each single-nucleotide polymorphism (SNP) is assessed to evaluate its potential effects on gene expression. When a SNP is found to be associated with nearby genes on the same chromosome, this gene–SNP pair is defined as cis-eQTL, while the mapping across the whole genome is termed trans-eQTL. eQTL mapping can help researchers gain deeper insights into genetic regulatory mechanisms governing the transcriptome, particularly in the context of diseases. For example, Edahiro et al. (2023) identified a risk allele at *IFNAR2* locus in classical monocytes in COVID-19 patients by performing single-cell eQTL mapping, revealing the host genetic contribution to COVID-19 pathophysiology. Robertson et al. (2021) highlighted a causal variant leading to *BACH2* expression upregulation in T cells in type I diabetes patients, providing potential targets for clinical trials.

eQTL mapping can also be performed at the single-cell level. Many single-cell eQTL studies aggregate/average cells per cell type/cluster per individual and use pseudo-bulk models. Van der Wijst et al. (2018) treated scRNA-seq data from human PBMC as 'bulk-like PBMC' and identified cell-type-specific cis-eQTLs and co-expression QTL based on the personalised co-expression networks. Aquino et al. (2023) explored the genetic variation in response to SARS-CoV-2 and identified *MMP1* as the strongest eQTL in myeloid cells, highlighting *MMP1* as a potential biomarker of severity. However, these pseudo-bulk strategies overlook the substantial heterogeneity even within relatively homogeneous cell populations (Schmiedel et al. 2022) and limit our understanding of the continuous transcriptional landscape (Nathan et al. 2022). To overcome these limitations and improve the power to detect subtle signals, several methods have been developed for single-cell transcriptome datasets without pseudo-bulk transformation. For instance, SCeQTL (Hu et al. 2020) employs zero-inflated negative binominal regression to model the single-cell transcriptome. StructLMM (Moore et al. 2019) uses linear mixed models (LMM) for eQTL statistical analysis. Its application on cell-context interaction analysis identified 46 previously unreported eQTL associated with autoimmune/infectious diseases and traits, such as one eQTL for *CTSW* in Crohn's disease. CellRegMap (Cuomo et al. 2022) aims to detect cell-type and cell-state eQTL in an unbiased way without priorly annotating cell types or cell states, potentially allowing the identification of continuous manner among cells.

15.2.2.3 Proteome and Cytokine Integration

Protein QTLs (pQTLs) allow the integration of genetics with proteome to measure the effect of genetic traits on protein/cytokine expression. Similar methods as described above can be useful to examine pQTL at steady state. One of the biggest pQTL study utilising proteomics and genetics datasets from more than 54,000 UK BioBank participants as a part of Pharma Proteomics Project has explored the genetic associations of 2923 proteins alongside ancestry-specific pQTLs. This extensive protein–genetics association study further using fine mapping and pathway analysis dissected the genetic regulation for both cis- and trans-QTLs. Such resources further aid in drug–target development, providing evidence for protein function in associated diseases (Sun et al. 2023). Another approach examined genetic associations to ratios of partially correlated proteins providing a hypothesis-free strength of association signals. This study showed stronger enrichment of genetic associations within known protein–protein interactions, thereby providing biological insights of genetic traits influencing the protein networks (Suhre 2024).

In addition, inter-individual variability in the production of cytokines upon stimulation with different inactivated viral or bacterial molecules has highlighted additional regulation driven through genetic traits, different from those at steady state (Li et al. 2016a, b). These approaches have previously been named cytokine quantitative trait loci (cQTL) studies. Consequently, the cQTLs examine the genetic traits associated with response to the stimulation or response QTL. Response QTL allows investigation of differences in the immune response regulation, thereby dissecting another layer of inter-individual variations. One of these studies explored the association between common genetic variants and cytokine production upon stimulation in a cohort of healthy individuals of European ancestry (Bakker et al. 2018), revealing multiple loci associated with monocyte- and T-cell-related function and their overall involvement in infectious diseases. More recently, new studies have examined the impact of cQTL on the healthy individuals of African ancestry, including a cohort of healthy individuals in Tanzania (Boahen et al. 2022) and a cohort of neonates in Guinea-Bissau (Boahen et al. 2023). These studies have highlighted the differences and similarities in regulating immune response between African and European genetic ancestries, emphasising the importance of multi-ancestry studies to fully understand host genetics' role in immune variation. Other recent studies, such as those conducted as part of the LymeProspect (Botey-Bataller et al. 2024) project, highlighted the first application of cQTL to study infectious diseases. The study pinpointed the significance of cQTL for Lyme borreliosis susceptibility and progression, revealing relevant pathways and the causal relationship between genetically driven variation in immune response and disease susceptibility. Other studies have also inspected the impact of cQTL on immune-mediated diseases, such as type 1 diabetes (T1D) (Chu et al. 2022). The mentioned study uncovered common and disease-specific regulations. Recent work also highlighted the impact of other immune modulators on genetically driven immune variation, such as the Bacillus Calmette–Guerin (BCG) vaccine (Koeken et al. 2020). Lastly, cQTL has been used to inspect immune responses in children, especially in the context of allergic asthma (Lin et al. 2022). In summary, while pQTLs are useful to identify genetic traits at steady state, cQTL analysis sheds light on genetic traits and their role in immune responses and diseases linked to immune response variabilities.

15.2.2.4 Metabolomics Integration

Assessing the genetic impact on metabolites' concentration offers essential insights for host-specific variation, particularly for studying metabolic diseases. Plasma metabolites may represent a sum of all metabolic processes in the system, similar to plasma proteome or cytokine concentrations. However, there are additional variabilities dependent on the diet, microbiome and ancestry. Studies examining metabolite QTLs (mQTLs) take advantage of methods, such as MR, co-localisation, etc., for biological interpretations and the likely function of these metabolites. A population-level cohort study examined genetic traits associated with plasma metabolites in a healthy cohort (Chen et al. 2023). In such analysis, to understand the biological relevance of the findings, multiple options are available, for example, summarising the metabolites based on their super-pathway or category, for example, amino acids, lipids and xenobiotics. The ratio of metabolites can be calculated using known databases, for example, KEGG, HMDB and PubChem. Since metabolites can be either substrates or products, the ratio of known pairs of metabolites can be evaluated for QTLs, essentially providing insights into the genetic determinants of metabolic flux. Additionally, multi-trait mapping of lipids has been shown to define novel loci in African ancestry (Kamiza et al. 2023). As metabolite concentrations are also affected by the microbiome, a cross-connection of genetics with metabolites and the microbiome abundance may provide further insights for various metabolic diseases (Liu et al. 2022). Metabolites can also play a role in the modulation of immune responses through multiple mechanisms. Chu et al. examined the genetic traits and the role of plasma metabolites in diseases by evaluating the impact of metabolites on immune cell types and cytokines. Among the other findings, they evaluated the role of locus overlapping *FADS2* gene in arachidonic metabolism and established causal effects of the locus in Crohn's disease (Chu et al. 2021a).

To further harness the power of other QTL and GWAS studies, co-localisation methods can be employed, for example, identifying overlapping eQTL, splicingQTL, cQTL or pQTLs from publicly available GWAS studies examining genetic determinants associated to diseases. Co-localisation analysis may also help in prioritising novel QTLs and genes in close proximity associated with a disease and therefore help in hypothesis building for the underlying mechanisms. To evaluate the statistical significance of the hypothesis, MR analysis can be conducted, for example, Chen et al. used MR to establish gene–metabolite associations to large GWAS of diseases predominantly influenced by either aging, metabolism or immune response (Chen et al. 2023). Liu et al. also showed the use of MR to examine the hypothesis evaluating the genetic trait–gut microbiome–blood metabolite causality in their study (Liu et al. 2022).

15.3 Multi-modal Integration to Model Dynamic Processes

Many transcriptomic studies aim to model time-dependent gene expression changes, for example, to follow disease dynamics, response to vaccination, dynamic changes during differentiation, etc. These datasets essentially provide a snapshot of the functional changes in the system, dependent on the condition examined. However, often, the changes observed at the transcriptome level are not replicated at the protein level, and assessing the gene function/role with metabolites is not straightforward either. Even though the central dogma governing the flow of information suggests transcriptome–proteome/metabolome, this is not always the case. For example, amino acids (measured in metabolites) are building blocks of proteins, and changes in amino acids captured may represent the changes and turn-around time of proteins/cytokines. Similarly, transcriptomic changes captured reflect the ongoing processes of regulation by transcription factor proteins induced by a stimulation or an infectious insult to the system. With these underlying complexities, the dissection of mechanism requires evaluation of datasets at the system level.

System-level integration of multi-omics layers may help in prioritising the mechanisms unique to examine the biological question at hand. In a previous study, Bakker et al. systematically investigated multi-omics datasets with deep immune phenotyping to evaluate the inter-individual variations in cytokine production in more than 500 healthy donors. In this work, the authors evaluated and examined the variations in genetics, transcriptomics, metabolomics and gut microbiome along with immunological parameters to establish the variabilities in cytokine production to different stimuli. Finally, using machine learning methods, the authors were able to show contribution of multi-omics in the prediction of cytokine responses, highlighting the role of different molecules in cytokine production (Bakker et al. 2018). In our recent work, we modelled system-level differences in vaccine response in elderly high-vaccine responders and low-vaccine responders to influenza vaccine in two different ways (Kumar et al. 2024). For each omics layer, we first identified the major differences within responders and non-responders. Instead of using differential gene expression, we summarised pathway-level differences using blood transcriptome modules (BTMs) in the two categories as well as cell-subset deconvolution for the different time points, pre- and post-vaccination. Metabolites were summarised based on their superfamily as amino acids, lipids, etc. To model the time-dependent differences, we associated the pathways, with changes in proteome and metabolome within each category. This allowed us to prioritise the transcriptomic pathways, proteins and metabolites that showed coordinated changes upon vaccination within responders. Analysing each omics layer separately allowed us to establish a mechanism where high pre-vaccination IL15 leading to higher NK cell proportions was linked to poor responder status, further confirmed with a mouse experimental model (Kumar et al. 2024).

Li et al. described building a Multiscale, Multifactorial Response Network (MMRN) to integrate system-level microarray-based transcriptome, targeted metabolome, flow cytometry and cytokine measurements examining Zostavax vaccination response (Li et al. 2017). The network essentially uses Partial Least-Squares Regression (PLSR) to establish associations across the different layers of information. The same lab also examined the loss of microbiome and its effects on antibody response to influenza vaccination upon antibiotic treatment. In this study, Hagan et al. integrated stool microbiome, blood transcriptome and metabolomics using MMRN to model vaccine dynamics in antibiotic-treated donors (Hagan et al. 2019). The above examples highlight some of the multi-omics integrative approaches that allow dissection of dynamical processes and therefore aid in hypothesis generation to address biological questions of interest.

15.4 Multi-modal Integration to Identify Common Variation Across Layers

With large cohort-based studies examining omics datasets in diseases and vaccination with additional clinical features being extensively generated and examined, great interest has been drawn in identifying the omics correlates or predictors of clinical outcomes. These correlates in certain cases may help in the stratification of patients, for example, disease prognosis, COVID-19 severity, etc. In such cases, integration and prioritisation of molecules from multiple modalities serve as a great tool for systematic examination of the cohort (Fig. 15.2). Apart

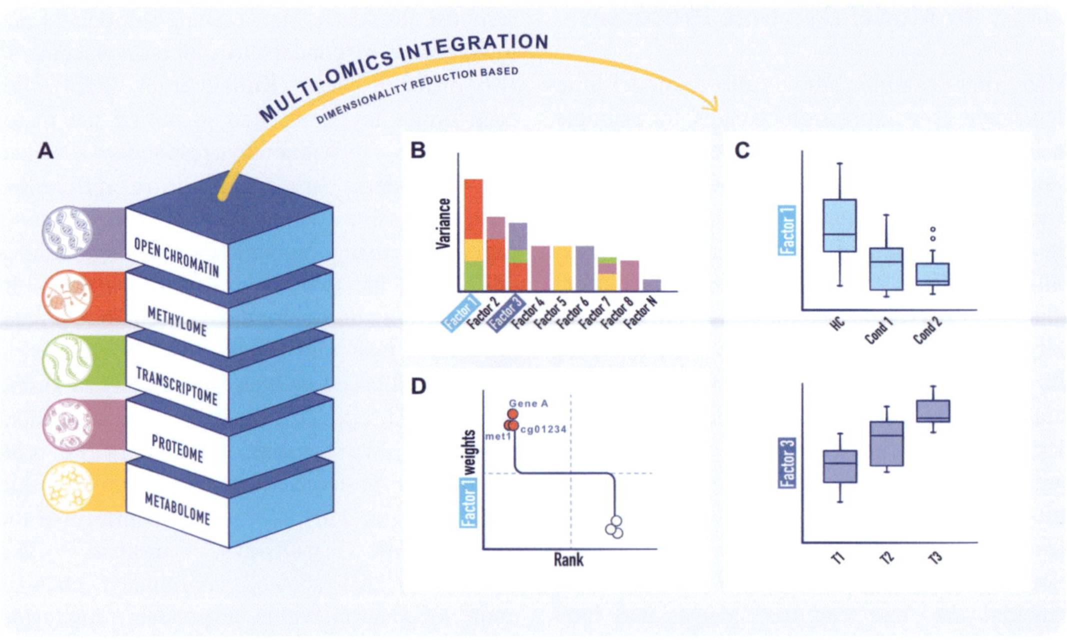

Fig. 15.2 Multi-omics integration schematic. (**a**) Multiple omics modalities from the same cohort; the tools can handle missingness of samples to a certain extent. (**b**) Latent factors identified, explaining the variance and contribution of each omics layer to the factor (colours match from A). (**c**) Factors explaining the variance between samples, e.g. Factor 1 captures the difference between Healthy Controls (HC) and Conditions, while Factor 3 captures time-dependence. (**d**) The top molecules for Factor 1 from the different modalities

from pairwise associations or regression methods, unsupervised dimensionality reduction methods across multiple layers of heterogenous datasets have been shown useful.

Multi-Omics Factor Analysis (MOFA) framework is one such unsupervised method that allows integration of multiple omics layers and infers latent factors capturing the common axis of variation across the layers (Argelaguet et al. 2020). MOFA can be applied with many different modalities (both single cell and bulk) and identifies the principal axis of heterogeneity across the samples, which may or may not be shared across the omics layers. In the elderly vaccination study, MOFA captured a common axis of variation in the form of a latent factor across all the three omics layers that separated elderly high responders from low responders (Kumar et al. 2024). Further, MOFA was able to capture the time dynamics of high responders in another latent factor. It also provides each omic layer and the top molecules contributing to the latent factor. Another study examining multi-omics dataset measuring up to five modalities from a large cohort of over 2000 people living with HIV showed great use of MOFA in dissecting the heterogeneity and identifying the latent factors associated to multiple comorbidities (Botey-Bataller et al. 2025). Botey-Bataller et al. were able to dissect the multi-omics profile associated with people with a history of comorbidity and inflammation during HIV infection. Another method, Multi-set Correlation and Factor Analysis (MCFA) offers unsupervised integration across omics layers using similar factor analysis concept. Brown et al. explored a large dataset of 614 samples to segregate the top factors capturing demographics across the omics layers of methylation, protein expression and RNA expression. The authors were able to find several factors enriched for GWAS hits, further integrating the genetic data to identify factors associated to metabolic diseases (Brown et al. 2023). Taken together, unsupervised factor analysis methods show great promise in dissecting the inter-individual heterogeneity and prioritising the molecular markers from different omics layers for each factor.

15.5 Single-Cell Integration Across Omics Layers

Single-cell technology advances have allowed to dissect the transcriptome and open chromatin heterogeneity at the cellular level, allowing the evaluation of gene expression and regulation within each cell at a high-throughput scale. Most commonly available datasets from these technologies capture the transcriptome (scRNA-seq), cell-surface proteome (CITE-Seq) and open chromatin regions (snATAC-seq) (Heumos et al. 2023). Whereas both transcriptome and open chromatin regions are measured from whole genome, surface proteome profiling via CITE-Seq is usually limited to 100–300 proteins per cell. Single-cell multiome technology offers further advancements allowing paired data from each cell, that is, scRNA-seq and surface protein quantification (CITE-Seq) or snRNA-seq + snATAC-seq (multiome). These technologies are currently producing a vast amount of data sets to study complex diseases, dissect heterogeneity and access disease relevant cellular states e.g. the cellular state linked to long COVID (https://www.nature.com/articles/s41590-025-02387-1).

Various data integration methods are widely available for the integration of at least two modalities at a time (Hao et al. 2021; Ko et al. 2023), with the methods taking into account the differences in the data structure of these modalities (Stuart et al. 2019; Korsunsky et al. 2019; Hao et al. 2021). Recently, another approach, 'bridge integration' using the principles of dictionary learning, allows the integration of all single-cell modalities generated from the same tissue (Hao et al. 2024). The tool requires a multiome dataset (e.g. snRNA-seq + snATAC-seq from the same cell), where each cell serves as an element of a 'dictionary'. This learned process then allows integration of unimodal datasets or multimodal data sets, common with either of the two modalities, thereby giving more power to the data analysis. In most cases of single-cell data integration, harmonisation of single-cell datasets from different batches and different modalities allows the increase in the size of datasets and may serve as a great resource for exploration of different biological questions. For example, protein velocity can be calculated from CITE-Seq datasets, along with RNA velocity, providing a more accurate representation of future cellular states. In this case, the future cellular state is modelled by combining information from unspliced RNA, spliced RNA and protein expression (Gorin et al. 2020). However, protein velocity is limited to the proteins sampled in the CITE-seq experiment (~100–300) proteins.

Integration of transcriptomics and open chromatin modalities at the single-cell level has been of great interest, particularly examining the gene regulation functional in each cell. scRNA-seq datasets had already seen a surge in studying and identifying the functional regulomes (transcription factors (TF) and target genes (TG) regulated) or gene regulatory networks (GRNs) within each cellular subset or cellular state. Open chromatin data adds an additional layer of evidence in deriving the gene regulatory networks of the datasets studied. Open chromatin data consists of regions within the genome that are open and thereby accessible for TF binding. This modality provides additional insights into the regulatory role of non-coding regions within the genome, also known as regulatory elements (RE), and their likely role in the expression of surrounding genes. Consequently, the two modalities combined allows fine-tuned exploration of GRNs (Badia-I-Mompel et al. 2023; Loers and Vermeirssen 2024). This TF-binding site within the RE provides evidence for TF function associated with the RE, while the closest gene to the RE provides evidence for the likely role of RE in gene expression. Combining the TF-RE-TG associations along with the TF-TG co-expression provides fine-tuned GRNs. These GRNs can further be used to examine cell-type-specific GRNs and GRNs involved in dynamic processes, for example, cellular differentiation (GRNs for pseudotime and GRN velocity) and perturbed GRNs in diseased versus healthy (Kamimoto et al. 2023; Badia-I-Mompel et al. 2023; Singh et al. 2024; Loers and Vermeirssen 2024).

Previously discussed single-cell eQTL integration associates the effect of genetic traits on gene expression in different cell subsets. Integrating the transcriptome with open chromatin and genetic traits may provide a mechanistic understanding of the regulation of gene expression dependent upon genetic traits. Emerging studies have further combined the transcriptome, open chromatin datasets from tissues and blood to integrate with GWAS studies, essentially dissecting the role of non-coding genetic traits in the regulation of cell-type-specific gene expression (Sakaue et al. 2024). One such example is demonstrated by Zhang et al., where they showed differential open chromatin accessibility of the same region dependent upon the genetic trait which was associated with the severity of COVID-19 and hospitalisation (Zhang et al. 2023). They showed differential open chromatin accessibility of the enhancer with the allele. The reduced enhancer accessibility was associated with CCR2 gene expression within monocytes and particularly enriched in hospitalised patients. As such, Zhang et al. showed the impact of genetic trait linked to the severity of COVID-19. Gupta et al. integrated transcriptome and open chromatin datasets from the same cells of a cell subset, to link the dynamic changing gene expression and open chromatin within the cell subset, for example, within CD4 T cells. Using multivariate Poisson model, they separated the open chromatin peaks from invariant open chromatin regions, for example, in CD4 T cells, seperated common celltype specific peaks from dynamic open chromatin regions, thereby identifying regions associated with changing gene expression within the CD4 T cells, identifying different cellular states. Further, they examined the 'disease heritability' within the two sets of open regions and found a greater enrichment of disease heritability within the dynamic open chromatin regions associated to changing gene expression in autoimmune diseases, for example, Rheumatoid Arthritis (RA) and Inflammatory Bowel Disease (IBD). Another study using the above approach showed tissue-specific enhancer–gene maps and their overlap in finding causal disease alleles (Gupta et al. 2023a; Sakaue et al. 2024). Although the above examples mostly focus on the integration of single-cell omics datasets, emerging studies have shown single-cell integration with datasets generated through cyTOF and flow cytometry technology. Schulte-Schrepping et al. conducted one of the first single-cell multi-omics studies of dysfunctional immune system in mild and severe COVID19 patients and determined changes in immune cell composition and activation in mild versus severe COVID-19 over time (Schulte-Schrepping et al. 2020). In summary, these examples show that integration of omics datasets at single cell resolution, and allows in-depth dissection of genetic and non-genetic regulation of unique and transient cellular states, aiding in mechanistic understanding of diseases and cellular processes.

15.6 Summary and Future Perspectives

The surge in availability of large omics datasets opens avenues to examine complex biological systems and perturbations thereof in complex diseases. In this chapter, we discussed different strategies, essential considerations and statistical methods with applications for various levels of data integration to investigate diseases, dynamical processes and inter-individual differences among the populations. While most of the statistical methods were designed for bulk studies, similar methods with additional considerations have also been used for single-cell omics-integrative studies. Further, use of machine learning-based methods shows promise in investigating dynamical systems. With the generation and implementation of these large omics datasets and integration approaches, new hypothesis can be generated, which can be further tested in model organisms. In the near future, the ever-increasing availability of computational power, machine learning and artificial intelligence methods will aid greatly in large-scale multi-omics integration, consequently aiding in translational research of complex diseases.

References

Abel HJ, Duncavage EJ (2013) Detection of structural DNA variation from next generation sequencing data: a review of informatic approaches. Cancer Genet 206(12):432–440. https://doi.org/10.1016/j.cancergen.2013.11.002

Aguirre-Gamboa R, Joosten I, Urbano PCM, van der Molen RG, van Rijssen E, van Cranenbroek B, Oosting M, Smeekens S, Jaeger M, Zorro M, Withoff S, van Herwaarden AE, Sweep FCGJ, Netea RT, Swertz MA, Franke L, Xavier RJ, Joosten LAB, Netea MG, Wijmenga C, Kumar V, Li Y, Koenen HJPM (2016) Differential effects of environmental and genetic factors on T and B cell immune traits. Cell Rep 17(9):2474–2487. https://doi.org/10.1016/j.celrep.2016.10.053

Aguirre-Gamboa R, De Klein N, Di Tommaso J, Claringbould A, Van Der Wijst MG, De Vries D, Brugge H, Oelen R, Võsa U, Zorro MM, Chu X, Bakker OB, Borek Z, Ricaño-Ponce I, Deelen P, Xu C-J, Swertz M, Jonkers I, Withoff S, Joosten I, Sanna S, Kumar V, Koenen HJPM, Joosten LAB, Netea MG, Wijmenga C, BIOS Consortium, Franke L, Li Y (2020) Deconvolution of bulk blood eQTL effects into immune cell subpopulations. BMC Bioinformatics 21(1):243. https://doi.org/10.1186/s12859-020-03576-5

Aquino Y, Bisiaux A, Li Z, O'Neill M, Mendoza-Revilla J, Merkling SH, Kerner G, Hasan M, Libri V, Bondet V, Smith N, de Cevins C, Ménager M, Luca F, Pique-Regi R, Barba-Spaeth G, Pietropaoli S, Schwartz O, Leroux-Roels G, Lee C-K, Leung K, Wu JT, Peiris M, Bruzzone R, Abel L, Casanova J-L, Valkenburg SA, Duffy D, Patin E, Rotival M, Quintana-Murci L (2023) Dissecting human population variation in single-cell responses to SARS-CoV-2. Nature 621(7977): 120–128. https://doi.org/10.1038/s41586-023-06422-9

Argelaguet R, Arnol D, Bredikhin D, Deloro Y, Velten B, Marioni JC, Stegle O (2020) MOFA+: a statistical framework for comprehensive integration of multi-modal single-cell data. Genome Biol 21(1):111. https://doi.org/10.1186/s13059-020-02015-1

Badia-I-Mompel P, Wessels L, Müller-Dott S, Trimbour R, Ramirez Flores RO, Argelaguet R, Saez-Rodriguez J (2023) Gene regulatory network inference in the era of single-cell multi-omics. Nat Rev Genet 24(11):739–754. https://doi.org/10.1038/s41576-023-00618-5

Bakker OB, Aguirre-Gamboa R, Sanna S, Oosting M, Smeekens SP, Jaeger M, Zorro M, Võsa U, Withoff S, Netea-Maier RT, Koenen HJPM, Joosten I, Xavier RJ, Franke L, Joosten LAB, Kumar V, Wijmenga C, Netea MG, Li Y (2018) Integration of multi-omics data and deep phenotyping enables prediction of cytokine responses. Nat Immunol 19(7):776–786. https://doi.org/10.1038/s41590-018-0121-3

Barreiro LB, Tailleux L, Pai AA, Gicquel B, Marioni JC, Gilad Y (2012) Deciphering the genetic architecture of variation in the immune response to mycobacterium tuberculosis infection. Proc Natl Acad Sci USA 109(4):1204–1209. https://doi.org/10.1073/pnas.1115761109

Battle A, Montgomery SB (2014) Determining causality and consequence of expression quantitative trait loci. Hum Genet 133(6):727–735. https://doi.org/10.1007/s00439-014-1446-0

Battram T, Yousefi P, Crawford G, Prince C, Sheikhali Babaei M, Sharp G, Hatcher C, Vega-Salas MJ, Khodabakhsh S, Whitehurst O, Langdon R, Mahoney L, Elliott HR, Mancano G, Lee MA, Watkins SH, Lay AC, Hemani G, Gaunt TR, Relton CL, Staley JR, Suderman M (2022) The EWAS catalog: a database of epigenome-wide association studies. Wellcome Open Res 7:41. https://doi.org/10.12688/wellcomeopenres.17598.2

Benner C, Spencer CCA, Havulinna AS, Salomaa V, Ripatti S, Pirinen M (2016) FINEMAP: efficient variable selection using summary data from genome-wide association studies. Bioinformatics 32(10):1493–1501. https://doi.org/10.1093/bioinformatics/btw018

Bird A (2002) DNA methylation patterns and epigenetic memory. Genes Dev 16(1):6–21. https://doi.org/10.1101/gad.947102

Boahen CK, Temba GS, Kullaya VI, et al (2022). A functional genomics approach in Tanzanian population identifies distinct genetic regulators of cytokine production compared to European population. Am J Hum Genet 109(3):471-485. https://doi.org/10.1016/j.ajhg.2022.01.014

Boahen CK, Moorlag SJCFM, Jensen KJ, Matzaraki V, Fanucchi S, Monteiro I, De Bree C, Fok ET, Mhlanga M, Joosten LAB, Aaby P, Benn CS, Netea MG, Kumar V (2023) Genetic regulators of cytokine responses upon BCG vaccination in children from West Africa. J Genet Genomics 50(6):434–446. https://doi.org/10.1016/j.jgg.2023.01.002

Botey-Bataller J, Vrijmoeth HD, Ursinus J, Kullberg B J, van den Wijngaard CC, Ter Hofstede H, Alaswad A, Gupta MK, Roesner LM, Huehn J, Werfel T, Schulz TF, Xu C-J, Netea MG, Hovius JW, Joosten LAB, Li Y (2024) A comprehensive genetic map of cytokine responses in Lyme borreliosis. Nat Commun 15(1): 3795. https://doi.org/10.1038/s41467-024-47505-z

Botey-Bataller J, van Unen N, Blaauw M, Vos WAJW, van Eekeren L, Vadaq N, Matzaraki V, Verbon A, Groenendijk AL, dos Santos JC, Cleophas MCP, Stalenhoef JE, Berrevoets MAH, Jiang X, Gupta MK, Nguyen N, Xu C-J, Joosten LAB, Netea MG, van der Ven AJAM, Li Y (2025) Integrative omics reveals the genetic and molecular landscape of immune responses and comorbidities in people living with HIV. Nat Med. https://doi.org/10.1038/s41591-025-03887-1

Brown BC, Wang C, Kasela S, Aguet F, Nachun DC, Taylor KD, Tracy RP, Durda P, Liu Y, Johnson WC, Van Den Berg D, Gupta N, Gabriel S, Smith JD, Gerzsten R, Clish C, Wong Q, Papanicolau G, Blackwell TW, Rotter JI, Rich SS, Barr RG, Ardlie KG, Knowles DA, Lappalainen T (2023) Multiset correlation and factor analysis enables exploration of multi-omics data. Cell Genom 3(8):100359. https://doi.org/10.1016/j.xgen.2023.100359

Burgess S, Davey Smith G, Davies NM, Dudbridge F, Gill D, Glymour MM, Hartwig FP, Kutalik Z, Holmes MV, Minelli C, Morrison JV, Pan W, Relton CL, Theodoratou E (2023) Guidelines for performing Mendelian randomization investigations: update for summer 2023. Wellcome Open Res 4:186. https://doi.org/10.12688/wellcomeopenres.15555.3

Byeon SK, Madugundu AK, Garapati K, Ramarajan MG, Saraswat M, Kumar-M P, Hughes T, Shah R, Patnaik MM, Chia N, Ashrafzadeh-Kian S, Yao JD, Pritt BS, Cattaneo R, Salama ME, Zenka RM, Kipp BR, Grebe SKG, Singh RJ, Sadighi Akha AA, Algeciras-Schimnich A, Dasari S, Olson JE, Walsh JR, Venkatakrishnan AJ, Jenkinson G, O'Horo JC, Badley AD, Pandey A (2022) Development of a multiomics model for identification of predictive biomarkers for COVID-19 severity: a retrospective cohort study. Lancet Digital Health 4(9):e632–e645. https://doi.org/10.1016/S2589-7500(22)00112-1

Chen H, Lareau C, Andreani T, Vinyard ME, Garcia SP, Clement K, Andrade-Navarro MA, Buenrostro JD, Pinello L (2019) Assessment of computational methods for the analysis of single-cell ATAC-seq data. Genome Biol 20(1):241. https://doi.org/10.1186/s13059-019-1854-5

Chen Y, Lu T, Pettersson-Kymmer U, Stewart ID, Butler-Laporte G, Nakanishi T, Cerani A, Liang KYH, Yoshiji S, Willett JDS, Su C-Y, Raina P, Greenwood CMT, Farjoun Y, Forgetta V, Langenberg C, Zhou S, Ohlsson C, Richards JB (2023) Genomic atlas of the plasma metabolome prioritizes metabolites implicated in human diseases. Nat Genet 55(1):44–53. https://doi.org/10.1038/s41588-022-01270-1

Chu X, Jaeger M, Beumer J, Bakker OB, Aguirre-Gamboa R, Oosting M, Smeekens SP, Moorlag S, Mourits VP, Koeken VACM, de Bree C, Jansen T, Mathews IT, Dao K, Najhawan M, Watrous JD, Joosten I, Sharma S, Koenen HJPM, Withoff S, Jonkers IH, Netea-Maier RT, Xavier RJ, Franke L, Xu C-J, Joosten LAB, Sanna S, Jain M, Kumar V, Clevers H, Wijmenga C, Netea MG, Li Y (2021a) Integration of metabolomics, genomics, and immune phenotypes reveals the causal roles of metabolites in disease. Genome Biol 22(1):198. https://doi.org/10.1186/s13059-021-02413-z

Chu X, Zhang B, Koeken VACM, Gupta MK, Li Y (2021b) Multi-omics approaches in immunological research. Front Immunol 12:668045. https://doi.org/10.3389/fimmu.2021.668045

Chu X, Janssen AWM, Koenen H, Chang L, He X, Joosten I, Stienstra R, Kuijpers Y, Wijmenga C, Xu C-J, Netea MG, Tack CJ, Li Y (2022) A genome-wide functional genomics approach uncovers genetic determinants of immune phenotypes in type 1 diabetes. elife 11:e73709. https://doi.org/10.7554/eLife.73709

Cui M, Cheng C, Zhang L (2022) High-throughput proteomics: a methodological mini-review. Lab Investig 102(11):1170–1181. https://doi.org/10.1038/s41374-022-00830-7

Cuomo ASE, Heinen T, Vagiaki D, Horta D, Marioni JC, Stegle O (2022) CellRegMap: a statistical framework for mapping context-specific regulatory variants using scRNA-seq. Mol Syst Biol 18(8):e10663. https://doi.org/10.15252/msb.202110663

Edahiro R, Shirai Y, Takeshima Y, Sakakibara S, Yamaguchi Y, Murakami T, Morita T, Kato Y, Liu Y-C, Motooka D, Naito Y, Takuwa A, Sugihara F, Tanaka K, Wing JB, Sonehara K, Tomofuji Y, Japan COVID-19 Task Force, Namkoong H, Tanaka H, Lee H, Fukunaga K, Hirata H, Takeda Y, Okuzaki D, Kumanogoh A, Okada Y (2023) Single-cell analyses and host genetics highlight the role of innate immune cells in COVID-19 severity. Nat Genet 55(5):753–767. https://doi.org/10.1038/s41588-023-01375-1

Ehrlich M, Gama-Sosa MA, Huang LH, Midgett RM, Kuo KC, McCune RA, Gehrke C (1982) Amount and distribution of 5-methylcytosine in human DNA from different types of tissues of cells. Nucleic Acids Res 10(8):2709–2721. https://doi.org/10.1093/nar/10.8.2709

Fairfax BP, Knight JC (2014) Genetics of gene expression in immunity to infection. Curr Opin Immunol 30:63–71. https://doi.org/10.1016/j.coi.2014.07.001

Filbin MR, Mehta A, Schneider AM, Kays KR, Guess JR, Gentili M, Fenyves BG, Charland NC, Gonye ALK, Gushterova I, Khanna HK, LaSalle TJ, Lavin-Parsons KM, Lilley BM, Lodenstein CL, Manakongtreecheep K, Margolin JD, McKaig BN, Rojas-Lopez M, Russo BC, Sharma N, Tantivit J, Thomas MF, Gerszten RE, Heimberg GS, Hoover PJ, Lieb DJ, Lin B, Ngo D, Pelka K, Reyes M, Smillie CS, Waghray A, Wood TE, Zajac AS, Jennings LL, Grundberg I, Bhattacharyya RP, Parry BA, Villani A-C, Sade-Feldman M, Hacohen N, Goldberg MB (2021) Longitudinal proteomic analysis of severe COVID-19 reveals survival-associated signatures, tissue-specific cell death, and cell-cell interactions. Cell Rep Med 2(5):100287. https://doi.org/10.1016/j.xcrm.2021.100287

Fu J, Zhu F, Xu C-J, Li Y (2023) Metabolomics meets systems immunology. EMBO Rep 24(4):e55747. https://doi.org/10.15252/embr.202255747

Gadd DA, Hillary RF, McCartney DL, Shi L, Stolicyn A, Robertson NA, Walker RM, McGeachan RI, Campbell A, Xueyi S, Barbu MC, Green C, Morris SW, Harris MA, Backhouse EV, Wardlaw JM, Steele JD, Oyarzún DA, Muniz-Terrera G, Ritchie C, Nevado-Holgado A, Chandra T, Hayward C, Evans KL, Porteous DJ, Cox SR, Whalley HC, McIntosh AM, Marioni RE (2022) Integrated methylome and phenome study of the circulating proteome reveals markers pertinent to brain health. Nat Commun 13(1):4670. https://doi.org/10.1038/s41467-022-32319-8

Goetz JJ, Trimarchi JM (2012) Transcriptome sequencing of single cells with Smart-Seq. Nat Biotechnol 30(8):763–765. https://doi.org/10.1038/nbt.2325

Gorin G, Svensson V, Pachter L (2020) Protein velocity and acceleration from single-cell multiomics experiments. Genome Biol 21(1):39. https://doi.org/10.1186/s13059-020-1945-3

Gruzieva O, Xu C-J, Yousefi P, Relton C, Merid SK, Breton CV, Gao L, Volk HE, Feinberg JI, Ladd-Acosta C, Bakulski K, Auffray C, Lemonnier N, Plusquin M, Ghantous A, Herceg Z, Nawrot TS, Pizzi C, Richiardi L, Rusconi F, Vineis P, Kogevinas M, Felix JF, Duijts L, den Dekker HT, Jaddoe VWV, Ruiz JL, Bustamante M, Antó JM, Sunyer J, Vrijheid M, Gutzkow KB, Grazuleviciene R, Hernandez-Ferrer C, Annesi-Maesano I, Lepeule J, Bousquet J, Bergström A, Kull I, Söderhäll C, Kere J, Gehring U, Brunekreef B, Just AC, Wright RJ, Peng C, Gold DR, Kloog I, DeMeo DL, Pershagen G, Koppelman GH, London SJ, Baccarelli AA, Melén E (2019) Prenatal particulate air pollution and DNA methylation in newborns: an Epigenome-wide meta-analysis. Environ Health Perspect 127(5):57012. https://doi.org/10.1289/EHP4522

Gupta A, Weinand K, Nathan A, Sakaue S, Zhang MJ, Accelerating Medicines Partnership RA/SLE Program and Network, Donlin L, Wei K, Price AL, Amariuta T, Raychaudhuri S (2023a) Dynamic regulatory elements in single-cell multimodal data implicate key immune cell states enriched for autoimmune disease heritability. Nat Genet 55(12):2200–2210. https://doi.org/10.1038/s41588-023-01577-7

Gupta MK, Peng H, Li Y, Xu C-J (2023b) The role of DNA methylation in personalized medicine for immune-related diseases. Pharmacol Ther 250:108508. https://doi.org/10.1016/j.pharmthera.2023.108508

Hagan T, Cortese M, Rouphael N, Boudreau C, Linde C, Maddur MS, Das J, Wang H, Guthmiller J, Zheng N-Y, Huang M, Uphadhyay AA, Gardinassi L, Petitdemange C, McCullough MP, Johnson SJ, Gill K, Cervasi B, Zou J, Bretin A, Hahn M, Gewirtz AT, Bosinger SE, Wilson PC, Li S, Alter G, Khurana S, Golding H, Pulendran B (2019) Antibiotics-driven gut microbiome perturbation alters immunity to vaccines in humans. Cell 178(6):1313–1328.e13. https://doi.org/10.1016/j.cell.2019.08.010

Hao Y, Hao S, Andersen-Nissen E, Mauck WM, Zheng S, Butler A, Lee MJ, Wilk AJ, Darby C, Zager M, Hoffman P, Stoeckius M, Papalexi E, Mimitou EP, Jain J, Srivastava A, Stuart T, Fleming LM, Yeung B, Rogers AJ, McElrath JM, Blish CA, Gottardo R, Smibert P, Satija R (2021) Integrated analysis of multimodal single-cell data. Cell 184(13):3573–3587.e29. https://doi.org/10.1016/j.cell.2021.04.048

Hao Y, Stuart T, Kowalski MH, Choudhary S, Hoffman P, Hartman A, Srivastava A, Molla G, Madad S, Fernandez-Granda C, Satija R (2024) Dictionary learning for integrative, multimodal and scalable single-cell analysis. Nat Biotechnol 42(2):293–304. https://doi.org/10.1038/s41587-023-01767-y

Hawe JS, Wilson R, Schmid KT, Zhou L, Lakshmanan LN, Lehne BC, Kühnel B, Scott WR, Wielscher M, Yew YW, Baumbach C, Lee DP, Marouli E, Bernard M, Pfeiffer L, Matías-García PR, Autio MI, Bourgeois S, Herder C, Karhunen V, Meitinger T, Prokisch H, Rathmann W, Roden M, Sebert S, Shin J, Strauch K, Zhang W, Tan WLW, Hauck SM, Merl-Pham J, Grallert H, Barbosa EGV, MuTHER Consortium, Illig T, Peters A, Paus T, Pausova Z, Deloukas P, Foo RSY, Jarvelin M-R, Kooner JS, Loh M, Heinig M, Gieger C, Waldenberger M, Chambers JC (2022) Genetic variation influencing DNA methylation provides insights into molecular mechanisms regulating genomic function. Nat Genet 54(1):18–29. https://doi.org/10.1038/s41588-021-00969-x

Hemani G, Zheng J, Elsworth B, Wade KH, Haberland V, Baird D, Laurin C, Burgess S, Bowden J, Langdon R, Tan VY, Yarmolinsky J, Shihab HA, Timpson NJ, Evans DM, Relton C, Martin RM, Davey Smith G, Gaunt TR, Haycock PC (2018) The MR-base platform supports systematic causal inference across the human phenome. eLife 7:e34408. https://doi.org/10.7554/eLife.34408

Heumos L, Schaar AC, Lance C, Litinetskaya A, Drost F, Zappia L, Lücken MD, Strobl DC, Henao J, Curion F, Single-cell Best Practices Consortium, Schiller HB, Theis FJ (2023) Best practices for single-cell analysis across modalities. Nat Rev Genet 24(8):550–572. https://doi.org/10.1038/s41576-023-00586-w

Holliday R, Pugh JE (1975) DNA modification mechanisms and gene activity during development: developmental clocks may depend on the enzymic modification of specific bases in repeated DNA sequences. Science 187(4173):226–232. https://doi.org/10.1126/science.187.4173.226

Hu Y, Xi X, Yang Q, Zhang X (2020) SCeQTL: an R package for identifying eQTL from single-cell parallel sequencing data. BMC Bioinformatics 21(1):184. https://doi.org/10.1186/s12859-020-3534-6

Huan T, Joehanes R, Song C, Peng F, Guo Y, Mendelson M, Yao C, Liu C, Ma J, Richard M, Agha G, Guan W, Almli LM, Conneely KN, Keefe J, Hwang S-J, Johnson AD, Fornage M, Liang L, Levy D (2019) Genome-wide identification of DNA methylation QTLs in whole blood highlights pathways for cardiovascular disease. Nat Commun 10(1):4267. https://doi.org/10.1038/s41467-019-12228-z

Irizarry RA, Ladd-Acosta C, Wen B, Wu Z, Montano C, Onyango P, Cui H, Gabo K, Rongione M, Webster M, Ji H, Potash J, Sabunciyan S, Feinberg AP (2009) The human colon cancer methylome shows similar hypo- and hypermethylation at conserved tissue-specific CpG Island shores. Nat Genet 41(2):178–186. https://doi.org/10.1038/ng.298

Joubert BR, Felix JF, Yousefi P, Bakulski KM, Just AC, Breton C, Reese SE, Markunas CA, Richmond RC, Xu C-J, Küpers LK, Oh SS, Hoyo C, Gruzieva O, Söderhäll C, Salas LA, Baïz N, Zhang H, Lepeule J, Ruiz C, Ligthart S, Wang T, Taylor JA, Duijts L, Sharp GC, Jankipersadsing SA, Nilsen RM, Vaez A, Fallin MD, Hu D, Litonjua AA, Fuemmeler BF, Huen K, Kere J, Kull I, Munthe-Kaas MC, Gehring U, Bustamante M, Saurel-Coubizolles MJ, Quraishi BM, Ren J, Tost J, Gonzalez JR, Peters MJ, Håberg SE, Xu Z, van Meurs JB, Gaunt TR, Kerkhof M, Corpeleijn E, Feinberg AP, Eng C, Baccarelli AA, Benjamin Neelon SE, Bradman A, Merid SK, Bergström A, Herceg Z, Hernandez-Vargas H, Brunekreef B, Pinart M, Heude B, Ewart S, Yao J, Lemonnier N, Franco OH, Wu MC, Hofman A, McArdle W, Van der Vlies P, Falahi F, Gillman MW, Barcellos LF, Kumar A, Wickman M, Guerra S, Charles M-A, Holloway J, Auffray C, Tiemeier HW, Smith GD, Postma D, Hivert M-F, Eskenazi B, Vrijheid M, Arshad H, Antó JM, Dehghan A, Karmaus W, Annesi-Maesano I, Sunyer J, Ghantous A, Pershagen G, Holland N, Murphy SK, DeMeo DL, Burchard EG, Ladd-Acosta C, Snieder H, Nystad W, Koppelman GH, Relton CL, Jaddoe VWV, Wilcox A, Melén E, London SJ (2016) DNA methylation in newborns and maternal smoking in pregnancy: genome-wide consortium meta-analysis. Am J Hum Genet 98(4):680–696. https://doi.org/10.1016/j.ajhg.2016.02.019

Kamimoto K, Stringa B, Hoffmann CM, Jindal K, Solnica-Krezel L, Morris SA (2023) Dissecting cell identity via network inference and in silico gene perturbation. Nature 614(7949):742–751. https://doi.org/10.1038/s41586-022-05688-9

Kamiza AB, Touré SM, Zhou F, Soremekun O, Cissé C, Wélé M, Touré AM, Nashiru O, Corpas M, Nyirenda M, Crampin A, Shaffer J, Doumbia S, Zeggini E, Morris AP, Asimit JL, Chikowore T, Fatumo S (2023) Multi-trait discovery and fine-mapping of lipid loci in 125,000 individuals of African ancestry. Nat Commun 14(1):5403. https://doi.org/10.1038/s41467-023-41271-0

Karjalainen MK, Karthikeyan S, Oliver-Williams C, Sliz E, Allara E, Fung WT, Surendran P, Zhang W, Jousilahti P, Kristiansson K, Salomaa V, Goodwin M, Hughes DA, Boehnke M, Fernandes Silva L, Yin X, Mahajan A, Neville MJ, Van Zuydam NR, De Mutsert R, Li-Gao R, Mook-Kanamori DO, Demirkan A, Liu J, Noordam R, Trompet S, Chen Z, Kartsonaki C, Li L, Lin K, Hagenbeek FA, Hottenga JJ, Pool R, Ikram MA, Van Meurs J, Haller T, Milaneschi Y, Kähönen M, Mishra PP, Joshi PK, Macdonald-Dunlop E, Mangino M, Zierer J, Acar IE, Hoyng CB, Lechanteur YTE, Franke L, Kurilshikov A, Zhernakova A, Beekman M, Van Den Akker EB, Kolcic I, Polasek O, Rudan I, Gieger C,

Waldenberger M, Asselbergs FW, China Kadoorie Biobank Collaborative Group, Estonian Biobank Research Team, FinnGen, Hayward C, Fu J, Den Hollander AI, Menni C, Spector TD, Wilson JF, Lehtimäki T, Raitakari OT, Penninx BWJH, Esko T, Walters RG, Jukema JW, Sattar N, Ghanbari M, Willems Van Dijk K, Karpe F, McCarthy MI, Laakso M, Järvelin M-R, Timpson NJ, Perola M, Kooner JS, Chambers JC, Van Duijn C, Slagboom PE, Boomsma DI, Danesh J, Ala-Korpela M, Butterworth AS, Kettunen J (2024) Genome-wide characterization of circulating metabolic biomarkers. Nature 628(8006):130–138. https://doi.org/10.1038/s41586-024-07148-y

Ko KD, Jiang K, Dell'Orso S, Sartorelli V (2023) Integrating single-cell transcriptomes, chromatin accessibility, and multiomics analysis of mesoderm-induced embryonic stem cells. STAR Protoc 4(2): 102307. https://doi.org/10.1016/j.xpro.2023.102307

Koeken VA, de Bree LCJ, Mourits VP, Moorlag SJ, Walk J, Cirovic B, Arts RJ, Jaeger M, Dijkstra H, Lemmers H, Joosten LA, Benn CS, van Crevel R, Netea MG (2020) BCG vaccination in humans inhibits systemic inflammation in a sex-dependent manner. J Clin Invest 130(10):5591–5602. https://doi.org/10.1172/JCI133935

Korsunsky I, Millard N, Fan J, Slowikowski K, Zhang F, Wei K, Baglaenko Y, Brenner M, Loh P-R, Raychaudhuri S (2019) Fast, sensitive and accurate integration of single-cell data with harmony. Nat Methods 16(12):1289–1296. https://doi.org/10.1038/s41592-019-0619-0

Kumar S, Zoodsma M, Nguyen N, Pedroso R, Trittel S, Riese P, Botey-Bataller J, Zhou L, Alaswad A, Arshad H, Netea MG, Xu C-J, Pessler F, Guzmán CA, Graca L, Li Y (2024) Systemic dysregulation and molecular insights into poor influenza vaccine response in the aging population. Sci Adv 10(39): eadq7006. https://doi.org/10.1126/sciadv.adq7006

Kurki MI, Karjalainen J, Palta P, Sipilä TP, Kristiansson K, Donner KM, Reeve MP, Laivuori H, Aavikko M, Kaunisto MA, Loukola A, Lahtela E, Mattsson H, Laiho P, Della Briotta Parolo P, Lehisto AA, Kanai M, Mars N, Rämö J, Kiiskinen T, Heyne HO, Veerapen K, Rüeger S, Lemmelä S, Zhou W, Ruotsalainen S, Pärn K, Hiekkalinna T, Koskelainen S, Paajanen T, Llorens V, Gracia-Tabuenca J, Siirtola H, Reis K, Elnahas AG, Sun B, Foley CN, Aalto-Setälä K, Alasoo K, Arvas M, Auro K, Biswas S, Bizaki-Vallaskangas A, Carpen O, Chen C-Y, Dada OA, Ding Z, Ehm MG, Eklund K, Färkkilä M, Finucane H, Ganna A, Ghazal A, Graham RR, Green EM, Hakanen A, Hautalahti M, Hedman ÅK, Hiltunen M, Hinttala R, Hovatta I, Hu X, Huertas-Vazquez A, Huilaja L, Hunkapiller J, Jacob H, Jensen J-N, Joensuu H, John S, Julkunen V, Jung M, Junttila J, Kaarniranta K, Kähönen M, Kajanne R, Kallio L, Kälviäinen R, Kaprio J, FinnGen, Kerimov N, Kettunen J, Kilpeläinen E, Kilpi T, Klinger K, Kosma V-M, Kuopio T, Kurra V, Laisk T, Laukkanen J, Lawless N, Liu A, Longerich S, Mägi R, Mäkelä J, Mäkitie A, Malarstig A, Mannermaa A, Maranville J, Matakidou A, Meretoja T, Mozaffari SV, Niemi MEK, Niemi M, Niiranen T, O'Donnell CJ, Obeidat M, Okafo G, Ollila HM, Palomäki A, Palotie T, Partanen J, Paul DS, Pelkonen M, Pendergrass RK, Petrovski S, Pitkäranta A, Platt A, Pulford D, Punkka E, Pussinen P, Raghavan N, Rahimov F, Rajpal D, Renaud NA, Riley-Gillis B, Rodosthenous R, Saarentaus E, Salminen A, Salminen E, Salomaa V, Schleutker J, Serpi R, Shen H, Siegel R, Silander K, Siltanen S, Soini S, Soininen H, Sul JH, Tachmazidou I, Tasanen K, Tienari P, Toppila-Salmi S, Tukiainen T, Tuomi T, Turunen JA, Ulirsch JC, Vaura F, Virolainen P, Waring J, Waterworth D, Yang R, Nelis M, Reigo A, Metspalu A, Milani L, Esko T, Fox C, Havulinna AS, Perola M, Ripatti S, Jalanko A, Laitinen T, Mäkelä TP, Plenge R, McCarthy M, Runz H, Daly MJ, Palotie A (2023) Author correction: FinnGen provides genetic insights from a well-phenotyped isolated population. Nature 615(7952):E19–E19. https://doi.org/10.1038/s41586-023-05837-8

Leonhardt H, Page AW, Weier HU, Bestor TH (1992) A targeting sequence directs DNA methyltransferase to sites of DNA replication in mammalian nuclei. Cell 71(5):865–873. https://doi.org/10.1016/0092-8674(92)90561-p

Li Y, Oosting M, Deelen P, Ricaño-Ponce I, Smeekens S, Jaeger M, Matzaraki V, Swertz MA, Xavier RJ, Franke L, Wijmenga C, Joosten LAB, Kumar V, Netea MG (2016a) Inter-individual variability and genetic influences on cytokine responses to bacteria and fungi. Nat Med 22(8):952–960. https://doi.org/10.1038/nm.4139

Li Y, Oosting M, Smeekens SP, Jaeger M, Aguirre-Gamboa R, Le KTT, Deelen P, Ricaño-Ponce I, Schoffelen T, Jansen AFM, Swertz MA, Withoff S, van de Vosse E, van Deuren M, van de Veerdonk F, Zhernakova A, van der Meer JWM, Xavier RJ, Franke L, Joosten LAB, Wijmenga C, Kumar V, Netea MG (2016b) A functional genomics approach to understand variation in cytokine production in humans. Cell 167(4):1099–1110.e14. https://doi.org/10.1016/j.cell.2016.10.017

Li S, Sullivan NL, Rouphael N, Yu T, Banton S, Maddur MS, McCausland M, Chiu C, Canniff J, Dubey S, Liu K, Tran V, Hagan T, Duraisingham S, Wieland A, Mehta AK, Whitaker JA, Subramaniam S, Jones DP, Sette A, Vora K, Weinberg A, Mulligan MJ, Nakaya HI, Levin M, Ahmed R, Pulendran B (2017) Metabolic phenotypes of response to vaccination in humans. Cell 169(5): 862–877.e17. https://doi.org/10.1016/j.cell.2017.04.026

Li Z, Schulz MH, Look T, Begemann M, Zenke M, Costa IG (2019) Identification of transcription factor binding sites using ATAC-seq. Genome Biol 20(1):45. https://doi.org/10.1186/s13059-019-1642-2

Lin L, Curtin JA, Regis E, Hirsman A, Howard R, Tutino M, Edwards MR, Prosperi M, Simpson A, Rattray M, Custovic A, Johnston SL (2022) A systems immunology approach to investigate cytokine responses to viruses and bacteria and their association with disease. Sci Rep 12(1):13463. https://doi.org/10.1038/s41598-022-16509-4

Liu Y, Aryee MJ, Padyukov L, Fallin MD, Hesselberg E, Runarsson A, Reinius L, Acevedo N, Taub M, Ronninger M, Shchetynsky K, Scheynius A, Kere J, Alfredsson L, Klareskog L, Ekström TJ, Feinberg AP (2013) Epigenome-wide association data implicate DNA methylation as an intermediary of genetic risk in rheumatoid arthritis. Nat Biotechnol 31(2):142–147. https://doi.org/10.1038/nbt.2487

Liu X, Tong X, Zou Y, Lin X, Zhao H, Tian L, Jie Z, Wang Q, Zhang Z, Lu H, Xiao L, Qiu X, Zi J, Wang R, Xu X, Yang H, Wang J, Zong Y, Liu W, Hou Y, Zhu S, Jia H, Zhang T (2022) Mendelian randomization analyses support causal relationships between blood metabolites and the gut microbiome. Nat Genet 54(1):52–61. https://doi.org/10.1038/s41588-021-00968-y

Loers JU, Vermeirssen V (2024) A single-cell multimodal view on gene regulatory network inference from transcriptomics and chromatin accessibility data. Brief Bioinform 25(5):bbae382. https://doi.org/10.1093/bib/bbae382

Lonsdale J, Thomas J, Salvatore M, Phillips R, Lo E, Shad S, Hasz R, Walters G, Garcia F, Young N, Foster B, Moser M, Karasik E, Gillard B, Ramsey K, Sullivan S, Bridge J, Magazine H, Syron J, Fleming J, Siminoff L, Traino H, Mosavel M, Barker L, Jewell S, Rohrer D, Maxim D, Filkins D, Harbach P, Cortadillo E, Berghuis B, Turner L, Hudson E, Feenstra K, Sobin L, Robb J, Branton P, Korzeniewski G, Shive C, Tabor D, Qi L, Groch K, Nampally S, Buia S, Zimmerman A, Smith A, Burges R, Robinson K, Valentino K, Bradbury D, Cosentino M, Diaz-Mayoral N, Kennedy M, Engel T, Williams P, Erickson K, Ardlie K, Winckler W, Getz G, DeLuca D, MacArthur D, Kellis M, Thomson A, Young T, Gelfand E, Donovan M, Meng Y, Grant G, Mash D, Marcus Y, Basile M, Liu J, Zhu J, Tu Z, Cox NJ, Nicolae DL, Gamazon ER, Im HK, Konkashbaev A, Pritchard J, Stevens M, Flutre T, Wen X, Dermitzakis ET, Lappalainen T, Guigo R, Monlong J, Sammeth M, Koller D, Battle A, Mostafavi S, McCarthy M, Rivas M, Maller J, Rusyn I, Nobel A, Wright F, Shabalin A, Feolo M, Sharopova N, Sturcke A, Paschal J, Anderson JM, Wilder EL, Derr LK, Green ED, Struewing JP, Temple G, Volpi S, Boyer JT, Thomson EJ, Guyer MS, Ng C, Abdallah A, Colantuoni D, Insel TR, Koester SE, Little AR, Bender PK, Lehner T, Yao Y, Compton CC, Vaught JB, Sawyer S, Lockhart NC, Demchok J, Moore HF (2013) The genotype-tissue expression (GTEx) project. Nat Genet 45(6):580–585. https://doi.org/10.1038/ng.2653

Merid SK, Novoloaca A, Sharp GC, Küpers LK, Kho AT, Roy R, Gao L, Annesi-Maesano I, Jain P, Plusquin M, Kogevinas M, Allard C, Vehmeijer FO, Kazmi N, Salas LA, Rezwan FI, Zhang H, Sebert S, Czamara D, Rifas-Shiman SL, Melton PE, Lawlor DA, Pershagen G, Breton CV, Huen K, Baiz N, Gagliardi L, Nawrot TS, Corpeleijn E, Perron P, Duijts L, Nohr EA, Bustamante M, Ewart SL, Karmaus W, Zhao S, Page CM, Herceg Z, Jarvelin M-R, Lahti J, Baccarelli AA, Anderson D, Kachroo P, Relton CL, Bergström A, Eskenazi B, Soomro MH, Vineis P, Snieder H, Bouchard L, Jaddoe VW, Sørensen TIA, Vrijheid M, Arshad SH, Holloway JW, Håberg SE, Magnus P, Dwyer T, Binder EB, DeMeo DL, Vonk JM, Newnham J, Tantisira KG, Kull I, Wiemels JL, Heude B, Sunyer J, Nystad W, Munthe-Kaas MC, Räikkönen K, Oken E, Huang R-C, Weiss ST, Antó JM, Bousquet J, Kumar A, Söderhäll C, Almqvist C, Cardenas A, Gruzieva O, Xu C-J, Reese SE, Kere J, Brodin P, Solomon O, Wielscher M, Holland N, Ghantous A, Hivert M-F, Felix JF, Koppelman GH, London SJ, Melén E (2020) Epigenome-wide meta-analysis of blood DNA methylation in newborns and children identifies numerous loci related to gestational age. Genome Med 12(1):25. https://doi.org/10.1186/s13073-020-0716-9

Min JL, Hemani G, Hannon E, Dekkers KF, Castillo-Fernandez J, Luijk R, Carnero-Montoro E, Lawson DJ, Burrows K, Suderman M, Bretherick AD, Richardson TG, Klughammer J, Iotchkova V, Sharp G, Al Khleifat A, Shatunov A, Iacoangeli A, McArdle WL, Ho KM, Kumar A, Söderhäll C, Soriano-Tárraga C, Giralt-Steinhauer E, Kazmi N, Mason D, McRae AF, Corcoran DL, Sugden K, Kasela S, Cardona A, Day FR, Cugliari G, Viberti C, Guarrera S, Lerro M, Gupta R, Bollepalli S, Mandaviya P, Zeng Y, Clarke T-K, Walker RM, Schmoll V, Czamara D, Ruiz Arenas C, Rezwan FI, Marioni RE, Lin T, Awaloff Y, Germain M, Aïssi D, Zwamborn R, van Eijk K, Dekker A, van Dongen J, Hottenga J-J, Willemsen G, Xu C-J, Barturen G, Català-Moll F, Kerick M, Wang C, Melton P, Elliott HR, Shin J, Bernard M, Yet I, Smart M, Gorrie-Stone T, BIOS Consortium, Shaw C, Al Chalabi A, Ring SM, Pershagen G, Melén E, Jiménez-Conde J, Roquer J, Lawlor DA, Wright J, Martin NG, Montgomery GW, Moffitt TE, Poulton R, Esko T, Milani L, Metspalu A, Perry JRB, Ong KK, Wareham NJ, Matullo G, Sacerdote C, Panico S, Caspi A, Arseneault L, Gagnon F, Ollikainen M, Kaprio J, Felix JF, Rivadeneira F, Tiemeier H, van IJzendoorn MH, Uitterlinden AG, Jaddoe VWV, Haley C, McIntosh AM, Evans KL, Murray A, Räikkönen K, Lahti J, Nohr EA, Sørensen TIA, Hansen T, Morgen CS, Binder EB, Lucae S, Gonzalez JR, Bustamante M, Sunyer J, Holloway JW, Karmaus W, Zhang H,

Deary IJ, Wray NR, Starr JM, Beekman M, van Heemst D, Slagboom PE, Morange P-E, Trégouët D-A, Veldink JH, Davies GE, de Geus EJC, Boomsma DI, Vonk JM, Brunekreef B, Koppelman GH, Alarcón-Riquelme ME, Huang R-C, Pennell CE, van Meurs J, Ikram MA, Hughes AD, Tillin T, Chaturvedi N, Pausova Z, Paus T, Spector TD, Kumari M, Schalkwyk LC, Visscher PM, Davey Smith G, Bock C, Gaunt TR, Bell JT, Heijmans BT, Mill J, Relton CL (2021) Genomic and phenotypic insights from an atlas of genetic effects on DNA methylation. Nat Genet 53(9): 1311–1321. https://doi.org/10.1038/s41588-021-00923-x

Moore LD, Le T, Fan G (2013) DNA methylation and its basic function. Neuropsychopharmacology 38(1): 23–38. https://doi.org/10.1038/npp.2012.112

Moore R, Casale FP, Jan Bonder M, Horta D, BIOS Consortium, Franke L, Barroso I, Stegle O (2019) A linear mixed-model approach to study multivariate gene-environment interactions. Nat Genet 51(1): 180–186. https://doi.org/10.1038/s41588-018-0271-0

Mortazavi A, Williams BA, McCue K, Schaeffer L, Wold B (2008) Mapping and quantifying mammalian transcriptomes by RNA-Seq. Nat Methods 5(7): 621–628. https://doi.org/10.1038/nmeth.1226

Nathan A, Asgari S, Ishigaki K, Valencia C, Amariuta T, Luo Y, Beynor JI, Baglaenko Y, Suliman S, Price AL, Lecca L, Murray MB, Moody DB, Raychaudhuri S (2022) Single-cell eQTL models reveal dynamic T cell state dependence of disease loci. Nature 606(7912):120–128. https://doi.org/10.1038/s41586-022-04713-1

Netea MG, Joosten LAB, Li Y, Kumar V, Oosting M, Smeekens S, Jaeger M, Ter Horst R, Schirmer M, Vlamakis H, Notebaart R, Pavelka N, Aguirre-Gamboa RR, Swertz MA, Tunjungputri RN, van de Heijden W, Franzosa EA, Ng A, Graham D, Lassen K, Schraa K, Netea-Maier R, Smit J, de Mast Q, van de Veerdonk F, Kullberg BJ, Tack C, van de Munckhof I, Rutten J, van der Graaf J, Franke L, Hofker M, Jonkers I, Platteel M, Maatman A, Fu J, Zhernakova A, van der Meer JWM, Dinarello CA, van der Ven A, Huttenhouwer C, Koenen H, Joosten I, Xavier RJ, Wijmenga C (2016) Understanding human immune function using the resources from the human functional genomics project. Nat Med 22(8):831–833. https://doi.org/10.1038/nm.4140

Nguyen N, Jennen D, Kleinjans J (2022) Omics technologies to understand drug toxicity mechanisms. Drug Discov Today 27(11):103348. https://doi.org/10.1016/j.drudis.2022.103348

Nørgaard M, Ehrenstein V, Vandenbroucke JP (2017) Confounding in observational studies based on large health care databases: problems and potential solutions – a primer for the clinician. Clin Epidemiol 9:185–193. https://doi.org/10.2147/CLEP.S129879

Okano M, Bell DW, Haber DA, Li E (1999) DNA methyltransferases Dnmt3a and Dnmt3b are essential for de novo methylation and mammalian development. Cell 99(3):247–257. https://doi.org/10.1016/s0092-8674(00)81656-6

Oliva M, Demanelis K, Lu Y, Chernoff M, Jasmine F, Ahsan H, Kibriya MG, Chen LS, Pierce BL (2023) DNA methylation QTL mapping across diverse human tissues provides molecular links between genetic variation and complex traits. Nat Genet 55(1):112–122. https://doi.org/10.1038/s41588-022-01248-z

Oltmanns C, Liu Z, Mischke J, Tauwaldt J, Mekonnen YA, Urbanek-Quaing M, Debarry J, Maasoumy B, Wedemeyer H, Kraft ARM, Xu C-J, Cornberg M (2023) Reverse inflammaging: long-term effects of HCV cure on biological age. J Hepatol 78(1):90–98. https://doi.org/10.1016/j.jhep.2022.08.042

Patel A, Ye T, Xue H, Lin Z, Xu S, Woolf B, Mason AM, Burgess S (2023) MendelianRandomization v0.9.0: updates to an R package for performing Mendelian randomization analyses using summarized data. Wellcome Open Res 8:449. https://doi.org/10.12688/wellcomeopenres.19995.1

Pourhoseingholi MA, Baghestani AR, Vahedi M (2012) How to control confounding effects by statistical analysis. Gastroenterol Hepatol Bed Bench 5(2):79–83

Qi C, Jiang Y, Yang IV, Forno E, Wang T, Vonk JM, Gehring U, Smit HA, Milanzi EB, Carpaij OA, Berg M, Hesse L, Brouwer S, Cardwell J, Vermeulen CJ, Acosta-Pérez E, Canino G, Boutaoui N, van den Berge M, Teichmann SA, Nawijn MC, Chen W, Celedón JC, Xu C-J, Koppelman GH (2020) Nasal DNA methylation profiling of asthma and rhinitis. J Allergy Clin Immunol 145(6):1655–1663. https://doi.org/10.1016/j.jaci.2019.12.911

Riggs AD (1975) X inactivation, differentiation, and DNA methylation. Cytogenet Cell Genet 14(1):9–25. https://doi.org/10.1159/000130315

Robertson CC, Inshaw JRJ, Onengut-Gumuscu S, Chen W-M, Santa Cruz DF, Yang H, Cutler AJ, Crouch DJM, Farber E, Bridges SL, Edberg JC, Kimberly RP, Buckner JH, Deloukas P, Divers J, Dabelea D, Lawrence JM, Marcovina S, Shah AS, Greenbaum CJ, Atkinson MA, Gregersen PK, Oksenberg JR, Pociot F, Rewers MJ, Steck AK, Dunger DB, Type 1 Diabetes Genetics Consortium, Wicker LS, Concannon P, Todd JA, Rich SS (2021) Fine-mapping, trans-ancestral and genomic analyses identify causal variants, cells, genes and drug targets for type 1 diabetes. Nat Genet 53(7):962–971. https://doi.org/10.1038/s41588-021-00880-5

Sager R, Kitchin R (1975) Selective silencing of eukaryotic DNA: a molecular basis is proposed for programmed inactivation or loss of eukaryotic DNA. Science 189(4201):426–433. https://doi.org/10.1126/science.189.4201.426

Sahoo K, Sundararajan V (2024) Methods in DNA methylation array dataset analysis: a review. Comput Struct Biotechnol J 23:2304–2325. https://doi.org/10.1016/j.csbj.2024.05.015

Sakaue S, Weinand K, Isaac S, Dey KK, Jagadeesh K, Kanai M, Watts GFM, Zhu Z, Accelerating Medicines

Partnership® RA/SLE Program and Network, Brenner MB, McDavid A, Donlin LT, Wei K, Price AL, Raychaudhuri S (2024) Tissue-specific enhancer-gene maps from multimodal single-cell data identify causal disease alleles. Nat Genet 56(4):615–626. https://doi.org/10.1038/s41588-024-01682-1

Schmiedel BJ, Gonzalez-Colin C, Fajardo V, Rocha J, Madrigal A, Ramírez-Suástegui C, Bhattacharyya S, Simon H, Greenbaum JA, Peters B, Seumois G, Ay F, Chandra V, Vijayanand P (2022) Single-cell eQTL analysis of activated T cell subsets reveals activation and cell type-dependent effects of disease-risk variants. Sci Immunol 7(68):eabm2508. https://doi.org/10.1126/sciimmunol.abm2508

Schulte-Schrepping J, Reusch N, Paclik D, Baßler K, Schlickeiser S, Zhang B, Krämer B, Krammer T, Brumhard S, Bonaguro L, De Domenico E, Wendisch D, Grasshoff M, Kapellos TS, Beckstette M, Pecht T, Saglam A, Dietrich O, Mei HE, Schulz AR, Conrad C, Kunkel D, Vafadarnejad E, Xu C-J, Horne A, Herbert M, Drews A, Thibeault C, Pfeiffer M, Hippenstiel S, Hocke A, Müller-Redetzky H, Heim K-M, Machleidt F, Uhrig A, Bosquillon de Jarcy L, Jürgens L, Stegemann M, Glösenkamp CR, Volk H-D, Goffinet C, Landthaler M, Wyler E, Georg P, Schneider M, Dang-Heine C, Neuwinger N, Kappert K, Tauber R, Corman V, Raabe J, Kaiser KM, Vinh MT, Rieke G, Meisel C, Ulas T, Becker M, Geffers R, Witzenrath M, Drosten C, Suttorp N, von Kalle C, Kurth F, Händler K, Schultze JL, Aschenbrenner AC, Li Y, Nattermann J, Sawitzki B, Saliba A-E, Sander LE, Deutsche COVID-19 OMICS Initiative (DeCOI) (2020) Severe COVID-19 is marked by a dysregulated myeloid cell compartment. Cell 182(6):1419–1440.e23. https://doi.org/10.1016/j.cell.2020.08.001

Shang L, Zhao W, Wang YZ, Li Z, Choi JJ, Kho M, Mosley TH, Kardia SLR, Smith JA, Zhou X (2023) meQTL mapping in the GENOA study reveals genetic determinants of DNA methylation in African Americans. Nat Commun 14(1):2711. https://doi.org/10.1038/s41467-023-37961-4

Singh R, Wu AP, Mudide A, Berger B (2024) Causal gene regulatory analysis with RNA velocity reveals an interplay between slow and fast transcription factors. Cell Syst 15(5):462–474.e5. https://doi.org/10.1016/j.cels.2024.04.005

Smith AK, Kilaru V, Kocak M, Almli LM, Mercer KB, Ressler KJ, Tylavsky FA, Conneely KN (2014) Methylation quantitative trait loci (meQTLs) are consistently detected across ancestry, developmental stage, and tissue type. BMC Genomics 15:145. https://doi.org/10.1186/1471-2164-15-145

Stuart T, Butler A, Hoffman P, Hafemeister C, Papalexi E, Mauck WM, Hao Y, Stoeckius M, Smibert P, Satija R (2019) Comprehensive integration of single-cell data. Cell 177(7):1888–1902.e21. https://doi.org/10.1016/j.cell.2019.05.031

Sudlow C, Gallacher J, Allen N, Beral V, Burton P, Danesh J, Downey P, Elliott P, Green J, Landray M, Liu B, Matthews P, Ong G, Pell J, Silman A, Young A, Sprosen T, Peakman T, Collins R (2015) UK Biobank: an open access resource for identifying the causes of a wide range of complex diseases of middle and old age. PLoS Med 12(3):e1001779. https://doi.org/10.1371/journal.pmed.1001779

Suhre K (2024) Genetic associations with ratios between protein levels detect new pQTLs and reveal protein-protein interactions. Cell Genomics 4(3):100506. https://doi.org/10.1016/j.xgen.2024.100506

Sun BB, Chiou J, Traylor M, Benner C, Hsu Y-H, Richardson TG, Surendran P, Mahajan A, Robins C, Vasquez-Grinnell SG, Hou L, Kvikstad EM, Burren OS, Davitte J, Ferber KL, Gillies CE, Hedman ÅK, Hu S, Lin T, Mikkilineni R, Pendergrass RK, Pickering C, Prins B, Baird D, Chen C-Y, Ward LD, Deaton AM, Welsh S, Willis CM, Lehner N, Arnold M, Wörheide MA, Suhre K, Kastenmüller G, Sethi A, Cule M, Raj A, Alnylam Human Genetics, AstraZeneca Genomics Initiative, Biogen Biobank Team, Bristol Myers Squibb, Genentech Human Genetics, GlaxoSmithKline Genomic Sciences, Pfizer Integrative Biology, Population Analytics of Janssen Data Sciences, Regeneron Genetics Center, Kang HM, Burkitt-Gray L, Melamud E, Black MH, Fauman EB, Howson JMM, Kang HM, McCarthy MI, Nioi P, Petrovski S, Scott RA, Smith EN, Szalma S, Waterworth DM, Mitnaul LJ, Szustakowski JD, Gibson BW, Miller MR, Whelan CD (2023) Plasma proteomic associations with genetics and health in the UK Biobank. Nature 622(7982):329–338. https://doi.org/10.1038/s41586-023-06592-6

Tang F, Barbacioru C, Wang Y, Nordman E, Lee C, Xu N, Wang X, Bodeau J, Tuch BB, Siddiqui A, Lao K, Surani MA (2009) mRNA-Seq whole-transcriptome analysis of a single cell. Nat Methods 6(5):377–382. https://doi.org/10.1038/nmeth.1315

Teo AYY, Squair JW, Courtine G, Skinnider MA (2024) Best practices for differential accessibility analysis in single-cell epigenomics. Nat Commun 15(1):8805. https://doi.org/10.1038/s41467-024-53089-5

Umans BD, Battle A, Gilad Y (2021) Where are the disease-associated eQTLs? Trends Genet 37(2):109–124. https://doi.org/10.1016/j.tig.2020.08.009

van der Wijst MGP, Brugge H, de Vries DH, Deelen P, Swertz MA, LifeLines Cohort Study, BIOS Consortium, Franke L (2018) Single-cell RNA sequencing identifies celltype-specific cis-eQTLs and co-expression QTLs. Nat Genet 50(4):493–497. https://doi.org/10.1038/s41588-018-0089-9

Verlouw JAM, Clemens E, de Vries JH, Zolk O, Verkerk AJMH, Am Zehnhoff-Dinnesen A, Medina-Gomez C, Lanvers-Kaminsky C, Rivadeneira F, Langer T, van Meurs JBJ, van den Heuvel-Eibrink MM, Uitterlinden AG, Broer L (2021) A comparison of genotyping arrays. Eur J Hum Genet 29(11):1611–1624. https://doi.org/10.1038/s41431-021-00917-7

Võsa U, Claringbould A, Westra H-J, Bonder MJ, Deelen P, Zeng B, Kirsten H, Saha A, Kreuzhuber R, Yazar S, Brugge H, Oelen R, De Vries DH, Van Der Wijst MGP, Kasela S, Pervjakova N, Alves I, Favé M-J, Agbessi M, Christiansen MW, Jansen R, Seppälä I, Tong L, Teumer A, Schramm K, Hemani G, Verlouw J, Yaghootkar H, Sönmez Flitman R, Brown A, Kukushkina V, Kalnapenkis A, Rüeger S, Porcu E, Kronberg J, Kettunen J, Lee B, Zhang F, Qi T, Hernandez JA, Arindrarto W, Beutner F, BIOS Consortium, 'T Hoen PAC, Van Meurs J, Van Dongen J, Van Iterson M, Swertz MA, i2QTL Consortium, Jan Bonder M, Dmitrieva J, Elansary M, Fairfax BP, Georges M, Heijmans BT, Hewitt AW, Kähönen M, Kim Y, Knight JC, Kovacs P, Krohn K, Li S, Loeffler M, Marigorta UM, Mei H, Momozawa Y, Müller-Nurasyid M, Nauck M, Nivard MG, Penninx BWJH, Pritchard JK, Raitakari OT, Rotzschke O, Slagboom EP, Stehouwer CDA, Stumvoll M, Sullivan P, 'T Hoen PAC, Thiery J, Tönjes A, Van Dongen J, Van Iterson M, Veldink JH, Völker U, Warmerdam R, Wijmenga C, Swertz M, Andiappan A, Montgomery GW, Ripatti S, Perola M, Kutalik Z, Dermitzakis E, Bergmann S, Frayling T, Van Meurs J, Prokisch H, Ahsan H, Pierce BL, Lehtimäki T, Boomsma DI, Psaty BM, Gharib SA, Awadalla P, Milani L, Ouwehand WH, Downes K, Stegle O, Battle A, Visscher PM, Yang J, Scholz M, Powell J, Gibson G, Esko T, Franke L (2021) Large-scale cis- and trans-eQTL analyses identify thousands of genetic loci and polygenic scores that regulate blood gene expression. Nat Genet 53(9): 1300–1310. https://doi.org/10.1038/s41588-021-00913-z

Wallace C (2021) A more accurate method for colocalisation analysis allowing for multiple causal variants. PLoS Genet 17(9):e1009440. https://doi.org/10.1371/journal.pgen.1009440

Wang G, Sarkar A, Carbonetto P, Stephens M (2020) A simple new approach to variable selection in regression, with application to genetic fine mapping. J R Stat Soc Ser B: Stat Methodol 82(5):1273–1300. https://doi.org/10.1111/rssb.12388

Wilhelm-Benartzi CS, Koestler DC, Karagas MR, Flanagan JM, Christensen BC, Kelsey KT, Marsit CJ, Houseman EA, Brown R (2013) Review of processing and analysis methods for DNA methylation array data. Br J Cancer 109(6):1394–1402. https://doi.org/10.1038/bjc.2013.496

Wimmers F, Burrell AR, Feng Y, Zheng H, Arunachalam PS, Hu M, Spranger S, Nyhoff LE, Joshi D, Trisal M, Awasthi M, Bellusci L, Ashraf U, Kowli S, Konvinse KC, Yang E, Blanco M, Pellegrini K, Tharp G, Hagan T, Chinthrajah RS, Nguyen TT, Grifoni A, Sette A, Nadeau KC, Haslam DB, Bosinger SE, Wrammert J, Maecker HT, Utz PJ, Wang TT, Khurana S, Khatri P, Staat MA, Pulendran B (2023) Multi-omics analysis of mucosal and systemic immunity to SARS-CoV-2 after birth. Cell 186(21): 4632–4651.e23. https://doi.org/10.1016/j.cell.2023.08.044

Wu X, Zhang Y (2017) TET-mediated active DNA demethylation: mechanism, function and beyond. Nat Rev Genet 18(9):517–534. https://doi.org/10.1038/nrg.2017.33

Xie S, Wang Z, Okano M, Nogami M, Li Y, He WW, Okumura K, Li E (1999) Cloning, expression and chromosome locations of the human DNMT3 gene family. Gene 236(1):87–95. https://doi.org/10.1016/s0378-1119(99)00252-8

Xu C-J, Bonder MJ, Söderhäll C, Bustamante M, Baïz N, Gehring U, Jankipersadsing SA, van der Vlies P, van Diemen CC, van Rijkom B, Just J, Kull I, Kere J, Antó JM, Bousquet J, Zhernakova A, Wijmenga C, Annesi-Maesano I, Sunyer J, Melén E, Li Y, Postma DS, Koppelman GH (2017) The emerging landscape of dynamic DNA methylation in early childhood. BMC Genomics 18(1):25. https://doi.org/10.1186/s12864-016-3452-1

Xu C-J, Söderhäll C, Bustamante M, Baïz N, Gruzieva O, Gehring U, Mason D, Chatzi L, Basterrechea M, Llop S, Torrent M, Forastiere F, Fantini MP, Carlsen KCL, Haahtela T, Morin A, Kerkhof M, Merid SK, van Rijkom B, Jankipersadsing SA, Bonder MJ, Ballereau S, Vermeulen CJ, Aguirre-Gamboa R, de Jongste JC, Smit HA, Kumar A, Pershagen G, Guerra S, Garcia-Aymerich J, Greco D, Reinius L, McEachan RRC, Azad R, Hovland V, Mowinckel P, Alenius H, Fyhrquist N, Lemonnier N, Pellet J, Auffray C, BIOS Consortium, van der Vlies P, van Diemen CC, Li Y, Wijmenga C, Netea MG, Moffatt MF, Cookson WOCM, Anto JM, Bousquet J, Laatikainen T, Laprise C, Carlsen K-H, Gori D, Porta D, Iñiguez C, Bilbao JR, Kogevinas M, Wright J, Brunekreef B, Kere J, Nawijn MC, Annesi-Maesano I, Sunyer J, Melén E, Koppelman GH (2018) DNA methylation in childhood asthma: an epigenome-wide meta-analysis. Lancet Respir Med 6(5):379–388. https://doi.org/10.1016/S2213-2600(18)30052-3

Yan F, Powell DR, Curtis DJ, Wong NC (2020) From reads to insight: a hitchhiker's guide to ATAC-seq data analysis. Genome Biol 21(1):22. https://doi.org/10.1186/s13059-020-1929-3

Zhang Z, Trypsteen W, Blaauw M, Chu X, Rutsaert S, Vandekerckhove L, Van Der Heijden W, Dos Santos JC, Xu C-J, Swertz MA, Van Der Ven A, Li Y (2021) IRF7 and RNH1 are modifying factors of HIV-1 reservoirs: a genome-wide association analysis. BMC Med 19(1):282. https://doi.org/10.1186/s12916-021-02156-5

Zhang B, Zhang Z, Koeken VACM, Kumar S, Aillaud M, Tsay H-C, Liu Z, Kraft ARM, Soon CF, Odak I, Bošnjak B, Vlot A, Deutsche COVID-19 OMICS Initiative (DeCOI), Swertz MA, Ohler U, Geffers R, Illig T, Huehn J, Saliba A-E, Sander LE, Förster R, Xu C-J, Cornberg M, Schulte LN, Li Y (2023) Altered and allele-specific open chromatin landscape reveals epigenetic and genetic regulators of innate immunity in

COVID-19. Cell Genom 3(2):100232. https://doi.org/10.1016/j.xgen.2022.100232

Zhao S, Jing W, Samuels DC, Sheng Q, Shyr Y, Guo Y (2018) Strategies for processing and quality control of Illumina genotyping arrays. Brief Bioinform 19(5): 765–775. https://doi.org/10.1093/bib/bbx012

Zheng GXY, Terry JM, Belgrader P, Ryvkin P, Bent ZW, Wilson R, Ziraldo SB, Wheeler TD, McDermott GP, Zhu J, Gregory MT, Shuga J, Montesclaros L, Underwood JG, Masquelier DA, Nishimura SY, Schnall-Levin M, Wyatt PW, Hindson CM, Bharadwaj R, Wong A, Ness KD, Beppu LW, Deeg HJ, McFarland C, Loeb KR, Valente WJ, Ericson NG, Stevens EA, Radich JP, Mikkelsen TS, Hindson BJ, Bielas JH (2017) Massively parallel digital transcriptional profiling of single cells. Nat Commun 8:14049. https://doi.org/10.1038/ncomms14049

Zhernakova A, Kurilshikov A, Bonder MJ, Tigchelaar EF, Schirmer M, Vatanen T, Mujagic Z, Vila AV, Falony G, Vieira-Silva S, Wang J, Imhann F, Brandsma E, Jankipersadsing SA, Joossens M, Cenit MC, Deelen P, Swertz MA, LifeLines cohort study, Weersma RK, Feskens EJM, Netea MG, Gevers D, Jonkers D, Franke L, Aulchenko YS, Huttenhower C, Raes J, Hofker MH, Xavier RJ, Wijmenga C, Fu J (2016) Population-based metagenomics analysis reveals markers for gut microbiome composition and diversity. Science 352(6285):565–569. https://doi.org/10.1126/science.aad3369

Zhou Q, Su X, Jing G, Chen S, Ning K (2018) RNA-QC-chain: comprehensive and fast quality control for RNA-Seq data. BMC Genomics 19(1):144. https://doi.org/10.1186/s12864-018-4503-6

Part III

Applications of Omics Approaches

Genomics in Health and Biomedicine 16

Maria Luís Cardoso, Hugo Martiniano, Luisa Mota-Vieira, and Astrid Moura Vicente

Abstract

Genomic information is rapidly becoming a cornerstone of personalized medicine, offering transformative potential for clinical practice. This chapter explores the critical role of genomics in enabling earlier diagnosis, precise treatment, risk prediction, and preventive healthcare strategies. Advances in sequencing technologies, data integration, and bioinformatics allow for tailored healthcare solutions based on an individual's genetic profile, combined with clinical, lifestyle, and environmental data. Integration with electronic health records, mHealth technologies, and artificial intelligence further enhances clinical decision-making. The chapter highlights current applications of genomic medicine in oncology, rare diseases, and pharmacogenomics and the growing relevance of polygenic risk scores in managing common chronic diseases. It also discusses the need for harmonized data governance, infrastructure development, professional training, and public engagement to ensure equitable and effective implementation. These developments are situated within the broader landscape of national and international initiatives—including ICPerMed, 1Million Genomes, and the Genome of Europe project—that aim to foster collaboration, standardization, and equitable access to genomic healthcare across populations. Clinical areas where genomics has already demonstrated substantial value are discussed while identifying key challenges and priorities for advancing the future of personalized medicine.

Keywords

Genomics · Personalized medicine · Rare diseases · Cancer genomics · Pharmacogenomics · Polygenic risk scores · Data governance

M. L. Cardoso · H. Martiniano · A. M. Vicente (✉)
Departamento de Promoção da Saúde e Prevenção de Doenças não Transmissíveis, Instituto Nacional de Saúde Doutor Ricardo Jorge, Lisboa, Portugal

BioISI – Biosystems and Integrative Sciences Institute, Faculdade de Ciências da Universidade de Lisboa, Lisboa, Portugal
e-mail: astrid.vicente@insa.min-saude.pt

L. Mota-Vieira
BioISI – Biosystems and Integrative Sciences Institute, Faculdade de Ciências da Universidade de Lisboa, Lisboa, Portugal

Serviço de Formação, Ensino e Investigação, Hospital Divino Espírito Santo de Ponta Delgada, Açores, Portugal

16.1 Genomic Information as Key Health Data for Personalized Medicine

Genomics is a dynamic and rapidly advancing field within the life sciences, driven by scientific and technological innovations that enable

M. Gama-Carvalho et al. (eds.), *Omics Approaches in Biomedicine and Biotechnology*, Advances in Experimental Medicine and Biology 1504, https://doi.org/10.1007/978-3-032-18966-0_16

progress across multiple disciplines. Its impact on medicine is particularly significant (Shendure et al. 2019). Genomic information has enormous potential in clinical settings, for earlier and more precise diagnoses, for personalized therapeutic decisions that minimize adverse drug reactions and optimize treatment effectiveness, and for disease risk estimation that can inform preventative actions. As a cornerstone of personalized medicine, genomic data enables tailored clinical strategies based on an individual's genetic makeup together with a broader health profile, including medical history and imaging and lifestyle and environmental exposure factors (Council of the European Union 2015; Vicente et al. 2020; Denny and Collins 2021).

The human genome remains unchanged throughout an individual's life, and therefore the information gained from sequencing the genome provides invaluable insights across the lifespan. Advances in genomic science and big data analytics increasingly enable the identification of genetic profiles associated with disease susceptibility, elucidating the biological pathways disrupted in disease, facilitating diagnostics, defining treatment targets that are adequate for each person, and paving the way for prevention strategies (Denny and Collins 2021). Importantly, even when current knowledge does not provide immediate clinical solutions, ongoing research and technological progress is constantly leading to discoveries that will offer answers in the future, rendering knowledge of the human genome useful during the whole life cycle of an individual.

In the medium term, integrating genomic data with electronic health records, emerging health-monitoring technologies such as mobile health (mHealth) devices for real-time health tracking, and environmental exposure assessment tools is expected to transform medical practice. Coupling this plethora of health information with artificial intelligence tools will enhance the ability to identify meaningful health data patterns, leading to improved healthcare decisions, better clinical care, and better patient outcomes. In this context, personalized medicine represents a paradigm shift in healthcare, leveraging technology and data to refine health systems (Vicente et al. 2020; Shen et al. 2023; Claussnitzer et al. 2020).

Currently, genomic medicine already plays a crucial role in diagnosing, predicting, managing risk, and treating diseases in fields such as oncology and rare diseases (Denny and Collins 2021; Claussnitzer et al. 2020; Green et al. 2020; Manolio et al. 2019). Whole Genome Sequencing is paving the way to lessen the so-called "diagnostic odyssey" for rare diseases, significantly reducing the time for etiological diagnosis through one single test for many conditions. This has important implications for timely patient care and for genetic counseling of family members. The use of genomic or gene panel information in oncology is key for the stratification of patients according to the underlying pathogenic mechanism of the tumors, which in turn directs the most effective treatment for each individual. For hereditary cancers running in families, it also allows the identification of relatives at risk, with implications for adequate personalized screening and surveillance and eventually preventive measures. Increasingly, information on pharmacogenes, which encode enzymes and other proteins involved in response to pharmacological treatments, is used to define drug therapies with a lower risk of adverse drug reactions or higher efficacy. The potential of genomics for preventing chronic illnesses is also driving significant expansion in biomedical research and clinical trials (Shendure et al. 2019; Vicente et al. 2020; Denny and Collins 2021; Claussnitzer et al. 2020; Green et al. 2020; Manolio et al. 2019). Chronic diseases, including cardiovascular disease, diabetes, cancer, and neurological or neuropsychiatric disorders, have a complex etiology, arising from a combination of genetic and environmental risk factors, including lifestyle choices like diet and tobacco use. The combination of variants in multiple genes, often common also in the general population, defines risk profiles that are the focus of intense research using a Polygenic Risk Score (PRS) approach. These conditions are often described as the great epidemic of the twenty-first century due to their high prevalence, mortality, and morbidity rates. Recognized as priority areas by the World Health

Organization (WHO), they are the focus of extensive research, innovation programs, and global investments, in the context of which genomics plays a key role.

The European Commission (EC) has been investing substantially in funding for research and implementation of genomics and personalized medicine. For instance, the International Consortium for Personalised Medicine (ICPerMed 2015), a platform to promote research and implementation of personalized medicine, supported an array of funded actions to develop clinical studies, promote personalized prevention, advance the development of economic models for personalized medicine, and foster personalized prevention. Importantly, ICPerMed is promoting international networking, bringing together programs across the world to support research and implementation. The recently launched European Partnership in Personalised Medicine (ICPerMed 2023) builds on these efforts as a platform for funding research and supporting innovation. In the European context, research funding in cancer and rare diseases places a major focus in personalized medicine, while progressively expanding to common noncommunicable diseases and emerging consideration given to personalized prevention.

For genomic research to have a tangible clinical impact, improve healthcare system efficiency, and benefit patients and citizens, several key areas must be addressed (Vicente et al. 2020; Green et al. 2020; Manolio et al. 2019; Fokstuen et al. 2016; Lopes et al. 2022). These include establishing robust data governance frameworks, developing infrastructure for clinical and genomic data management, and creating standardized guidelines, quality protocols, and policies for genomic testing. Making data interoperable and harmonizing on a global scale is key for the high quality and comparability of genomic findings. Given the fast-paced nature of technological advancements, regular review of protocols and regulatory frameworks are crucial to accelerating the clinical adoption of new genomic tests and treatments. Equally important is incorporating patient perspectives and improving public literacy in genomics, empowering individuals to make informed healthcare decisions. Health systems must adapt through new processes, infrastructure, and tools that effectively support clinical practice. In this context, advanced training for professionals across disciplines is also essential (Genomics England 2021). Multidisciplinary teams are needed, with professionals proficient in genomic-related disciplines, that not only excel in their areas of expertise, but also maintain a functional understanding of related fields to ensure seamless communication and collaboration. The education of professionals is thus key to addressing the complexities of clinical genomics, genetic counseling, genetic laboratory testing, pharmacogenomics, genetic variant interpretation, and bioinformatics, among others.

A close integration between clinical practice and research, bolstered by industry partnerships, is crucial for translational medicine. On one hand, real-world clinical data can drive new discoveries, while research findings should be rapidly incorporated into clinical applications to maximize patient benefits. Open communication and a shared understanding of terminology and concepts are essential to bridging research and clinical practice, ensuring that clinical data informs innovation and vice versa, in a continuous collaboration between research and clinical applications for patient-centered advancements.

Crucial to advancing genomic medicine is the secure and controlled access to genomic information for research and for healthcare. The 1 Million Genomes Initiative (1+MG) (European Commission 2024a, b) includes signatory member states that closely cooperate to provide access to genomes across Europe, for research and healthcare. In the context of this initiative, a federated European network will allow secure access to genomic information from millions of European citizens. In parallel to the technical aspects of data access, the initiative is establishing a legal and ethical framework for genomic data sharing, developing quality control guidelines, and elaborating evidence-based recommendations for implementation of genomics in healthcare, including health economic models to support public health decision-making. The resulting framework will make genomic data sharing truly effective and sustainable, for the benefit of citizens and patients.

The United Kingdom has been a pioneer in genomic medicine with programs like Genomic England and the 100,000 Genomes Project (100,000 Genomes Project Pilot Investigators 2021), as well as the UK Biobank (Backman et al. 2021), and is currently driving efforts for the implementation of genomic medicine in the UK National Health Service (NHS) (Friedrich et al. 2024). Several other countries also have well-advanced genomic programs, including France (Abadie et al. 2025), Denmark (Danish National Genome Center 2025), Finland (Kurki et al. 2023), Estonia (Milani et al. 2025), and Spain (IMPaCT 2025). The 1+MG has been a stimulus for countries to establish genomic programs at the national level, which will contribute genomes from research projects as well as from healthcare. These programs will contribute to the Genome of Europe (GoE) project (Genome of Europe 2025), which will analyze whole-genome sequence data from 100,000 Europeans, building a reference genome to advance research and facilitate healthcare. Because genomes will be provided by European countries in proportion to their population size in Europe, a main contribution of this project is to generate genomic information on European citizens of variable ancestry, promoting a better understanding of the true genetic variability across Europe. Considering that the genomic knowledge available globally is still largely based on Caucasians of European descent, this will be a major step toward equitable access to personalized healthcare, independently of ancestry, for European citizens. Complementing these efforts at the global level, there is a growing number of genomic initiatives in all continents (Howley et al. 2025), including the Americas, namely in the USA (Jeong et al. 2025; All of Us Research Program Investigators 2019) and Canada (Canadian Institutes of Health Research 2022), Africa and the Middle East (Gooden 2024; Ateia et al. 2023; Mbarek et al. 2022), Asia (Wong et al. 2023), Australia (Australian Genomics 2025), and multiple other countries. The global interest stresses the need for an effort in aligning ethical and legal frameworks, policies, regulation, quality of data, as well as shared infrastructure. Work being carried out by the 1+MG and ICPerMed, as well as by the Global Alliance for Genomic Health (Rehm et al. 2021) and the related National Initiatives Forum (GA4GH 2025), is crucial to this effort and testifies to the value of genomics in personalized medicine.

This chapter will focus three main areas of medicine where genomics is well advanced, exemplifying successes and benefits for patients and citizens, but also highlighting the challenges and the needs moving forward for the future of personalized healthcare.

16.2 Advancing the Diagnosis of Rare Diseases Through Genomics

Rare diseases (RD), as defined by the European Union, affect fewer than 1 in 2000 individuals (EURORDIS 2025). Despite their rarity, they collectively pose a substantial health burden, impacting approximately 350 million people globally. Over 10,000 rare diseases have been identified, with 80% having a genetic origin, nearly 70% manifesting in childhood, and approximately 95% lacking approved treatments (Global Genes 2025). Although monogenic, the genetic basis of more than half of all identified Mendelian diseases remains unclear (Rahit and Tarailo-Graovac 2020). Diagnosing these diseases is particularly challenging due to their clinical heterogeneity, low frequency, and overlap with more common disorders. The process often entails a prolonged and complex "diagnostic odyssey", marked by years of inconclusive tests and clinical uncertainty.

On average, an accurate diagnosis is close to 5 years (Faye et al. 2024). However, the development and use of genome sequencing, using next-generation sequencing (NGS) or other genetic testing methods, enable earlier and more precise diagnoses (Hartley et al. 2020; Kernohan and Boycott 2024; Wojcik et al. 2024; Schobers et al. 2024). Integrated with artificial intelligence (AI) and multi-omic analyses, these cutting-edge technologies also enhance our understanding of the molecular mechanisms underlying rare diseases, facilitating the development of personalized and effective treatments (Abdallah et al. 2023). Despite these advances, challenges remain, including the need for more integrated analysis and clinical interpretation of complex datasets (phenotyping, omics, and/and imaging) (Hou et al. 2020; Schmidt et al. 2024) and for equitable access to sequencing tools across diverse populations and healthcare systems (Cohen et al. 2024; Cavaller-Bellaubi et al. 2023). Here, we explore the genomic approaches for diagnosing rare diseases, highlighting those to solve undiagnosed cases, as well as for newborn screening to enhance the early diagnosis of treatable rare diseases.

16.2.1 Genomic Testing for Rare Diseases

Genomic tests for rare disease investigation are based on NGS due to their ability to detect genome sequence alterations (genetic variants) associated with the disease (Wojcik et al. 2024). These alterations include (i) relatively simple variants, such as single-nucleotide variants (SNVs) or short (up to several dozen base pairs) insertions or deletions (indels), and (ii) more complex variants, such as complex chromosomal rearrangements (structural variants, SVs) or copy number variants (CNVs) (Eichler 2019). The diagnostic yield of NGS in RD varies significantly, ranging from 24% to 68%, and depends mainly on the sequencing technology and bioinformatic tools for variant analysis, combined with the disease features and patient phenotype assessment (age, clinical presentation, family structure, etc.). Three main NGS approaches are frequently prescribed in clinical practice: gene panel testing (GPT), clinical and whole exome sequencing (CES and WES), and whole genome sequencing (WGS) (Chen et al. 2024; Klein et al. 2014). Sometimes, NGS can be preceded or completed by chromosomal microarray analysis or other auxiliary methods (Iyshwarya and Veerabathiran 2024; Schuler et al. 2022). Despite these approaches, phenotype assessment and clinical suspicion remain crucial steps in the diagnostic process.

Gene panel testing —single gene or multi-genes—are focused on targeting specific gene (s) or coding gene regions known to be associated with a disease or disease group. For this reason, GPT is being applied in the diagnosis of various rare diseases, such as hypertrophic cardiomyopathy (Duarte et al. 2025), skeletal dysplasia (MacCarrick et al. 2024), and neuromuscular disorders (Ng et al. 2022). Compared to exome and genome sequencing, targeted gene panels allow higher-depth sequencing, reduced costs, shorter turnaround times, and a lower risk of incidental findings or variants of uncertain significance (VUS). However, these panels require regular updates to remain current due to the finding of new disease-associated genes. Additionally, gene panels have limitations in detecting structural variants or deep intronic variants (both resolved by WGS). Despite these constraints, GPT remains a valuable tool in genetic diagnostics, offering a balance between comprehensive analysis and practical implementation.

CES and WES are targeted approaches that focus mainly on sequencing the protein-coding regions (exons), but they differ in their scope and application. CES examines around 5000 disease-associated genes and their flanking splice regions. It is especially valuable for the diagnosis of Mendelian diseases in patients with suspected genetic conditions or when the phenotype is complex (Gorukmez et al. 2023; Lee et al. 2014). Many studies are showing its benefits, such as the identification of seven pathogenic variants in the *NPHS1*, *COL2A1*, *OCRL*, *SHOC2*, *TPRV4*, *MTM1*, and *STAC3* genes in an infant cohort from South Africa (Campbell et al. 2023), as well as the diagnostic utility of CES in a prospective cohort study of patients with suspected rare genetic diseases (Hartley et al. 2024). WES is designed to sequence all ~20,000 protein-coding genes (1–2%) of the human genome, in which over 95% of disease-causing variants are located, but it can also capture some noncoding regions, such as long noncoding RNAs or small regulatory RNAs. This approach has the potential to identify variants in genes not previously associated with human diseases, thus expanding the understanding of genotype–phenotype relationships (Glotov et al. 2023). CES and WES data can be analyzed using a strategy based on a phenotype-driven virtual gene panel, known to be associated with the patient's disease (Alvarez-Mora et al. 2023; Benedetti et al. 2022; Wang et al. 2019). WES data can also be analyzed through an exome-wide screening for all rare and potentially damaging variants, followed by a comparative analysis of the phenotypes associated with these genes against the patient's clinical presentation. This latter strategy enhances the potential for discovering novel disease-associated genes, making it an invaluable tool in clinical diagnostics and research settings (Sáenz et al. 2021). Despite its utility, WES has technical limitations. These include incomplete coverage of certain genomic regions and difficulties in detecting CNVs and SVs. However, advances in sequencing technology and bioinformatics algorithms are significantly improving exon coverage and extending variant detection outside the exons, including known disease-causing intronic variants. CES and WES are valuable tools in genetic diagnostics, with the choice between them often depending on the suspected genetic condition and the balance between clinically oriented testing and a broad analysis.

WGS allows the sequencing and analysis of the entire human genome (~3.2 billion base pairs), offering the potential to identify novel genes, which can then be considered in the design of new gene panels, and providing insights into understanding complex inheritance patterns, such as the non-Mendelian inheritance of some rare diseases (Ross 2024). Compared to other sequencing approaches, WGS detects a broader spectrum of genetic variations, including SNVs, indels, SVs, CNVs, and variants in deep intronic regions (Jiao et al. 2019). Two recent studies illustrate these advantages. One identified CNVs in three clinical cases, after negative results with other genetic testing (Bowman et al. 2025); the other study described the identification of a novel pathogenic GAA repeat expansion in *FGF14* gene in unsolved individuals with adult-onset of ataxia (Rafehi et al. 2023). Despite its advantages, the widespread adoption of WGS faces several challenges, such as limited equipment availability, high costs, and genomic complexity, particularly the clinical interpretation of noncoding variants. Together, genome and exome sequencing also introduce new concerns for clinicians and patients due to the identification of incidental findings and VUS. Nevertheless, these approaches, combined with novel genomic technologies such as long-read WGS (AlAbdi et al. 2023; Ohori et al. 2021), significantly advance the clinical diagnosis and research of rare diseases. To this end, the re-analysis of genomic sequencing data, supported by evolving bioinformatics tools, variant interpretation guidelines, and genetic databases (Hartley et al. 2023; Kernohan and Boycott 2024), is also contributing.

16.2.2 Early Detection of Rare Diseases by Genomic Newborn Screening

Prenatal screening tests for rare diseases are entering a new phase with the launch of genomic newborn screening (gNBS). These programs have the potential to enhance the early diagnosis of treatable rare diseases and provide lifelong health benefits (Bick et al. 2022; Jiang et al. 2023; Stark and Scott 2023). Several initiatives are exploring the use of gNBS at the population scale. One is the Generation Study, a UK initiative led by Genomics England in partnership with NHS England (Lewis et al. 2024). This study aims to screen up to 100,000 newborns for more than 200 rare genetic conditions using WGS. Blood samples are collected from the umbilical cord at birth in approximately 40 participating hospitals. Expectant parents receive study information during pregnancy. If they choose to enroll, a research midwife ensures informed consent. To date, over 1000 participants have provided consent. Genetic results are provided within 28 days if a condition is detected, or within a few months otherwise. If the screening result is positive, families receive additional NHS testing, clinical support, and treatment. This pioneering initiative represents a major advance in gNBS and personalized medicine for rare diseases. Moreover, the Generation Study complements, but does not replace, the NHS blood spot screening (heel prick test), which remains essential for detecting nine rare conditions in all newborns.

16.2.3 Opportunities, Challenges and Future Directions

Despite significant advancements in genomic technologies, approximately 30% of rare disease cases remain unresolved, highlighting the need for continuous innovation (van Karnebeek et al. 2024). Emerging approaches such as long-read sequencing enhance the resolution of complex genomic regions, including structural rearrangements and repetitive sequences (Nakamichi et al. 2024; Steyaert et al. 2024). Additionally, the integration of artificial intelligence and machine learning into genomic analysis is already used in clinical practice to predict variant pathogenicity, revolutionizing the diagnosis, understanding, and treatment of rare genetic diseases (Abdallah et al. 2023; Hong et al. 2024). However, challenges persist, such as the interpretation of rare variants, the complexity of noncoding regions, the development of functional assays for variant analysis, tools for predicting phenotype–genotype associations, and the integration of multi-omics data into clinical-genomic databases (Braconi et al. 2025; IGVF Consortium 2024; Roberts et al. 2024; Wright et al. 2024). These challenges also introduce ethical concerns in clinical practice, including incidental findings, data privacy, and patient access and autonomy (Adachi et al. 2023; Quintal et al. 2023). Moreover, access to genomic diagnostics remains unequal due to high costs, limited infrastructure in low-resource settings, and the underrepresentation of diverse populations in genomic databases (Cohen et al. 2024; Delaye et al. 2022; Marshall et al. 2023). Addressing these gaps requires collaborative projects, research, innovation, training, education, and investment, as they create opportunities for diagnosis and provide new perspectives for clinical trials (Denton et al. 2022; Jobanputra et al. 2024; Mishra and Venkatesh 2024; Ronald and Gui 2024; Schmidt et al. 2024). These efforts are shaping a future where all patients can access timely and accurate diagnoses, enhancing their quality of life and driving advancements in innovative therapies and personalized medicine for rare diseases.

16.3 Cancer Genomics and Precision Oncology: Transforming Cancer Care

The evolution of genomics has been particularly beneficial for cancer care (Hoeben et al. 2021; Desai et al. 2023). The possibility of using genomic profiles to accurately classify tumor types enabled the development of targeted therapies, improving the efficacy of cancer treatment and transforming patient outcomes. Furthermore, genomic insights catalyzed the advancement of immunotherapy, ushering in a new era of precision oncology. Genomics established a rapid pace of progress in cancer research and translation into clinical practice, with breakthrough therapies like next-generation immunotherapies. For this, collaboration between research, clinical settings, and the pharma industry was essential, demonstrating the value of synergies for the benefit of patients and healthcare systems.

16.3.1 Diagnosis, Screening, and Implications for Therapeutic Choices and Preventive Measures

The understanding of the molecular biology underlying particular cancers dates back from the 1970s, when the first mutations in tumor DNA were identified, revealing biological pathways that could be selectively targeted by novel therapies (Leon et al. 1977). This triggered a change in standard-of-care practices to personalized strategies, which are vastly enhancing treatment efficacy.

While standard-of-care practices used to rely on histopathology and tumor morphology, the availability of genomic and molecular profiling allowed molecular subtyping of tumors, with therapeutic implications (Johnson et al. 2021). For example, precise classification of breast cancer as HER2-positive, hormone receptor-positive, and triple-negative breast cancer enabled targeted treatment according to subtypes. HER2-targeted therapies such as *trastuzumab* and *pertuzumab* have dramatically improved survival and reduced recurrence rates for HER2-positive breast cancer patients. Similarly, lung cancers are now characterized by genetic drivers like EGFR mutations, ALK rearrangements, and KRAS mutations, guiding the use of targeted agents such as *osimertinib* for EGFR-mutant tumors or *alectinib* for ALK-rearranged cancers (Li et al. 2023). Other notable examples include the use of BRAF inhibitors (*vemurafenib and dabrafenib*) for BRAF-mutated melanoma (Castellani et al. 2023).

Genomic screening has reshaped cancer prevention by identifying individuals at elevated risk for hereditary cancers (Garutti et al. 2023). Mutations in genes such as *BRCA1*, *BRCA2*, or mismatch repair genes linked to Lynch syndrome, drive higher risks of breast, ovarian, colorectal, and other cancers. This knowledge enables preventive strategies like enhanced surveillance, prophylactic surgery, and risk-reducing therapies.

Genomic companion diagnostics are nowadays used to identify actionable mutations in cancer patient tumor before the start of treatment, ensuring that patients receive therapies specifically suited to their tumor's genetic profile, improving treatment outcomes and minimizing unnecessary side effects. Concurrently, innovations in genomic and epigenomic technologies have led to the development of multi-cancer early detection (MCED) assays capable of identifying malignancies at presymptomatic stages (Imai et al. 2025). Collectively, these advances hold significant potential to transform cancer-screening paradigms, enabling earlier therapeutic interventions and enhancing patient outcomes.

16.3.2 Cancer Immunotherapy Advances

Genomic testing has also contributed to the rise of cancer immunotherapy treatment by identifying patients with tumors that are more vulnerable to immune system dysfunction. Such patients can truly benefit from immunotherapy treatments that are costly and prone to eliciting very serious adverse reactions (Esfahani et al. 2020; Mukherjee 2019). For instance, genomic testing is able to identify molecular characteristics in the tumor, such as a high Tumor Mutational Burden (TMB), by detecting neoantigens on the tumor surface. The neoantigens are more recognizable to the immune system, and therefore immune checkpoint inhibitors like *pembrolizumab* and *nivolumab* are more effective in treating patients with TMB-high tumors (Marabelle et al. 2020). Similarly, cancer cells with defects in their DNA repair mechanisms, for instance with high microsatellite instability, accumulate mutations very quickly, rendering them more visible to the immune system and more responsive to immune checkpoint inhibitors.

Genomic testing can also detect inflammation in the tumor environment, which correlates better with improved response to immunotherapy. High genomic expression of PD-L1 can determine which patients respond better to anti-PD-1or anti-PD-L1 antibodies (Le et al. 2017). Genomic testing, therefore, can help identify patients who are more likely to benefit from immunotherapy, expanding treatment options beyond conventional therapy and radiotherapy. Clinical studies like the KEYNOTE-522 or the NIAGARA Trial increasingly demonstrate how patients can benefit from innovative treatment approaches that significantly improve life expectancy for cancers that are difficult to treat, such as triple-negative breast cancer (Schmid et al. 2020), melanoma, or bladder cancer (Bellmunt et al. 2021). For such cancers, treatment options were very limited in the past, and these novel approaches provide precious insights regarding the occurrence of drug toxicity and the benefits for patients' survival.

16.3.3 Leveraging Genomics for Therapy: Antibody–Drug Conjugates

Research introducing antibody–drug conjugates (ADCs) has been equally prolific (Beck et al. 2017). ADCs deliver chemotherapy by coupling an antibody designed to recognize and bind a specific antigen, at the tumor cell surface, with a chemotherapy drug, forming a complex molecular structure that is endocyted by the cell. This complex delivers their payload directly inside the specifically recognized cells, with reduced toxicity and potentially higher antitumor activity.

Genetic testing can improve the precision of targeting, again by identifying the patients overexpressing the tumor antigen. For instance, very effective treatment for HER-2-positive cancer cells uses *Trastuzumab deruxtecan* (trastuzumab, the same antibody as Herceptin, plus a chemo payload) (Modi et al. 2020). However, the treatment requires genetic testing for HER-2, because it is specific for this subtype. Also, tumor cells with genetic variants that hinder internalization or drug metabolism may resist ADC therapy. Finally, some gene variations, for instance, in cytochrome P450 (CYP450) liver enzymes, could predict higher risk for ADC toxicity, requiring drug adjustment and close monitoring (Beck et al. 2017).

16.3.4 Liquid Biopsy and Real-Time Tumor Monitoring

The future holds promising strategies for cancer screening, early detection, and surveillance. Liquid biopsy technologies are enabling major advances by minimally invasive detection of circulating tumor DNA (ctDNA) in blood samples (Wan et al. 2017; Diaz and Bardelli 2014). Liquid biopsy avoids traditional invasive and costly tumor biopsies, enabling easier diagnosis, particularly for hard-to-reach cancers. Furthermore, detection of ctDNA allows real-time tracking of tumor evolution, facilitating treatment monitoring and detection of resistance mutations, with the possibility of fast adjustment of therapies in response to genetic changes (Oxnard et al. 2016). Tumor cells can frequently acquire mutations during disease progression, with a large percentage of the tumor being eventually more responsive to a different drug. ctDNA testing is expected to allow routine and frequent monitoring of these changes over time, at reasonable costs, opening the options for more precise treatment over the course of the illness (Bartolomucci et al. 2025).

The UK is currently offering ctDNA testing in the National Health System (NHS) as a pilot for particular cancers, to demonstrate the feasibility of integrating liquid biopsy into routine clinical care. These pilots are paving the way for widespread use in longitudinal monitoring of the disease, including surveillance for signs of progression at the earliest onset, but also identifying actionable markers that could indicate the need to change a treatment or accurately predict its efficacy (NHS England 2024). This has the potential for monitoring the status of an illness in a cost-effective way and modifying therapy if needed (Bartolomucci et al. 2025).

16.3.5 Tumor-Agnostic Therapies: A Paradigm Change

Research is also focusing on tumor agnostic drugs, where a particular drug can be effective to treat tumors arising in different tissue types by targeting a specific genetic mutation, biomarker, or molecular feature, thus treating the genetic driver of the cancer independently of what tissue the tumor affects (Gouda et al. 2023). Examples include therapies with immune checkpoint therapies like *Pembrolizumab* or with *Larotrectinib*, which treats solid tumors showing NTRK gene fusions (Drilon et al. 2018). Such therapies can be particularly valuable for rare cancers, where comprehensive genomic profiling is essential for identifying actionable targets and expanding treatment options.

16.3.6 Challenges and Opportunities

Despite the significant progress, integrating genomics into cancer care comes with challenges (Zalis et al. 2024). The complexity of cancer genomes, tumor heterogeneity, and the emergence of resistance mechanisms necessitate continuous research and refinement of genomic-based strategies. However, as research and technology continue to evolve, genomics provides huge opportunities for the future of cancer care, which will be increasingly personalized, improving the survival rates and quality of life for patients worldwide.

16.4 Pharmacogenomics

Over the last decade, scientific advances in genomics have raised expectations about their impact on healthcare, particularly in disease prevention, diagnosis, and treatment. Integrating research outputs in the field of genomics into clinical practice has become the new challenge. Some specific areas are already seeing high levels of success in implementation, improving diagnosis rates and treatment outcomes through the selection of targeted therapies. One particularly successful area is the use of genomics for the identification of individuals with a high risk of developing adverse drug reactions or nonresponding to pharmacological treatment, therefore guiding more personalized, safe, and effective treatments.

16.4.1 Pharmacogenomics and Related Concepts

The aim of pharmacogenetics and pharmacogenomics (PGx) is to study how genetic inheritance contributes to variations in individual drug responses. Sometimes, these biologically unexpected behaviors can manifest as extreme phenotypes including severe, life-threatening adverse drug reactions or a complete lack of therapeutic effectiveness (Weinshilboum and Wang 2006). Pharmacogenomics has also been integrated into the drug development and approval process by medicines' regulatory agencies (EMA 2011). PGx can be described as "*the study of genomic technologies to enable the discovery and development of novel drugs, and the optimization of drug dose and choice in individual patients to maximize efficacy and minimize toxicity*" (Pirmohamed 2023).

Pharmacogenes are a set of genes that encode enzymes and transporters playing key roles in pharmacokinetics, covering drug absorption, distribution, metabolism, and elimination, as well as in pharmacodynamics, affecting the drug targets or modulating biological pathways responsible for the patient's sensitivity to medication (Katara and Yadav 2019). Particular attention has long been given to CYP450 enzymes, due to their central role in pharmacokinetics, as they are responsible for the metabolism of a wide range of drugs. This superfamily of microsomal heme-thiolate enzymes catalyzes oxidation, peroxidation, and reduction reactions, processing both endogenous compounds (e.g., steroids, prostaglandins, and fatty acids) and exogenous substances (e.g., drugs, environmental pollutants, and agrochemicals). CYP450 enzymes mediate phase I metabolism for over 90% of commonly prescribed drugs, excluding biologics and vaccines, significantly contributing to inter-individual variability in drug response, toxicity, and adverse effects (Danielson 2002; Guengerich 2017; Zanger and Schwab 2013).

Genetic sequence or structural variants in the genes encoding phase I enzymes, such as cytochromes P450, lead to distinct phenotypes associated with drug metabolism, enabling to group individuals into metabolism rate groups. It is expectable that individuals who inherit two copies of the wild-type allele of a CYP450 exhibit standard enzymatic activity, classified as extensive (or normal) metabolizers (EM). However, as genetic variants can lead to reduced or absent enzyme activity, this may result in intermediate metabolizers (IM), and poor metabolizers (PM) who face an increased risk of drug accumulation and toxicity, and may also impair the activation of prodrugs, potentially resulting in therapeutic failure. Conversely, individuals with multiple copies of functional alleles exhibit increased enzymatic activity and are classified as rapid (RM) or ultra-rapid metabolizers (UM), resulting in therapeutic failure due to accelerated inactivation and rapid drug clearance or, in the case of prodrugs, rapid conversion to active metabolites, raising the risk of toxicity and adverse effects (Salyakina et al. 2019). Note that one person can be a PM of a drug and a RM for a different one depending on the genes and variants involved.

Variations in phase II enzymes, which mediate conjugation reactions, as well as drug transporters can also influence drug metabolism and clearance, impacting treatment safety and efficacy (Yiannakopoulou 2013; Babalola et al. 2023). Evidence supports genotyping for phase II enzymes TPMT and UGT1A1, with recommendations incorporated into clinical guidelines (DPWG 2021; CPIC 2024). In addition, PGx of *NAT2* is gaining interest due to its potential to optimize tuberculosis treatment (Ridolfi et al. 2024; Mahajan and Tyagi 2024); this is particularly relevant given that tuberculosis remains one of the leading causes of death globally.

For membrane drug transporters, the evidence remains limited with a few exceptions like SLCO1B1, a solute carrier transporter, for which PGx guidelines have been established to guide statin therapy reducing the risk of statin-induced myopathy (CPIC 2024; DPWG 2021). Among ATP-binding cassette (ABC) transporters, while some studies support their strong influence on drug clearance and clinical use (Loscocco et al. 2021; Lalatović et al. 2023), others emphasize the need for further research in the field (Bruckmueller and Cascorbi 2021; Kasten and Cascorbi 2024). Genetic variants can also affect drug target genes which are involved in pharmacodynamics. Despite some progress that has been made, namely in neuropsychiatric medications (Hertz et al. 2021; Gross and Daniel 2018), the evidence supporting their testing is weaker than that for genes involved in pharmacokinetics.

16.4.2 Clinical Implementation

For many years, PGx was confined to academic research and clinical trials (Ji et al. 2012). In healthcare settings, it was used only reactively, as a diagnostic tool, to identify the cause of severe adverse drug reactions (ADRs). However, more recently, PGx has emerged as a key element of personalized medicine with significant impact in therapeutic outcomes. Today, it stands as one of the strongest examples of how genomic information can drive the development of more precise and personalized treatments, influencing clinical practice. PGx is an additional tool that clinicians can use for treatment selection and personalization, shifting from a "one-size-fits-all" approach to a more tailored strategy that considers individual genetic variability (Padmanabhan et al. 2023).

Abacavir hypersensitivity is a very good example of the success of translational PGx (Fig. 16.1). This antiretroviral drug is used in combination with other HIV medications to treat human immunodeficiency virus (HIV) infection. However, abacavir can cause a severe, potentially life-threatening hypersensitivity reaction in approximately 5% of treated patients, which has been strongly linked to the presence of the HLA-B*57:01 allele (Hetherington et al. 2001). This association was confirmed by multiple studies worldwide. Pretreatment genetic testing for HLA-B57:01 has demonstrated clear clinical utility in identifying individuals at risk (Chaponda and Pirmohamed 2011). In consequence, drug regulatory agencies across the globe recommend HLA-B*57:01 genotyping before prescribing abacavir. Over the years, the routine implementation of this genetic screening has nearly eliminated abacavir hypersensitivity, as long as genotyping prior to abacavir administration became standard practice (Hughes et al. 2008; Peter et al. 2019). The widespread adoption of HLA-B*5701 pre-emptive testing represents a milestone in the application of genetic testing in clinical settings to improve drug safety (Phillips and Mallal 2007).

Following the success of *abacavir* PGx screening, there has been growing interest in the potential of PGX to improve patient outcomes, supporting its integration into clinical practice.

Other breakthroughs in the integration of PGx into hospital care came with two landmark studies from the Netherlands, which played a pivotal role in establishing genotyping of the dihydropyrimidine dehydrogenase gene *(DPYD)* before fluoropyrimidine therapy (namely with *capecitabine* or *5-fluorouracil*). *DPYD* encodes the enzyme dihidropyrimidine dehydrogenase

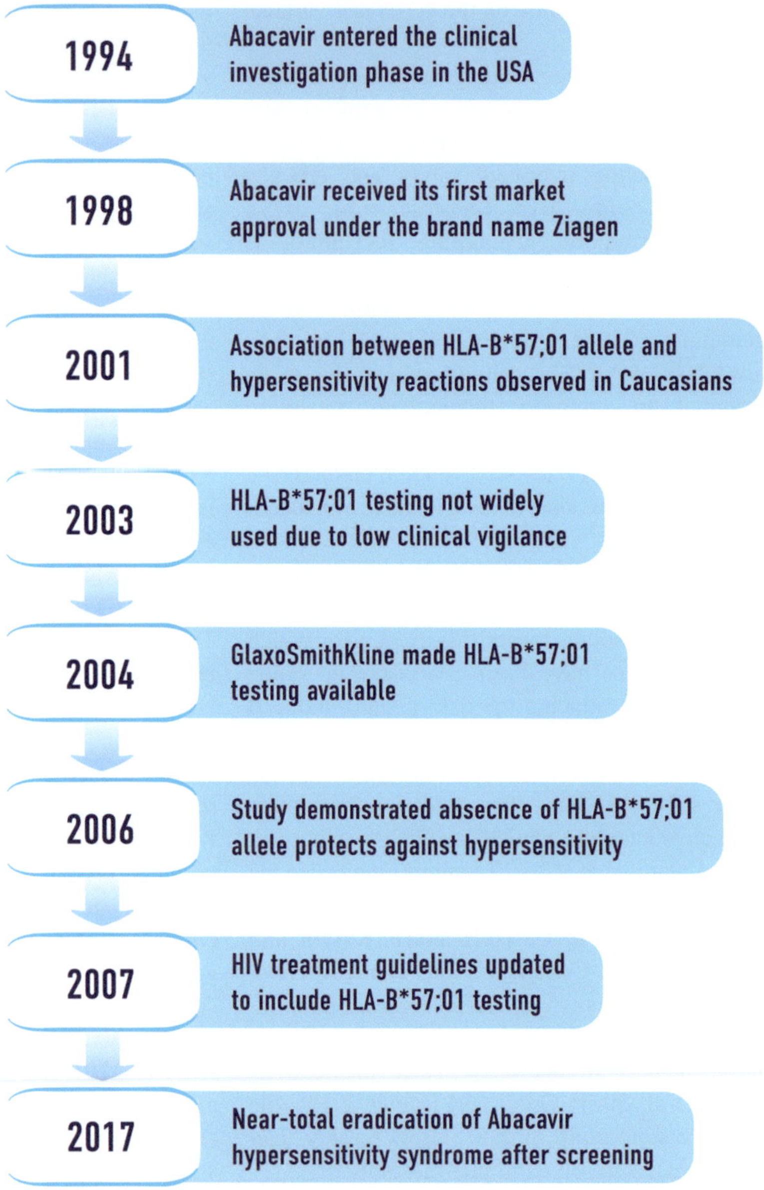

Fig. 16.1 Preventing adverse drug reactions: The success story of HLA-B*57:01 testing for abacavir

(DPD; EC 1.3.1.2), which catalyzes a rate-limiting step in the catabolism of 5-fluoropyrimidine metabolism. Specific gene sequence variants are known to decrease the activity of the enzyme (Henderson 2019) and can confer an increased risk of severe or even fatal toxicity, when patients harboring one or more copies of these variants are treated with fluoropyrimidines. A study on the prospective *DPYD* genotyping to reduce the risk of fluoropyrimidine-induced severe toxicity (Lunenburg et al. 2016) and the prospective safety study about *DPYD* genotype-guided dose individualization of fluoropyrimidines therapy in cancer patients (Henricks et al. 2018) laid the foundation for the subsequent incorporation of *DPYD* genotyping into clinical protocol.

In early 2018, the French agency in charge of medicinal product safety ruled in favor of routine screening for dihydropyrimidine dehydrogenase

(DPD) deficiency before initiating treatment with fluoropyrimidines (INCa and HAS 2018), following strong advocacy from patient associations, with recommendations favoring phenotyping primarily by plasma uracil measurement, complemented by *DPYD* genotyping for specific alleles, and advice on adjusting the drug dosage accordingly (Loriot et al. 2018). In 2020, the European Medicines Agency (EMA) recommended the pre-emptive use of *DPYD* genotyping to prevent serious, even fatal adverse reactions to fluoropyrimidines (EMA 2020). This recommendation was rapidly adopted by EU national regulatory agencies and implemented across European countries (de With et al. 2023). In 2024, a Health Impact Assessment (HIA) was conducted in three European countries—Finland, Italy, and Portugal—concluding on the positive impact of going beyond recommendations and issuing a mandatory public health policy for *DPYD* testing (data not published). Modeling the impact of a policy for systematic *DPYD* genotyping predicts a reduction of severe toxicity and mortality; most important, it would impact the equity of access to *DPYD* testing.

Testing individuals on a drug-by-drug, gene-by-gene basis presents significant costs and logistical challenges as each patient would require multiple tests for different medications. Many patients take more than one medication, and several drugs share metabolic pathways involving common detoxification enzymes. As knowledge of the allele frequencies and understanding of their impact on drug pharmacokinetics and pharmacodynamics expands (alongside advancements in genotyping technologies, increased accessibility, and reduced costs), the possibility of using panel-based genotyping has emerged (UPGX 2016; Chenchula et al. 2024). The shift from reactive single-gene testing to comprehensive pre-emptive genotyping panels offers significant promise for enhancing drug-prescribing practices (Peruzzi et al. 2025; Mai et al. 2024). This approach allows for the simultaneous testing of multiple variants across several pharmacogenes, thereby optimizing therapy and enhancing patient safety (Swen et al. 2023; Chenchula et al. 2024).

A key study that advanced this perspective was the PREPARE (**Pre-**emptive **P**harmacogenomic Testing for **P**reventing **A**dverse Drug **Re**actions) clinical study carried out by the U-PGx (Ubiquitous Pharmacogenomics) consortium. It enrolled 18 institutions from seven European countries, which implemented and evaluated the impact of PGx testing on therapy outcomes (UPGx website). The findings demonstrated that implementing pre-emptive pharmacogenetic testing significantly reduced the frequency of ADRs in patients receiving genotype-guided therapy, compared to those following standard care. Additionally, the study highlighted the feasibility and clinical utility of panel-based genotyping as an effective approach to optimize drug therapy and enhance patient safety (Swen et al. 2023).

Another ongoing implementation study, the PROGRESS (**P**harmacogenetics **R**oll **O**ut - **G**auging **Res**ponse to **S**ervice) study in the UK (NHS 2023), aims to evaluate the feasibility of providing an NHS-wide diagnostic service and identify challenges in implementing PGx into primary care. The study will also collect information on patient and healthcare professional attitudes regarding this service. In this clinical trial, a panel of genes with known implications in the efficacy and safety of a range of commonly prescribed medicines (including statins, proton pump inhibitors, selective serotonin reuptake inhibitors, and tricyclic antidepressants) was selected and is currently being tested. An informatics tool was developed to help guide the prescription.

Whether through individual gene–drug pairs or via genotyping panels, PGx is gaining traction among reference hospitals in regions where there is a strong focus on genomic medicine.

16.4.3 Guidelines

The clinical significance of PGx spans various medical fields, contributing to enhanced treatment effectiveness and the prevention of ADRs in many specialities. Beyond oncology (Leong et al. 2025), PGx can impact fields like cardiology

(Chatelin et al. 2017; Mauriello et al. 2023; Inshutiyimana et al. 2024), psychiatry (Lo et al. 2024; Adiukwu et al. 2023; Salatin et al. 2025; Dhieb and Bastaki 2025), gastroenterology (Coelho et al. 2025; Stanton et al. 2025; Appell et al. 2025), infectious diseases (Carolino et al. 2017; Mahajan and Tyagi 2024; Amorim et al. 2025; Covington et al. 2025), nephrology (Adams et al. 2018), transplantation (Wang et al. 2025; Jackson et al. 2025), and pediatrics (Coelho et al. 2025; Kennedy et al. 2024; Jackson et al. 2025), among others.

Throughout the last decade, several international scientific consortia and some national health agencies have published guidelines, reporting and reinforcing numerous associations between genetic variants and drug responses. Most relevant are those from the Dutch Pharmacogenetics Working Group (DPWG), the Clinical Pharmacogenetics Implementation Consortium (CPIC), the Canadian Pharmacogenomics Network for Drug Safety (CPNDS), and the French RNPGx (Abdullah-Koolmees et al. 2021; DPWG 2021; CPIC 2024). These guidelines are periodically updated and provide support for the implementation of PGx in clinical practice, recommending evidence-based interventions for specific genotypes, which can enhance treatment precision and efficacy (Cecchin and Stocco 2020). However, unlike the case of fluoropyrimidines and *DPYD*, most countries have not yet widely adopted these guidelines. There are some academic and specialized hospitals, which have strategically invested in this field, but the implementation is still not uniform across Europe. Moreover, as most PGx guidelines focus on single drug–gene(s) interactions, it is essential to advance research further and develop tools, including artificial intelligence, that can predict multiple drug–drug–gene interactions, incorporating other clinical variables such as renal function, especially in elderly patients with comorbidities (Božina et al. 2020; Bobo et al. 2022; Dhieb and Bastaki 2025).

16.4.4 Main Factors Influencing the Adoption of PGx in Clinical Care

Despite advances in genotyping and sequencing technologies that have lowered costs and improved accessibility, broad implementation has yet to occur.

According to several experts (Relling and Evans 2015; Asiimwe et al. 2021; Adiukwu et al. 2023; Apellaniz-Ruiz et al. 2024), PGx's successful implementation depends on:

(i) The results of randomized and nonrandomized clinical trials evidencing clinical utility
(ii) Clear clinical guidelines developed by expert groups from international consortia
(iii) Recommendations from regulatory entities or endorsement by scientific societies to facilitate the adoption of such guidelines
(iv) The availability of effective alternative treatments for individuals with "high-risk" genetic profiles
(v) Healthcare systems' preparedness to integrate new prescribing protocols (e.g., algorithms)
(vi) The implementation of electronic clinical decision support (CDS) tools which are crucial for integrating genetic data into clinical workflows and ensuring appropriate, genetics-guided medication prescription
(vii) An engaged and well-trained health workforce

The successful integration of PGx into routine clinical practice will require a multidisciplinary and collaborative approach, engaging a broad range of professionals, including physicians and pharmacists in both primary and secondary care, as well as clinical laboratory geneticists responsible for genetic testing. Additionally, software engineers, informatics specialists, healthcare researchers, and policymakers will play key roles. Citizens and patients must be central to this effort to ensure the effective and equitable integration of PGx into healthcare (Turner et al. 2022).

Several critical questions remain unresolved and need further investigation. Although PGx has the potential to optimize treatment outcomes and reduce adverse drug reactions (ADRs), data on the cost-effectiveness of implementing pre-emptive multigene testing across different clinical areas remains limited (Morris et al. 2022; Jameson et al. 2024).

Another key issue is assessing the applicability of current PGx guidelines to patients of non-European descent and determining the impact of rare PGx variants that are not detected by current genotyping methods (Peruzzi et al. 2025; Mosch et al. 2025). The integration of whole-genome sequencing into routine clinical care can help to bridge this gap (Balciuniene et al. 2023; Hsu et al. 2024; Youn et al. 2024; Sherman et al. 2024). This challenge will be also addressed in the ongoing GoE project, which aims to establish a reference genome for Europe. GoE will analyze whole-genome sequences from 100,000 European citizens, with country representation according to the relative dimension of its population in Europe, thus capturing the ancestry variability that is widely expected given the recent migration waves (European Commission 2024a, b). This will allow a clearer depiction of pharmacogene allele frequencies present in European citizens, with important impact on the quality of PGx healthcare.

Another issue is genotype–phenotype mismatches which often result from drug–drug interactions. Understanding the influence of polypharmacy, including the potential of drug-induced phenoconversion, requires further research into the mechanisms of enzyme inhibition and induction in patients taking multiple concurrent medications (Stingl and Viviani 2025; Sarömba et al. 2025).

To date, drug dosage guidelines have largely relied on single-gene variants. However, drug pharmacokinetics and pharmacodynamics involve multiple proteins, encoded and regulated by a diverse set of genes. Polygenic risk scores (PRS) offer a promising approach by incorporating the polygenic nature of a patient's genetic makeup, potentially improving predictive accuracy beyond single-gene assessments (Jayasinghe et al. 2024; Simona et al. 2023). Ongoing research and large-scale collaborations are expected to drive progress in this field, ultimately enabling more personalized, safer, and more effective pharmacotherapy in coming years (Zhai et al. 2023; Singh et al. 2024).

16.5 Polygenic Risk Scores

Predicting an individual's susceptibility to disease has long been a goal of clinical medicine. Family history has traditionally served this purpose, but its predictive ability is limited by several factors, including incomplete penetrance and complex nature of disease etiologies (Chatterjee et al. 2016; Tada et al. 2016). In recent years, Polygenic Risk Scores (PRS) have emerged as a tool for estimating an individual's genetic susceptibility to a disease or trait. Despite methodological, ethical, and efficacy concerns, PRS hold significant promise for improving healthcare through more targeted approaches to early detection, prevention, and treatment strategies (Torkamani et al. 2018; Lewis and Vassos 2020; Abu-El-Haija et al. 2023).

The most common noncommunicable complex disorders are polygenic, influenced by the interplay of many genetic variants, each contributing with a small individual effect (Chatterjee et al. 2016; Ge et al. 2017; Lewis and Vassos 2020). Large-scale genome-wide association studies (GWAS) have identified these variants, providing the foundation for PRS construction. Early PRS focused on genome-wide significant variants (Dudbridge 2013). However, methodological advances now allow the inclusion of variants with sub-genome-wide significance (i.e., those that do not reach the stringent threshold for genome-wide significance but still contribute to risk), leading to more comprehensive PRS (Lloyd-Jones et al. 2019; Mak et al. 2017a, b; Privé et al. 2021). This evolution reflects the increasingly sophisticated methods used to capture the complex polygenic architecture of common diseases (Wray et al. 2021).

In the last few years, PRS have been gaining attention for their ability to refine clinical decision-making by enabling personalized risk stratification (Wray et al. 2021). PRS offer a mechanism to identify individuals across a spectrum of genetic risk, facilitating earlier detection, targeted prevention, and personalized treatment. This ability to stratify risk is expected to improve outcomes for patients and advance personalized medicine.

PRS are statistical tools used for risk stratification, not for diagnostic assays. They provide a probabilistic estimate of relative genetic risk rather than a definitive prediction. In the clinical application of PRS, it is important to distinguish between clinical validity, which PRS have shown by effectively stratifying risk, and clinical utility, which requires robust validation in diverse clinical settings (Koch et al. 2023; Kumuthini et al. 2022). The cost-effectiveness of PRS implementation also requires further evaluation. This section explores the potential applications of PRS in contemporary clinical practice while also identifying some of the challenges that have to be addressed before their widespread clinical adoption (Koch et al. 2023; Kumuthini et al. 2022).

16.5.1 Calculation and Interpretation of PRS

16.5.1.1 Calculation

The calculation of PRS involves a weighted sum of an individual's risk alleles at PRS-specific variants, where the weights are the effect size (for binary traits) or the beta-coefficient (for continuous traits), estimated from GWAS summary statistics. The outcome is a risk score for each individual for a disease or continuous trait (Lewis and Vassos 2020; Wray et al. 2021). Several software tools such as LDPred (Vilhjálmsson et al. 2015), lassosum (Mak et al. 2017a, b), and PRSice-2 (Choi and O'Reilly 2019) automate the process of building PRS from GWAS summary data.

16.5.1.2 Interpretation

The interpretation of PRS values depends on the distribution of the score within the reference population used for its derivation. An individual's PRS is interpreted as a percentile rank within the reference population's distribution, reflecting their relative genetic risk. Since PRS quantify relative, not absolute, risk, standard clinical risk assessment and vigilance remain essential, regardless of the PRS result (Wray et al. 2021).

16.5.2 Clinical Applications of PRS

16.5.2.1 Risk Stratification for Personalized Prevention

PRS enable risk stratification, supporting personalized prevention strategies tailored to individual genetic profiles (Polygenic Risk Score Task Force of the International Common Disease Alliance 2021; Abu-El-Haija et al. 2023). By identifying individuals across the spectrum of genetic risk, PRS can guide targeted preventive interventions. In familial cancer clinics, PRS helps to refine risk assessment in high-risk families with uninformative germline results, leading to more precise prevention plans. Multiple studies have shown that PRS can predict breast cancer (BC) risk even in women from high-risk families without *BRCA1/2* mutations (Evans et al. 2017; Shieh et al. 2017; Mavaddat et al. 2019; Choudhury et al. 2021; Lin et al. 2025). A widely cited BC PRS with 313 SNPs (Mavaddat et al. 2019) was shown to accurately stratify BC patients of European ancestry, with odds-ratio (OR) of 1.61 (95% CI 1.57–1.65), an area under the curve (AUC) of 0.630 (95% CI 0.628–0.651), and a 32.6% lifetime risk of overall BC in the top 1% of the PRS. This PRS has been used for informing recommendations such as intensified lifestyle changes or chemoprevention. Vince Jr et al. (2025) report a modest performance for prostate cancer PRS in identifying patients at a higher risk of developing this disease. Tools such as CanRisk, which incorporate PRS, are being evaluated for their impact on personalized BC prevention (Koch et al. 2023). PRSs have also demonstrated discriminatory power in diabetes: they can differentiate between (i) type 1 and healthy controls, (ii) between type 1 and type 2 diabetes, (iii) type 1 and monogenic diabetes, and (iv) between type 2 diabetes patients and healthy people (Padilla-Martínez et al. 2020; Ritchie et al. 2024).

16.5.2.2 Treatment Decision-Making

PRS hold promise for informing treatment decisions, enabling personalized therapy based on individual genetic risk and predicted treatment response (Läll et al. 2017). More specifically, PRSs can identify patients who are more likely to benefit from, or experience adverse effects with, certain therapies. Coronary artery disease patients with high PRS derived greater benefit from the PCSK9 inhibitor *alirocumab* (Damask et al. 2021). Similarly, Marston et al. (2020) reported that patients with atherosclerotic cardiovascular disease with elevated genetic risk, as determined by PRS, showed enhanced benefit from *evolocumab* (another PCSK9 inhibitor), regardless of clinical risk. These studies highlight the potential of PRS to guide personalized therapeutic strategies in cardiovascular disease.

16.5.2.3 Population-Level Screening and Prognosis

PRS can improve population-level screening and disease prognosis, enabling more targeted public health interventions and efficient resource allocation (Läll et al. 2017). In screening programs, PRS refine risk stratification, allowing for more focused interventions on individuals at a higher genetic risk. For example, Forgetta et al. (2020) showed that PRS-based screening for low bone mineral density could reduce the need for bone densitometry tests by approximately40% without compromising sensitivity or specificity. In BCr screening, integrated models combining PRS, questionnaire data, and mammographic density have been shown to improve the identification of women at elevated 10-year risk. The BOADICEA v5 model (implemented in the CanRisk tool), which incorporates the 313-variant BC PRS discussed above (Mavaddat et al. 2019), has already been used in clinical settings to inform population-based breast cancer risk assessments and guide prevention strategies. Ongoing trials like WISDOM and PERSPECTIVE are evaluating PRS-guided mammographic screening strategies (Abu-El-Haija et al. 2023). A recent review concluded that a majority of studies indicated that polygenic risk-informed cancer screening was likely to be more cost-effective than alternatives highlighting, however, the uncertainties in the cost of conducting PRS at the population level (Dixon et al. 2022). These examples illustrate the growing role of PRS in refining risk assessment and improving the efficiency of public health initiatives.

16.5.3 Challenges and Considerations for Clinical Translation of PRSs

Despite the potential applications of PRS outlined in Sect. 16.3, critical appraisals underscore the need for more robust evidence of clinical utility before widespread clinical adoption (Koch et al. 2023; Kumuthini et al. 2022; Abu-El-Haija et al. 2023). While PRS demonstrate clinical validity through accurate risk stratification, evidence supporting improved patient outcomes remains limited. Kumuthini et al. (2022) classified some studies as showing "near evidence" of clinical utility, where benefit is plausible, but reported no studies demonstrating unequivocal improvements in patient health outcomes. This limitation is partly due to the moderate discriminative power of current PRS (Koch et al. 2023) and their inability to serve as standalone predictors (Wray et al. 2021).

16.5.3.1 Diagnostic and Prognostic Performance

The current utility of PRS for many common, complex diseases is limited by their moderate discriminative ability (Sud et al. 2023). Koch et al. (2023) report AUC values often ranging from 0.6 to 0.7 when PRS are used alone, indicating the limited ability to distinguish between cases and controls. This modest diagnostic and prognostic performance, also noted by Wray et al. (2021), calls for cautious clinical integration and discourages over-reliance on PRS as standalone predictors. Moreover, the incremental predictive value gained from incorporating PRS into established clinical risk models is often small, further limiting their immediate clinical utility.

16.5.3.2 Predictive Performance in Diverse Populations

The diminished predictive accuracy of PRS in non-European populations, stemming from the strong European bias in GWAS, represents a critical barrier to equitable clinical implementation (Morales et al. 2018; Martin et al. 2019; Koch et al. 2023; Kumuthini et al. 2022; Wray et al. 2021). This limitation is particularly pronounced in individuals of African ancestry (Wray et al. 2021). PRS accuracy is maximized when applied to individuals whose ancestry closely matches that of the original GWAS cohort, due to population-specific variations in allele frequencies and linkage disequilibrium patterns. Application across significantly different ancestries requires careful validation to account for these differences. For instance, while Ji et al. (2021) report an improvement in the accuracy of polygenic risk prediction of BC among East Asian populations through the use of European GWAS data, such cross-population generalizability is not consistently observed, particularly in BC risk prediction. This disparity can be addressed through various approaches, such as developing ancestry-specific PRSs or the inclusion of diverse ancestries directly into PRS development and validation. Ongoing research efforts, including expanding GWAS sample collection across diverse global populations (Wray et al. 2021), aim to enhance PRS performance and promote equitable access to their benefits, irrespective of ancestry. The recently launched Genome of Europe project is expected to make a major contribution to tackle this challenge, by providing a thorough, in-depth view of the genetic diversity of the European population, including principal component analysis of genetic variation across regions, reference panels for ancestry-specific imputation, and population distributions for polygenic risk scores.

16.5.3.3 Interpretation and Communication

Communicating the probabilistic nature of PRS to both clinicians and patients remains a significant barrier for effective clinical integration. While experts recognize that PRS provide estimates of risk rather than definitive diagnoses, translating these probabilities into actionable clinical guidance is complex, as highlighted by recent systematic reviews (Koch et al. 2023; Kumuthini et al. 2022).

Clear guidelines for interpreting PRS results in a clinically meaningful way are still lacking (Kumuthini et al. 2022), hindering informed decision-making. A persistent challenge is the misconception that genetic tests provide deterministic diagnostic accuracy (Wray et al. 2021). Improving genetic literacy among both patients and healthcare providers can help to clarify the probabilistic nature of PRS and support appropriate clinical use. The American College of Medical Genetics and Genomics (ACMG) recommends prioritizing the development of robust, standardized pre-test counseling protocols and educational resources to ensure accurate interpretation and management of expectations surrounding PRS testing (Abu-El-Haija et al. 2023). These resources should focus on conveying the probabilistic, nondeterministic nature of PRS, and provide clear frameworks for integrating PRS information with other clinical and environmental risk factors.

16.5.4 Future Directions: Realizing the Clinical Promise of PRS

The field of PRS is rapidly evolving, with ongoing research focused on enhancing accuracy, generalizability, and critically demonstrating clinical utility and impact on patient outcomes. Future research should prioritize several key directions to fully realize the clinical promise of PRS.

16.5.4.1 Developing PRS for More Diverse Populations

Expanding GWAS and PRS development to encompass diverse populations is crucial to improve accuracy and generalizability across ancestries, ensuring equitable access and benefit for all populations (Peterson et al. 2019). This imperative extends beyond scientific advancement to directly address health equity and ensure PRS benefit all communities. To achieve this, sustained efforts to increase GWAS sample collection across global population groups are essential (Wray et al. 2021).

16.5.4.2 Establishing Clinical Utility

As clinical validity becomes increasingly established, the next step is to demonstrate unequivocal clinical utility through rigorous prospective studies. Randomized controlled trials (RCTs) are considered essential to provide the robust evidence base needed for responsible clinical translation (Lewis and Green 2021; Koch et al. 2023; Kumuthini et al. 2022). The integration of PRS into clinical guidelines must be guided by robust evidence of improved patient health outcomes (Abu-El-Haija et al. 2023). Future research must prioritize the responsible translation of PRS into clinical practice and the demonstration of measurable improvements in patient health outcomes (Koch et al. 2023; Kumuthini et al. 2022). Furthermore, future trials should also incorporate health economic evaluations to assess the cost-effectiveness of PRS-guided strategies.

16.5.4.3 Developing Enhanced Communication and Education Strategies

The creation of effective communication strategies and educational resources is essential to ensure patients and healthcare providers accurately understand PRS results, their probabilistic nature, and their appropriate application in clinical decision-making (Kumuthini et al. 2022). This effort should include research into optimal communication modalities and tools to effectively convey the nuances of PRS results, prevent misinterpretations, support informed consent and shared decision-making, and improve genetic risk literacy to ensure appropriate interpretation and utilization of PRS (Wray et al. 2021).

PRS hold transformative potential for healthcare by enabling personalized risk assessments and tailored interventions. Ongoing research focusing on these future directions is essential to confirm their clinical value and secure their role in the future of genomic medicine. As evidence grows, PRS are likely to play an increasingly important and beneficial role in future healthcare ushering in a new era of genomic medicine. Personalized risk prediction may contribute to improved patient outcomes and more effective public health strategies.

References

100,000 Genomes Project Pilot Investigators (2021) 100,000 genomes pilot on rare-disease diagnosis in health care — preliminary report. N Engl J Med 385: 1868–1880. https://doi.org/10.1056/NEJMoa2035790

Abadie C et al (2025) PFMG2025–integrating genomic medicine into the national healthcare system in France. Lancet Reg Health—Europe 50:101183. https://doi.org/10.1016/j.lanepe.2024.101183

Abdallah S, Sharifa M, I Kh Almadhoun MK et al (2023) The impact of artificial intelligence on optimizing diagnosis and treatment plans for ra re genetic disorders. Cureus 15:e46860. https://doi.org/10.7759/cureus.46860

Abdullah Koolmees H, van Keulen AM, Nijenhuis M et al (2021) Pharmacogenetics guidelines: overview and comparison of the DPWG, CPIC, CPN DS, and RNPGx guidelines. Front Pharmacol 11:595219. https://doi.org/10.3389/fphar.2020.595219

Abu-El-Haija A, Reddi HV, Wand H et al (2023) The clinical application of polygenic risk scores: a points to consider statement of the American College of Medical Genetics and Genomics (ACMG). Genet Med 25(5):100803. https://doi.org/10.1016/j.gim.2023.100803

Adachi T, El-Hattab AW, Jain R et al (2023) Enhancing equitable access to rare disease diagnosis and treatment around the world: a review of evidence, policies, and challenges. Int J Environ Res Public Health 20:4732. https://doi.org/10.3390/ijerph20064732

Adams SM, Crisamore KR, Empey PE (2018) Clinical pharmacogenomics: applications in nephrology. Clin J Am Soc Nephrol (CJASN) 13(10):1561–1571. https://doi.org/10.2215/CJN.02730218

Adiukwu F, Adesokun O, Essien E et al (2023) Pharmacogenetic testing in psychiatry: perspective on clinical utility. Asian J Psychiatr 86:103674. https://doi.org/10.1016/j.ajp.2023.103674

AlAbdi L, Shamseldin HE, Khouj E et al (2023) Beyond the exome: utility of long-read whole genome sequencing in exome-negative autosomal recessive diseases. Genome Med 15:114. https://doi.org/10.1186/s13073-023-01270-8

Alvarez-Mora MI, Rodríguez-Revenga L et al (2023) Implementation of exome sequencing in clinical practice for neurological disorders. Genes 14:813. https://doi.org/10.3390/genes14040813

Amorim G, Jaworski J, Yang J et al (2025) Pharmacogenetics of tuberculosis treatment toxicity and effectiveness in a large Brazilian cohort. Pharmacogenet Genomics 35(2):55–64. https://doi.org/10.1097/FPC.0000000000000552

Apellaniz-Ruiz M, Barrachina J, Castro-Sanchez P et al (2024) Status of the implementation of pharmacogenetics in clinical practice in Spain: from regional to national initiatives. Drug Metab Pers Ther 39(4):183–199. https://doi.org/10.1515/dmpt-2024-0042

Appell ML, Hindorf U, Almer S et al (2025) Response to azathioprine treatment in autoimmune hepatitis is dependent on glutathione transferase genotypes. Dig Liver Dis. https://doi.org/10.1016/j.dld.2024.12.026

Asiimwe IG, Zhang EJ, Osanlou R et al (2021) Warfarin dosing algorithms: a systematic review. Br J Clin Pharmacol 87(4):1717–1729. https://doi.org/10.1111/bcp.14608

Ateia H, Ogrodzki P, Wilson HV et al (2023) Population genome programs across the Middle East and North Africa: successes, challenges, and future directions. Biomed Hub 8(1):60–71. https://doi.org/10.1159/000530619

Australian Genomics (2025) www.australiangenomics.org.au. Assessed 15 Apr 2025

Babalola CP, Kotila O, Iwuchukwu OF (2023) The expanding role of gene-based prescribing for phase II drug-metabolizing enzymes. Minirev Pharmacol Toxicol 2(1):1–14. https://ajpps.org/view-pdf/?article=3b72a58055a88db31ae06ec526a70dbb61cBykGZECQ=

Backman JD, Li AH, Marcketta A et al (2021) Exome sequencing and analysis of 454,787 UK biobank participants. Nature 599(7886):628–634. https://doi.org/10.1038/s41586-021-04103-z

Balciuniene J, Liu R, Bean L, Guo F et al (2023) At-risk genomic findings for pediatric-onset disorders from genome sequencing vs medically actionable gene pane l in proactive screening of newborns and children. JAMA Netw Open 6(7):e2326445. https://doi.org/10.1001/jamanetworkopen.2023.26445

Bartolomucci A, Nobrega M, Ferrier T et al (2025) Circulating tumor DNA to monitor treatment response in solid tumors and advance precision o ncology. NPJ Precis Oncol 9:84. https://doi.org/10.1038/s41698-025-00876-y

Beck A et al (2017) Strategies and challenges for the next generation of antibody–drug conjugates. Nat Rev Drug Discov 16(5):315–337. https://doi.org/10.1038/nrd.2016.268

Bellmunt J et al (2021) Enfortumab vedotin in previously treated advanced urothelial carcinoma. N Engl J Med 384(13):1125–1135. https://doi.org/10.1056/NEJMoa2035807

Benedetti V, Canzoneri R, Perrelli A et al (2022) Next-generation sequencing advances the genetic diagnosis of cerebral cavernous malformation (CCM). Antioxidants (Basel) 11:1294. https://doi.org/10.3390/antiox11071294

Bick D, Ahmed A, Deen D et al (2022) Newborn screening by genomic sequencing: opportunities and challenges. Int J Neonatal Screen 8:40. https://doi.org/10.3390/ijns8030040

Bobo WV, Van Ommeren B, Athreya AP (2022) Machine learning, pharmacogenomics, and clinical psychiatry: predicting antidepressant response in patients with major depressive disorder. Expert Rev Clin Pharmacol 15(8):927–944. https://doi.org/10.1080/17512433.2022.2112949

Bowman P, Grimes H, Dallosso AR et al (2025) Whole genome sequencing for copy number variant detection to improve diagnosis and management of rare diseases. Dev Med Child Neurol 67:126–131. https://doi.org/10.1111/dmcn.15985

Božina N, Vrkić Kirhmajer M, Šimičević L et al (2020) Use of pharmacogenomics in elderly patients treated for cardiovascular diseases. Croat Med J 61(2):147–158. https://doi.org/10.3325/cmj.2020.61.147

Braconi D, Nadwa H, Bernardini G, Santucci A (2025) Omics and rare diseases: challenges, applications, and future perspectives. Expert Rev Proteomics 20:1–16. https://doi.org/10.1080/14789450.2025.2468300

Bruckmueller H, Cascorbi I (2021) ABCB1, ABCG2, ABCC1, ABCC2, and ABCC3 drug transporter polymorphisms and their impact on drug bioavailabil ity: what is our current understanding? Expert Opin Drug Metab Toxicol 17(4):369–396. https://doi.org/10.1080/17425255.2021.1876661

Campbell L, Fredericks J, Mathivha K et al (2023) The implementation and utility of clinical exome sequencing in a south African infant cohort. Front Genet 14:1277948. https://doi.org/10.3389/fgene.2023.1277948

Canadian Institutes of Health Research (2022) Sequencing our future: Institute of Genetics strategic plan 2022–2027. https://cihr-irscgcca/e/52972html. Assessed 29 Jan 2025

Carolino F, Santos N, Piñeiro C et al (2017) Prevalence of Abacavir-associated hypersensitivity syndrome and HLA-B*5701 allele in a Portuguese HIV-positive population. Porto Biomed J 2(2):59–62. https://doi.org/10.1016/j.pbj.2016.12.004

Castellani G, Buccarelli M, Arasi MB et al (2023) BRAF mutations in melanoma: biological aspects, therapeutic implications, and circulating biomarkers. Cancer

15(16):4026. https://doi.org/10.3390/cancers15164026

Cavaller-Bellaubi M, Hughes-Wilson W, Kubinová Š et al (2023) Patients, payers and developers of orphan medicinal products: lessons learned from 10 years' multi-stakeholder dialogue on improving access in Europe via MoCA. Orphanet J Rare Dis 18:144. https://doi.org/10.1186/s13023-023-02774-7

Cecchin E, Stocco G (2020) Pharmacogenomics and personalized medicine. Genes 11(6):679. https://doi.org/10.3390/genes11060679

Chaponda M, Pirmohamed M (2011) Hypersensitivity reactions to HIV therapy. Br J Clin Pharmacol 71(5): 659–671. https://doi.org/10.1111/j.1365-2125.2010.03784.x

Chatelin J, Stathopoulou MG, Arguinano AA, Xie T, Visvikis-Siest S (2017) Pharmacogenomic challenges in cardiovascular diseases: examples of D rugs and considerations for future integration in clinical practice. Curr Pharm Biotechnol 18(3):231–241. https://doi.org/10.2174/1389201018666170123153626

Chatterjee N, Shi J, García-Closas M (2016) Developing and evaluating polygenic risk prediction models for stratified disease prevention. Nat Rev Genet 17(7): 392–406. https://doi.org/10.1038/nrg.2016.27

Chen F, Ahimaz P, Nguyen QM et al (2024) Phenotype driven molecular genetic test recommendation for diagnosing pediatric rare disorders. NPJ Digit Med 7: 333. https://doi.org/10.1038/s41746-024-01331-1

Chenchula S, Atal S, Uppugunduri CRS (2024) A review of real-world evidence on preemptive pharmacogenomic testing for preventing adverse drug reactions: a reality for future health care. Pharmacogenomics J 24(2):9. https://doi.org/10.1038/s41397-024-00326-1

Choi SW, O'Reilly PF (2019) PRSice-2: polygenic risk score software for Biobank-scale data. Gigascience 8(7):giz082. https://doi.org/10.1093/gigascience/giz082

Choudhury PP, Brook MN, Hurson AN et al (2021) Comparative validation of the BOADICEA and Tyrer-Cuzick breast cancer risk models incorporating classical risk factors and polygenic risk in a population-based prospective cohort of women of European ancestry. Breast Cancer Res 23(1):22. https://doi.org/10.1186/s13058-021-01399-7

Claussnitzer M, Cho JH, Collin R et al (2020) A brief history of human disease genetics. Nature 577(7789): 179–189. https://doi.org/10.1038/s41586-019-1879-7

Coelho T, Cheng G, Lewis S et al (2025) Pharmacogenomic assessment of genes implicated in Thiopurine metabolism and toxicity in a UK cohort of pediatric patients with inflammatory bowel disease. Inflamm Bowel Dis 31(2):362–375. https://doi.org/10.1093/ibd/izae126

Cohen ASA, Berrios CD, Zion TN et al (2024) Genomic answers for kids: toward more equitable access to genomic testing for rare diseases in rural populations. Am J Hum Genet 111:825–832. https://doi.org/10.1016/j.ajhg.2024.03.016

Council of the European Union (2015) Council conclusions on personalised medicine for patients (2015/C 421/03). Off J Eur Union, C421:17–23. https://eur-lex.europa.eu/legal-content/EN/TXT/PDF/?uri=CELEX:52015XG1217(01)

Covington N, Luetkemeyer AF, Imperial MZ et al (2025) Pharmacogenetics of plasma dolutegravir exposure during 1-month rifapentine/isoniazid treatment of latent tuberculosis. Pharmacogenet Genomics. https://doi.org/10.1097/FPC.0000000000000562

CPIC – Clinical Pharmacogenetics Implementation Consortium guidelines (2024) https://cpicpgx.org/guidelines/. Assessed 15 Nov 2024

Damask A, Steg PG, Schwartz GG et al (2021) Patients With high genome-wide polygenic risk scores for coronary artery disease may receive greater clinical benefit from Alirocumab treatment in the ODYSSEY OUTCOMES trial. Circulation 141(8):624–636. https://doi.org/10.1161/CIRCULATIONAHA.119.044434

Danielson PB (2002) The cytochrome P450 superfamily: biochemistry, evolution and drug metabolism in humans. Curr Drug Metab 3(6):561–597. https://doi.org/10.2174/1389200023337054

Danish National Genome Center (2025) https://www.eng.ngc.dk. Assessed 13 Feb 2025

de With M, Sadlon A, Cecchin E et al (2023) Implementation of dihydropyrimidine dehydrogenase deficiency testing in Europe. ESMO Open 8(2):101197. https://doi.org/10.1016/jesmoop2023101197

Delaye J, Cacciatore P, Kole A (2022) Valuing the "burden" and impact of rare diseases: a scoping review. Front Pharmacol 13:914338. https://doi.org/10.3389/fphar.2022.914338

Denny JC, Collins FS (2021) Precision medicine in 2030-seven ways to transform healthcare. Cell 184(6): 1415–1419. https://doi.org/10.1016/j.cell.2021.01.015

Denton N, Mulberg AE, Molloy M et al (2022) Sharing is caring: a call for a new era of rare disease research and development. Orphanet J Rare Dis 17:389. https://doi.org/10.1186/s13023-022-02529-w

Desai A, Reddy NK, Subbiah V (2023) Top advances of the year: precision oncology. Cancer 129(11): 1634–1642. https://doi.org/10.1002/cncr.34743

Dhieb D, Bastaki K (2025) Pharmaco-multiomics: a new frontier in precision psychiatry. Int J Mol Sci 26(3):10:82. https://doi.org/10.3390/ijms26031082

Diaz LA, Bardelli A (2014) Liquid biopsies: genotyping circulating tumor DNA. J Clin Oncol 32:579–586. https://doi.org/10.1200/jco.2012.45.2011

Dixon P, Keeney E, Taylor JC et al (2022) Can polygenic risk scores contribute to cost-effective cancer screening? A system atic review. Genet Med 24(8): 1604–1617. https://doi.org/10.1016/j.gim.2022.04.020

DPWG – Dutch Pharmacogenetics Working Group guideline recommendations (2021) https://www.knmp.nl/

dossiers/farmacogenetica/pharmacogenetics Assessed 15 Nov 2024

Drilon A, Laetsch TW, Kummar S et al (2018) Efficacy of Larotrectinib in TRK fusion–positive cancers in adults and children. N Engl J Med 378(8):731–739. https://doi.org/10.1056/NEJMoa1714448

Duarte F, Oliveira L, Baixia M et al (2025) Unique genetic profiles in hypertrophic cardiomyopathy patients from São Miguel Island (Azores, Portugal). Clin Genet 107: 434–440. https://doi.org/10.1111/cge.14656

Dudbridge F (2013) Power and predictive accuracy of polygenic risk scores. PLoS Genet 9(3):e1003348. https://doi.org/10.1371/journal.pgen.1003348

Eichler EE (2019) Genetic variation, comparative genomics, and the diagnosis of disease. N Engl J Med 4(381):64–74. https://doi.org/10.1056/nejmra1809315

EMA – European Medicines Agency – Committee for Medicinal Products for Human Use (CHMP) (2011) Guideline on the use of pharmacogenetic methodologies in the pharmacokinetic evaluation of medicinal products. EMA/CHMP/37646/2009. https://www.tga.gov.au/sites/default/files/2024-08/guideline_on_the_use_of_methodologies_in_the_pharmacokinetic_evaluation_of_medicinal_products.pdf. Accessed 10 Dec 2024

EMA – European Medicines Agency (2020) EMA recommendations on DPD testing prior to treatment with fluorouracil, capecitabine, tegafur and flucytosine. Eur Med Agency 2020;31(April):3. https://www.fagg.be/sites/default/files/content/dhpc_fluorouracil_nl_-_website.pdf

Esfahani K, Roudaia L, Buhlaiga N et al (2020) A review of cancer immunotherapy: from the past, to the present, to the future. Curr Oncol 27(Suppl 2):S87–S97. https://doi.org/10.3747/co.27.5223

European Commission (2024a) Genome of Europe project launched: first step towards a European reference genome. Digital Strategy. https://digital-strategy.ec.europa.eu/en/news/genome-europe-project-launched-first-step-towards-european-reference-genome. Assessed 25 Mar 2025

European Commission (2024b) European '1+ Million Genomes' Initiative. https://digital-strategy.ec.europa.eu/en/policies/1-million-genomes. Accessed 10 Mar 2025

EURORDIS Rare Disease Europe (2025) What is a rare disease? https://www.eurordis.org/information-support/what-is-a-rare-disease/. Accessed 9 Mar 2025

Evans DG, Brentnall A, Byers H et al (2017) The impact of a panel of 18 SNPs on breast cancer risk in women attending a UK familial screening clinic: a case-control study. J Med Genet 54(2):111–113. https://doi.org/10.1136/jmedgenet-2016-104125

Faye F, Crocione C, Anido de Peña R et al (2024) Time to diagnosis and determinants of diagnostic delays of people living with a rare disease: results of a rare barometer retrospective patient survey. Eur J Hum Genet 32:1116–1126. https://doi.org/10.1038/s41431-024-01604-z

Fokstuen S, Makrythanasis P, Hammar E et al (2016) Experience of a multidisciplinary task force with exome sequencing for Mendelian disorders. Hum Genomics 10(1):24. https://doi.org/10.1186/s40246-016-0080-4

Forgetta V, Keller-Baruch J, Forest M et al (2020) Development of a polygenic risk score to improve screening for fracture risk: a genetic risk prediction study. PLoS Med 17(7):e1003152. https://doi.org/10.1371/journal.pmed.1003152

Friedrich B, Vindrola-Padros C, Lucassen AM et al (2024) "A very big challenge": a qualitative study to explore the early barriers and enablers to implementing a national genomic medicine service in England. Front Genet 14:1282034. https://doi.org/10.3389/fgene.2023.1282034

GA4GH (2025) National Initiatives Forum. https://wwwga4ghorg/what-we-do/national-initiatives forum/. Assessed 10 Apr 2025

Garutti M, Foffano L, Mazzeo R et al (2023) Hereditary cancer syndromes: a comprehensive review with a visual tool. Genes 14(5):1025. https://doi.org/10.3390/genes14051025

Ge T, Chen CY, Neale BM et al (2017) Phenome-wide heritability analysis of the UK biobank. PLoS Genet 13(4):e1006711. https://doi.org/10.1371/journal.pgen.1006711

Genome of Europe (2025) https://digital-strategy.ec.europa.eu/en/news/genome-europe-project-launched-first-step-towards-european-reference-genome. Assessed 9 Apr 2025

Genomics England (2021) What we do and why we do it: genomics England's Innovation Showcase 21. Available at: https://www.genomicsengland.co.uk/news/innovation-showcase-21. Accessed 5 Feb 2025

Global Genes (2025) Rare disease facts. https://globalgenes.org/rare-disease-facts/. Accessed 9 Mar 2025

Glotov OS, Chernov AN, Glotov AS (2023) Human exome sequencing and prospects for predictive medicine: analysis of international data and own experience. J Pers Med 13:1236. https://doi.org/10.3390/jpm13081236

Gooden A (2024) A pathway to strengthening open science: comments on the draft South African Ethics in Health Research Guidelines. Front Pharm acol 15: 1304950. https://doi.org/10.3389/fphar.2024.1304950

Gorukmez O, Gorukmez O, Topak A (2023) Clinical exome sequencing findings in 1589 patients. Am J Med Genet A 191:1557–1564. https://doi.org/10.1002/ajmg.a.63190

Gouda MA, Nelson BE, Buschhorn L, Wahida A, Subbiah V (2023) Tumor-agnostic precision medicine from the AACR GENIE database: clinical implications. Clin Cancer Res 29(15):2753–2760. https://doi.org/10.1158/1078-0432.ccr-23-0090

Green ED, Gunter C, Biesecker LG et al (2020) Strategic vision for improving human health at the forefront of genomics. Nature 586(7831):683–692. https://doi.org/10.1038/s41586-020-2817-4

Gross T, Daniel J (2018) Overview of pharmacogenomic testing in clinical practice. Ment Health Clin 8(5): 235–241. https://doi.org/10.9740/mhc.2018.09.235

Guengerich FP (2017) Intersection of the roles of cytochrome P450 enzymes with xenobiotic and endogenous substrates: relevance to toxicity and drug interactions. Chem Res Toxicol 30(1):2–12. https://doi.org/10.1021/acs.chemrestox.6b00226

Hartley T, Lemire G, Kernohan KD et al (2020) New diagnostic approaches for undiagnosed rare genetic diseases. Annu Rev Genomics Hum Genet 21:351–372. https://doi.org/10.1002/ajmg.a.63190

Hartley T, Soubry É, Acker M et al (2023) Bridging clinical care and research in Ontario, Canada: maximizing diagnoses from reanalysis of clinical exome sequencing data. Clin Genet 103:288–300. https://doi.org/10.1111/cge.14262

Hartley T, Marshall D, Acker M et al (2024) Evaluation of the diagnostic accuracy of exome sequencing and its impact on diagnostic thinking for patients with rare disease in a publicly funded health care system: a prospective cohort study. Genet Med 26:101012. https://doi.org/10.1016/j.gim.2023.101012

Henderson S (2019) 100,000 genomes project: validation and reporting guidance for DPYD variants. Genomics England.10.06.2019. Available from https://www.genomicsengland.co.uk/assets/forms/100000-Genomes-Project-Validation-and-Reporting-Guidance-for-DPYD-variants-v1.0-signed.pdf

Henricks LM, Lunenburg CATC, de Man FM et al (2018) DPYD genotype-guided dose individualisation of fluoropyrimidine therapy in patients with cancer: a prospective safety analysis. Lancet Oncol 19(11): 1459–1467. https://doi.org/10.1016/S1470-2045(18)30686-7

Hertz DL, Ramsey LB, Gopalakrishnan M, Leeder JS, Van Driest SL (2021) Analysis approaches to identify pharmacogenetic associations With pharmacodynamics. Clin Pharmacol Ther 110(3):589–594. https://doi.org/10.1002/cpt.2312

Hetherington S, McGuirk S, Powell G et al (2001) Hypersensitivity reactions during therapy with the nucleoside reverse transcriptase inhibitor abacavir. Clin Ther 23(10):1603–1614. https://doi.org/10.1016/s0149-2918(01)80132-6

Hoeben A, Joosten EAJ, van den Beuken-van Everdingen MHJ (2021) Personalized medicine: recent Progress in cancer therapy. Cancer 13(2):242. https://doi.org/10.3390/cancers13020242

Hong J, Lee D, Hwang A et al (2024) Rare disease genomics and precision medicine. Genomics Inform 22:28. https://doi.org/10.1186/s44342-024-00032-1

Hou YC, Yu HC, Martin R et al (2020) Precision medicine integrating whole-genome sequencing, comprehensive metabolomics, and advanced imaging. Proc Natl Acad Sci USA 117:3053–3062. https://doi.org/10.1073/pnas.1909378117

Howley C, Haas MA, Al Muftah WA et al (2025) The expanding global genomics landscape: converging priorities from national genomics programs. Am J Hum Genet 112(4):751–763. https://doi.org/10.1016/j.ajhg.2025.02.008

Hsu JS, Wu DC, Shih SH et al (2024) Complete genomic profiles of 1496 Taiwanese reveal curated medical insights. J Adv Res 66:197–207. https://doi.org/10.1016/j.jare.2023.12.018

Hughes AR, Spreen WR, Mosteller M et al (2008) Pharmacogenetics of hypersensitivity to abacavir: from PGx hypothesis to confirmation to clinical utility. Pharmacogenomics J 8(6):365–374. https://doi.org/10.1038/tpj.2008.3

ICPerMed – International Consortium for Personalised Medicine (2015) https://www.icpermed.eu/. Accessed 15 Feb 2025

ICPerMed – International Consortium for Personalised Medicine (2023) Guide for the next generation of researchers: EP PerMed Training and Edu cation Framework. Version 2.0. [pdf]. https://www.icpermed.eu/wp-content/uploads/EP_PerMed_guide_V2.0.pdf. Accessed 1 Feb 2025

IGVF Consortium (2024) Deciphering the impact of genomic variation on function. Nature 633:47–57. https://doi.org/10.1038/s41586-024-07510-0

Imai M, Nakamura Y, Yoshino T (2025) Transforming cancer screening: the potential of multi-cancer early detection (MCED) technologies. Int J Clin Oncol 30(2):180–193. https://doi.org/10.1007/s10147-025-02694-5

IMPaCT (2025) Precision Medicine Infrastructure associated with Science and Technology. https://impact.isciii.es/en/. Assessed 10 Feb 2025

INCa & HAS (2018) Recherche de déficit en dihydropyrimidine déshydrogenase en vue de prévenir certaines toxicités sévères survenant sous traitement comportant des fluoropyrimidines (5-fluorouracile), Recommandations et référentiels INCa, HAS, Dec 2018. https://www.has-sante.fr/upload/docs/application/pdf/2018-12/recherche_dun_deficit_en_dihydropyrimidine_deshydrogenase_visant_a_prevenir_certaines_toxicites_severes_associees_aux_traite.pdf. Accessed 1 Feb 2025

Inshutiyimana S, Ramadan N, Razzak RA et al (2024) Pharmacogenomics revolutionizing cardiovascular therapeutics: a narrative review. Health Sci Rep 7(10):e70139. https://doi.org/10.1002/hsr2.70139

Iyshwarya BK, Veerabathiran R (2024) Techniques for the diagnosis of rare genetic disorders. In: Umair M, Rafeeq M, Alam Q (eds) Rare genetic disorder s. Springer, Singapore, pp 35–60. https://doi.org/10.1007/978-981-99-9323-9_2

Jackson RL, Heyrend C, Bucher B et al (2025) Impact of pharmacogenomic testing in pediatric heart and kidney transplant. Pediatr Transplant 29(2):e70044. https://doi.org/10.1111/petr.70044

Jameson A, Tomlinson J, Medlinskiene K et al (2024) Normal ising the implementation of Pharmacogenomic

(PGx) testing in adult mental health settings: a theory-based systematic review. J Pers Med 14(10):1032. https://doi.org/10.3390/jpm14101032

Jayasinghe D, Eshetie S, Beckmann K et al (2024) Advancements and limitations in polygenic risk score methods for genomic prediction: a scoping review. Hum Genet 143(12):1401–1431. https://doi.org/10.1007/s00439-024-02716-8

Jeong H, Roghanizad AR, Master H et al (2025) Data from the all of us research program reinforces existence of activity inequality. NPJ Digit Med 8(1):8. https://doi.org/10.1038/s41746-024-01358-4

Ji Y, Biernacka J, Snyder K, Drews M et al (2012) Catechol O-methyltransferase pharmacogenomics and selective serotonin reuptake inhibitor response. Pharmacogenomics J 12(1):78–85. https://doi.org/10.1038/tpj.2010.69

Ji Y, Long J, Kweon SS et al (2021) Incorporating European GWAS findings improve polygenic risk prediction accuracy of breast cancer among Ea st Asians. Genet Epidemiol 45(5):471–484. https://doi.org/10.1002/gepi.22382

Jiang S, Wang H, Gu Y (2023) Genome sequencing for newborn screening – an effective approach for tackling rare diseases. JAMA New Open 6(e2331):141. https://doi.org/10.1001/jamanetworkopen.2023.31141

Jiao Q, Sun H, Zhang H et al (2019) The combination of whole-exome sequencing and copy number variation sequencing enables the diagnosis of rare neurological disorders. Clin Genet 96:140–150. https://doi.org/10.1111/cge.13548

Jobanputra V, Schroeder B, Rehm HL et al (2024) Advancing access to genome sequencing for rare genetic disorders: recent progress and call to action. NPJ Genom Med 9:23. https://doi.org/10.1038/s41525-024-00410-2

Johnson KS, Conant EF, Soo MS (2021) Molecular subtypes of breast cancer: a review for breast radiologists. J Breast Imaging 3(1):12–24. https://doi.org/10.1093/jbi/wbaa110

Kasten A, Cascorbi I (2024) Understanding the impact of ABCG2 polymorphisms on drug pharmacokinetics: focus on rosuvastatin and allopurinol. Expert Opin Drug Metab Toxicol 20(6):519–528. https://doi.org/10.1080/17425255.2024.2362184

Katara P, Yadav A (2019) Pharmacogenes (PGx-genes): current understanding and future directions. Gene 718:144050. https://doi.org/10.1016/j.gene.2019.144050

Kennedy AM, Griffiths AM, Muise AM et al (2024) Landscape of TPMT and NUDT15 pharmacogenetic variation in a cohort of Canadian pediatric inflammatory bowel disease patients. Inflamm Bowel Dis 30(12):2418–2427. https://doi.org/10.1093/ibd/izae109

Kernohan KD, Boycott KM (2024) The expanding diagnostic toolbox for rare genetic diseases. Nat Rev Genet 25:401–415. https://doi.org/10.1038/s41576-023-00683-w

Klein HG, Bauer P, Hambuch T (2014) Whole genome sequencing (WGS), whole exome sequencing (WES) and clinical exome sequencing (CES) in patient care. J Lab Med 38:221–230. https://doi.org/10.1515/labmed-2014-0025

Koch S, Schmidtke J, Krawczak M et al (2023) Clinical utility of polygenic risk scores: a critical 2023 appraisal. J Community Genet 14(5):471–487. https://doi.org/10.1007/s12687-023-00645-z

Kumuthini J, Zick B, Balasopoulou A et al (2022) The clinical utility of polygenic risk scores in genomic medicine practices: a systematic re view. Hum Genet 141(11):1697–1704. https://doi.org/10.1007/s00439-022-02452-x

Kurki MI, Karjalainen J, Palta P et al (2023) FinnGen provides genetic insights from a well-phenotyped isolated population. Nature 613(7944):508–518. https://doi.org/10.1038/s41586-022-05473-8

Lalatović N, Ždralević M, Antunović T et al (2023) Genetic polymorphisms in ABCB1 are correlated with the increased risk of atorvastati n-induced muscle side effects: a cross-sectional study. Sci Rep 13(1): 17895. https://doi.org/10.1038/s41598-023-44792-2

Läll K, Mägi R, Morris A et al (2017) Personalized risk prediction for type 2 diabetes: the potential of genetic risk scores. Genet Med 19(3):322–329. https://doi.org/10.1038/gim.2016.103

Le DT et al (2017) Mismatch repair deficiency predicts response of solid tumors to PD-1 blockade. Science 357(6349):409–413. https://doi.org/10.1126/science.aan6733

Lee H, Deignan JL, Dorrani N et al (2014) Clinical exome sequencing for genetic identification of rare Mendelian disorders. JAMA 312:1880–1887. https://doi.org/10.1001/jama.2014.14604

Leon SA, Shapiro B, Sklaroff DM, Yaros MJ (1977) Free DNA in the serum of cancer patients and the effect of therapy. Cancer Res 37:646–650

Leong IUS, Cabrera CP, Cipriani V et al (2025) Large-scale pharmacogenomics Analy sis of patients With cancer within the 100,000 genomes project combining whole-genome sequencing and medical records to inform clinical practice. J Clin Oncol 43(6):682–693. https://doi.org/10.1200/JCO.23.02761

Lewis ACF, Green RC (2021) Polygenic risk scores in the clinic: new perspectives needed on familiar ethical issues. Genome Med 13(1):14. https://doi.org/10.1186/s13073-021-00829-7

Lewis CM, Vassos E (2020) Polygenic risk scores: from research tools to clinical instruments. Genome Med 12:44. https://doi.org/10.1186/s13073-020-00742-5

Lewis C, Boardman F, Buchanan J et al (2024) Exploring the feasibility, acceptability and impact of genomic newborn screening for rare diseases in England: a study protocol for the generation study – process and impact evaluation. medRxiv [Preprint]. https://doi.org/10.1101/2024.05.14.24307295

Li Y, Yan B, He S (2023) Advances and challenges in the treatment of lung cancer. Biomed Pharmacother 169: 115891. https://doi.org/10.1016/j.biopha.2023.115891

Lin YH, Hung CC, Lin GC et al (2025) Utilizing polygenic risk score for breast cancer risk prediction in a Taiwanese population. Cancer Epidemiol 94:102701. https://doi.org/10.1016/j.canep.2024.102701

Lloyd-Jones LR, Zeng J, Sidorenko J et al (2019) Improved polygenic prediction by Bayesian multiple regression on summary statistics. Nat Commun 10(1): 5086. https://doi.org/10.1038/s41467-019-12653-0

Lo EAG, Wan YM, Ho CSH (2024) The promise and challenges of pharmacogenomics in psychiatry. Ann Acad Med Singapore 53(1 2):710–712. https://doi.org/10.47102/annals-acadmedsg.2024367

Lopes MF, Merchant A, Costa A et al (2022) Genomics in Healthcare issues for implementation. Policy Brief. Available at: https://b1mg-project.eu/images/pdf/Policy_Brief_Genomics_in_Healthcare_2022.pdf. Accessed 15 Apr 2025

Loriot MA, Ciccolini J, Thomas F et al (2018) Dépistage du défic it en dihydropyrimidine déshydrogénase (DPD) et sécurisation des chimiothérapies à base de fluoropyrimidines: mise au point et recommandations nationales du GPCO-Unicancer et du RNPGx [Dihydropyrimidine déhydrogenase (DPD) deficiency screening and securing of fluoropyrimidine-based c hemotherapies: update and recommendations of the French GPCO-Unicancer and RNPGx networks]. Bull Cancer 105(4):397–407. https://doi.org/10.1016/j.bulcan.2018.02.001

Loscocco F, Visani G, Ruzzo A et al (2021) Clinical relevance of ABCB1, ABCG2, and ABCC2 gene polymorphisms in chronic myeloid leukemia patients treated With Nilotinib. Front Oncol 11:672287. https://doi.org/10.3389/fonc.2021.672287

Lunenburg CATC, Henricks LM, Guchelaar HJ et al (2016) Prospective DPYD genotyping to reduce the risk of fluoropyrimidine-induced severe toxicity. ready for prime time. Eur J Cancer 54:40–48. https://doi.org/10.1016/j.ejca.2015.11.008

MacCarrick G, Aradhya S, Bailey M et al (2024) Clinical utility of comprehensive gene panel testing for common and rare causes of skeletal dysplasia and other skeletal disorders: results from the largest cohort to date. Am J Med Genet A 194:e63646. https://doi.org/10.1002/ajmg.a.63646

Mahajan R, Tyagi AK (2024) Pharmacogenomic insights into tuberculosis treatment shows the NAT2 genetic variants linked to hepatotoxicity risk: a systematic review and meta-analysis. BMC Genomic Data 25(1): 103. https://doi.org/10.1186/s12863-024-01286-y

Mai CW, Sridhar SB, Karattuthodi MS et al (2024) Scoping review of enablers and challenges of implementing pharmacogenomics testing in the primary care settings. BMJ Open 14(11):e087064. https://doi.org/10.1136/bmjopen-2024-087064

Mak TSH, Porsch RM, Choi SW, Zhou X, Sham PC (2017a) Polygenic scores via penalized regression on summary statistics. Genet Epidemiol 41(6):469–480. https://doi.org/10.1002/gepi.22050

Mak TSH, Porsch RM, Choi SW et al (2017b) Polygenic scores via penalized regression on summary statistics. Genet Epidemiol 41(6):469–480. https://doi.org/10.1002/gepi.22050

Manolio TA, Rowley R, Williams MS et al (2019) Opportunities, resources, and techniques for implementing genomics in clinical care. Lancet (London, England) 394(10197):511–520. https://doi.org/10.1016/S0140-6736(19)31140-7

Marabelle A et al (2020) Pembrolizumab for patients with TMB-high cancers: updated KEYNOTE-158 study results. J Clin Oncol 38(15_suppl):3116–3116. https://doi.org/10.1200/JCO.2020.38.15_suppl.3116

Marshall DA, Gerber B, Lorenzetti DL et al (2023) Are we capturing the socioeconomic burden of rare genetic disease? A scoping review of economic evaluations and cost-of-illness studies. Pharmacoeconomics 41: 1563–1588. https://doi.org/10.1007/s40273-023-01308-0

Marston NA, Kamanu FK, Nordio F et al (2020) Predicting benefit from evolocumab therapy in patients With atherosclerotic disease using a genetic risk score: results from the FOURIER trial. Circulation 141(8): 616–623. https://doi.org/10.1161/CIRCULATIONAHA.119.043805

Martin AR, Kanai M, Kamatani Y et al (2019) Clinical use of current polygenic risk scores may exacerbate health disparities. Nat Genet 51(4):584–591. https://doi.org/10.1038/s41588-019-0379-x

Mauriello A, Ascrizzi A, Molinari R et al (2023) Pharmacogenomics of Cardiovascular Drugs for Atherothro mbotic. Thromboembolic and atherosclerotic risk. Genes 14(11):2057. https://doi.org/10.3390/genes14112057

Mavaddat N, Michailidou K, Dennis J et al (2019) Polygenic risk scores for prediction of breast cancer and breast cancer subtypes. Am J Hum Genet 104(1): 21–34. https://doi.org/10.1016/j.ajhg.2018.11.002

Mbarek H, Gandhi GH, Selvaraj S et al (2022) Qatar genome: insights on genomics from the Middle East. Hum Mutat 43(4):499–510. https://doi.org/10.1002/humu.24336

Milani L, Alver M, Laur S et al (2025) The Estonian Biobank's journey from biobanking to personalized medicine. Nat Commun 16(1):3270. https://doi.org/10.1038/s41467-025-58465-3

Mishra S, Venkatesh MP (2024) Rare disease clinical trials in the European Union: navigating regulatory and clinical challenges. Orphanet J Rare Dis 19:285. https://link.springer.com/article/10.1186/s13023-024-03146-5

Modi S et al (2020) Trastuzumab deruxtecan in previously treated HER2-positive breast cancer. N Engl J Med 382(7):610–621. https://doi.org/10.1056/NEJMoa1914510

Morales J, Welter D, Bowler EH et al (2018) A standardized framework for representation of ancestry data in genomics studies, with applicati on to the NHGRI-EBI GWAS catalog. Genome Biol 19(1):21. https://doi.org/10.1186/s13059-018-1396-2

Morris SA, Alsaidi AT, Verbyla A et al (2022) Cost effectiveness of Pharmacogenetic testing for drugs with clinical Pharmacogenetics implementation consortium (CPIC) guidelines: a systematic review. Clin Pharmacol Ther 112(6):1318–1328. https://doi.org/10.1002/cpt.2754

Mosch R, van der Lee M, Guchelaar HJ et al (2025) Pharmacogenetic panel testing: a review of current practice and potential for clinical implementation. Annu Rev Pharmacol Toxicol 65(1):91–109. https://doi.org/10.1146/annurev-pharmtox-061724-080935

Mukherjee S (2019) Genomics-guided immunotherapy for precision medicine in cancer. Cancer Biother Radiopharm 34(8):487–497. https://doi.org/10.1089/cbr.2018.2758

Nakamichi K, Huey J, Sangermano R et al (2024) Targeted long-read sequencing enriches disease-relevant genomic regions of interest to provide complete Mendelian disease diagnostics. JCI Insight 9: e183902. https://doi.org/10.1172/jci.insight.183902

Ng KWP, Chin HL, Chin AXY et al (2022) Using gene panels in the diagnosis of neuromuscular disorders: a mini-review. Front Neurol 13:997551. https://doi.org/10.3389/fneur.2022.997551

NHS England (2024) Thousands more lung cancer patients to get innovative blood test as part of NHS pilot. https://www.england.nhs.uk/2024/03/thousands-more-lung-cancer-patients-to-get-innovative-blood-test-as-part-of-nhs-pilot/. Accessed 16 Apr 2025

NHS, Health Research Authority (2023) Pharmacogenetics Roll Out – Gauging Response to Service (PROGRESS). https://www.hra.nhs.uk/planning-and-improving-research/application-summaries/research-summaries/pharmacogenetics-roll-out-gauging-response-to-service-progress/. Assessed 15 Mar 2023

Ohori S, Tsuburaya RS, Kinoshita M et al (2021) Long-read whole-genome sequencing identified a partial MBD5 deletion in an exome-negative patient with neurodevelopmental disorder. J Hum Genet 66:697–705. https://doi.org/10.1038/s10038-020-00893-8

Oxnard GR, Paweletz CP, Kuanf Y et al (2016) Noninvasive detection of response and resistance in EGFR-mutant lung cancer using circulating tumor DNA profiling. Clin Cancer Res 22(4):961–970. https://pmc.ncbi.nlm.nih.gov/articles/PMC3959249/

Padilla-Martínez F, Collin F, Kwasniewski M, Kretowski A (2020) Systematic review of polygenic risk scores for type 1 and type 2 diabetes. Int J Mol Sci 21(5): 1703. https://doi.org/10.3390/ijms21051703

Padmanabhan S, du Toit C, Dominiczak AF (2023) Cardiovascular precision medicine – a pharmacogenomic perspective. Camb Prisms Precis Med 1:e28. https://doi.org/10.1017/pcm.2023.17

Peruzzi E, Roncato R, De Mattia E et al (2025) Implementation of pre-emptive testing of a pharmacogenomic panel in clinical practice: where do we stand? Br J Clin Pharmacol 91(2):270–282. https://doi.org/10.1111/bcp.15956

Peter J, Choshi P, Lehloenya RJ (2019) Drug hypersensitivity in HIV infection. Curr Opin Allergy Clin Immunol 19(4):272–282. https://doi.org/10.1097/ACI.0000000000000545

Peterson RE, Kuchenbaecker K, Walters RK et al (2019) Genome-wide association studies in ancestrally diverse populations: opportunities, methods, pitfalls, and recommendations. Cell 179(3):589–603. https://doi.org/10.1016/j.cell.2019.08.051

Phillips E, Mallal S (2007) Drug hypersensitivity in HIV. Curr Opin Allergy Clin Immunol 7(4):324–330. https://doi.org/10.1097/aci.0b013e32825ea68a

Pirmohamed M (2023) Pharmacogenomics: current status and future perspectives. Nat Rev Genet 24(6): 350–362. https://doi.org/10.1038/s41576 022 00572 8

Polygenic Risk Score Task Force of the International Common Disease Alliance (2021). Responsible use of polygenic risk scores in the clinic: potential benefits, risks and gaps. Nature medicine, 27(11):1876–1884. https://doi.org/10.1038/s41591-021-01549-6

Privé F, Arbel J, Vilhjálmsson BJ (2021) LDpred2: better, faster, stronger. Bioinf (Oxf) 36(22–23):5424–5431. https://doi.org/10.1093/bioinformatics/btaa1029

Quintal A, Carreau I, Grenier AD et al (2023) An ethics action plan for rare disease care: participatory action research approach. J Particip Med 15:e46607. https://doi.org/10.2196/46607

Rafehi H, Read J, Szmulewicz DJ et al (2023) An intronic GAA repeat expansion in FGF14 causes the autosomal-dominant adult-onset ataxia SCA27B / ATX-FGF14. Am J Hum Genet 110:105–119. https://doi.org/10.1016/j.ajhg.2022.11.015

Rahit KMTH, Tarailo-Graovac M (2020) Genetic modifiers and rare Mendelian disease. Genes (Basel) 11:239. https://doi.org/10.3390/genes11030239

Rehm HL, Page AJH, Smith L et al (2021) GA4GH: international policies and standards for data sharing across genomic research and healthcare. Cell Genom 1(2):100029. https://doi.org/10.1016/j.xgen.2021.100029

Relling M, Evans W (2015) Pharmacogenomics in the clinic. Nature 526:343–350. https://www.nature.com/articles/nature15817

Ridolfi F, Amorim G, Peetluk LS et al (2024) Regional prospective observational research in tuberculosis (RePORT)–Brazil consortium prediction models for adverse drug reactions during tuberculosis treatment in Brazil. J Infect Dis 229(3):813–823. https://doi.org/10.1093/infdis/jiae025

Ritchie SC, Taylor HJ, Liang Y et al (2024) Integrated clinical risk prediction of type 2 diabetes with a multifactorial polygenic risk score. MedRxiv. https://doi.org/10.1101/2024.08.22.24312440

Roberts AM, DiStefano MT, Riggs ER et al (2024) Toward robust clinical genome interpretation: developing a consistent terminology to characterize

Mendelian disease-gene relationships-allelic requirement, inheritance modes, and disease mechanisms. Genet Med 26:101029. https://doi.org/10.1016/j.gim.2023.101029

Ronald A, Gui A (2024) The potential and translational application of infant genetic research. Nat Genet 56:1346–1354. https://doi.org/10.1038/s41588-024-01822-7

Ross L (2024) From Mendel's laws to non-Mendelian inheritance. Nat Rev Genet 25:677. https://doi.org/10.1038/s41576-024-00753-7

Sáenz SS, Arias B, Hosomichi K, Romero VI (2021) The limits of clinical findings in similar phenotypes, from carpenter to ATRX syndrome using a whole exome sequencing approach: a case review. Hum Genomics 15:49. https://doi.org/10.1186/s40246-021-00348-x

Salatin S, Shafiee-Kandjani AR, Hamidi S et al (2025) Individualized psychiatric care: integration of therapeutic drug monitoring, pharmacogenomics, and biomarkers. Personal Med 22(1):29–44. https://doi.org/10.1080/17410541.2024.2442897

Salyakina D, Roy S, Wang et al (2019) Results and challenges of Cytochrome P450 2D6 (CYP2D6) testing in an ethnically diverse South Florida population. Mol Genet Genomic Med 7(9):e922. https://doi.org/10.1002/mgg3.922

Sarömba JA, Müller JP, Tupiec J et al (2025) Solanidine-derived CYP2D6 phenotyping elucidates phenoconversion in multimedicated geriatric patients. Br J Clin Pharmacol. https://doi.org/10.1002/bcp.70004

Schmid P, Adams S, Rugo HS et al (2020) Atezolizumab and nab-paclitaxel in advanced triple-negative breast cancer (IMpassion130). N Engl J Med 379(22):2108–2212. https://doi.org/10.1056/NEJMoa1809615

Schmidt A, Danyel M, Grundmann K et al (2024) Next-generation phenotyping integrated in a national framework for patients with ultrarare disorders improves genetic diagnostics and yields new molecular findings. Nat Genet 56:1644–1653. https://doi.org/10.1038/s41588-024-01836-1

Schobers G, Derks R, den Ouden A et al (2024) Genome sequencing as a generic diagnostic strategy for rare disease. Genome Med 16:32. https://doi.org/10.1186/s13073-024-01301-y

Schuler BA, Nelson ET, Koziura M et al (2022) Lessons learned: next-generation sequencing applied to undiagnosed genetic diseases. J Clin Invest 132:e154942. https://doi.org/10.1172/jci154942

Shen X, Kellogg R, Panyard DJ et al (2023) Multi-omics microsampling for the profiling of lifestyle-associated changes in health. Nat Biomed Eng. https://doi.org/10.1038/s41551-022-00999-8

Shendure J, Findlay GM, Snyder MW (2019) Genomic medicine-progress, pitfalls, and promise. Cell 177(1):45–57. https://doi.org/10.1016/j.cell.2019.02.003

Sherman CA, Claw KG, Lee SB (2024) Pharmacogenetic analysis of structural variation in the 1000 genomes project using whole genome sequences. Sci Rep 14(1):22774. https://doi.org/10.1038/s41598-024-73748-3

Shieh Y, Hu D, Ma L et al (2017) Joint relative risks for estrogen receptor-positive breast cancer from a clinical model, polygenic risk score, and sex hormones. Breast Cancer Res Treat 166(2):603–612. https://doi.org/10.1007/s10549-017-4430-2

Simona A, Song W, Bates DW et al (2023) Polygenic risk scores in pharmacogenomics: opportunities and challenges-a mini review. Front Genet 14:1217049. https://doi.org/10.3390/ijns8030040

Singh S, Stocco G, Theken KN (2024) Pharmacogenomics polygenic risk score: ready or not f or prime time? Clin Transl Sci 17(8):e13893. https://doi.org/10.1111/cts.13893

Stanton MB, Solinski MA, Hanauer SB (2025) Genetic polymorphisms impacting clinical pharmacology of drugs used to treat inflammatory bowel disease: a precursor to multi-omics approach to precision medicine. Expert Rev Clin Immunol:1–12. https://doi.org/10.1080/1744666X.2025.2461584

Stark Z, Scott RH (2023) Genomic newborn screening for rare diseases. Nat Rev Genet 24:755–766. https://doi.org/10.1038/s41576-023-00621-w

Steyaert W, Sagath L, Demidov G et al (2024) Unravelling undiagnosed rare disease cases by HiFi long-read genome sequencing. medRxiv [Preprint]. https://doi.org/10.1101/2024.05.03.24305331

Stingl JC, Viviani R (2025) Pharmacogenetic guided drug therapy – how to deal with phenoconversion in polypharmacy. Expert Opin Drug Metab Toxicol:1–9. https://doi.org/10.1080/17425255.2025.2451440

Sud A, Horton RH, Hingorani AD et al (2023) Realistic expectations are key to realising the benefits of polygenic scores. BMJ 380:e073149. https://doi.org/10.1136/bmj-2022-073149

Swen JJ, van der Wouden CH, Manson LE et al (2023) A 12-gene pharmacogenetic panel to prevent adverse drug reactions: an open-label, multicentre, controlled, cluster-randomised crossover implementation study. Lancet 401(10374):347–356. https://doi.org/10.1016/S0140-6736(22)01841-4

Tada H, Melander O, Louie JZ et al (2016) Risk prediction by genetic risk scores for coronary heart disease is independent of self-reported family history. Eur Heart J 37(6):561–567. https://doi.org/10.1093/eurheartj/ehv462

The All of Us Research Program Investigators, Denny JC, Rutter JL et al (2019) The "All of Us" research program. N Engl J Med 381(7):668–676. https://doi.org/10.1056/NEJMsr1809937

Torkamani A, Wineinger NE, Topol EJ (2018) The personal and clinical utility of polygenic risk scores. Nat Rev Genet 19(9):581–590. https://doi.org/10.1038/s41576-018-0018-x

Turner RM, Magavern EF, Pirmohamed M (2022) Pharmacogenomics: relevance and opportunities for clinical pharmacology. Br J Clin Pharmacol 88(9):3943–3946. https://doi.org/10.1111/bcp.15329

UPGX, U-PGx: Ubiiquitous Pharmacogenomics (2016) https://upgx.eu/. Assessed 10 Feb 2025

van Karnebeek CDM, O'Donnell-Luria A, Baynam G et al (2024) Leaving no patient behind! Expert recommendation in the use of innovative technologies for diagnosing rare diseases. Orphanet J Rare Dis 19:357. https://doi.org/10.1186/s13023-024-03361-0

Vicente AM, Ballensiefen W, Jönsson JI (2020) How personalised medicine will transform healthcare by 2030: the ICPerMed vision. J Transl Med 18(1):180. https://doi.org/10.1186/s12967-020-02316-w

Vilhjálmsson BJ, Yang J, Finucane HK et al (2015) Modeling linkage disequilibrium increases accuracy of polygenic risk scores. Am J Hum Genet 97(4): 576–592. https://doi.org/10.1016/j.ajhg.2015.09.001

Vince RA Jr, Sun H, Singhal U et al (2025) Assessing the clinical utility of published prostate cancer polygenic risk scores in a large Biobank data set. Eur Urol Oncol 8(1):47–55. https://doi.org/10.1016/j.euo.2024.04.017

Wan JCM, Massie C, Garcia-Corbacho J et al (2017) Liquid biopsies come of age: towards implementation of circulating tumour DNA. Nat Rev Cancer 17(4): 223–238. https://doi.org/10.1038/nrc.2017.7

Wang X, Shen X, Fang F et al (2019) Phenotype-driven virtual panel is an effective method to analyze WES data of neurological disease. Front Pharmacol 9:1529. https://doi.org/10.3389/fphar.2018.01529

Wang J, Pasternak AL, Maggo S et al (2025) CYP3A5 pharmacogenetic testing for tacrolimus in pediatric heart transplant patients: a budget impact analysis. Pharmacogenet Genomics 35(2):81–86. https://doi.org/10.1097/FPC.0000000000000549

Weinshilboum RM, Wang L (2006) Pharmacogenetics and pharmacogenomics: development, science, and translation. Annu Rev Genomics Hum Genet 7:223–245. https://doi.org/10.1146/annurev.genom.6.080604.162315

Wojcik MH, Lemire G, Berger E et al (2024) A genome sequencing for diagnosing rare diseases. N Engl J Med 390:1985–1997. https://doi.org/10.1056/nejmoa2314761

Wong E, Bertin N, Hebrard M et al (2023) The Singapore national precision medicine strategy. Nat Genet 55(2): 178–186. https://doi.org/10.1038/s41588-022-01274-x

Wray NR, Lin T, Austin J, et al. From Basic Science to Clinical Application of Polygenic Risk Scores: A Primer. JAMA Psychiatry. 2021;78(1): 101–109. https://doi.org/10.1001/jamapsychiatry.2020.3049

Wright MW, Thaxton CL, Nelson T et al (2024) Generating clinical-grade gene-disease validity classifications through the ClinGen data platforms. Annu Rev Biomed Data Sci 7:31–50. https://doi.org/10.1146/annurev-biodatasci-102423-112456

Yiannakopoulou EC (2013) Pharmacogenomics of phase II metabolizing enzymes and drug transporters: clinical implications. Pharmacogenomics J 13(2):105 109. https://doi.org/10.1038/tpj.2012.42

Youn MS, Ahn SH, Kim JH (2024) Pharmacogenomic profiling of the South Korean population: insights and implications for personalized medicine. Front Pharmacol 15:1476765. https://doi.org/10.3389/fphar.2024.1476765

Zalis M, Viana Veloso GG, Aguiar PN Jr et al (2024) Next-generation sequencing impact on cancer care: applications, challenges, and future directions. Front Genet 15:1420190. https://doi.org/10.3389/fgene.2024.1420190

Zanger UM, Schwab M (2013) Cytochrome P450 enzymes in drug metabolism: regulation of gene expression, enzyme activities, and impact of genetic variation. Pharmacol Ther 138(1):103–141. https://doi.org/10.1016/j.pharmthera.2012.12.007

Zhai S, Mehrotra DV, Shen J (2023) Applying polygenic risk score methods to pharmacogenomics GWAS: challenges and opportunities. Brief Bioinform 25(1): bbad470. https://doi.org/10.1093/bib/bbad470

Genome Sequencing in Infectious Disease Outbreaks

17

João Paulo Gomes

Abstract

Genome sequencing has become a crucial tool in the management and understanding of infectious disease outbreaks. By decoding the entire genetic material of pathogens, this technology allows scientists to track the spread and evolution of infectious agents with unprecedented precision. During an outbreak, one of the key applications of genome sequencing is in tracing the transmission pathways of the disease. By comparing the genetic sequences of pathogens from different patients and sources, epidemiologists can determine the source of the outbreak and how the disease is spreading through populations. In addition, genome sequencing has the potential to identify mutations that may affect transmissibility, virulence, or resistance to treatment, providing critical insights for public health responses, for instance enabling targeted interventions, such as specific treatments or vaccines. During the last years, this was particularly evident for multiple internationally occurring severe outbreaks caused by either bacteria or virus, with special emphasis during the COVID-19 pandemic, where genome sequencing played a vital role in monitoring the emergence of new variants and informing vaccine strategies. As sequencing technologies continue to advance and associated costs decline, their integration into public health strategies will be essential for more effective control and prevention of infectious disease outbreaks in the future.

Keywords

Infectious diseases · Outbreak · Next-generation sequencing · Genome sequencing · Infectious source · Virulence · Antibiotic resistance · Vaccine

17.1 1 Monitoring Outbreaks Before the Next-Generation Sequencing (NGS) Era

Before the advent of whole genome sequencing (WGS) of microbial pathogens, mainly through the next-generation sequencing (NGS) approaches, the monitoring and management of infectious disease outbreaks relied on a combination of traditional microbiological techniques, epidemiological investigations, and more rudimentary genetic methods. These approaches, while effective to some extent, were often limited in their ability to rapidly identify pathogens, track their spread, and understand their evolution

J. P. Gomes (✉)
Department of Infectious Diseases, Genomics and Bioinformatics Unit, National Institute of Health Doutor Ricardo Jorge (INSA), Lisbon, Portugal

Veterinary and Animal Research Centre (CECAV), Faculty of Veterinary Medicine, Lusófona University, Lisbon, Portugal
e-mail: j.paulo.gomes@insa.min-saude.pt

M. Gama-Carvalho et al. (eds.), *Omics Approaches in Biomedicine and Biotechnology*, Advances in Experimental Medicine and Biology 1504, https://doi.org/10.1007/978-3-032-18966-0_17

during an outbreak. Nevertheless, they laid the foundation for the more advanced technologies we use today.

The identification of pathogens during an outbreak was primarily achieved through some phenotypic characteristics yielded by culture-based methods, which still constitute, nowadays, a fundamental first step in the investigation of outbreaks caused by some non-viral pathogens (Koser et al. 2014; Singh et al. 2006). These techniques involve growing the suspected pathogens in laboratory conditions, often using selective media that favor the growth of specific organisms. Once isolated, these pathogens can be identified based on their morphology, staining characteristics, antimicrobial susceptibility, and biochemical properties. For example, during bacterial outbreaks, such as those caused by *Salmonella* spp. or *Escherichia coli*, clinical samples from patients (e.g., blood, stool) are cultured on selective agar plates. The colonies that grow are subjected to a series of tests, including Gram staining, catalase and oxidase tests, and carbohydrate fermentation profiles, to identify the specific bacterial species responsible for the outbreak. While culture-based methods are important for pathogen identification, they have significant limitations. First, they are time-consuming, often requiring days to weeks to obtain results. This delay can be critical in the context of an outbreak, where rapid identification is necessary to implement control measures. Second, this approach is completely dependent on the viability of the pathogens, requiring appropriate transport and storage conditions of the collected sample. Third, some pathogens are difficult or impossible to culture under standard laboratory conditions, leading to potential misdiagnosis or failure to identify the causative agent. Additionally, these methods lack the resolution needed to distinguish between closely related strains of a pathogen, making it difficult to track the precise source and transmission routes during an outbreak (El Houmami et al. 2016; Singh et al. 2006).

To overcome some of these limitations, additional phenotypic techniques like serotyping and phage typing were developed (Diep et al. 2019; Schlichting et al. 1993), providing more detailed information about specific strains. Serotyping involves identifying specific antigens present on the surface of pathogens. For bacteria like *Salmonella* and *Escherichia coli*, serotyping is used to classify strains based on their O (lipopolysaccharide) and H (flagellar) antigens. This method allows epidemiologists to differentiate between strains within a species, aiding in the tracking of outbreaks. However, serotyping is still relatively limited in its discriminatory power and cannot provide insights into the genetic relationships between strains. Phage typing, another method used for strain differentiation, involves exposing bacterial isolates to a set of bacteriophages (viruses that infect bacteria) to determine their susceptibility (Holmberg et al. 1984; Schlichting et al. 1993). Different strains of bacteria show varying susceptibility patterns to the phages, allowing for their classification. Phage typing was particularly useful during outbreaks of *Salmonella* spp. and *Staphylococcus aureus*. However, like serotyping, phage typing has limitations in its discriminatory power and is labor-intensive and time-consuming.

One of the most significant advances in molecular epidemiology before the NGS era was the development of pulsed-field gel electrophoresis (PFGE) in the 1980s (Finney 1993; Schwartz et al. 1984). PFGE provides a higher resolution method for differentiating bacterial strains based on their DNA. It involves the digestion of bacterial DNA with restriction enzymes that cut the DNA at specific sites, resulting in large fragments of varying sizes. These fragments are then separated using an electric field that periodically changes direction (pulsed-field), allowing for the separation of large DNA fragments. The resulting banding pattern, or "DNA fingerprint," can be used to compare strains and track the spread of a pathogen during an outbreak. PFGE became the gold standard for outbreak investigations, particularly for foodborne pathogens like *Listeria monocytogenes*, *Escherichia coli* O157, and *Salmonella* spp. (Hunt et al. 2021; Shi et al. 2015). It allowed for the identification of outbreak clusters and provided crucial evidence in linking cases to a common source. However, PFGE also has its limitations. It depends on the culture isolation of the bacteria, it requires specialized equipment and technical expertise, and the interpretation of results can be subjective. Additionally, while PFGE reveals a considerable capacity to differentiate between strains (although not at nucleotide level), it does not provide information about genes or mutations associated with differences in specific traits such as in pathogenicity or antibiotic resistance.

Among the various genotyping methods relying on the use of the polymerase chain reaction (PCR) for the amplification of multiple genomic regions, the multi-locus sequence typing (MLST) was certainly the most widely used (Maiden et al. 1998; Pérez-Losada 2013; Maiden et al. 2013). The principle behind MLST involves the PCR amplification and sequencing of specific gene loci (typically seven) that are conserved across strains of a species (i.e., housekeeping genes). Variations in these sequences are used to assign each isolate a unique sequence type (ST), allowing for a quite precise identification of strains and the tracking of their spread during an outbreak, making it a valuable tool for epidemiological investigations. However, as MLST requires multiple PCR and Sanger sequencing reactions *per* strain, it is labor-intensive and expensive compared to other typing methods. Additionally, MLST relies heavily on existing databases of sequence types for comparison, which may not always be comprehensive. Also, it provides limited information about the pathogen genetic makeup, hampering a full resolution power and hindering efforts to understand its evolutionary dynamics. Despite these challenges, MLST (and MLST-like approaches) remains a crucial tool in the field of microbial genomics and public health for managing and understanding outbreaks, namely for laboratories lacking the capacity to perform whole genome sequencing and for pathogens for which culture isolation (a convenient first step before sequencing) is still defiant (e.g., *Legionella pneumophila*) (Halbedel et al. 2023).

In conclusion, before the advent of next-generation sequencing, outbreak monitoring relied on a combination of traditional microbiological techniques, such as serotyping, phage typing, PFGE, and MLST, coupled with epidemiological investigations. These methods, while effective in many cases, had significant limitations in terms of speed, cost, resolution, and the ability to provide detailed genetic information. Nevertheless, the principles of microbiology, epidemiology, and molecular biology that guided pre-NGS outbreak investigations continue to be relevant, even as we embrace more advanced technologies. The lessons learned from traditional methods continue to inform and enhance modern approaches to infectious disease surveillance.

17.2 The Technological Transition: Development, Advantages, and Challenges of NGS Approaches

The limitations of traditional outbreak monitoring methods underscored the need for more advanced tools capable of providing faster, more detailed, and more accurate information about pathogens during an outbreak. The development of NGS technologies in the early 2000s marked a turning point in the field of infectious disease surveillance by enabling comprehensive, high-resolution genomic analyses that facilitate a more detailed understanding of pathogen transmission and evolution. NGS refers to a suite of technologies that allow for the rapid sequencing of large amounts of DNA. Unlike traditional Sanger sequencing ("first generation of sequencing"), which sequences DNA fragments one at a time, NGS can process millions of sequences from multiple microbes in parallel in a single run followed by the use of computational tools to assemble the sequences into complete genomes.

The development of NGS methodologies has been rapid and transformative. The first NGS platforms, such as 454 pyrosequencing, Ion Torrent, and Illumina, marked a significant departure from Sanger sequencing by introducing massively parallel sequencing. The principle behind these "second generation of sequencing" platforms involves fragmenting the DNA of interest before amplifying these fragments through polymerase chain reaction methods and sequencing them by synthesis (McCombie et al. 2019; Mandlik et al. 2024), yielding what is commonly designated by "short reads" of approximately 150–300 nucleotides. It soon became evident that genomes from multiple microbial species are highly complex with long repetitive elements, copy number alterations of genes and mobile elements, and structural variations that are important to evolution, adaptation, and disease. As the short-read technologies hardly resolved these issues, long-read sequencing was also developed through platforms like PacBio and Oxford Nanopore ("third generation of sequencing") (Clarke et al. 2009). These approaches differ from short-read approaches in that they do not rely on populations of amplified small DNA fragments because non-fragmented DNA is directly sequenced. Such long reads (frequently >1 kb) can span complex or repetitive regions with a single continuous read, thus eliminating ambiguity in the size, number, or positions of genomic elements (Goodwin et al. 2016). Throughout the last 10 years, the available NGS technologies have become more advanced, offering increasingly faster, cheaper, and more accurate sequencing capabilities. These advancements have facilitated the routine use of whole-genome sequencing in clinical and public health laboratories, making it possible to quickly sequence and analyze the genomes of pathogens involved in outbreaks.

17.3 Application of Genome Sequencing Approaches to Monitor Outbreaks

Over the past two decades, several infectious disease outbreaks have had a significant impact on public health worldwide. These outbreaks, which spanned various pathogens, including viruses and bacteria, have challenged healthcare systems and revealed the need for a more efficient tool to monitor and control the spread of diseases. Such transformative tool has been the NGS, which has revolutionized the surveillance and management of outbreaks by providing real-time genomic data on pathogens, and providing critical insights into transmission dynamics, evolution, and resistance patterns, thus facilitating more informed decision-making for outbreak control.

Whereas NGS is now part of the routine in several countries to monitor outbreaks, there are a handful of examples in the last 10–15 years that perfectly illustrate its incomparable potential to guide investigations and influence public health measures. One of such examples is the cholera outbreak in Haiti that occurred in 2010, following a major earthquake. Cholera, which is caused by the bacterium *Vibrio cholerae*, is a disease of poverty affecting people with inadequate access to safe water and basic sanitation and is characterized by an acute diarrhea that can kill within hours if left untreated. It is estimated that each year there are 1.3 to 4.0 million cases, and 21,000 to 143,000 deaths worldwide due to cholera (Ali 2015). The outbreak in Haiti, which is believed to have caused up to 200,000 deaths (the official numbers are a matter of some dispute) (Kolbe 2010), tremendously affected Haiti's infrastructures by exacerbating the poor sanitation and inadequate access to clean water, thus contributing to the massive spread of cholera. It is believed that more over 820,000 persons have been sickened from the outbreak, and more than 10,000 have died (https://www.cdc.gov/orr/responses/haiti-cholera-outbreak.html). During laboratorial and epidemiological investigations, NGS played a pivotal role in tracing the origin of the cholera strain responsible for the outbreak. By sequencing the genome of the *V. cholerae* strain that was spreading in Haiti and by comparing it with the available sequences in public databases, researchers presented convincing evidences that the strain had been introduced into Haiti from South Asia, namely from Bangladesh or Nepal (Chen-Shan Chin 2011; Hendriksen 2011). In particular, it has been suggested that it had been introduced by United Nations peacekeepers from Nepal, where cholera was endemic (Hendriksen 2011). The theory of an external origin of this massive outbreak was supported by the fact that no cases of cholera had been reported in Haiti during the last century. In addition to identifying the source of the outbreak,

NGS allowed researchers to monitor the spread of cholera across Haiti and neighboring countries. Sequencing data helped track the evolution of the strain, providing insights into how it was adapting to local conditions and how future outbreaks might be prevented. The ability to definitively trace the source of the outbreak through genome sequencing had significant public health and political implications, as it highlighted the importance of international health protocols and sanitation measures.

Another important example of NGS in outbreak management is the 2011 outbreak of Shiga toxin-producing *Escherichia coli* (STEC) O104 in Germany, which caused around 4000 cases of bloody diarrhea, 850 cases of hemolytic uremic syndrome (HUS), and 50 deaths (Grad 2012). Traditional diagnostic methods struggled to quickly identify the strain responsible for the outbreak. However, NGS allowed researchers to rapidly sequence the genome of the outbreak strain and identify its unique combination of virulence factors. Within days of the outbreak being identified, the full genome of the *E. coli* strain was sequenced and made publicly available. Genome analysis revealed a Shiga toxin-producing *E. coli* of serotype O104:H4, which possessed a plasmid, pAA, characteristic of enteroaggregative *E. coli*, as well as a plasmid encoding an extended-spectrum β-lactamase, making it particularly dangerous due to its ability to cause both severe diarrhea and HUS, a condition that leads to kidney failure (Scheutz 2011; Grad 2012). The genome analysis of *E. coli* isolates from multiple food sources also helped researchers identify the origin of the strain and track its spread. In fact, this outbreak, which spread across several European countries (Gault 2011), was linked to a 15,000 kg contaminated fenugreek sprouts seed shipment from Egypt that arrived in Germany in December 2009. NGS also enabled the development of specific diagnostic tests to detect this strain in patients and food products, significantly speeding up the identification of contaminated food sources, thus helping public health authorities to contain the outbreak more efficiently and prevent further cases.

One of the most significant and deadly outbreaks in recent history was the Ebola virus outbreak in West Africa that occurred between 2014 and 2015. Outbreaks caused by viruses of the genera *Ebolavirus* represent a major public health issue in sub-Saharan Africa and are associated with a case fatality rate of 30–90%, depending on the virus species (Feldmann 2011). This outbreak affected Guinea, Liberia, and Sierra Leone, leading to more than 28,000 cases and over 11,000 deaths. Prior to this outbreak, Ebola outbreaks had been relatively contained due to their location in more rural areas. However, the 2014 outbreak quickly spiraled out of control due to the virus's spread into urban areas and the delayed international response. NGS played a crucial role in understanding the spread of the Ebola virus by providing vital insights into the origins and transmission of the virus. The genomic data revealed that the outbreak originated from a single zoonotic spillover event, most likely from a bat to a two-year-old boy in December 2013 (Baize 2014), and then continued spreading through human-to-human transmission with no evidence of additional zoonotic sources (Gire 2014). Despite the difficulties in performing robust epidemiological analysis, which may be unreliable owing to the difficulties of obtaining patient information, the identification of specific viral lineages helped track the movement of the virus across different regions, allowing researchers to map out transmission chains more precisely than ever before. Genomic sequencing also revealed that the virus was mutating during the course of the outbreak (Baize 2014; Gire 2014). These mutations, most of them altering protein sequences, provided insights into how the virus was evolving in response to transmission pressures and helped monitor the effectiveness of potential treatments and vaccines.

In 2015–2016, Brazil experienced an unprecedented epidemic of Zika virus. It is usually transmitted by the bite of infected mosquitoes and infections in humans were sporadic for half a century before emerging in the Pacific and the Americas (Musso 2016). Transmission of Zika virus in the Americas was first confirmed in

May 2015 in northeast Brazil and, by December 2016, more than 200,000 cases had already been reported in more than 45 countries in the Americas (Faria 2017). This Zika virus outbreak, particularly in Brazil, raised global concerns due to its unprecedented widespread transmission and association with microcephaly and other birth defects in newborns, posing a significant public health challenge, especially for pregnant women. Moreover, the rampant spread of Zika virus infections in Brazil posed challenges for public health preparedness and surveillance for the Olympics and Paralympics, which was held in Rio de Janeiro in August 2016. Considering all this, in February 2016, the World Health Organization declared Zika virus infection a public health emergency of international concern (Petersen 2016). NGS and subsequent phylogenetic and molecular clock analyses showed a single introduction of a likely Asian genotype lineage of Zika virus into the Americas, which is estimated to have occurred by late 2013, more than 12 months prior to the detection of the virus in Brazil (Faria 2016). Although apparently surprising, these lag times are not unreasonable, considering not only the time required for a significant number of infections to be observed, but also the fact that the most obvious effects are observed in babies born several months after their mothers have been infected. By tracking the evolution of the virus, NGS helped map its geographic spread and provided insights into how the virus was evolving as it moved through different populations. Finally, NGS was used to track Zika virus in mosquito populations, providing critical information for vector control efforts.

17.4 The Hallmark Example of COVID-19

The COVID-19 pandemic, caused by the severe acute respiratory syndrome coronavirus 2 (SARS-CoV-2), began in December 2019 in China through the identification of clusters of patients with pneumonia of unknown cause, epidemiologically linked to a seafood and wet animal wholesale market in Wuhan, Hubei Province, China. It rapidly spread across the globe, resulting in an unprecedented public health crisis (Zhu 2020). Declared a pandemic by the World Health Organization (WHO) in March 2020, COVID-19 has led so far to more than 779 million cases and more than seven million deaths (https://data.who.int/dashboards/covid19/data, by February 20, 2026), overburdened healthcare systems, and severe economic and social disruptions. The highly transmissible virus, which primarily spreads through respiratory droplets, presented unique challenges for containment due to frequent asymptomatic transmission, its ability to evolve quickly, and the lack of early therapeutic interventions. From the onset, NGS has been pivotal in identifying and characterizing the virus, tracking its global spread, monitoring its evolution, and informing public health responses, including vaccine development. These following sections explore how NGS has been applied to different facets of SARS-CoV-2 surveillance and control.

17.4.1 Identifying the First Genome Sequence of SARS-CoV-2

The identification of the first genome sequence of SARS-CoV-2 was a pivotal moment in the global response to the emerging pandemic. In early January 2020, Chinese researchers used both metagenomic RNA sequencing of a sample of bronchoalveolar lavage fluid (Wu 2020) and human airway epithelial cell virus isolates (Zhu 2020) to obtain the complete genome of the causative agent of what later became a pandemic. They identified a viral genome of about 30,000 nucleotides and subsequent phylogenetic analysis revealed that the virus was most closely related to a group of SARS-like coronaviruses (genus *Betacoronavirus*, subgenus *Sarbecovirus*) that had previously been found in bats in China (Hu 2018), with notable genome similarity particularly in the receptor-binding domain (Drosten 2003). Initially named 2019-nCoV ("2019-novel coronavirus"), it was later designated SARS-CoV-2 (naming the coronavirus disease

(COVID-19) and the virus that causes it). Deep genome analysis also allowed the establishment of theories about SARS-CoV-2 origins, suggesting that it would be improbable that SARS-CoV-2 emerged through laboratory manipulation of a related SARS-CoV-like coronavirus. Instead, it has been proposed that it would be highly plausible the existence of natural selection in an intermediary animal host (e.g., the pangolin, also sold in the Chinese markets) before zoonotic transfer (Andersen 2020; Lu 2020). The SARS-CoV-2 genome sequence was rapidly shared with the global scientific community, allowing for the immediate development of diagnostic tests and forming the foundation for subsequent research on vaccines and therapeutics. Also, the rapid publication of the genome sequence in public databases and the free access to developed user-friendly bioinformatics platforms (https://nextstrain.org/sars-cov-2) (Hadfield 2018) allowed laboratories worldwide to track the virus's spread and evolution from the very beginning, and thus contribute to an informed public health decision-making process.

17.4.2 Understanding How COVID-19 Started in Specific Countries: Identifying Introductions

As COVID-19 spread from its origin in China to the rest of the world, NGS became instrumental in identifying the points of introduction in different countries. By sequencing viral genomes from patients in various countries, researchers could trace the origin of SARS-CoV-2 strains and determine whether infections were linked to local transmission or multiple international introductions. For instance, genome sequencing efforts in Europe and the United States revealed that the virus had multiple entry points, with different strains circulating concurrently (Lemieux 2021; Worobey 2020; Alteri 2021). That was the case of Italy, where early sequencing efforts showed the presence of seven viral lineages in Lombardy (one of the most affected areas in Europe), and the presence of local transmission clusters, revealing that sustained community transmission was underway before the first COVID-19 case had been detected (Alteri 2021). In a similar study performed in Portugal, about 300 independent SARS-CoV-2 introductions, mostly from European countries, were estimated to have occurred during early March 2020 (Borges 2022). Curiously, it has been observed that a single introduction from Italy, of a lineage that was circulating in Lombardy, represented about 20% of all sampled genomes collected until the end of March in Portugal (Borges 2020, 2022). In Boston, North America, early sequencing also showed numerous introductions of the virus, and the existence of two superspreading events with dissimilar health impact. Whereas one of them occurred in nursing facility with vulnerable population, leading to rapid transmission and significant mortality but little broader spread, the other was at an international business conference and resulted in extensive regional, national, and international spread (Lemieux 2021). Globally, NGS has provided detailed maps of SARS-CoV-2 dissemination. By continuously sequencing viral genomes from various countries, researchers have been able to create phylogenetic trees that show how different lineages of the virus are related and how they spread over time. These efforts were crucial for public health authorities in developing strategies for border control and quarantine, as they helped clarify how the virus was entering new regions and how it was spreading once introduced. This information informed decisions about lockdowns, travel restrictions, testing protocols, and vaccination campaigns, contributing to the management of early outbreaks.

17.4.3 Identifying Local Outbreaks and Monitoring the Local Dissemination of Lineages

The use of viral genomics has also been critical in identifying local outbreaks and tracing the spread of SARS-CoV-2 lineages. By comparing viral genomes from different patients within a specific region, public health officials have been able to determine whether cases are part of a single outbreak or if multiple independent introductions are occurring. High-resolution genetic sequencing of viral samples allowed public health officials to pinpoint transmission pathways within settings like nursing homes, schools, and other community environments, where identifying the source of infection could inform targeted interventions to control the spread (Arons 2020; Moline 2022; van Hensbergen 2021; Nickbakhsh 2022). For example, in a nursing facility in Washington, USA, the introduction of SARS-CoV-2 led to a rapid and widespread transmission affecting 64% of residents. As more than half of residents with positive test results were asymptomatic, it was concluded that infection-control strategies focused solely on symptomatic residents were not sufficient to prevent transmission after SARS-CoV-2 introduction into this facility (Arons 2020). In the Netherlands, viral genomics coupled with epidemiological data has virtually shown that the first long-term care facility COVID-19 outbreak, which affected 41% of the tested residents, was caused by cross-border introduction from Germany (van Hensbergen 2021). In Portugal, while monitoring a nosocomial outbreak of SARS-CoV-2 in a non-COVID-19 hospital ward that yielded a mortality among patients of 57%, virus genome sequencing was a pivotal tool to unveil a scenario of silent SARS-CoV-2 dissemination, mostly driven by the close contact within healthcare workers and between these and patients (Borges 2021). In educational settings, for example, a large cluster of 519 COVID-19 cases was identified in a university of Scotland, and using genomic data, researchers estimated at least 11 independent introductions of SARS-CoV-2 into the student population, with four comprising the majority of detected cases and consistent with separate outbreaks (Nickbakhsh 2022). Even in sports, the use of virus genomics was essential to understand the viral introductions and transmission dynamics. This is well illustrated by the study of a COVID-19 outbreak among a university's men's and women's soccer teams in Chicago, USA, in the Summer of 2020, where, after a probable single source introduction, numerous social gatherings with limited mask use or social distancing resulted in multiple laboratory-confirmed COVID-19 cases across both teams (Teran 2020).

17.4.4 Monitoring the Emergence and Global Dissemination of SARS-CoV-2 Variants

One of the most critical applications of NGS during the COVID-19 pandemic has been in monitoring the emergence of SARS-CoV-2 variants. As the virus replicates, mutations naturally accumulate in its genome, and some of these mutations may confer advantages, such as increased transmissibility or resistance to neutralizing antibodies. NGS has allowed researchers to track these mutations in real time, providing crucial insights into how the virus is evolving. On behalf of this, during the pandemics, the World Health Organization (WHO) classified several SARS-CoV-2 variants as "variants of concern" (VOC) as they presented mutations that impact the virus's transmission, severity, or resistance to neutralization by immunity (https://www.who.int/activities/tracking-SARS-CoV-2-variants). Criteria for designating a VOC include evidence of increased transmissibility, changes in clinical disease presentation, or reduced effectiveness of public health measures, diagnostics, vaccines, and therapeutics. WHO also defined "variants of interest" and "variants under monitoring" whenever some evidences pointed to the possibility of these virus might fulfil the above criteria for a VOC in a near future. Variants such as Alpha, Beta, Gamma, Delta, and Omicron met these criteria and have been designated VOCs due to their notable effects on global case rates and healthcare systems. The VOC Alpha (B.1.1.7) was the first of the highly publicized variants. Alpha first appeared in Great Britain in November 2020 and soon surfaced around the world becoming dominant (Volz 2021). It was followed by Beta (B.1.351) (which was identified in South Africa at the end of 2020) and Gamma (P.1) (identified in Brazil also at the end of 2020 (Tegally 2021; Faria 2021). None of these two reached global predominance but experts had been concerned about their several mutations potentially associated with immune evasion. Delta (B.1.617.2) was first identified in India in late 2020, and became predominant worldwide some months later (Mlcochova 2021), being also associated with more severe disease than other variants in people who were not vaccinated (Twohig 2022). Finally, Omicron (B.1.1.529) emerged in late November of 2021 in Botswana and South Africa and has been dominating the world epidemiological scenario since then through its dozens of evolving lineages (Karim 2021). It is characterized by an abnormal number of mutations (illustrating an "evolutionary jump") essentially on the virus's spike protein, the part that attaches to human cells, and several of those are believed to increase the probability of infection (Tegally 2022). Throughout the COVID-19 pandemics, monitoring these variants through genomic sequencing has been essential for adjusting public health measures, therapeutic interventions, and vaccine strategies. Regarding the latter, multiple adaptations have been made on vaccine formulations to better suit the rapid emergence of viral variants and lineages with novel mutational combinations associated with immune evasion (Lundstrom 2024; Zhao 2024). This includes updating vaccines, implementing booster campaigns, and developing new treatments that are more effective against variants with immune evasion capabilities. Although staying ahead of the virus has not been possible due to the rapid emergence of new viral lineages, the real-time monitoring of these mutations has been essential and ensuring that vaccines and treatments remain effective.

17.5 The Ongoing Mpox Multi-Country Outbreak

Mpox, also known as monkeypox, is an infectious disease caused by the monkeypox virus (MPXV), primarily endemic to central and western African regions and infections outside these endemic areas are rarely reported. The disease is generally transmitted through close contact with infected animals, particularly rodents, but human-to-human transmission can also occur, often through close or intimate contact with lesions, respiratory droplets, or contaminated materials. The clinical manifestations of human monkeypox infection exhibit many similarities to those of smallpox but are typically much milder in nature (Elsayed 2022).

On May 7, 2022, WHO was informed of a confirmed case of mpox in an individual who traveled from the United Kingdom to Nigeria and subsequently returned to the United Kingdom. Two weeks later, about 100 confirmed cases had already been reported in multiple non-endemic countries, especially in the UK, Portugal, and Spain (https://www.who.int/emergencies/disease-outbreak-news/item/2022-DON385) and it soon became evident that the outbreak was being mainly transmitted in defined gay, bisexual, or MSM (i.e., men who have sex with men) sexual networks. In July 2022, WHO declared mpox a public health emergency of international concern, on account of the unprecedented global spread of the disease outside endemic countries (Mitjà 2023). By November 2024, about 102,000 mpox cases caused by the viral strain of the 2022 outbreak had already been reported in more than 110 countries, with more than 220 reported deaths (https://www.cdc.gov/mpox/outbreaks/2022/index-1.html).

The virus is classified into two main genetic clades: Clade I (formerly the Congo Basin clade) and Clade II (formerly the West African clade), which is less virulent than the former. Since the beginning of the 2022 global outbreak, the use of NGS has proved to be an essential tool for the outbreak characterization. In fact, a new lineage B.1, classified as clade IIb for its close relationship to clade II, has been identified in this outbreak (Happi 2022). Also, phylogenetic analysis revealed that the lineage B.1 formed a divergent branch descendant from a branch with viruses associated with the exportation of MPXV in 2018 and 2019 from an endemic country (Nigeria) to the United Kingdom, Israel, and Singapore, with genetic linkage to a large outbreak occurring in Nigeria in 2017–2018 (Isidro 2022; Yinka-Ogunleye 2019). This led to the hypothesis that the MPXV detected in 2022 potentially represented the continuous circulation and evolution of the virus that caused the 2017–2018 Nigeria outbreak. Deeper genomic analysis of the first sequenced virus also highlighted a process of accelerated evolution revealed by a (hyper)mutation signature likely resulting from the potential action of APOBEC3 (Apolipoprotein B mRNA-editing enzyme catalytic polypeptide-like 3) deaminases, host enzymes with antiviral function (Isidro 2022). Additional analysis of signatures of editing by APOBEC3 deaminase led researchers to estimate the existence of MPXV-sustained transmission in the human population since at least 2016. This presents a fundamental shift to the paradigm of mpox epidemiology as a zoonosis and highlights the need for revising public health messaging around mpox outbreak management and control (O'Toole 2023).

During outbreak management, with the aid of viral genomics coupled with epidemiological data, it was also possible to identify travel-associated introductions and autochthonous transmissions. For example, in a large study performed in Portugal (one of the first countries to detect mpox cases), it was observed that the emergence of some outbreak sublineages, or their early global dissemination, most probably occurred in that country, suggesting that Portugal's epidemics played an important role in the early and widespread dissemination of mpox worldwide (Borges 2023). That integrative genomic epidemiologic study also consolidated the key role of events or venues with superspreader potential and sexual networks in mpox transmission. For instance, in that study it was observed that 11 Portuguese viral genetic subclusters, 8 of which including international sequences, were originated in a specific sauna in the Lisbon area.

Contrarily to the scenario of clade IIb MPXV multi-country outbreak observed since 2022, the genetically distinct clade I strains, primarily endemic to central and western Africa, are responsible for the majority of cases in Africa. In fact, the greatest historical burden of mpox globally has occurred within the Democratic Republic of Congo (DRC), where the number of suspected mpox cases has been doubling each year in the last years (Kinganda-Lusamaki 2024). Within this clade, the recently (late 2023) identified clade Ib lineage has emerged as particularly significant due to its association with sexual transmission and is predominantly spreading in the eastern provinces of the DRC, and neighboring countries (Vakaniaki 2024). According to the Africa CDC Epidemic Intelligence Report issued on November 25, 2024 (https://africacdc.org/download/africa-cdc-weekly-event-based-surveillance-report-november-2024/), over 56,000 mpox cases have been reported from 20 African Union Member States so far in 2024, including 12,586 confirmed cases and 1142 deaths (case fatality of 2.01%). Taken together, these events have led to the declaration of a second Public Health Emergency of International Concern for mpox by the WHO director-general on August 14, 2024 (https://www.who.int/news/item/14-08-2024-who-director-general-declares-mpox-outbreak-a-public-health-emergency-of-international-concern). Genomic sequencing allowed researchers to develop rapid PCR assays to distinguish between clade Ia and the recently disseminated clade Ib, facilitating a better understanding of transmission patterns. For instance, signatures of APOBEC3-mediated mutation were observed in clade Ib strains, which suggests human-to-human transmission, supported by epidemiological data indicating sexual contact as a frequent route. On the contrary, clade Ia retains its zoonotic and household transmission routes (Vakaniaki 2024; Kibungu 2024). Also, NGS is being used by non-endemic countries to confirm clade Ib introductions. To date, multiple countries outside of Africa have already detected clade Ib MPXV, such as Sweden, Thailand, Germany, the United States of America, Canada, Pakistan, India, Belgium, the United Kingdom, France, Ireland and Portugal. According to the available epidemiological information, although most of these countries detected cases among travelers from affected countries in eastern and central Africa, some of them already reported community transmission and are conducting public health measures to avoid the dissemination of the clade Ib strains (https://worldhealthorg.shinyapps.io/mpx_global/).

17.6 The Role of Epidemiological Investigations to Complement Genomic-Based Monitoring of Outbreaks

As described throughout this chapter, it is unquestionable that microbial genomic data are particularly powerful for monitoring outbreaks; however, they require integration with robust epidemiological data. Indeed, epidemiological data play a critical role in complementing genomic-based monitoring of outbreaks by contextualizing genomic findings through the identification of demographic, temporal, and geographic patterns of disease spread, thus offering a holistic understanding of disease transmission and facilitating effective public health interventions. For example, during the 2014–2016 West African Ebola virus outbreak, genomic sequencing revealed multiple introductions of the virus into human populations and provided insights into its evolution. Epidemiological investigations complemented this by tracking transmission chains, identifying spreading events, and identifying imported cases in non-endemic countries, thus helping to target containment efforts effectively (Gire 2014; Caroll 2015; Lopaz 2015). In another illustrative example, while monitoring the 2011 *E. coli* O104:H4 outbreak in Germany, a combination of genomic sequencing and epidemiological analysis identified fenugreek sprouts as the source of infection. Genomic data confirmed the unique hybrid strain responsible, while epidemiological studies traced the supply chain, allowing for rapid removal of contaminated products from the market (Grad 2012; Gault 2011). Finally, no better example than COVID-19 to demonstrate the usefulness of joining efforts between the genomic and public health fields, as well demonstrated in thousands of scientific publications (Arons 2020; Moline 2022; van Hensbergen 2021; Nickbakhsh 2022; Borges 2021). Besides its plentiful applications and implications, a simple but powerful example of its combined use relies on the tremendous effort to identify the emergence and dissemination of highly transmissible SARS-CoV-2 VOC in high-risk populations such as in elderly and in hospitalized patients towards the interruption of transmission chains in order to reduce morbidity and mortality. Globally, by integrating genomic and epidemiological datasets, public health authorities can implement targeted interventions, evaluate the effectiveness of control measures, and predict future outbreak dynamics. This synergy, which should be also applied at the cross-border level, underscores the importance of maintaining robust epidemiological and genomic surveillance systems to enhance outbreak response capabilities.

17.7 Conclusion

The application of NGS approaches has significantly transformed the investigation, monitoring, and control of infectious disease outbreaks. This arises from the fact that the genomes of microbial pathogens accumulate mutations, enabling researchers to track their genetic changes over time and space. This capability allows the understanding of evolutionary dynamics, the identification of emerging lineages/variants, the identification of transmission sources, and the detailed resolution of transmission chains, which, coupled with available epidemiological data, are crucial for outbreak response. From the 2011 *E. coli* outbreak in Germany, to the 2014 Ebola outbreak in West Africa, or the 2019 COVID-19 pandemic, NGS has proven its value as an indispensable tool for public health authorities, allowing them to respond more effectively to emerging threats. Nevertheless, while NGS has proven to be a powerful tool in the monitoring and control of infectious disease outbreaks, it is not without its challenges. One of the primary hurdles is the need for robust bioinformatics infrastructure to analyze and interpret the vast amounts of data generated by sequencing efforts. Not all countries or regions have the capacity to implement NGS-based surveillance at the same level, which can lead to disparities in outbreak response and control. Additionally, the cost of sequencing, while decreasing, can still be prohibitive for some low-resource settings. However, as technology continues to advance, and as the global health community invests in capacity-building efforts, these challenges are gradually being addressed and the role of NGS in safeguarding global health security will only become more critical. There is no turning back. The investigation of outbreaks in public health has definitively entered the era of microbial genomics.

References

Ali M, Nelson AR, Lopez AL, Sack DA (2015) Updated global burden of cholera in endemic countries. PLoS Negl Trop Dis 9(6):e0003832. https://doi.org/10.1371/journal.pntd.0003832

Alteri C, Cento V, Piralla A et al (2021) Genomic epidemiology of SARS-CoV-2 reveals multiple lineages and early spread of SARS-CoV-2 infections in Lombardy, Italy. Nat Commun. 12(1):434. Published 2021 Jan 19. https://doi.org/10.1038/s41467-020-20688-x

Andersen KG, Rambaut A, Lipkin WI, Holmes EC, Garry RF (2020) The proximal origin of SARS-CoV-2. Nat Med 26(4):450–452. https://doi.org/10.1038/s41591-020-0820-9

Arons MM, Hatfield KM, Reddy SC et al (2020) Presymptomatic SARS-CoV-2 infections and transmission in a skilled nursing facility. N Engl J Med 382(22):2081–2090. https://doi.org/10.1056/NEJMoa2008457

Baize S, Pannetier D, Oestereich L, Rieger T, Koivogui L, Magassouba N, Soropogui B, Sow MS, Keïta S, De Clerck H, Tiffany A, Dominguez G, Loua M, Traoré A, Kolié M, Malano ER, Heleze E, Bocquin A, Mély S, Raoul H, Caro V, Cadar D, Gabriel M, Pahlmann M, Tappe D, Schmidt-Chanasit J, Impouma B, Diallo AK, Formenty P, Van Herp M, Günther S (2014) Emergence of Zaire Ebola virus disease in Guinea. N Engl J Med 371(15):1418–1425. https://doi.org/10.1056/NEJMoa1404505

Borges V, Isidro J, Cortes-Martins H et al (2020) Massive dissemination of a SARS-CoV-2 spike Y839 variant in Portugal. Emerg Microbes Infect 9(1):2488–2496. https://doi.org/10.1080/22221751.2020.1844552

Borges V, Isidro J, Macedo F et al (2021) Nosocomial outbreak of SARS-CoV-2 in a "non-COVID-19" hospital Ward: virus genome sequencing as a key tool to understand cryptic transmission. Viruses 13(4):604. Published 2021 Apr 1. https://doi.org/10.3390/v13040604

Borges V, Isidro J, Trovão NS et al (2022) SARS-CoV-2 introductions and early dynamics of the epidemic in Portugal. Commun Med (Lond). 2:10. Published 2022 Jan 28. https://doi.org/10.1038/s43856-022-00072-0

Borges V, Duque MP, Martins JV et al (2023) Viral genetic clustering and transmission dynamics of the 2022 mpox outbreak in Portugal. Nat Med 29(10):2509–2517. https://doi.org/10.1038/s41591-023-02542-x

Carroll MW, Matthews DA, Hiscox JA et al (2015) Temporal and spatial analysis of the 2014-2015 Ebola virus outbreak in West Africa. Nature 524(7563):97–101. https://doi.org/10.1038/nature14594

Chin CS, Sorenson J, Harris JB, Robins WP, Charles RC, Jean-Charles RR, Bullard J, Webster DR, Kasarskis A, Peluso P, Paxinos EE, Yamaichi Y, Calderwood SB, Mekalanos JJ, Schadt EE, Waldor MK (2011) The origin of the Haitian cholera outbreak strain. N Engl J Med 364(1):33–42. https://doi.org/10.1056/NEJMoa1012928

Clarke J, Wu HC, Jayasinghe L, Patel A, Reid S, Bayley H (2009) Continuous base identification for single-molecule nanopore DNA sequencing. Nat Nanotechnol 4(4):265–270. https://doi.org/10.1038/nnano.2009.12

Diep B, Barretto C, Portmann AC, Fournier C, Karczmarek A, Voets G, Li S, Deng X, Klijn A (2019) Salmonella serotyping; comparison of the traditional method to a microarray-based method and an in silico platform using whole genome sequencing data. Front Microbiol 10:2554. https://doi.org/10.3389/fmicb.2019.02554

Drosten C, Günther S, Preiser W et al (2003) Identification of a novel coronavirus in patients with severe acute respiratory syndrome. N Engl J Med 348(20):1967–1976. https://doi.org/10.1056/NEJMoa030747

El Houmami N, Minodier P, Dubourg G, Mirand A, Jouve JL, Basmaci R, Charrel R, Bonacorsi S, Yagupsky P, Raoult D, Fournier PE (2016) Patterns of Kingella kingae disease outbreaks. Pediatr Infect Dis J 35(3):340–346. https://doi.org/10.1097/INF.0000000000001010

Elsayed S, Bondy L, Hanage WP (2022) Monkeypox virus infections in humans. Clin Microbiol Rev 35(4):e0009222. https://doi.org/10.1128/cmr.00092-22

Faria NR, Azevedo RDSDS, Kraemer MUG et al (2016) Zika virus in the Americas: early epidemiological and genetic findings. Science 352(6283):345–349. https://doi.org/10.1126/science.aaf5036

Faria NR, Quick J, Claro IM et al (2017) Establishment and cryptic transmission of Zika virus in Brazil and the Americas. Nature 546(7658):406–410. https://doi.org/10.1038/nature22401

Faria NR, Mellan TA, Whittaker C et al (2021) Genomics and epidemiology of the P.1 SARS-CoV-2 lineage in Manaus, Brazil. Science 372(6544):815–821. https://doi.org/10.1126/science.abh2644

Feldmann H, Geisbert TW (2011) Ebola haemorrhagic fever. Lancet 377(9768):849–862. https://doi.org/10.1016/S0140-6736(10)60667-8

Finney M (1993) Pulsed-field gel electrophoresis. In: Ausubel FM, Brent R, Kingston RE, Moore DD,

Seidman JG, Smith JA, Struhl K (eds) Current protocols in molecular biology, vol 1. Greene-Wiley, New York, N.Y, pp 2.5.9–2.5.17

Gault G, Weill FX, Mariani-Kurkdjian P, Jourdan-da Silva N, King L, Aldabe B, Charron M, Ong N, Castor C, Mace M, Bingen E, Noel H, Vaillant V, Bone A, Vendrely B, Delmas Y, Combe C, Bercion R, d'Andigne E, Desjardin M, de Valk H, Rolland P (2011) Outbreak of haemolytic uraemic syndrome and bloody diarrhoea due to Escherichia coli O104:H4, south-west France, June 2011. Euro Surveill 16(26):19905. https://doi.org/10.2807/ese.16.26.19905-en

Gire SK, Goba A, Andersen KG, Sealfon RS, Park DJ, Kanneh L, Jalloh S, Momoh M, Fullah M, Dudas G, Wohl S, Moses LM, Yozwiak NL, Winnicki S, Matranga CB, Malboeuf CM, Qu J, Gladden AD, Schaffner SF, Yang X, Jiang PP, Nekoui M, Colubri A, Coomber MR, Fonnie M, Moigboi A, Gbakie M, Kamara FK, Tucker V, Konuwa E, Saffa S, Sellu J, Jalloh AA, Kovoma A, Koninga J, Mustapha I, Kargbo K, Foday M, Yillah M, Kanneh F, Robert W, Massally JL, Chapman SB, Bochicchio J, Murphy C, Nusbaum C, Young S, Birren BW, Grant DS, Scheiffelin JS, Lander ES, Happi C, Gevao SM, Gnirke A, Rambaut A, Garry RF, Khan SH, Sabeti PC (2014) Genomic surveillance elucidates Ebola virus origin and transmission during the 2014 outbreak. Science 345(6202):1369–1372. https://doi.org/10.1126/science.1259657

Goodwin S, McPherson JD, McCombie WR (2016) Coming of age: ten years of next-generation sequencing technologies. Nat Rev Genet 17(6):333–351. https://doi.org/10.1038/nrg.2016.49

Grad YH, Lipsitch M, Feldgarden M, Arachchi HM, Cerqueira GC, Fitzgerald M, Godfrey P, Haas BJ, Murphy CI, Russ C, Sykes S, Walker BJ, Wortman JR, Young S, Zeng Q, Abouelleil A, Bochicchio J, Chauvin S, Desmet T, Gujja S, McCowan C, Montmayeur A, Steelman S, Frimodt-Møller J, Petersen AM, Struve C, Krogfelt KA, Bingen E, Weill FX, Lander ES, Nusbaum C, Birren BW, Hung DT, Hanage WP (2012) Genomic epidemiology of the Escherichia coli O104:H4 outbreaks in Europe, 2011. Proc Natl Acad Sci USA 109(8):3065–3070. https://doi.org/10.1073/pnas.1121491109

Hadfield J, Megill C, Bell SM et al (2018) Nextstrain: real-time tracking of pathogen evolution. Bioinformatics 34(23):4121–4123. https://doi.org/10.1093/bioinformatics/bty407

Halbedel S, Sperle I, Lachmann R, Kleta S, Fischer MA, Wamp S, Holzer A, Lüth S, Murr L, Freitag C, Espenhain L, Stephan R, Pietzka A, Schjørring S, Bloemberg G, Wenning M, Al Dahouk S, Wilking H, Flieger A (2023) Large multicountry outbreak of invasive Listeriosis by a listeria monocytogenes ST394 clone linked to smoked rainbow trout, 2020 to 2021. Microbiol Spectrum 11(3):e0352022. https://doi.org/10.1128/spectrum.03520-22

Happi C, Adetifa I, Mbala P et al (2022) Urgent need for a non-discriminatory and non-stigmatizing nomenclature for monkeypox virus. PLoS Biol. 20(8): e3001769. Published 2022 Aug 23. https://doi.org/10.1371/journal.pbio.3001769

Hendriksen RS, Price LB, Schupp JM, Gillece JD, Kaas RS, Engelthaler DM, Bortolaia V, Pearson T, Waters AE, Upadhyay BP, Shrestha SD, Adhikari S, Shakya G, Keim PS, Aarestrup FM (2011) Population genetics of Vibrio cholerae from Nepal in 2010: evidence on the origin of the Haitian outbreak. MBio 2(4): e00157–e00111. https://doi.org/10.1128/mBio.00157-11

Holmberg SD, Wachsmuth IK, Hickman-Brenner FW, Cohen ML (1984) Comparison of plasmid profile analysis, phage typing, and antimicrobial susceptibility testing in characterizing Salmonella typhimurium isolates from outbreaks. J Clin Microbiol 19(2): 100–104. https://doi.org/10.1128/jcm.19.2.100-104.1984

Hu D, Zhu C, Ai L et al (2018) Genomic characterization and infectivity of a novel SARS-like coronavirus in Chinese bats [published correction appears in Emerg Microbes Infect. 2020 Dec;9(1):2727. doi:101080/22221751202018757048]. Emerg Microbes Infect 7(1):154. Published 2018 Sep 12. https://doi.org/10.1038/s41426-018-0155-5

Hunt K, Jordan K (2021) Pulsed-field gel electrophoresis (PFGE) analysis of listeria monocytogenes. Methods in molecular biology (Clifton, N.J.) 2220:79–88. https://doi.org/10.1007/978-1-0716-0982-8_6

Isidro J, Borges V, Pinto M et al (2022) Phylogenomic characterization and signs of microevolution in the 2022 multi-country outbreak of monkeypox virus [published correction appears in Nat Med. 2022 28(10):2220–2221. doi:10.1038/s41591-022-02036-2]. Nat Med 28(8):1569–1572. https://doi.org/10.1038/s41591-022-01907-y

Karim SSA, Karim QA (2021) Omicron SARS-CoV-2 variant: a new chapter in the COVID-19 pandemic. Lancet 398(10317):2126–2128. https://doi.org/10.1016/S0140-6736(21)02758-6. Epub 2021 Dec 3. Erratum in: Lancet. 2022 Jan 8;399(10320): 142. doi:10.1016/S0140-6736(21)02876-2

Kibungu EM, Vakaniaki EH, Kinganda-Lusamaki E, Kalonji-Mukendi T, Pukuta E, Hoff NA et al (2024) Clade I–associated mpox cases associated with sexual contact, the democratic Republic of the Congo. Emerg Infect Dis 30(1):172–176. https://doi.org/10.3201/eid3001.231164

Kinganda-Lusamaki E, Amuri-Aziza A, Fernandez-Nuñez N et al (2024) Clade I mpox virus genomic diversity in The Democratic Republic of the Congo, 2018–2024: predominance of zoonotic transmission. Cell. Published online October 24. https://doi.org/10.1016/j.cell.2024.10.017

Kolbe AR, Hutson RA, Shannon H, Trzcinski E, Miles B, Levitz N et al (2010) Mortality, crime and access to basic needs before and after the Haiti earthquake: a

random survey of Port-au-Prince households. Med Confl Surviv 26(4):281–297. https://doi.org/10.1080/13623699.2010.535279
Köser CU, Ellington MJ, Peacock SJ (2014) Whole-genome sequencing to control antimicrobial resistance. Trends Genet 30(9):401–407
Lemieux JE, Siddle KJ, Shaw BM et al (2021) Phylogenetic analysis of SARS-CoV-2 in Boston highlights the impact of superspreading events. Science 371(6529): eabe3261. https://doi.org/10.1126/science.abe3261
Lopaz MA, Amela C, Ordobas M et al (2015) First secondary case of Ebola outside Africa: epidemiological characteristics and contact monitoring, Spain, September to November 2014. Euro Surveill 20(1): 21003. Published 2015 Jan 8. https://doi.org/10.2807/1560-7917.es2015.20.1.21003
Lu R, Zhao X, Li J et al (2020) Genomic characterisation and epidemiology of 2019 novel coronavirus: implications for virus origins and receptor binding. Lancet 395(10224):565–574. https://doi.org/10.1016/S0140-6736(20)30251-8
Lundstrom K (2024) COVID-19 vaccines: where did we stand at the end of 2023? Viruses 16(2):203. Published 2024 Jan 29. https://doi.org/10.3390/v16020203
Maiden MC, Bygraves JA, Feil E, Morelli G, Russell JE, Urwin R, Zhang Q, Zhou J, Zurth K, Caugant DA, Feavers IM, Achtman M, Spratt BG (1998) Multilocus sequence typing: a portable approach to the identification of clones within populations of pathogenic microorganisms. Proc Natl Acad Sci USA 95(6): 3140–3145. https://doi.org/10.1073/pnas.95.6.3140
Maiden MC, Jansen van Rensburg MJ, Bray JE, Earle SG, Ford SA, Jolley KA, McCarthy ND (2013) MLST revisited: the gene-by-gene approach to bacterial genomics. Nat Rev Microbiol 11(10):728–736. https://doi.org/10.1038/nrmicro3093
Mandlik JS, Patil AS, Singh S (2024) Next-generation sequencing (NGS): platforms and applications. J Pharm Bioallied Sci 16(Suppl 1):S41–S45. https://doi.org/10.4103/jpbs.jpbs_838_23
McCombie WR, McPherson JD, Mardis ER (2019) Next-generation sequencing technologies. Cold Spring Harb Perspect Med 9(11):a036798. https://doi.org/10.1101/cshperspect.a036798
Mitjà O, Ogoina D, Titanji BK et al (2023) Monkeypox [published correction appears in Lancet. 2022;400(10367):1926. doi:10.1016/S0140-6736(22)02414-X]. Lancet 401(10370):60–74. https://doi.org/10.1016/S0140-6736(22)02075-X
Mlcochova P, Kemp SA, Dhar MS et al (2021) SARS-CoV-2 B.1.617.2 Delta variant replication and immune evasion. Nature 599(7883):114–119. https://doi.org/10.1038/s41586-021-03944-y
Moline HL, Keaton A, Rice W et al (2022) Effectiveness of coronavirus disease 2019 (COVID-19) mRNA vaccines against infection during an outbreak of severe acute respiratory syndrome coronavirus 2 (SARS-CoV-2) Beta (B.1.351) variant in a skilled nursing facility: Virginia, March–April 2021. Clin Infect Dis 75(Suppl 2):S155–S158. https://doi.org/10.1093/cid/ciac526
Musso D, Gubler DJ (2016) Zika Virus. Clin Microbiol Rev 29(3):487–524. https://doi.org/10.1128/CMR.00072-15. PMID: 27029595; PMCID: PMC4861986
Nickbakhsh S, Hughes J, Christofidis N et al (2022) Genomic epidemiology of SARS-CoV-2 in a university outbreak setting and implications for public health planning. Sci Rep. 12(1):11735. Published 2022 Jul 19. https://doi.org/10.1038/s41598-022-15661-1
O'Toole Á, Neher RA, Ndodo N et al (2023) APOBEC3 deaminase editing in mpox virus as evidence for sustained human transmission since at least 2016. Science 382(6670):595–600. https://doi.org/10.1126/science.adg8116
Pérez-Losada M, Cabezas P, Castro-Nallar E, Crandall KA (2013) Pathogen typing in the genomics era: MLST and the future of molecular epidemiology. Infect Genet Evol 16:38–53. https://doi.org/10.1016/j.meegid.2013.01.009
Petersen E, Wilson ME, Touch S et al (2016) Rapid spread of Zika virus in the Americas—implications for public health preparedness for mass gatherings at the 2016 Brazil Olympic Games. Int J Infect Dis 44:11–15. https://doi.org/10.1016/j.ijid.2016.02.001
Scheutz F, Nielsen EM, Frimodt-Møller J, Boisen N, Morabito S, Tozzoli R, Nataro JP, Caprioli A (2011) Characteristics of the enteroaggregative Shiga toxin/verotoxin-producing Escherichia coli O104:H4 strain causing the outbreak of haemolytic uraemic syndrome in Germany, May to June 2011. Euro Surveill 16(24): 19889. https://doi.org/10.2807/ese.16.24.19889-en
Schlichting C, Branger C, Fournier JM, Witte W, Boutonnier A, Wolz C, Goullet P, Döring G (1993) Typing of Staphylococcus aureus by pulsed-field gel electrophoresis, zymotyping, capsular typing, and phage typing: resolution of clonal relationships. J Clin Microbiol 31(2):227–232. https://doi.org/10.1128/jcm.31.2.227-232.1993
Schwartz DC, Cantor CR (1984) Separation of yeast chromosome-sized DNAs by pulsed field gradient gel electrophoresis. Cell 37(1):67–75. https://doi.org/10.1016/0092-8674(84)90301-5
Shi C, Singh P, Ranieri ML, Wiedmann M, Moreno Switt AI (2015) Molecular methods for serovar determination of Salmonella. Crit Rev Microbiol 41(3):309–325. https://doi.org/10.3109/1040841X.2013.837862
Singh A, Goering RV, Simjee S, Foley SL, Zervos MJ (2006) Application of molecular techniques to the study of hospital infection. Clin Microbiol Rev 19(3): 512–530. https://doi.org/10.1128/CMR.00025-05
Tegally H, Wilkinson E, Giovanetti M et al (2021) Detection of a SARS-CoV-2 variant of concern in South Africa. Nature 592(7854):438–443. https://doi.org/10.1038/s41586-021-03402-9
Tegally H, Moir M, Everatt J et al (2022) Emergence of SARS-CoV-2 omicron lineages BA.4 and BA.5 in South Africa. Nat Med 28(9):1785–1790. https://doi.org/10.1038/s41591-022-01911-2

Teran RA, Ghinai I, Gretsch S et al (2020) COVID-19 outbreak among a University's Men's and Women's Soccer Teams—Chicago, Illinois, July–August 2020. MMWR Morb Mortal Wkly Rep. 69(43):1591–1594. Published 2020 Oct 30. https://doi.org/10.15585/mmwr.mm6943e5

Twohig KA, Nyberg T, Zaidi A et al (2022) Hospital admission and emergency care attendance risk for SARS-CoV-2 delta (B.1.617.2) compared with alpha (B.1.1.7) variants of concern: a cohort study. Lancet Infect Dis 22(1):35–42. https://doi.org/10.1016/S1473-3099(21)00475-8

Vakaniaki EH, Kacita C, Kinganda-Lusamaki E et al (2024) Sustained human outbreak of a new MPXV clade I lineage in eastern Democratic Republic of the Congo. Nat Med 30:2791–2795. https://doi.org/10.1038/s41591-024-03130-3

van Hensbergen M, den Heijer CDJ, Wolffs P et al (2021) COVID-19: first long-term care facility outbreak in the Netherlands following cross-border introduction from Germany, March 2020. BMC Infect Dis 21(1):418. Published 2021 May 4. https://doi.org/10.1186/s12879-021-06093-9

Volz E, Mishra S, Chand M et al (2021) Assessing transmissibility of SARS-CoV-2 lineage B.1.1.7 in England. Nature 593(7858):266–269. https://doi.org/10.1038/s41586-021-03470-x

Worobey M, Pekar J, Larsen BB et al (2020) The emergence of SARS-CoV-2 in Europe and North America. Science 370(6516):564–570. https://doi.org/10.1126/science.abc8169

Wu F, Zhao S, Yu B, Chen YM, Wang W, Song ZG, Hu Y, Tao ZW, Tian JH, Pei YY, Yuan ML, Zhang YL, Dai FH, Liu Y, Wang QM, Zheng JJ, Xu L, Holmes EC, Zhang YZ (2020) A new coronavirus associated with human respiratory disease in China. Nature 579(7798):265–269. https://doi.org/10.1038/s41586-020-2008-3

Yinka-Ogunleye A, Aruna O, Dalhat M et al (2019) Outbreak of human monkeypox in Nigeria in 2017–2018: a clinical and epidemiological report. Lancet Infect Dis 19(8):872–879. https://doi.org/10.1016/S1473-3099(19)30294-4

Zhao Z, Bashiri S, Ziora ZM, Toth I, Skwarczynski M (2024) COVID-19 variants and vaccine development. Viruses. 16(5):757. Published 2024 May 10. https://doi.org/10.3390/v16050757

Zhu N, Zhang D, Wang W, Li X, Yang B, Song J, Zhao X, Huang B, Shi W, Lu R, Niu P, Zhan F, Ma X, Wang D, Xu W, Wu G, Gao GF, Tan W, China novel coronavirus investigating and research team (2020) A novel coronavirus from patients with pneumonia in China, 2019. N Engl J Med 382(8):727–733. https://doi.org/10.1056/NEJMoa2001017

A Journey into Plant-Microbe Interactions: From Lab to Field for a Sustainable Agriculture

18

Raffaella Balestrini, Ivan Bevilacqua, Manuela Campa, Walter Chitarra, Francesco Favaretto, Micol Guaschino, Loredana Moffa, Anna Narduzzo, Luca Nerva, Marina Niero, Giuseppe Paradiso, Andrea Rossetto, Alberto Spada, Filippo Villano, and Elisa Zizzamia

Abstract

The relationship between plants and microorganisms has ancient roots. However, only recently we have begun to appreciate the true complexity of the holobiont and its intricate interactions. Fungi, bacteria and viruses residing within or around plants play a crucial role in enhancing their resilience to biotic and abiotic stressful factors. The impact of these interactions is not limited to stress resilience but are strictly correlated with the plant's life cycle. The recognition of specific effects on plant growth and resilience has driven research towards utilizing microorganisms as sustainable solutions for agricultural challenges. The application of selected microorganisms, combined with the development of SynCom consortia, marks the beginning of a promising strategy for achieving sustainability in agriculture. This chapter explores the hidden world of plant-associated microorganisms, emphasizing their diverse beneficial impacts and providing evidence to support their potential applications in modern agriculture.

Keywords

Holobiont · Plant microbiome · Endophytes · Plant growth-promoting microorganisms · Synthetic microbial communities

Balestrini Raffaella, Bevilacqua Ivan, Campa Manuela, Chitarra Walter, Favaretto Francesco, Guaschino Micol, Moffa Loredana, Narduzzo Anna, Nerva Luca, Niero Marina, Paradiso Giuseppe, Rossetto Andrea, Spada Alberto, Villano Filippo and Zizzamia Elisa contributed equally with all other contributors.

R. Balestrini
Institute of Biosciences and Bioresources, National Research Council (CNR-IBBR), Bari, Italy

I. Bevilacqua · F. Favaretto · A. Narduzzo · A. Spada
Council for Agricultural Research and Economics—Research Centre for Viticulture and Enology (CREA-VE), Conegliano, TV, Italy

Department of Agronomy, Food, Natural Resources, Animals and Environment, University of Padua, Legnaro, PD, Italy

M. Campa
Department of Genetics, Faculty of Agrisciences, Stellenbosch University, Stellenbosch, South Africa

W. Chitarra (✉) · L. Nerva (✉)
Council for Agricultural Research and Economics—Research Centre for Viticulture and Enology (CREA-VE), Conegliano, TV, Italy

National Research Council of Italy—Institute for Sustainable Plant Protection (CNR-IPSP), Torino, TO, Italy
e-mail: walter.chitarra@crea.gov.it; luca.nerva@crea.gov.it

M. Guaschino · L. Moffa · M. Niero · G. Paradiso · A. Rossetto · F. Villano · E. Zizzamia
Council for Agricultural Research and Economics—Research Centre for Viticulture and Enology (CREA-VE), Conegliano, TV, Italy

M. Gama-Carvalho et al. (eds.), *Omics Approaches in Biomedicine and Biotechnology*, Advances in Experimental Medicine and Biology 1504, https://doi.org/10.1007/978-3-032-18966-0_18

18.1 Understanding the Holobiont Concept: An Overview

The symbiotic relationship between plants and microorganisms is widely acknowledged to have ancient origins. However, it was not until 2002 that research revealed a significantly greater diversity of microorganisms colonizing plant roots, highlighting a previous gap in our understanding of root-associated microorganisms (Vandenkoornhuyse et al. 2002).

In 1943, the holobiont theory was empirically formulated by the German theoretical biologist Adolf Meyer-Abich, but it was then in 1991 that the famous symbiosis researcher Lynn Margulis defined the holobiont as a gain, with the holobiont representing an association of partners (bionts) throughout a significant portion of the host life history (Margulis and Fester 1991).

Nowadays, microbiome research confirmed not only the holobiont theory, but also how it substantially contributed to better insights into host-microbiota relationships and metabolite interplay (Berg et al. 2020; Cordovez et al. 2019). Analysing the plant microbiome involves the integration of microbial ecology with the biology and functioning of the host plant. Microorganisms within the microbiome serve as a reservoir of additional genes and functions, improving the host plant capabilities (Bulgarelli et al. 2013). Although these interactions might initially seem asymptomatic, the benefits provided by the plant-associated microbiome are crucial. They in fact significantly extend the plant ability to adapt to various environmental conditions and stresses (Bulgarelli et al. 2013), which is of primary importance in view of the sessile lifestyle of plants (Fig. 18.1).

Nevertheless, our understanding of the role of the plant microbiome remains limited, indeed we are only just beginning to comprehend its ecological functions (Kiers et al. 2011).

The upshot of this better understanding will have substantial impacts on a variety of research investigations and applications, such as possible innovations for crop production (Kiers et al. 2011; Werner et al. 2014).

Nowadays, grasping the holobiont concept is crucial not only for unravelling the coevolution of microorganisms and hosts, but also for enhancing breeding technologies that boost plant resistance (Giudice et al. 2021). With global climate change increasing the frequency of biotic and abiotic stresses, developing new and sustainable crop protection strategies is urgently needed to enhance agricultural resilience. In this context, uncovering plant-microbiome interactions is vital to identifying key factors in the recruitment of beneficial microbes by plants (Giudice et al. 2021, 2022). An example of which are beneficial endophytes, such as plant growth-promoting microorganisms (including actinomycetes and arbuscular mycorrhizal fungi (AMF)), that colonize plant tissues and play a crucial role in protecting plants from stressful factors (Carrión et al. 2019).

Insight into the interactions between plants and endophytic microorganisms can enable the cultivation of well-characterized microbial strains and the development of synthetic microbial communities (SynComs) that mimic natural microbiome function (Carlström et al. 2019; Sandrini et al. 2022a, b). By studying these SynComs through multi- and interdisciplinary approaches, it is possible to better understand plant microbiome assembly and its effects on plant performance(s) under a variety of environmental conditions (Vannier et al. 2019).

In addition, SynComs not only facilitate basic research into plant-microorganism interactions and the holobiont concept, but also provide practical applications for enhancing plant growth and improving other beneficial traits in host plants (Ma et al. 2021).

Drawing from the concepts explored in this section, the holobiont emerges as a pivotal framework reshaping our understanding of biological systems, emphasizing the intricate relationships between hosts and their microbiomes (Vandenkoornhuyse et al. 2015). Recognizing microorganisms as active participants in host health and resilience, this perspective not only expands our comprehension of evolutionary dynamics, but also opens new pathways for

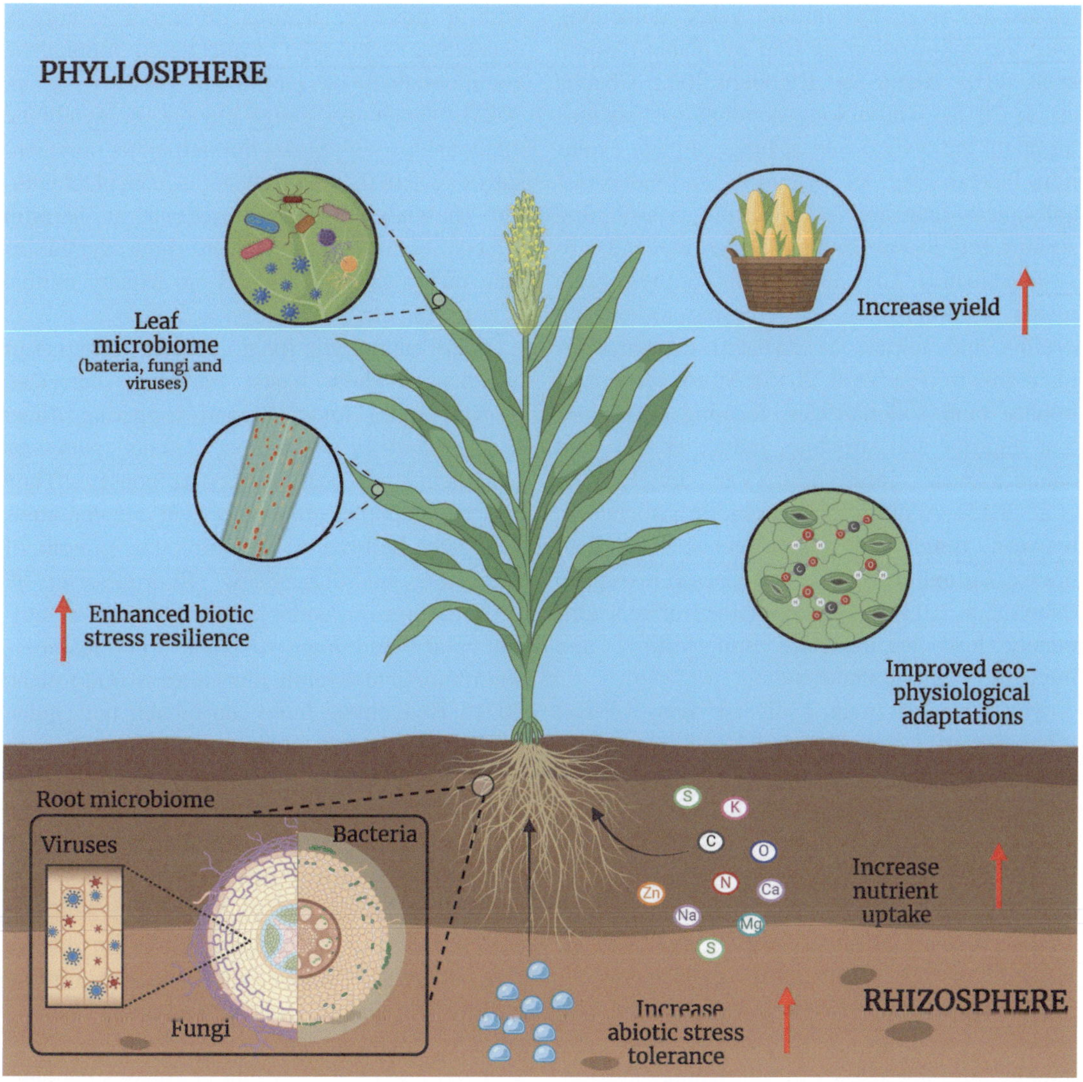

Fig. 18.1 Overview of the main plant's beneficial effects derived from the interactions with the associated microbiome. Plant-associated microbiome can be divided into leaf-associated microbes, forming the so-called phyllosphere, and root-associated microbes, forming the rhizosphere. Overall the interactions between plant and its associated beneficial microbes improve the holobiont fitness, nutritional status and resilience to both biotic and abiotic stresses

enhancing agricultural sustainability, and environmental conservation efforts. In the future, continued exploration and application of the holobiont concept will revolutionize our approach to investigate and harness biological complexity.

18.2 Plants-Fungi Relationships: A Brief Overview

Fossil evidence suggests that the symbiotic relationship between plants and fungi, particularly with endophytes and mycorrhizal fungi, have existed for over 400 million years, coinciding with the colonization of land by plants (Anjum et al. 2019; Baron and Rigobelo 2022; Chadha et al. 2015). These long-standing associations highlight the pivotal role of fungi in plant evolution. Indeed, fungi are ubiquitous and indispensable microorganisms in agriculture, contributing significantly to soil health, plant growth and ecosystem balance. They are essential for the organic matter decomposition, contributing to nutrient cycling and release of essential elements for plant growth (Frąc et al. 2015). Moreover, fungal hyphae bind soil particles, forming aggregates that improve soil structure, reduce erosion and enhance water infiltration (Van der Heijden et al. 2015). Additionally, symbiotic associations between fungi and plants enhance nutrient uptake, drought tolerance and disease resistance (Newsham 2011). These combined effects ultimately lead to increased soil fertility and improved plant productivity.

Plants interact with a diverse array of soil fungi, fostering both beneficial and detrimental associations (Thoms et al. 2021). Pathogenic fungi interact negatively with plants, using this connection to their exclusive advantage by infecting both green and woody tissues, causing disease or even death of the host plant. Fungi exhibit diverse lifestyles, which can also be switched during plant infection. Necrotrophs, such as *Alternaria alternata* and *Botrytis cinerea*, are pathogenic agents for several crops and directly kill plant tissues for nutrient acquisition. Biotrophs, such as *Blumeria graminis*, have developed strategies to be capable of weakening the plant's immune system, thus altering the plant cells metabolism, to extract nutrients without causing the plants' death. Finally, Hemibiotrophs, such as *Colletotrichum higginsianum*, well known to colonize leaf tissues of several *Brassica* and *Raphanus* species, initially behave like biotrophs, developing hyphae inside leaf epidermal cells, but at a late stage it shifts its behaviour to a secondary necrotrophic phase, penetrating host cells cytoplasm and feeding on both living and dead cells (Yan et al. 2018). Conversely, symbiosis refers to any type of close and long-term interactions between different organisms with a positive, neutral or negative impact, depending on the type of interaction: mutualism, commensalism or parasitism (Martins et al. 2023). Generally, fungi engage in symbiotic relationships with plants through either commensalism, where the fungus benefits from plant without showing apparent beneficial or harmful effects, or mutualism, where both symbionts gain reciprocal benefits and are usually mutual (Casadevall et al. 2000).

Many endophytic fungi can offer a myriad of benefits to their hosts, including enhanced growth, stress tolerance and triggering direct (e.g. antibiosis) or indirect defence pathways (e.g. activating host defence responses). They can, for example, produce bioactive compounds that deter herbivores or inhibit the growth of competing microorganisms (Rodriguez et al. 2009). Moreover, endophytes can facilitate nutrient uptake and improve plant water relations, contributing to overall plant fitness (Guo et al. 2021). Endophytic fungi can inhabit both green and woody tissues, and can exhibit both beneficial and pathogenic lifestyles, influenced by a complex interplay of biotic and environmental conditions (Redman et al. 2001). The endophytic fungus *Trichoderma* (teleomorph *Hypocrea*) has garnered significant attention due to its multifaceted interactions with plants. Found in diverse ecosystems, *Trichoderma* species exhibit a strong antagonistic and mycoparasitic nature, enabling them to suppress plant pathogens in the soil and mitigate disease severity. Furthermore, it has been elucidated the ability of certain *Trichoderma* strains to establish direct interactions with plant roots, leading to enhanced plant growth and disease resistance (Hermosa et al. 2012).

Arbuscular mycorrhizal fungi (AMF), belonging to *phylum Mucoromycota* and *subphylum Glomeromycotina* (Zou et al. 2021), form symbiotic associations with the roots of about 80% of terrestrial plants. They create an extensive hyphal network that efficiently absorbs water and minerals from a large soil volume in exchange of carbohydrates and lipids (Keymer et al. 2017). This enables AMF to provide plants with phosphorus and other mineral nutrients, enhance water uptake, improve leaf photosynthesis and increase the hydraulic conductivity of roots (Othman et al. 2024; Balestrini et al. 2020; Jacott et al. 2017). These benefits translate to improved plant tolerance towards abiotic stresses, allowing them to thrive under harsh environmental conditions (Porcel et al. 2015; Khaliq et al. 2022). The interactions between soil microorganisms, including AMF and plant growth-promoting rhizobacteria (PGPR), offer a promising avenue for developing a more sustainable agriculture. In fact, several studies report synergistic effects of AMF and PGPR in promoting plant growth and protection against pathogens (Nanjundappa et al. 2019). *Rhizobia*, known for nitrogen fixation in the soil, share signalling pathways with AMF, facilitating their association with plants (Primieri et al. 2021). Recent research findings highlight the potential existence of a positive correlation between root AMF colonization and soil microbial diversity (Ferreira et al. 2021). Dissecting and exploiting the interactions between different microorganisms offer potential for developing commercial products, such as biostimulants. Microbial biostimulants can represent sustainable and eco-friendly alternatives to mineral fertilizers. They enhance metabolic and enzymatic processes in plants, improving yield, productivity, crop quality and resistance to abiotic stresses (Sun and Shahrajabian 2023).

Further research is required to unravel these inter-kingdom communications, moving from lab-controlled to real-field conditions. By delving deeper into the ancient alliance between plants and microbes, we can harness its power to cultivate a more resilient and productive agriculture, fostering a healthier planet.

18.3 Plant-Bacteria Interactions in Agriculture: From Quorum Sensing to SynComs Development

Plant-associated bacteria can be classified by their interaction with their host, which can be beneficial, deleterious or neutral (Santos et al. 2021). Most of them reside in the rhizosphere, attracted by root exudates that provide energy and nutrients for their reproduction (Kuzyakov and Domanski 2000). Several of these bacteria are categorized as plant growth-promoting bacteria (PGPB) and contribute to multiple abiotic stress tolerances in plants through various mechanisms. These mechanisms include the production of phytohormones such as auxins, cytokinins and gibberellins, which regulate plant growth processes (Maheshwari et al. 2015). PGPB can also solubilize nutrients like phosphorus and iron, thereby enhancing their availability to plants (Etesami et al. 2020). Additionally, under stressful conditions, PGPB synthesize and release stress-induced metabolites, including osmolytes, proline and antioxidant enzymes, which assist plants in coping with abiotic stress (Subiramani et al. 2020). Beyond abiotic stress tolerance, PGPB also play a crucial role in biotic stress management. Prominent genera of PGPB, like *Pseudomonas*, *Bacillus*, *Azospirillum* and *Streptomyces*, can behave as biocontrol agents against pathogens by utilizing various mechanisms (Wang et al. 2021). These include direct competition for resources, antibiosis through the production of antibiotics, enzymatic lysis of pathogens, signal interference and the induction of host plant defences. Furthermore, PGPB can trigger a systemic defence response in plants, known as induced systemic resistance (ISR) (Bukhat et al. 2020). This involves the activation of signalling pathways mediated by jasmonate and ethylene, leading to the production of defence-related enzymes and metabolites, as well as morphological changes that enhance plant resistance to a broad spectrum of pathogens.

All of bacteria traits are ineffective when carried out by single bacteria alone, rather they have the ability to work together in multicellular assemblies. To regulate these group activities, bacteria utilize an interspecific and intraspecific communication process between cells called quorum sensing (QS) (Mukherjee et al. 2019). The most studied is the N-acyl-homoserine lactone (AHL) quorum sensing systems, common in many Gram-negative bacteria (Fuqua et al. 1994). When a critical bacterial concentration, or quorum, is reached, the AHL biosynthesis is induced, leading to the activation of a set of genes essential as nutrient polymer degradation, virulence induction, synthesis of siderophores and biofilm formation (Harmann et al. 2024). Plants, in an as-yet-unidentified mechanism can perceive these bacterial signals, and undergo an AHL-mediated effect. Although the mechanisms by which plants perceive bacterial signalling molecules are not yet fully understood, the impacts of certain AHLs has been documented on several occasions. These impacts include changes in the expression of various genes, alterations in proteomes and effects on root development (Mathesius et al. 2003; Schuhegger et al. 2006; Schenk et al. 2012; Schikora et al. 2011). The bacteria interaction with the complex microbial biofilms decides about success or failure of beneficial microbes to support plant health, or of pathogens to develop disease. In this context, it is evident that not all bacteria exhibit synergistic effects that are beneficial to the plant. In some cases, pathogens may interact with the plant's native microbiome, thereby increasing their virulence. A notable example is the olive knot disease caused by *Pseudomonas savastanoi* pv. *Savastanoi* (Psv). The disease occurs with tumours (knot) along the branches and twigs within which there is a high concentration of the non-pathogenic bacterium *Erwinia toletana* (Buonaurio et al. 2015). This endophyte produces the same AHLs as *P. savastanoi*, enabling them to exchange signals (Hosni et al. 2011). Co-inoculation experiments between a *P. savastanoi* AHL synthase mutant, which can not produce AHLs, and the wild-type strain of *E. toletana* showed that the *P. savastanoi* mutant could induce knot formation in olive plants similar to the wild-type strain. This suggests that the two strains form a consortium where *Et* rescues the virulence of *P. savastanoi* by providing AHLs (Hosni et al. 2011; Buonaurio et al. 2015). Another example, reported by Kim et al. 2020, demonstrated that the plant pathogen *Burkholderia glumae* employs a specific type-6 secretion system (T6SS) to interact with rice endophytic microbes. This interaction results in a decreased population of beneficial bacterial genera, such as *Luteibacter* and *Dyella*, which are known to promote plant growth and protect against pathogenic bacteria (Mannaa et al. 2021).

Despite this, the shared lifestyle of bacteria has prompted the development of synthetic microbial communities (SynComs), composed of mixed PGPB species, with the aim of enhancing disease suppression, plant growth promotion and stress resilience through synergistic interactions (Nerva et al. 2022a, b). In this context, the compatibility between PGPB strains, including factors such as synergistic metabolic activities and complementary modes of action, emerges as a crucial determinant of overall SynCom performance and stability. To fully harness the potential of PGPB-based biocontrol strategies, a comprehensive understanding of the complex interactions within SynComs and their impact on the plant microbiome is imperative. Future research should prioritize unravelling these intricate relationships to optimize SynCom formulations and enhance their effectiveness in promoting sustainable agricultural practices.

18.4 Size Doesn't Matter! Can Viral Entities Control the Holobiont?

The viral communities (viromes) associated with organisms exhibit remarkable diversity and complexity. Viruses can infect all cells across all kingdoms of life, with these infections resulting in a spectrum of effects on the host, ranging from beneficial or neutral to detrimental (Roossinck 2005, 2011). However, the scientific interest around viruses was historically due to their pathogenic effects that result in disease, epidemics and pandemics in different kingdoms from plants to animals. Virology research has recently been strengthened thanks to the new sequencing techniques which allow one to identify the majority of the viruses present in a cell (Fadiji et al. 2022). The technological advancement opened the door for identifying and appreciating the true complexity of viromes in various organisms, including plants like grapevine (Nuzzo et al. 2022; Shvets et al. 2022), and wheat (Redila et al. 2021). Recently, Rivarez et al. (2023)

analysed the virome of healthy and diseased tomato plants, discovering a total of 79 novel viruses. Only 50% of those viruses were associated with the diseases, while the other 50% were either exclusively detected in asymptomatic plants or present in both groups confirming that viruses do not always have pathogenic effects.

Due to the new vision of plants as a holobiont entity (Vandenkoornhuyse et al. 2015), limiting the studies to the plant virome leads inevitably to missing lots of information. Indeed, it was proven that the intimate relationship between plant and endophytes, especially fungi, led to an exchange of viral entities between the two kingdoms that has been defined cross-kingdom viral transmission (Dolja et al. 2020; Sun et al. 2020). This cross-kingdom effect seems to be ancient since plant and fungal viruses shared evolutionary origins (Roossinck 2019). For instance, it was suggested that the *Ourmiaviruses* infecting plant derived from the *Narnaviruses* infecting fungal mitochondria after the acquisition of the capsid and the movement proteins (Lefeuvre et al. 2019). The close relationship between plant and fungal viruses was initially supported by the discovery that some mycoviruses can replicate within plant cells (Nerva et al. 2017). This finding was further strengthened by evidence suggesting that specific plant movement proteins may facilitate the spread of these mycoviruses throughout plant tissues (Andika et al. 2023; Deom et al. 1990).

Following the holobiont point of view it was recently published that the first study comprehensively analysed the whole endophytic virome of a maize plant (Fadiji et al. 2022). They monitored the variability of the root endophytes virome in maize plants cultivated with different agricultural practices, observing that organic farming also increases the virome biodiversity. The main limitations of these whole virome studies are the lacking information about the functional contributions that these viruses have to the holobiont. Currently, the two main literature-supported beneficial activities of the viruses to the holobiont are drought stress tolerance and hypovirulence (Roossinck 2015). Additionally, viruses can participate in tripartite interactions, influencing both the host and the superhost as described by Márquez et al. (2007).

Pagliarani et al. (2022) showed that the pathogenic *Tomato yellow leaf curl Sardinia virus* (*TYLCSV*) encodes a specific C4 protein able to induce stress tolerance in tomato plants. The drought tolerance promotion passes through an ABA independent mechanism; indeed, the accumulation of ABA is four times less in C4 expressing plants than in controls. Another interesting discovery was the protein 6 K2 from the *Turnip mosaic virus* (TuMV) (Prakash et al. 2023). 6 K2 induces drought resistance in *Nicotiana benthamiana* and *Arabidopsis thaliana* plants by promoting salicylic acid accumulation (Miura et al. 2013). An even different pathway was discovered by Wu et al. (2024) on the common bean (*Phaseolus vulgaris*). In this specific case a 21 nucleotides vsiRNA produced from the *Cowpea mild mottle virus* (*CPMMV*) leads to the degradation of the *PvERD15* genes that positively regulates stomatal aperture, ending with an improved drought tolerance.

Hypovirulence is the ability of an infecting virus to reduce host virulence, representing the most studied aspect of mycoviruses (Wagemans et al. 2022). The first interesting application of hypovirulence was made against *Cryphonectria parasitica*, the causative agent of the chestnut blight. The infection by *Cryphonectria parasitica hypovirus 1* (*CHV1*) strongly reduces the symptoms of the disease and has been effectively used to treat chestnut blight (Prospero and Rigling 2016; Rigling and Prospero 2018; Turina and Rostagno 2007). Lemus-Minor et al. (2015) discovered a *Fusarium oxysporum f.* sp. *Dianthi* infected with the Fusarium oxysporum f. sp. Dianthi mycovirus 1 (FodV1) that showed reduced growth and reduced the virulence to carnation compared to the uninfected (Lemus-Minor et al. 2018). This activity was then confirmed in Torres-Trenas et al. (2019) where they microscopically assess the reduction of the growth and virulence in plant.

Moreover, an interesting hypovirus was found infecting *Sclerotinia sclerotium*, one of the most widespread pathogens with a great number of hosts (Bolton et al. 2006). The identified

Slcerotinia sclerotium hypovirulence associated DNA virus 1 (*SsHADV-1*) was able to induce strong hypovirulence effect on the host and moreover was able to modify the host vegetative incompatibility to favour this spread though normally incompatible hosts (Yu et al. 2010). Recently Zhang et al. (2020) analysing the effect of *SsHADV-1* strain infected in *Brassica napus* discovered that the virus has much more than just a hypovirulent effect to the host. Their molecular analysis showed that the mycovirus infection reduces all the molecular pathogenic pathways in *S. sclerotium* such as the cell-wall degrading enzymes and all the effectors able to inhibit plant PTI. TEM (transmission electron microscopy) analysis proved that the infected *S. sclerotium* strain has an endophytic growth without causing any damage. These effects result in an enhancement of plant growth and increased resistance to other fungal pathogens due to the priming effect, concluding that *SsHADV-1* is able to transform a parasitic host into a beneficial associated organism.

Viruses present exciting opportunities for researchers to discover and harness new beneficial functions within holobionts revealing key roles in host physiology and functionality. However, our understanding of complexity and their precise roles remains in their early stages, highlighting the need for further research.

18.5 Integrating OMICs Approaches to Understand Plant-Pathogen Interactions and Microbial Diversity

The advent of OMICs technologies has revolutionized our understanding of plant-pathogen interactions and microbial diversity, offering novel strategies for sustainable agriculture. Genomics, transcriptomics, proteomics and metabolomics provide an in-depth perspective on host-microbe interactions, aiding in the identification of key microbial players involved in plant resilience (Nerva et al. 2022a, b).

18.5.1 OMICs in Plant-Pathogen Interactions

Plant-pathogen interactions involve complex signalling networks that determine disease resistance or susceptibility. Genomics and transcriptomics have been instrumental in identifying resistance genes and regulatory pathways activated in response to pathogen attacks (Sandrini et al. 2022a, b). For instance, transcriptomic studies have revealed differential gene expression patterns in plants inoculated with beneficial microbes, highlighting the activation of defence-related pathways. Similarly, metagenomics approaches have provided insights into microbial community shifts upon pathogen infection, enabling the identification of microbial consortia with biocontrol potential (Nerva et al. 2022a, b). In the context of plant-pathogen interactions, OMICs data have been instrumental in unravelling the intricate immune responses of plants. For instance, Jiang et al. demonstrated that Arabidopsis exhibits distinct immune responses to different pathogens, such as *Golovinomyces orontii* and *Botrytis cinerea*, highlighting the role of specific proteins in mediating these responses (Jiang et al. 2016). Similarly, Abuqamar et al. (2016) emphasized the importance of high-throughput sequencing technologies in understanding the genomic variability and pathogenic diversity of *B. cinerea*, which can inform strategies for enhancing plant resistance. These studies underscore the potential of OMICs to provide a systems biology perspective on plant immunity, allowing for the identification of key genes and pathways that can be targeted for crop improvement.

Proteomic and metabolomic analyses complement these findings by characterizing protein expression changes and metabolite shifts in infected plants. Recent studies have demonstrated that beneficial microbes can induce systemic resistance through the production of volatile organic compounds and secondary metabolites that modulate host immunity. Furthermore, multi-OMICs approaches integrating metabolomic and transcriptomic data have been applied to uncover plant biochemical responses to fungal infections, elucidating key pathways involved in disease suppression (Mendes et al. 2013; Sandrini et al. 2022a, b and references therein), enabling the development of predictive models for managing plant diseases (Velasquez-Camacho et al. 2022).

18.5.2 The Role of SynCom in Enhancing Microbial Diversity and Plant Health

As cited above, the SynCom approach further enhances our understanding of plant-microbe interactions by allowing researchers to manipulate and study specific microbial consortia in controlled environments. Vorholt et al. (2017), highlighted the potential of SynComs to establish causality in plant-microbe interactions, providing insights into how specific microbial members contribute to plant health. This approach not only aids in identifying beneficial microbes but also facilitates the development of microbial inoculants that can be used to promote plant growth and disease resistance in agricultural settings. However, selecting compatible microbial strains that can effectively colonize the plant rhizosphere and confer growth benefits remains a challenge. OMICs tools have been pivotal in optimizing SynCom compositions by characterizing microbial interactions at both the taxonomic and functional levels (Nerva et al. 2022a, b).

Metagenomics and metabarcoding have been extensively used to assess microbial diversity in SynCom applications. These approaches allow for the identification of microbial species present in complex soil and root-associated microbiomes, enabling the design of SynComs that mimic natural microbial communities (Sandrini et al. 2022b). Meta-transcriptomics and meta-proteomics further enhance this understanding by providing functional insights into the metabolic activities of SynCom members, allowing researchers to optimize community composition for improved plant performance (Yin et al. 2021; Nerva et al. 2022b).

18.5.3 Advancing OMICs Integration in Microbial-Based Solutions: What's Next?

Despite significant advances, several challenges remain in the application of OMICs to plant-microbe interactions. One major limitation is the difficulty in integrating multi-OMICs datasets to generate holistic models of microbial function and plant response (Nerva et al. 2022a, b). The development of standardized bioinformatics pipelines and machine-learning algorithms is essential for extracting meaningful insights from OMICs data (Yin et al. 2021; Sandrini et al. 2022b; Crossa et al. 2025). Additionally, future research should focus on applying SynCom formulations under field conditions to validate their effectiveness beyond controlled environments. Combining high-throughput sequencing with metabolomics will allow for the identification of key microbial metabolites that drive plant growth and stress resilience, paving the way for next-generation microbial-based biofertilizers. The integration of OMICs technologies with microbial diversity assessment and SynCom applications holds great potential for advancing sustainable agriculture. By unravelling the intricate mechanisms governing plant-microbe interactions, researchers can develop targeted microbial solutions to enhance plant resilience against biotic and abiotic stresses (Sandrini et al. 2022b).

18.6 Conclusions and Future Prospects

The intricate interactions between plants and their microbiomes, alongside the complex communication networks between plants and microorganisms, as well as within the microbiome itself, present significant challenges in studying and understanding these systems. However, recent technological advances have enabled the identification and characterization of specific interactions within the holobiont, unveiling the beneficial effects that can arise from these relationships. Only a few beneficial fungi and a limited number of bacterial consortia are commercially available for agricultural use, as the complexity of the plant-associated microbiome limits their effectiveness. Fortunately, researchers worldwide are developing and implementing innovative techniques, including microbial-based solutions, to address agricultural challenges on a larger scale. SynComs, in particular, show great promise for enhancing sustainability. Future research should aim to expand the applicability of these green, bio-based technologies.

In addition to bacterial and fungal microorganisms, the often-overlooked viral entities, particularly mycoviruses (viruses that infect fungi), offer potential for pest control and broader agricultural applications.

Continued research is essential to deepen our understanding of the complex interactions within the holobiont. Fortunately, many promising initiatives are emerging from various research groups, with several poised to translate their findings into practical applications for farmers and the agricultural industry.

Acknowledgements This work has been supported by LegnoSano (Funding ID: 2023.0094–53737) project funded by Cariverona Foundation and by CIRCOVINO (CUP C27G22000030001) funded by the Italian Ministry of Environment and Energy Security EC-DEC 86-2023.

References

Abu Qamar SF, Moustafa K, Tran LSP (2016) 'Omics' and plant responses to Botrytis cinerea. Front Plant Sci 7:1658

Andika IB, Tian M, Bian R, Cao X, Luo M, Kondo H, Sun L (2023) Cross-kingdom interactions between plant and fungal viruses. Annu Rev Virol 10:119–138. https://doi.org/10.1146/annurev-virology-111821-122539

Anjum R, Afzal M, Baber R, Khan MAJ, Kanwal W, Sajid W, Raheel A (2019) Endophytes: as potential biocontrol agent—review and future prospects. JAS 11:113. https://doi.org/10.5539/jas.v11n4p113

Balestrini R, Brunetti C, Chitarra W, Nerva L (2020) Photosynthetic traits and nitrogen uptake in crops: which is the role of arbuscular mycorrhizal fungi? Plants 9:1105. https://doi.org/10.3390/plants9091105

Baron NC, Rigobelo EC (2022) Endophytic fungi: a tool for plant growth promotion and sustainable agriculture. Mycology 13:39–55. https://doi.org/10.1080/21501203.2021.1945699

Berg G, Rybakova D, Fischer D, Cernava T, Vergès M-CC, Charles T, Chen X, Cocolin L, Eversole K, Corral GH, Kazou M, Kinkel L, Lange L, Lima N, Loy A, Macklin JA, Maguin E, Mauchline T, McClure R, Mitter B, Ryan M, Sarand I, Smidt H, Schelkle B, Roume H, Kiran GS, Selvin J, de Souza RSC, van Overbeek L, Singh BK, Wagner M, Walsh A, Sessitsch A, Schloter M (2020) Microbiome definition re-visited: old concepts and new challenges. Microbiome 8:103. https://doi.org/10.1186/s40168-020-00875-0

Bolton MD, Thomma BPHJ, Nelson BD (2006) Sclerotinia sclerotiorum (Lib.) de Bary: biology and molecular traits of a cosmopolitan pathogen. Mol Plant Pathol 7:1–16. https://doi.org/10.1111/j.1364-3703.2005.00316.x

Bukhat S, Imran A, Javaid S, Shahid M, Majeed A, Naqqash T (2020) Communication of plants with microbial world: exploring the regulatory networks for PGPR mediated defense signaling. Microbiol Res 238:126486

Bulgarelli D, Schlaeppi K, Spaepen S, Ver Loren, van Themaat E, Schulze-Lefert P (2013) Structure and functions of the bacterial microbiota of plants. Annu Rev Plant Biol 64:807–838. https://doi.org/10.1146/annurev-arplant-050312-120106

Buonaurio R, Moretti C, Passos da Silva D, Cortese C, Ramos C, Venturi V (2015) The olive knot disease as a model to study the role of interspecies bacterial communities in plant disease. Front Plant Sci 6:434

Carlström CI, Field CM, Bortfeld-Miller M, Müller B, Sunagawa S, Vorholt JA (2019) Synthetic microbiota reveal priority effects and keystone strains in the Arabidopsis phyllosphere. Nat Ecol Evol 3:1445–1454. https://doi.org/10.1038/s41559-019-0994-z

Carrión VJ, Perez-Jaramillo J, Cordovez V, Tracanna V, de Hollander M, Ruiz-Buck D, Mendes LW, van Ijcken WFJ, Gomez-Exposito R, Elsayed SS, Mohanraju P, Arifah A, van der Oost J, Paulson JN, Mendes R, van Wezel GP, Medema MH, Raaijmakers JM (2019) Pathogen-induced activation of disease-suppressive functions in the endophytic root microbiome. Science 366:606–612. https://doi.org/10.1126/science.aaw9285

Casadevall A, Pirofski LA (2000) Host-pathogen interactions: basic concepts of microbial commensalism, colonization, infection, and disease. Infect Immun 68. https://doi.org/10.1128/iai.68.12.6511-6518.2000

Chadha N, Mishra M, Rajpal K, Bajaj R, Choudhary DK, Varma A (2015) An ecological role of fungal endophytes to ameliorate plants under biotic stress. Arch Microbiol 197:869–881. https://doi.org/10.1007/s00203-015-1130-3

Cordovez V, Dini-Andreote F, Carrión VJ, Raaijmakers JM (2019) Ecology and evolution of plant microbiomes. Ann Rev Microbiol 73:69–88. https://doi.org/10.1146/annurev-micro-090817-062524

Crossa J, Montesinos-Lopez OA, Costa-Neto G, Vitale P, Martini JWR, Runcie D, Fritsche-Neto R, Montesinos-Lopez A, Pérez-Rodríguez P, Gerard G, Dreisigacker S, Crespo-Herrera L, Pierre CS, Lillemo M, Cuevas J, Bentley A, Ortiz R (2025) Machine learning algorithms translate big data into predictive breeding accuracy. Trends Plant Sci 30(2): 167 184. https://doi.org/10.1016/j.tplants.2024.09.011

Deom CM, Schubert KR, Wolf S, Holt CA, Lucas WJ, Beachy RN (1990) Molecular characterization and biological function of the movement protein of tobacco mosaic virus in transgenic plants. Proc Nat Acad Sci 87:3284–3288. https://doi.org/10.1073/pnas.87.9.3284

Dolja VV, Krupovic M, Koonin EV (2020) Deep roots and splendid boughs of the global plant Virome. Annu Rev Phytopathol 58:23–53. https://doi.org/10.1146/annurev-phyto-030320-041346

Etesami H, Adl SM (2020) Plant growth-promoting rhizobacteria (PGPR) and their action mechanisms in availability of nutrients to plants. In: Phyto-microbiome in stress regulation, pp 147–203

Fadiji AE, Galeemelwe O, Babalola OO (2022) Unravelling the endophytic virome inhabiting maize plant. Agronomy 12:1867

Ferreira DA, Da Silva TF, Pylro VS, Salles JF, Andreote FD, Dini-Andreote F (2021) Soil microbial diversity affects the plant-root colonization by Arbuscular Mycorrhizal fungi. Microb Ecol 82:100–103. https://doi.org/10.1007/s00248-020-01502-z

Frąc M, Łabanowski J, Olejnik J (2015) Fungal biodiversity and their role in soil health. Front Microbiol 6:707

Fuqua WC, Winans SC, Greenberg EP (1994) Quorum sensing in bacteria: the LuxR-LuxI family of cell density-responsive transcriptional regulators. J Bacteriol 176(2):269–275

Giudice G, Moffa L, Varotto S, Cardone MF, Bergamini C, De Lorenzis G, Velasco R, Nerva L, Chitarra W (2021) Novel and emerging biotechnological crop protection approaches. Plant Biotechnol J 19: 1495–1510. https://doi.org/10.1111/pbi.13605

Giudice G, Moffa L, Niero M, Duso C, Sandrini M, Vazzoler LF, Luison M, Pasini E, Chitarra W, Nerva L (2022) Novel sustainable strategies to control Plasmopara viticola in grapevine unveil new insights on priming responses and arthropods ecology. Pest Manag Sci 78:2342–2356. https://doi.org/10.1002/ps.6860

Guo Y, Liu X, Li Y, Zhang H, Li G (2021) Endophytic fungi: a promising resource for sustainable agriculture and environmental protection. Microbiol Mol Biol Rev 85(3):e00024–e00021

Hartmann A, Binder T, Rothballer M (2024) Quorum sensing-related activities of beneficial and pathogenic bacteria have important implications for plant and human health. FEMS Microbiol Ecol 100:6

Hermosa R, Viterbo A, Chet I, Monte E (2012) Plant-beneficial effects of Trichoderma and of its genes. Microbiology. https://doi.org/10.1099/mic.0.052274-0

Hosni T, Moretti C, Devescovi G, Suarez-Moreno ZR, Barek FM, Guarnaccia C, Pongor S, Onofri A, Buonaurio R, Venturi V (2011) Sharing of quorum-sensing signals and role of interspecies communities in a bacterial plant disease. ISMEJ 5:1857–1870

Jacott C, Murray J, Ridout C (2017) Trade-offs in Arbuscular Mycorrhizal symbiosis: disease resistance, growth responses and perspectives for crop breeding. Agronomy 7:75. https://doi.org/10.3390/agronomy7040075

Jiang Z, Dong X, Zhang Z (2016) Network-based comparative analysis of Arabidopsis immune responses to Golovinomyces orontii and Botrytis cinerea infections. Sci Rep 6(1):19149

Keymer A, Pimprikar P, Wewer V, Huber C, Brands M, Bucerius SL, Delaux P-M, Klingl V, Röpenack-Lahaye EV, Wang TL, Eisenreich W, Dörmann P, Parniske M, Gutjahr C (2017) Lipid transfer from plants to arbuscular mycorrhiza fungi. eLife 6: e29107. https://doi.org/10.7554/eLife.29107

Khaliq A, Perveen S, Alamer KH, Zia Ul Haq M, Rafique Z, Alsudays IM, Althobaiti AT, Saleh MA, Hussain S, Attia H (2022) Arbuscular mycorrhizal fungi symbiosis to enhance plant–soil interaction. Sustainability 14:7840. https://doi.org/10.3390/su14137840

Kiers ET, Duhamel M, Beesetty Y, Mensah JA, Franken O, Verbruggen E, Fellbaum CR, Kowalchuk GA, Hart MM, Bago A, Palmer TM, West SA, Vandenkoornhuyse P, Jansa J, Bücking H (2011) Reciprocal rewards stabilize cooperation in the mycorrhizal symbiosis. Science 333:880–882. https://doi.org/10.1126/science.1208473

Kim N, Kim JJ, Kim I, Mannaa M, Park J, Kim J, Lee HH, Lee SB, Park DS, Sul WJ, Seo Y-S (2020) Type VI secretion systems of plant pathogenic Burkholderia glumae BGR1 play a functionally distinct role in interspecies interactions and virulence. Mol Plant Pathol 21:1055–1069

Kuzyakov Y, Domanski G (2000) Carbon input by plants into the soil. Review J Plant Nutr Soil Sci 163(4): 421–431

Lefeuvre P, Martin DP, Elena SF, Shepherd DN, Roumagnac P, Varsani A (2019) Evolution and ecology of plant viruses. Nat Rev Microbiol 17:632–644

Lemus-Minor CG, Cañizares MC, García-Pedrajas MD, Pérez-Artés E (2015) Complete genome sequence of a novel dsRNA mycovirus isolated from the phytopathogenic fungus Fusarium oxysporum f. sp. dianthi. Arch Virol 160:2375–2379. https://doi.org/10.1007/s00705-015-2509-9

Lemus-Minor CG, Cañizares MC, García-Pedrajas MD, Pérez-Artés E (2018) Fusarium oxysporum f. sp.

dianthi virus 1 accumulation is correlated with changes in virulence and other phenotypic traits of its fungal host. Phytopathology 108:957–963. https://doi.org/10.1094/PHYTO-06-17-0200-R

Ma K-W, Niu Y, Jia Y, Ordon J, Copeland C, Emonet A, Geldner N, Guan R, Stolze SC, Nakagami H, Garrido-Oter R, Schulze-Lefert P (2021) Coordination of microbe–host homeostasis by crosstalk with plant innate immunity. Nat Plants 7:814–825. https://doi.org/10.1038/s41477-021-00920-2

Maheshwari DK, Dheeman S, Agarwal M (2015) Phytohormone-producing PGPR for sustainable agriculture. Bacterial metabolites in sustainable agroecosystem:159–182

Mannaa M, Seo Y-S (2021) Plants under the attack of allies: moving towards the plant pathobiome paradigm. Plants 10:125

Margulis L, Fester R (1991) Symbiosis as a source of evolutionary innovation: speciation and morphogenesis. MIT Press

Márquez LM, Redman RS, Rodriguez RJ, Roossinck MJ (2007) A virus in a fungus in a plant: three-way symbiosis required for thermal tolerance. Science 315:513–515. https://doi.org/10.1126/science.1136237

Martins SJ, Pasche J, Silva HAO, Selten G, Savastano N, Abreu LM, Bais HP, Garrett KA, Kraisitudomsook N, Pieterse CMJ, Cernava T (2023) The use of synthetic microbial communities to improve plant health. Phytopathology 113(8):1369–1379. https://doi.org/10.1094/PHYTO-01-23-0016-IA

Mathesius U, Mulders S, Gao M, Teplitski M, Caetano-Anollés G, Rolfe GB, Bauer DW (2003) Extensive and specific responses of an eucaryote to bacterial quorum sensing-signal. Proc Natl Acad Sci USA 100:1444–1449

Mendes R, Garbeva P, Raaijmakers JM (2013) The rhizosphere microbiome: significance of plant beneficial, plant pathogenic, and human pathogenic microorganisms. FEMS Microbiol Rev 37(5):634–663

Miura K, Okamoto H, Okuma E, Shiba H, Kamada H, Hasegawa PM, Murata Y (2013) SIZ1 deficiency causes reduced stomatal aperture and enhanced drought tolerance via controlling salicylic acid-induced accumulation of reactive oxygen species in Arabidopsis. Plant J 73:91–104. https://doi.org/10.1111/tpj.12014

Mukherjee S, Bassler BL (2019) Bacterial quorum sensing in complex and dynamically changing environments. Nat Rev Microbiol 17:371–382

Nanjundappa A, Bagyaraj DJ, Saxena AK, Kumar M, Chakdar H (2019) Interaction between arbuscular mycorrhizal fungi and Bacillus spp. in soil enhancing growth of crop plants. Fungal Biol Biotechnol 6:23. https://doi.org/10.1186/s40694-019-0086-5

Nerva L, Varese GC, Falk BW, Turina M (2017) Mycoviruses of an endophytic fungus can replicate in plant cells: evolutionary implications. Sci Rep 7:1908

Nerva L, Garcia JF, Favaretto F, Giudice G, Moffa L, Sandrini M, Cantu D, Zanzotto A, Gardiman M, Velasco R, Gambino G, Chitarra W (2022a) The hidden world within plants: metatranscriptomics unveils the complexity of wood microbiomes. J Exp Bot 73(8):2682–2697. https://doi.org/10.1093/jxb/erac032

Nerva L, Sandrini M, Moffa L, Velasco R, Balestrini R, Chitarra W (2022b) Breeding toward improved ecological plant–microbiome interactions. Trends Plant Sci 27(11):1134–1143

Newsham KK (2011) Arbuscular mycorrhizas: an introduction. In: Bucking H, van der Heijden MGA (eds) Arbuscular mycorrhizas in the rhizosphere. Springer, Berlin, Heidelberg, pp 1–17

Nuzzo F, Moine A, Nerva L, Pagliarani C, Perrone I, Boccacci P, Gribaudo I, Chitarra W, Gambino G (2022) Grapevine virome and production of healthy plants by somatic embryogenesis. Microb Biotechnol 15:1357–1373. https://doi.org/10.1111/1751-7915.14011

Othman YA, Alananbeh KM, Tahat MM (2024) Can arbuscular mycorrhizal fungi enhance crop productivity and quality in hydroponics? a meta-analysis. Sustainability 16:3662. https://doi.org/10.3390/su16093662

Pagliarani C, Moine A, Chitarra W, Nerva L, Catoni M, Tavazza R, Matić S, Vallino M, Secchi F, Noris E (2022) The C4 protein of tomato yellow leaf curl Sardinia virus primes drought tolerance in tomato through morphological adjustments. Hortic Res 9: uhac164

Porcel R, Redondo-Gómez S, Mateos-Naranjo E, Aroca R, Garcia R, Ruiz-Lozano JM (2015) Arbuscular mycorrhizal symbiosis ameliorates the optimum quantum yield of photosystem II and reduces non-photochemical quenching in rice plants subjected to salt stress. J Plant Physiol 185:75–83. https://doi.org/10.1016/j.jplph.2015.07.006

Prakash V, Nihranz CT, Casteel CL (2023) The Potyviral protein 6K2 from turnip mosaic virus increases plant resilience to drought. Mol Plant-Microbe Interact 36: 189–197. https://doi.org/10.1094/MPMI-09-22-0183-R

Primieri S, Santos JCP, Antunes PM (2021) Nodule-associated bacteria alter the mutualism between arbuscular mycorrhizal fungi and N2 fixing bacteria. Soil Biol Biochem 154:108149. https://doi.org/10.1016/j.soilbio.2021.108149

Prospero S, Rigling D (2016) Using molecular markers to assess the establishment and spread of a mycovirus applied as a biological control agent against chestnut blight. BioControl 61:313–323. https://doi.org/10.1007/s10526-015-9713-0

Redila CD, Prakash V, Nouri S (2021) Metagenomics analysis of the wheat virome identifies novel plant and fungal-associated viral sequences. Viruses 13: 2457

Redman RS, Dunigan DD, Rodriguez RJ (2001) Fungal symbiosis from mutualism to parasitism: who controls the outcome, host or invader? New Phytol 151:705–

716. https://doi.org/10.1046/j.0028-646x.2001.00210.x

Rigling D, Prospero S (2018) Cryphonectria parasitica, the causal agent of chestnut blight: invasion history, population biology and disease control. Mol Plant Pathol 19:7–20. https://doi.org/10.1111/mpp.12542

Rivarez MPS, Pecman A, Bačnik K, Maksimović O, Vučurović A, Seljak G, Mehle N, Gutiérrez-Aguirre I, Ravnikar M, Kutnjak D (2023) In-depth study of tomato and weed viromes reveals undiscovered plant virus diversity in an agroecosystem. Microbiome 11:60. https://doi.org/10.1186/s40168-023-01500-6

Rodriguez RJ, White JF, Arnold AE, Redman RS (2009) Fungal endophytes: symbiotic associations with plants. New Phytol 182(3):337–363

Roossinck MJ (2005) Symbiosis versus competition in plant virus evolution. Nat Rev Microbiol 3:917–924

Roossinck MJ (2011) The good viruses: viral mutualistic symbioses. Nat Rev Microbiol 9:99–108

Roossinck MJ (2015) A new look at plant viruses and their potential beneficial roles in crops. Mol Plant Pathol 16: 331

Roossinck MJ (2019) Evolutionary and ecological links between plant and fungal viruses. New Phytol 221:86–92. https://doi.org/10.1111/nph.15364

Sandrini M, Moffa L, Velasco R, Balestrini R, Chitarra W, Nerva L (2022a) Microbe-assisted crop improvement: a sustainable weapon to restore holobiont functionality and resilience. Hort Res 9:uhac160. https://doi.org/10.1093/hr/uhac160

Sandrini M, Nerva L, Sillo F, Balestrini R, Chitarra W, Zampieri E (2022b) Abiotic stress and belowground microbiome: the potential of omics approaches. Int J Mol Sci 23(3):1091

Santos LF, Olivares FL (2021) Plant microbiome structure and benefits for sustainable agriculture. Curr Plant Biol 26:100198

Schenk ST, Stein E, Kogel K-H, Schikora A (2012) Arabidopsis growth and defense are modulated by bacterial quorum sensing molecules. Plant Signal Behav 7:178–181

Schikora A, Schenk ST, Stein E, Molitor A, Zuccaro A, Kogel KH (2011) N-acyl-homoserine lactone confers resistance toward biotrophic and hemibiotrophic pathogens via altered activation of AtMPK6. Plant Physiol 157:1407–1418

Schuhegger R, Ihring A, Gantner S, Bahnweg G, Knappe C, Vogg G, Hutzler P, Schmid M, Breusegem VF, Eberl L, Hartmann A, Langebartels C (2006) Induction of systemic resistance in tomato plants by N- ac ylhomoserine lactone-producing rhizosphere bacteria. Plant Cell Environ 29:909–918

Shvets D, Porotikova E, Sandomirsky K, Vinogradova S (2022) Virome of grapevine germplasm from the Anapa Ampelographic collection (Russia). Viruses 14:1314

Subiramani S, Ramalingam S, Muthu T, Nile SH, Venkidasamy B (2020) Development of abiotic stress tolerance in crops by plant growth-promoting rhizobacteria (PGPR). Phyto-microbiome in stress regulation, pp 125–145

Sun W, Shahrajabian MH (2023) The application of arbuscular mycorrhizal fungi as microbial biostimulant, sustainable approaches in modern agriculture. Plants 12:3101. https://doi.org/10.3390/plants12173101

Sun L, Kondo H, Bagus AI (2020) Cross-kingdom virus infection. In: Encyclopedia of virology: vol 1–5, 4th edn. Elsevier, pp 443–449

Thoms D, Liang Y, Haney CH (2021) Maintaining symbiotic homeostasis: how do plants engage with beneficial microorganisms while at the same time restricting pathogens? MPMI 34(5):462–469. https://doi.org/10.1094/MPMI-11-20-0318-FI

Torres-Trenas A, Prieto P, Cañizares MC, García-Pedrajas MD, Pérez-Artés E (2019) Mycovirus Fusarium oxysporum f. sp. dianthi virus 1 decreases the colonizing efficiency of its fungal host. Front Cell Infect Microbiol 9:51

Turina M, Rostagno L (2007) Virus-induced hypovirulence in Cryphonectria parasitica: still an unresolved conundrum. J Plant Pathol:165–178

Van der Heijden MG, Martin F, Selosse MA, Sanders IR (2015) Mycorrhizal ecology and evolution: the past, the present, and the future. New Phytol 205(3): 299–323

Vandenkoornhuyse P, Baldauf SL, Leyval C, Straczek J, Young JPW (2002) Extensive fungal diversity in plant roots. Science 295:2051. https://doi.org/10.1126/science.295.5562.2051

Vandenkoornhuyse P, Quaiser A, Duhamel M, Le Van A, Dufresne A (2015) The importance of the microbiome of the plant holobiont. New Phytol 206:1196–1206. https://doi.org/10.1111/nph.13312

Vannier N, Agler M, Hacquard S (2019) Microbiota-mediated disease resistance in plants. PLoS Pathog 15:e1007740. https://doi.org/10.1371/journal.ppat.1007740

Velasquez-Camacho L, Otero M, Basile B, Pijuan J, Corrado G (2022) Current trends and perspectives on predictive models for mildew diseases in vineyards. Microorganisms 11(1):73

Vorholt JA, Vogel C, Carlström CI, Müller DB (2017) Establishing causality: opportunities of synthetic communities for plant microbiome research. Cell Host Microbe 22(2):142–155

Wagemans J, Holtappels D, Vainio E, Rabiey M, Marzachì C, Herrero S, Ravanbakhsh M, Tebbe CC, Ogliastro M, Ayllón MA, Turina M (2022) Going viral: virus-based biological control agents for plant protection. Annu Rev Phytopathol 60:21–42. https://doi.org/10.1146/annurev-phyto-021621-114208

Wang H, Liu R, You MP, Barbetti MJ, Chen Y (2021) Pathogen biocontrol using plant growth-promoting bacteria (PGPR): role of bacterial diversity. Microorganisms 9(9):1988

Werner GDA, Strassmann JE, Ivens ABF, Engelmoer DJP, Verbruggen E, Queller DC, Noë R, Johnson NC, Hammerstein P, Kiers ET (2014) Evolution of microbial markets. Proc Natl Acad Sci USA 111: 1237–1244. https://doi.org/10.1073/pnas.1315980111

Wu X, Chen S, Zhang Z, Zhou W, Sun T, Ning K, Xu M, Ke X, Xu P (2024) A viral small interfering RNA-host plant mRNA pathway modulates virus-induced drought tolerance by enhancing autophagy. Plant Cell 36:koae158

Yan Y, Yuan Q, Tang J, Huang J, Hsiang T, Wei Y, Zheng L (2018) Colletotrichum higginsianum as a model for understanding host-pathogen interactions: a review. Int J Mol Sci 19(7):2142. https://doi.org/10.3390/ijms19072142

Yin C, Casa Vargas JM, Schlatter DC, Hagerty CH, Hulbert SH, Paulitz TC (2021) Rhizosphere community selection reveals bacteria associated with reduced root disease. Microbiome 9:1–18

Yu X, Li B, Fu Y, Jiang D, Ghabrial SA, Li G, Peng Y, Xie J, Cheng J, Huang J, Yi X (2010) A geminivirus-related DNA mycovirus that confers hypovirulence to a plant pathogenic fungus. Proc Natl Acad Sci 107: 8387–8392. https://doi.org/10.1073/pnas.0913535107

Zhang H, Xie J, Fu Y, Cheng J, Qu Z, Zhao Z, Cheng S, Chen T, Li B, Wang Q (2020) A 2-kb mycovirus converts a pathogenic fungus into a beneficial endophyte for Brassica protection and yield enhancement. Mol Plant 13:1420–1433

Zou Y-N, Wu Q-S, Kuča K (2021) Unravelling the role of arbuscular mycorrhizal fungi in mitigating the oxidative burst of plants under drought stress. Plant Biol J 23:50–57. https://doi.org/10.1111/plb.13161

Decoding Plant Immunity: Proteomics as a Biotechnological Tool for Crop Improvement

19

Rita B. Santos, Catarina Paiva-Silva, and Andreia Figueiredo

Abstract

Proteomics is an already established transformative tool for dissecting the molecular mechanisms underlying plant immunity and pathogen virulence. As crop losses due to plant diseases intensify under climate change, understanding these interactions at the protein level has become essential for sustainable crop improvement. Advances in mass spectrometry, bioinformatics, and quantitative proteomics have expanded the capacity to characterize dynamic changes in host and pathogen proteomes, enabling the identification of defense-related proteins, signaling networks, effector targets, and pathogenicity determinants. This chapter synthesizes recent proteomic insights across major crops as soybean, tomato, rice, grapevine, maize, and wheat, highlighting how proteome profiling has revealed key regulators of pathogen-associated molecular pattern (PAMP)-triggered and effector-triggered immunity, pathogen infection strategies, and the impact of biocontrol agents and priming compounds. Case studies illustrate how proteomics uncovers cultivar-specific responses, subcellular reprogramming, and protein networks critical for resistance. We further explore how proteomics informs biotechnology, from identifying candidate genes for breeding and genome editing to supporting safety assessments of engineered crops.

Keywords

Mass spectrometry · Plant–pathogen interactions · Soybean · Tomato · Rice · Grapevine · Wheat

19.1 Introduction

Proteomics, defined as the comprehensive study of proteins, namely protein composition, structure, function, and interactions within a biological system (Jorrin Novo 2021), became a pivotal approach in the field of biotechnology. Proteomic methods have evolved from simple gel-based approaches, being the most used traditional two-dimensional gel electrophoresis (2-DE), which separates proteins by their isoelectric point and molecular weight (Scheele 1975; Klose 1975) coupled to mass spectrometry for protein identification. For a long time, 2-DE was the preferred technique for analyzing plant proteomes, mainly using descriptive or comparative methods to study its modulation (Jorrin Novo 2021). The integration of computational tools and bioinformatics has further enhanced the capacity

R. B. Santos · C. Paiva-Silva · A. Figueiredo (✉)
Grapevine Pathogen Systems Lab, Biosystems and Integrative Sciences Institute, Faculdade de Ciências, Universidade de Lisboa, Lisbon, Portugal
e-mail: absantos@ciencias.ulisboa.pt; cmdsilva@ciencias.ulisboa.pt; aafigueiredo@ciencias.ulisboa.pt

M. Gama-Carvalho et al. (eds.), *Omics Approaches in Biomedicine and Biotechnology*, Advances in Experimental Medicine and Biology 1504, https://doi.org/10.1007/978-3-032-18966-0_19

to analyze complex proteomic data, providing insights into protein networks and pathways. Recent technological advancements in mass spectrometry (MS)-based proteomics, including differential gel electrophoresis (DIGE) and the development of various gel-free methods using liquid chromatography, have significantly sped up and enhanced the characterization of protein dynamics in plant–pathogen interactions (Balotf et al. 2022). Additionally, shotgun analysis has emerged as the predominant platform for studying proteomics in plant–pathogen interactions (Balotf et al. 2022). However, it is important to recognize that understanding these complex interactions requires a holistic view of the cellular 'tug-of-war,' where regulatory hubs at the DNA, transcript, protein, and metabolite levels determine the final outcome of the infection. While each of these layers provides critical insights into host resilience—ranging from the epigenetic reprogramming and transcriptional shifts (Pereira et al. 2022, 2025) to the metabolic biomarkers and spatial localization of defense compounds (Maia et al. 2020)—this chapter focuses specifically on the role of proteomics. By narrowing the scope to the proteome, we aim to bridge the gap between upstream genomic regulation and the final phenotypic defense response that defines plant immunity.

Crop losses due to plant diseases pose a major threat to food security worldwide, particularly in a climate change scenario. Each pathogen employs distinct infection strategies and triggers specific host responses. In the Omics era, proteomics arose as a powerful tool to understand the molecular mechanisms underlying plant–pathogen interactions. Proteomic approaches are being used to elucidate both sides of the interaction, characterizing host–pathogen communication, pathogenicity, defense-related proteins, and the dynamics and networks that define the outcome of the interaction. The identification of key components of the interaction is paramount for the definition of new targets for biotechnological approaches towards disease management. Despite these advancements, achieving a comprehensive understanding of the molecular intricacies of plant–pathogen interactions remains elusive (Jain et al. 2021).

In this chapter we explore the pivotal role of proteomics in understanding plant–pathogen interactions and its potential applications in biotechnology, such as the engineering of disease-resistant plants and the development of novel antimicrobial agents. Real-world applications and ongoing research efforts will be highlighted to illustrate the transformative potential of proteomics in agriculture and biotechnology.

Furthermore, we will address the challenges and future directions in this field. Despite significant advancements, several technical and methodological limitations remain, such as the need for better quantification techniques and the integration of multi-omics data. We will explore emerging technologies and propose future research areas that could pave the way for groundbreaking developments in plant protection and crop improvement.

19.2 Plant–Pathogen Interactions

Plants are continuously challenged by many organisms (virus, bacteria, fungi, oomycetes, nematodes, among others) as for them plants are simply tempting as food and living space providers. Though some interactions are beneficial to the host, such as symbionts, most are pathogenic and an arms race between the host and pathogens occurs. How pathogens gain access to host tissues depends on their lifestyle. Plant pathogens can be either biotrophs, keeping the cells alive or necrotrophs that degrade host tissues to perform colonization (Santos and Figueiredo 2021). Hemibiotroph lifestyle (characteristic of many *Colletotrichum* species) combines a short biotrophic phase and a necrotrophic phase at later stages of infection (Fig. 19.1). At the molecular level, pathogens employ an array of strategies to infect and colonize their plant hosts. Pathogens may secrete effector proteins to suppress or manipulate the host's defense responses (Toruño et al. 2016; Gouveia et al. 2024; Paiva-Silva et al. 2025). For instance, bacteria might inject effector proteins into plant cells via type III secretion systems, disrupting normal cellular processes to favor infection (Hann and Rathjen 2010).

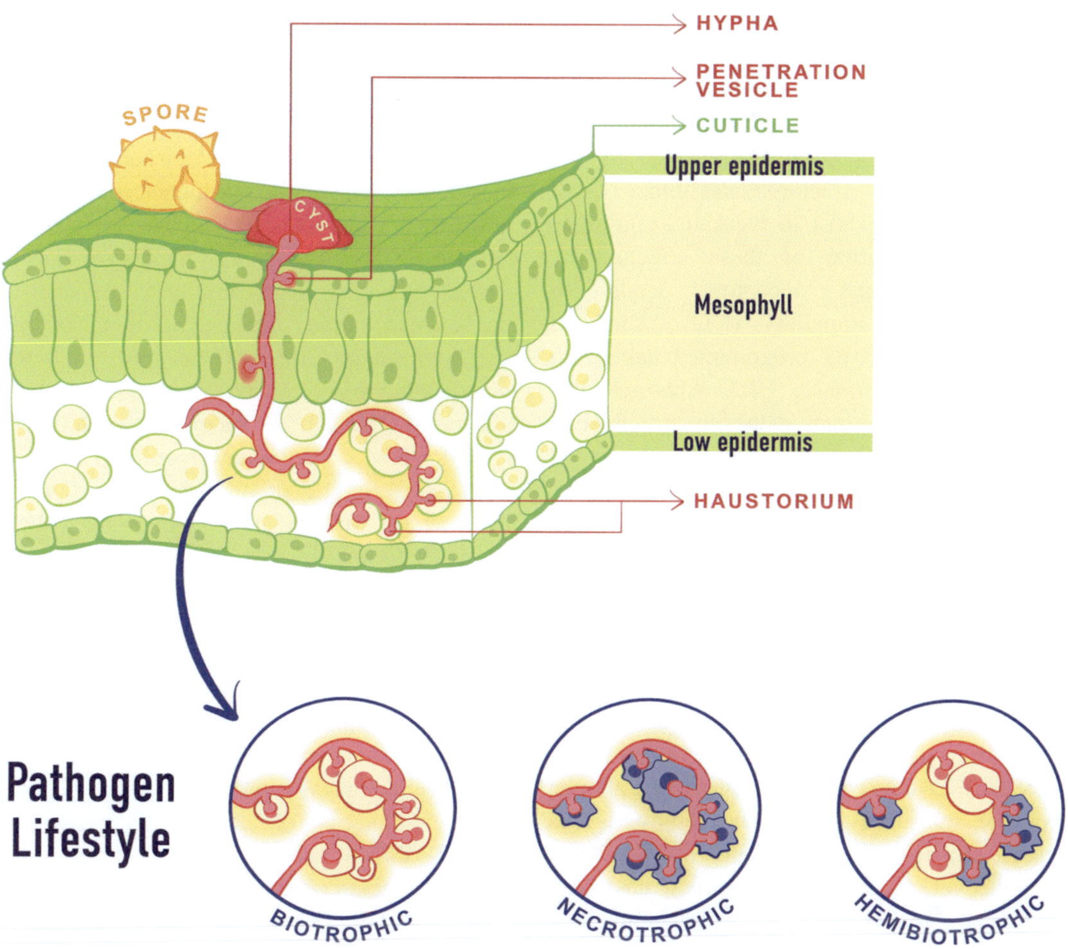

Fig. 19.1 Pathogen lifestyles and penetration strategies in plant tissues. The diagram illustrates how pathogens initiate infection by breaching the plant cuticle and epidermis using a penetration structure and developing intracellular feeding hyphae. After germination, the spore forms a cyst and penetration vesicle that enable entry through the upper epidermis, followed by hyphal growth and haustorium formation within mesophyll cells. Pathogens adopt distinct nutritional lifestyles that shape their mode of colonization. Biotrophic pathogens maintain host cells alive and extract nutrients through specialized structures such as haustoria. Necrotrophic pathogens kill host cells and degrade tissues to obtain nutrients. Hemibiotrophic pathogens begin infection with a brief biotrophic phase before transitioning to necrotrophy at later stages

Viruses, on the other hand, hijack the host's replication machinery to reproduce and spread within the plant (Patarroyo et al. 2013). In response, plants have developed sophisticated mechanisms to detect these effectors and mount appropriate defense responses, involving the activation of signaling cascades and the production of antimicrobial compounds (Dong and Ronald 2019).

In-depth analysis of plant–pathogen interactions reveals that pathogens must initially overcome physical barriers (as trichomes or cell walls). Once these barriers are breached, a complex multi-layered defense response is initiated, schematized in the *zig-zag model* (Jones and Dangl 2006). This model describes how microbe or pathogen-associated molecular patterns (PAMPs or MAMPs) and damage-associated molecular patterns (DAMPs) from the host are recognized by pattern recognition receptors (PRRs) on the plasma membrane, triggering PAMP-triggered immunity (PTI). To circumvent PTI, pathogens secrete proteins (such as Vir proteins or Avr effectors) and other small molecules into the host to weaken its immune defenses. Recognition of these effectors by specialized R-gene receptors leads to effector-triggered immunity (ETI), which is often accompanied by programmed cell death (hypersensitive response, HR). Although PTI activation often restrains pathogen growth, ETI typically activates a stronger and more sustained defense response. The establishment of PTI, ETI, and HR involves communication within and between cells, with several plant subcellular compartments playing specialized roles in defense (Hossain et al. 2012; Figueiredo et al. 2023a).

Proteins play a crucial role in plant defense mechanisms. They act as receptors, signaling molecules, and as defense compounds (e.g., proteases), enabling plants to perceive pathogen presence, transduce signals, and deploy various defense strategies (Bednarek and Osbourn 2009; Schreiber et al. 2021). Proteomics has been instrumental in identifying key proteins involved in these processes, such as PTI and ETI, shedding light on the complex interplay between plants and their pathogens. For example, receptor-like kinases (RLKs) play crucial roles in signal perception and transduction in PTI. Similarly, R proteins, which typically contain nucleotide-binding sites and leucine-rich repeat domains, are central to ETI (Figueiredo et al. 2022). Proteomics plays not only a critical role in unraveling these complex molecular interactions but may also be a key approach to identify potential biotechnological targets.

In the next sections we will explore recent case studies where a sound contribution of proteomics was given to characterize host and pathogen proteins crucial for the definition of the outcome of the interaction. We will focus on crops economically relevant worldwide such as *Glycine max* (soybean), *Solanum lycopersicum* (tomato), *Oryza sativa* (rice), *Vitis vinifera* (grapevine), *Zea mays* (maize), and *Triticum aestivum* (wheat).

19.2.1 *Glycine max* (L.) Merr

Soybean is globally valued for its high protein and oil content. This crop is extensively used in animal feed, human food products, and industrial applications, playing a significant role in both agricultural economies and dietary regimes. Its cultivation faces significant challenges due to various pathogens that cause diseases such as the soybean sudden death syndrome, soybean rust, soybean root rot, soybean cyst, and root-knot, leading to severe yield losses and economic constraints. Through 2D electrophoresis followed by ESI-MS/MS, 18 unique proteins in soybean roots that were differentially accumulated during the soybean sudden death syndrome (SDS), caused by *Fusarium virguliforme* were identified. Among these proteins, the GmRLK18-1, a receptor-like kinase located at the Rhg1Rfs2 locus emerged as a key player in soybean resistance against SDS. This kinase is crucial for the activation of signaling pathways that trigger the defense response, and it may be used as a target for biotechnological approaches such as genome editing (Iqbal et al. 2016). By using an LC-ESI-MS/MS approach five *F. virguliforme* secreted proteins were also identified in soybean plants xylem sap (Abeysekara and Bhattacharyya 2014). One of those proteins presented high similarity with cerato-platanin, a known phytotoxin, suggesting that multiple host-selective toxins may be produced by *F. virguliforme*, contributing to SDS. The detection of 112 soybean proteins in the xylem sap, including four pathogenicity-related defense proteins specifically in infected plants, also highlights the plant's response to infection. These authors have shown that *F. virguliforme* proteins can travel through the xylem, playing a role in disease progression, and lays the groundwork for further investigation into the molecular mechanisms of SDS development and soybean defense strategies.

Pereira et al. (2014) used 2-DE followed by MS to pinpoint changes in protein levels in soybean leaves during its interaction with the fungus *Phakopsora pachyrhizi*, which causes the Asian soybean rust (Pereira et al. 2013). The study identified soybean proteins such as carbonic anhydrase, 1-deoxy-D-xylulose-5-phosphate reductoisomerase, fructose-bisphosphate aldolase, and glutamine synthetase. The involvement of carbonic anhydrase and other proteins in plant defense responses highlights the biochemical adjustments occurring in the host during the interaction. 2-DE was also used to characterize the host responses after *Phytophthora sojae* infection, causing agent of soybean stem and root rot. Eighty-three host proteins were identified and by comparing the protein accumulation levels in resistant and susceptible genotypes, the study illustrated that reactive oxygen species (ROS) and salicylic acid (SA) pathways as pivotal in defense against *P. sojae* (Jing et al. 2015).

Nematodes are one of the most devastating pests for the soybean crop yields (e.g., soybean cyst nematode, *Heterodera glycines*, and the root-knot nematode, *Meloidogyne incognita*). A differentially labeled proteins by isobaric tags (iTRAQ) approach was used to explore the potential of the biocontrol agent *Sinorhizobium frediiv* strain Sneb183 as a tool for soybean cyst nematode control (Wang et al. 2020). This nitrogen-fixing bacterium enhances soybean resistance through isoflavonoid biosynthesis modulation. Key proteins such as GmPAL (phenylalanine ammonia-lyase), GmCHR (chalcone reductase), GmCHS (chalcone synthase), and GmIFS (isoflavone synthase) showed higher expression levels in treated samples. Another study using label-free quantification through MS identified pivotal proteins mediating constitutive defenses against root-knot nematodes in both susceptible and tolerant soybean genotypes. The phenylpropanoid pathway emerged as a crucial component in plant defense against this biotic stress, since it limits nematode reproduction (Arraes et al. 2022). The results from these studies highlighted the importance of an oxidative burst triggered during soybean–nematode interactions, involving reactive oxygen species (ROS) and compounds like H_2O_2 and nitric oxide (NO) to control the disease.

19.2.2 *Solanum lycopersicum* L.

Tomato is one of the most cultivated and consumed fruit crops around the world, revered for their versatility in culinary uses, from fresh salads to sauces and soups. Diseases like tomato late blight, wilt, and tobacco virus can devastate tomato crops, leading to substantial economic losses. Recent advances in proteomic and multi-omics studies have provided valuable insights into tomato's defense mechanisms against various pathogens. Fan et al. explored the modulation in the tomato proteome during infection by the hemibiotrophic oomycete *Phytophthora infestans*, which causes late blight disease, using a data-independent acquisition (DIA) proteomics approach (ref). This approach was recently developed as a global MS-based proteomics strategy where the precursor ions are isolated into pre-defined isolation windows and fragmented. All the fragmented ions are analyzed in a high-resolution manner (Li et al. 2021), which results in high reproducibility and accuracy. In this study, it was shown that during the biotrophic stage, proteins involved in glycolysis, fatty acid/lipid biosynthesis, and abscisic acid (ABA) signaling were accumulated while those involved in direct defense, redox homeostasis, and JA signaling were down accumulated. During the necrotrophic stage, resistance-related proteins, including those with direct defense functions, immune regulators mediating hypersensitive response (HR) and programmed cell death (PCD), and biosynthetic enzymes of defensive secondary metabolites, were also accumulated. Key proteins such as PR-5, ENO1, TPL3, PIP1, and CHLH/ABAR, were identified as key targets of *P. infestans* to suppress host defense/immunity via transcriptional or translational/post-translational regulation during different infection stages. The identification of these targets pose new ways to manipulate the interaction to favor the host through gene editing approaches.

The vascular wilt of tomato, caused by *Fusarium oxysporum* is another relevant disease, affecting most regions where this plant is grown. In 2024, Šimkovicová et al. explored the proteome modulation in resistant rootstocks (that harbors different R gene-mediated resistance) that were shown to protect a susceptible scion from fusarium wilt (Šimkovicová et al. 2024). Comparative proteomics by LC-MS showed distinct PR protein profiles in the host xylem sap. It was pinpointed that despite the structural diversity of the immune receptors from the resistant rootstock lines, there was an accumulation of a common set of four PR proteins, highlighting them as prime candidates for disease development prevention and therefore targets for biotechnological approaches. Another type of pathogens affecting tomato plants are viruses, such as the tomato spotted wilt virus (TSWV). Gupta et al. (2020) used a tandem mass tag (TMT)-based quantitative proteomics approach to identify TSWV-responsive proteins in susceptible and resistant cultivars (Gupta et al. 2020). The main advantage of this technique is to enable the simultaneous identification and quantification of proteins in multiple samples in a single experiment, increasing the accuracy and reducing experimental variability. In this study, the authors identified 1022 proteins with significant changes in response to TSWV infection. These proteins were mainly involved in secondary metabolism, ROS detoxification, MAP kinase signaling, calcium signaling, and jasmonate biosynthesis. A notable difference was observed between the susceptible and resistant cultivars. This approach enabled the identification of a larger number or proteins that were modulated in this pathosystem, enabling a deeper comprehension of the system.

Another strategy that is being explored as an alternative to conventional pesticides is the use of biocontrol agents. The study of three-way interactions between plants, pathogens, and biocontrol agents provides a more comprehensive understanding of disease suppression mechanisms compared to traditional two-way interactions. Paranthaman et al. (2023) used comparative proteomics (2D-MS) to identify key functional proteins involved in these interactions (Paranthaman et al. 2023). The research revealed significant changes in the proteome of tomato plants when treated with both the pathogen *Fusarium oxysporum* and the biocontrol agent *Pseudomonas fluorescens* (TPf12). *Pseudomonas fluorescens* application through seed treatment and soil drenching effectively prevents vascular wilt disease. This is the first study to investigate triadic protein expression in tomatoes in response to *Fusarium oxysporum* f. sp. solani infection under the antagonistic effects of *P. fluorescens*. The interaction between tomato plants, the biocontrol agent, and the pathogen revealed that proteins such as phenylalanine ammonia-lyase and lipid transfer-like proteins are upregulated in TPf12-treated roots, indicating their role in enhancing resistance to wilt disease. The complexity of the system, which requires the simultaneous actions of multiple players, indicates that omics tools and computational biology are essential to fully visualize the interactions of biocontrol agents. Understanding these triadic interactions is crucial for developing reliable biopesticides that meet field and market demands. This knowledge can help in selecting effective biological agents and their application methods for disease management strategies.

19.2.3 *Oryza sativa* L.

Rice is a primary food source for more than half of the global population, particularly in Asia. Diseases like rice blast, bacterial blight, and sheath blight affect the entire production. Proteomics also enabled gaining more knowledge into rice–pathogen interactions and into the application of priming agents. In 2022, Zhou et al. combined a two-dimensional liquid chromatography with the matrix-assisted laser desorption/ionization time-of-flight mass spectrometry to compare and identify differentially labeled proteins by isobaric tags (iTRAQ) between blast resistant and susceptible cultivars, with and without infection by *Magnaporthe grisea* and with and without application of salicylic acid (SA). SA was previously pointed out to have potential to increase rice resistance to this pathogen. Overall, 139 differentially accumulated proteins (DAP)s were identified mainly associated with energy, signal transduction, redox homeostasis, and protein hydrolysis. Comparing the SA-treated and untreated cultivars, the authors showed that the accumulation patterns tend to return to normal (recovery) after SA application in the susceptible genotype, whereas in the resistant, SA-mediated pathways seemed to be constitutively activated (Zhou et al. 2022).

The fungus *Rhizoctonia solani*, causing the rice sheath blight, is another relevant rice pathogen. Feng et al. (2022) also used an iTRAQ based quantitative proteomics approach to examine the defense responses triggered by infection in rice in a resistant cultivar (Feng et al. 2022). The study identified 319 DAPs involved in defense-related pathways, such as cell redox homeostasis, carbohydrate metabolism, and phenylpropanoid biosynthesis. Proteins related to photosynthesis and chlorophyll biosynthesis were downregulated upon infection, suggesting pathogen strategies to facilitate colonization. The findings provide valuable insights into the host proteins involved in rice resistance against *R. solani*, enhancing our understanding of the rice–pathogen interaction dynamics.

Rice bacterial blight disease is caused by *Xanthomonas oryzae* pv. *Oryzae* (Xoo). To gain more insights into this interaction, Wu et al. (2021) employed 2-DE comparative proteomics to identify host proteins that affect pathogen virulence (Wu et al. 2021). Five proteins from Xoo were significantly modulated in susceptible rice leaf extract treatment. Four host-induced proteins were shown to affect virulence and two novel host-induced virulence factors XanA and Imp were identified. These virulence factors were shown to induce significant changes in the global gene expression of susceptible rice genotypes. Overall, this study enabled the identification of pathogen key virulence factors that may be targets of biotechnological approaches. Another study used an LC-MS/MS approach to analyze proteome-level differences between incompatible and compatible interactions in rice infected with Xoo (Zhang et al. 2020). A total of 374 differentially abundant rice proteins and 117 Xoo proteins were identified, highlighting key regulatory pathways in the defense responses of rice against Xoo. This study revealed that during an incompatible interaction, rice plants prevent pathogen invasion and initiate multi-component defense responses that inhibit disease development. Most of Xoo DAPs were related to pathogen virulence and less abundant in the incompatible interaction (e.g., type III secretion systems proteins, TonB-dependent receptors). Rice DAPs were involved mainly in phenylpropanoid and flavonoid metabolisms, also SA-related DAPs were more accumulated in the incompatible interaction.

19.2.4 *Vitis vinifera* L.

Grapevine is an economically pivotal crop worldwide, primarily cultivated for wine production, fresh fruit, raisins, and juice. The cultivated grapevine species, *Vitis vinifera*, is susceptible to several diseases that severely impact health, fruit quality, and yield. Grapevine pathogens such as *Plasmopara viticola* (downy mildew), *Botrytis cinerea* (gray mold), among others present unique challenges to viticulture (Santos and Figueiredo 2023).

Several studies have focused on the early onset of the infection, exploiting the whole leaf proteome modulation and targeting subcellular compartments as the extracellular space. The extracellular space (i.e., apoplast) is the space outside of the plasma membrane, including all interfibrillar and intermicellar cell wall spaces, the fluid in the intercellular spaces, as well as all spaces and matrices generated between the plant plasma membrane and pathogen cell wall during infection (Farvardin et al. 2020). In plant–pathogen interactions, the apoplast is the main contact point of host and pathogen secretomes. The outcome of an interaction between a plant and a pathogen is often determined in the apoplast (Figueiredo et al. 2023b). A gel free LC-MS/MS approach was followed to characterize the proteome modulation in the apoplast and in the whole leaf at 6 and 12 h of the interaction between tolerant grapevine cultivar, 'Regent,' and *Plasmopara viticola* (Figueiredo et al. 2017, 2022). Santos et al. (2020) pinpointed that defense-related proteins were negatively modulated at 6 h post-inoculation but accumulated at 12 h post-inoculation (Santos et al. 2020). Indicators of both effector-triggered susceptibility (ETS) and effector-triggered immunity (ETI) were observed, suggesting that these processes co-occur and that the defense process is highly plastic. This study provided insights into defense strategies of the tolerant grapevine cultivar against *P. viticola*, identifying potential resistance-associated candidates for breeding or biotechnological based approaches. In 2023, the early cellular communication dynamics established an incompatible interaction between grapevine and *P. viticola*, showcasing upregulated defense-related proteins involved in pathogen recognition and signal transduction pathways in the plant apoplast (Figueiredo et al. 2023b). Upon inoculation with this pathogen, the cultivar 'Regent' activates mechanisms such as oomycete recognition, modifications to plant cell walls, ROS signaling, and secretion of proteins aimed at disrupting the oomycete structures. While at 6 h post-inoculation, proteins related to development and virulence are predominant in the *P. viticola* proteome within the plant (Figueiredo et al. 2023b). Another study exploited the *P. viticola* proteome during its interaction with susceptible and resistant grapevine hosts, revealing distinct infection strategies in the pathosystem (Figueiredo et al. 2022). *P. viticola* was shown to manipulate the SA pathway and isoprenoid biosynthesis to facilitate plant colonization in the compatible interaction. In contrast, in the resistant grapevine cultivar, development-associated proteins increased, and oxidoreductases protect *P. viticola* from ROS-associated plant defenses. Pathogen effector proteins with the conserved RXLR-EER motif were also found during the interaction with resistance cultivar 'Regent' (Figueiredo et al. 2022). Furthermore, the role of proteases in plant defense, including their contribution to hypersensitive response, priming, and immunity, was emphasized, although their specific involvement in grapevine defense remains underexplored (Santos and Figueiredo 2021).

Liu et al. (2021) performed a comparative proteomic analysis of the resistant genotype *V. davidii* "LiuBa-8" and susceptible *V. vinifera* 'Pinot Noir' (PN) 12 h after inoculation with *P. viticola* through an iTRAQ approach. Several DAPs were involved in photosynthesis, metabolism, stress, and redox homeostasis, providing insights into early infection mechanisms. Enzymes like glucose-6-phosphate 1-dehydrogenase and 6-phosphogluconate dehydrogenase were induced in resistant genotypes, aiding in defense through phenolic compound production. Pyruvate decarboxylase strongly increased in the resistant genotype, indicating enhanced disease resistance. HSP70 and HSP90, along with small heat shock proteins like HSP17.8, were shown to play crucial roles in pathogen resistance, particularly in resistant genotypes (Liu et al. 2021a).

One of the strategies to manage the downy mildew disease is the use of crossing hybrids, developed in breeding programs that present one of more resistance loci to *P. viticola*. In that sense, proteomics was also used to understand the host response to pathogen inoculation and to identify potential targets for protection strategies. By studying a pyramided crossing hybrid, with two resistance loci—Rpv1/Rpv3, the authors showed that significant proteomic changes occurred at 48 and 96 hours post-infection with *P. viticola*, mainly associated to antioxidant metabolism and defense responses, highlighting the systemic response to infection and identifying potential resistance-associated candidates (Nascimento-Gavioli et al. 2017). Another approach to manage grapevine diseases is the use of priming agents, on that sense, 2-DE was used to study the onset of laminarin compounds application (sulfated (PS3) and the shortened (H13) forms of laminarin are both known to elicit defense responses whereas only PS3 induces resistance against downy mildew) (Lemaître-Guillier et al. 2017). This comparative proteomics approach enabled the cross comparison of the proteomes obtained from the two treatments upon elicitation and pathogenic challenge stages. The two treatments presented distinct proteome modulation, the PS3 provoked a major shift in protein accumulation, compatible with defense priming agents, whereas H13 induced weak proteome modulation. After inoculation, PS3 was shown to enhance carbohydrate metabolism, both JA and SA metabolism, and to accumulate defense-related proteins "Arsenite resistance protein" (homolog to Serrate protein) and "OPR 12-oxophytodienoate reductase 11." These proteins were pointed as candidates in grapevine resistance against *P. viticola*.

Similarly, grapevine red blotch virus infection was shown by shotgun nLC-MS/MS to increase flavonoid biosynthesis enzymes, indicating an activated defense mechanism (Buchs et al. 2018). Pierce's disease, caused by *Xylella fastidiosa*, induced antioxidant strategies, phytoalexins, and cell wall modifications in grapevines (Zaini et al. 2018). Botryosphaeria dieback affected different cultivars differently, with increased pathogenesis-related proteins, antioxidants, and phenolics in infected wood across cultivars, but varying impacts on primary metabolism (Spagnolo et al. 2014). Advances in proteomic techniques, such as deep learning neural network predictions, have enabled the identification of a broader range of proteins associated with cellular responses to oxidative stress in grapevines infected with pathogens like *Xylella fastidiosa* (Helena Duarte Sagawa et al. 2020). These studies highlight common defense responses in grapevines, including upregulation of pathogenesis-related proteins, antioxidants, and cell wall modifications, while also revealing pathogen-specific and cultivar-specific variations in response to different diseases.

Lastly, as a perennial crop, studies in the grapevine plant model are restrained to the growing season, therefore new approaches as cell culture pose a promising biotechnological tool to study grapevine–pathogen interactions. 2-DE comparative proteomics was used to understand cell response to *Botrytis cinerea* infection (Dadakova et al. 2015). Thirty proteins with increased abundance after infection that were identified, including PR10 proteins, peroxidase, ascorbate peroxidase, glutathione S-transferases and Hsc proteins, had previously been found to be modulated by other pathogens in *V. vinifera* leaves and stems but not in berries, suggesting similar response of cells and green vegetative parts. This work opens new insights into the use of grapevine cell culture as a proxy to the green parts of the plant model. This presents high potential for biotechnological approaches.

19.2.5 *Zea mays* L.

Maize is a staple crop that serves as a vital food source and a key agricultural commodity worldwide. Its relevance extends beyond basic nutrition; maize is crucial for economies as it is used in animal feed, biofuel production, and as a raw material in various industries. Diseases caused by fungi, bacteria, and viruses, such as corn smut, blight, and maize dwarf mosaic virus, can reduce photosynthesis, hinder growth, and diminish the crop's commercial value.

The host–pathogen interaction of maize with *Ustilago maydis*, a pathogen causing tumor formation in maize, was studied by using in vivo proximity labeling and co-immunoprecipitation techniques. Molecular characterization of cell-specific mechanisms using cell type–specific approaches at the proteome level can provide new biological insights. To overcome the barriers, subcellular and cell-specific proteomics, in vivo proteomic labeling with proximity-dependent biotinylation of cytosolic proteins using biotin ligase TurboID, coupled with mass spectrometry (MS) of labeled proteins, emerged as a powerful strategy for cell type–specific proteomics in the native state of cells without the need for cellular isolation. This approach enabled the identification of the host target network of the tumor-inducing pathogen effector UmSee1, which is crucial for tumor formation in maize leaves during infection which is necessary for the pathogen to induce tumors in the bundle sheath cells of maize. This effector interacts with maize SGT1 and affects the ubiquitin-proteasome pathway, namely ZmSIP1, ZmSIP2, and ZmSIP3, which are involved in the regulation of the cell cycle. This study shed light on how UmSee1 manipulates host processes to induce tumor formation during *U. maydis–Z. mays* interaction (Shi et al. 2023).

Considering subcellular proteomics, Hafiz et al. (2022), through LC-MS/MS, identified specific proteins in the maize stem apoplast that were modulated in response to the infection with *Fusarium verticillioides*, the fungus producing the mycotoxin fumonisin, and the causal agent of rotting of maize plants (Hafiz et al. 2022; Omotayo and Babalola 2023). Hundred and twenty proteins were shown to be differentially accumulated after infection, being mainly related to the reinforcement of the cell wall, to plant immune response and signaling pathways activation. Another study explored the complementation of transcriptomics and proteomics to further disclose maize's response to *F. verticillioides* infection (Zhang et al. 2023). It identified differentially accumulated proteins and genes involved in glutathione metabolism, starch and sucrose metabolism, and phenylpropanoid biosynthesis. Key defense-related genes like *ZmCTA1*, *ZmWIP1*, and *ZmLOX2* were highlighted, with *ZmLOX2* playing a role in the defense response against fungal invasion via the SA-dependent pathway. These findings offer insights into maize's defense mechanisms against *F. verticillioides*, providing valuable information for potential breeding strategies targeting new genes for enhanced resistance.

Another disease affecting maize plants is the northern corn leaf blight disease, caused by the fungus *Setosphaeria turcica* (Liu et al. 2021b). A comparative proteomic analysis provided insights into the molecular mechanisms underlying the interaction between maize and the pathogen. By analyzing the proteomic changes in maize upon infection, the researchers were able to identify key proteins involved in the plant's response to *S. turcica*. The analysis revealed that proteins involved in photosynthesis were significantly reduced in infected plants, suggesting that the infection disrupts the plant's ability to produce energy efficiently. Proteins related to stress responses and defense mechanisms, such as those involved in the production of ROS and pathogenesis-related proteins, were found to be increased in infected plants, highlighting the plant's effort to combat the infection. The study also found that proteins associated with cell wall reinforcement were upregulated, indicating that maize strengthens its cell walls as a physical barrier against the fungal invasion.

19.2.6 *Triticum aestivum* L.

Wheat stands as one of the world's most crucial food crops, vital for global food security due to its role as a staple for over a third of the global population. It is a primary source of calories and protein in human diets across many cultures, particularly in the form of breads, pastas, and cereals. Diseases such as rust, smut, and blight not only lead to reduced crop productivity but also impact the nutritional value of the harvested grain.

Fusarium graminearum, a major pathogen causing Fusarium head blight (FHB) and Fusarium crown rot, interacts with wheat through a variety of cell wall proteins (CWPs) that play crucial roles in virulence and host interaction. A comprehensive proteomic analysis identified 1373 CWPs in *F. graminearum*, with specific proteins like FCA7 being critical for virulence on wheat (Moon et al. 2024). The accumulation of FCA7 and Cpd1 in planta, and the confirmation of their localization predominantly in the cell wall and periplasmic space, and the disruption of FCA7 lead to reduced virulence on wheat.

In the case of *Fusarium pseudograminearum*, causing Fusarium crown rot, TMT quantitative proteomics was used to evaluate the response of a disease-susceptible and a disease-tolerant wheat cultivar inoculated with the pathogen. Unique pathways were identified in both cultivars. In the susceptible cultivar, UC1110, the pathways were mainly related to linoleic acid metabolism, plant hormone signal transduction, mitogen activated protein kinase (MAPK) signaling pathway, and ubiquinone biosynthesis, while in the disease-tolerant cultivar these pathways were mainly related to photosynthesis, carbon fixation, flavone and flavonol biosynthesis, glyoxylate and dicarboxylate metabolism. This approach provides a perspective for wheat genetic improvement and breeding (Qiao et al. 2021). Furthermore, a gel-free LC-MS/MS approach was applied to study the interaction between three wheat cultivars with different susceptibility degrees and three *F. graminearum* strains of varying aggressiveness. Through comparative proteomics analysis the authors highlighted that most proteome adjustments in wheat were shared across cultivars, and cultivar distinction being mainly due to basal abundance differences, especially in chloroplast-associated functions (Fabre et al. 2021).

iTRAQ was used to exploit the differences of wheat spikes to *Tilletia controversa* and *Tilletia foetida*, causing dwarf bunt and common bunt, respectively. The DAP identified were mainly associated with metabolic process, catalytic activity, transferase activity, photosynthetic membrane, extracellular region, and oxidoreductase activity. Pathogenesis-related proteins were also accumulated. The results expanded our understanding of the proteome in wheat spikes in response to *T. controversa* and *T. foetida* infection and provide a basis for further investigation for improving the defense mechanism of the wheat crops (Li et al. 2018). Pathogenesis-related proteins were also accumulated in wheat apoplastic fluid after *Puccinia triticina* infection, mainly peroxidases and chitinases (Rampitsch et al. 2019). A label-free LC-MS/MS approach was conducted to investigate temporal changes in the abundance of the apoplastic fluid proteome of resistant and susceptible wheat leaves infected with *P. triticina*. Most of the host proteins altered their abundance after 72 h of infection, being the PR proteins the main modulated proteins. Five days after inoculation with the susceptible apoplast, 150 *P. triticina* proteins were detected and 117 host proteins. The latter represents potential targets of pathogen effectors (Rampitsch et al. 2019).

The response of wheat to stripe rust and powdery mildew, caused by *Puccinia striiformis* f. sp. *tritici* and *Blumeria graminis* f. sp. *tritici*, respectively, was studied by TMT proteomics coupled to LC-MS/MS. This approach enabled the identification of candidates that play a key role in powdery mildew resistance but not to stripe rust (Wang et al. 2022). Lastly, the proteomic analysis of *Pyrenophora tritici-repentis*, causing tan spot, identified secreted proteins that interact directly with the wheat host, providing insights into the pathogen's infection mechanisms (Moffat et al. 2018).

Proteomics was also used to study the biocontrol potential of *Bacillus velezensis* to the *Gaeumannomyces graminis* var. *tritici* pathogen (wheat leaf rust caused) showing that wheat prioritizes defense against the pathogen, with SA-dependent resistance being crucial for combating the necrotrophic pathogen (Kang et al. 2019). Proteomic analysis of wheat bunt fungi, including *Tilletia controversa*, *T. caries*, and *T. foetida*, identified differentially expressed proteins that could serve as markers for pathogen detection and monitoring (Li et al. 2018).

19.3 Proteomics as a Tool for Advancing Plant Biotechnology

Much of our knowledge in plant–pathogen interactions was achieved through genomic and transcriptomic approaches. Nonetheless correlation between transcripts and protein levels is often low. Through detailed proteomic approaches, researchers have identified key proteins and pathways that are crucial for the outcome of a plant–pathogen interaction. Their quantification through mass spectrometry enables a deeper understanding of their role within complex signaling pathways (e.g., proteins involved in the production of antimicrobial compounds or cell wall reinforcement are often upregulated in response to pathogens). By identifying these proteins and understanding their function, scientists can engineer plants to enhance the expression of specific proteins or pathways, thereby boosting the plant's innate immunity (Mahmood et al. 2022).

Additionally, proteomics can reveal the effects of genetic engineering at the proteome level, ensuring that engineered plants maintain a balanced defense response without compromising important agronomic traits. The CRISPR-Cas system has become the most widely used for gene-editing approaches due to its accuracy and efficiency; however, off-targets and insertion of

Cas genes in gene-edited crops may occur, potentially generating new transcripts or proteins. Safety assessments are also being carried out through proteomics (Xiao-Jing et al. 2023). Xiao-Jing et al. (2023) have used a TMT-MS proteomics combined to a transcriptomics approach to compare the effects of CRISPR-Cas9 rice-edited plants with WT plants, stating that rice-edited plants presented little effect on the transcription levels and that no new proteins were generated in gene-edited rice lines (Xiao-Jing et al. 2023). Proteomic studies were also conducted to understand the alterations caused by CRISPR/Cas9 gene editing in rice–*M. oryzae* interaction. Nawaz et al. (2020) performed a comparative proteomics profiling between mutant homozygous rice lines and wild type to evaluate the gene-editing effect targeting the Pi21 gene (broad-spectrum resistance that encodes a proline-rich protein that includes a putative heavy metal–binding domain and putative protein–protein interaction motifs). They have identified 372 DAPs and provided a comprehensive overview of Pi21 loss of function mutations. The homozygous mutant lines will be used for rice blast resistance improvement (Nawaz et al. 2020).

Proteomics approaches also contribute to plant breeding programs by providing novel targets that might be used in the selection process. One significant application is marker-assisted selection (MAS), which utilizes DNA markers identified through proteomic analyses to expedite the breeding of plants with desirable traits (Eldakak et al. 2013). This approach allows breeders to screen for specific genetic markers associated with enhanced resistance or productivity, thus streamlining the breeding process. Moreover, proteomics offers comprehensive insights into the functional dynamics of plant–pathogen interactions, highlighting potential genetic modifications that could improve crop resilience (Liu et al. 2019). By integrating proteomic data with traditional breeding techniques, the efficiency and effectiveness of plant breeding programs are greatly enhanced.

The use of selected reaction monitoring (SRM) methodology in plant proteomics aids in biomarker validation, moving beyond the exploratory phase to practical applications in crop improvement (Hu et al. 2015). SRM methodology plays a crucial role in plant proteomics by providing precise and reliable quantification of specific proteins involved in plant–pathogen interactions. This targeted mass spectrometry technique allows for the validation of potential biomarkers linked to disease resistance, facilitating the development of robust, disease-resistant crop varieties. By focusing on key proteins and pathways, SRM helps researchers move beyond the exploratory phase of proteomics to practical applications in crop improvement, such as targeted breeding programs that select for naturally resistant plants. Furthermore, SRM's integration with other omics approaches, such as genomics and metabolomics, provides a comprehensive understanding of the regulatory networks governing plant defense mechanisms. This holistic view enables the identification of new targets for genetic engineering and breeding, contributing to the development of crops that can better withstand evolving pathogens and changing environmental conditions. The cost-effectiveness and high-throughput capability of SRM make it particularly valuable in large-scale breeding programs, ensuring the sustainable development of resilient crop varieties.

19.4 Future Directions in Proteomics for Plant–Pathogen Interactions

Future work in the field of plant–pathogen interactions, particularly in proteomics, will likely focus on several key areas. One priority will be the deeper integration of proteomics with other omics technologies, such as genomics, transcriptomics, and metabolomics, to create a comprehensive view of the molecular mechanisms governing plant defenses (Elmore et al. 2021). This systems-level approach could provide insights into complex signaling pathways and regulatory networks, offering more precise targets for genetic engineering and breeding programs. Another area of focus will be

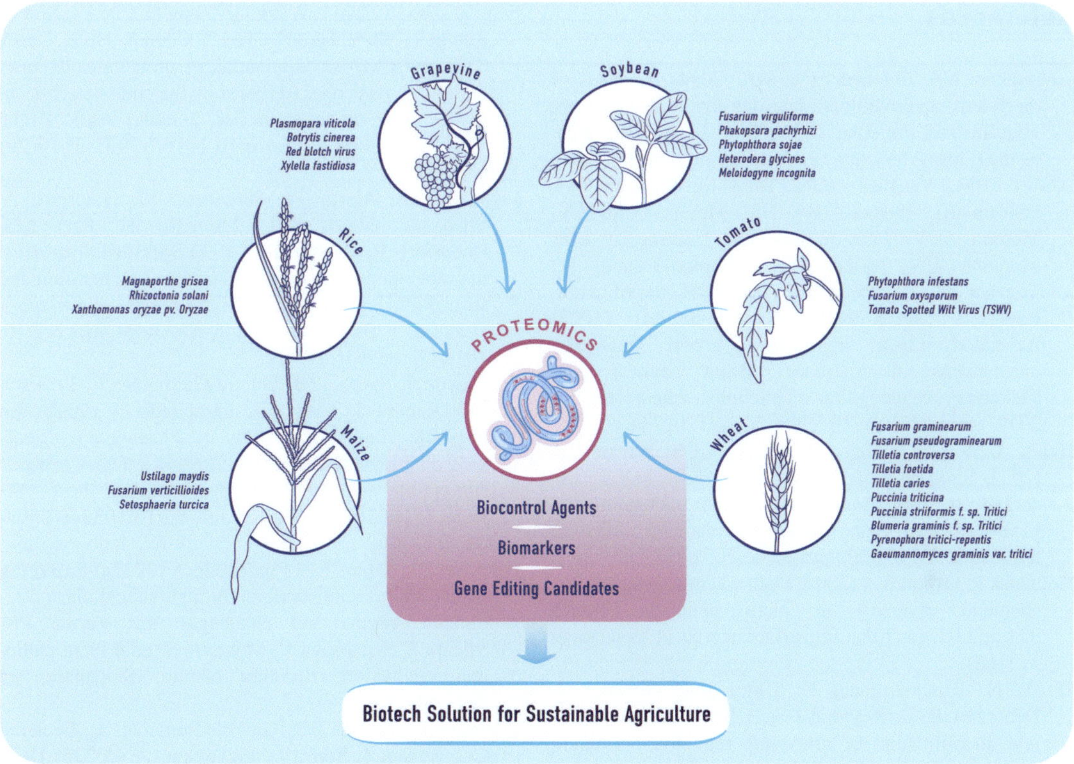

Fig. 19.2 Future directions for proteomics in plant–pathogen interactions. The figure highlights how proteomics will increasingly integrate with other omics, advanced mass spectrometry, CRISPR-based editing, and AI-driven data analysis to identify biomarkers and targets for crop resistance. As climate change intensifies stress on plants, proteomics will also help ensure that engineered traits remain effective and resilient across diverse environmental conditions

improving the accuracy and efficiency of proteomic tools, such as mass spectrometry, to increase the sensitivity of detecting low-abundance proteins, which are often critical in early defense responses or in specific plant subcellular locations such as the apoplast. As CRISPR-Cas technology continues to evolve, future studies should assess the long-term effects of gene editing at the proteome level to ensure both efficacy and safety, particularly in commercial crop development. Additionally, there is potential for the use of artificial intelligence and machine learning to analyze large-scale proteomic data, which could significantly speed up the identification of key biomarkers and targets for enhancing crop resistance (Tran et al. 2023). Finally, as environmental conditions continue to shift due to climate change, proteomic research should also investigate how these stressors impact plant–pathogen interactions and the stability of engineered traits, ensuring that future crops are not only disease-resistant but also resilient to a broader range of abiotic stresses (Fig. 19.2).

Acknowledgments RBS was funded through the FCT CEEC program (2021.00795.CEECIND; https://doi.org/10.54499/2021.00795.CEECIND/CP1654/CT0005) and CP-S was funded through an FCT PhD fellowship (2025.02061.BD). This work was supported by UID/04046/2025—Biosystems and Integrative Sciences Institute Centre grant from FCT, Portugal.

References

Abeysekara NS, Bhattacharyya MK (2014) Analyses of the xylem sap proteomes identified candidate Fusarium virguliforme Proteinacious toxins. PLoS One 9(5): e93667. https://doi.org/10.1371/journal.pone.0093667

Arraes FBM, Vasquez DDN, Tahir M, Pinheiro DH, Faheem M, Freitas-Alves NS, Moreira-Pinto CE, Moreira VJV, Paes-de-Melo B, Lisei-de-Sa ME, Morgante CV, Mota APZ, Lourenço-Tessutti IT, Togawa RC, Grynberg P, Fragoso RR, de Almeida-Engler J, Larsen MR, Grossi-de-Sa MF (2022) Integrated Omic approaches reveal molecular mechanisms of tolerance during soybean and Meloidogyne incognita interactions. Plants 11(20): 2744. https://doi.org/10.3390/plants11202744

Balotf S, Wilson R, Tegg RS, Nichols DS, Wilson CR (2022) Shotgun proteomics as a powerful tool for the study of the proteomes of plants, their pathogens, and plant–pathogen interactions. Proteomes 10(1):5. https://doi.org/10.3390/proteomes10010005

Bednarek P, Osbourn A (2009) Plant-microbe interactions: chemical diversity in plant defense. Science 324(5928):746–748. https://doi.org/10.1126/science.1171661

Buchs N, Braga-Lagache S, Uldry A-C, Brodard J, Debonneville C, Reynard J-S, Heller M (2018) Absolute quantification of grapevine red blotch virus in grapevine leaf and petiole tissues by proteomics. Front Plant Sci 9. https://doi.org/10.3389/fpls.2018.01735

Dadakova K, Havelkova M, Kurkova B, Tlolkova I, Kasparovsky T, Zdrahal Z, Lochman J (2015) Proteome and transcript analysis of *Vitis vinifera* cell cultures subjected to *Botrytis cinerea* infection. J Proteome 119:143–153. https://doi.org/10.1016/j.jprot.2015.02.001

Dong OX, Ronald PC (2019) Genetic engineering for disease resistance in plants: recent Progress and future perspectives. Plant Physiol 180(1):26–38. https://doi.org/10.1104/pp.18.01224

Eldakak M, Milad SI, Nawar AI, Rohila JS (2013) Proteomics: a biotechnology tool for crop improvement. Front Plant Sci 4. https://doi.org/10.3389/fpls.2013.00035

Elmore JM, Griffin BD, Walley JW (2021) Advances in functional proteomics to study plant-pathogen interactions. Curr Opin Plant Biol 63:102061. https://doi.org/10.1016/j.pbi.2021.102061

Fabre F, Urbach S, Roche S, Langin T, Bonhomme L (2021) Proteomics-based data integration of wheat cultivars facing Fusarium graminearum strains revealed a core-responsive pattern controlling Fusarium head blight. Front Plant Sci 12. https://doi.org/10.3389/fpls.2021.644810

Farvardin A, González-Hernández AI, Llorens E, García-Agustín P, Scalschi L, Vicedo B (2020) The Apoplast: a key player in plant survival. Antioxidants 9(7):604. https://doi.org/10.3390/antiox9070604

Feng Z, Gao P, Zhao J, Wang G, Zhang H, Cao W, Xue X, Zhang Y, Ma Y, Hua R, Chen Z, Chen X, Hu K, Zuo S (2022) iTRAQ-based quantitative proteomics analysis of defense responses triggered by the pathogen *Rhizoctonia solani* infection in rice. J Integr Agric 21(1): 139–152. https://doi.org/10.1016/S2095-3119(20)63499-2

Figueiredo A, Martins J, Sebastiana M, Guerreiro A, Silva A, Matos AR, Monteiro F, Pais MS, Roepstorff P, Coelho AV (2017) Specific adjustments in grapevine leaf proteome discriminating resistant and susceptible grapevine genotypes to *Plasmopara viticola*. J Proteome 152:48–57. https://doi.org/10.1016/j.jprot.2016.10.012

Figueiredo J, Santos RB, Guerra-Guimarães L, Leclercq CC, Renaut J, Malhó R, Figueiredo A (2022) An in-planta comparative study of Plasmopara viticola proteome reveals different infection strategies towards susceptible and Rpv3-mediated resistance hosts. Sci Rep 12(1):20794. https://doi.org/10.1038/s41598-022-25164-8

Figueiredo J, Santos RB, Figueiredo A (2023a) Exploring plant-pathogen interactions through subcellular proteomics: insights and challenges. In: Verma PK, Mishra S, Srivastava V, Mehrotra S (eds) Plant pathogen interaction. Springer Nature, Singapore, pp 287–310

Figueiredo J, Santos RB, Guerra-Guimarães L, Leclercq CC, Renaut J, Sousa L, Figueiredo A (2023b) Deep into the Apoplast: grapevine and Plasmopara viticola proteomes reveal the secret beneath host and pathogen communication at 6 h after contact. Phytopathology 113(5):893–903. https://doi.org/10.1094/PHYTO-09-22-0340-FI

Gouveia C, Santos RB, Paiva-Silva C, Buchholz G, Malhó R, Figueiredo A (2024) The pathogenicity of Plasmopara viticola: a review of evolutionary dynamics, infection strategies and effector molecules. BMC Plant Biol 24(1):327. https://doi.org/10.1186/s12870-024-05037-0

Gupta R, Min CW, Kim SW, Yoo JS, Moon A R, Shin A-Y, Kwon S-Y, Kim ST (2020) A TMT-based quantitative proteome analysis to elucidate the TSWV induced signaling Cascade in susceptible and resistant cultivars of Solanum lycopersicum. Plants 9(3):290. https://doi.org/10.3390/plants9030290

Hafiz AH, Jun Z, Yu-shuang G, Mei-xu G, Wei G (2022) Proteomic analysis of pathogen-responsive proteins from maize stem apoplast triggered by *Fusarium verticillioides*. J Integr Agric 21(2):446–459. https://doi.org/10.1016/S2095-3119(21)63657-2

Hann DR, Rathjen JP (2010) The long and winding road: virulence effector proteins of plant pathogenic bacteria. Cell Mol Life Sci 67(20):3425–3434. https://doi.org/10.1007/s00018-010-0428-1

Helena Duarte Sagawa C, Zaini PA, de A. B. Assis R, Saxe H, Salemi M, Jacobson A, Wilmarth PA, Phinney BS, M. Dandekar A (2020) Deep learning neural network prediction method improves proteome profiling of vascular sap of grapevines during Pierce's disease development. Biology (Basel) 9(9):261. https://doi.org/10.3390/biology9090261

Hossain Z, Nouri M-Z, Komatsu S (2012) Plant cell organelle proteomics in response to abiotic stress. J Proteome Res 11(1):37–48. https://doi.org/10.1021/pr200863r

Hu J, Rampitsch C, Bykova NV (2015) Advances in plant proteomics toward improvement of crop productivity and stress resistance. Front Plant Sci 6. https://doi.org/10.3389/fpls.2015.00209

Iqbal MJ, Majeed M, Humayun M, Lightfoot DA, Afzal AJ (2016) Proteomic profiling and the predicted Interactome of host proteins in compatible and incompatible interactions between soybean and Fusarium virguliforme. Appl Biochem Biotechnol 180(8): 1657–1674. https://doi.org/10.1007/s12010-016-2194-5

Jain A, Singh HB, Das S (2021) Deciphering plant-microbe crosstalk through proteomics studies. Microbiol Res 242:126590. https://doi.org/10.1016/j.micres.2020.126590

Jing M, Ma H, Li H, Guo B, Zhang X, Ye W, Wang H, Wang Q, Wang Y (2015) Differential regulation of defense-related proteins in soybean during compatible and incompatible interactions between Phytophthora sojae and soybean by comparative proteomic analysis. Plant Cell Rep 34(7):1263–1280. https://doi.org/10.1007/s00299-015-1786-9

Jones JDG, Dangl JL (2006) The plant immune system. Nature 444(7117):323–329. https://doi.org/10.1038/nature05286

Jorrin Novo JV (2021) Proteomics and plant biology: contributions to date and a look towards the next decade. Expert Rev Proteomics 18(2):93–103. https://doi.org/10.1080/14789450.2021.1910028

Kang X, Wang L, Guo Y, ul Arifeen MZ, Cai X, Xue Y, Bu Y, Wang G, Liu C (2019) A comparative transcriptomic and proteomic analysis of Hexaploid wheat's responses to colonization by bacillus velezensis and Gaeumannomyces graminis, Both Separately and Combined. MPMI 32(10):1336–1347. https://doi.org/10.1094/MPMI-03-19-0066-R

Klose J (1975) Protein mapping by combined isoelectric focusing and electrophoresis of mouse tissues. Hum Genet 26(3):231–243. https://doi.org/10.1007/BF00281458

Lemaître-Guillier C, Hovasse A, Schaeffer-Reiss C, Recorbet G, Poinssot B, Trouvelot S, Daire X, Adrian M, Héloir M-C (2017) Proteomics towards the understanding of elicitor induced resistance of grapevine against downy mildew. J Proteome 156:113–125. https://doi.org/10.1016/j.jprot.2017.01.016

Li C, Wei X, Gao L, Chen W, Liu T, Liu B (2018) iTRAQ-based proteomic analysis of wheat bunt fungi Tilletia controversa, T. caries, and T. foetida. Curr Microbiol 75(8):1103–1107. https://doi.org/10.1007/s00284-018-1490-4

Li J, Smith LS, Zhu H-J (2021) Data-independent acquisition (DIA): an emerging proteomics technology for analysis of drug-metabolizing enzymes and transporters. Drug Discov Today Technol 39:49–56. https://doi.org/10.1016/j.ddtec.2021.06.006

Liu Y, Lu S, Liu K, Wang S, Huang L, Guo L (2019) Proteomics: a powerful tool to study plant responses to biotic stress. Plant Methods 15(1):135. https://doi.org/10.1186/s13007-019-0515-8

Liu G-T, Wang B-B, Lecourieux D, Li M-J, Liu M-B, Liu R-Q, Shang B-X, Yin X, Wang L-J, Lecourieux F, Xu Y (2021a) Proteomic analysis of early-stage incompatible and compatible interactions between grapevine and P. viticola. Horticulture Research 8:100. https://doi.org/10.1038/s41438-021-00533-y

Liu Y, Gong X, Zhou Q, Liu Y, Liu Z, Han J, Dong J, Gu S (2021b) Comparative proteomic analysis reveals insights into the dynamic responses of maize (*Zea mays* L.) to *Setosphaeria turcica* infection. Plant Sci 304:110811. https://doi.org/10.1016/j.plantsci.2020.110811

Mahmood MA, Naqvi RZ, Mansoor S (2022) Engineering crop resistance by manipulating disease susceptibility genes. Mol Plant 15(10):1511–1513. https://doi.org/10.1016/j.molp.2022.09.010

Maia M, Ferreira AEN, Nascimento R et al (2020) Integrating metabolomics and targeted gene expression to uncover potential biomarkers of fungal/oomycetes-associated disease susceptibility in grapevine. Sci Rep 10:15688. https://doi.org/10.1038/s41598-020-72781-2

Moffat CS, Stoll T, Moolhuijzen P (2018) Proteomics of the wheat tan spot pathogen Pyrenophora tritici-repentis. BMC Res Notes 11(1):846. https://doi.org/10.1186/s13104-018-3936-y

Moon H, Min K, Winarto J, Shin S, Jeon H, Song D-G, Son H (2024) Proteomic analysis of cell wall proteins with various linkages in Fusarium graminearum. J Agric Food Chem 72(11):6028–6039. https://doi.org/10.1021/acs.jafc.3c07746

Nascimento-Gavioli MCA, Agapito-Tenfen SZ, Nodari RO, Welter LJ, Sanchez Mora FD, Saifert L, da Silva AL, Guerra MP (2017) Proteome of *Plasmopara viticola*-infected *Vitis vinifera* provides insights into grapevine *Rpv1/Rpv3* pyramided resistance to downy mildew. J Proteome 151:264–274. https://doi.org/10.1016/j.jprot.2016.05.024

Nawaz G, Usman B, Peng H, Zhao N, Yuan R, Liu Y, Li R (2020) Knockout of Pi21 by CRISPR/Cas9 and iTRAQ-based proteomic analysis of mutants revealed new insights into M. oryzae Resistance in Elite Rice Line. Genes 11(7):735. https://doi.org/10.3390/genes11070735

Omotayo OP, Babalola OO (2023) Fusarium verticillioides of maize plant: potentials of propitious phytomicrobiome as biocontrol agents. Front Fungal Biol 4. https://doi.org/10.3389/ffunb.2023.1095765

Paiva-Silva C, Proença Pereira J, Marcolino F, Figueiredo A, Santos RB (2025) Silenced cutters: mechanisms and effects of protease inhibition in plant-pathogen interactions. J Exp Bot 76(15): 4248–4261. https://doi.org/10.1093/jxb/eraf156

Paranthaman L, Seethapathy P, Pandita D, Gopalakrishnan C, Sankaralingam S, Venkatesh S,

Malaisamy A, Pandita A, Casini R, Alataway A, Dewidar AZ, Almutairi KF, Elansary HO (2023) Levering proteomic analysis of Pseudomonas fluorescens mediated resistance responses in tomato during pathogenicity of Fusarium oxysporum f sp oxysporum. Front Sustain Food Syst 7. https://doi.org/10.3389/fsufs.2023.1157575

Patarroyo C, Laliberté J-F, Zheng H (2013) Hijack it, change it: how do plant viruses utilize the host secretory pathway for efficient viral replication and spread? Front Plant Sci 3. https://doi.org/10.3389/fpls.2012.00308

Pereira MR, Gouvêa BC, Marcelino-Guimarães FC, de Ramos HJO, Moreira MA, de Barros EG (2013) Proteomic analysis of soybean leaves in a compatible and an incompatible interaction with Phakopsora pachyrhizi. Organelles Proteomics 1 (1): 16-27. https://doi.org/10.2478/orpr-2013-0004

Pereira G, Pereira J, Santos RB, Figueiredo A (2022) Uncovering the role of DNA methyltransferases in grapevine-plasmopara viticola interaction: from genome-wide characterization to global methylation patterns. Gene 837:146693

Pereira JP, Bevilacqua I, Santos RB, Varotto S, Chitarra W, Nerva L, Figueiredo A (2025) Epigenetic regulation and beyond in grapevine-pathogen interactions: a biotechnological perspective. Physiol Plant 177(2):e70216. https://doi.org/10.1111/ppl.70216

Qiao F, Yang X, Xu F, Huang Y, Zhang J, Song M, Zhou S, Zhang M, He D (2021) TMT-based quantitative proteomic analysis reveals defense mechanism of wheat against the crown rot pathogen Fusarium pseudograminearum. BMC Plant Biol 21(1):82. https://doi.org/10.1186/s12870-021-02853-6

Rampitsch C, Huang M, Djuric-Cignaovic S, Wang X, Fernando U (2019) Temporal quantitative changes in the resistant and susceptible wheat leaf Apoplastic proteome during infection by wheat leaf rust (Puccinia triticina). Front Plant Sci 10. https://doi.org/10.3389/fpls.2019.01291

Santos RB and Figueiredo A (2021) Two sides of the same story in grapevine–pathogen interactions. J Exp Bot 72(9):3367–3380. https://doi.org/10.1093/JXB/ERAB091

Santos RB and Figueiredo A (2023) Biotic and abiotic stress Management in Grapevine: recent advances and major breakthroughs. Agronomy 13(6):1584. https://doi.org/10.3390/agronomy13061584

Santos RB, Nascimento R, V. Coelho A, Figueiredo A (2020) Grapevine–Downy Mildew Rendezvous: proteome analysis of the first hours of an incompatible interaction. Plants 9(11):1498. https://doi.org/10.3390/plants9111498

Scheele GA (1975) Two-dimensional gel analysis of soluble proteins. Characterization of Guinea pig exocrine pancreatic proteins. J Biol Chem 250(14):5375–5385. https://doi.org/10.1016/S0021-9258(19)41192-7

Schreiber KJ, Chau-Ly IJ, Lewis JD (2021) What the wild things do: mechanisms of plant host manipulation by bacterial type III-secreted effector proteins. Microorganisms 9(5):1029. https://doi.org/10.3390/microorganisms9051029

Shi W, Stolze SC, Nakagami H, Misas Villamil JC, Saur IML, Doehlemann G (2023) Combination of in vivo proximity labeling and co-immunoprecipitation identifies the host target network of a tumor-inducing effector in the fungal maize pathogen Ustilago maydis. J Exp Bot 74(15):4736–4750. https://doi.org/10.1093/jxb/erad188

Šimkovicová M, Kramer G, Rep M, Takken FLW (2024) Tomato R-gene-mediated resistance against Fusarium wilt originates in roots and extends to shoots via xylem to limit pathogen colonization. Front Plant Sci 15. https://doi.org/10.3389/fpls.2024.1384431

Spagnolo A, Magnin-Robert M, Alayi TD, Cilindre C, Schaeffer-Reiss C, Van Dorsselaer A, Clément C, Larignon P, Ramirez-Suero M, Chong J, Bertsch C, Abou-Mansour E, Fontaine F (2014) Differential responses of three grapevine cultivars to Botryosphaeria dieback. Phytopathology 104(10): 1021–1035. https://doi.org/10.1094/PHYTO-01-14-0007-R

Toruño TY, Stergiopoulos I, Coaker G (2016) Plant-pathogen effectors: cellular probes interfering with plant defenses in spatial and temporal manners. Annu Rev Phytopathol 54:419–441. https://doi.org/10.1146/annurev-phyto-080615-100204

Tran JC, Campuzano IDG, Baker ES, Wysocki VH (2023) Introducing the high-throughput in mass spectrometry special focus issue. J Am Soc Mass Spectrom 34(8): 1543–1544. https://doi.org/10.1021/jasms.3c00253

Wang Y, Yang R, Feng Y, Sikandar A, Zhu X, Fan H, Liu X, Chen L, Duan Y (2020) iTRAQ-based proteomic analysis reveals the role of the biological control agent, Sinorhizobium fredii strain Sneb183, in enhancing soybean resistance against the soybean cyst nematode. Front Plant Sci 11. https://doi.org/10.3389/fpls.2020.597819

Wang Q, Guo J, Jin P, Ying GM, Guo J, Peng C, Li Q, Wang B (2022) Proteomics and functional analysis revealed TaGSTU6/TaCBSX3 enhances wheat resistance to powdery mildew. 2022.01.31.478538

Wu G, Zhang Y, Wang B, Li K, Lou Y, Zhao Y, Liu F (2021) Proteomic and transcriptomic analyses provide novel insights into the crucial roles of host-induced carbohydrate metabolism enzymes in Xanthomonas oryzae pv. Oryzae virulence and Rice-Xoo interaction. Rice 14(1):57. https://doi.org/10.1186/s12284-021-00503-x

Xiao-Jing L, Bao X, Meng-Yu W, Xiao-Man L, Xu-Jing W, Zhi-Xing W (2023) No obvious unintended effects was found in gene editing rice through transcriptional and proteomic analysis. GM Crops Food 14(1):1–16. https://doi.org/10.1080/21645698.2023.2229927

Zaini PA, Nascimento R, Gouran H, Cantu D, Chakraborty S, Phu M, Goulart LR, Dandekar AM (2018) Molecular profiling of Pierce's disease outlines the response circuitry of Vitis vinifera to Xylella fastidiosa infection. Front Plant Sci 9. https://doi.org/10.3389/fpls.2018.00771

Zhang F, Zhang F, Huang L, Zeng D, Cruz CV, Li Z, Zhou Y (2020) Comparative proteomic analysis reveals novel insights into the interaction between rice and Xanthomonas oryzae pv. oryzae. BMC Plant Biol 20(1):563. https://doi.org/10.1186/s12870-020-02769-7

Zhang L, Hou M, Zhang X, Cao Y, Sun S, Zhu Z, Han S, Chen Y, Ku L, Duan C (2023) Integrative transcriptome and proteome analysis reveals maize responses to Fusarium verticillioides infection inside the stalks. Mol Plant Pathol 24(7):693–710. https://doi.org/10.1111/mpp.13317

Zhou H, Hwarari D, Zhang Y, Mo X, Luo Y, Ma H (2022) Proteomic analysis reveals salicylic acid as a pivotal signal molecule in Rice response to blast disease infection. Plants 11(13):1702. https://doi.org/10.3390/plants11131702

Clinical Proteomics Applied to Food Allergy

20

Robert Stryński, Ana G. Abril, Jesús Mateos, and Mónica Carrera

Abstract

Food allergy (FA) is an immune-related condition, mainly characterized by hypersensitivity to food allergens in children and adults. Clinical proteomics is among the first medical disciplines supported by novel technologies, on the basis of which it becomes possible to understand FA from a therapeutic point of view. Within this chapter, three subdisciplines of proteomics will be covered: discovery clinical proteomics, targeted clinical proteomics, and clinical proteomics-based systems biology. Discovery clinical proteomics refers to the analysis of protein expression aiming at biomarker identification and knowledge about disease mechanisms which bring out completely new proteins connected to FAs. In targeted clinical proteomics, proteins that were identified in the exploratory studies are followed up so as to increase sensitivity for biomarkers detection in diagnostics. Proteomic-based systems biology is a means by which proteomics data can be integrated with other "omics" technologies for simulating immune response pathways governing biological systems and identifying therapeutic targets. Over the years, it has been possible to study what causes diseases, find new markers, and customize drug prescriptions according to protein results obtained from patients. Clinical proteomics has provided insight on immune responses, allergen diversity, and treatment possibilities in relation to FAs. It provides a detailed mapping of protein alterations as a result of allergens that are responsible for essential interactions with the immune system. In turn, clinical proteomics helps individualized medicine through identifying specific proteins signatures in patients enabling customized treatment approaches that will enhance its efficacy while mitigating any ill-effect. In this way, basic research is supposed to be united with clinical practice to offer novel opportunities for diagnosing, monitoring, and managing food allergic conditions that impact positively on outcomes leading to better life quality for patients.

R. Stryński
Department of Biochemistry, Faculty of Biology and Biotechnology, University of Warmia and Mazury in Olsztyn, Olsztyn, Poland

A. G. Abril
Department of Food Technology, Institute of Marine Research (IIM-CSIC), Spanish National Research Council (CSIC), Vigo, Spain

Department of Functional Biology and Health Sciences, Faculty of Biology, University of Vigo, Vigo, Spain

J. Mateos
Clinical Pharmacology Group, Health Research Institute of Santiago de Compostela (IDIS), Santiago de Compostela, A Coruña, Spain

M. Carrera (✉)
Department of Food Technology, Institute of Marine Research (IIM-CSIC), Spanish National Research Council (CSIC), Vigo, Spain
e-mail: mcarrera@iim.csic.es

M. Gama-Carvalho et al. (eds.), *Omics Approaches in Biomedicine and Biotechnology*, Advances in Experimental Medicine and Biology 1504, https://doi.org/10.1007/978-3-032-18966-0_20

Keywords

Food allergy · Clinical proteomics · Mass spectrometry · Discovery proteomics · Targeted proteomics · Proteomics-based systems biology

20.1 Introduction

Food allergy (FA) represents an immune-mediated, hypersensitive reaction to specific food allergens and affects 6–8% of children and 3–4% of adults (Burks et al. 2012; Warren et al. 2020). This process involves the identification of food antigens by IgE antibodies in a response mediated primarily with adaptive immunity. Through this interaction, a cascade of events is triggered that results in the release of histamines and other mediators from mast cells and basophils causing allergic symptoms (Anvari et al. 2019).

Symptoms can be mild like skin reactions, nausea, vomiting, and diarrhea as well to more severe life-threatening anaphylaxis. The European Union (EU) and most of the rest of the world's countries have set out 14 different types for food allergens to be identified on a label when they are part of any food product. This group has been called the "Big 14 foods" that are: cereals containing gluten, crustaceans, eggs, fish, mollusk, peanut, soybeans, milk, nuts, celery, mustard, sesame seeds, sulfur dioxide, sulfites, and lupin (European Parliament and Council of the European Union, Regulation No 1169/2011). The most appropriate and successful treatment is to stop using the specific food antigens or anything that contains them.

Omics technologies, such as genomics, proteomics, and metabolomics, have the potential to offer valuable tools for comprehending and addressing FAs (Dhondalay et al. 2018). Through the collaboration of data across these various disciplines, professionals in the research and medical fields can work toward creating improved methods for diagnosis, treatment, and prevention, leading to enhanced outcomes for patients and advancements in public health.

Clinical proteomics is a sophisticated area at the interface of proteomics principles and clinical applications with an insight into the root causes of a variety of diseases, including FA (Gillette et al. 2024). This methodology allows for the discovery of possible diagnostic biomarkers and mechanisms of allergic reactions and the development of new therapeutic approaches. Application of proteomics in FA research makes use of mass spectrometry (MS), chromatography, and bioinformatics tools to analyze protein levels, post-translational modifications (PTMs), and interactions (Birhanu 2023).

An important application of proteomics in FA is in diagnostic biomarker identification. The biomarkers would enhance the accuracy of FA diagnoses, which are otherwise based on historical information, skin prick tests, and serum-specific IgE tests. Specific protein profiles related to allergic reactions have been identified through proteomic techniques of discovery proteomics (Reisdorph et al. 2009). Moreover, proteomic techniques are capable of identifying a number of allergens at a time, thereby offering overall and more accurate diagnosis (Bianco et al. 2023). The identification of such biomarkers is likely to lead to the development of more accurate diagnostic tools. Targeted proteomics deals with monitoring and quantification of certain proteins of interest after the discovery phase. This approach increases the accuracy and reliability of detection of the target biomarker, which in the long term could have advantages for developing diagnostic tests and monitoring devices. Levels of specific IgE-binding proteins, for example, can be measured to achieve improved diagnosis and management practices for FA (Kattan and Sicherer 2015).

Proteomics can also provide insight into the mechanisms of allergic reactions. Through the proteome of immune cells and changes after allergen challenge, researchers can find out which pathways are involved in the allergic response (Li Xu et al. 2022). For instance, studies may examine the proteomic profile of dendritic cells in the presence of the major allergens, looking for substantial changes in proteins related to antigen processing and presentation that underline the role of dendritic cells in FA sensitization (Humeniuk et al. 2017). Researching protein profiles in individuals with FAs allows one to detect changes in protein abundance and PTMs taking place during allergic reactions (Carrera et al. 2021). All this knowledge is useful in comprehending the pathophysiology of FA and in the development of targeted therapies.

Recent studies indicate a very important role of gut microbiota in modulating immune responses to food allergens. Proteomic techniques targeting gut microbiota and metabolites have shed light on how microbial dysbiosis may drive FAs (Abril et al. 2023b). A good example is microbial metabolites, which have been described to modulate the activity of regulatory T cells essential for maintaining immune tolerance to food allergens (Luu et al. 2020). Further, this will be realized through metaproteomics and shotgun proteomics, allowing for in-depth analysis of proteins expressed by microbial communities (Abril et al. 2024). That is, the protein content of gut microbiota can reveal how changes in this microbiome contribute to the development of FA. This would be important for the development of microbiome-targeted therapies to help alleviate allergic reactions.

Proteomic profiling can enable personalized medicine treatment by personalizing therapies according to protein expression profiles of a given individual. In this respect, patients at risk could be sorted out to have higher chances of severe allergic reactions, and then treatments could be designed in a more personalized way. This facilitates the understanding of protein interactions and pathways involved in FAs, thus helping researchers identify new therapeutic targets. Proteomics could help in the identification of proteins that are key players in the allergic response and which novel drugs or therapies could target to attenuate allergic reactions (Nony et al. 2016). Allergen-specific immunotherapy (AIT) is one such treatment given for repeated exposure to minute portions of allergens to desensitize the immune system from their action. Proteomic techniques could thus make valuable contributions in the development of AIT by identifying the exact proteins and epitopes inducing tolerance (Nye-Wood et al. 2021).

The chapter further details a step-by-step workflow on how to set up proteomic tools for the design of FA vaccines. This workflow is constructed based on the characterization of B-cell epitopes specific for a given food allergen and includes the following five steps (Carrera et al. 2019). This workflow brings together techniques from proteomics, bioinformatics, and immunology to promote the development of vaccines against FAs (Carrera and Magadán 2022).

The future of clinical proteomics in FA research is bright. Proteomic data integrated with genomic, transcriptomic, and metabolomic information give almost a full view of allergic diseases. Besides, personalized medicine approaches informed by proteomic profiles may lead to tailored treatments and dietary recommendations for patients with FAs.

20.2 Proteomics Workflows: Discovery, Targeted, and Proteomics-Based Systems Biology

Proteomics is the systemic analysis of a cell, tissue, or organism for its protein component with the purpose of providing information on protein expression, function, and interactions (Aebersold and Mann 2016). In relation to clinical FA, proteomics presents an opportunity for new diagnostic biomarker identification and mechanisms of allergic reactions, as well as targeted therapies. Figure 20.1 illustrates three strategies of clinical proteomics in FA.

Discovery proteomics is defined as the exploratory approach that detects and quantitates as many different proteins as possible in a biological sample (Paik et al. 2008). This unbiased technique is key to the identification of novel biomarkers and to the understanding of the mechanisms of diseases. Preparing the biological samples is the first step of discovery proteomics, and these samples are generally sources from blood, urine, cerebrospinal fluid, or tissue

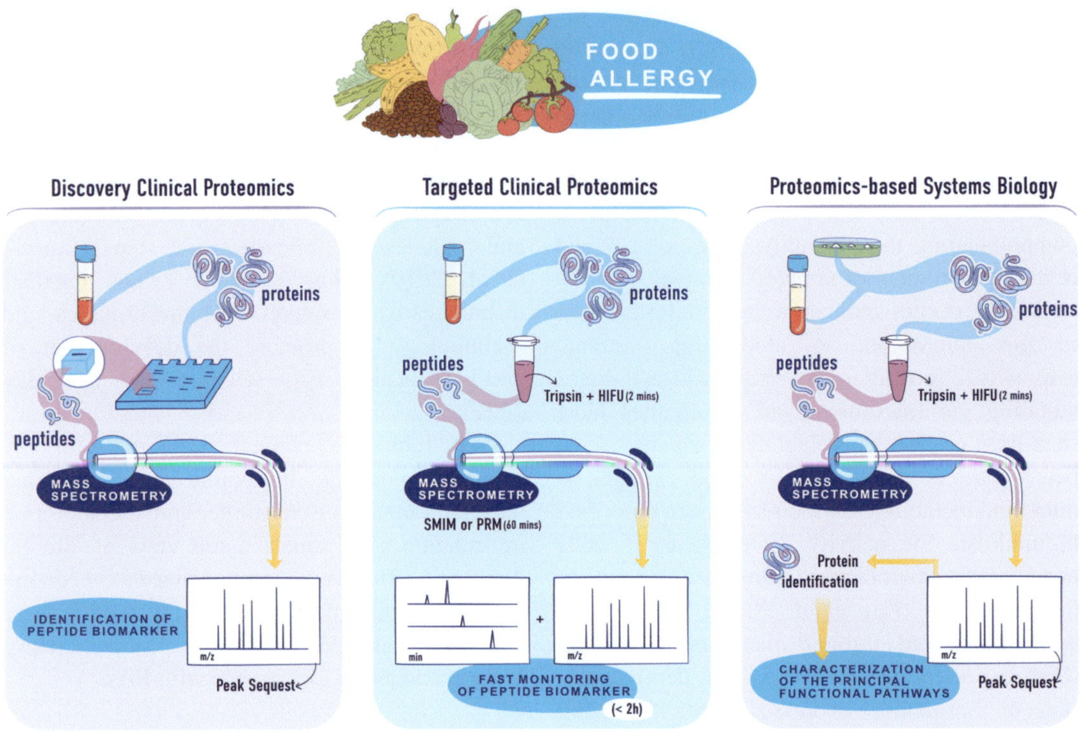

Fig. 20.1 Workflows of clinical proteomics strategies in food allergy. The figure presents three major approaches used in clinical proteomics. In discovery proteomics, proteins from biological samples are digested into peptides and analyzed by mass spectrometry to identify potential peptide biomarkers. In targeted proteomics, proteins are rapidly digested and selected peptides are monitored by MS using SRM or PRM methods, enabling fast and sensitive quantification of known biomarkers in less than two hours. Finally, proteomics-based systems biology integrates protein identification with functional pathway analysis, providing insights into the principal molecular mechanisms underlying food allergy. Together, these strategies complement one another, supporting both biomarker discovery and mechanistic understanding of molecular processes

biopsies. Extraction conditions should be optimized with respect to the type of sample used to achieve maximum protein yield with minimum degradation. Detergent-based lysis, sonication, and centrifugation are methods employed in the disruption of cells during protein extraction. Techniques like two-dimensional gel electrophoresis (2-DE) and liquid chromatography (LC) are normally employed in separating extracted proteins after extraction (De Angelis et al. 2010). With these techniques, complex protein mixtures can be resolved. Besides, automated sample preparation methods for MS-based clinical proteomics were also published (Müller et al. 2023). These proteins are then identified and quantified by MS methods. High-resolution MS techniques, specifically tandem MS (MS/MS), gave a new face to protein identification, which was earlier done on the available methods of low sensitivity and accuracy. The bulk data generated from MS techniques requires robust bioinformatics tools to analyze and sort out data. Protein identification from peptide mass fingerprints and matching against databases, for example, UniProt or NCBI, are performed via software platforms such as Proteome Discover, Mascot, and MaxQuant (Eng et al. 2008). Following this, statistical testing concerning differentially regulated proteins, for instance, between allergic and non-allergic individuals is performed—that is, using, for example, Percolator algorithm—according to Käll et al. (2007). De novo sequencing of the peptide is required with the help of tools such as PEAKS when a protein of interest is not in the database (Ma and Johnson 2012). Protein quantification is often necessary since many biological variations are not due to a protein/peptide biomarker but rather to variations in their levels. These techniques include, among others, stable isotope labeling by amino acids in cell culture (i.e., SILAC), isotope tagging via chemical reactions (i.e., iTRAQ, TMT), stable isotope incorporation via enzymatic reactions (i.e., 18O), and label-free quantitation (LFQ), where the peptide intensity at the MS level can be used (Rozanova et al. 2021). There are a number of studies that have used discovery proteomics in the identification of potential FA biomarkers. For example, proteomics studies on peanut allergy have revealed differentially expressed proteins related to immune responses and inflammation (Czolk et al. 2020). Such findings can therefore form a basis for the development of diagnostic tests and understanding the molecular mechanisms behind FAs.

Targeted proteomics is the analysis targeting of predefined groups of proteins (Mendes and Dittmar 2022). Such methods are very useful in validating candidate biomarkers detected in discovery studies and monitoring specific proteins involved in diseases, such as particular proteins involved in allergic reactions (Wenk et al. 2024). The target proteins will be selected according to previously acquired knowledge from discovery proteomics, literature reviews, and biological relevance to FA. Such common targets include allergens, cytokines, and proteins involved in the immune signaling pathways. Thereafter, techniques like selected MS/MS ion monitoring—SMIM, according to Jorge et al. (2007)—or MS-based selected/multiple/parallel reaction monitoring, SRM/MRM/PRM are used (Carrera et al. 2016). These methods have been reported with very high sensitivity and specificity, thus allowing for the accurate monitoring and quantification of target proteins in complex biological samples. Isotope-labeled standards are often used to ensure accuracy and reproducibility. Moreover, data-independent acquisition methods have been applied according to Gillet et al. (2012), which involves sequential window acquisition of all theoretical mass MS that identified and quantified large protein sets without a priori specification (SWATH). Targeted proteomics has been applied in clinical studies on the validation of biomarkers for different diseases, including diagnosis and prognosis related to FA (Devonshire et al. 2023). Additionally, targeted proteomics will further help in monitoring the efficacy of immunotherapy and other therapeutic interventions.

Proteomics-based systems biology is the integration of proteomics data with other omics datasets for constructing models on biological systems comprehensively (Gstaiger and Aebersold 2009). Independently, this holistic approach takes a deeper understanding of the complicated networks and pathways involved in diseases. To do this, integration of proteomics data with genomics, transcriptomics, metabolomics, and epigenomics data lies at the heart of systems biology. This multi-omics approach allows the identification of key regulatory nodes and interactions driving the disease process. For example, integrating proteomics with transcriptomics can answer how changes in gene expression are translated to protein level alterations. Proteins are projected onto biological pathways and interaction networks using bioinformatics tools and databases such as STRING, KEGG, and Reactome (Fabregat et al. 2017). Network analysis identifies hub proteins and key pathways that, when dysregulated in diseases, offer potential targets for therapeutic intervention. These techniques use mathematical and computational modeling methodologies, such as differential equation modeling and machine learning, toward modeling the biological process and predicting perturbation outcomes (Wang et al. 2019). Such models may identify critical points in the disease pathways and suggest strategies for intervention. In the case of FA, systems biology and the massive proteomics methodology have very high potential for clinical translation (Cao et al. 2024). For example, a part of the approaches to systems biology can be utilized in building predictive models for allergy severity and treatment response. Furthermore, novel drug targets and biomarkers for personalized medicine can be identified by integrative analyses (Chen et al. 2012).

20.3 Discovery Proteomics in Food Allergy

In this part of the chapter, we will focus on the discovery proteomics approach, which aims to identify proteins in a single cell up to a whole organism. This often involves bottom-up ("shotgun") and top-down proteomics (Carbonara et al. 2021). Bottom-up approach combines LC and MS/MS to analyze peptide mixtures obtained from total proteolytic digests of proteome extracts. This method enables the identification and quantification of proteins in complex mixtures by reconstructing the original protein sequences from the detected peptide fragments. The identified peptides are then matched to a corresponding canonical protein (i.e., amino acid sequence) by searching protein sequence databases (Matallana-Surget et al. 2010; Marcus et al. 2020; Carbonara et al. 2021). Top-down approaches can be divided into integrative MS and MS-intensive proteome analysis methods (Toby et al. 2016). Integrative MS involves the separation of intact protein species via 2-DE prior to peptide MS. Additionally, spots of high abundance or areas at the pH extremes and unresolved small peptides in the migrating front can be further subjected to third electrophoretic separations. MS-intensive involves separation of intact protein species, currently mainly via gel-eluted liquid fraction entrapment electrophoresis (GELFrEE), prior to intact protein MS. GELFrEE is single dimension continuous elution tube gel electrophoresis providing low resolution separation of proteins by molecular weight (MW) (Tran and Doucette 2008) (Fig. 20.2). In recent years, the bottom-up method has become the dominant method in the field of proteomics due to its high throughput, its ability to generate large amounts of raw data, its high sensitivity, as well as the claim that it is "faster" and "simpler."

In general, discovery proteomics has become a powerful method for profiling clinical samples to identify biomarkers such as allergens and applying this knowledge in the fields of allergy, asthma, and general immunology, in particular to characterize inflammatory reactions and improve the quality of life of patients suffering from them.

Clinical proteomics for the purpose of biomarker discovery generally focuses on the two mentioned above, basic workflows: bottom-up and top-down. These proteomics techniques have been used to discover novel biomarkers in biofluids and tissues from people with a variety of diseases, including allergies, asthma, atopic dermatitis, inflammatory diseases including parasitic infections, chronic obstructive pulmonary disease, and others (McDonald and Yates 2002; Aebersold and Mann 2003; Rioux et al. 2008; Reisdorph et al. 2009; Silberring and Ciborowski 2010; Herrero et al. 2012; Morphew et al. 2014; Carrera et al. 2016, 2024; Stryiński et al. 2019, 2020, 2022; Ahmed et al. 2019; Raposo de Magalhães et al. 2020).

20.3.1 Identifying Allergens

Food allergies are considered by the World Health Organization (WHO) to be the fourth most important public health problem. About 75% of allergic reactions among children in Europe are due to eggs, peanuts, cows' milk, and fish (Tanno and Demoly 2022). The European Union has identified 14 common food allergens, but new allergens are likely to emerge as dietary habits continue to change. Plant-based proteins are of great importance as sources of protein both in non-developed countries due to their low cost and in developed countries due to the growing number of vegetarians. The main allergic plant-based foods are peanuts, soy, tree nuts, and wheat (Tanno and Demoly 2022).

The identification of new allergens in the different type of food matrix can be performed with the conventional gel electrophoresis followed by Western immunoblotting with sera from allergic patients and MS analysis to identify new allergens (Grishina et al. 2017). MS is a promising alternative to conventional antibody-

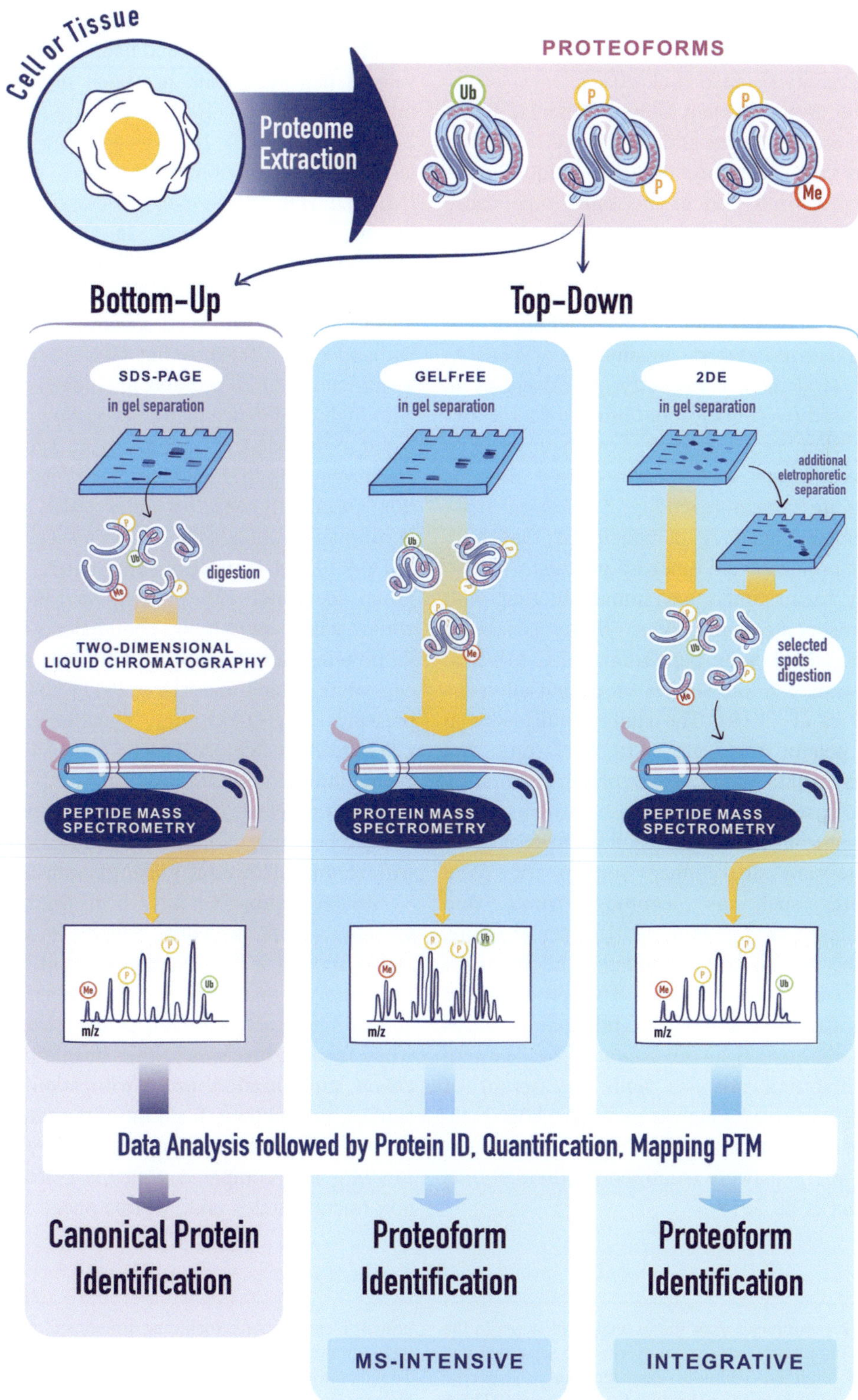

Fig. 20.2 Top-down and bottom-up proteomics workflows. This schematic illustrates the complementary strategies used in proteomic studies. Both approaches rely on mass spectrometry for final identifications yet differ in

based assays to quantify multiple allergenic proteins in complex matrices with high sensitivity and selectivity (Bianco et al. 2023).

In the group of plant allergy proteins, gluten proteins are a complex group of 70–100 abundant proteins that give wheat flour its unique visco-elastic properties and at the same time cause serious health problems in humans (Battais et al. 2005). Among others, there are omega-5 gliadins, a group of important sensitizing allergens in a severe form of FA, the so-called wheat-dependent, exercise-induced anaphylaxis (WDEIA) (Morita et al. 2003). Recently, Altenbach et al. (2018) used two-dimensional immunoblot analysis of flour proteins probed with sera from WDEIA patients and identified two proteins with strong immunoreactivity. Subsequent mass spectrometry analysis demonstrated that both reactive proteins were omega-5 gliadins.

2-DE-based proteomics, immunoblotting with allergen-specific antibodies from sensitized patients, and shotgun peptidomics (LC-MS/MS) was also used in the studies on peanut allergens (Prodic et al. 2018). The study confirmed that major peanut allergens, Ara h 2 and Ara h 6, remain intact after peanut grain digestion by oral and gastric phase enzymes, highlighting their important roles in allergic reactions to peanut.

There are also other sources of plant allergens, such as legumes, fruits, and vegetables. Cardona et al. (2018), for example, identified of 12 low abundance allergens of mango (*Mangifera indica*). The proteome of mango was extracted and the presence of traces of the allergens was investigated by Western immunoblotting analysis with the serum of patients sensitive to mango proteins. After reactive spot elution and MS analyses, four major mango allergens were identified for the first time (Cardona et al. 2018).

Food allergens from animal foods come mainly from eggs, milk, and fish. Proteomic analysis of hen egg white has been performed by many laboratories. Raikos et al. (2006) used 2-DE to separate the proteins and matrix-assisted laser desorption/ionization-time of flight (MALDI-TOF)-based peptide mass fingerprinting to analyze the spots. In a further study, 16 proteins were identified using 2-DE, peptide mass fingerprinting and LC-MS/MS with a quadrupole(Q)-TOF mass spectrometer (Guérin-Dubiard et al. 2006). Mann (2007) reported the identification of 78 proteins in chicken egg white with high confidence using a workflow that combined sodium dodecyl sulfate–polyacrylamide gel electrophoresis (SDS-PAGE) for protein separation with LC-MS/MS and MS_3 analysis performed on a linear ion trap-Fourier transform (LTQ-FT) mass spectrometer. Later, the same group identified 158 proteins in chicken egg white with very high confidence, including 79 proteins identified for the first time in chicken egg white (Mann and Mann 2011), using a dual pressure LTQ-Orbitrap instrument, the LTQ-Orbitrap Velos. Later, using 1-DE and 2-DE immunoblotting and MALDI-TOF/TOF MS, Martos et al. (2013) found that two minor allergic proteins, ovoinhibitor and clusterin, react with serum IgE from egg allergic individuals.

Similar approaches have been used to study the cow's milk proteome. Nissen et al. (2012, 2013) applied a fractionation method to bovine colostrum or mature milk that resulted in a cell and fat-free fraction, a cell pellet fraction and a whey fraction that was further treated by acidification, ultrafiltration, or centrifugation. In these studies, the proteins from the different fractions were digested with trypsin, analyzed by 2-D-LC-MS/MS, and compared with the corresponding non-fractionated milk proteome. Recently,

Fig. 20.2 (continued) their initial analytical steps. In the top-down approach, intact proteoforms are separated and analyzed directly, enabling identification of specific proteoforms and post-translational modifications (PTMs). In contrast, the bottom-up approach involves enzymatic digestion into peptides prior to MS, leading to canonical protein identifications, including amino acid variants, but with limited resolution of individual proteoforms. Together, these strategies highlight the value of advanced, high-resolution MS technologies for comprehensive proteome characterization

Vincent et al. (2016) using bottom-up proteomics by nanoHPLC-electrospray ionization-tandem mass spectrometry (nLC-ESI-MS/MS) identified 186 different proteins, and highlighted breed differences between the proteins in milk of Jersey and Holstein-Friesian cows. Lately, Bavaro et al. (2019) investigated the effect of thermal treatment like baking of cow's milk proteins that were employed as ingredients in the preparation of muffins. A proteomic workflow was applied to the analysis of the protein bands highlighted along the SDS gel followed by western blot analyses with sera of milk allergic children.

Fish allergies are mainly triggered by parvalbumins, which are calcium-binding proteins that exist in two isoforms: α-parvalbumins, which are non-allergenic, and β-parvalbumins, which are IgE-reactive (Fernandes et al. 2015). The proteomics data published by Carrera et al. (2010) enabled the comprehensive de novo MS sequencing of 41 novel β-PRVB isoforms from the Merlucciidae family through a combination of bottom-up proteomics, precise molecular mass measurement by Fourier transform ion cyclotron resonance (FTICR-MS), and selected MS/MS ion monitoring. With the aim of identifying a potential peptide vaccine, shotgun proteomics was also used to characterize B-cell epitopes of β-parvalbumins. This strategy could be used for the development of new potential immunotherapies (Carrera et al. 2019).

20.3.2 Characterizing Allergenic Responses

Nowadays, animal models for allergy studies are standard in laboratories all over the world. There are a number of different allergy animal models, including mice, rats, guinea pigs, dogs, pigs, and monkeys (Bøgh et al. 2016). As an indicator, the immediate response of the animal can be assessed both in vivo and in vitro after sacrifice. Mouse models are most commonly used to study the development of allergic sensitization, elicitation, and the potential of immunotherapeutic measures. However, the genetic background of the different mouse strains has an impact on the development of allergic symptoms. BALB/c mice develop high IgE levels after allergen sensitization. In contrast, C57BL/6 mice are known to react with medium or low IgE levels. However, the results obtained in an animal model must be interpreted with caution and may not be comparable to immune responses in humans (Bøgh et al. 2016). Thus, the human models of FA are of great importance to close the gap between in vivo animal models and clinical studies with human patients (Humeniuk et al. 2017).

In addition to in vivo animal models and in vitro models from various human cell lines (DCs, THP-1, keratinocytes, etc.), scientists are using tissues directly from patients. Bronchial biopsies taken from asthmatic patients are sources of fully differentiated lung tissue that contain the cellular repertoire required to study allergen-specific responses and when these samples are exposed to allergen in culture, they can reflect the allergic response of asthmatic lung tissue (Wood et al. 2013). Similarly, intestinal biopsy samples were used a long time ago for short-term cultures to investigate FAs (Bischoff et al. 1997). When we add high-throughput proteomics, it is possible to obtain information about the molecular mechanisms of allergic reactions. By analyzing the changes in the proteome of immune cells, specific organoids, or individuals after exposure to an allergen, scientists can learn which signaling pathways are involved in the allergic reaction and how exactly it occurs (Li Xu et al. 2022). Below you will find some examples of the use of proteomics in FA research.

To investigate the organ-specific proteomic alterations of the allergy responses, the spleen and intestine of mice sensitized with high (shrimp and clam) and weak (fish) allergenic tropomyosins were analyzed using SWATH-MS-based proteomics (Li Xu et al. 2022). In the intestine, tropomyosin allergy concurred with the downregulation of 35 proteins involved in different metabolic pathways, absorption processes, and ammonium ion responses. These data can provide new functional biomarkers of tropomyosin-induced immune responses as well as candidate targets for intervention.

Authors of different study, in order to investigate molecular alterations in intestinal tissues during cow's milk protein allergy occurrence, analyzed the jejunal tissue of β-lactoglobulin (BLG)-sensitized mice through quantitative tandem mass tag (TMT)-labeled proteomics (Li et al. 2024). A total of 94 differentially expressed proteins (65 up-regulated, 29 down-regulated) were identified with corresponding pathways enriched during the immune response to BLG.

The study on the relation between the human milk proteome and allergy of both mother and child also used proteomic approach. Dekker et al. (2022) analyzed proteins in milk samples from a subset of 300 mother-child dyads selected based on maternal and child allergy phenotypes. Authors observed a trend that milk for infants who develop an allergy by 3 years of age contains higher abundances of immunoglobulin chains, irrespective of the allergy status of the mother. Furthermore, network analysis showed overall increased connectivity of proteins in the milk of allergic mothers and milk for infants who ultimately develop an allergy. Together, these findings give new insights into how the human milk proteome, through differences in the abundance of individual proteins and protein-protein associations, relates to the allergy status of mother and child (Dekker et al. 2022).

Allergic dermatitis (AD) is the predominant form of immune response in humans and intensive work is underway to identify the major metabolic pathways involved in the response to specific allergens. The foods that are commonly associated with protein contact dermatitis include raw seafood, eggs, and flour, and less commonly, meat, vegetables, fruits, and spices (Amado and Jacob 2007). For example, Goleva et al. (2020) used skin strip samples to identify pathways associated with allergen polysensitization in children with AD and FA. The authors suggested that altered expression of certain skin proteins may support allergen penetration through a damaged barrier and the subsequent type 2 inflammatory response that occurs after penetration, suggesting potential new targets for future early interventions to prevent AD and FA.

The next step in the study of FA is the use of human-derived organoids (Stanton and Grabrucker 2023). These are complex, three-dimensional structures that consist of multiple cells and serve as representative models for certain organs. Organoids are more physiologically relevant than immortalized cell lines, and in the case of animal models, the results of studies performed on organoids do not need to be validated as they may not be comparable to human immune responses. As they are also easy to handle, these "mini-organs" could serve as an effective tool for research into the etiology of FA and new treatments.

In addition to trace metals, vitamins, and large nutritional components, organoid models have been used to identify potential effects from various nutraceuticals and food additives in recent years (Stanton and Grabrucker 2023). Currently, there are not many studies on human-derived organoids as a model for FA, but interest in this approach is growing and some studies using organoids in the context of FA, such as celiac disease, are already being conducted (Freire et al. 2019). The application of proteomics methods to this model of FA is only a matter of time.

20.3.3 Microbiome Analysis and Dysbiosis in Food Allergy

Research on the association of FAs with the microbiome is underway (Bunyavanich and Berin 2019). Development and regulation of the immune system are highly dependent on the gut microbiome, whose dysbiosis may initiate the development of allergies, including FAs (Zubeldia-Varela et al. 2022). The impact of this community during early gut microbiome, more precisely during infants, seems to affect development. It can also be driven by antibiotic use, diet, and mode of delivery, whether vaginal or cesarean. From the studies, it emerged that the gut microbiota of persons with FAs might differ in terms of diversity and composition. The risk for developing FAs appears to be higher in a low-diversity microbiome. Moreover, metabolites from the gut microbiota—for example, short-chain fatty acids—may sustain

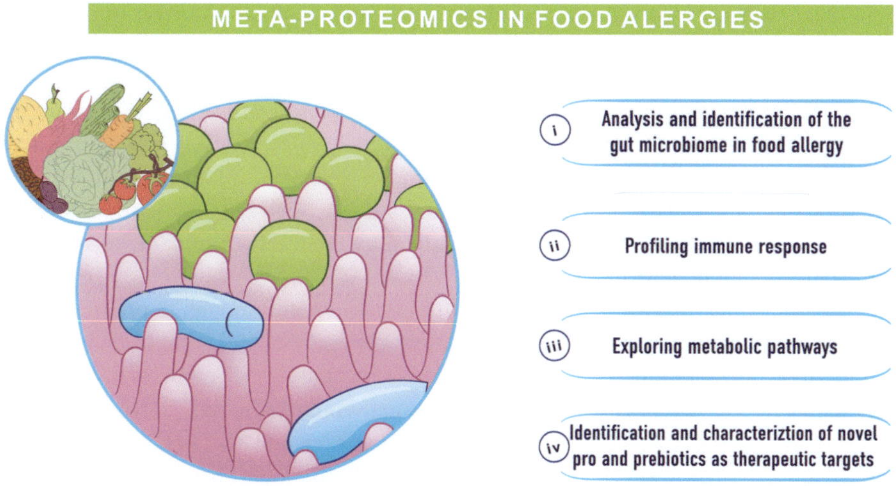

Fig. 20.3 Meta-proteomic insights in the study of gut microbiome in FAs. This schematic highlights the applications of meta-proteomics in understanding food allergy mechanisms through the gut microbiome. Meta-proteomic approaches enable (i) analysis and identification of gut microbial communities associated with allergic responses, (ii) profiling of host immune responses to allergenic proteins in presence of specific microbiota (eubiosis or dysbiosis), (iii) exploration of microbial and host metabolic pathways influenced by dietary allergens (e.g., fish, peanuts, processed foods), and (iv) identification and characterization of novel probiotics and prebiotics as potential therapeutic targets. Together, these insights provide a systems-level understanding of gut microbiome function in food allergy and support the development of microbiome-based interventions

immunological tolerance and modulate immune responses while exerting anti-inflammatory effects.

A proteomics approach known as "meta-proteomics" detects and evaluates proteins within a complex environmental microbial community (Xiao et al. 2017). This approach generally applies a method of discovery proteomics where complex peptide/protein samples are analyzed with high-throughput liquid chromatography coupled with tandem mass spectrometry (LC-MS/MS). The term "meta-proteomics" was first used by P. Wilmes and P. L. Bond in 2004 (Wilmes and Bond 2004). Since concepts of metaproteome and meta-proteomics emerged back in the year 2004, interest in analyzing overall protein profiles of microbial communities in different samples has grown. This approach gives details on metabolic pathways, microbial activities, and the signal transduction or functions present at any given time in a community. It allows understanding the interactions between a host microbiome and health or dysbiosis and disorder phenotypes (Abril et al. 2023b). Metaproteomic investigations of the gut microbiota, including peptide/protein biomarker identification, may reveal the molecular underpinning of the molecular causes of many diseases. Microbial metaproteomics has been applied to many environments, such as soil, marine ecosystems, or fecal simples (Abril et al. 2023b).

Changes in the gut microbiome have been implicated in a range of disorders, including FA-associated diseases (Zubeldia-Varela et al. 2022). Thus, profiling the proteome of the microbiome in patients with FAs may define protein patterns characteristic of an allergic response. Metaproteomics can identify specific microbial proteins or pathways influencing the immune system's response to food proteins. Understanding proteomic interactions between microbiota and dietary proteins might shed more light on the development of FA. The proteomic links between the microbiome and host cells can be further explored to gain insights into how the immune system works in response to these food antigens (Fig. 20.3). This can elaborate the mechanisms on how the microbiota contributes to the induction or regulation of FAs. Other metaproteomic applications concern the gut microbiota processing or modifying ingested food proteins. This information would be useful in explaining which protein modifications or breakdown products contribute to allergies.

For metaproteomics, the final data analysis is the biggest challenge. Identification of peptides and proteins and taxonomic and functional studies requires genome and protein databases for microorganisms, most of which are uncultivable; however, this has considerably changed with genome and metagenome sequencing collections. A single genome catalog contains more than 200,000 reference genomes from the human gut microbiome. Metaproteogenomics links proteomics to genomics, thus enabling gene prediction errors to be corrected and protein products to be identified from genome regions not previously recognized as coding regions. In addition, various programs and algorithms have been developed and implemented in different microbiome studies, allowing data analysis to occur.

Samples for microbiome studies can be obtained from the following microbial biomass sources, according to Xiao et al. (2017): fecal samples, ileum, cecum, small and large intestines; internal gut samples like the mucus layer, according to van der Post and Arike (2021) and Zhang et al. (2018). This thereby calls for microbial enrichment and isolation methods, chemical, mechanical, and thermal cell lysis procedures in the presence of diverse pollutants and contaminants in these simples (Abril et al. 2023b). The meta-omics fields, genomics, transcriptomics, proteomics, and metabolomics, have probably stated a relation between microbiomes and inflammatory diseases, including inflammatory bowel disease (Presley et al. 2012; Zhang et al. 2018), Crohn's disease (Mills et al. 2019), allergies (Kolmeder et al. 2016; Kingkaw et al. 2020; Mok et al. 2021). Further research in the direction is required, and studies on FAs are currently at process. Other studies have shown differences in the gut microbiome between children with FAs and healthy controls (Ling et al. 2014; Azad et al. 2015; Inoue et al. 2017; Berni Canani et al. 2018; Dong et al. 2018; Kourosh et al. 2018; Abdel-Gadir et al. 2019; Mauras et al. 2019; Kingkaw et al. 2020; Goldberg et al. 2020; Yamagishi et al. 2021).

A metaproteomic study conducted in Thailand identified differences in the metabolic composition of reporter proteins between the microbiomes of infants with atopic dermatitis and those of healthy children (Kingkaw et al. 2020). Using LC-MS/MS, eight proteins were identified as biomarkers for atopic dermatitis through stool sample analysis. All eight biomarkers were linked to metabolic processes. Among these, dimethylmenaquinone methyltransferase from *Bacteroides* and triosephosphate isomerase produced by the bifidobacterium *Alloscardovia* were particularly notable, as they were only expressed in infants with the disease (Kolmeder et al. 2016).

Moreover, the development of host diseases is significantly influenced by the gastro-intestinal (GI) tract mycobiome, which consists of fungi (Pettersen et al. 2022). Mok et al. (2021) investigated the fungal microbiota of 9- to 12-month-old infants diagnosed with atopic dermatitis, dividing the patients into two groups based on the severity of their symptoms (recovered or ongoing). Using both metagenomic and metaproteomic methods (LC-MS/MS), they concluded that the group still affected by the disease had a more varied mycobiome. While the healthy group displayed significantly greater levels of *Wickerhamomyces* and *Kodamaea* species, the infants undergoing GI dysbiosis showed a decrease in the *Ascomycota*/*Basidiomycota* ratio and an increase in *Rhodotorula* species. Surprisingly, bacteria belonging to the genera *Acremonium* and *Rhizopus* were elevated in the recovered group. Furthermore, the researchers discovered that the ongoing disease cohort produced more fungal proteins than the other groups, with *Rhodotorula* species being the main producer. Using a metaproteomic approach, they identified five fungi from each sample group as biomarkers. Two unique proteins produced by *Rhodotorula* species, glycerol kinase and RAN-binding protein 1, were only observed in infants suffering from atopic dermatitis, indicating that they may play a role in the development of this condition (Mok et al. 2021).

In a multi-omics investigation involving metagenomics, metatranscriptomics, and metametabolomics, infants with atopic dermatitis who were sensitized to allergens were assessed and compared with healthy controls. The study identified potential microbial pathways associated with the early onset of infant atopic dermatitis and described an abnormal developmental trajectory of the gut microbiota and stool metabolome. These modifications included a reduction in the number of bacteria that produce butyrate and propionate. However, the impact of the subjects' genetic backgrounds was not evaluated, preventing the integration of the data with the host's genetic profile (Fazlollahi et al. 2018).

A different study assessed the relationship between microbe-associated metabolites and allergic and non-allergic childhood asthma, revealing distinct microbial populations in individuals with these conditions (Zheng et al. 2022). The group with allergic asthma displayed 42 metabolites produced by microbes, while the group with non-allergic asthma displayed 58 metabolites. These findings suggest that controlling metabolite suppression or modification may be a helpful therapeutic strategy for reducing asthma symptoms.

Recent studies have implicated early-life alterations in gut microbiota in the development of FA. Hendrickx et al. (2023) investigated the composition of the fecal microbiota by 16S rRNA gene amplicon sequencing and its functional properties through metaproteomics in a cohort of 40 neonates diagnosed with cow's milk allergy (CMA) at baseline. While some infants grew out of CMA by one year, others did not. Tryptic digest peptides from stool samples were analyzed by LC-MS/MS and obtained protein sequences were cross-referenced versus a target at the genus level database. A 16S rRNA sequencing study showed that the lower relative abundance of *Lachnospiraceae* and *Bacteroidaceae* is a characteristic feature of persistence of CMA. Besides, metaproteomics findings showed marked differences in human and microbial protein classes between the allergy groups (Hendrickx et al. 2023).

Additional research using multi-omics techniques reveals that individuals with atopic dermatitis (AD) and FAs have non-lesional skin surfaces with specific features, including a unique combination of skin proteins, lipids, and RNA transcripts, due to an immature skin barrier and type 2 immune activation. The study assessed the skin microbiota, transcriptome, proteome, and stratum corneum composition of patients with AD and FA, patients with AD without FA, and healthy controls, based on transepidermal water loss. These findings suggest that the outermost layer of non-lesional skin in AD patients with FA may have distinct characteristics associated with an immature skin barrier and type 2 immune activation (Leung et al. 2019).

20.4 Targeted Clinical Proteomics in Food Allergy

With advancements in proteomics, particularly targeted clinical proteomics, there are new opportunities to understand the mechanisms underlying FAs, identify biomarkers for monitoring, and develop therapeutic targets (Sicherer and Sampson 2018). This section delves into targeted clinical proteomics in FA, focusing on biomarkers monitoring and therapeutic targets.

20.4.1 Biomarkers Monitoring

Protein biomarkers are specific proteins whose expression, absence, or altered levels can signify a particular biological condition, such as an allergic reaction. In the case of FAs, protein biomarkers can offer essential information about immune responses and pathophysiological mechanisms (Muraro and Arasi 2018). The identification and monitoring of these biomarkers by targeted proteomics would thus be very imperative for the betterment of clinical management in FAs. Protein biomarker monitoring is the measurement of specific proteins that are supposed to be continuously or repeatedly monitored so as to establish any alteration over time. This method is crucial due to a number of reasons (Sindher et al. 2018): it provides for the differential diagnosis of allergic reactions against other adverse food reactions in the diagnosis of FAs; an estimation is made with regard to the severity and progression of FAs; gives insights into the activity of disease and effectiveness of interventions monitored over time; and finally, biomarker profiling helps in personalizing treatment to the individual patient, hence bringing better outcomes and fewer side effects (Hoffmann-Sommergruber 2016).

Several types of protein biomarkers are recognized in FAs (Sahiner et al. 2023): allergen-specific IgE antibodies, such as those specific to food allergens, serve as primary markers for diagnosing FAs; markers like CD63 and CD203c indicate basophil activation in response to allergens; and signaling molecules, including cytokines and chemokines such as IL-4, IL-5, IL-13, and TNF-α, play crucial roles in allergic inflammation.

Accurate diagnosis of FAs is important in effective management. Traditional diagnostic tools, such as skin prick tests and oral food challenge, have obvious limitations in regard to risk of severe reactions and subjectivity. Thus, protein biomarker monitoring can become a more accurate and less invasive alternative (Lozano-Ojalvo and Benedé 2023). However, sensitization alone is not specific for a clinical allergy, and additional biomarkers are warranted to confirm the allergy. The basophil activation test (BAT) gives information on the degree of activation of the basophils through allergens, which is defined by markers such as CD63 and CD203c (Santos et al. 2021). It represents functional proof of an allergic reaction and hence supplements the measurements of IgE. Therefore, BAT should be especially useful in cases where the determination of IgE is low or ambiguous. Component-resolved diagnostics can define which allergenic components are present in a given food allergen, providing further detail on the allergenic profile and thus being potentially useful for differentiation between primary sensitization and cross-reactivity (Calamelli et al. 2019).

Protein biomarkers could become indicators of prognosis, risk of severe allergic reactions, and likelihood of outgrowing an allergy (Nony et al. 2016). Basophil activation markers and certain cytokines have been current biomarkers that may point to the possible severity of an allergic reaction (Hemmings et al. 2018). Higher levels of specific IgE or strong basophil responses often correspond to more serious reactions. Finally, the changes in the biomarkers over time inform on the natural history of the FAs. For example, a decrease in the degree of specific IgE or alterations in cytokine profiles may predict tolerance. Longitudinal studies have established that those children whose milk or egg protein IgE levels are declining have an excellent prognosis for outgrowing these allergies (Lee et al. 2024). Cytokines and chemokines, such as IL-4, IL-5, IL-13, and TNF-α, are involved in the inflammatory process of FAs and might, therefore, provide information about ongoing inflammation and the efficacy of anti-inflammatory therapies when monitored (Muraro and Arasi 2018). Eosinophils have long been considered one of the hallmarks of allergic inflammation; on the other hand, some biomarkers, such as eosinophil cationic protein and eosinophil-derived neurotoxin, may express eosinophil activation, which is to be used as an indicator for follow-up on the activity of eosinophil-related diseases such as eosinophilic esophagitis, a possible complication of FA (Cianferoni 2020). Such predictive models can be constructed with the integration of multiple biomarkers. For instance, in severe allergic reaction, there may be the integration of specific IgE levels with basophil activation markers and cytokine profiles in a model that gives a more diffuse assessment of risk.

20.4.2 Therapeutic Targets

The efficacy of therapeutic interventions, such as specific allergen immunotherapy, pharmacological treatments, and dietary management can also be monitored by these protein biomarkers (Keet and Wood 2014). Immunotherapy, aimed at inducing tolerance to particular allergens, may utilize biomarkers such as specific IgE, IgG4, and cytokine levels to monitor the immune response against the treatment (Virkud et al. 2019). Successful immunotherapy is mostly marked by increased IgG4 with a concomitant decrease in the level of specific IgE. Dietary avoidance of the allergen is a common management strategy in FA, and protein biomarker monitoring can serve for assessment of compliance and inadvertent exposure. Specific IgE levels, for example, can reflect an ongoing sensitization or new sensitization to other allergens (Santos et al. 2023). Symptomatic medications in FAs include antihistamines and corticosteroids, while biologics are used for the management of symptoms and inflammation. Such protein biomarkers will be helpful in monitoring the treatment for its efficiency by follow-up on the changes in inflammatory markers and immune responses. Biomarker profiling can tailor treatment in patients with an optimized outcome and minimal side effects. This biomarker profile is therefore useful in stratifying patients at risk based on the severity of reactions, probability of tolerance, and response to treatments; it hence can allow more personalized management. Further, the biomarkers may provide a basis upon which precision immunotherapy approaches—for example, the appropriate allergen dose and schedule—are based for each patient (Virkud et al. 2019). Such a personalized approach might further improve on the safety and efficacy of immunotherapy.

Allergen immunotherapy relies on a process of desensitization against allergens, achieved through incremental exposure to the allergen. This approach of targeted proteomics will enable the identification and monitoring of all proteins participating in the allergic response, acting as a guide on how to develop relevant, effective immunotherapy protocols. Peanut oral immunotherapy is based on the gradual introduction of peanut proteins for patient desensitization. It was applied to targeted proteomics measurements of individual IgE and IgG4 against peanut allergens Ara h 1, Ara h 2, and Ara h 3 as surrogate markers of the immune response for monitoring of the oral immunotherapy (OIT) process (Reinwald et al. 2022). Changes in their levels could define successful desensitization, which would be very valuable in adjusting treatment protocols. Milk OIT realizes the desensitization of patients against milk proteins. Targeted proteomics techniques can thus estimate milk allergens-specific IgE and IgG4 at a time when an immune response takes place (D'Auria et al. 2018). The frequent reduction in specific IgE levels and increase in IgG4 indicative of success regarding desensitization could hence be a possible tool to monitor the effectiveness of treatment. Proteomic measure of IgE before and after the treatment may act as a measure of effectiveness of the therapy. Some promising results in reducing IgE levels and alleviating symptoms come from using anti-IgE mAb called omalizumab for the treatment of FA (Virkud et al. 2018). Several mAbs targeted to IL-4, IL-5, and IL-13 have been generated to reduce allergic inflammation. Such targeted proteomics quantifies the set of cytokines underpinning development and monitoring of anti-cytokine therapies. For example, dupilumab is an anti-IL-4/IL-13 mAb that proved efficacy in the treatment of allergic diseases, including FA. Small molecule inhibitors specifically inhibit certain proteins or pathways in the allergic response (Dodson and Lio 2022). Such targets could be identified, and efficiency of such inhibitors quantified by targeted proteomics. TNF-α is a cytokine possessing a pro-inflammatory role in serious allergic reactions. It can be quantified via target proteomics, measuring the levels of TNF-α to quantify the efficiency of such inhibitors driving along drug development processes. Small molecule inhibitors targeted against TNF-α would decrease inflammation and reduce symptoms (Jang et al. 2021). Measurements of such protein levels would also be very helpful in targeted proteomics, which would be useful in proving the efficacy of these inhibitors and eventually in developing them as drugs.

Once allergenic proteins are identified, the next step is to map the epitopes responsible for IgE binding. Epitope mapping involves determining the specific regions of the allergenic protein that interact with IgE antibodies. Proteomics techniques can be used in combination with bioinformatics analysis in this process (Carrera et al. 2019). By understanding the structure of these epitopes, researchers can design hypoallergenic variants of these proteins that can be used in vaccines to desensitize the immune system without triggering severe allergic reactions.

Peptide-based vaccines are one of the promising approaches in FA immunotherapy (Carrera and Magadán 2022). These vaccines use short synthetic peptides that represent specific epitopes of allergenic proteins. By administering these peptides in a controlled manner, the immune system can be trained to tolerate the allergen without eliciting a full-blown allergic reaction. Proteomics techniques are essential for identifying the most relevant epitopes and designing peptides that can effectively induce immune tolerance (Carrera and Magadán 2022).

Organoids combined with proteomics offer significant potential for advancing FA research and treatment. Organoids provide accurate models of human organs and immune responses, accelerating therapy development and enhancing our understanding of FAs (Hung et al. 2019). Gut organoids, mimicking the human intestine's structure and cell diversity, are particularly valuable for studying the initial interaction of food allergens with the body. These models allow researchers to introduce allergens and study immune responses, offering insights beyond simpler models. Immune organoids, containing key immune cells, enable the study of allergen processing and IgE antibody production. Proteomics provides detailed protein analysis, helping uncover the molecular and cellular mechanisms behind FAs. This combined approach improves understanding and paves the way for innovative treatments.

Integrating proteomics data with other omics datasets, such as genomics, transcriptomics, and metabolomics, can provide a more comprehensive understanding of FAs. This multi-omics approach can reveal new therapeutic targets and biomarkers, facilitating the development of personalized medicine.

20.5 Proteomics-Based Systems Biology

Proteomics-based systems biology offers a promising avenue for understanding the complex interactions and mechanisms underlying FAs (Carrera et al. 2018). By integrating proteomic data with systems biology frameworks (Malmström et al. 2007), researchers can gain deeper insights into the molecular basis of allergic reactions, identify potential biomarkers, and develop novel therapeutic strategies.

One of the approaches that systems biology takes toward the analysis of interactions between different biological components is network analysis. In the case of FA, this may involve the construction of a protein-protein interaction (PPI) network to identify key players involved in allergic responses. Through these networks, one can find crucial nodes and pathways that potentially become targets of intervention (Ruiz Amores and Martínez-Antonio 2022). Integration of proteomic data with other omics data, for example, genomics, transcriptomics, metabolomics, may contribute to a better understanding of FAs (Babu and Snyder 2023). For example, integration of proteomic data with genomic data can highlight how genetic variation may influence protein expression and allergenicity. Computational models and their corresponding simulations lie at the heart of systems biology, with eventual clinical implications (Cai 2014). Such computational models would yield insight into the behavior of biological systems under different conditions and thus offer potential points of intervention. In the field of FA research, modeling allergic responses is instrumental in understanding IgE-binding dynamics, release of mediators, and the manifestation of symptoms.

The mechanisms by which intracellular signaling in the immune system contributes to susceptibility to FAs remain poorly understood. For B and T lymphocytes, the B-cell receptor (BCR) and T-cell receptor (TCR) signaling pathways are known to be essential (Maggi 1998). Understanding how these pathways interact with others could be crucial for discovering new methods of preventing and treating FAs. Developing a comprehensive model that represents B and T-cell signaling may provide researchers with deeper insights into the mechanisms driving food allergic reactions. This could pave the way for more effective strategies to manage and mitigate FAs. To address this, our research group is employing proteomics-based systems biology approaches to investigate splenocytes in the context of fish allergy (Abril et al. 2023a). This involves studying changes in protein abundance and the dynamics protein-protein interactions at various time points following stimuli of immune cells. These datasets will be integrated into functional and dynamic networks using computational systems biology methods. Integrating this data into a holistic immune-cell model for FA will provide a valuable signaling map, enhancing our understanding of the mechanisms of action of specific allergens.

Single-cell proteomics is a game-changing tool in FA research that permits unprecedented understanding of the immune response at a cellular level (Perkel 2021). This technique could, therefore, allow the examination of the proteomic profile of single cells to obtain information on cellular behavior, protein abundance, and its modifications against allergens. Thus, it enables detailed profiling of the immune cells, like B and T lymphocytes, involved in food allergic reactions. The application of single-cell proteomics in research into FA is just beginning but holds great promise. Further studies will be conducted to refine techniques, enhance sensitivity and accuracy, and apply them to bigger and more diverse populations to derive comprehensive insights into FAs.

Computational modeling and simulation are powerful tools in systems biology. Models can simulate allergic responses, predict the behavior of biological systems, and identify potential intervention points. In FA research, models can simulate IgE-binding dynamics, mediator release, and symptom progression. These simulations aid in understanding the complex interactions underlying allergic reactions and guide the development of therapeutic strategies.

Although proteomics-based systems biology has huge potential for FA research, there are a number of issues that must be brought out: (i) the integration with and interpretation of large-scale proteomic data in conjunction with multi-omics will require sophisticated computational tools and expertise to apply them. In this respect, standardized protocols in the analysis and sharing of data can help facilitate its full harnessing. (ii) FAs exhibit important inter-individual variability with respect to the sensitivity to allergens or immune responses directed against them. Diagnostic and therapeutic strategies have therefore to be individually tailored according to genetic and epigenetic determinants and also environmental factors. (iii) Rigorous validation and testing are the challenges for translation from results in proteomics-based systems biology into clinical application. This will succeed only with appropriate collaboration among researchers, clinicians, and regulatory agencies to facilitate the translation of novel, safe, and effective diagnostic and therapeutic strategies.

20.5.1 Personalized Medicine

Personalized medicine tailors medical treatment to individual characteristics, such as genetic makeup and proteomic profiles. In FA, personalized approaches can consider individual variability in allergen sensitivity and immune responses. Proteomic and systems biology data can inform personalized diagnostic tests, therapeutic strategies, and dietary recommendations, improving patient outcomes.

Current diagnostic tools, like the prick test and measurement of specific IgE, have sensitivity and specificity limitations (Cuomo et al. 2017). Proteomic-based systems biology holds great promise for improved diagnostics with novel biomarkers and precise assays. Multiplexed assays allow for the simultaneous detection of multiple allergenic proteins that can define an overall sensitization to the allergen.

This will be possible with the guide of proteomics-based systems biology in the development of hypoallergenic food variants for the subcutaneous immunotherapy of FA (Zuidmeer-Jongejan et al. 2015). The application identifies and modifies the allergenic epitopes present in the proteins of foods to make them less likely to trigger allergic reactions. In this way, it can decrease the risks due to accidental exposure to allergens, enhancing the quality of life of patients who suffer from FAs. AIT will further be aimed at desensitizing people to allergens and decreasing the symptoms of allergies (Durham and Shamji 2023). Proteomics can also help in optimizing AIT by characterization of the allergen extracts used in the course of therapy and monitor immune responses to therapy. Given the molecular changes induced by AIT, its efficacy and safety will be greatly improved, hence making it a viable therapy for FAs.

The translation of the results from proteomics-based systems biology into practice will need rigorous validation. Consequently, interaction among researchers, clinicians, and regulatory agencies becomes important to ensure the safety and efficacy of new diagnostic and therapeutic approaches. Two important approaches for validating the clinical utility of proteomic biomarkers and targeted therapies are clinical trials and real-world studies.

20.6 Concluding Remarks and New Perspectives

Future trends in clinical proteomics for FA will focus on bridging the gap between research and clinical application.

Such proteomic discoveries could allow the development of new diagnostic tests with improved sensitivity and specificity compared with techniques available today. New diagnostic tests could be developed that are capable of detecting, early on, sensitization to specific foods and thus allow for early intervention. Furthermore, there is a high degree of genetic variability in how the immune response is mounted toward food allergens. Hence, researching different populations may enable the discovery of the genetic factors of FAs, consequently leading to therapies that will work for everybody. FA rates are also greatly influenced by dietary habits and cultural practices. In this regard, considering these factors in proteomic studies may be better placed to understand how different diets and environments affect allergic responses. In part, this need to expand proteomic research to underserved and underrepresented populations underlies health disparities. Future trends likely will be toward global health initiatives that can promote inclusive research and ensure benefits for all from proteomic advancements.

On the other hand, proteomics can also identify new therapeutic targets by uncovering the proteins and pathways involved in allergic reactions. This would probably involve future research in the development of drugs against these proteins, which would give new treatment possibilities for FA patients. Clinical proteomics will therefore revolutionize research into FA and treatment. This can thus be achieved by multi-omics approaches, progress in personalized medicine, enhanced technology innovation, discovery of novel biomarkers, translation to practice, extension into research populations, increased collaboration and sharing, and attention to ethical considerations.

One of the key elements in implementing proteomic technologies in clinical settings lies in the collaboration between researchers, clinicians, and regulatory bodies. The trends under development will thus be oriented toward developing protocols with uniform guidelines on how to apply proteomic tools in clinical practice, so wide acceptance and diffusion are achieved. It is possible through collaborative networks including researchers, clinicians, and industrial partners that give a fast track in discovery and innovation. These allow sharing of facilities, expertise, and data that eventually lead to more solid and far-reaching research. Notably, very important are the ethical frameworks for proteomic studies, to ensure that research is responsibly carried out while guarding the patient rights. Future trends will involve developing guidelines on the ethical use of proteomic data, including privacy and security requirements.

References

Abdel-Gadir A, Stephen-Victor E, Gerber GK, Noval Rivas M, Wang S, Harb H, Wang L, Li N, Crestani E, Spielman S, Secor W, Biehl H, DiBenedetto N, Dong X, Umetsu DT, Bry L, Rachid R, Chatila TA (2019) Microbiota therapy acts via a regulatory T cell MyD88/RORγt pathway to suppress food allergy. Nat Med 25:1164–1174. https://doi.org/10.1038/s41591-019-0461-z

Abril AG, Villa T, Freire J, Magadán S, Pazos M, Carrera M (2023a) Fish allergy sensitization and gut microbiome: a metagenomic approach in a mouse model. In: In: 3rd Food Chemistry Conference, Dresden

Abril AG, Carrera M, Sánchez-Pérez Á, Villa TG (2023b) Gut microbiome proteomics in food allergies. Int J Mol Sci 24:2234. https://doi.org/10.3390/ijms24032234

Abril AG, Calo-Mata P, Barros-Velázquez J, Villa TG, Pazos M, Carrera M (2024) Metaproteomic analysis of microbial communities affecting fishery products. Methods Mol Biol 2820:89–98. https://doi.org/10.1007/978-1-0716-3910-8_9

Aebersold R, Mann M (2003) Mass spectrometry-based proteomics. Nature 422:198–207. https://doi.org/10.1038/nature01511

Aebersold R, Mann M (2016) Mass-spectrometric exploration of proteome structure and function. Nature 537:347–355. https://doi.org/10.1038/nature19949

Ahmed F, Kumar G, Soliman FM, Adly MA, Soliman HAM, El-Matbouli M, Saleh M (2019) Proteomics for understanding pathogenesis, immune modulation and host pathogen interactions in aquaculture. Comp Biochem Physiol Part D Genomics Proteomics 32:1–15. https://doi.org/10.1016/j.cbd.2019.100625

Altenbach SB, Chang H-C, Simon-Buss A, Jang Y-R, Denery-Papini S, Pineau F, Gu YQ, Huo N, Lim S-H, Kang C-S, Lee J-Y (2018) Towards reducing the immunogenic potential of wheat flour: omega gliadins encoded by the D genome of hexaploid wheat may also harbor epitopes for the serious food allergy WDEIA. BMC Plant Biol 18:291. https://doi.org/10.1186/s12870-018-1506-z

Amado A, Jacob SE (2007) Dermatitis de contacto por alimentos. Actas Dermosifiliogr 98:452–458. https://doi.org/10.1016/S0001-7310(07)70107-2

Anvari S, Miller J, Yeh C-Y, Davis CM (2019) IgE-mediated food allergy. Clin Rev Allergy Immunol 57:244–260. https://doi.org/10.1007/s12016-018-8710-3

Azad MB, Konya T, Guttman DS, Field CJ, Sears MR, HayGlass KT, Mandhane PJ, Turvey SE, Subbarao P, Becker AB, Scott JA, Kozyrskyj AL, Investigators the CS (2015) Infant gut microbiota and food sensitization: associations in the first year of life. Clin Exp Allergy 45:632–643. https://doi.org/10.1111/cea.12487

Babu M, Snyder M (2023) Multi-omics profiling for health. Mol Cell Proteomics 22:100561. https://doi.org/10.1016/j.mcpro.2023.100561

Battais F, Courcoux P, Popineau Y, Kanny G, Moneret-Vautrin DA, Denery-Papini S (2005) Food allergy to wheat: differences in immunoglobulin E-binding proteins as a function of age or symptoms. J Cereal Sci 42:109–117. https://doi.org/10.1016/j.jcs.2005.01.004

Bavaro SL, De Angelis E, Barni S, Pilolli R, Mori F, Novembre Elio M, Monaci L (2019) Modulation of milk allergenicity by baking milk in foods: a proteomic investigation. Nutrients 11:1536. https://doi.org/10.3390/nu11071536

Berni Canani R, De Filippis F, Nocerino R, Paparo L, Di Scala C, Cosenza L, Della Gatta G, Calignano A, De Caro C, Laiola M, Gilbert JA, Ercolini D (2018) Gut microbiota composition and butyrate production in children affected by non-IgE-mediated cow's milk allergy. Sci Rep 8:12500. https://doi.org/10.1038/s41598-018-30428-3

Bianco M, Ventura G, Calvano CD, Losito I, Cataldi TRI (2023) Food allergen detection by mass spectrometry: from common to novel protein ingredients. Proteomics 23. https://doi.org/10.1002/pmic.202200427

Birhanu AG (2023) Mass spectrometry-based proteomics as an emerging tool in clinical laboratories. Clin Proteomics 20:32. https://doi.org/10.1186/s12014-023-09424-x

Bischoff SC, Mayer J, Wedemeyer J, Meier PN, Zeck-Kapp G, Wedi B, Kapp A, Cetin Y, Gebel M, Manns MP (1997) Colonoscopic allergen provocation (COLAP): a new diagnostic approach for gastrointestinal food allergy. Gut 40:745–753. https://doi.org/10.1136/gut.40.6.745

Bøgh KL, van Bilsen J, Głogowski R, López-Expósito I, Bouchaud G, Blanchard C, Bodinier M, Smit J, Pieters R, Bastiaan-Net S, de Wit N, Untersmayr E, Adel-Patient K, Knippels L, Epstein MM, Noti M, Nygaard UC, Kimber I, Verhoeckx K, O'Mahony L (2016) Current challenges facing the assessment of the allergenic capacity of food allergens in animal models. Clin Transl Allergy 6:21. https://doi.org/10.1186/s13601-016-0110-2

Bunyavanich S, Berin MC (2019) Food allergy and the microbiome: current understandings and future directions. J Allergy Clin Immunol 144:1468–1477. https://doi.org/10.1016/j.jaci.2019.10.019

Burks AW, Tang M, Sicherer S, Muraro A, Eigenmann PA, Ebisawa M, Fiocchi A, Chiang W, Beyer K, Wood R, Hourihane J, Jones SM, Lack G, Sampson HA (2012) ICON: food allergy. J Allergy Clin Immunol 129:906–920. https://doi.org/10.1016/j.jaci.2012.02.001

Cai Y (2014) Computational proteomics, systems biology and clinical implications. Biochim Biophys Acta 1844:

163–164. https://doi.org/10.1016/j.bbapap.2013.10.001

Calamelli E, Liotti L, Beghetti I, Piccinno V, Serra L, Bottau P (2019) Component-resolved diagnosis in food allergies. Medicina (Kaunas) 55:498. https://doi.org/10.3390/medicina55080498

Cao Z, Li Q, Li Y, Wu J (2024) Identification of plasma protein markers of allergic disease risk: a mendelian randomization approach to proteomic analysis. BMC Genomics 25:503. https://doi.org/10.1186/s12864-024-10412-0

Carbonara K, Andonovski M, Coorssen JR (2021) Proteomes are of Proteoforms: embracing the complexity. Proteomes 9:38. https://doi.org/10.3390/proteomes9030038

Cardona EEG, Heathcote K, Teran LM, Righetti PG, Boschetti E, D'Amato A (2018) Novel low abundance allergens from mango via combinatorial peptide libraries treatment: a proteomics study. Food Chem 269:652–660. https://doi.org/10.1016/j.foodchem.2018.06.113

Carrera M, Magadán S (2022) Proteomics for development of food allergy vaccines. Methods Mol Biol 2410:673–689. https://doi.org/10.1007/978-1-0716-1884-4_36

Carrera M, Cañas B, Vázquez J, Gallardo JM (2010) Extensive De Novo sequencing of new Parvalbumin isoforms using a novel combination of bottom-up proteomics, accurate molecular mass measurement by FTICR−MS, and selected MS/MS ion monitoring. J Proteome Res 9:4393–4406. https://doi.org/10.1021/pr100163e

Carrera M, Gallardo JM, Pascual S, González ÁF, Medina I (2016) Protein biomarker discovery and fast monitoring for the identification and detection of Anisakids by parallel reaction monitoring (PRM) mass spectrometry. J Proteome 142:130–137. https://doi.org/10.1016/j.jprot.2016.05.012

Carrera M, Cañas B, Gallardo JM (2018) Advanced proteomics and systems biology applied to study food allergy. Curr Opin Food Sci 2019 16. https://doi.org/10.1016/j.cofs.2017.12.001

Carrera M, González-Fernández Á, Magadán S, Mateos J, Pedrós L, Medina I, Gallardo JM (2019) Molecular characterization of B-cell epitopes for the major fish allergen, parvalbumin, by shotgun proteomics, protein-based bioinformatics and IgE-reactive approaches. J Proteome 200:123–133. https://doi.org/10.1016/j.jprot.2019.04.005

Carrera M, Cañas B, Lopez-Ferrer D (2021) Rapid shotgun phosphoproteomics analysis. Methods Mol Biol 2259:259–268. https://doi.org/10.1007/978-1-0716-1178-4_17

Carrera M, Abril AG, Pazos M, Calo-Mata P, Villa TG, Barros-Velázquez J (2024) Proteins and peptides: proteomics approaches for food authentication and allergen profiling. Curr Opin Food Sci 57:101172. https://doi.org/10.1016/j.cofs.2024.101172

Chen R, Mias GI, Li-Pook-Than J, Jiang L, Lam HYK, Chen R, Miriami E, Karczewski KJ, Hariharan M, Dewey FE, Cheng Y, Clark MJ, Im H, Habegger L, Balasubramanian S, O'Huallachain M, Dudley JT, Hillenmeyer S, Haraksingh R, Sharon D, Euskirchen G, Lacroute P, Bettinger K, Boyle AP, Kasowski M, Grubert F, Seki S, Garcia M, Whirl-Carrillo M, Gallardo M, Blasco MA, Greenberg PL, Snyder P, Klein TE, Altman RB, Butte AJ, Ashley EA, Gerstein M, Nadeau KC, Tang H, Snyder M (2012) Personal omics profiling reveals dynamic molecular and medical phenotypes. Cell 148:1293–1307. https://doi.org/10.1016/j.cell.2012.02.009

Cianferoni A (2020) Non-IgE mediated food allergy. Curr Pediatr Rev 16:95–105. https://doi.org/10.2174/1573396315666191031103714

Cuomo B, Indirli GC, Bianchi A, Arasi S, Caimmi D, Dondi A, La Grutta S, Panetta V, Verga MC, Calvani M (2017) Specific IgE and skin prick tests to diagnose allergy to fresh and baked cow's milk according to age: a systematic review. Ital J Pediatr 43. https://doi.org/10.1186/s13052-017-0410-8

Czolk R, Klueber J, Sørensen M, Wilmes P, Codreanu-Morel F, Skov PS, Hilger C, Bindslev-Jensen C, Ollert M, Kuehn A (2020) IgE-mediated peanut allergy: current and novel predictive biomarkers for clinical phenotypes using multi-omics approaches. Front Immunol 11:594350. https://doi.org/10.3389/fimmu.2020.594350

D'Auria E, Mameli C, Piras C, Cococcioni L, Urbani A, Zuccotti GV, Roncada P (2018) Precision medicine in cow's milk allergy: proteomics perspectives from allergens to patients. J Proteome. https://doi.org/10.1016/j.jprot.2018.01.018

De Angelis M, Di Cagno R, Minervini F, Rizzello CG, Gobbetti M (2010) Two-dimensional electrophoresis and IgE-mediated food allergy. Electrophoresis 31:2126–2136. https://doi.org/10.1002/elps.201000101

Dekker PM, Azad MB, Boeren S, Mandhane PJ, Moraes TJ, Simons E, Subbarao P, Turvey SE, Saccenti E, Hettinga KA (2022) The human milk proteome and allergy of mother and child: exploring associations with protein abundances and protein network connectivity. Front Immunol 13. https://doi.org/10.3389/fimmu.2022.977470

Devonshire A, Gautam Y, Johansson E, Mersha TB (2023) Multi-omics profiling approach in food allergy. World Allergy Organ J 16:100777. https://doi.org/10.1016/j.waojou.2023.100777

Dhondalay GK, Rael E, Acharya S, Zhang W, Sampath V, Galli SJ, Tibshirani R, Boyd SD, Maecker H, Nadeau KC, Andorf S (2018) Food allergy and omics. J Allergy Clin Immunol 141:20–29. https://doi.org/10.1016/j.jaci.2017.11.007

Dodson J, Lio PA (2022) Biologics and small molecule inhibitors: an update in therapies for allergic and immunologic skin diseases. Curr Allergy Asthma Rep 22:183–193. https://doi.org/10.1007/s11882-022-01047-w

Dong P, Feng J-J, Yan D-Y, Lyu Y-J, Xu X (2018) Early-life gut microbiome and cow's milk allergy—a prospective case—control 6-month follow-up study. Saudi J Biol Sci 25:875–880. https://doi.org/10.1016/j.sjbs.2017.11.051

Durham SR, Shamji MH (2023) Allergen immunotherapy: past, present and future. Nat Rev Immunol 23:317–328. https://doi.org/10.1038/s41577-022-00786-1

Eng JK, Fischer B, Grossmann J, Maccoss MJ (2008) A fast SEQUEST cross correlation algorithm. J Proteome Res 7:4598–4602. https://doi.org/10.1021/pr800420s

European Parliament and Council of the European Union (2011) Regulation (EU) No 1169/2011 of the European Parliament and of the Council of 25 October 2011 on the provision of food information to consumers. Off J Eur Union https://doi.org/10.3000/19770677.L_2011.304

Fabregat A, Sidiropoulos K, Viteri G, Forner O, Marin-Garcia P, Arnau V, D'Eustachio P, Stein L, Hermjakob H (2017) Reactome pathway analysis: a high-performance in-memory approach. BMC Bioinformatics 18:142. https://doi.org/10.1186/s12859-017-1559-2

Fazlollahi M, Chun Y, Grishin A, Wood RA, Burks AW, Dawson P, Jones SM, Leung DYM, Sampson HA, Sicherer SH, Bunyavanich S (2018) Early-life gut microbiome and egg allergy. Allergy 73:1515–1524. https://doi.org/10.1111/all.13389

Fernandes TJR, Costa J, Oliveira MBPP, Mafra I (2015) An overview on fish and shellfish allergens and current methods of detection. Food Agric Immunol 26:848–869. https://doi.org/10.1080/09540105.2015.1039497

Freire R, Ingano L, Serena G, Cetinbas M, Anselmo A, Sapone A, Sadreyev RI, Fasano A, Senger S (2019) Human gut derived-organoids provide model to study gluten response and effects of microbiota-derived molecules in celiac disease. Sci Rep 9:7029. https://doi.org/10.1038/s41598-019-43426-w

Gillet LC, Navarro P, Tate S, Röst H, Selevsek N, Reiter L, Bonner R, Aebersold R (2012) Targeted data extraction of the MS/MS spectra generated by data-independent acquisition: a new concept for consistent and accurate proteome analysis. Mol Cell Proteomics 11(O111):016717. https://doi.org/10.1074/mcp.O111.016717

Gillette MA, Jimenez CR, Carr SA (2024) Clinical proteomics: a promise becoming reality. Mol Cell Proteomics 23:100688. https://doi.org/10.1016/j.mcpro.2023.100688

Goldberg MR, Mor H, Magid Neriya D, Magzal F, Muller E, Appel MY, Nachshon L, Borenstein E, Tamir S, Louzoun Y, Youngster I, Elizur A, Koren O (2020) Microbial signature in IgE-mediated food allergies. Genome Med 12:92. https://doi.org/10.1186/s13073-020-00789-4

Goleva E, Calatroni A, LeBeau P, Berdyshev E, Taylor P, Kreimer S, Cole RN, Leung DYM (2020) Skin tape proteomics identifies pathways associated with transepidermal water loss and allergen polysensitization in atopic dermatitis. J Allergy Clin Immunol 146:1367–1378. https://doi.org/10.1016/j.jaci.2020.04.022

Grishina G, Bardina L, Grishin A (2017) 2D-electrophoresis and immunoblotting in food. Allergy:59–69

Gstaiger M, Aebersold R (2009) Applying mass spectrometry-based proteomics to genetics, genomics and network biology. Nat Rev Genet 10:617–627. https://doi.org/10.1038/nrg2633

Guérin-Dubiard C, Pasco M, Mollé D, Désert C, Croguennec T, Nau F (2006) Proteomic analysis of hen egg white. J Agric Food Chem 54:3901–3910. https://doi.org/10.1021/jf0529969

Hemmings O, Kwok M, McKendry R, Santos AF (2018) Basophil activation test: old and new applications in allergy. Curr Allergy Asthma Rep 18:77. https://doi.org/10.1007/s11882-018-0831-5

Hendrickx DM, An R, Boeren S, Mutte SK, Chatchatee P, Nowak-Wegrzyn A, Lange L, Benjaponpitak S, Chong KW, Sangsupawanich P, van Ampting MTJ, Oude Nijhuis MM, Harthoorn LF, Langford JE, Knol J, Knipping K, Garssen J, Trendelenburg V, Pesek R, Davis CM, Muraro A, Erlewyn-Lajeunesse M, Fox AT, Michaelis LJ, Beyer K, Noimark L, Stiefel G, Schauer U, Hamelmann E, Peroni D, Boner A, Lambert JM, Belzer C (2023) Assessment of infant outgrowth of cow's milk allergy in relation to the faecal microbiome and metaproteome. Sci Rep 13:12029. https://doi.org/10.1038/s41598-023-39260-w

Herrero M, Simó C, García-Cañas V, Ibáñez E, Cifuentes A (2012) Foodomics: MS-based strategies in modern food science and nutrition. Mass Spectrom Rev 31:49–69. https://doi.org/10.1002/mas.20335

Hoffmann-Sommergruber K (2016) Proteomics and its impact on food allergy diagnosis. EuPA Open Proteom 12:10–12. https://doi.org/10.1016/j.euprot.2016.03.016

Humeniuk P, Dubiela P, Hoffmann-Sommergruber K (2017) Dendritic cells and their role in allergy: uptake, proteolytic processing and presentation of allergens. Int J Mol Sci 18:1491. https://doi.org/10.3390/ijms18071491

Hung L, Obernolte H, Sewald K, Eiwegger T (2019) Human ex vivo and in vitro disease models to study food allergy. Asia Pac Allergy 9:e4. https://doi.org/10.5415/apallergy.2019.9.e4

Inoue R, Sawai T, Sawai C, Nakatani M, Romero-Pérez GA, Ozeki M, Nonomura K, Tsukahara T (2017) A preliminary study of gut dysbiosis in children with food allergy. Biosci Biotechnol Biochem 81:2396–2399. https://doi.org/10.1080/09168451.2017.1383849

Jang D, Lee A-H, Shin H-Y, Song H-R, Park J-H, Kang T-B, Lee S-R, Yang S-H (2021) The role of tumor necrosis factor alpha (TNF-α) in autoimmune disease and current TNF-α inhibitors in therapeutics. Int J Mol Sci 22:2719. https://doi.org/10.3390/ijms22052719

Jorge I, Casas EM, Villar M, Ortega-Pérez I, López-Ferrer D, Martínez-Ruiz A, Carrera M, Marina A, Martínez P, Serrano H, Cañas B, Were F, Gallardo

JM, Lamas S, Redondo JM, García-Dorado D, Vázquez J (2007) High-sensitivity analysis of specific peptides in complex samples by selected MS/MS ion monitoring and linear ion trap mass spectrometry: application to biological studies. J Mass Spectrom 42: 1391–1403. https://doi.org/10.1002/jms.1314

Käll L, Canterbury JD, Weston J, Noble WS, MacCoss MJ (2007) Semi-supervised learning for peptide identification from shotgun proteomics datasets. Nat Methods 4: 923–925. https://doi.org/10.1038/nmeth1113

Kattan JD, Sicherer SH (2015) Optimizing the diagnosis of food allergy. Immunol Allergy Clin N Am 35:61–76. https://doi.org/10.1016/j.iac.2014.09.009

Keet CA, Wood RA (2014) Emerging therapies for food allergy. J Clin Invest 124:1880–1886. https://doi.org/10.1172/JCI72061

Kingkaw A, Nakphaichit M, Suratannon N, Nitisinprasert S, Wongoutong C, Chatchatee P, Krobthong S, Charoenlappanit S, Roytrakul S, Vongsangnak W (2020) Analysis of the infant gut microbiome reveals metabolic functional roles associated with healthy infants and infants with atopic dermatitis using metaproteomics. PeerJ 8:e9988. https://doi.org/10.7717/peerj.9988

Kolmeder CA, Salojärvi J, Ritari J, de Been M, Raes J, Falony G, Vieira-Silva S, Kekkonen RA, Corthals GL, Palva A, Salonen A, de Vos WM (2016) Faecal metaproteomic analysis reveals a personalized and stable functional microbiome and limited effects of a probiotic intervention in adults. PLoS One 11: e0153294. https://doi.org/10.1371/journal.pone.0153294

Kourosh A, Luna RA, Balderas M, Nance C, Anagnostou A, Devaraj S, Davis CM (2018) Fecal microbiome signatures are different in food-allergic children compared to siblings and healthy children. Pediatr Allergy Immunol 29:545–554. https://doi.org/10.1111/pai.12904

Lee ECK, Trogen B, Brady K, Ford LS, Wang J (2024) The natural history and risk factors for the development of food allergies in children and adults. Curr Allergy Asthma Rep 24:121–131. https://doi.org/10.1007/s11882-024-01131-3

Leung DYM, Calatroni A, Zaramela LS, LeBeau PK, Dyjack N, Brar K, David G, Johnson K, Leung S, Ramirez-Gama M, Liang B, Rios C, Montgomery MT, Richers BN, Hall CF, Norquest KA, Jung J, Bronova I, Kreimer S, Conover Talbot C, Crumrine D, Cole RN, Elias P, Zengler K, Seibold MA, Berdyshev E, Goleva E (2019) The nonlesional skin surface distinguishes atopic dermatitis with food allergy as a unique endotype. Sci Transl Med 11. https://doi.org/10.1126/scitranslmed.aav2685

Li Xu L, Wei Zhang H, Lin H, Mei Zhang X, Qi Wen Y, Long Zhao J, Xing Li Z, Gasset M (2022) SWATH-MS-based proteomics reveals functional biomarkers of Th1/Th2 responses of tropomyosin allergy in mouse models. Food Chem 383:132474. https://doi.org/10.1016/j.foodchem.2022.132474

Li Q, Deng Y, Xu Z, Zhou H (2024) Combined transcriptomics and TMT-proteomics reveal abnormal complement and coagulation cascades in cow's milk protein allergy. Int Immunopharmacol 131:111806. https://doi.org/10.1016/j.intimp.2024.111806

Ling Z, Li Z, Liu X, Cheng Y, Luo Y, Tong X, Yuan L, Wang Y, Sun J, Li L, Xiang C (2014) Altered fecal microbiota composition associated with food allergy in infants. Appl Environ Microbiol 80:2546–2554. https://doi.org/10.1128/AEM.00003-14

Lozano-Ojalvo D, Benedé S (2023) Molecular approaches for food protein allergenicity assessment and the diagnosis and treatment of food allergies. Foods 12:1205. https://doi.org/10.3390/foods12061205

Luu M, Monning H, Visekruna A (2020) Exploring the molecular mechanisms underlying the protective effects of microbial SCFAs on intestinal tolerance and food allergy. Front Immunol 11. https://doi.org/10.3389/fimmu.2020.01225

Ma B, Johnson R (2012) De Novo sequencing and homology searching. Mol Cell Proteomics 11(O111):014902. https://doi.org/10.1074/mcp.O111.014902

Maggi E (1998) The TH1/TH2 paradigm in allergy. Immunotechnology 3:233–244. https://doi.org/10.1016/S1380-2933(97)10005-7

Malmström J, Lee H, Aebersold R (2007) Advances in proteomic workflows for systems biology. Curr Opin Biotechnol 18:378–384. https://doi.org/10.1016/j.copbio.2007.07.005

Mann K (2007) The chicken egg white proteome. Proteomics 7:3558–3568. https://doi.org/10.1002/pmic.200700397

Mann K, Mann M (2011) In-depth analysis of the chicken egg white proteome using an LTQ Orbitrap Velos. Proteome Sci 9:7. https://doi.org/10.1186/1477-5956-9-7

Marcus K, Lelong C, Rabilloud T (2020) What room for two-dimensional gel-based proteomics in a shotgun proteomics world? Proteomes 8:17. https://doi.org/10.3390/proteomes8030017

Martos G, López-Fandiño R, Molina E (2013) Immunoreactivity of hen egg allergens: influence on in vitro gastrointestinal digestion of the presence of other egg white proteins and of egg yolk. Food Chem 136:775–781. https://doi.org/10.1016/j.foodchem.2012.07.106

Matallana-Surget S, Leroy B, Wattiez R (2010) Shotgun proteomics: concept, key points and data mining. Expert Rev Proteomics 7:5–7. https://doi.org/10.1586/epr.09.101

Mauras A, Wopereis H, Yeop I, Esber N, Delannoy J, Labellie C, Reygner J, Kapel N, Slump R, van Eijndthoven T, Rutten L, Knol J, Garssen J, Harthoorn LF, Butel M, Bajaj-Elliott M, Hartog A, Waligora-Dupriet A (2019) Gut microbiota from infant with cow's milk allergy promotes clinical and immune features of atopy in a murine model. Allergy 74: 1790–1793. https://doi.org/10.1111/all.13787

McDonald WH, Yates JR (2002) Shotgun proteomics and biomarker discovery. Dis Markers 18:99–105. https://doi.org/10.1155/2002/505397

Mendes ML, Dittmar G (2022) Targeted proteomics on its way to discovery. Proteomics 22. https://doi.org/10.1002/pmic.202100330

Mills RH, Vázquez-Baeza Y, Zhu Q, Jiang L, Gaffney J, Humphrey G, Smarr L, Knight R, Gonzalez DJ (2019) Evaluating metagenomic prediction of the metaproteome in a 4.5-year study of a patient with Crohn's disease. mSystems 4. https://doi.org/10.1128/mSystems.00337-18

Mok K, Suratanon N, Roytrakul S, Charoenlappanit S, Patumcharoenpol P, Chatchatee P, Vongsangnak W, Nakphaichit M (2021) ITS2 sequencing and targeted meta-proteomics of infant gut mycobiome reveal the functional role of Rhodotorula sp. during atopic dermatitis manifestation. J Fungi 7:748. https://doi.org/10.3390/jof7090748

Morita E, Matsuo H, Mihara S, Morimoto K, Savage AWJ, Tatham AS (2003) Fast ω-gliadin is a major allergen in wheat-dependent exercise-induced anaphylaxis. J Dermatol Sci 33:99–104. https://doi.org/10.1016/S0923-1811(03)00156-7

Morphew RM, MacKintosh N, Hart EH, Prescott M, LaCourse EJ, Brophy PM (2014) In vitro biomarker discovery in the parasitic flatworm Fasciola hepatica for monitoring chemotherapeutic treatment. EuPA Open Proteom 3:85–99. https://doi.org/10.1016/j.euprot.2014.02.013

Müller T, Cremonini MA, Kliewer G, Krijgsveld J (2023) Automated sample preparation for mass spectrometry-based clinical proteomics. Methods Mol Biol 2718: 181–211. https://doi.org/10.1007/978-1-0716-3457-8_11

Muraro A, Arasi S (2018) Biomarkers in food allergy. Curr Allergy Asthma Rep 18:64. https://doi.org/10.1007/s11882-018-0816-4

Nissen A, Bendixen E, Ingvartsen KL, Røntved CM (2012) In-depth analysis of low abundant proteins in bovine colostrum using different fractionation techniques. Proteomics 12:2866–2878. https://doi.org/10.1002/pmic.201200231

Nissen A, Bendixen E, Ingvartsen KL, Røntved CM (2013) Expanding the bovine milk proteome through extensive fractionation. J Dairy Sci 96:7854–7866. https://doi.org/10.3168/jds.2013-7106

Nony E, Le Mignon M, Brier S, Martelet A, Moingeon P (2016) Proteomics for allergy: from proteins to the patients. Curr Allergy Asthma Rep 16:64. https://doi.org/10.1007/s11882-016-0642-5

Nye-Wood MG, Juhász A, Bose U, Colgrave ML (2021) Proteome analysis and epitope mapping in a commercial reduced-gluten wheat product. Front Nutr 8. https://doi.org/10.3389/fnut.2021.705822

Paik Y-K, Kim H, Lee E-Y, Kwon M-S, Cho SY (2008) Overview and introduction to clinical proteomics. In: Methods in molecular biologyTM. Humana Press, Totowa, pp 1–31

Perkel JM (2021) Single-cell proteomics takes centre stage. Nature 597:580–582. https://doi.org/10.1038/d41586-021-02530-6

Pettersen VK, Dufour A, Arrieta M-C (2022) Metaproteomic profiling of fungal gut colonization in gnotobiotic mice. Anim Microbiome 4:14. https://doi.org/10.1186/s42523-022-00163-2

Presley LL, Ye J, Li X, LeBlanc J, Zhang Z, Ruegger PM, Allard J, McGovern D, Ippoliti A, Roth B, Cui X, Jeske DR, Elashoff D, Goodglick L, Braun J, Borneman J (2012) Host–microbe relationships in inflammatory bowel disease detected by bacterial and metaproteomic analysis of the mucosal–luminal interface. Inflamm Bowel Dis 18:409–417. https://doi.org/10.1002/ibd.21793

Prodic I, Stanic-Vucinic D, Apostolovic D, Mihailovic J, Radibratovic M, Radosavljevic J, Burazer L, Milcic M, Smiljanic K, van Hage M, Cirkovic Velickovic T (2018) Influence of peanut matrix on stability of allergens in gastric-simulated digesta: 2S albumins are main contributors to the IgE reactivity of short digestion-resistant peptides. Clin Exp Allergy 48: 731–740. https://doi.org/10.1111/cea.13113

Raikos V, Hansen R, Campbell L, Euston SR (2006) Separation and identification of hen egg protein isoforms using SDS–PAGE and 2D gel electrophoresis with MALDI-TOF mass spectrometry. Food Chem 99: 702–710. https://doi.org/10.1016/j.foodchem.2005.08.047

Raposo de Magalhães CSF, Cerqueira MAC, Schrama D, Moreira MJV, Boonanuntanasarn S, Rodrigues PML (2020) A proteomics and other omics approach in the context of farmed fish welfare and biomarker discovery. Rev Aquac 12:122–144. https://doi.org/10.1111/raq.12308

Reinwald S, Rolland JM, O'Hehir RE, van Zelm MC (2022) Peanut oral immunotherapy: current trends in clinical trials. Immunother Adv 2. https://doi.org/10.1093/immadv/ltac004

Reisdorph NA, Reisdorph R, Bowler R, Broccardo C (2009) Proteomics methods and applications for the practicing clinician. Ann Allergy Asthma Immunol 102:523–529. https://doi.org/10.1016/S1081-1206(10)60128-7

Rioux M-C, Carmona C, Acosta D, Ward B, Ndao M, Gibbs BF, Bennett HP, Spithill TW (2008) Discovery and validation of serum biomarkers expressed over the first twelve weeks of Fasciola hepatica infection in sheep. Int J Parasitol 38:123–136. https://doi.org/10.1016/j.ijpara.2007.07.017

Rozanova S, Barkovits K, Nikolov M, Schmidt C, Urlaub H, Marcus K (2021) Quantitative mass spectrometry-based proteomics: an overview. Methods Mol Biol 2228:85–116. https://doi.org/10.1007/978-1-0716-1024-4_8

Ruiz Amores G, Martínez-Antonio A (2022) Basics on network theory to analyze biological systems: a hands-on outlook. Funct Integr Genomics 22:1433–1448. https://doi.org/10.1007/s10142-022-00907-y

Sahiner UM, Giovannini M, Escribese MM, Paoletti G, Heffler E, Alvaro Lozano M, Barber D, Canonica GW, Pfaar O (2023) Mechanisms of allergen immunotherapy and potential biomarkers for clinical evaluation. J Pers Med 13:845. https://doi.org/10.3390/jpm13050845

Santos AF, Alpan O, Hoffmann H (2021) Basophil activation test: mechanisms and considerations for use in clinical trials and clinical practice. Allergy 76: 2420–2432. https://doi.org/10.1111/all.14747

Santos AF, Riggioni C, Agache I, Akdis CA, Akdis M, Alvarez-Perea A, Alvaro-Lozano M, Ballmer-Weber B, Barni S, Beyer K, Bindslev-Jensen C, Brough HA, Buyuktiryaki B, Chu D, Del Giacco S, Dunn-Galvin A, Eberlein B, Ebisawa M, Eigenmann P, Eiwegger T, Feeney M, Fernandez-Rivas M, Fisher HR, Fleischer DM, Giovannini M, Gray C, Hoffmann-Sommergruber K, Halken S, Hourihane JO, Jones CJ, Jutel M, Knol E, Konstantinou GN, Lack G, Lau S, Marques Mejias A, Marchisotto MJ, Meyer R, Mortz CG, Moya B, Muraro A, Nilsson C, Lopes de Oliveira LC, O'Mahony L, Papadopoulos NG, Perrett K, Peters RL, Podesta M, Poulsen LK, Roberts G, Sampson HA, Schwarze J, Smith P, Tham EH, Untersmayr E, Van Ree R, Venter C, Vickery BP, Vlieg-Boerstra B, Werfel T, Worm M, Du Toit G, Skypala I (2023) EAACI guidelines on the diagnosis of IgE-mediated food allergy. Allergy 78:3057–3076. https://doi.org/10.1111/all.15902

Sicherer SH, Sampson HA (2018) Food allergy: a review and update on epidemiology, pathogenesis, diagnosis, prevention, and management. J Allergy Clin Immunol 141:41–58. https://doi.org/10.1016/j.jaci.2017.11.003

Silberring J, Ciborowski P (2010) Biomarker discovery and clinical proteomics. TrAC Trends Anal Chem 29: 128–140. https://doi.org/10.1016/j.trac.2009.11.007

Sindher SB, Long A, Acharya S, Sampath V, Nadeau KC (2018) The use of biomarkers to predict aero-allergen and food immunotherapy responses. Clin Rev Allergy Immunol 55:190–204. https://doi.org/10.1007/s12016-018-8678-z

Stanton JE, Grabrucker AM (2023) The use of organoids in food research. Curr Opin Food Sci 49:100977. https://doi.org/10.1016/j.cofs.2022.100977

Stryiński R, Mateos J, Pascual S, González ÁF, Gallardo JM, Łopieńska-Biernat E, Medina I, Carrera M (2019) Proteome profiling of L3 and L4 Anisakis simplex development stages by TMT-based quantitative proteomics. J Proteome 201:1–11. https://doi.org/10.1016/j.jprot.2019.04.006

Stryiński R, Łopieńska-Biernat E, Carrera M (2020) Proteomic insights into the biology of the most important foodborne parasites in Europe. Foods 9:1403. https://doi.org/10.3390/foods9101403

Stryiński R, Łopieńska-Biernat E, Carrera M (2022) Proteomic advances in seafood and aquaculture. In: Pedrouso M, Ruiz D, Lorenzo J (eds) Food proteomics: technological advances, current applications and future perspectives, 1st edn. Academic Press, pp 113–150

Tanno LK, Demoly P (2022) Food allergy in the World Health Organization's International Classification of Diseases (ICD)-11. Pediatr Allergy Immunol 33. https://doi.org/10.1111/pai.13882

Toby TK, Fornelli L, Kelleher NL (2016) Progress in top-down proteomics and the analysis of Proteoforms. Annu Rev Anal Chem 9:499–519. https://doi.org/10.1146/annurev-anchem-071015-041550

Tran JC, Doucette AA (2008) Gel-eluted liquid fraction entrapment electrophoresis: an electrophoretic method for broad molecular weight range proteome separation. Anal Chem 80:1568–1573. https://doi.org/10.1021/ac702197w

van der Post S, Arike L (2021) Metaproteomics analysis of host-microbiota interfaces. Methods Mol Biol 2259: 167–179. https://doi.org/10.1007/978-1-0716-1178-4_10

Vincent D, Ezernieks V, Elkins A, Nguyen N, Moate PJ, Cocks BG, Rochfort S (2016) Milk bottom-up proteomics: method optimization. Front Genet 6. https://doi.org/10.3389/fgene.2015.00360

Virkud YV, Wang J, Shreffler WG (2018) Enhancing the safety and efficacy of food allergy immunotherapy: a review of adjunctive therapies. Clin Rev Allergy Immunol 55:172–189. https://doi.org/10.1007/s12016-018-8694-z

Virkud YV, Kelly RS, Wood C, Lasky-Su JA (2019) The nuts and bolts of omics for the clinical allergist. Ann Allergy Asthma Immunol 123:558–563. https://doi.org/10.1016/j.anai.2019.09.017

Wang Y, Zhang X-S, Chen L (2019) Systems biology intertwines with single cell and AI. BMC Bioinformatics 20:204. https://doi.org/10.1186/s12859-019-2731-7

Warren CM, Jiang J, Gupta RS (2020) Epidemiology and burden of food allergy. Curr Allergy Asthma Rep 20:6. https://doi.org/10.1007/s11882-020-0898-7

Wenk D, Zuo C, Kislinger T, Sepiashvili L (2024) Recent developments in mass-spectrometry-based targeted proteomics of clinical cancer biomarkers. Clin Proteomics 21:6. https://doi.org/10.1186/s12014-024-09452-1

Wilmes P, Bond PL (2004) The application of two-dimensional polyacrylamide gel electrophoresis and downstream analyses to a mixed community of prokaryotic microorganisms. Environ Microbiol 6: 911–920. https://doi.org/10.1111/j.1462-2920.2004.00687.x

Wood RA, Sicherer SH, Burks AW, Grishin A, Henning AK, Lindblad R, Stablein D, Sampson HA (2013) A phase 1 study of heat/phenol-killed, E. coli -encapsulated, recombinant modified peanut proteins Ara h 1, Ara h 2, and Ara h 3 (EMP-123) for the treatment of peanut allergy. Allergy 68:803–808. https://doi.org/10.1111/all.12158

Xiao M, Yang J, Feng Y, Zhu Y, Chai X, Wang Y (2017) Metaproteomic strategies and applications for gut microbial research. Appl Microbiol Biotechnol 101:

3077–3088. https://doi.org/10.1007/s00253-017-8215-7

Yamagishi M, Akagawa S, Akagawa Y, Nakai Y, Yamanouchi S, Kimata T, Hashiyada M, Akane A, Tsuji S, Kaneko K (2021) Decreased butyric acid-producing bacteria in gut microbiota of children with egg allergy. Allergy 76:2279–2282. https://doi.org/10.1111/all.14795

Zhang X, Deeke SA, Ning Z, Starr AE, Butcher J, Li J, Mayne J, Cheng K, Liao B, Li L, Singleton R, Mack D, Stintzi A, Figeys D (2018) Metaproteomics reveals associations between microbiome and intestinal extracellular vesicle proteins in pediatric inflammatory bowel disease. Nat Commun 9:2873. https://doi.org/10.1038/s41467-018-05357-4

Zheng P, Zhang K, Lv X, Liu C, Wang Q, Bai X (2022) Gut microbiome and metabolomics profiles of allergic and non-allergic childhood asthma. J Asthma Allergy 15:419–435. https://doi.org/10.2147/JAA.S354870

Zubeldia-Varela E, Barker-Tejeda T, Obeso D, Villaseñor A, Barber D, Pérez-Gordo M (2022) Microbiome and allergy: new insights and perspectives. J Investig Allergol Clin Immunol 32: 327–344. https://doi.org/10.18176/jiaci.0852

Zuidmeer-Jongejan L, Huber H, Swoboda I, Rigby N, Versteeg SA, Jensen BM, Quaak S, Akkerdaas JH, Blom L, Asturias J, Bindslev-Jensen C, Bernardi ML, Clausen M, Ferrara R, Hauer M, Heyse J, Kopp S, Kowalski ML, Lewandowska-Polak A, Linhart B, Maderegger B, Maillere B, Mari A, Martinez A, Mills ENC, Neubauer A, Nicoletti C, Papadopoulos NG, Portoles A, Ranta-Panula V, Santos-Magadan S, Schnoor HJ, Sigurdardottir ST, Stahl-Skov P, Stavroulakis G, Stegfellner G, Vázquez-Cortés S, Witten M, Stolz F, Poulsen LK, Fernandez-Rivas M, Valenta R, van Ree R (2015) Development of a hypoallergenic recombinant Parvalbumin for first-in-man subcutaneous immunotherapy of fish allergy. Int Arch Allergy Immunol 166:41–51. https://doi.org/10.1159/000371657

Applications of Metabolomics in the Development of Biomedicine and Biotechnology

21

Eric H. Rosenn, Allison Walker-Elders, Murugesh Padmanarayana, Aditya Upadhayay, Aditya Kumar Sharma, Neha Dubey, Yogita Soni, and Vijay Soni

Abstract

Metabolomics has emerged as a powerful discipline for capturing dynamic biochemical states and linking molecular processes to human health, disease, and biotechnology. This chapter explores the evolution of metabolomics from early profiling efforts to its current integration with advanced analytical platforms, computational biology, and precision medicine. We highlight how targeted and untargeted metabolomic strategies have advanced biomarker discovery, clarified disease mechanisms, and informed the development of personalized therapies in cancer, cardiovascular disease, diabetes, and other conditions. The role of pharmacometabolomics in refining drug response predictions and guiding therapeutic decisions is also examined. Beyond biomedicine, metabolomics drives innovation in biotechnology, synthetic biology, and bio-based production, while emerging tools such as single-cell metabolomics and machine learning continue to expand its scope. By directly tracing metabolic fluxes and engineering new pathways, metabolomics is becoming a practical engine for designing microbes, optimizing bioprocesses, and accelerating drug discovery.

Keywords

Metabolome · Biomarkers · Personalized medicine · Pharmacometabolomics · Biosynthetic pathways · Synthetic biology · Host-pathogen relationships · Drug-target interactions · Oncometabolites

E. H. Rosenn
Department of Biomedical Engineering, New York University, Tandon School of Engineering, New York, NY, USA

A. Walker-Elders
College of Osteopathic Medicine, Touro University, New York, NY, USA

M. Padmanarayana
Department of Biochemistry, Stanford University, Beckman Center for Molecular and Genetic Medicine, Stanford, Palo Alto, CA, USA

A. Upadhayay
Hospital for Special Surgery, New York, NY, USA

A. K. Sharma
Department of Pathology, College of Medicine, University of Illinois at Chicago, Chicago, IL, USA

N. Dubey
Department of Molecular Microbiology, Center for Women's Infectious Disease Research, Washington University School of Medicine, St. Louis, MO, USA

Y. Soni
Department of Chemical and Biological Engineering, Tufts University, Medford, MA, USA

V. Soni (✉)
Division of Infectious Diseases, Weill Department of Medicine, Weill Cornell Medicine, New York, NY, USA
e-mail: vis2032@med.cornell.edu

M. Gama-Carvalho et al. (eds.), *Omics Approaches in Biomedicine and Biotechnology*, Advances in Experimental Medicine and Biology 1504, https://doi.org/10.1007/978-3-032-18966-0_21

21.1 Introduction

Metabolomics is the systematic study of the metabolome—the complete set of small molecules (below 1.5 kDa) involved in cellular biochemical processes. Because the metabolome is dynamic, fluctuations in metabolite levels can reveal patterns in metabolism and define characteristic metabolic states. In the late 1940s, Roger Williams proposed the concept of unique metabolic statuses in individuals, a notion he confirmed by identifying compounds in the saliva of patients with schizophrenia using paper chromatography (Lokhov et al. 2021). This was a significant step in the early development of metabolomics. However, it wasn't until 1971 that Horning first used the term "metabolomic profile" to describe metabolite measurements found in saliva and urine samples by gas chromatography (Nalbantoğlu 2019). Around that time, Nuclear Magnetic Resonance (NMR) also began to be used for "metabolic profiling," and in 1984, Professor Jeremy Nicholson demonstrated the clinical efficacy of metabolomics by using NMR to diagnose diabetes mellitus (Yang et al. 2019).

While this "snapshot" of metabolic information holds great promise for discovery, interpretation has long been hindered by the difficulty of obtaining reliable absolute measurements. As with many technologies, however, progress over the past two decades has accelerated rapidly. In 2005, the Metabolomics Society set up the "Metabolomics Standards Initiative (MSI)" committee that has established rules to standardize metabolomics systems (Fiehn et al. 2007). The Human Metabolome project completed its first phase in 2007, establishing the Human Metabolome Database (HMDB) which originally contained data on around 2500 metabolites, but its newest version, released in 2022 contains over 220,000 entries (visit https://hmdb.ca/ for updates) (Hartman and Lees 2023). METLIN, the first tandem mass spectrometry database for characterizing human metabolites, was established in 2005, and now contains data on over a million molecules identified through tandem mass spectrometry (MS/MS) (Tautenhahn et al. 2012; Chen et al. 2022a) (https://metlin.scripps.edu/). Among other databases and repositories, many tools have emerged to utilize this data. MetaboAnalyst is the most prominent platform for metabolite identification and analysis (Pang et al. 2024). Among its many functions, the program can automatically process raw MS data and facilitate biomarker identification. Interfacing with metabolite data banks, such as the HMDB, various analysis methods can be performed on metabolites of interest by inputting their associated identifier codes.

These technological advances have enabled the characterization of a wide range of biomarkers through metabolomic techniques. Such biomarkers may arise from endogenous metabolites, dietary components, or environmental exposures, and many have been linked to specific health and disease states. Importantly, groups of biomarkers forming distinct metabolic profiles have been developed into reliable diagnostic indicators for conditions including diabetes, cardiovascular disease, cancer, and numerous other disorders (Di Minno et al. 2022).

A comprehensive analysis of the metabolome cannot yet be achieved with a single analytic method. To overcome this limitation, hyphenated approaches, such as LC-NMR-MS, are being developed to provide broader coverage for global metabolite profiling (McCullagh and Probert 2024). Integrating these complex technologies remains a challenge, yet they are expected to evolve into critical quantitative platforms that will help define the field (Pinu et al. 2019). The rapid expansion of computational technology, and programs to analyze the output of such platforms, has been a major factor driving the development of metabolomics applications.

Major metabolomics approaches include targeted metabolomics, untargeted metabolomics, fluxomics, and metabolite imaging (Karnovsky and Li 2020). *Targeted metabolomics* aims to identify and quantify a small subset of metabolites and is ideal for biomarker detection. *Untargeted metabolomics* attempts to characterize as many metabolites as possible within a biological system, most commonly by LC-MS

(Sindelar and Patti 2020). *Fluxomics* is a branch of targeted metabolomics that tracks the movement of isotopic labels in metabolic intermediates, providing a measurement of metabolite reaction rates. *Metabolite imaging* is an emerging field of metabolomics that involves detecting and visualizing metabolites in tissues (Saoi and Britz-McKibbin 2021).

While untargeted metabolomics provides a full scan of the metabolome associated with a particular state, sometimes referred to as a "metabolic fingerprint," targeted metabolomics aims to track specific metabolite levels and the activity of biochemical pathways (Nalbantoğlu 2019). Untargeted metabolomics can thus generate data that reveals important patterns and is considered "hypothesis-generating." (Schrimpe-Rutledge et al. 2016) Standard methods for screening metabolites in untargeted, or "global," approaches include selection based on the statistical significance or fold change. The XCMS tool is an online platform that offers an additional level of candidate selection through comparative analysis of pathway maps (Guijas et al. 2018a). In contrast, targeted metabolomics focuses on a defined subset of metabolites of interest and is considered "hypothesis confirming."

In the current era of technological evolution, a central goal is to harness human health through precision medicine, which seeks to predict, prevent, and treat disease by considering each individual's genetics, environment, and lifestyle in contrast to the traditional "one-size-fits-all" reactive model of care (Babu and Snyder 2023). By using a quantitative approach to healthcare, metabolomics exemplifies how precision medicine draws on mathematics and analytics to predict the course of pathology more accurately. Promoting the translation of omics research to in-clinic point-of-care methods, metabolomics supports a growing holistic approach in medical care that views each patient as a distinct combination of biochemical, physiological, and environmental interactions (Ashrafian et al. 2021) (Fig. 21.1).

In this chapter, we will examine how various methods of metabolomics analysis are being applied across a wide range of scientific fields and industries. We will highlight important recent advances in biomedicine and personalized health, and also explore the advantages of metabolomics relative to other omics platforms. Throughout our discussion, we will clarify basic concepts underlying broad topics such as analytic instrumentation, computational biology, and drug development.

21.2 Applications of Metabolomics in Biomedicine

Biomedicine encompasses a diverse range of disciplines focused on understanding health and disease through molecular and genetic insights. Among these, the omics sciences—genomics, proteomics, and metabolomics—stand out for their transformative impact. Genomics provides insights into inherited traits and disease susceptibility, which are crucial for personalized medicine (Lander 2011). Proteomics investigates the complete set of proteins within cells, essential for understanding cellular functions and identifying therapeutic targets (Aebersold and Mann 2016).

Metabolomics, a rapidly expanding field within biomedicine, centers on metabolites—small molecules that reflect genetic activity and respond to environmental influences. This approach, known as metabolic profiling and phenotyping, captures real-time snapshots of physiological states (Nicholson and Lindon 2008). Analyzing these profiles enables the identification of disease biomarkers from genetic pathways or environmental exposures, revealing disease mechanisms and facilitating early interventions. Advances in metabolomic technologies have revolutionized drug discovery, development, and therapeutic strategies, ushering in a new era of precision medicine tailored to individual metabolite profiles (Patti et al. 2012; Wishart 2016).

21.2.1 Metabolomics: Unveiling Small Molecules' Role in Disease and Personalized Medicine

The central dogma of molecular biology has long provided foundational insights into the molecular

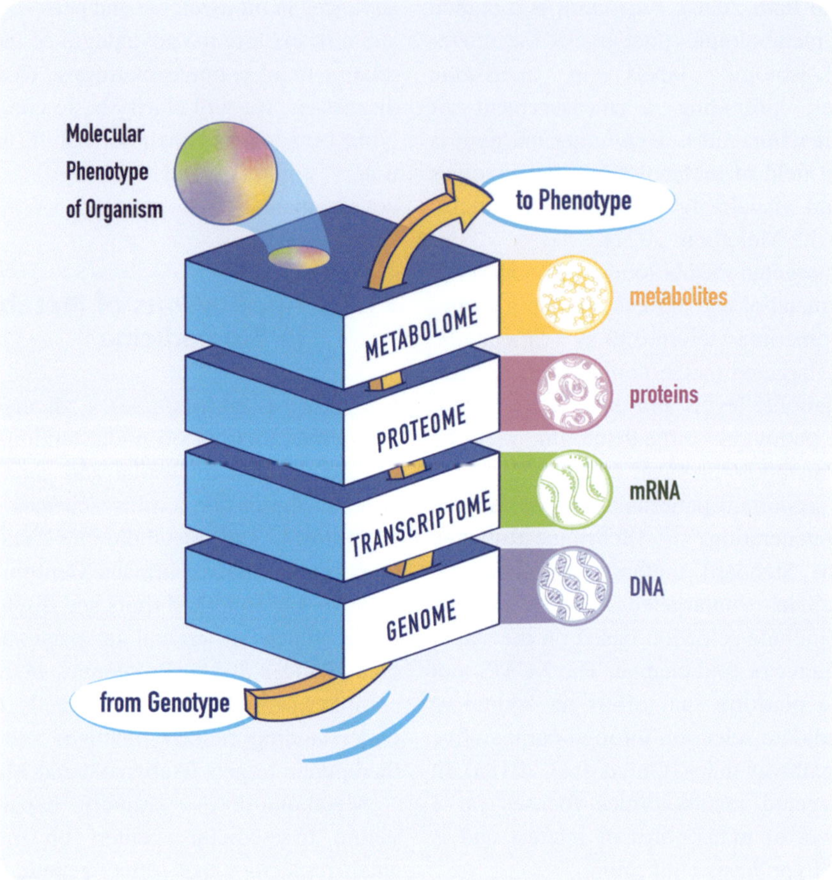

Fig. 21.1 Multi-level interconnection of "-omes" forming the metabolome (Balashova et al. 2020)

basis of life, human susceptibility to diseases, and the varying responses to therapeutic drugs. While genes, transcripts, and proteins have historically dominated biomedical research, metabolomics offers a new perspective centered on the role of small molecules in cellular function (Beger et al. 2016). By adopting a small molecule-centric view encompassing metabolites from both internal cellular processes and external environmental influences, metabolomics has unveiled unexpected contributors to chronic diseases such as atherosclerosis, diabetes, and cancer (Qiu et al. 2023; Wild et al. 2013). This approach probes complex biochemical interactions at cellular and organismal levels, illuminating the critical role of metabolites in disease development, cellular signaling, and physiological regulation. This unique capability makes metabolomics a powerful tool for identifying biomarkers, understanding disease mechanisms, and developing early diagnostic strategies.

21.2.2 Metabolomics in Biomarker Discovery

The metabolome, which represents the final product of the omics hierarchy, is closely linked to the phenotype (Patti et al. 2012). This proximity allows metabolomics to be a powerful tool in disease diagnosis by identifying specific biomarkers. These biomarkers reflect normal biological activities, disease processes, or responses to environmental exposures and treatments. Often, metabolomic changes can be detected before clinical symptoms arise, enabling early risk assessment, screening, diagnosis, prognosis, therapeutic response prediction, and ongoing disease monitoring (Holmes et al. 2008; Ou et al. 2021). This early detection capacity positions metabolomics as an asset in disease diagnosis. We will discuss the role of metabolomics in the discovery of various disorders by biomarkers.

21.2.2.1 Cancer

In the 1950s, Otto Warburg proposed that cancer cells exhibit a distinct metabolic phenotype compared to normal differentiated adult cells. Cancer cells demonstrate heightened glucose uptake and preferentially produce lactate through (aerobic) glycolysis, even in the presence of sufficient oxygen, a process known as the Warburg effect (Hanahan and Weinberg 2011; Warburg 1956). This altered metabolism supports rapid cell growth and division by generating Adenosine triphosphate (ATP) and providing essential substrates. The regulation of this metabolic reprogramming involves oncogenes and tumor suppressor genes, highlighting cancer as a metabolic disorder rather than solely a genetic one (Levine and Puzio-Kuter 2010). Metabolites crucial for sustaining tumor growth and metastasis are often termed "oncometabolites," acting as signaling molecules or regulators that influence key cellular processes such as cell division (Yang et al. 2013).

For instance, studies comparing colorectal tumor biopsies with healthy colon tissue have identified potential biomarkers linked to increased tissue hypoxia, glycolysis, nucleotide biosynthesis, lipid metabolism, inflammation, and steroid metabolism. These findings underscore significant metabolic differences between cancerous and normal tissues (Cheng et al. 2012).

Furthermore, analysis of serum from pancreatic cancer patients has revealed distinct metabolic profiles compared to healthy individuals. Pancreatic cancer is characterized by reduced levels of metabolites like 3-hydroxybutyrate, lactate, and trimethylamine N-oxide, alongside elevated concentrations of isoleucine, triglycerides, leucine, and creatinine (Tognetti et al. 2025). These metabolic alterations offer critical insights into the biochemical landscape of pancreatic cancer, presenting opportunities for early detection strategies and personalized treatment approaches (Urayama et al. 2010).

21.2.2.2 Atherosclerosis

Atherosclerosis, traditionally associated with elevated cholesterol levels, has undergone a paradigm shift following the discoveries of Stanley Hazen and his team (Wang et al. 2011a; Lusis 2012). They elucidated a significant link between diet, the microbiome, and trimethylamine N-oxide (TMAO), a compound implicated in plaque formation. Subsequent studies in both animal models and humans have consistently shown strong correlations, indicating that elevated TMAO levels in plasma are closely associated with increased risks of adverse myocardial events. This research highlights TMAO as a promising biomarker not only for assessing atherosclerosis progression but also for predicting subsequent myocardial infarction events (Wang et al. 2014; Koeth et al. 2013; Gregory et al. 2015).

21.2.2.3 Diabetes

Metabolites, particularly amino acids like branched-chain amino acids (BCAAs) and aromatic amino acids (AAAs), have emerged as valuable biomarkers for predicting the risk of developing type 2 diabetes. Research utilizing metabolomic techniques has shown that elevated levels of isoleucine, leucine, valine and phenylalanine, tyrosine, as well as amino-adipic acid, can precede the onset of diabetes by up to 15 years (Wurtz et al. 2013; Wang et al. 2011b; Li et al. 2013). These metabolites are not only influenced by dietary intake but also by the gut microbiome, which plays a role in their abundance (Adel et al. 2025). Importantly BCAAs, especially leucine, activate mTORC1 signaling and can influence insulin signaling. Chronic elevation of these amino acids can lead to insulin resistance, a hallmark of type 2 diabetes (Wang et al. 2011b; Chace and Spitzer 2011).

Metabolomics offers a more predictive approach than traditional genetic studies (GWAS) in identifying individuals at risk for diabetes (Barrett et al. 2024; Lauber et al. 2022). By detecting early metabolic changes associated with amino acid dysregulation, metabolomic profiling provides a window into potential disease development long before clinical symptoms manifest. This ability to identify at-risk individuals early enables targeted interventions, such as dietary modifications or microbiome-targeted therapies, emphasizing the clinical utility of metabolites as biomarkers in diabetes prevention and management (Wang et al. 2011b).

21.2.3 Metabolomics for Clinical Technologies and Prognostic Tools

As previously mentioned, biomarkers can serve as a strong diagnostic tool as the method to obtain them is relatively fast and non-invasive and greatly decreases in price year by year (Kroemer et al. 2024). These biomarkers should thus be considered for their predictive value and the potential of including these measurements in regular screening (Olivier et al. 2019; Everett et al. 2019) (Table 21.1). Samples can be collected in a clinical setting or at the point of care, making metabolomics a convenient tool. Unlike other prognostic tools, such as genomic testing, metabolomic procedures rely on currently existing laboratory and clinical infrastructure. This makes its use cost-efficient as well as relatively quick (Resurreccion and Fong 2022). As a tool for preventive medicine, metabolomics has even greater implications for cost-savings—assessing a patient's risk for developing a given disease enables the patient and provider to invest in lifestyle modifications and other preventive measures that lower the likelihood of developing said disease, thereby saving healthcare costs in treatment downstream (Singh et al. 2023a).

There is some evidence supporting the clinical efficacy of this (Table 21.2). For instance, particular and characteristic gut enterotypes are related to polyphenol metabolism and the likely severity of Alzheimer's development. A separate study examining plasma metabolites identified a minimum signature of nine metabolites that predicted dementia due to Alzheimer's disease more than 3 years prior to the onset of cognitive decline (Tremblay-Franco et al. 2024; Soares Dias Portela et al. 2024; Liu et al. 2025; Fletcher et al. 2025). An assessment of how rapid and at what age Alzheimer's onset is expected to occur has many practical implications and issues. However, focusing on the positive aspects of early detection can allow patients and caretakers to prepare for a future diagnosis. The application of metabolomics in clinical prognostics and preventive medicine has further implications for personalized treatment and precision care (Azad and Shulaev 2019).

Detection of metabolites through metabolomics has shown effectiveness in the realm of predicting the development of cancer as well. Metabolites present in the urea cycle are elevated in patients with prostate cancer relative to those with normal prostates (Resurreccion and

Table 21.1 Metabolomic alterations prognostic for diseases

Condition	Study outcome	Citation
Prostate cancer	Patients with elevated sarcosine levels had a more aggressive disease progression than those with normal levels	Sreekumar et al. (2009)
Breast cancer	Ten metabolites were associated with an increased risk of breast cancer in a group of premenopausal women	Jobard et al. (2021)
Alzheimer's disease	Combined aqueous and lipid metabolomics predicted cognitive decline	Tremblay-Franco et al. (2024)
Type II diabetes mellitus	Individuals with altered levels of glycine, lysophosphatidylcholine, and acetylcarnitine were at increased risk of developing type II diabetes within 7 years	Wang-Sattler et al. (2012)
Myocardial infarction	A machine-learning model which combined metabolites and clinical scenarios identified acetylcarnitine, pyruvate, and lathosterol as significant contributors to risk of MI	Yu et al. (2024)

Table 21.2 Select experiments displaying metabolomics methods in bioprocess engineering

Organism	Design	Outcome	Citation
E. coli	Ensemble model incorporating 148 metabolites, using time-resolved metabolite and protein measurements	Identified pathways with high influence on product yield for cell-free protein synthesis	Horvath et al. (2017)
C. cellulovorans	Dynamic model, which integrated experimental proteomics, metabolomics, and fermentation kinetics	Identified key regulators and guide metabolicengineering of n-butanol biosynthesis	Ou et al. (2020)
S. cerevisiae	Metabolic and proteomic flux analysis, pull-push-block engineering of carbon flux, and cofactor balancing	Identified and modified genes affecting upstream targets regulating the production of fatty alcohols by fatty acid reductases	d'Espaux et al. (2017)

Fong 2022). Further research into metabolomics has revealed myriad metabolites that have the potential to positively predict whether a patient will develop prostate cancer. For instance, the dysregulated synthesis of fatty acids and cholesterol increases the risk for prostate cancer development, as most prostate cancer does not rely solely on the Warburg effect (Gómez-Cebrián et al. 2019). Higher levels of metabolites from de novo lipogenesis have the potential to not only predict a patient's risk for developing prostate cancer but also offer a pathway to preventive medicine through diet and pharmaceutical interventions.

The next phase of metabolomics as a tool for translational medicine is in clinical application to guide therapeutics (Castelli et al. 2022). By applying precision therapeutics, metabolomics could be used to optimize drug doses. Pathways associated with different medications can be compared to inform side effects and negative sequelae (Schmidt et al. 2021). As an example, serum bile acid levels are positively associated with increased mortality from acetaminophen-associated liver failure (AALF) (Woolbright et al. 2014), while glycodeoxycholic acid predicted survival at a statistically significant rate. Measuring metabolite biomarkers of patients with AALF has the potential to inform the course of illness from acetaminophen toxicity.

21.2.4 Metabolomics in Precision Medicine

Precision or personalized medicine aims to tailor medical treatments based on individual biomarker profiles derived from advanced omics testing. One of the most notable successes in this field has been the application of metabolomics in newborn screening (Chace and Spitzer 2011; Lehotay et al. 2011; Grebe and Singh 2011). Initially introduced over 50 years ago for detecting phenylketonuria using a bacterial inhibition assay on blood spots from newborns, the practice expanded in the 1970s to include screening for congenital hypothyroidism. Recent advancements in analytical instruments such as electrospray ionization and tandem mass spectrometry have enabled rapid, high-throughput analysis of samples extracted from dried blood spots (DBS) (Chace et al. 1995; Chace et al. 1993). Pioneers like Millington, Chace, and others have significantly contributed by developing methods for screening multiple inherited metabolic diseases using targeted metabolomics (Chace et al. 1997; Millington et al. 1992). Today, these techniques not only facilitate early disease detection but also optimize therapies such as enzyme replacement or dietary restrictions and monitor therapeutic efficacy through personalized dose adjustments (Cossu et al. 2023).

Moreover, metabolomics and clinical chemistry assays have revolutionized the identification and treatment of conditions like pre-diabetes (using glucose levels), hemochromatosis (using iron levels), and dietary deficiencies (using vitamins and micronutrients) (Xie et al. 2025; Stechemesser et al. 2017). Significant cost reductions in gene sequencing, Single Nucleotide Polymorphism (SNP) profiling, and proteomics are facilitating the adaptation of other omics technologies into precision and personalized medicine. This has led pioneers such as Mike

Snyder and Leroy Hood to deploy "integrated personal omics profiling" (iPOP) as part of their efforts to establish "predictive, preventive, personalized, and participatory" (P4) medicine (Chen et al. 2012). These initiatives gather multi-omics data—including genomic, transcriptomic, proteomic, and metabolomics data—from large cohorts over extended periods. This data-rich approach not only assesses individual health statuses comprehensively but also identifies illnesses early and guides tailored medical interventions (Flores et al. 2013).

Metabolomics is becoming increasingly important in drug response and therapeutic monitoring, areas that pose significant challenges because of the wide variability in individual treatment outcomes (Wilkinson 2005). Reports from 1998 indicated a staggering two million cases of adverse drug reactions (ADR) annually in the United States, resulting in approximately 100,000 deaths, figures that have since continued to rise (Lazarou et al. 1998; Shepherd et al. 2012). While personalized medicine initially focused on pharmacogenomics—where genetic factors influence drug responses—the anticipated impact on patient care has been somewhat limited (Roden et al. 2011; McInnes et al. 2021; Pirmohamed 2001; Mancinelli et al. 2000). A key reason for this limitation is that genetics alone does not consider crucial contextual factors such as alcohol consumption, age, diet, concurrent medications, gut microbiota composition, physical activity, occupation, stress levels, and overall health status, all of which significantly influence how individuals respond to drug treatments. Genetics typically explains only 20 to 40% of the variability in drug metabolism and response among individuals (Ventola 2013). Consequently, the field of pharmacometabolomics has emerged to elucidate the underlying mechanisms of drug effects or xenobiotic metabolism, enhancing our ability to predict individual variations in drug response phenotypes. This approach leverages baseline metabolic profiles before treatment initiation and monitors metabolic changes over time during drug therapy, thereby providing a more comprehensive understanding of personalized drug responses (Clayton et al. 2006).

For example, immunosuppressant drugs have a narrow therapeutic window and can lead to severe adverse effects such as leukopenia, thrombosis, and anemia (Sallustio 2010). Accordingly, rapid mass spectrometry (MS)-based monitoring of immunosuppressant drugs and their metabolites is now employed to optimize dosing on an individual basis. This method is also crucial for patients with liver disease, where dosing and potential byproduct buildup are a concern (Stewart et al. 2011). Pharmacometabolomics has likewise been pivotal in investigating responses to statin therapy in cardiovascular disease. Plasma levels of several microbial metabolites—including bile acids such as lithocholic acid, taurolithocholic acid, and glycolithocholic acid, as well as microbially derived 2-hydroxyvaleric acid—have been identified as markers of favorable statin response, suggesting that specific microbiome compositions may influence treatment outcomes (Krauss et al. 2013).

In the context of aspirin therapy for cardiovascular disease prevention, serotonin levels have been shown to influence platelet reactivity, illustrating how pharmacometabolomics provides mechanistic insights that extend beyond those obtainable through genotyping alone (Krauss et al. 2013; Zhu et al. 2013).

Recent advancements in pharmacometabolomics leverage large-scale GWAS and metabolome-wide association studies (MWAS) to explore SNP-induced variations in "resting" metabolite levels among healthy individuals (Prakash 2011). These studies not only elucidate individual differences in metabolite levels but also highlight the potential of combining endogenous metabolites with SNP-typing to predict or enhance understanding of responses to drugs or therapies (Shin et al. 2014). For example, in studies of statin therapy, GWAS have identified significant SNPs such as rs4149056 in the SLCO1B1 gene, impacting statin uptake into the liver and influencing LDL cholesterol levels as well as treatment response. This insight highlights that effective statin personalization requires balancing genetic determinants with metabolomic factors, including liver enzyme activity, to improve outcomes and reduce adverse effects (Stewart 2013). Similarly, investigations into antipsychotic medications have uncovered SNPs in genes like DRD2 (rs1799978) and COMT (rs4680), affecting dopamine metabolism pathways (Taheri et al. 2023). Variants in these genes can alter receptor function, dopamine signaling efficiency, and neurotransmitter availability, influencing treatment response and side effect profiles in psychiatric care. Another study highlighted variations in patient responses to the antihypertensive drug atenolol, where specific intronic SNPs (rs9652472 in hepatic lipase gene and rs7250148 in phospholipase A2 gene) were identified through metabolomics data, rationalizing individual and racial differences in atenolol response (Wikoff et al. 2013).

The future of metabolomics lies in the continuous refinement of technologies and methodologies to further uncover the complexity of metabolic networks and their interactions with other omics. Integrating metabolomics with other omics data and clinical information will enhance our ability to develop more precise diagnostic tools, identify new therapeutic targets, and personalized treatment strategies (Nicholson and Lindon 2008; Holmes et al. 2011). One promising area is the application of metabolomics in understanding the role of the gut microbiome in health and disease. The microbiome's influence on host metabolism is profound, and metabolomics offers a powerful tool to elucidate these interactions. Future research may focus on microbial metabolites contribute to disease mechanisms and how altering the microbiome could lead to novel therapeutic strategies (Holmes et al. 2011; Nicholson and Wilson 2003).

Another exciting development is the use of single-cell metabolomics, which allows the study of metabolic heterogeneity within cell populations. This approach can provide insights into how individual cells contribute to overall tissue function and disease pathology, offering new avenues for targeted therapies (Zenobi 2013). Finally, advances in computational tools and machine learning algorithms will play a crucial role in analyzing the vast amounts of data generated by metabolomics studies. These tools will enable the integration of multi-omics data, providing a holistic view of biological systems and facilitating the discovery of new biomarkers and therapeutic targets (Chong et al. 2019; Weissler et al. 2021). Metabolomics is revolutionizing our understanding of disease mechanisms, offering new diagnostic and therapeutic opportunities. Its ability to provide a comprehensive snapshot of metabolic states makes it invaluable tool in precision medicine, enabling tailored treatments based on individual metabolic profiles.

21.3 Application of Metabolomics in Biotechnology

Metabolomics has been described as a compendium of technologies involved in metabolic analysis. Many technologies have contributed to the development of metabolomics, and there has indeed been a coevolution between the field and analytical and data processing platforms. We will delineate here between technologies that have explicitly developed in conjunction with metabolomics and those that have developed because of the discipline. For instance, some technologies applied in metabolomics are generally used in multi-omics, while other technologies have emerged specifically to address needs in metabolomics. We will also discuss technologies that are not central to metabolomics but have been devised or improved based on understandings founded in metabolomics.

21.3.1 Increased Automation of Hyphenated Techniques

Metabolomics has led to the establishment of tried-and-true analysis techniques. Still, it has co-opted the development of some modalities for its own needs, thus stimulating their functional development. As a result, the underlying technology has seen increased accuracy, automation, and user-centric features that are facilitating translation to clinical use.

Mass spectrometry (MS) can provide qualitative information on compound structure and quantitative information like relative abundance;

if appropriately conducted, absolute quantitation can provide metabolite concentrations (Xiao et al. 2012). The workings of various MS methods can be challenging to conceptualize.

Historical Context

Over the past century, seven Nobel Prizes have been awarded for breakthroughs that advanced mass spectrometry. Among the laureates was John Fenn, recognized for developing electrospray ionization (ESI), a technique that transformed biomolecular analysis (Grayson 2011). Fenn is said to have drawn inspiration from the principles of jet propulsion that he studied while working on a rocket development project during World War II. Despite the profound impact of his technique, controversy followed—he was accused of downplaying the significance of his research while at Yale University and ultimately lost a lawsuit to the institution, exceeding one million dollars.

MS is generally coupled to a separation process. Gas chromatography (GC) separates compounds from a volatile sample in the gaseous state, while liquid chromatography (LC) separates compounds from a non-volatile sample in the liquid state based on interactions with the stationary phase such as polarity, hydrophobicity, and charge. Ultra-performance LC (UPLC)-MS is a standard in biomedical research (Alsaleh et al. 2019). GC-MS is also widely used, especially for volatile and derivatized metabolites, and benefits from reproducible retention indices and electron ionization spectra that can be matched to large libraries (Sumner et al. 2007; Fiehn 2016). In LC-MS workflows, substantial information can be obtained from accurate mass, isotopic patterns, and retention time without fragmentation, although secure metabolite identification typically relies on MS/MS to capture diagnostic product ions (Schymanski et al. 2014; Kind et al. 2009; Witting and Böcker 2020). The separation ideally allows one molecule at a time to enter MS for analysis. A compound is ionized in a typical MS process, and the mass-to-charge ratio (*m/z*) of resulting ions is measured. Subsequently, the compound is fractured, producing a distinct spectrum. The spectrum is screened against spectra of known compounds for identification. This process is completely automated and involves complex statistical analysis for dimensional reduction, as data is highly collinear, which can increase false positive identifications. Absolute quantitative data is needed for clinical translation use; thus, robust internal standards are required, and extensive planning is done before every LC-MS experiment (Rischke et al. 2023).

NMR is based on irradiating a compound with an electromagnetic pulse. The pulse has a specific frequency, and when the source is turned off, the compound releases the quantized energy in a wave that forms a characteristic spectrum (Paris et al. 2018). Though NMR demonstrates the number of advantages over LC-MS, the lack of NMR data in metabolite databases has classically limited NMR and LC-MS has become the most popular platform for metabolomics (Wishart et al. 2022a). However, NMR databases have expanded considerably in recent years. The field has also benefited from the unique strengths and advances in NMR technologies, along with continued developments in computational tools to analyze complex metabolite mixtures. As a result, several automated, quantitative NMR platforms have emerged that can analyze samples faster and more reliably than LC-MS, at just minutes per sample (Nagana Gowda and Raftery 2021). Another advantage of NMR is its superior ability to detect compounds that do not readily ionize, such as sugars, organic acids, alcohols, polyols, and other highly polar compounds (Emwas et al. 2019).

Until recently, these methods were used purely in research methodologies; however, automation and user-based designing are allowing them to be translated into clinical applications. However, advances in electronics are making NMR instruments smaller, cheaper, and more clinically feasible (Wishart et al. 2022a). For instance, using magnetic resonance imaging (MRI) scanners, NMR-based in vivo metabolomics profiling can potentially be utilized in the clinic. Additionally, NMR is easily referenced to metabolic flux and metabolic imaging studies, making it ideally suited for probing living cells, tissues, and organs (Moco 2022).

Metabolite imaging is another area of technological advancement being catalyzed by metabolomics. Metabolite imaging involves the in vivo detection of metabolites within tissues using NMR, magnetic resonance spectroscopy (MRS), or positron emission tomography (PET), among other methods (Wishart 2016). Improvements in NMR pulse sequences, magnetic field strength, and computer processing speed allow various compounds to be detected at lower concentrations through MRS. As a result, NMR has become a clinically utilized and even preferred route for analyzing plasma lipoprotein and cholesterol classes (Crook and Powers 2020). One outcome of these advances has been more accurate tumor phenotyping, which can be important for diagnostics and staging (Corona et al. 2021).

Very recently, a considerable dilemma involving LC-MS quantification has been brought up. For many years, it has been thought that there are a vast number of unknown metabolites, mostly small molecules. The low concentrations and high noise produced by this supposed pool of compounds made their isolation and identification by MS virtually impossible (Mahieu and Patti 2017; Chi et al. 2025). Many recent and significant works have posited that this background noise is actually just an instrumentation artifact and may be the result of compound fragmentation occurring during ionization, resulting in a multitude of electrospray ionization (ESI) shadow peaks called features (Chi et al. 2025; Giera et al. 2024a; Kuhl et al. 2012). If true, an estimated 70% of LC-MS/MS data could result from in-source fragment ions (ISF) and not unique metabolites (Roca et al. 2021; Wishart et al. 2022b; Giera et al. 2024b; Guo et al. 2022). While this reduces the number of theorized unknown metabolites, it is agreed that current coverage for all metabolites represents only a small percentage. New tools will likely provide answers in the near future; for example, the METLIN database has incorporated an AI feature, "fragment similarity search" to match feature peaks with fragments of known metabolites (Guijas et al. 2018b).

Another analytical method that has emerged for metabolomics applications has been extracellular flux analysis (EFA) (Aderemi et al. 2021). EFA can greatly help scientists who need to extract metabolic data but are not highly skilled in other quantitative and computational techniques (Yoo et al. 2024). Instead of detecting specific biomarkers, EFA strives to detect overall changes in energy state and energy usage or changes in cell metabolic activity. Briefly, EFA measures primary bioenergetic pathway activity, such as the mitochondrial oxidative phosphorylation and glycolysis processes (Ryan and Lee 2025). Shifts in these pathways denote "characteristic metabolic phenotypes." This can further characterize the baseline metabolic state of the cells being studied and highlight bioenergetic and biosynthetic pathways likely to be of primary focus for the study (Voss et al. 2021).

21.3.2 Advances in Metabolic Engineering and Biochemical Production

Many chemicals have intrinsic value, but those central to industrial processes are particularly subject to research on faster, cost-effective means of production, with an increasing concern being placed on their environmental impact as well. With advances in genetic engineering, industrial microbiology has emerged as a major source of many compounds. As opposed to mining resources from natural reserves, or production by synthetic processes, large quantities of a desired compound can be produced by harnessing the capacity of microorganisms. Today, microorganisms can be genetically altered and cultivated to produce anything from perfume to food ingredients (Galanie et al. 2020). The field of industrial biosynthesis aims to improve production by altering microbial metabolism and growth conditions.

Metabolic engineering has emerged as a distinct discipline with a methodologic approach to improving microbial strains to optimize chemical production (Xu et al. 2020). In short, natural metabolites are produced by microbes that can be cultivated and isolated, thus making these compounds at industrial scales in green, sustainable, and cost-effective manners (Wei et al. 2020). The production of biofuels, particularly via microbial fermentation, is environmentally sustainable and has the potential to disrupt reliance on fossil fuels (Liu et al. 2015). Metabolic engineering aims to improve strain performance and make bioprocess production competitive with petrochemical development processes. A deeper understanding of the underlying metabolic mechanisms of the host organisms can decrease the need for repeated trial-and-error development cycles. More efficient processes while raising yield, also aim to lower costs and reduce labor. Metabolomics capabilities are thus a key driver of technologies in the industry (Table 21.1).

Escherichia coli has been the most utilized strain in industrial bioprocess production and metabolic engineering due to the extensive informational coverage on every aspect of the organism (Pontrelli et al. 2018). Even where *E. coli* is not the best strain to produce a material, it is usually preferred because our knowledge of the organism's physiology and biochemistry allows for significant phenotype construction and alterations that make it more suited to produce most compounds. In this way, *E. coli* has been engineered for the industrial production of myriad organic and nonorganic materials, including proteins, DNA, alcohol, synthetic sugars and amino compounds, and biofuels (Keasling et al. 2021). While alterations in rates of metabolite formation and cell turnover affecting substance production are optimized using a number of omics techniques, the ability to produce a variety of materials stems from the capacity to introduce variation in enzymes that result in the desired material. The discovery of novel enzyme functions that are the basis for producing many compounds relies on metabolomics strategies (Notebaart et al. 2018).

As an example of this, many studies are currently focusing on increasing *E. coli*'s natural capacity to produce hydrogen and ethanol from glycerol. This essentially converts a fossil fuel byproduct into a clean biofuel. Valle et al. (2023). showed a large increase in H_2 yield by promoting a shift in the carbon metabolic pathway resulting from an alteration in the central metabolic enzyme citrate lyase. [Similar to Le Chatelier's principle, the enzyme is chosen based on metabolite levels and the desired shift in direction for the metabolic pathway, in this case, by overexpressing the enzyme.]

Microbial strains with greater commercial and industrial implications have been developed rapidly through systems of metabolic engineering (Choi et al. 2019). This field, which combines systems and synthetic biology with metabolic engineering, facilitates the design of discrete metabolic pathways that result in the desired substance (see Fig. 21.2). This novel process is termed "metabolic pathway reconstruction" and contrasts with the traditional process of pre-existing endogenous pathways to produce chemicals (Choi et al. 2019). Metabolomics helps identify rate-limiting steps and prioritizes metabolic engineering targets (Liu and Nielsen 2019).

For example, the alteration of type 1 fatty acid synthase in *Saccharomyces cerevisiae* results in a remarkable variation in oleochemicals (Zhou et al. 2016). By deleting certain genes, researchers discovered that a free fatty acid derivative could be produced. Commercially, these could be used to produce chemicals ranging from food additive flavorings and aromas to medications, lubricants, and the main components of biodiesels.

A mathematical approach to study metabolism and rational design of *S. cerevisiae* as a cell factory uses reaction kinetics to model time-dependent dynamics. A more recent review on the development of kinetics models for *S. cerevisiae* (Ting et al. 2023) suggests that the metabolomics datasets can be used to re-fit parameter values or validate kinetic model simulations. For instance, optimization and risk

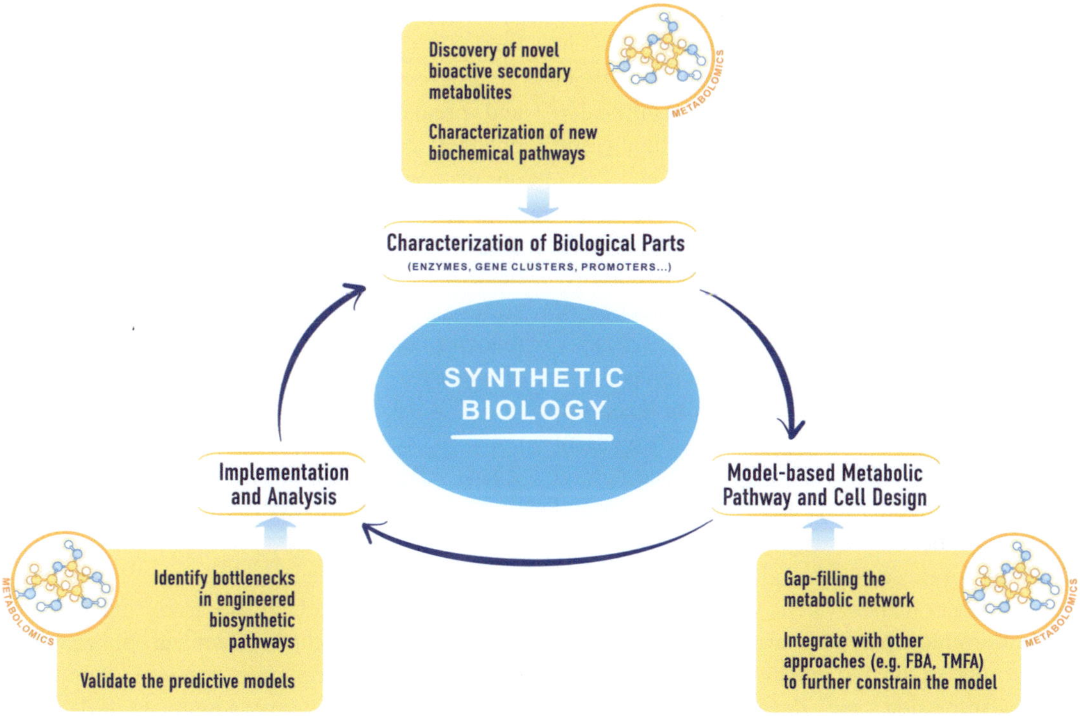

Fig. 21.2 Role of metabolomics in synthetic biology (Nguyen et al. 2012): This schematic shows metabolomics enabling each stage of synthetic biology. Discovery data defines biological parts, constraints-based model pathway and cell design through network gap filling and integration with FBA and TMFA, and then validates and optimizes implementations by identifying bottlenecks in engineered biosynthetic pathways

analysis of complex living entities (ORCALE) framework that utilized metabolomics with *metabolic control analysis* (MCA) can be used to create kinetic models that are more effective in predicting outputs. These models give important information about bioprocess reactions such as the rate of substrate consumption, biotechnologically important metabolite synthesis, or the growth of *S. cerevisiae* (Ting et al. 2023; Xenios et al. 2022).

Genome-scale metabolic modeling/model (GSMM or GEM) is a concept in systems biology heavily tied to metabolomics. This mathematical and computational method of metabolic modeling imposes constraints to predict a range of possible phenotypes based on protein-gene relationships (Ye et al. 2022). The first whole-cell metabolic model to be developed was of the human RBC in 1989 after two decades of research. In 1995, the first whole bacterial genome, that of *Haemophilus influenzae*, had been sequenced, and by 1999, software for simulating cellular environments based on GEMs had emerged. Today, advanced software accounting for thousands of variables is used to generate phenotypic predictions to produce strains used in industrial biosynthesis processes (Sheikh et al. 2024; Blaise et al. 2021). These same mathematical models are used for simulating biological phenomena such as interactions between nutrients or drugs during digestion.

Another technological advancement that has been propelled by its potential utility in industrial bioengineering is the ability to collect real-time metabolic data. The ability to observe and analyze reactions and intermediates occurring in biosynthetic processes in real-time can generate more data than a random or periodic sampling of media (Singh et al. 2024; Chaisupa and Wright 2024). Furthermore, this enables real-time adjustments to reaction parameters. The first real-time single-cell metabolic profiling was accomplished in 2015, and since then, flow-injection time-of-flight mass spectrometry (TOF-MS) has been applied to analyze biosynthetic reactions in real time (Link et al. 2015).

Metabolomics plays a central role in the Design-Build-Test cycle (DBTLc) of synthetic biology, particularly in the applications of discovery, diversification, and optimized production of natural products (Hollywood et al. 2018)—including "new-to-nature metabolites" or molecules containing nonbiological chemical elements (Haas and Nikel 2023). An important outcome of the ability to easily produce natural compounds via microbial culturing advances and genetic engineering strategies has been a revival of the use of these compounds in pharmaceutical development (Atanasov et al. 2021). Historically, natural products played a large role in drug discovery, especially for infectious diseases. However, exploration and development of these products has declined since the 1990s due to the relatively high cost involved in screening, purification, and optimization. With new biosynthetic capabilities and the rising threat of multi-drug-resistant infections, there is hope that a robust pipeline for new antibiotics can be established (Mei et al. 2025).

The next step in advancing synthetic biology will be in integrating novel technologies to automate the DBTLc (Gurdo et al. 2023). Automation routines in biotechnology are supported by robotics, DNA sequencing, data processing, and artificial intelligence (AI). Computer-aided design and analysis are termed "design automation," *and this concept is extended to* synthetic biology with the term "*bio-design automation*" (BDA). These developments have enabled multi-part, high-throughput assembly on a scale that would not have been possible otherwise. The concept of automated laboratories that will result from the compilation of these technologies have been termed "biofoundries," and such automated facilities are already being developed (Szymanski et al. 2023). The exploration of metabolomics in silico that will be entailed in this process is discussed in the upcoming research directions section.

In terms of bioprocess engineering, analytic and predictive platforms assist with focusing potential lead compounds by parameters for scaling up possibility and narrowing down the number of potential natural compound candidates for drug discovery from large libraries. AI programs have the potential to further predict issues in scale-up, propose workarounds, and compare reaction conditions for maximal yield and efficiency.

21.4 Application of Metabolomics in Research and Development in General

Metabolomics stands apart from other omics fields by capturing the biochemical consequences of genomic, transcriptomic, and proteomic variation. Because these measurements reflect the actual intermediates and end products of cellular processes, they offer a direct window into phenotype and its molecular underpinnings (Perakakis et al. 2018).

Metabolomics research spans metabolite identification, pathway annotation, and network reconstruction (Nguyen et al. 2024). Analyses operate at three levels. At the metabolite level, we characterize identity, chemical properties, reactivity, and spatial or temporal concentration profiles. At the pathway level, metabolites are viewed as reactants, products, or cofactors within defined reactions; two metabolites detected in a sample may be linked as intermediates in the same pathway or connected through a shared catalytic function. At the network level, pathways are integrated at common nodes to form graphs of metabolic reactions that include metabolites, enzymes, transporters, signaling factors, non-endogenous compounds, and other proteins. These association networks are constructed using both data-driven correlations and prior biochemical knowledge (Fig. 21.3). The resulting structures generate high-volume information that clarifies enzyme function and reveals how

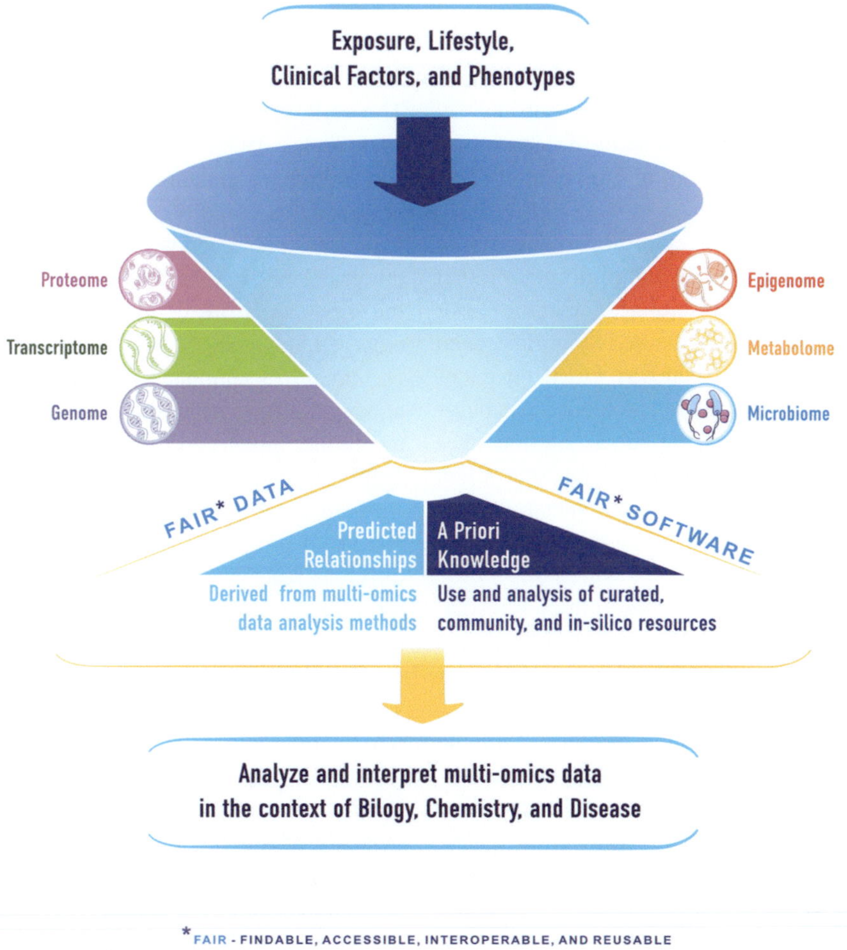

Fig. 21.3 Integration of metabolomics with multi-omics workflows (Eicher et al. 2020): The way in which various omics data are analyzed affects how results can be interpreted and involves complex guidelines. The funnel shows exposures, lifestyle, and clinical factors generating layered omics datasets, including genome, proteome, microbiome, epigenome, transcriptome, and metabolome. These data feed two complementary streams—data-driven predicted relationships and curated a priori knowledge—both supported by FAIR data and software, which are then combined to interpret biology, chemistry, and disease

perturbations in one process propagate to other physiological functions (Battiston et al. 2020; Methods in Molecular Biology 2018). In this context, gene-to-metabolite regulatory paths can be traced as the series of relationships through which a gene, via its encoded enzyme or regulator, influences metabolite levels.

Metabolomics has the unique advantage of application at the delta of environmental factors, genetic inheritance, microbiome, and exposome. Through translational medical applications, metabolomics studies put research that connects biomarker identification to diagnosis and prognosis at the forefront (Babu and Snyder 2023). Advances in identifying biomarkers associated with disease states serve as diagnostic factors and provide further information on disease pathogenesis. Metabolic information has been incorporated into making comprehensive models of disease pathogenesis that allow us further insight into in vivo interactions. A combination of metabolomics study and multivariate data analysis provides a broader systematic view of metabolic pathways that are perturbed in a particular disease state (Guijas et al. 2018a; Perakakis et al. 2018). Changes in biomarkers can identify previously unknown involvement of certain biochemical pathways, indicating directions for further research.

Metabolomics is the epitome of phenotypic expression and provides us with information that other multi-omics disciplines cannot. For instance, we can predict the activation and even steps in a biochemical pathway based on genomics and transcriptomics, but metabolomics helps to identify and quantify the actual intermediates and products being formed (Jendoubi 2021; Chen et al. 2022b). Thus, only by using metabolomics can we gain a full view of the interaction between intrinsic and extrinsic factors within the body. This "bird's eye" view of multi-omics information reveals greater revelations about fundamental causes of diseases than could be realized by analyzing single pathways.

In this section, we will discuss innovations in gathering and interpreting this vital information. New computational techniques and technology allow us to make research advances in the understanding of physiological processes beyond other omics such as functions of the microbiome and host-pathogen relations, as well as subsequent disease pathogenesis. These insights impact basic research at a fundamental level and translate into actionable information for drug and therapeutic design among other biotechnologies.

21.4.1 Metabolomics in the Big Data Era

The earliest metabolic simulations can be dated back to the 1980s. These were two-component systems where an annotated genome, or sequence of genes that could be associated with functions, was cross-referenced with related metabolites. By the year 1999, computational platforms that were the first tools in computational biology had emerged, using the combination of *H. influenza*'s annotated genome and known biochemical information to establish the organism's metabolic phenotype (Edwards and Palsson 1999).

With today's computational power and complexity of program architecture (Table 21.3), the different reactions and environmental pressures on a single molecule moving through digestion is dependent on the states of single cells in whole organs and organ systems (Perakakis et al. 2018). This involves thousands to billions of factors, or what are termed "high-dimensional input variables," that result in "big data"—data with dimensions in the order of trillions or more. An example of this is the use of metabolomics data gained from tumor biopsies to predict the nonlinear behavior of tumor cells, predict the growth of particular cell lines, and possible response to treatment (Miller et al. 2022).

Massive computational resources are needed to store, organize, and access the large amounts of data being generated by increasingly efficient

Table 21.3 Mathematical techniques used to verify key metabolites altered in pathologies

Condition	Abnormal metabolites	Analytic technique	Citation
Preeclampsia	Hydroxyhexanoylcarnitine, alanine, phenylalanine, and glutamate	Logistic regression model	Odibo et al. (2011)
Postpartum epression	Glutathione-disulfide, adenylosuccinate, ATP and seven others	Partial least squares—discriminant analysis (MetaboAnalyst software)	Papadopoulou et al. (2019)
Myocardial infarction	Acetylcarnitine, pyruvate, and lathosterol	Generative adversarial model	Yu et al. (2024)
Colon and other tumors	Microenvironment-specific Warburg metabolism of lactate	Feedforward neural networks and nearest neighbor analysis	Hsieh et al. (2022)

high-throughput omics platforms (Grapov et al. 2018). Moreover, complex computational platforms are needed to analyze this data, and so it can be integrated into platforms for molecular genetic mapping, reaction modeling, computer-aided drug design, and many other applications. The amount and variety of data being processed is more than can be handled by the largest research initiatives and surpasses the abilities of traditional statistical modeling to identify trends in data. Many of these purposes thus rely on AI programs that can handle Big Data.

Deep phenotyping integrates longitudinal metabolomics data with other multi-omics information to fully understand perturbants' long-term effects on expression pathways (Jendoubi 2021; Wörheide et al. 2021). Such analysis has yielded data on the changes in the human body that astronauts in space experience (Manna et al. 2024), while metabolites in saliva can act as a measure of immune response to vaccination (Reicher et al. 2025; Wright and Herzberg 2021). The 100 K Wellness Project demonstrated that analyzing the blood, saliva, and stool composition of a large sample led to very accurate predictions for the emergence of illness (Babu and Snyder 2023). The Human Phenotype Project (2025) is a large deep-phenotype prospective cohort of about 28,000 enrollees that integrates longitudinal lifestyle, clinical, imaging, and multi-omics data, and its multimodal AI trained on diet and continuous glucose monitoring has outperformed existing methods for predicting disease onset (Reicher et al. 2025).

Studies to date have shown that many basic and life sciences researchers are unfamiliar with simple tenets of AI and matters traditionally considered the domain of Computer Science (Soares Dias Portela et al. 2024). This is an issue in the field as there is a disconnect between engineers creating omics and other scientific software and the researchers or physicians who are using the products. An analogy for this would be a microbiologist not understanding the operating principles of a microscope. On the other hand, in very recent years, there has been a massive push toward establishing integrative courses of study and focus on fields such as Systems Biology, Biomedical Engineering, Computational Biology, Bio/chemoinformatics, and other applied Data Science subjects.

21.4.2 Metabolomics Information Integrated to Simulate In Vivo Interactions

In their 1999 paper, Varner and Ramkrishna (1999) make important observations about a turning point in Omics development and define a paradigm of in vivo reaction prediction we are still striving to achieve today. First, they describe the necessity of a dynamic metabolic model that not only includes the snapshot of metabolites provided by metabolomics but incorporates system kinetics data.

The other interesting point is that they considered themselves on the cusp of new massively parallel sequencing technology and predicted what we now call the era of Big Data. Their observation over 25 years ago was: "In the coming 'high-information' era, mathematical models will most certainly play a central role."

At that time applying concepts of linear algebra for phenotypic mapping was a marvel. What they called "massively parallel" was orders of magnitude slower than sequencing techniques that are today referred to as "high throughput." However, today, we may just be entering the era of Big Data as new repositories begin to store information mapping the trajectory of single atoms through metabolic pathways (Starke and Wegner 2022). Our mathematical concepts have advanced and now form the algorithms that underlie in silico platforms for predicting in vivo dynamics. However, with the ability to account for more parameters, experimental design also has more hurdles. For this reason, in their 2021 paper, Lakrisenko and Weindl (2021) describe dynamic models as an underutilized nascent technology.

When information from metabolomics analysis is combined with other omics data such as genomic, proteomic, transcriptomic, and epigenomic data, we can gain a highly comprehensive understanding of a biological system

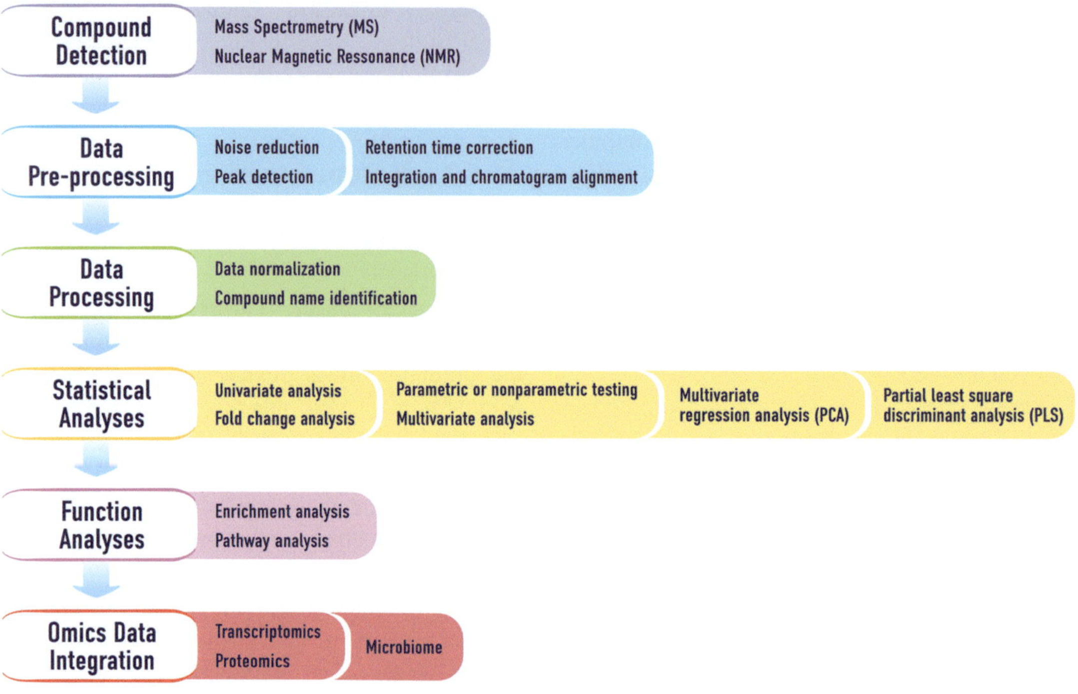

Fig. 21.4 Metabolomics process yields complex information (Chen et al. 2022b): (1) Compound detection starts with sample analysis using analytical techniques; (2) Processing of resultant data such as NMR spectrum is required to remove artifacts and obtain a clearer reading; (3) Further processing identifies metabolites in an automated fashion; (4) Advanced statistical and AI processes are applied to match identified structures with known structures in metabolic databases; this may involve matching a fragment to a parent molecule; (5) Further computational analysis places identified compounds in the context of established metabolic and gene expression networks; (6) The resulting data is integrated with other omics data to give a better understanding of a disease state or greater predictive power regarding metabolic interactions

(Fig. 21.4). Here a systems biology view becomes useful; this looks at the body as a steady system and through understanding all physiologic parameters the response of the system to some perturbation can be predicted (Helmy et al. 2020). Integrating varying omics data and creating such a complex model accurately requires highly advanced AI programs.

When such omics data is available modern computational methods can create extraordinary models that predict biological phenomena (Chi et al. 2024). In this manner in silico models can be thought to simulate in vivo interactions. Dynamic models simulate the change in global metabolite concentrations that would occur in response to some pressure or perturbance exerted on an organism. Such a model can be used to predict a patient's response to a particular drug, and further select an optimal candidate for the desired outcome (Biswas and Chakrabarti 2020). AI programs like this are now being widely applied in the pharmaceutical industry to identify therapeutic targets as well as select and optimize drug compounds with favorable metabolic characteristics.

Another modeling approach is 3D molecular modeling. This can essentially simulate the structure of a protein as well as that of an enzyme and predict the kinetics of their reaction (Helmy et al. 2020). In this manner it is possible to generate experimental data on biological phenomena based only on simulations; however, this is currently done in a mostly exploratory capacity. Another way this technology is used is to take the results of an in vitro experiment and more closely fit the data to in vivo parameters: called in vitro to in vivo extrapolation (IVIVE) (Benet and Sodhi 2020).

Aside from the heterogeneous omics data being highly complex and difficult to integrate, there are a number of other hurdles involved in the use of such systems. The curse of dimensionality describes one of the largest setbacks in AI training in that the amount of data required to predict a variable is far exceeded by the number of variables; this is highly apparent where large omics datasets are available but from very few individuals. A related problem is that there is often model overfitting where many collinear variables exist. Lastly, the black box problem describes the inverse relationship between an AI program's complexity and interpretability (Flores et al. 2023). While more advanced programs can perform more complicated functions, the code behind the function becomes unintelligible to the point where a program designer cannot discern the operating logic, making the program impossible to troubleshoot.

21.4.3 Insights into the Human Microbiome and Host-Pathogen Relationships

Metabolomics has also helped to reveal the importance of the microbiome (and its major internal influences) on human health and disease. Its unique ability to probe complex biochemistry at both a cellular and an organismal level has already helped to identify unexpected chemical causes for several important chronic and complex diseases, including atherosclerosis, cancer, and diabetes. Such studies are revealing that metabolites (both exogenous and endogenous) have a far more central role in disease pathogenesis and physiologic response than recognized.

The metabolome closely corresponds with phenotype and gives a snapshot of intracellular changes in response to antibiotics and adaptations that can confer resistance. Primary and secondary metabolites excreted by an organism form the exometabolome which provides more information about metabolic processes occurring under various conditions. For instance, microbial metabolic pathways may have distinct primary metabolites resulting from unique biochemical reactions (Patra et al. 2023). These primary metabolites can be distinguished from secondary metabolites, which may both result from and contribute to other chemical pathways. Without accounting for both primary and secondary metabolites, only a partial glimpse of the production cycle is obtained. A holistic approach must be taken to identify opportunities for engineering the microbiome, such as through untargeted metabolomics (Di Minno et al. 2021). While engineering multiple pathways to account for multiple primary and secondary metabolites is time-intensive, it is a more efficient strategy than identifying a potentially ineffective target based on incomplete information.

The gut microbiome, which represents the greatest number of microbes in the human body has proven to be a rich ecosystem for metabolite research. In a study comparing samples taken from the small intestine, colon, and an ileostomy, different respective biochemical pathways were noted in each location (Zoetendal et al. 2012). Furthermore, the observed metabolites provided clues not only to the pathophysiology behind digestion and the gut microbiome, but also to the evolutionary advantages gained through adapting carbohydrate digestion in the small intestines based on dietary availability. Metabolomic evaluation of the gut microbiome has the potential to therefore predict aspects of a patient's digestive health, which influences conditions including type II diabetes mellitus, atherosclerosis, hepatic steatosis, and colorectal cancer (de Vos et al. 2022).

Shifts in metabolism underlie stages of pathogenesis and mechanisms of immune invasion that may confer resistance. Bacteria can utilize diversity in their metabolic pathways to combat environmental pressures (Kost et al. 2023). For instance, many bacteria can switch from a high-energy state utilizing aerobic respiration to an anaerobic state when resources are scarce or depleted. Though a method of conservation, this can also be a strategy for evasion of host defenses, as this switch confers resistance to aminoglycosides. While in a "dormant" state, antimicrobials that utilize active cellular processes for uptake and as targets are less effective. Likewise, antibiotics such as β-lactams show increased effects in conjunction with higher bacterial respiration.

Outside endogenous microbiome pathology, metabolomics has the potential to evaluate patients with microbial infection (Wang et al. 2024). Research conducted on biomarkers associated with tuberculosis infection (Conde et al. 2022) has revealed six biomarkers associated with *Mycobacterium tuberculosis*. Additionally, patients with active tuberculosis infection had altered levels of 2-aminobutyrate and mannose compared to patients with latent tuberculosis (TB). Further research has revealed a more comprehensive diagnostic profile differentiating the metabolomic profile of patients with active TB infection versus those with latent infection (Albors-Vaquer et al. 2020).

Identification of drug-resistant strains of TB has also shown promise. In a study using ultra-high-performance liquid chromatography and mass spectrometry, Huang et al. identified four primary metabolites associated with extensively drug-resistant tuberculosis (TB-XDR) (Huang et al. 2020). The research team then constructed a differential diagnostic model based on these results with over 85% sensitivity and specificity in identifying TB-XDR. Additional research has shown that metabolic profiles of pre-XDR and TB-XDR can be distinguished with 100% sensitivity and 100% specificity (Chaiyachat et al. 2023). Tests using metabolomics to distinguish between drug-resistant strains of tuberculosis have the potential to greatly inform patient diagnosis and management. In combination with medicinal chemistry, these insights have yielded antitubercular leads such as coumarin-NO donor hybrids that show promising activity (Kareem et al. 2024).

21.4.4 Metabolomics in Pharmaceutical Development

With the advent of high throughput technologies, the last decade has seen the emergence of a streamlined drug development process termed "rational drug design" (RDD) (Doytchinova 2022). Broadly, these steps include target identification, the hit-to-lead process (H2L) or lead generation, profiling, and then clinical testing before marketing approval. Briefly, rational design starts with information on physiologic mechanisms to select a component—such as a protein—whose actions can mediate disease pathogenesis, and which has a reactive site where molecule binding can alter its function. Such a site on a protein of interest is a druggable target. A large library of compounds is screened for structural compatibility or binding activity at the reactive site, generating a number of suitable compounds called hits. The most promising hits are optimized by making changes to the base structure that enhance desirable pharmaceutical properties, namely the absorption, distribution, metabolism, and excretion—often referred to as a compound's ADME (Tian et al. 2024). An optimized hit is called a lead compound and is the finalized structure for a drug candidate. For example, natural-product screening has identified acetohydroxyacid synthase inhibitors as anti-tuberculosis lead compounds (Niu et al. 2025). The pharmacokinetics and pharmacodynamic properties of the lead are extensively studied, mainly using computational methods to define quantitative structure-activity relationships (QSAR). All the steps above represent the pre-clinical portion of drug development. In clinical trials, the compound is assessed for safety first in animals and then in groups of humans, and the data from these trials is evaluated by the FDA when applying for drug approval (Singh et al. 2023b).

Every step of the modern drug design process involves some element of metabolomics; however, this discussion is limited to a few key points impacted by the rapid expansion of metabolomics technologies. Despite being high throughput, screening for potential hits can actually be very time-consuming and produce many reactive candidates that must be ruled out. By analyzing metabolite profiles, much about drug-target relationships can be predicted, which can be used to narrow down suitable candidates (Astarita et al. 2023). Library screening can either be fully replaced by a high throughput metabolomics approach or metabolomics information can be integrated into screening parameters, both approaches greatly reduce the time involved in generating quality hits.

A breakthrough in revealing drug-target interactions was made by Holbrook-Smith et al. (2022)., involving both a novel analysis method and the use of an emerging profiling technology. The basis of their theory is that the metabolic profile resulting from a drug interaction and that of an organism where the corresponding gene is mutated are ideally highly similar. In this method, a metabolic profile of an organism's post-drug interaction is generated for each compound, which is screened against that of the mutant. This can be done quickly using flow injection analysis time-of-flight mass spectrometry (FIA-TOF-MS) (Fuhrer et al. 2011), which can process MS samples in under a minute—as fast as seconds—compared to classic methods that may take an hour per sample. In the experiment by Holbrook-Smith et al. metabolic profiles for specimens treated with a library of 1280 compounds were screened against profiles compiled from *S. cerevisiae* strains overexpressing genes of interest under synthetic inducible promoters. The identified hits displayed some known drug reactions, a planned element that acted as a positive standard, and revealed some new interactions that were then experimentally verified. Further, samples of treated organisms were collected again and analyzed using LC-MS, yielding highly similar results to the FIA-TOF-MS, thus displaying the robustness of the method. While there are significant limitations to this method, with such high throughput processing, an almost real-time readout of global metabolite levels can be accomplished.

Reichling et al. (2023). demonstrated the capacity of rapid metabolome profiling by using FIA-MS to analyze the response to rapamycin in S. *cerevisiae* by analyzing multiple samples taken at various time intervals post-administration. The purpose of the experiment was to compile a longitudinal metabolome measurement covering the key period of dynamic perturbation, which was predicted to be a more accurate reflection of phenotypic response than would a sample collected at a steady state. They were able to correlate the metabolomics data with functional gene distances, indicating the involvement of several known genes as well as several previously non-characterized loci.

Dynamic metabolomic responses to drugs can further elucidate a compound's mode of action (MOA) (Gonzalez-Covarrubias et al. 2022). Information on compounds' MOA is useful at multiple stages of development as it can assist in validating hits, predicting pharmacodynamic properties such as dose response, and identifying possible off-target effects. As has been discussed regarding personalized medicine, metabolomics is being explored as a potential method to explain intra-individual variations in drug response (Phapale 2021).

21.5 Conclusion

Metabolomics is revolutionizing our understanding of disease mechanisms, offering new diagnostic and therapeutic opportunities. Its ability to provide a comprehensive snapshot of metabolic states makes it an invaluable tool in precision medicine, enabling tailored treatments based on individual metabolic profiles. As technologies and methodologies continue to evolve, the potential for metabolomics to transform biomedicine is immense, promising new insights into disease mechanisms and paving the way for personalized healthcare solutions.

In biomedicine, metabolomics plays an important role in biomarker discovery which has enhanced our understanding of many pathologies including cancer, atherosclerosis, and diabetes.

The identification of oncometabolites, for instance, has highlighted metabolic alterations in cancer cells that are a potential avenue for early diagnosis and targeted therapies. Similarly, the discovery of characteristic altered metabolite levels in cardiovascular diseases and diabetes underscores the potential of metabolomics in predicting disease risk and tailoring preventive strategies.

In the era of biotechnology, metabolomics has driven the development of new technologies and methodologies for metabolic analysis. The automation and refinement of hyphenated techniques like LC-NMR-MS have improved the accuracy and efficiency of metabolomics studies, facilitating translation into clinical applications. In conjunction with these advanced computational tools, machine learning algorithms have emerged that can handle the vast amounts of data generated by metabolomics studies, allowing for higher-quality analyses of larger data with more coverage.

Looking forward, the future of metabolomics involves further integration with other omics disciplines as well as exploring applications in emerging fields such as single-cell metabolomics and microbiome research. Understanding the metabolic interactions between host metabolism and the microbiome will provide deeper insights into disease mechanisms and identify novel therapeutic targets. Understanding metabolic profiles will likewise streamline the discovery of drug-target interactions; and provide more info in analyzing compounds' therapeutic MOAs.

In conclusion, the field of metabolomics stands at the forefront of biomedical and biotechnological research, offering powerful insights that are reshaping our understanding of health and disease. Its continued evolution and integration with other scientific disciplines hold great promise for advancing precision medicine, improving disease diagnostics, and developing innovative therapeutic strategies. As we harness the full potential of metabolomics, we move closer to a future where personalized healthcare becomes the norm, ultimately improving patient outcomes and advancing public health.

References

Adel M et al (2025) Changes in gut microbial diversity and correlation with clinical outcome in children with acute myeloid leukemia receiving induction chemotherapy. Children 12:1176

Aderemi AV, Ayeleso AO, Oyedapo OO, Mukwevho E (2021) Metabolomics: a scoping review of its role as a tool for disease biomarker discovery in selected non-communicable diseases. Metabolites 11. https://doi.org/10.3390/metabo11070418

Aebersold R, Mann M (2016) Mass-spectrometric exploration of proteome structure and function. Nature 537:347–355. https://doi.org/10.1038/nature19949

Albors-Vaquer A et al (2020) Active and prospective latent tuberculosis are associated with different metabolomic profiles: clinical potential for the identification of rapid and non-invasive biomarkers. Emerg Microbes Infect 9:1131–1139. https://doi.org/10.1080/22221751.2020.1760734

Alsaleh M et al (2019) Mass spectrometry: a guide for the clinician. J Clin Exp Hepatol 9:597–606. https://doi.org/10.1016/j.jceh.2019.04.053

Ashrafian H et al (2021) Metabolomics: the stethoscope for the twenty-first century. Med Princ Pract 30:301–310. https://doi.org/10.1159/000513545

Astarita G, Kelly RS, Lasky-Su J (2023) Metabolomics and lipidomics strategies in modern drug discovery and development. Drug Discov Today 28:103751. https://doi.org/10.1016/j.drudis.2023.103751

Atanasov AG et al (2021) Natural products in drug discovery: advances and opportunities. Nat Rev Drug Discov 20:200–216. https://doi.org/10.1038/s41573-020-00114-z

Azad RK, Shulaev V (2019) Metabolomics technology and bioinformatics for precision medicine. Brief Bioinform 20:1957–1971. https://doi.org/10.1093/bib/bbx170

Babu M, Snyder M (2023) Multi-omics profiling for health. Mol Cell Proteomics 22:100561. https://doi.org/10.1016/j.mcpro.2023.100561

Balashova EE, Maslov DL, Trifonova OP (2020) Metabolomics Community in Russia: history of development, key participants, and results. Biotech 9:20

Barrett JC et al (2024) Metabolomic and genomic prediction of common diseases in 700,217 participants in three national biobanks. Nat Commun 15:10092. https://doi.org/10.1038/s41467-024-54357-0

Battiston F et al (2020) Networks beyond pairwise interactions: structure and dynamics. Phys Rep 874:1–92. https://doi.org/10.1016/j.physrep.2020.05.004

Beger RD et al (2016) Metabolomics enables precision medicine: "a white paper, community perspective". Metabolomics 12:149. https://doi.org/10.1007/s11306-016-1094-6

Benet LZ, Sodhi JK (2020) Investigating the theoretical basis for in vitro-in vivo extrapolation (IVIVE) in predicting drug metabolic clearance and proposing

future experimental pathways. AAPS J 22:120. https://doi.org/10.1208/s12248-020-00501-9
Biswas N, Chakrabarti S (2020) Artificial intelligence (AI)-based systems biology approaches in multi-omics data analysis of cancer. Front Oncol 10. https://doi.org/10.3389/fonc.2020.588221
Blaise BJ et al (2021) Statistical analysis in metabolic phenotyping. Nat Protoc 16:4299–4326. https://doi.org/10.1038/s41596-021-00579-1
Castelli FA et al (2022) Metabolomics for personalized medicine: the input of analytical chemistry from biomarker discovery to point-of-care tests. Anal Bioanal Chem 414:759–789. https://doi.org/10.1007/s00216-021-03586-z
Chace DH, Spitzer AR (2011) Altered metabolism and newborn screening using tandem mass spectrometry: lessons learned from the bench to bedside. Curr Pharm Biotechnol 12:965–975. https://doi.org/10.2174/138920111795909104
Chace DH et al (1993) Rapid diagnosis of phenylketonuria by quantitative analysis for phenylalanine and tyrosine in neonatal blood spots by tandem mass spectrometry. Clin Chem 39:66–71
Chace DH et al (1995) Rapid diagnosis of maple syrup urine disease in blood spots from newborns by tandem mass spectrometry. Clin Chem 41:62–68
Chace DH, Hillman SL, Van Hove JL, Naylor EW (1997) Rapid diagnosis of MCAD deficiency: quantitative analysis of octanoylcarnitine and other acylcarnitines in newborn blood spots by tandem mass spectrometry. Clin Chem 43:2106–2113
Chaisupa P, Wright RC (2024) State-of-the-art in engineering small molecule biosensors and their applications in metabolic engineering. SLAS Technol 29:100113. https://doi.org/10.1016/j.slast.2023.10.005
Chaiyachat P et al (2023) Metabolomic analysis of Mycobacterium tuberculosis reveals metabolic profiles for identification of drug-resistant tuberculosis. Sci Rep 13:8655. https://doi.org/10.1038/s41598-023-35882-2
Chen R et al (2012) Personal omics profiling reveals dynamic molecular and medical phenotypes. Cell 148:1293–1307. https://doi.org/10.1016/j.cell.2012.02.009
Chen Y, Li E-M, Xu L-Y (2022a) Guide to metabolomics analysis: a bioinformatics workflow. Metabolites 12:357
Chen Y, Li EM, Xu LY (2022b) Guide to metabolomics analysis: a bioinformatics workflow. Metabolites 12. https://doi.org/10.3390/metabo12040357
Cheng Y et al (2012) Distinct urinary metabolic profile of human colorectal cancer. J Proteome Res 11:1354–1363. https://doi.org/10.1021/pr201001a
Chi J et al (2024) Artificial intelligence in metabolomics: a current review. TrAC Trends Anal Chem 178:117852. https://doi.org/10.1016/j.trac.2024.117852
Chi Y, Mitchell JM, Zheng S, Li S (2025) Systematic pre-annotation explains the "dark matter" in LC-MS metabolomics. bioRxiv. https://doi.org/10.1101/2025.02.04.636472
Choi KR et al (2019) Systems metabolic engineering strategies: integrating systems and synthetic biology with metabolic engineering. Trends Biotechnol 37:817–837. 10.1016/j.tibtech.2019.01.003
Chong J, Wishart DS, Xia J (2019) Using MetaboAnalyst 4.0 for comprehensive and integrative metabolomics data analysis. Curr Protoc Bioinformatics 68:e86. https://doi.org/10.1002/cpbi.86
Clayton TA et al (2006) Pharmaco-metabonomic phenotyping and personalized drug treatment. Nature 440:1073–1077. https://doi.org/10.1038/nature04648
Conde R et al (2022) Discovery of serum biomarkers for diagnosis of tuberculosis by NMR metabolomics including cross-validation with a second cohort. Biomed J 45:654–664. https://doi.org/10.1016/j.bj.2021.07.006
Corona G et al (2021) (1)H-NMR plasma lipoproteins profile analysis reveals lipid metabolism alterations in HER2-positive breast cancer patients. Cancers (Basel) 13. https://doi.org/10.3390/cancers13225845
Cossu M et al (2023) Metabolomic studies in inborn errors of metabolism: last years and future perspectives. Metabolites 13:447
Crook AA, Powers R (2020) Quantitative NMR-based biomedical metabolomics: current status and applications. Molecules 25. https://doi.org/10.3390/molecules25215128
d'Espaux L et al (2017) Engineering high-level production of fatty alcohols by Saccharomyces cerevisiae from lignocellulosic feedstocks. Metab Eng 42:115–125. https://doi.org/10.1016/j.ymben.2017.06.004
de Vos WM, Tilg H, Van Hul M, Cani PD (2022) Gut microbiome and health: mechanistic insights. Gut 71:1020. https://doi.org/10.1136/gutjnl-2021-326789
Di Minno A, Gelzo M, Stornaiuolo M, Ruoppolo M, Castaldo G (2021) The evolving landscape of untargeted metabolomics. Nutr Metab Cardiovasc Dis 31:1645–1652. https://doi.org/10.1016/j.numecd.2021.01.008
Di Minno A et al (2022) Challenges in metabolomics-based tests, biomarkers revealed by metabolomic analysis, and the promise of the application of metabolomics in precision medicine. Int J Mol Sci 23. https://doi.org/10.3390/ijms23095213
Doytchinova I (2022) Drug design-past, present, future. Molecules 27. https://doi.org/10.3390/molecules27051496
Edwards JS, Palsson BO (1999) Systems properties of the Haemophilus influenzae Rd metabolic genotype. J Biol Chem 274:17410–17416. https://doi.org/10.1074/jbc.274.25.17410
Eicher T et al (2020) Metabolomics and multi-omics integration: a survey of computational methods and resources. Metabolites 10:202
Emwas A-H et al (2019) NMR spectroscopy for metabolomics research. Metabolites 9:123
Everett JR, Holmes E, Veselkov KA, Lindon JC, Nicholson JK (2019) A unified conceptual framework for metabolic phenotyping in diagnosis and prognosis.

Trends Pharmacol Sci 40:763–773. https://doi.org/10.1016/j.tips.2019.08.004

Fiehn O (2016) Metabolomics by gas chromatography–mass spectrometry: combined targeted and untargeted profiling. Curr Protoc Mol Biol 114:10.1002/0471142727.mb3004s114

Fiehn O et al (2007) The metabolomics standards initiative (MSI). Metabolomics 3:175–178. https://doi.org/10.1007/s11306-007-0070-6

Fletcher E et al (2025) Modeling the importance of life exposure factors on memory performance in diverse older adults: a machine learning approach. Alzheimers Dement 21:e70428. https://doi.org/10.1002/alz.70428

Flores M, Glusman G, Brogaard K, Price ND, Hood L (2013) P4 medicine: how systems medicine will transform the healthcare sector and society. Per Med 10:565–576. https://doi.org/10.2217/pme.13.57

Flores JE et al (2023) Missing data in multi omics integration: recent advances through artificial intelligence. Front Artif Intell 6:1098308. https://doi.org/10.3389/frai.2023.1098308

Fuhrer T, Heer D, Begemann B, Zamboni N (2011) High-throughput, accurate mass metabolome profiling of cellular extracts by flow injection-time-of-flight mass spectrometry. Anal Chem 83:7074–7080. https://doi.org/10.1021/ac201267k

Galanie S, Entwistle D, Lalonde J (2020) Engineering biosynthetic enzymes for industrial natural product synthesis. Nat Prod Rep 37:1122–1143. https://doi.org/10.1039/C9NP00071B

Giera M, Aisporna A, Uritboonthai W, Siuzdak G (2024a) The hidden impact of in-source fragmentation in metabolic and chemical mass spectrometry data interpretation. Nat Metab. https://doi.org/10.1038/s42255-024-01076-x

Giera M, Aisporna A, Uritboonthai W, Siuzdak G (2024b) The hidden impact of in-source fragmentation in metabolic and chemical mass spectrometry data interpretation. Nat Metab 6:1647–1648. https://doi.org/10.1038/s42255-024-01076-x

Gómez-Cebrián N et al (2019) Metabolomics contributions to the discovery of prostate cancer biomarkers. Metabolites 9. https://doi.org/10.3390/metabo9030048

Gonzalez-Covarrubias V, Martínez-Martínez E, Del Bosque-Plata L (2022) The potential of metabolomics in biomedical applications. Metabolites 12. https://doi.org/10.3390/metabo12020194

Grapov D, Fahrmann J, Wanichthanarak K, Khoomrung S (2018) Rise of deep learning for genomic, proteomic, and metabolomic data integration in precision medicine. OMICS 22:630–636. https://doi.org/10.1089/omi.2018.0097

Grayson MA (2011) John Bennett Fenn: a curious road to the prize. J Am Soc Mass Spectrom 22:1301–1308. https://doi.org/10.1007/s13361-011-0136-6

Grebe SK, Singh RJ (2011) LC-MS/MS in the clinical laboratory - where to from here? Clin Biochem Rev 32:5–31

Gregory JC et al (2015) Transmission of atherosclerosis susceptibility with gut microbial transplantation. J Biol Chem 290:5647–5660. https://doi.org/10.1074/jbc.M114.618249

Guijas C, Montenegro-Burke JR, Warth B, Spilker ME, Siuzdak G (2018a) Metabolomics activity screening for identifying metabolites that modulate phenotype. Nat Biotechnol 36:316–320. https://doi.org/10.1038/nbt.4101

Guijas C et al (2018b) METLIN: a technology platform for identifying knowns and unknowns. Anal Chem 90:3156–3164. https://doi.org/10.1021/acs.analchem.7b04424

Guo J, Yu H, Xing S, Huan T (2022) Addressing big data challenges in mass spectrometry-based metabolomics. Chem Commun 58:9979–9990. 10.1039/d2cc03598g

Gurdo N, Volke DC, McCloskey D, Nikel PI (2023) Automating the design build test learn cycle towards next-generation bacterial cell factories. New Biotechnol 74:1–15. 10.1016/j.nbt.2023.01.002

Haas R, Nikel PI (2023) Challenges and opportunities in bringing nonbiological atoms to life with synthetic metabolism. Trends Biotechnol 41:27–45. https://doi.org/10.1016/j.tibtech.2022.06.004

Hanahan D, Weinberg RA (2011) Hallmarks of cancer: the next generation. Cell 144:646–674. https://doi.org/10.1016/j.cell.2011.02.013

Hartman TE, Lees HJ (2023) In: Soni V, Hartman TE (eds) Metabolomics: recent advances and future applications. Springer International Publishing, pp 1–37

Helmy M, Smith D, Selvarajoo K (2020) Systems biology approaches integrated with artificial intelligence for optimized metabolic engineering. Metab Eng Commun 11:e00149. https://doi.org/10.1016/j.mec.2020.e00149

Holbrook-Smith D, Durot S, Sauer U (2022) High-throughput metabolomics predicts drug-target relationships for eukaryotic proteins. Mol Syst Biol 18:e10767. https://doi.org/10.15252/msb.202110767

Hollywood KA, Schmidt K, Takano E, Breitling R (2018) Metabolomics tools for the synthetic biology of natural products. Curr Opin Biotechnol 54:114–120. 10.1016/j.copbio.2018.02.015

Holmes E, Wilson ID, Nicholson JK (2008) Metabolic phenotyping in health and disease. Cell 134:714–717. https://doi.org/10.1016/j.cell.2008.08.026

Holmes E, Li JV, Athanasiou T, Ashrafian H, Nicholson JK (2011) Understanding the role of gut microbiome-host metabolic signal disruption in health and disease. Trends Microbiol 19:349–359. https://doi.org/10.1016/j.tim.2011.05.006

Horvath N et al (2017) Toward a genome scale sequence specific dynamic model of cell-free protein synthesis in Escherichia coli. Metab Eng Commun 10:e00113

Hsieh WC et al (2022) Spatial multi-omics analyses of the tumor immune microenvironment. J Biomed Sci 29:96. https://doi.org/10.1186/s12929-022-00879-y

Huang H et al (2020) The novel potential biomarkers for multidrug-resistance tuberculosis using UPLC-Q-

TOF-MS. Exp Biol Med (Maywood) 245:501–511. https://doi.org/10.1177/1535370220903464

Jendoubi T (2021) Approaches to integrating metabolomics and multi-omics data: a primer. Metabolites 11. https://doi.org/10.3390/metabo11030184

Jobard E et al (2021) Investigation of circulating metabolites associated with breast cancer risk by untargeted metabolomics: a case-control study nested within the French E3N cohort. Br J Cancer 124:1734–1743. https://doi.org/10.1038/s41416-021-01304-1

Kareem AI, Malan SF, Kapp E, Shamido S, Joubert J (2024) Synthesis, characterization, and biological evaluation of coumarin-nitric oxide donor hybrids as antitubercular agents. European Journal of Medicinal Chemistry Reports 12:100211. https://doi.org/10.1016/j.ejmcr.2024.100211

Karnovsky A, Li S (2020) In: Li S (ed) Computational methods and data analysis for metabolomics. Springer US, pp 387–400

Keasling J et al (2021) Microbial production of advanced biofuels. Nat Rev Microbiol 19:701–715. https://doi.org/10.1038/s41579-021-00577-w

Kind T et al (2009) FiehnLib: mass spectral and retention index libraries for metabolomics based on quadrupole and time-of-flight gas chromatography/mass spectrometry. Anal Chem 81:10038–10048. https://doi.org/10.1021/ac9019522

Koeth RA et al (2013) Intestinal microbiota metabolism of L-carnitine, a nutrient in red meat, promotes atherosclerosis. Nat Med 19:576–585. https://doi.org/10.1038/nm.3145

Kost C, Patil KR, Friedman J, Garcia SL, Ralser M (2023) Metabolic exchanges are ubiquitous in natural microbial communities. Nat Microbiol 8:2244–2252. https://doi.org/10.1038/s41564-023-01511-x

Krauss RM, Zhu H, Kaddurah-Daouk R (2013) Pharmacometabolomics of statin response. Clin Pharmacol Ther 94:562–565. https://doi.org/10.1038/clpt.2013.164

Kroemer G, Chan TA, Eggermont AMM, Galluzzi L (2024) Immunosurveillance in clinical cancer management. CA Cancer J Clin 74:187–202. https://doi.org/10.3322/caac.21818

Kuhl C, Tautenhahn R, Böttcher C, Larson TR, Neumann S (2012) CAMERA: an integrated strategy for compound spectra extraction and annotation of liquid chromatography/mass spectrometry data sets. Anal Chem 84:283–289. https://doi.org/10.1021/ac202450g

Lakrisenko P, Weindl D (2021) Dynamic models for metabolomics data integration. Curr Opin Syst Biol 28:100358

Lander ES (2011) Initial impact of the sequencing of the human genome. Nature 470:187–197. https://doi.org/10.1038/nature09792

Lauber C et al (2022) Lipidomic risk scores are independent of polygenic risk scores and can predict incidence of diabetes and cardiovascular disease in a large population cohort. PLoS Biol 20:e3001561. https://doi.org/10.1371/journal.pbio.3001561

Lazarou J, Pomeranz BH, Corey PN (1998) Incidence of adverse drug reactions in hospitalized patients: a meta-analysis of prospective studies. JAMA 279:1200–1205. https://doi.org/10.1001/jama.279.15.1200

Lehotay DC et al (2011) LC-MS/MS progress in newborn screening. Clin Biochem 44:21–31. https://doi.org/10.1016/j.clinbiochem.2010.08.007

Levine AJ, Puzio-Kuter AM (2010) The control of the metabolic switch in cancers by oncogenes and tumor suppressor genes. Science 330:1340–1344. https://doi.org/10.1126/science.1193494

Li X et al (2013) Chronic leucine supplementation increases body weight and insulin sensitivity in rats on high-fat diet likely by promoting insulin signaling in insulin-target tissues. Mol Nutr Food Res 57:1067–1079. https://doi.org/10.1002/mnfr.201200311

Link H, Fuhrer T, Gerosa L, Zamboni N, Sauer U (2015) Real-time metabolome profiling of the metabolic switch between starvation and growth. Nat Methods 12:1091–1097. https://doi.org/10.1038/nmeth.3584

Liu Y, Nielsen J (2019) Recent trends in metabolic engineering of microbial chemical factories. Curr Opin Biotechnol 60:188–197. 10.1016/j.copbio.2019.05.010

Liu Y, Shin H-D, Li J, Liu L (2015) Toward metabolic engineering in the context of system biology and synthetic biology: advances and prospects. Appl Microbiol Biotechnol 99:1109–1118. https://doi.org/10.1007/s00253-014-6298-y

Liu Z et al (2025) Single-cell multiregion epigenomic rewiring in Alzheimer's disease progression and cognitive resilience. Cell 188:4980–5002.e4929. https://doi.org/10.1016/j.cell.2025.06.031

Lokhov PG, Balashova EE, Trifonova OP, Maslov DL, Archakov AI (2021) A decade of Russian metabolomics: the history of development and achievements. Biochemistry (Moscow) Suppl B: Biomed Chem 15:1–15. https://doi.org/10.1134/S1990750821010042

Lusis AJ (2012) Genetics of atherosclerosis. Trends Genet 28:267–275. https://doi.org/10.1016/j.tig.2012.03.001

Mahieu NG, Patti GJ (2017) Systems-level annotation of a metabolomics data set reduces 25 000 features to fewer than 1000 unique metabolites. Anal Chem 89:10397–10406. https://doi.org/10.1021/acs.analchem.7b02380

Mancinelli L, Cronin M, Sadée W (2000) Pharmacogenomics: the promise of personalized medicine. AAPS PharmSci 2:E4. https://doi.org/10.1208/ps020104

Manna OM et al (2024) Microgravity and human body: unraveling the potential role of heat-shock proteins in spaceflight and future space missions. Biology 13:921

McCullagh J, Probert F (2024) New analytical methods focusing on polar metabolite analysis in mass spectrometry and NMR-based metabolomics. Curr Opin Chem Biol 80:102466. https://doi.org/10.1016/j.cbpa.2024.102466

McInnes G, Yee SW, Pershad Y, Altman RB (2021) Genomewide association studies in pharmacogenomics. Clin Pharmacol Ther 110:637–648. https://doi.org/10.1002/cpt.2349

Mei Y et al (2025) Engineered Mycobacterium smegmatis expressing anti-PD-L1/IL-15 immunocytokine induces and activates specific antitumor immunity. J Immunother Cancer 13:e010118. https://doi.org/10.1136/jitc-2024-010118

Miller HA, Lowengrub J, Frieboes HB (2022) Modeling of tumor growth with input from patient-specific metabolomic data. Ann Biomed Eng 50:314–329. https://doi.org/10.1007/s10439-022-02904-5

Millington DS et al (1992) The role of tandem mass spectrometry in the diagnosis of fatty acid oxidation disorders. Prog Clin Biol Res 375:339–354

Moco S (2022) Studying metabolism by NMR-based metabolomics. Front Mol Biosci. https://doi.org/10.3389/fmolb.2022.882487

Nagana Gowda GA, Raftery D (2021) In: Shen Hu (ed) Cancer metabolomics: methods and applications. Springer International Publishing, pp 19–37

Nalbantoğlu S (2019) Metabolomics: basic principles and strategies. Mol Med 10(1)

Nguyen Q-T et al (2012) Metabolomics methods for the synthetic biology of secondary metabolism. FEBS Lett 586:2177–2183. 10.1016/j.febslet.2012.02.008

Nguyen Q-H, Nguyen H, Oh EC, Nguyen T (2024) Current approaches and outstanding challenges of functional annotation of metabolites: a comprehensive review. Brief Bioinform **25**:bbae498. https://doi.org/10.1093/bib/bbae498

Nicholson JK, Lindon JC (2008) Systems biology: metabonomics. Nature 455:1054–1056. https://doi.org/10.1038/4551054a

Nicholson JK, Wilson ID (2003) Opinion: understanding 'global' systems biology: metabonomics and the continuum of metabolism. Nat Rev Drug Discov 2:668–676. https://doi.org/10.1038/nrd1157

Niu Y et al (2025) Discovery of acetohydroxyacid synthase inhibitors as anti-tuberculosis lead compounds from natural products. Bioorg Med Chem 118:118041. https://doi.org/10.1016/j.bmc.2024.118041

Notebaart RA, Kintses B, Feist AM, Papp B (2018) Underground metabolism: network-level perspective and biotechnological potential. Curr Opin Biotechnol 49:108–114. 10.1016/j.copbio.2017.07.015

Odibo AO et al (2011) First-trimester prediction of preeclampsia using metabolomic biomarkers: a discovery phase study. Prenat Diagn 31:990–994. 10.1002/pd.2822

Olivier M, Asmis R, Hawkins GA, Howard TD, Cox LA (2019) The need for multi-omics biomarker signatures in precision medicine. Int J Mol Sci 20:4781

Ou J et al (2020) Intracellular metabolism analysis of Clostridium cellulovorans via modeling integrating proteomics, metabolomics and fermentation. Process Biochem 89:9–19

Ou FS, Michiels S, Shyr Y, Adjei AA, Oberg AL (2021) Biomarker discovery and validation: statistical considerations. J Thorac Oncol 16:537–545. https://doi.org/10.1016/j.jtho.2021.01.1616

Pang Z et al (2024) MetaboAnalyst 6.0: towards a unified platform for metabolomics data processing, analysis and interpretation. Nucleic Acids Res 52:W398–w406. https://doi.org/10.1093/nar/gkae253

Papadopoulou Z et al (2019) Unraveling the serum metabolomic profile of post-partum depression. Front Neurosci 13:833. https://doi.org/10.3389/fnins.2019.00833

Paris D, Maniscalco M, Motta A (2018) Nuclear magnetic resonance-based metabolomics in respiratory medicine. Eur Respir J 52. https://doi.org/10.1183/13993003.01107-2018

Patra P, B.R D, Kundu P, Das M, Ghosh A (2023) Recent advances in machine learning applications in metabolic engineering. Biotechnol Adv 62:108069. https://doi.org/10.1016/j.biotechadv.2022.108069

Patti GJ, Yanes O, Siuzdak G (2012) Innovation: metabolomics: the apogee of the omics trilogy. Nat Rev Mol Cell Biol 13:263–269. https://doi.org/10.1038/nrm3314

Perakakis N, Yazdani A, Karniadakis GE, Mantzoros C (2018) Omics, big data and machine learning as tools to propel understanding of biological mechanisms and to discover novel diagnostics and therapeutics. Metabolism 87:A1–a9. https://doi.org/10.1016/j.metabol.2018.08.002

Phapale P (2021) Pharmaco-metabolomics opportunities in drug development and clinical research. Anal Sci Adv 2:611–616. https://doi.org/10.1002/ansa.202000178

Pinu FR et al (2019) Systems biology and multi-omics integration: viewpoints from the metabolomics research community. Metabolites 9. https://doi.org/10.3390/metabo9040076

Pirmohamed M (2001) Pharmacogenetics and pharmacogenomics. Br J Clin Pharmacol 52:345–347. 10.1046/j.0306-5251.2001.01498.x

Pontrelli S et al (2018) Escherichia coli as a host for metabolic engineering. Metab Eng 50:16–46. https://doi.org/10.1016/j.ymben.2018.04.008

Prakash S (2011) Human metabolic individuality in biomedical and pharmaceutical research. Circ Cardiovasc Genet 4:714–715. https://doi.org/10.1161/circgenetics.111.962175

Qiu S et al (2023) Small molecule metabolites: discovery of biomarkers and therapeutic targets. Signal Transduct Target Ther 8:132. https://doi.org/10.1038/s41392-023-01399-3

Reicher L et al (2025) Deep phenotyping of health–disease continuum in the human phenotype project. Nat Med. https://doi.org/10.1038/s41591-025-03790-9

Reichling S et al (2023) Dynamic metabolome profiling uncovers potential TOR signaling genes. elife 12. https://doi.org/10.7554/eLife.84295

Resurreccion EP, Fong KW (2022) The integration of metabolomics with other omics: insights into understanding prostate cancer. Metabolites 12. https://doi.org/10.3390/metabo12060488

Rischke S et al (2023) Small molecule biomarker discovery: proposed workflow for LC-MS-based clinical research projects. J Mass Spectrom Adv Clin Lab 28: 47–55. https://doi.org/10.1016/j.jmsacl.2023.02.003

Roca M, Alcoriza MI, Garcia-Cañaveras JC, Lahoz A (2021) Reviewing the metabolome coverage provided by LC-MS: focus on sample preparation and chromatography-A tutorial. Anal Chim Acta 1147: 38–55. https://doi.org/10.1016/j.aca.2020.12.025

Roden DM, Wilke RA, Kroemer HK, Stein CM (2011) Pharmacogenomics: the genetics of variable drug responses. Circulation 123:1661–1670. https://doi.org/10.1161/CIRCULATIONAHA.109.914820

Ryan P, Lee J (2025) In vitro senescence and senolytic functional assays. Biomater Sci 13:3509–3531. https://doi.org/10.1039/d4bm01684j

Sallustio BC (2010) LC–MS/MS for immunosuppressant therapeutic drug monitoring. Bioanalysis 2:1141–1153. https://doi.org/10.4155/bio.10.58

Saoi M, Britz-McKibbin P (2021) New advances in tissue metabolomics: a review. Metabolites 11:672. https://doi.org/10.3390/metabo11100672

Schmidt JC et al (2021) Metabolomics as a truly translational tool for precision medicine. Int J Toxicol 40: 413–426. https://doi.org/10.1177/10915818211039436

Schrimpe-Rutledge AC, Codreanu SG, Sherrod SD, McLean JA (2016) Untargeted metabolomics strategies-challenges and emerging directions. J Am Soc Mass Spectrom 27:1897–1905. https://doi.org/10.1007/s13361-016-1469-y

Schymanski EL et al (2014) Identifying small molecules via high resolution mass spectrometry: communicating confidence. Environ Sci Technol 48:2097–2098. https://doi.org/10.1021/es5002105

Sheikh M et al (2024) Integrating artificial intelligence and high-throughput phenotyping for crop improvement. J Integr Agric 23:1787–1802. https://doi.org/10.1016/j.jia.2023.10.019

Shepherd G, Mohorn P, Yacoub K, May DW (2012) Adverse drug reaction deaths reported in United States vital statistics, 1999-2006. Ann Pharmacother 46:169–175. https://doi.org/10.1345/aph.1P592

Shin SY et al (2014) An atlas of genetic influences on human blood metabolites. Nat Genet 46:543–550. https://doi.org/10.1038/ng.2982

Sindelar M, Patti GJ (2020) Chemical discovery in the era of metabolomics. J Am Chem Soc 142:9097–9105. https://doi.org/10.1021/jacs.9b13198

Singh S, Sarma DK, Verma V, Nagpal R, Kumar M (2023a) Unveiling the future of metabolic medicine: omics technologies driving personalized solutions for precision treatment of metabolic disorders. Biochem Biophys Res Commun 682:1–20. 10.1016/j.bbrc.2023.09.064

Singh N et al (2023b) Drug discovery and development: introduction to the general public and patient groups. Front Drug Discov 3. https://doi.org/10.3389/fddsv.2023.1201419

Singh NB, Kumar B, Usman UL, Susan MABH (2024) Nano revolution: exploring the frontiers of nanomaterials in science, technology, and society. Nano-Structures & Nano-Objects 39:101299. https://doi.org/10.1016/j.nanoso.2024.101299

Soares Dias Portela A et al (2024) Role of artificial intelligence in multinomial decisions and preventative nutrition in Alzheimer's disease. Mol Nutr Food Res 68: e2300605. https://doi.org/10.1002/mnfr.202300605

Sreekumar A et al (2009) Metabolomic profiles delineate potential role for sarcosine in prostate cancer progression. Nature 457:910–914. https://doi.org/10.1038/nature07762

Starke C, Wegner A (2022) MetAMDB: metabolic atom mapping database. Metabolites 12. https://doi.org/10.3390/metabo12020122

Stechemesser L et al (2017) Metabolomic profiling identifies potential pathways involved in the interaction of iron homeostasis with glucose metabolism. Mol Metab 6:38–47. https://doi.org/10.1016/j.molmet.2016.10.006

Stewart A (2013) SLCO1B1 polymorphisms and statin-induced myopathy. PLoS Curr 5. https://doi.org/10.1371/currents.eogt.d21e7f0c58463571bb0d9d3a19b82203

Stewart NA, Buch SC, Conrads TP, Branch RA (2011) A UPLC-MS/MS assay of the "Pittsburgh cocktail": six CYP probe-drug/metabolites from human plasma and urine using stable isotope dilution. Analyst 136:605–612. https://doi.org/10.1039/c0an00643b

Sumner LW et al (2007) Proposed minimum reporting standards for chemical analysis. Metabolomics 3: 211–221. https://doi.org/10.1007/s11306-007-0082-2

Szymanski NJ et al (2023) An autonomous laboratory for the accelerated synthesis of novel materials. Nature 624:86–91. https://doi.org/10.1038/s41586-023-06734-w

Taheri N, Pirboveiri R, Sayyah M, Bijanzadeh M, Ghandil P (2023) Association of DRD2, DRD4 and COMT genes variants and their gene-gene interactions with antipsychotic treatment response in patients with schizophrenia. BMC Psychiatry 23:781. https://doi.org/10.1186/s12888-023-05292-9

Tautenhahn R et al (2012) An accelerated workflow for untargeted metabolomics using the METLIN database. Nat Biotechnol 30:826–828. https://doi.org/10.1038/nbt.2348

Tian T et al (2024) Benchmarking compound activity prediction for real-world drug discovery applications. Commun Chem 7:127. https://doi.org/10.1038/s42004-024-01204-4

Ting T-Y, Li Y, Bunawan H, Ramzi AB, Goh H-H (2023) Current advancements in systems and synthetic biology studies of Saccharomyces cerevisiae. J Biosci Bioeng 135:259–265. 10.1016/j.jbiosc.2023.01.010

Tognetti M et al (2025) Serum proteomics reveals survival-associated biomarkers in pancreatic cancer patients treated with chemoimmunotherapy. iScience 28:112230. https://doi.org/10.1016/j.isci.2025.112230

Tremblay-Franco M et al (2024) Integrative multimodal metabolomics to early predict cognitive decline among amyloid positive community-dwelling older adults. J Gerontol A Biol Sci Med Sci 79. https://doi.org/10.1093/gerona/glae077

Urayama S, Zou W, Brooks K, Tolstikov V (2010) Comprehensive mass spectrometry based metabolic profiling of blood plasma reveals potent discriminatory classifiers of pancreatic cancer. Rapid Commun Mass Spectrom 24:613–620. https://doi.org/10.1002/rcm.4420

Valle A et al (2023) Metabolomics of Escherichia coli for disclosing novel metabolic engineering strategies for enhancing hydrogen and ethanol production. Int J Mol Sci 24. https://doi.org/10.3390/ijms241411619

Varner J, Ramkrishna D (1999) Metabolic engineering from a cybernetic perspective: aspartate family of amino acids. Metab Eng 1:88–116. https://doi.org/10.1006/mben.1998.0104

Ventola CL (2013) Role of pharmacogenomic biomarkers in predicting and improving drug response: part 1: the clinical significance of pharmacogenetic variants. P T 38:545–560

Voss K et al (2021) A guide to interrogating immunometabolism. Nat Rev Immunol 21:637–652. https://doi.org/10.1038/s41577-021-00529-8

Walker JM (2018) In: Methods in Molecular Biology. Springer New York. Print ISSN 1064-3745

Wang Z et al (2011a) Gut flora metabolism of phosphatidylcholine promotes cardiovascular disease. Nature 472:57–63. https://doi.org/10.1038/nature09922

Wang TJ et al (2011b) Metabolite profiles and the risk of developing diabetes. Nat Med 17:448–453. https://doi.org/10.1038/nm.2307

Wang Z et al (2014) Prognostic value of choline and betaine depends on intestinal microbiota-generated metabolite trimethylamine-N-oxide. Eur Heart J 35:904–910. https://doi.org/10.1093/eurheartj/ehu002

Wang Z et al (2024) Antimicrobial resistance: factors to findings. Springer International Publishing, pp 387–413

Wang-Sattler R et al (2012) Novel biomarkers for pre-diabetes identified by metabolomics. Mol Syst Biol 8:615. https://doi.org/10.1038/msb.2012.43

Warburg O (1956) On the origin of cancer cells. Science 123:309–314. https://doi.org/10.1126/science.123.3191.309

Wei X, Meng D, You C (2020) In: Liu L, Du G, Liu Y (eds) Systems and synthetic metabolic engineering. Academic Press, pp 183–206

Weissler EH et al (2021) The role of machine learning in clinical research: transforming the future of evidence generation. Trials 22:537. https://doi.org/10.1186/s13063-021-05489-x

Wikoff WR et al (2013) Pharmacometabolomics reveals racial differences in response to atenolol treatment. PLoS One 8:e57639. https://doi.org/10.1371/journal.pone.0057639

Wild CP, Scalbert A, Herceg Z (2013) Measuring the exposome: a powerful basis for evaluating environmental exposures and cancer risk. Environ Mol Mutagen 54:480–499. https://doi.org/10.1002/em.21777

Wilkinson GR (2005) Drug metabolism and variability among patients in drug response. N Engl J Med 352:2211–2221. https://doi.org/10.1056/NEJMra032424

Wishart DS (2016) Emerging applications of metabolomics in drug discovery and precision medicine. Nat Rev Drug Discov 15:473–484. https://doi.org/10.1038/nrd.2016.32

Wishart DS et al (2022a) NMR and metabolomics-A roadmap for the future. Metabolites 12. https://doi.org/10.3390/metabo12080678

Wishart DS et al (2022b) HMDB 5.0: the human metabolome database for 2022. Nucleic Acids Res 50:D622–d631. https://doi.org/10.1093/nar/gkab1062

Witting M, Böcker S (2020) Current status of retention time prediction in metabolite identification. J Sep Sci 43:1746–1754. https://doi.org/10.1002/jssc.202000060

Woolbright BL et al (2014) Glycodeoxycholic acid levels as prognostic biomarker in acetaminophen-induced acute liver failure patients. Toxicol Sci 142:436–444. https://doi.org/10.1093/toxsci/kfu195

Wörheide MA, Krumsiek J, Kastenmüller G, Arnold M (2021) Multi-omics integration in biomedical research—A metabolomics-centric review. Anal Chim Acta 1141:144–162. https://doi.org/10.1016/j.aca.2020.10.038

Wright JT, Herzberg MC (2021) Science for the next century: deep phenotyping. J Dent Res 100:785–789. https://doi.org/10.1177/00220345211001850

Wurtz P et al (2013) Branched-chain and aromatic amino acids are predictors of insulin resistance in young adults. Diabetes Care 36:648–655. https://doi.org/10.2337/dc12-0895

Xenios S, Weilandt D, Vasilis H, Miskovic L, Kokosis A (2022) In: Montastruc L, Negny S (eds) Computer Aided Chemical Engineering, vol 51. Elsevier, pp 541–546

Xiao JF, Zhou B, Ressom HW (2012) Metabolite identification and quantitation in LC-MS/MS-based metabolomics. Trends Analyt Chem 32:1–14. https://doi.org/10.1016/j.trac.2011.08.009

Xie Y et al (2025) Natural products as modulators of iron metabolism and ferroptosis in diabetes and its complications. Nutrients 17:2714

Xu X, Liu Y, Du G, Liu L (2020) In: Liu l, Du G, Liu Y (eds) Systems and synthetic metabolic engineering. Academic Press, pp 1–31

Yang M, Soga T, Pollard PJ (2013) Oncometabolites: linking altered metabolism with cancer. J Clin Invest 123:3652–3658. https://doi.org/10.1172/JCI67228

Yang Q et al (2019) Metabolomics biotechnology, applications, and future trends: a systematic review. RSC Adv 9:37245–37257. https://doi.org/10.1039/c9ra06697g

Ye C et al (2022) Genome-scale metabolic network models: from first-generation to next-generation. Appl Microbiol Biotechnol 106:4907–4920. https://doi.org/10.1007/s00253-022-12066-y

Yoo I, Ahn I, Lee J, Lee N (2024) Extracellular flux assay (Seahorse assay): diverse applications in metabolic research across biological disciplines. Mol Cells 47: 100095. https://doi.org/10.1016/j.mocell.2024.100095

Yu S et al (2024) Prediction of myocardial infarction using a combined generative adversarial network model and feature-enhanced loss function. Metabolites 14:258

Zenobi R (2013) Single-cell metabolomics: analytical and biological perspectives. Science 342:1243259. https://doi.org/10.1126/science.1243259

Zhou YJ et al (2016) Production of fatty acid-derived oleochemicals and biofuels by synthetic yeast cell factories. Nat Commun 7:11709. https://doi.org/10.1038/ncomms11709

Zhu H et al (2013) Pharmacometabolomics of response to sertraline and to placebo in major depressive disorder - possible role for methoxyindole pathway. PLoS One 8: e68283. https://doi.org/10.1371/journal.pone.0068283

Zoetendal EG et al (2012) The human small intestinal microbiota is driven by rapid uptake and conversion of simple carbohydrates. ISME J 6:1415–1426. https://doi.org/10.1038/ismej.2011.212

MIX
Papier aus verantwortungsvollen Quellen
Paper from responsible sources
FSC® C105338

If you have any concerns about our products,
you can contact us on
ProductSafety@springernature.com

In case Publisher is established outside the EU,
the EU authorized representative is:
Springer Nature Customer Service Center GmbH
Europaplatz 3, 69115 Heidelberg, Germany

Printed by Libri Plureos GmbH
in Hamburg, Germany